Conversion of units
(See Appendix B for a more complete list.)

Acre (U.S. survey)	4.047×10^3 m²
Angstrom (Å)	1.000×10^{-10} m
Astronomical unit (AU)	1.496×10^{11} m
Atmosphere (standard)	1.013×10^5 N/m²
Barrel (42 gal)	1.590×10^{-1} m³
British thermal unit (Btu)	1.054×10^3 J
Calorie	4.182 J
Day (mean solar)	8.640×10^4 s
Dyne	1.000×10^{-7} N
Erg	1.000×10^{-7} J
Foot (ft)	$3.048^* \times 10^{-1}$ m
Gallon (U.S. liquid)	3.785×10^{-3} m³
Gauss	$1.000^* \times 10^{-4}$ T
Horsepower	7.457×10^2 W
Inch	$2.540^* \times 10^{-2}$ m
Light year	9.461×10^{15} m
Liter† (l)	$1.000^* \times 10^{-3}$ m³
Mile (statute)	1.609×10^3 m
Mile (international nautical)	1.852×10^3 m
Ounce (avoirdupois)	2.835×10^{-2} kg
Pound (mass, lbm)	4.536×10^{-1} kg
Pound (force, lbf)	4.448 N
Slug	1.459×10^1 kg
Ton (metric), tonne	$1.000^* \times 10^3$ kg
Ton (short, 2000)	9.072×10^2 kg
Torr (mmHg, 0°C)	1.333×10^2 N/m²

*Indicates an exact equivalence.

INTRODUCTORY
PHYSICS

MASHURI L. WARREN

SAN FRANCISCO STATE UNIVERSITY
and
LAWRENCE BERKELEY LABORATORY

W. H. FREEMAN AND COMPANY
San Francisco

Sponsoring Editor: Peter Renz; *Manuscript Editor:* Kevin Gleason; *Designer:* Robert Ishi; *Production Coordinator:* M.Y. Mim; *Illustration Coordinator:* Batyah Janowski; *Artist:* Eric Hieber; *Compositor:* York Graphic Services; *Printer and Binder:* Kingsport Press.

Credits for front cover and part opening illustrations

Front cover: Courtesy of Marriott's Great America®
Part I: Courtesy of the Hale Observatories
Part II: Courtesy of Dr. Harold Edgerton, Massachusetts Institute of Technology
Part III: Courtesy of Eugene S. Ferguson, the Hagley Museum
Part IV: Courtesy of S. A. Prentice
Part V: Courtesy of the Hale Observatories
Part VI: Courtesy of Jeffrey Richardson, University of California, Lawrence Livermore Laboratory
Part VII: Courtesy of the High Altitude Observatory

Library of Congress Cataloging in Publication Data

Warren, Mashuri L., 1940–
   Introductory physics.

   Includes bibliographies and index.
   1. Physics. I. Title.
QC21.2.W38   530   78-22089
ISBN 0-7167-1008-0

9 8 7 6 5 4 3 2

To
Ivana and Harris

Contents

Preface

Physics is a source of continual change in our lives. Changes flow not only from the spectacular discoveries of the twentieth century but from classical physics as well. Modern physics makes possible such innovations as computer electronics, television, the lasers used in ophthalmology, and nuclear medicine; new applications of classical physics give us such diverse benefits as contact lenses and ultrasonic techniques in medical diagnosis, and even new and better roller coasters. Today we can begin to see the physical principles that will be important in the emerging twenty-first century technology. This book was written with such principles in mind.

This text is for the one-year introductory course, without calculus, for students of biology, for those preparing to study medicine, and for those who plan to major in other sciences. The array of topics chosen gives a comprehensive introduction to basic physics. Many of those in the book's intended audience share a common trait. To wit, even students who have a command of the basic algebra and trigonometry prerequisite to this text often are unsure in applying mathematics to physics. For such students, and for those who need review in mathematics, the mathematical concepts used are reviewed and illustrated by examples. Moreover, the text presents systematic problem-solving techniques that the students are encouraged to adopt and thereby develop their confidence. Appendixes provide a review of basic mathematics, and a discussion of problem solving using a calculator. Preference is given to SI units throughout the text; however, other units that remain in common use are also employed where appropriate.

Each chapter begins by introducing the fundamental ideas related to a set of basic physical phenomena and principles, and develops them with concrete examples. To illustrate the physics and to emphasize the quantitative aspects, these ideas are applied to examples from diverse areas of experience, including sports, transportation, and daily life, as well as biology and modern medical practice. Where appropriate, historical material is used to introduce a topic and to show how scientific discoveries are made. Where possible, current scientific research is given attention. Each chapter is structured to give the instructor flexibility in adapting the text to his or her classroom needs. The instructor may choose to emphasize the later portions of a chapter, to assign them as additional reading, or to omit them altogether as the organization and pace of the course require.

At the end of each chapter are a review list of important terms, suggestions for further reading, and numerous problems arranged to follow the sequence of topics in that chapter. To interest and challenge students, many chapters contain additional problems of greater difficulty or which require the application of several concepts. The answers to the odd-numbered problems are found at the back of the book.

The text follows the traditional sequence of topics, with one exception: Part I concludes with a discussion of special relativity (Chapter 6) as an extension of mechanics. My experience has been that introducing relativity early stimulates student interest in physics. However, at the instructor's discretion this material can be deferred until after treatment of electromagnetic waves (Chapter 18) or wave optics (Chapter 20). The text's seven parts represent major divisions of physics. The material is presented in ways that meet the needs of the students within the constraints of a two-semester or three-quarter introductory course. The suggested division of topics in the adacemic year is shown below.

1st quarter	PART I / MECHANICS	
	PART II / MOTION	1st semester
2nd quarter	PART III / THERMAL PHYSICS	
	PART IV / ELECTRICITY AND MAGNETISM	
3rd quarter	PART V / OPTICS	
	PART VI / THE ATOM	2nd semester
	PART VII / THE NUCLEUS	

To emphasize the procedures of the experimental physicist, I consistently give attention to the observational bases of ideas in physics. For example, before discussing speed and velocity I discuss in some detail the methods by which distance and time are measured, and the process by which physicists go from observations to mathematical analysis. This process is familiar to the instructor but new to most students. I discuss also measurement of speed of light, for the observed constancy of the speed of light in different reference frames is a cornerstone of the theory of special relativity.

Throughout the text physics is related to contemporary concerns. Thermal physics, for example, is related to home energy loss, home insulation, the energy balance in the environment. The discussions of electricity and magnetism not only offer basic knowledge of electric circuits and electric power but also familiarize students with the working principles of radio and TV. In discussing uranium fission I describe also the development of nuclear power and the consequent problems of radioactive wastes. The text concludes on a positive note, however, with a discussion of fusion, energy production in stars, and the prospect for practical uses of fusion power on earth.

Throughout the text the physics is illustrated by numerous examples of relevance to the biological and medical sciences. Angular motion, for example, is given extended application to blood pressure and the mechanical work done by the heart. Electric signals are discussed with respect to their production by nerve cells, the brain, and heart. A chapter is devoted to discussing the safe use of electrical devices and instruments in modern hospitals. The optics of human vision and the resolving power of the modern microscope are treated. Measurement of radiation is applied to nuclear medicine and to environmental health and safety.

ACKNOWLEDGMENTS

The text evolved from classroom experience in teaching introductory physics at California State University, Hayward, and at San Francisco State University. I thank my students for their curiosity, their eagerness to learn, their persistent and insistent questioning, and their patience with new material. It is my hope that this text will be valuable in introducing students to the world of physics.

It is impossible to produce a comprehensive text without the help of many persons, whose assistance I gratefully acknowledge. I thank my teaching colleagues for their suggestions and encouragement; many persons provided information, technical details, and photographs and illustrations for the many examples. The reviewers' thoughtful readings of the manuscript yielded valuable suggestions that guided me in refining the text. In particular I thank Ann Birge, California State University, Hayward; John S. Blair, University of Washington; Norman Goldstein, California State University, Hayward; Arthur Huffman, Colorado College; Robert Karplus, Lawrence Hall of Science, Berkeley; Harold Metcalf, State University of New York, Stony Brook; Partick Argos, Southern Illinois University; and Evans W. Paschal, Stanford Electronics Laboratories. Special thanks go to my publisher and its staff for their encouragement and assistance in bringing this project to successful completion—in particular: Arthur Bartlett, Peter Renz, Fred Raab, Robert Ishi, and Kevin Gleason. I thank also Dean Bocobo and Christopher Hodges for their help with the problem solutions. Any mistakes that appear in the text are my own, and I welcome comments and suggestions for corrections and improvements.

Finally, I give thanks to God for grace and blessings through this long effort.

December 1978   *Mashuri L. Warren*

INTRODUCTORY
PHYSICS

I
SPACE, TIME, AND MATTER

1

Space and Time

Plato and Aristotle . . . Pythagoras, Heraclitus, Diogenes, Archimedes . . . provided
the intellectual basis for what we call modern science. Hellenic culture
gave us faith in the belief that rational thinking and
verifiable facts constitute the stuff of science and
that scientific knowledge helps man elevate
his life above brutish existence.

René Dubos, Reason Awake

Since ancient times, humankind has observed the coming and going of the ocean tides, the rising and setting of the sun, moon, and stars, and the seasonal progression of the constellations of the zodiac through the heavens. We have witnessed the birth of animals and of our own children. We have watched the growth, flowering, and decay of plants. From very earliest times our myths, legends, and religions have given this flux of events and of life a perspective and structure. Lesser and greater gods caused the sun to shine and the thunder to roll. A few believed that the power of the mind could make sense of it all. Plato, Aristotle, Socrates, Euclid, and Pythagoras all shared the belief that the cosmos and its workings could be explained rationally. This was the beginning of natural philosophy, or what since the nineteenth century has been called science.

The study of nature is roughly divided between the biological sciences, which study living systems, and the physical sciences, which include astronomy, chemistry, geology, and physics. Physics is devoted to the study of inanimate bodies that do not undergo changes in chemical structure. We begin our study of physics by examining the simplest ideas, namely how we locate objects in space and time.

MEASUREMENT OF SIZE AND DISTANCE

Our experience of size and distance is visual. We judge an object's size by its appearance close up. Our notion of large distances is conditioned by how far we can see on a clear day or how far we can walk in an afternoon. For objects too small for the naked eye or distances too great to be immediately observed, we must rely on instruments to extend our senses.

Our direct perceptions of size and distance are inescapably subjective. They are like so many fish stories (the ones that got away always seem larger than the ones we caught—and usually get larger as time passes). To communicate our experiences of size and space we must develop mutually agreed upon units of measurement. Whereas a day's journey by automobile can mean different distances to different people, a distance of, say, 375 miles or 600 kilometers is unambiguous, even though it tells nothing about the type of terrain or about our experiences.

Units of measure seem to date from prehistoric times, and ancient civilizations had theirs as well. But the age of the exact sciences and the flourishing of industry and commerce, both beginning roughly in the seventeenth century, required ever-greater precision and uniformity.

The Metric System and SI

In 1790 the French Academy of Sciences formulated a system of units meant to be suitable for adoption throughout the civilized world. This system was based on the **meter** (abbreviated m; spelled metre in Europe), defined as one ten-millionth of the distance between the equator and the North Pole along the meridian through Paris, all other units in the **metric system** being multiples or divisions of the meter by ten. A centimeter is $\frac{1}{100}$ meter; a millimeter is $\frac{1}{1000}$ meter; a kilometer is 1000 meters. Table 1-1 shows different metric units of length.

In 1867, however, it was discovered that the earth's circumference was quite different from what it had been thought to be, so 26 nations convened in Paris in 1872. In 1875 an international treaty, the Metric Convention, was signed by 18 nations, including the United States and

TABLE 1-1
Common metric units of length.

1 kilometer(km)	= 1 000.	m = 10^3 m
1 hectometer(hm)	= 100.	m = 10^2 m
1 dekameter(dam)	= 10.	m = 10^1 m
1 meter(m)	= 1.000	m = 10^0 m
1 decimeter(dm)	= 0.100	m = 10^{-1} m
1 centimeter(cm)	= 0.010	m = 10^{-2} m
1 millimeter(mm)	= 0.001	m = 10^{-3} m
1 micrometer(m)	= 0.000 001	m = 10^{-6} m
1 nanometer(nm)	= 0.000 000 001	m = 10^{-9} m
1 Ångstrom(Å)	= 0.000 000 000 100 m	= 10^{-10} m

Russia, and the International Bureau of Weights and Measures was established near Paris. A General Conference was also to meet every six years in Paris to handle all international matters concerning the metric system. In 1887 French instrument makers produced 30 prototype meters and 40 prototype kilograms to serve as standards of length and mass, respectively. To be uniformly accurate the meter bars were to be compared at 0° Celsius and each was accompanied by a precise thermometer.

As technology became more sophisticated, measurements could be made to an accuracy greater than the finest line scratched on a metal bar. In 1960 the standard for the meter was defined as 1 650 763.73 wavelengths[1] in vacuum of a characteristic orange light emitted by an atom of **krypton–86.** (From this primary standard the foot is defined as 0.304 800 000 meters.) This same general conference also established the **International System of Units** (*Système International d'Unités*), universally abbreviated **SI.** This metric system of units is gradually replacing other systems for both scientific and practical measurement. Although in this book we introduce various units of measurement in common usage, we shall strongly emphasize SI units, anticipating their universal acceptance in both commerce and industry.

The English System of Measurement

Although the metric system has become the predominant system of weights and measures in science and commerce throughout the world, one other system remains in common use: the **English system.** The system used in this country today is nearly the same as that brought here in the eighteenth century, when it was developed as the standard for British colonial commerce. In the United States we are introduced as children to the intricacies of the foot system: 1 foot = 12 inches, 1 mile = 5280 feet, etc.; see Table 1-2. This system is the remnant of a more elaborate set of units, some of which are still used in

---

[1] For greater ease of reading, digits are placed in groups of three to the left and right of the decimal point. Commas between groups are omitted. This style, recommended by the American Society for Testing and Materials, avoids confusion caused by the European use of commas to express decimal points and vice versa. If there are four digits to the left or right of the decimal point, the spacing is optional.

TABLE 1-2
English units of length.

Unit	Equivalent
1 foot	12 inches
1 yard	3 feet
1 mile(statute)	5280 feet
1 mile(US nautical)	6076 feet
1 furlong	1/8 mile
1 fathom	6 feet
1 league	3 miles
1 pica(printer's)	1/6 inch
1 point(printer's)	0.0138 inch
1 barley corn	1/3 inch
1 mil	1/1000 inch

FIGURE 1-1
Prototype kilogram no. 20 is the primary standard for the measure of mass in the United States. Courtesy of the National Bureau of Standards.

special applications—for example, the rod and chain in surveying; and furlongs in horse racing.

Yet, paradoxically, the United States had early been involved with the metric system. In 1805 our first Superintendent of Weights and Measures brought our first standard meter to this country. In 1875 the United States signed the Metric Convention. In 1893 the U.S. Secretary of the Treasury formally recognized meter no. 27 and kilogram no. 20 as the basis of all standards of weight and length in the United States. Since 1901 the National Bureau of Standards has had primary responsibility for standards of measurement in U.S. commerce and industry. Figures 1-1 and 1-2 show, respectively, the prototype kilogram and the prototype meter now kept by the National Bureau of Standards.

Over the past hundred years adoption of the metric system as the principal system of measurement in this country has been much discussed. But in practice our system remains English-based and only partially metric. With the great influx of metrically engineered foreign cars, for example, our automotive service industries have "tooled up" accordingly—for example, most mechanics have a 14-millimeter wrench that fits their 3/8-inch socket drive. But only recently did our aerospace industry cease to measure accelerations in feet per second squared and distance in nautical miles. Only Britain, Canada, and the United States have long conducted their commerce in metric units.

FIGURE 1-2
The prototype meter was the original standard for the measure of length. Courtesy of the National Bureau of Standards.

Our understanding of units is conditioned by our experience. By repetition and reinforcement we get to know the approximate size of units: The maximum highway speed in the United States is 55 miles per hour; a race or football field is 100 yards long; lumber is measured in feet or inches (our houses are built with "two-by-fours"—which most of us know measure 1.5 by 3.5 inches!); our height is 5' 8" or 6' 1" or whatever.

The metric system is no more complicated than the English units. It is only less familiar. To become familiar with metric units, you should obtain your own metric ruler or tape measure and measure common objects. How wide is a quarter? How long is a dollar bill? "One thin dime" has a thickness of about 1 millimeter (mm). Your own body can also provide useful metric measurements. How wide is your thumb? the span of your hand from thumb to pinky? the distance from nose to outstretched fingertips? Knowing these dimensions will help you visualize metric lengths.

Scientific Notation

Because numbers associated with natural phenomena range from very large—millions of billions—to very small—millionths of billionths—we use **scientific notation** based on powers of ten: A thousand = 10^3; a million = 10^6; a billion = 10^9. Perhaps the largest number to have a name is 10^{100}: a googol, invented by the nephew of the American mathematician Edward Kasner.

In scientific notation a number is represented by a value between 1.0000 and 9.999 multiplied by 10 raised to some power:

$$0.040 = 4.0 \times 10^{-2}; \quad 7500 = 7.5 \times 10^3$$

In this notation multiplication or division can be accomplished by manipulating the small numbers and the powers of ten separately. For instance, in multiplication:

$$7500 \times 0.040 = (7.5 \times 10^3) \times (4 \times 10^{-2})$$
$$= (7.5 \times 4) \times 10^{3-2}$$
$$= 30 \times 10^1 = 3 \times 10^2$$

In division:

$$\frac{7500}{0.040} = \frac{7.5 \times 10^3}{4.0 \times 10^{-2}} = 1.875 \times 10^{3+2}$$
$$= 1.875 \times 10^5$$

A very large number can be converted to scientific notation by shifting the decimal point to the left and increasing the power of ten by one for each decimal place shifted.

EXAMPLE

The speed of light c is 299 792 458. meters per second (m/s).

$$299\ 792\ 458 \text{ m/s} = 299\ 792\ 458. \times 10^0 \text{ m/s}$$
$$= 29\ 979\ 245.8 \times 10^1 \text{ m/s}$$
$$= 2\ 997\ 924.58 \times 10^2 \text{ m/s}$$
$$\text{and so on to}$$
$$= 2.99\ 792\ 458 \times 10^8 \text{ m/s}$$

Rounded to three significant figures, the speed of light $c = 3.00 \times 10^8$ m/s.

A very small number can be converted to scientific notation by shifting the decimal point to the right and decreasing the power of ten by one for each decimal place shifted.

EXAMPLE

The wavelength of light from a sodium vapor lamp is 0.000 000 589 4 meters (m). This can be written as

$$0.000\ 000\ 589\ 4 \times 10^0 \text{ m}$$

or $\quad 0.000\ 005\ 894 \quad \times 10^{-1} \text{ m}$

or $\quad 0.000\ 058\ 94 \quad \times 10^{-2} \text{ m}$

and so on to

$$5.894 \quad \times 10^{-7} \text{ m}$$

TABLE 1-3
Prefixes adopted for use with SI units.

Prefix	Symbol	Value	Example
exa-	E-	10^{18}	
peta	P-	10^{15}	
tera-	T-	10^{12}	terawatt(TW)
giga-	G-	10^{9}	gigahertz(GHz)
mega-	M-	10^{6}	megajoule(MJ)
kilo-	k-	10^{3}	kilogram(kg)
hecto-	h-	10^{2}	hectometer(hm)
deka-	da-	10^{1}	dekameter(dam)
deci-	d-	10^{-1}	decibel(dB)
centi-	c-	10^{-2}	centimeter(cm)
milli-	m-	10^{-3}	milliampere(mA)
micro-	μ-	10^{-6}	microvolt(μV)
nano-	n-	10^{-9}	nanometer(nm)
pico-	p-	10^{-12}	picofarad(pf)
femto-	f-	10^{-15}	femtometer(fm)
atto-	a-	10^{-18}	

Prefixes used with SI units provide a shorthand method of expressing powers of ten: A centimeter (cm) is 10^{-2} meter; a megawatt (MW), 10^6 watts; a kilometer (km), 10^3 meter, and so on. These and other prefixes, all of which can be used with any of the metric units, are listed in Table 1-3.

The meter is a linear unit representing length or distance. The SI unit of area is the square meter (m^2). The common metric unit of land area is the hectare, which is a square hectometer, or an area 100 meters on each side. One hectare is 10^4 m^2. Other metric-based units of area are the square kilometer (1 $km^2 = 10^6$ m^2), the square centimeter (1 $cm^2 = 10^{-4}$ m^2), and the square millimeter (1 $mm^2 = 10^{-6}$ m^2).

The SI unit of volume is the cubic meter (m^3). The common metric unit for volume is the liter. One liter is a cubic decimeter (dm^3), or the volume of a cube 0.1 meters on a side. Thus, one liter is a volume of 10^{-3} m^3 or 10^3 cm^3. One cubic centimeter (cm^3) is one milliliter (ml).

Significant Figures

Whenever we write down a number there is an implied accuracy and an implied uncertainty. To say that the national debt in a particular year was $500 billion is less accurate than to say that it was $461.1 billion. The difference is in the number of **significant figures**: the number of digits that contain information. Often, the last signifi-

cant digit to the left of a decimal point is underscored to indicate this. In the first case, only the 5 of the 5̲00 billion contained information, for we had rounded off the number to only one significant digit. In the second case, all four digits contain information, so the number has four significant figures. Asked to estimate the width of a street or the distance from a golf ball to the cup on the putting green, we might say about 20 feet (ft). This could mean any distance between, say, 15 ft and 25 ft. If we were to carefully pace off the distance, we might get within, say, 6 inches (in), 20 ft $\pm$ ¹/₂ ft. On the other hand, with a surveyor's tape measure, we could measure the distance to be 20.14 ft $\pm$ 0.005 ft.

In the first case, only the digit 2 of the 20 conveys information: We really have only one significant figure. The distance is 2̲0 ft, or in scientific notation 2×10^1 ft. In the second case, there are two significant figures, 20̲ ft or 2.0×10^1 ft. In the third case, there are four significant figures, 20.14 or 2.014×10^1 ft. The number of digits indicates the accuracy of a measurement. By convention, the uncertainty in measurements is usually assumed to be plus or minus about one-half the smallest division on the scale.

For numbers smaller than 1.0, leading zeros to the right of the decimal point are not considered significant digits. However, zeros to the right of significant digits are themselves significant and indicate the precision of the number:

	significant figures:
0.0010	two
0.000 003 45	three
0.3400	four

In doing calculations, we must be aware of the accuracy of our information. The accuracy of our conclusions is no greater than that of our least accurate information. For example, the circumference of a circle is given by

$$C = \pi d$$

where d is the diameter. The value π can be written to any accuracy required:

$$\pi = 3.141\ 592\ 653\ 59$$

$$\pi = 3.141\ 6$$

$$\pi = 3.14$$

having 3, 5, or even 12 significant figures. If we estimate the diameter of a Pacific Coast redwood tree to be 12 ft (a number with only two significant figures), it is pointless to use a value of π to 12 significant figures.

The result of any multiplication or division should contain no more significant figures than did the numbers that went into it. We can be misled by our pocket calculators, which give us 8 or 10 digits at the press of a button. In doing a series of calculations it is usually appropriate to keep one extra digit, so that we do not accumulate errors in rounding off intermediate results. We should always round off our final results to the proper number of significant digits.

In this book we will be working to three significant figures unless greater accuracy is required for special purposes. For a number of the order of 1.00 ± 0.005 this is 0.5% accuracy. For a number of the order of 5.00 ± 0.005 this is 0.1% accuracy. Slide rules are inherently accurate to about 0.3%, or about three significant figures. Properly used, the slide rule will not introduce

uncertainties into your calculations. If you use a calculator or do longhand calculations, you must maintain constant vigilance lest you start believing all those insignificant digits.

EXAMPLE

A box is 18.2 inches (in) long, 12.1 in wide, and 2.5 in thick. What is its volume?

$$V = xyz$$

$$V = (18.2\ \text{in})(12.1\ \text{in})(2.5\ \text{in})$$

$$V = 550.55\ \text{in}^3 = 5\underline{5}0\ \text{in}^3$$

to two significant figures.

Conversion of Units

As long as either English or metric units are used, one unit is converted to another simply by multiplying or dividing it by some whole number. However, conversion from metric to English units or vice versa is not so simple (see Table 1-4). Do you multiply by 0.3048? or divide by

TABLE 1-4
Conversion of metric and English units.

Metric Units	English Units
1 mm	0.03937 in
1 cm	0.3937 in
1 m	39.37 in
1 m	3.2808 ft
1 km	0.62137 mi (statute)
0.00254 cm	1 mil
2.54 cm	1 in
0.3048 m	1 ft
0.9144 m	1 yd
1.609 km	1 mi (statute)
1.852 km	1 mi (nautical)

0.3048? or 39.37? or 1.609? Because we later encounter a variety of units for energy, force, power, electric current, magnetic field, etc., it is well at the outset to develop a system for converting units.

There are two basic rules for converting units. Rule one: A quantity multiplied by unity (that is, by one), does not change. Rule two: Units cancel just as any other algebraic quantity. To multiply by unity we use the definition of the various units.

$$1 = \frac{1.0000 \text{ ft}}{0.3048 \text{ m}} \qquad 1 = \frac{5280 \text{ ft}}{1 \text{ mile}}$$

EXAMPLE

How many kilometers (km) in one statute mile?

$$l = 1.0000 \text{ mi}$$

$$l = 1.0000 \text{ mi} \times \frac{5280 \text{ ft}}{1.0000 \text{ mi}} \times \frac{0.3048 \text{ m}}{1.0000 \text{ ft}}$$

$$l = 5280 \times 0.3048 \frac{\cancel{\text{mi}} \times \cancel{\text{ft}} \times \text{m}}{\cancel{\text{mi}} \times \cancel{\text{ft}}}$$

$$l = 1609.344 \cancel{\text{m}} \times \frac{1.000 \text{ km}}{1\,000 \cancel{\text{m}}}$$

$$l = 1.609\,344 \text{ km}$$

Since, by definition, 1 ft = 0.3048000 m exactly and 1 mi = 5280.0000 ft, our number is good to seven significant figures. For practical work we usually round this back to three or four significant figures: 1 mi = 1.609 km.

THE MICROSCOPIC AND THE ASTRONOMIC

The Microscopic

We can make direct measurements with our hands and observations with our eyes only over a limited range. We can directly measure distances as large as a few tens of kilometers and sizes as small as that of a speck—about 0.1 mm or 10^{-4} m. (We can see light reflected from even smaller objects but we cannot directly measure them.)

As our understanding of nature and our technology alike have improved, new instruments have extended our ability to see and to measure. The invention of the microscope opened to scientists the new world of biological structure. In his *Micrographia*, published in 1665, Robert Hooke reported on his observations through a microscope, and his drawings recorded fundamental discoveries in the living world: the compound eye of the fly, the structure of feathers, details of the flea, the stinger of the bee and the nettle, and many others. As the microscope was refined, the world of cellular organisms and cellular biology was opened to inquiry.

Once Max von Laue discovered the wave properties of x-rays, in 1912, the nature of crystals could be probed. In 1913 Sir William Bragg was able (as discussed in Chapter 22) to describe the interaction of x-ray waves with the structure of crystals, and was able to accurately measure the spacing of atoms in crystals, a space of the order of 10^{-10} m. In 1953 the refinement of x-ray crystallography reached its culmination, at the Cavendish laboratory at Cambridge University in England, with the discovery of the structure of the DNA molecule—the carrier of genetic codes.

At this same Cavendish laboratory in 1913 another important discovery was made. Ernest Rutherford, then director of the laboratory, and assisted by Hans Geiger and Ernest Marsden, had been probing the atom with energetic alpha rays. (This work is also discussed in Chapter 22.) From results of experiments they deduced that the atom was not, as had been thought, a solid sphere of matter. Instead, most of the matter was concentrated in a small center—the nucleus. Rutherford calculated the diameter of the nucleus as of the order of 10^{-14} m. From this discovery came further understanding of the atom (discussed in Chapter 23) and exploration of the nucleus (discussed in Chapter 24). These developments in modern physics have brought us to the Alice in Wonderland world of charmed particles and quarks, the benefits of civilian nuclear power, and the perils of nuclear armaments. It has given us deep understanding of the structure of the material world and has also raised the stakes in the game of human civilization. Table 1-5 shows the range of small and microscopic sizes of physical phenomena.

TABLE 1-5
The small and the microscopic.

Object	Size	Other units
The human heart	9–13 $\times 10^{-2}$ m	9–13 cm
Smallest measurable speck	10^{-4} m	0.1 mm
Human red blood cell	10^{-5} m	10 μm
Human nerve axon	0.3–14 $\times 10^{-6}$ m	0.3–14 μm
Wavelength of sodium yellow light	5.894 $\times 10^{-7}$ m	589.4 nm
Cell walls and cell organelles	7.5 $\times 10^{-9}$ m	7.5 nm
Sodium chloride crystal spacing	2.81 $\times 10^{-10}$ m	0.281 nm
Bohr radius of hydrogen atom	5.28 $\times 10^{-11}$ m	0.528 Å
Wavelength 0.51 MeV gamma ray	2.4 $\times 10^{-12}$ m	0.024 Å
Radius of uranium-238 nucleus	7.4 $\times 10^{-15}$ m	7.4 fm
Radius of the proton	1.2 $\times 10^{-15}$ m	1.2 fm

The Astronomic

Human inquiry into distances has extended from early attempts to measure the earth's circumference to our contemplation of the furthest reaches of the universe. The ancients observed the motions of the sun, moon, and planets against the background of the stars. They used these motions to regulate their lives in harmony with the seasons, but they were unable to fathom the reaches of space. The earth's size was first accurately measured by the Alexandrian astronomer Eratosthenes in about 200 B.C. At the summer solstice, Eratosthenes noted the position of the noontime sun with respect to the zenith in two cities a known distance apart. From this information, he used simple trigonometry to determine the earth's radius. (See Figure 1-3.)

In about A.D. 140, Ptolemy was the first to accurately measure the distance from the earth to the moon (see Figure 1-4). Using trigonometry, he determined that the moon's orbit was at a distance of 59 earth radii. Toward the end of the sixteenth century the Danish astronomer Tycho Brahe made voluminous observations and measurements of the stars with the naked eye that have been surpassed in accuracy only by the telescope. Unable to observe any relative motion, or parallax, of the stars, Tycho concluded that the earth must be at rest. He did not imagine that the stars were so far away.

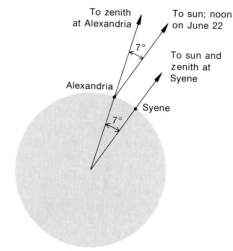

FIGURE 1-3
Eratosthenes' method for measuring the size of the earth.

In 1609 Galileo Galilei heard rumor of a Dutch device for examining distant objects. Using his knowledge of Greek optics, he quickly constructed his own device,

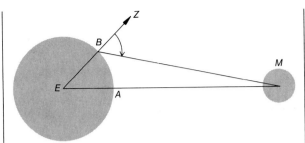

Ptolemy's method for measuring the distance to the moon is essentially identical to the method used today.

which we now call the telescope. His observations, published in 1610, in a pamphlet, *Message from the Stars,* were spectacular. Venus has phases similar to the moon's. The moon has craters and mountains not unlike those on the earth. He discovered that Jupiter has four moons, which, on a moonless night, when Jupiter is up, are visible through ordinary binoculars. See Figure 1-5. Galileo also observed the Milky Way and found that, rather than being wispy clouds floating in the heavens, it consisted of thousands upon thousands of points of light, millions of stars never before seen as stars.

In 1613 Galileo published his *Letters on the Sunspots,* reporting his observations of moving spots on the solar disk, and entered the open fight for the Copernican theory that the earth and planets revolve about the sun. This led to Galileo's confrontation with the Roman Catholic Church. Some of Galileo's contemporaries refused to look through his telescope lest they become confused and be persuaded to what was heresy against medieval theology. Nevertheless, a beginning had been made. Instruments could amplify our senses and extend our ability to describe and to measure.

However, not until the nineteenth century did the true vastness of space begin to be revealed. Telescopes became sufficiently refined to measure the parallax of nearby stars against the background of distant stars as the earth makes its annual pilgrimage around the sun. The first successful measurement of parallax, made in the 1830s, placed the star system Alpha Centauri a distance of some 4×10^{16} meters (m) from the earth.

As more becomes known about the universe, the bits and pieces come together to make a whole. From our models and explanations a picture of the universe emerges with the earth in orbit around the sun at a distance of

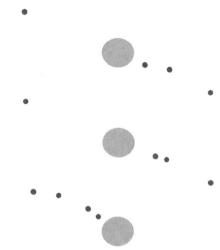

FIGURE 1-5
The moons of Jupiter, discovered by Galileo: "On the Seventh day of January in this present year 1610 . . . I perceived that beside the planet Jupiter there were three starlets . . . on January eighth—led by what, I do not know—I found a very different arrangement . . ."

some 3×10^{11} m. Our sun is one of the thousands upon thousands of stars in our galaxy. Our concept of space has become so vast that we cannot measure it in miles or kilometers. We need a larger unit of length, and we have one, which is based on the speed of light (measurement of which is discussed in Chapter 6).

The speed of light c is some 186 000 miles per second (mi/s), or some 3.0×10^8 meters per second (m/s). Light takes about 1.26 seconds to travel between the earth and the moon, 8 minutes from the sun to the earth, and 5 hours from the sun to the outermost planet, Pluto. So great is the distance to even the star nearest our solar system, Alpha Centauri, that its light must travel for years to reach us. One **light year** is the distance that light, traveling at 2.9979×10^8 m/s, covers in one mean solar year of 3.156×10^7 s:

$$1 \text{ light-yr} = 9.46 \times 10^{15} \text{ m} \cong 10^{16} \text{ m}$$

The distance to Alpha Centauri is 4.3 light-yr. The distance from earth to the center of our galaxy is some 25 000 light-yr. Our galaxy has a diameter of about 10^5 light-yr.

TABLE 1-6
The large and the astronomical.

Object	Size	Other Units
Height of a person	1.8×10^0 m	6.0 ft
1 mile	1.609×10^3 m	1.609 km
Radius of earth	6.37×10^6 m	6 370 km
Distance light travels in 1 s	3.00×10^8 m	1.000 light-s
Distance to moon	3.78×10^8 m	1.26 light-s
Distance to sun from earth	1.50×10^{11} m	8.3 light-min
Distance from sun to Pluto	5.9×10^{12} m	5.0 light-hr
Distance light travels in 1 yr	9.46×10^{15} m	1.000 light-yr
Distance to Alpha-Centauri	4.1×10^{16} m	4.3 light-yr
Distance to galactic center	2.4×10^{20} m	25 000 light-yr
Approximate diameter of galaxy	10^{21} m	10^5 light-yr
Distance to Andromeda galaxy	1.4×10^{22} m	1.5 million light-yr
Estimated size of universe	10^{26} m	10 billion light-yr

During the early 1920s, the American astronomer Edwin Hubble studied the light coming from variable stars in the nebula of Andromeda, near the constellation of Pegasus, which is visible in the evening sky in the fall. Andromeda is a faint wispy cloud just barely visible through binoculars. Until that time nebulae, gaseous bodies in the heavens, were thought to be planetary systems forming about very young stars. Using the large telescope at the Mount Wilson Observatory, Hubble was able to determine the approximate distance of certain stars. To his own and everyone else's astonishment, the Andromeda nebula (Figure 1-6) was discovered to be 1.5 million light-yr away, far beyond the limit of our own galaxy. Since then, the great cluster of galaxies in the constellation of Coma Berenices has been found to be some 300 million light-yr away. The dimensions of the universe are truly beyond our ability to comprehend. The estimated size of the universe is some 10 billion light years! The scale of large and astronomical distances is shown in Table 1-6. For large distances our notions of linear distance are inextricably connected to our ideas about time.

Human vision has reached out beyond the earth, the solar system, and this galaxy to contemplate the mysteries of creation. From the medieval cosmology that placed a perishable and transient world, the earth, at the center of a perfect creation, our knowledge has expanded to address the universe as a dynamic and seemingly infinite evolution. Some four billion human beings inhabit this planet of a minor star, among the 100 million or so stars in our galaxy, among uncountable galaxies.

TIME

Time is perceived as the flow and evolution of our lives. Different scales of measuring time—which date back to prehistoric ages—reflect different human needs. From time immemorial, for example, the farmer has had to sow and harvest as the seasons decree. Thus it was that units of time originally reflected recurring natural events: Days were the recurrent coming and going of the sun; months noted the phases of the moon; the year denoted a complete revolution of the zodiac in the night sky.

The early Romans kept time with a 355-day lunar calendar, which eventually was completely out of phase with the seasons of the year. The time was out of joint and Julius Caesar decreed that the year 46 B.C. (fittingly known as the year of confusion) was to have 445 days! He decreed that thereafter the year was to be 365 days with an

FIGURE 1-6
The great spiral nebula in Andromeda.
Courtesy of Hale Observatories.

extra day in February every fourth (or leap) year. This calendar, known as the Julian calendar, was in common use throughout the European world for the next 1600 years. Pope Gregory XIII, after consultation with his astronomers, determined that the calendar had fallen behind by 10 days and decreed that 1583 be shortened by 10 days and that only one out of four century years (years, like 1900 and 2000, that mark the beginning of a century) be a leap year. This established the Gregorian calendar, which was not adopted in the American colonies until 1752.

Clocks

Over the centuries, methods of keeping time have become more sophisticated. According to legend, Galileo was the first to understand the motion of the pendulum, an object suspended at the end of a cord that is free to swing. While attending mass in the cathedral at Pisa one day, Galileo, then a medical student, fixed his attention on a swinging chandelier. Timing the swing with his own pulse, Galileo observed that the period of motion (the time required for a complete swing) seemed the same for a small or a large swing. This observation contradicted the wisdom of the day, which was: the wider the pendulum's swing, the longer the duration, or period. (Observe a pendulum for yourself: If the pendulum's length is kept constant, does a wider swing take longer?)

Galileo constructed a pendulum to be used for measuring the pulse rate in patients. While Galileo's observation was the basis for development of the accurate clock, he himself did not pursue it. In 1656 a Dutchman, Christian Huygens, invented the first successful pendulum clock by counting out the swing of a pendulum of fixed length. From then on the precision and accuracy of clocks rapidly increased.

Any phenomenon that is repeated with regularity can be used as a more or less accurate **clock,** be it a pendulum or the oscillation of the balance wheel of a wristwatch. In tuning fork watches, a small tuning fork vibrates at 360 times a second and a mechanical linkage advances a gear to record the advance of time. The accuracy of these watches depends on the mechanical properties of the tuning fork, which is sensitive to temperature.

FIGURE 1-7
The quartz crystal watch. Electric vibration of the quartz crystal controls the vibration of the tuning fork, which in turn drives the hands of the dial. The canister at the left encloses the tiny quartz crystal and microelectronics to drive the tuning fork at exactly 341.333 Hz. Courtesy of Bulova Watch Company, Inc.

In a typical quartz watch (as seen in Figure 1-7) the **quartz crystal** vibrates at exactly 32 768 cycles per second. The SI unit for frequency in cycles per second is the **hertz** (Hz), after the scientist Heinrich Hertz, who studied the propagation of electromagnetic waves at the end of the nineteenth century. The quartz crystal is much less sensitive to changes in temperature than the mechanical tuning fork and is therefore more accurate. Using a tiny electronic circuit, the quartz vibrating frequency drives a tuning fork, at 341.333 Hz, which drives the sweep second hand. You can hear the tuning fork humming faintly. Modern digital watches also work from the vibration of a quartz crystal.

International Atomic Time

The U.S. Naval Observatory in Washington, D.C., the Royal Greenwich Observatory in England, and other laboratories around the world regularly observe passage of the stars to precisely determine time according to the earth's motion. Universal Time (UT), the mean solar time at the meridian of Greenwich, England, has been measured since 1925. The Bureau International de l'Heure has coordinated time determinations among different observatories since 1928.

The rotational speed of the earth is not uniform, as was first discovered by S. Newcomb in 1874 from astronomical observations and was confirmed in the 1930s, when the exceedingly accurate quartz crystal clocks came into use. The speed of the earth's rotation varies only a few parts in 10^8. The direction of the earth's axis of rotation in space changes slightly from year to year, wobbling through a 25 000 year cycle. And seasonal changes in the mass of the polar ice caps also slightly change the earth's rate of rotation. A uniform time scale based on astronomical data requires that polar motion and seasonal variations be corrected for.

By international agreement at the General Conference for Weights and Measures in 1956, the second was defined as a precise unit of time in terms of the mean solar year independent of the rotation of the earth, because the year is more precisely defined: The **mean solar year 1900** was defined to be exactly 31 556 925. 974 7 seconds long (365.2422 days). By 1958 scientists at the U.S. Naval Observatory determined the precise radiation frequency of a **cesium-133** atom which was insensitive to external influences such as electric and magnetic fields. Since 1 January 1958, at 0 hr, 0 min, 0 s, the time scale known as the **International Atomic Time** has been in effect, and is determined by comparison of atomic clocks at different laboratories throughout the world.

The **atomic clock** is an extremely stable primary standard for the measurement of time. At the thirteenth General Conference of Weights and Measures in 1964, the radiation frequency of cesium was defined to be exactly 9 192 631 770.0000 Hz (cycles per second). The vibration of a quartz crystal is stabilized by reference to a fundamental atomic vibration of the atom. Modern portable ultrastable cesium clocks can be aligned independently of one another and will agree in frequency within a few parts in 10^{12}.

Our clock time is defined by the International Atomic Time. Over the past eight years the length of the mean solar day has been 86 400.0027 seconds, or about 0.027 seconds longer than the clock day. The difference between solar time and clock time after a year is 365×0.0027 s or 0.9855 s. To keep atomic time synchronized with the earth's motion about the sun, "leap seconds" were introduced. Leap seconds have been added once a year since 1 July 1972. Whenever needed, one leap second is added just after 23 hrs, 59 min, 60 s, and just before 0 hr, 0 min, and 0 s on 1 July or 1 January.

Duration

Our perception of short intervals of time is limited by our reaction times for operating switches, which is about 0.2 seconds, and our ability to discriminate two flashes of light separated by about 0.02 seconds. Observation of brief events required new technology. (For example, the high-speed motion picture camera takes 1000 frames/s for short stretches of time.)

Electronic instrumentation and the oscilloscope make it possible to display and record events of a few millionths, or even billionths, of a second in duration. This requires the use of small units of time:

$$1 \text{ millisecond } = 10^{-3} \text{ s } = 1 \text{ ms}$$

$$1 \text{ microsecond } = 10^{-6} \text{ s } = 1 \text{ } \mu\text{s}$$

$$1 \text{ nanosecond } = 10^{-9} \text{ s } = 1 \text{ ns}$$

$$1 \text{ picosecond } = 10^{-12} \text{ s } = 1 \text{ ps}$$

Because contemporary storage oscilloscopes are capable of observing and storing a single electronic event lasting less than a **microsecond** or only a few **nanoseconds,** high-speed computers have been developed that can execute arithmetic operations in times as short as 4 ns.[2]

As Table 1-7 shows, the time scale of physical phenomena extends from the very short to the very long. Modern physics has established methods for determining approximate times in the distant past. In 1949 Dr. Willard Libby, then at the University of Chicago, determined a way to date the age of objects containing carbon by

---

[2]The Cray-1, at present the world's fastest computer, can perform 138 million operations/s for sustained periods and 250 million/s in short bursts.

Conventional carbon-14 dates in carbon-14 years

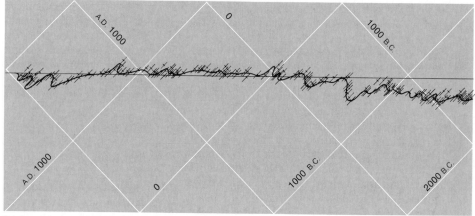

Bristlecone-pine dates in calendar years

TABLE 1-7
Duration in time.

Event	Time	Time in seconds
Time for light to go one meter	3.33 ns	3.33×10^{-9} s
Lifetime of an excited atom	10 ns	$1. \times 10^{-8}$ s
Time for a high-speed CDC 7600 computer to do an operation	27 ns	2.7×10^{-8} s
Potential spike on human nerve axon	1 ms	$1. \times 10^{-3}$ s
Human reaction time	0.2 s	$2. \times 10^{-1}$ s
1 hour	3 600 s	3.600×10^{3} s
1 day	86 400 s	8.64×10^{4} s
1 mean solar year	365.2422 days	3.156×10^{7} s
1 century	100 yr	3.156×10^{9} s
Time since birth of Christ	1979 yr	6.2×10^{10} s
Bristle cone pine chronology	8200 yr	2.6×10^{11} s
Close of last ice age	10 000 yr	$3. \times 10^{11}$ s
Light from Andromeda to reach us	1.5 million yr	4.7×10^{13} s
Oldest Precambrian rocks on earth	3.6 billion yr	1.1×10^{17} s
Estimated age of universe	10 billion yr	$3. \times 10^{17}$ s

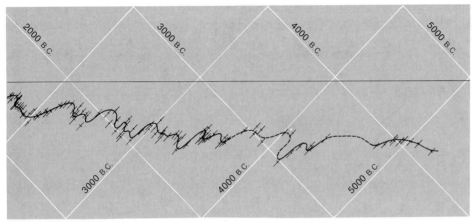

FIGURE 1-8
Dates yielded by carbon-14 dating have been compared against the ages of bristlecone pine back to about 5000 B.C. From Colin Renfrew, "Carbon 14 and the Prehistory of Europe." Copyright © 1971 by Scientific American, Inc. All rights reserved.

measuring the amount of a specific isotope, **carbon-14,** present. Testing fragments of wood from trees known to have grown from 1031 to 928 B.C. gave dates that are accurate within 200 years. For his discovery, Professor Libby received the Nobel Prize in chemistry in 1960.

The Carbon-14 chronology has been calibrated with measurements of the age of samples of the bristlecone pine (see Figure 1-8). From the rings of such trees, the late Dr. Edmund Schulman and Dr. Charles Wesley Ferguson, at the Laboratory of Tree-Ring Research at the University of Arizona, compiled a continuous absolute chronology reaching back nearly 8200 years.

With the discovery of naturally radioactive elements that decay over periods of millions of years, it became possible to date rocks from the crust of the earth and meteorites from outer space. (Natural radioactivity is discussed in Chapter 24.) The oldest measured rocks on the earth are of the order of 3.6 billion years old. Stony meteorites show a maximum age of the order of 4.5 billion years. Furthermore, the existence of lead-207 in association with natural uranium indicates that these heavy elements must have been formed sometime within the last seven billion years. Cosmological models estimate the age of the universe to be between 7 and 20 billion years.

Review

Give the significance of each of the following terms. Where appropriate give two examples.

meter
metric system
krypton-86
International System of Units
SI
English system
scientific notation

prefix
significant figures
light year
clock
quartz crystal
hertz
mean solar year 1900

cesium-133
International Atomic Time
atomic clock
microsecond
nanosecond
carbon-14

Further Readings

"Brief History of Measurement Systems," U.S. Department of Commerce, National Bureau of
   Standards, Special Publication 304A. An introduction to metric available from the U.S.
   Government Printing Office, Washington, D.C.
John Cohen, "Psychological Time," *Scientific American* (November 1964).
Edward S. Deevey, Jr., "Radioactive Dating," *Scientific American* (February 1952).
P. M. Hurley, "Radioactivity and Time," *Scientific American* (August 1949).
Colin Renfrew, "Carbon-14 and the Prehistory of Europe," *Scientific American* (October 1971).

Problems

MEASUREMENT OF LENGTH AND DISTANCE

1 The gross national product for the United States in fiscal year 1972–1973 was $1,151 800 000 000.
(a) Write this in scientific notation.
(b) Express this number in megadollars and giga-dollars. (M$, G$)

2 The Bohr radius for the electron surrounding the hydrogen atom is 0.000 000 000 052 8 meters.
(a) Write this in scientific notation.
(b) Express this number in micrometer (μm), nanometer (nm), and Angstrom (Å).

3 A circle with a radius R has a circumference $C = 2\pi R$ and an area $A = \pi R^2$. Calculate the circumference and cross-sectional area of the following circles. Express your answer in scientific notation to the correct number of significant figures.
(a) A frisbee of radius 9.0 in.

(b) A brass rod with a diameter of 2.54 cm.
(c) A wire with a diameter of 20 mils
   (1 mil = 0.001 in).
(d) A great circle on the earth with a radius of
   3964 mi, or 6378 km.

4 A sphere of radius R has a surface area $S = 4\pi R^2$ and a volume $V = \frac{4}{3}\pi R^3$. Calculate the surface area and volume to the correct number of significant figures and express your answers in scientific notation.
(a) A golf ball of radius 2.2 cm.
(b) A basketball of radius 24.5 cm, or 9.65 in.
(c) A spherical balloon of diameter 12 in.
(d) A spherical tank of diameter 25 ft.
(e) The moon, radius 1080 mi.
(f) The earth, radius 6378 km, 3964 mi.

5 How many golf balls could you place in a suitcase? Hint: How big is the suitcase? How big is a golf ball? Estimate the accuracy of your answer.

6 How many hairs are there on your head? Hint: How far apart are the hairs on your head? How much of your head is covered with hair?

7 How many grains are there in a bucket of sand?

8 One mole of water has a mass of 18 g, occupies a volume of about 18 cm³, and has about 6×10^{23} molecules.
(a) How many molecules of water are in a bucket of water?
(b) How many molecules are there in a bathtub full of water?
(c) How close are your answers?

9 One mole of a gas will fill a volume of 22 400 cm³ at standard temperature and pressure, and will have about 6×10^{23} molecules.
(a) How many molecules are in an inflated balloon?
(b) How many molecules are in your physics classroom?

10 How many people can you get on a football field?

11 Convert each of the following to meters.
(a) Basketball player Wilt Chamberlin is 7 ft 1 in tall.
(b) The Washington Monument is 555 ft tall.
(c) The Empire State Building, in New York City, is 1250 ft tall.
(d) Mount Everest is 29 028 ft tall.
(e) The mean distance to the moon is 238 857 miles.

12 Convert each of the following into English units.
(a) Eddie Carmel, one of the world's tallest circus giants, measured 2.31 m.
(b) *La Tour Eiffel*, built in 1889 in Paris, measures 300 m high.
(c) The deepest recorded place, the Challenger deep in the Pacific Ocean, is about 11 km deep.
(d) The shore of the Dead Sea is 394 m below sea level.

13 Which race is longer?
(a) The 100 yard or the 100 meter race?
(b) The 1500 meter or the mile race?
(c) The 400 meter or the 440 yard race?

14 A 12 fluid ounce can of frozen orange juice is cylindrical with a height of 12 cm and a diameter of 6.5 cm.
(a) What is the volume of the can in cm³? in liters?
(b) What is the approximate volume of one fluid ounce expressed in cm³?

15 Two walking steps make a pace. Walk 10 paces and measure this distance in meters. How long is your pace? Using a wristwatch, count the number of paces you normally walk in one minute. How long will it take you to walk a distance of one kilometer? one mile?

THE MICROSCOPIC AND THE ASTRONOMICAL

16 Express each of the following in meters, using scientific notation:
(a) Helium-neon laser light of wavelength 632.8 nm.
(b) A proton of radius 1.2 fm.
(c) A red blood cell 10 μm in diameter.

17 The brightest star in the sky, Sirius, is 8.7 light-yr from the earth. The mean distance from the earth to the sun is equal to one astronomical unit (AU). 1 Au = 1.49×10^8 km.

(a) How far is Sirius in meters?
(b) How far is Sirius in AU?

TIME

18 The frequency of our standard of time, cesium-133, is defined to be 9 192 631 700.0000 Hz.
   (a) How many significant figures are in this number?
   (b) Write this number in scientific notation to three and six significant figures.

19 In an experiment, Galileo sent a flash of light to an assistant about a mile away. How long does it take for light to travel a distance of one mile and back?

20 In one experiment, a beam of light is sent across a room and reflected back by a mirror. The distance from the light source to the mirror is 4.0 m. How long does the round trip take in s? μs? ns?

2

Motion

POSITION IN TIME

We are all aware of motion. We move; things move—an automobile, a bicycle, the second hand on a clock, a flying bird, or children at play. Our lives are surrounded by continual motion. But what is the nature of motion? A basic task of physics is to observe, measure, and analyze motion.

Yet much motion takes place so rapidly that observers of a particular event can disagree as to what happened and when and in what sequence. For instance, if you drop a coin and a baseball from eye level, which one strikes the ground first? If you drop a baseball and a feather the answer is obvious. Early natural philosophers concluded that "obviously" a heavier object must fall faster than a lighter object. These early philosophers also concluded that the natural state of objects was rest. After all, if something is pushed it quickly comes to rest. They also concluded that an object hurled through space must have some motive force or angel to push it along to keep it in motion. In this chapter we describe motion, kinematics; in Chapter 3 we discuss forces.

Much of the work of science is to analyze phenomena. Observations and measurements yield data from which we can do computations to determine averages, differences, correlations, etc. The important question of physics is, What is the functional relationship between observed quantities? If one quantity changes, what happens to the others?

In experiments involving motion we can neither stop nor reverse time. Time is thus an independent variable. The location of a moving object depends at least partly on time: Location is a dependent variable. To analyze an object's motion we must observe and measure its position at different instants in time. Either we must choose kinds of motion that occur slowly enough for precise recording of positions at different instants, or we must use instruments that accurately record measurements of quick changes of position. Today the latter alternative is quite simple: We can analyze motion pictures, as Figure 2-1 shows. And stroboscopic photography shows quick successions of instants by multiple flashes (as seen in the photograph on the opening page of Part II.

The Commuter

Our primary data for the study of motion are the positions of an object at successive instants of time. With this information we can answer such questions as, How fast is it going at a particular instant? How long does it take to complete a given distance?

Suppose that we live in a suburb and must drive a distance of some 40.7 miles to work. We must drive through the suburb, take a freeway, drive through an urban area, take another freeway, and finally drive the remaining distance to the parking lot at work. The time and mileage at the beginning and end of each stretch are our primary measurements, our raw data. From calculated data we can glean information by comparing how the numbers change from reading to reading: From the time and odometer readings we can compute for any observation (1, 2, . . . , 6) the elapsed time (t_1, t_2, . . . , t_6) and distance traveled (s_1, s_2, . . . , s_6). Raw and calculated data are recorded in Table 2-1. By scanning down the columns

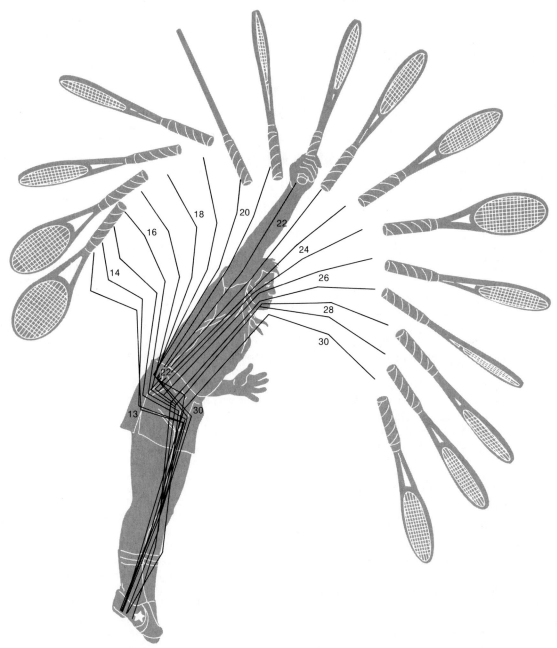

FIGURE 2-1
Tennis service by Ken Rosewall.
The drawing is made from motion pictures taken at 50 frames per second. The tennis racket is 67.5 cm.
Impact with the ball occurs between frames 22 and 23. Courtesy of Stanley Plagenhoef. From *Patterns
of Human Motion: A Cinematographic Analysis,* © 1971. Reprinted by permission of Prentice-Hall Inc.,
Englewood Cliffs, New Jersey.

TABLE 2-1
The commuter. Time and distance going to work.

Observation number	Time	Odometer reading	Elapsed time minutes	Distance traveled miles	Time interval minutes	Distance interval miles	Average speed mi/hr
i		miles	t_i	s_i	Δt	Δs	
1	7:58:30	27277.1	0.0	0.0			
					6.0	2.5	25.0
2	8:04:30	27279.6	6.0	2.5			
					9.8	9.4	57.6
3	8:14:20	27289.0	15.8	11.0			
					8.2	3.4	24.9
4	8:22:30	27292.4	24.0	15.3			
					21.5	21.3	59.4
5	8:44:00	27313.7	45.5	36.6			
					10.3	4.1	23.9
6	8:54:20	27317.8	55.8	40.7			

of numbers (t_i, s_i) we can see that as elapsed time increases, distance traveled also increases.

Average Speed

From the observational data in Table 2-1 we can calculate the average speed over any time interval. The **distance interval** Δs is the difference between any two observations of the distance traveled (throughout the text the delta, Δ, represents difference in value or change in a quantity).

$$\Delta s = s_5 - s_4$$

is the distance interval between observation 5 and 4. The corresponding **time interval** Δt is

$$\Delta t = t_5 - t_4$$

From our tabulated data we can calculate the time and distance intervals for each segment of the trip, also shown in Table 2-1. The **average speed** is defined as the distance per unit time interval.

$$v_{av} = \frac{\Delta s}{\Delta t} \tag{2-1}$$

For any average speed we must specify the time interval over which we are doing our calculation.

EXAMPLE

Calculate the average speed in the interval between observations 4 and 5 in Table 2-1.
The time interval is given by

$$\Delta t = t_5 - t_4$$
$$\Delta t = 21.5 \text{ min}$$

The distance interval is

$$\Delta s = s_5 - s_4$$
$$\Delta s = 21.3 \text{ mi}$$

The average speed in this interval is

$$v_{av} = \frac{\Delta s}{\Delta t} = \frac{21.3 \text{ mi}}{21.5 \text{ min}}$$

$$v_{av} = 0.991 \frac{\text{mi}}{\text{min}} \times \frac{60 \text{ min}}{1 \text{ hr}}$$

$$v_{av} = 59.4 \text{ mi/hr}$$

This stretch involves freeway driving with only moderate traffic. During the final segment of the commute the average speed was only 23.9 mi/hr. As Table 2-1 shows, the average speed in each segment is different.

Calculate the average speed over the entire trip. The entire time interval is

$$\Delta t = t_6 - t_1 = 55.8 \text{ min}$$

The total distance traveled is

$$\Delta s = s_6 - s_1 = 40.7 \text{ mi}$$

The average speed is

$$v_{av} = \frac{40.7 \text{ mi}}{55.8 \text{ min}} = 43.8 \text{ mi/hr}$$

However, the car is virtually never going at exactly this speed.

The data in Table 2-1 can be given much more visual immediacy if we make a graph, as in Figure 2-2.

Graphical Analysis

Each observation consists of paired readings of elapsed time and distance traveled (t_i, s_i). Each pair of values can be represented as a point on a graph. Time is the independent variable. Customarily the independent variable is plotted on the horizontal axis (the *abscissa*). The other variable, the dependent variable—in this case the distance traveled—is plotted with reference to the vertical axis (the *ordinate*). The graph of our data for the commuter is shown in Figure 2-2. Often one glance at a graph discloses the general nature of the motion and permits some deductions. In this case it is apparent that the speed is not uniform.

Suppose a bowling ball is rolled down the center of a bowling alley lane and travels at almost constant speed until it strikes the pins. A motion picture or a multiple-flash exposure of the moving ball would show its position as a function of time. The ball travels at a speed of about 24 ft/s and traverses the 60 ft from the foul line to the

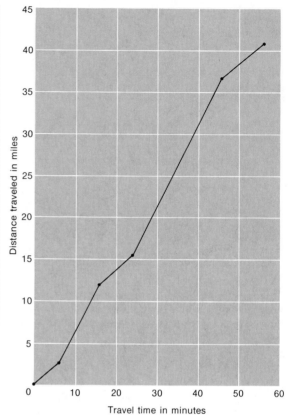

FIGURE 2-2
The commuter. The graph shows the distance traveled as a function of elapsed time.

head pin in about 2.3 s. Figure 2-3 shows a graph of the hypothetical data.

We can approximate the ball's uniform advance as a function of time by fitting a straight line to our data. We can write the equation that describes the motion as

$$s = mt + b$$

The constant b is the intercept on the vertical axis and is the value of the distance s_0 at the initial time $t_0 = 0$; s_0 corresponds to the initial position of the ball had it been traveling at uniform speed for its entire path. In this case

$$s_0 = b = 4.8 \text{ ft}$$

The constant m is the slope of the straight line.

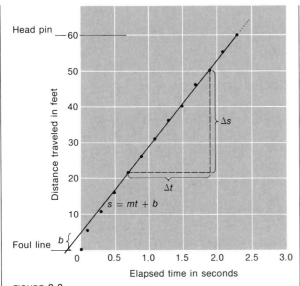

FIGURE 2-3
The bowling ball.
The graph shows the position of a bowling ball as a function of time. The slope of the graph gives the average speed of the ball over time interval Δt.

$$m = \frac{\Delta s}{\Delta t} = v_{av}$$

The slope is just the average speed over the interval Δt.

We can evaluate (express in numbers) the slope, and thus the average speed, by choosing representative points lying on the line as shown in Figure 2-3.

$$v_{av} = \frac{28.7 \text{ ft}}{1.2 \text{ s}} = 24 \text{ ft/s}$$

We can also write a general relationship, or equation, for the distance traveled as a function of elapsed time. The intercept b is equal to s_0. The slope m is equal to the average velocity v_{av}. The equation of uniform motion is

$$s = v_{av}t + s_0 \qquad (2\text{-}2)$$

where t is the elapsed time, since the initial time $t_0 = 0$. This can also be written

$$s = s_0 + v_{av}(t - t_0)$$

This expression can be rearranged to give our original definition of average speed:

$$v_{av} = \frac{s - s_0}{t - t_0} = \frac{\Delta s}{\Delta t}$$

A simple graph is a valuable tool in the preliminary analysis of data. We plot one variable against another. Any relationship between variables will be revealed graphically as a straight line, a parabola, a hyperbola, or some other mathematical curve. If there is a linear dependence, the relationship is simply a straight line. We then have direct evidence of a functional relationship between two variables.

In this book we will pursue relationships between various measured and deduced quantities. Once established, these relationships can be combined to form new relationships. The language we use is algebra. All of these relationships ultimately represent a best fit to someone's observations of the behavior of objects. Thus, in our analysis we move from observation and measurement to establishing empirical relationships that are generalizations from our observations. These in turn will lead, in Chapter 3, to a theory of motion.

MOTION IN THREE DIMENSIONS

So far we have discussed motion in only one dimension, yet we live in a world of three-dimensional space. Our binocular vision allows us to perceive distance and the arrangement of things in space. To describe the location of a point in space we must establish a **frame of reference,** *an arbitrary set of axes, usually at right angles, from which positions are measured.* The **point of origin,** in a reference frame, *is an arbitrary point in space to which all measurements are referred.* The points of the compass—north, south, east, west—provide a geographic frame of reference for direction, as Figure 2-4 shows. To make our description complete, we must also give the angle from the horizon. For instance, a 747 jetliner may approach Dulles International Airport on a flight path from a direction 45° east of north and may at a given instant be 5 km from the control tower at an approach angle of 3° above

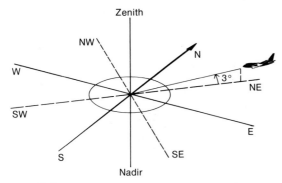

FIGURE 2-4
A geographic coordinate system. Compass points, zenith, and nadir can indicate direction and spatial position.

the horizon. This description uniquely defines the aircraft's position, for it gives the point of origin (the control tower), the direction in space, and the distance from the point of origin (5 km).

With the frame of reference called the **Cartesian coordinate system** (x, y, z), we can locate a point in space by specifying the coordinates of the point, that is, the distance in the x, y, and z directions from a point of origin. Figure 2-5 shows a representation of a three-dimensional Cartesian coordinate system. The distance from the origin to point (x, y, z) is given, by the Pythagorean theorem, as

$$R^2 = r^2 + z^2$$

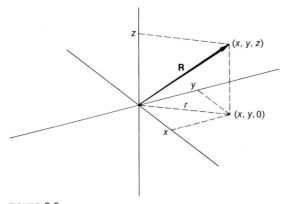

FIGURE 2-5
The Cartesian coordinate system. The position of any point in space is represented by the coordinates (x, y, z).

where r is the distance from the origin in the x-y plane. The distance r is given by

$$r^2 = x^2 + y^2$$

The total distance from the origin to point (x, y, z) is then

$$R^2 = x^2 + y^2 + z^2$$

Three dimensions can be difficult to represent in a two-dimensional picture. Because many motions take place in a two-dimensional plane, a two-dimensional coordinate system is often quite sufficient.

Vectors

A **vector** *is a mathematical way of representing a physical quantity that has a* **magnitude** *and a* **direction** *in space.* The direction in space given by a vector is independent of any particular coordinate system used to describe that space. The simplest vector is the **position vector, R**, which points from the origin of our reference system to a point in space. Vector quantities appear in **boldface (R, v)** when we wish to emphasize the vector property. (When a vector quantity is written on the blackboard, a little arrow $\vec{R}$ is often placed over it to remind us of its vector nature.)

The **displacement, ΔR,** *from one position to another is the difference between two position vectors* (in Figure 2-5, R_1 and R_2). The displacement ΔR is also a vector:

$$\Delta \mathbf{R} = \mathbf{R}_2 - \mathbf{R}_1$$

Velocity *is a vector: It is a quantity describing a body's rate of change in position and the body's direction of motion.* Average velocity (also a vector) is the rate of change of the displacement per unit time. The average speed relates the distance traveled to unit of time. The concept of speed does not incorporate direction of motion. That is, speed is a **scalar** *quantity: It has only magnitude.* The magnitude is a numerical value expressed in appropriate units: 100 km/hr, 88 ft/s, etc.

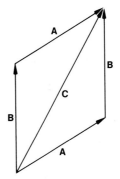

FIGURE 2-6
The head-to-tail method of vector addition.

The average velocity is the displacement per unit time.

$$\mathbf{v}_{av} = \frac{\Delta \mathbf{R}}{\Delta t}$$

At different instants a moving object may be going in different directions. Velocity is a vector quantity because it incorporates both the speed and direction of motion. The direction of travel may be specified according to any of several references: gravity (up and down); compass points (north, southeast, etc); some angle from the horizontal, or some coordinate system (x-direction, y-direction). We specify the positive direction of motion in our reference frame, and negative velocity is motion in the opposite direction.

If vectors are multiplied or divided by a scalar quantity, the result is still a vector. When a vector is multiplied by a scalar, the magnitude of the vector changes but its direction does not:

$$\mathbf{p} = m\,\mathbf{v}$$

The multiplication of a vector by a vector involves special rules and definitions, which we shall discuss later.

Vector Addition

Two vectors that describe the same physical quantity can be added to one another. The sum of two vectors is also a vector and is independent of the order of addition. This is the commutative law. If **A** and **B** are both vectors, then the sum **C** can be written:

$$\mathbf{C} = \mathbf{A} + \mathbf{B} = \mathbf{B} + \mathbf{A}$$

We can represent any vector as a line pointing in the appropriate direction in space, whose length is proportional to the magnitude of the quantity. The intersection of any two lines in space defines a two-dimensional plane. Consequently we can draw these two vectors on a piece of paper, and add them graphically using the **head-to-tail method:** As Figure 2-6 shows, the tail of the second vector **B** starts at the head of the first, **A**. The resultant vector **C** points from the starting point to the head of the second vector. The order of addition of vectors does not affect the result.

EXAMPLE

A treasure map tells a pirate to start at the old tree stump and walk 15 paces to the southeast (**A**). He then must walk a distance of 25 paces to the north (**B**) to find the spot where the treasure is buried. Figure 2-7 shows the graphical addition of these two vectors. With a ruler and protractor we can read off the magnitude and direction of the resultant vector **C**.

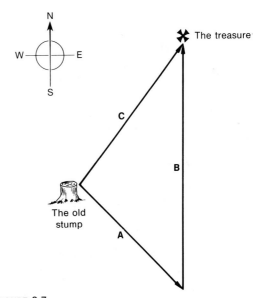

FIGURE 2-7
The treasure map. The resultant vector **C** can be represented as the graphical sum of vectors **A** and **B**.

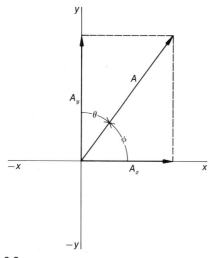

FIGURE 2-8
Any vector can be resolved into perpendicular components. The magnitude of vector **A** is A and the magnitudes of its components are A_x and A_y.

Components of a Vector

Any vector can be resolved uniquely into mutually perpendicular (orthogonal) **components.** (See Figure 2-8.) A vector **A** in the x–y plane can be resolved into a component A_x in the x-direction and a component A_y in the y-direction. We can do this using the definitions of the sine and cosine. If A represents the magnitude of the vector, then the components are given by

$$A_x = A \sin \theta = A \cos \phi$$
$$A_y = A \cos \theta = A \sin \phi$$

The magnitude of the vector is related to the components of the vector by the Pythagorean theorem:

$$A^2 = A_x{}^2 + A_y{}^2$$
$$A^2 = (A \sin \theta)^2 + (A \cos \theta)^2$$
$$A^2 = A^2 (\sin^2 \theta + \cos^2 \theta) = A^2$$

since

$$\sin^2 \theta + \cos^2 \theta = 1$$

The components of a vector **A** can be determined from either the angle θ or from the complementary angle ϕ. The two angles are related by

$$\theta + \phi = 90°$$

Whether to use the sine or the cosine to determine the component using a particular angle must be decided by careful inspection in each case. Remembering the definitions of the sine and cosine functions and reasoning it out each time is better than memorizing a formula for each special case.

In a right triangle as shown in Figure 2-9 the side opposite the 90° angle is the hypotenuse (hyp). The side

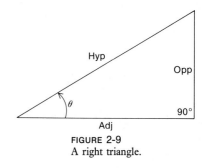

FIGURE 2-9
A right triangle.

next to the angle θ is the side adjacent (adj), and the side across from the angle θ is the side opposite (opp). The sine and cosine are defined by

$$\sin \theta = \frac{\text{opp}}{\text{hyp}}$$

and

$$\cos \theta = \frac{\text{adj}}{\text{hyp}}$$

The sine and cosine of the angles formed by these common triangles are shown in Table 2-2. Either memorize this table or become familiar with the proportions of the triangles upon which these angles are based. A more complete table of sine and cosine values is given in Appendix B.

TABLE 2-2
*The value of the sine and cosine function
of common angles.*

angle θ degrees	$\sin \theta$	$\cos \theta$
0	0.000	1.000
30	0.500	0.866
37	0.60	0.80
45	0.707	0.707
53	0.80	0.60
60	0.866	0.500
90	1.000	0.000

EXAMPLE

London, England, is 710 km from Paris, France, in a direction of about 13° west of north. How far west and north of Paris is London? We can resolve the displacement into two components

$A_\mathrm{w} = A \sin \theta = A \sin 13°$ $A_\mathrm{n} = A \cos \theta = A \cos 13°$

$A_\mathrm{w} = (710 \,\mathrm{km})(0.226)$ $A_\mathrm{n} = (710 \,\mathrm{km})(0.975)$

$A_\mathrm{w} = 160 \,\mathrm{km}$ west $A_\mathrm{n} = 692 \,\mathrm{km}$ north

The Component Method

With the **component method** we add vectors by resolving each vector into components and then adding the components:

$$C = A + B$$

In the x-direction we obtain

$$C_x = A_x + B_x$$

In the y-direction we obtain

$$C_y = A_y + B_y$$

To resolve the components we choose whatever coordinate system is most convenient for the problem.

EXAMPLE

An aircraft carrier travels a distance of 50 km on a course 37° west of north (**A**). It then travels 150 km on a course 37° east of north (**B**). How far has the ship gone?

If we consider the positive x-direction as east, and the positive y-direction as north. The graphical solution as well as the resolution of components is shown in Figure 2-10.

Vector **A** and **B** can be resolved into x and y components. The magnitude of vector **A** is just $A = 50$ km. The direction is $\theta_A = -37°$.

$A_x = A \sin \theta_A$ $A_y = A \cos \theta_A$

$A_x = (50 \,\mathrm{km})(-0.60)$ $A_y = (50 \,\mathrm{km})(0.80)$

$A_x = -30 \,\mathrm{km}$ $A_y = +40 \,\mathrm{km}$

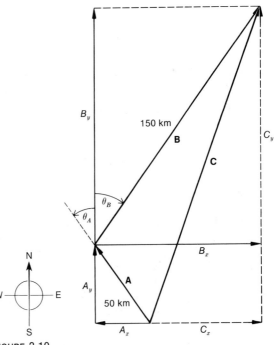

FIGURE 2-10
Vectors can be added by resolving into components.

The magnitude of vector **B** is just $B = 150$ km. The direction is $\theta_B = +37°$. The components of vector **B** are then

$$B_x = B \sin \theta_B \qquad\qquad B_y = B \cos \theta_B$$

$$B_x = (150 \text{ km})(0.60) \qquad B_y = (150 \text{ km})(0.80)$$

$$B_x = 90 \text{ km} \qquad\qquad B_y = 120 \text{ km}$$

The resultant vector components can then be found:

$$C_x = A_x + B_x \qquad\qquad C_y = A_y + B_y$$

$$C_x = -30 \text{ km} + 90 \text{ km} \qquad C_y = 40 \text{ km} + 120 \text{ km}$$

$$C_x = 60 \text{ km} \qquad\qquad C_y = 160 \text{ km}$$

These components are shown on Figure 2-10. By the Pythagorean theorem the magnitude of the resultant vector is

$$C = (C_x{}^2 + C_y{}^2)^{1/2}$$

$$C = [(60 \text{ km})^2 + (160 \text{ km})^2]^{1/2}$$

$$C = 172 \text{ km}$$

The angle of the resultant vector from the y axis can be found from

$$\sin \theta_C = \frac{C_x}{C} = \frac{60 \text{ km}}{172 \text{ km}} = 0.35$$

$$\theta_C \cong 20° \text{ to the east of north}$$

Perhaps one of the simplest examples of addition of velocities (which are vector quantities) is a ship's motion on the water or an airplane through the air, for the water and air are also in motion. The actual velocity of the ship or plane will be the sum of its velocity relative to the medium, **relative velocity,** plus the velocity of the medium.

EXAMPLE

A commercial airplane is flying at an altitude of 7000 m and encounters a prevailing wind from the

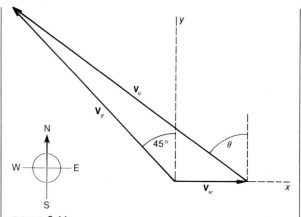

FIGURE 2-11
The ground velocity is the sum of the wind velocity and the velocity relative to the air.

west with a speed of 200 km/hr, $\mathbf{v}_w$. The pilot wishes to fly northwest with a ground velocity $\mathbf{v}_g = 650$ km/hr. What are the air velocity and the direction of flight?

The actual velocity of the aircraft is the ground velocity $\mathbf{v}_g$.

$$\mathbf{v}_g = \mathbf{v}_a + \mathbf{v}_w$$

In this case it is the air velocity $\mathbf{v}_a$ which is unknown. The air velocity is shown in Figure 2-11 and is given by

$$\mathbf{v}_a = \mathbf{v}_g - \mathbf{v}_w$$

We can solve for the air velocity by the component method assuming that east is the positive x-direction and north the positive y-direction. The components of the wind velocity are

$$v_{wx} = +200 \text{ km/hr} \qquad v_{wy} = 0 \text{ km/hr}$$

The components of the ground velocity are

$$v_{gx} = -460 \text{ km/hr} \qquad v_{gy} = +460 \text{ km/hr}$$

The components of the air velocity are given by

$$v_{ax} = v_{gx} - v_{wx} \qquad v_{ay} = v_{gy} - v_{wy}$$

$$v_{ax} = -660 \text{ km/hr} \qquad v_{ay} = 460 \text{ km/hr}$$

The magnitude of the air velocity is 804 km/hr. The angle of flight is given by

$$\sin \theta = \frac{v_{ax}}{v_a} = \frac{-660 \text{ km/hr}}{804 \text{ km/hr}} = -0.82$$

$$\theta = -55°$$

The direction of the aircraft is 55° to the west of north.

GRAVITY AND ACCELERATION

So far we have discussed uniform motion, that is, motion in which average speed or velocity is constant. However, when an object falls because of the earth's **gravity,** it is accelerated; that is, its velocity increases with time. To describe the motion of an object in one dimension under the influence of gravity, we shall develop a set of equations of motion.

Galileo's Experiment

If we try to perform an experiment such as timing a dropped ball, the experiment is over before we have had a chance to react. A ball dropped from shoulder height, 1.60 m, takes only 0.57 s to reach the ground. We would have difficulty timing this motion with a stopwatch, since our own reaction time is 0.1 or 0.2 s. To "dilute gravity" we can roll a ball down an inclined plane. Galileo was the first to systematically perform this experiment.

If we take a board 2 m long, elevate one end by 5 cm, and roll a ball down the plane, we can mark its position each second to about the nearest centimeter. For more precision we could have a strobe light flash at half-second intervals and take a photograph of the successive positions of the ball.

Table 2-3 lists the position of the ball as a function of time for a set of hypothetical measurements. If we make a graph of distance traveled as a function of time, we find that it is not a straight line, as is seen in Figure 2-12. The slope of a line through the points is continuously increasing, indicating that the velocity along the plane is continuously increasing.

TABLE 2-3

Galileo's experiment. An inclined plane 200 cm long is raised by 5 cm at one end. The distance down the plane is recorded as a function of the elapsed time. Velocities are calculated from alternate measurements.

Measurement number	Elapsed time s	Distance cm	Velocity cm/s
i	t_i	y_i	
1	0.0	0	
2	0.5	2	
3	1.0	9	18
4	1.5	20	
5	2.0	35	35
6	2.5	55	
7	3.0	79	52
8	3.5	107	
9	4.0	140	70
10	4.5	177	

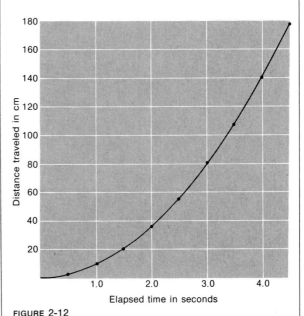

FIGURE 2-12
Galileo's experiment.
The graph shows the distance traveled as a function of time for a ball rolling down an incline that drops 5 cm in 200 cm.

From our data in Table 2-3 we can compute the average velocity. To determine the average velocity at the time t_3 we should use measurements just before and after. In the interval between 0.5 s and 1.5 s the displacement increases from 2 cm to 20 cm. If y represents the distance down the plane, the average velocity down the plane is given by

$$v_{av} = \frac{y_4 - y_2}{t_4 - t_2}$$

$$v_{av} = \frac{20 \text{ cm} - 2 \text{ cm}}{1.5 \text{ s} - 0.5 \text{ s}}$$

$$v_{av} = 18 \text{ cm/s}$$

Since we have calculated the average velocity in the time interval between t_4 and t_2, we should record this average velocity at the average time

$$t_{av} = \frac{t_4 + t_2}{2} = 1.0 \text{ s}$$

Table 2-3 shows the velocity calculated from alternate measurements.

Equations of Motion

The average velocity in the above example is increasing with time. The ball is going faster and faster; that is, it is accelerating along the direction of motion. A graph of the average velocity as a function of time shows that it increases uniformly, as is shown in Figure 2-13. A straight line can be drawn through the points to give a good fit to the data. Thus the velocity v can be represented as a straight line.

$$v = at + b$$

The constant b represents the velocity v_0 at $t_0 = 0$, so that

$$b = v_0$$

The velocity can be simply related to the straight line

$$v = at + v_0$$

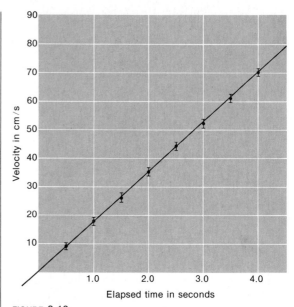

FIGURE 2-13
The average velocity over 1 s intervals as a function of time for a ball rolling down an incline.

The constant a, the slope of the velocity graph, gives us the **acceleration** down the plane.

$$a = \frac{\Delta v}{\Delta t}$$

If the acceleration is uniform throughout the duration of motion, then the graph of velocity as a function of time is a straight line. Because the acceleration has both magnitude and direction, it is a vector.

If the velocity down the plane changes from v_0 to v in the time interval between t_0 and t, then there is an average acceleration down the plane over the interval given by

$$a = \frac{\Delta v}{\Delta t} = \frac{v - v_0}{t - t_0}$$

This can also be written as

$$v - v_0 = a(t - t_0)$$

Setting $t_0 = 0$, we get an equation of motion for velocity with **uniform acceleration.**

$$v = v_0 + at \qquad (2\text{-}3)$$

In this example the velocity of the ball rolling down the inclined plane increases uniformly from $t_o = 0$ s to $t = 4.0$ s without changing direction. The average of the initial velocity v_o and the final velocity v is given by

$$v_{av} = \frac{(v + v_o)}{2} = \frac{(70 \text{ cm/s} + 0 \text{ cm/s})}{2}$$

$$v_{av} = 35 \text{ cm/s}$$

If we calculate the average velocity over the entire interval, we have

$$v_{av} = \frac{\Delta y}{\Delta t} = \frac{140 \text{ cm} - 0 \text{ cm}}{4.0 \text{ s} - 0.0 \text{ s}}$$

$$v_{av} = 35 \text{ cm/s}$$

which is the same result.

We have thus far established two linear relationships involving velocity and displacement as a function of time, equation 2-2:

$$y = y_o + v_{av}t$$

and equation 2-3:

$$v = v_o + at$$

where the average velocity can be written as

$$v_{av} = \frac{v_o + v}{2}$$

In uniform acceleration the velocity increases at a uniform rate. We can calculate the average velocity over any interval from $t_o = 0$ to time t by combining the last two relationships:

$$v_{av} = \frac{v_o + v}{2}$$

$$v_{av} = \frac{v_o + (v_o + at)}{2}$$

$$v_{av} = v_o + \tfrac{1}{2}at$$

We can now use this value for the average velocity over the interval to calculate the total displacement:

$$y = y_o + v_{av}t$$

$$y = y_o + (v_o + \tfrac{1}{2}at)t$$

$$y = y_o + v_o t + \tfrac{1}{2}at^2 \qquad (2\text{-}4)$$

This is the equation of motion, or relationship for the total displacement in the case of uniformly accelerated motion.

In our example of the ball rolling from rest down an inclined plane, the ball's initial position and velocity are zero $y_o = 0$, $v_o = 0$. The relationship between velocity and time takes the form of a straight line through the origin:

$$v = at$$

The relationship between the displacement and time takes the form

$$y = \tfrac{1}{2}at^2$$

which is a parabola. We can obtain the acceleration from the slope of the velocity graph:

$$a = \frac{\Delta v}{\Delta t}$$

$$a = \frac{70 \text{ cm/s}}{4 \text{ s}}$$

$$a = 17.5 \text{ cm/s}^2$$

We have used the inclined plane to "dilute gravity." If, on the other hand, we drop or throw an object, the earth's gravity acts on it, producing a downward acceleration. Near the earth's surface the acceleration due to gravity g is more or less constant: 9.780 39 m/s² at sea level on the equator, increasing slightly with increasing latitude to 9.801 71 m/s² at sea level at 40° North latitude. For most practical purposes we shall assume that g is

$$g = 9.80 \text{ m/s}^2 = 980 \text{ cm/s}^2 = 32 \text{ ft/s}^2$$

Instantaneous Velocity

Thus far we have calculated only the average velocity over a given time or distance interval. In accelerated or decelerated motion, the velocity of an object is changing continuously. How can we determine the speed of an object at a particular instant in time?

Imagine a speed trap beside a highway. The simplest version—the one used for highway speed surveillance by aircraft in remote stretches of our interstate highways—could be two markers stretched across the road at a given distance. The time for an automobile to traverse this interval could be measured with a stopwatch. If, for instance, the markers were 250 m apart, at 100 km/hr it would take 9.0 s for a car to complete the distance; at 120 km/hr it would take only 7.5 s. This time interval is clearly measurable with a stopwatch. Suppose, however, that a speeding driver spotted the speed trap as he entered it. He could slow down, and the measured time to complete the distance would give only his average velocity over the interval.

His velocity could be more precisely determined by measuring his travel over much shorter distance and time intervals. If the two distance markers were only 30 m apart, then at 100 km/hr and 120 km/hr it would take 1.08 s and 0.90 s, respectively, to complete the distance. Measured with a hand stopwatch, this time difference would be largely obscured by the human reaction time of 0.1 to 0.2 s. However, using electronics accurate to milliseconds, we could accurately determine the speed before the offending driver had a chance to slow down. At 100 km/hr an automobile takes about 11 milliseconds to go a distance of 0.3 m.

The shorter the time interval of measurement, Δt, the more closely the measured velocity approximates the actual velocity of the object at a given instant. If $\Delta \mathbf{y}$ is the displacement in a time Δt the **instantaneous velocity** is defined as the limit where the time interval approaches zero, $\Delta t \to 0$:

$$\mathbf{v} = \lim_{\Delta t \to 0} \frac{\Delta \mathbf{y}}{\Delta t} \qquad (2\text{-}5)$$

For uniformly accelerated motion the average velocity over a given time interval is equal to the instantaneous velocity at the midpoint of the time interval. If we can assume that the acceleration is constant, we can choose a time interval suitable to our measuring equipment and still determine the velocity of an object as a function of time.

PROBLEM SOLVING

One purpose of a physics course is to develop skills in solving problems that use numerical and functional relationships. Many of the basic concepts of physics can be introduced descriptively. However, the power of these concepts does not begin to show until they are applied to concrete situations. Many students approach problem solving in the wrong way—as a mysterious process wherein answers are arrived at by divine revelation or by some miracle; others try to solve problems in their heads, writing nothing down but their answers; still others become hopelessly confused, their attempts to solve problems degenerating to grasping at straws.

Problem solving serves a valuable function, but only if approached systematically. The series of steps that follow should help you solve virtually any physics problem you encounter.

Systematic Problem Solving

1. Read text Read (do not just skim) the relevant text material *before* attempting to solve the problems. Each problem in this text is an application, extension, or illustration of one or more concepts. Wherever possible the examples are drawn from life experience or from biology and medicine. Each chapter develops new concepts and new techniques of analysis. So do yourself a favor and *read the material*. Review what you have read. Determine what you understand, and what you don't understand. Reread as necessary; ask questions. Make use of the professor; find out!

2. Picture Read the problem and try to picture in your mind the circumstances it depicts. What is happening? Get an intuitive feeling for what the outcome might be. Have you ever seen or been in such a situation?

3. Sketch Systematic problem solving involves learning to think on paper. Drawing a sketch of the situation is the first step.

4. Given? List the information given in the sketch, by marking off distances, drawing little arrows next to velocities, etc. This is the primary information with which you work.

5. Find? List clearly what it is that you are being asked to find. It may be useful to ask yourself, Why am I being asked to find that? What is the point of the problem?

6. Relationships Write down the relevant relationships between the various known and unknown quantities. This step is critical in learning to solve problems on paper. If you are not sure which of two relationships to use, write them both down. Then you can verify that each relationship is written correctly, and can critically examine each to see whether it is the proper relationship to use.

7. Solve Solve the problem. Using algebra, rearrange the relevant equation or equations to give a relationship for the unknown quantity. Substitute the numerical values and proper units in the expression. Reduce both the numbers and the units to the simplest form to give the answer. Round off the answer to the appropriate number of significant figures.

Plug-in Problems

The simplest problems—often called plug-in problems: you plug in the numbers and turn the crank—require the selection of only a single relationship, and solving for only one unknown. Once it is established which relationship is proper to use, we can solve for the unknown quantity. The numerical values with units can be plugged in, and the result reduced to its simplest form.

EXAMPLE

A man stands on a bridge and drops a stone from rest into the stream below. The stone is accelerated down-

ward at 9.8 m/s², and takes 1.2 s to reach the stream. How far is it to the stream? Analysis:
(1) Read text material on motion.
(2) Read problem and picture the situation.
(3) Sketch.
(4) Given? List the information given in the sketch (Figure 2-14).

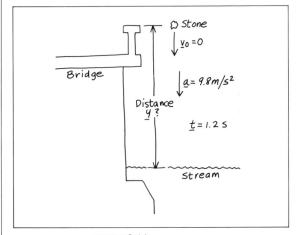

FIGURE 2-14
A sketch: dropping a stone.

(5) Find? What is being asked for? You are being asked to find the distance y to the stream.
(6) Relevant relationships?

(a) $v = v_0 + at$ (?)   No: v not asked for.

(b) $y = v_0 t + \frac{1}{2}at^2$ (?)   Yes: a, t, and v_0 given.

The first relationship (a) will tell you how fast the stone was going. This is a fair question, but it was not asked. The second relationship (b) is the proper one to use to determine the distance traveled.
(7) Solve the problem.

The stone was initially at rest, $v_0 = 0$. If in our reference system we choose distance downward from the point of origin as positive, then the downward velocity is also positive. The acceleration due to gravity $a = +9.8 \text{ m/s}^2$ is also positive. The elapsed time is $t = 1.2$ s. We have one equation (b) and only one unknown variable, y. We can solve by substitution.

$$y = v_o t + \tfrac{1}{2}at^2$$

$$y = (0 \text{ m/s})(1.2 \text{ s}) + (0.5)(9.8 \text{ m/s}^2)(1.2 \text{ s})^2$$

$$y = 0 + (0.5)(9.8)(1.2)^2 \frac{\text{m-s}^2}{\text{s}^2}$$

Doing the arithmetic we get 7.056, which to two significant figures is 7.1. The units simplify to meters (m). Thus, the final answer is

$$y = 7.1 \text{ m}$$

EXAMPLE

A Porsche 911 Carrera sportscar will accelerate from 0 to 96.5 km/hr in 8.2 s. What is the average acceleration in m/s²? Analysis:
(1) Read the material on acceleration.
(2) Read the problem and picture it.
(3) Draw a sketch (shown in Figure 2-15).

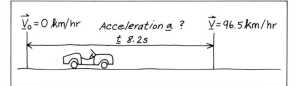

FIGURE 2-15
A sketch: an accelerating automobile.

(4) List information given.
(5) What is being asked for? The unknown is the average acceleration a.
(6) Relevant relationships.

(a) $y = v_o t + \tfrac{1}{2}at^2$ (?)   No: Distance not given or asked for.

(b) $v = v_o + at$ (?)   Yes: v, v_o, and t given.

(c) $a = \dfrac{\Delta v}{\Delta t}$ (?)   Yes: Δv and Δt given.

In this case, either of two valid relationships, (b) or (c), could be used to solve for the one unknown, a.
(7) Solve. The relationship in (c) can be rewritten:

$$a = \frac{v - v_o}{t}$$

$$a = \frac{96.5 \text{ km/hr} - 0 \text{ km/hr}}{8.2 \text{ s}}$$

$$a = 11.8 \text{ km/hr-s} = 12 \text{ km/hr-s}$$

That is, the speed of the automobile is increasing by 12 km/hr each passing second.

To determine the acceleration in m/s² we can either convert v and v_o into m/s and rework the problem, or we can convert the answer.

$$a = 11.8 \frac{\text{km}}{\text{hr-s}} \times \frac{1000 \text{ m}}{1 \text{ km}} \times \frac{1 \text{ hr}}{3600 \text{ s}}$$

$$a = \frac{(11.8)(1000) \; \cancel{\text{km}} - \text{m} - \cancel{\text{hr}}}{(3600) \quad \cancel{\text{hr}}\text{-s} - \cancel{\text{km}}\text{-s}}$$

$$a = 3.3 \text{ m/s}^2$$

Quadratic Forms

Some problems involving a single unknown are solved by a quadratic equation:

$$ax^2 + bx + c = 0 \qquad (2\text{-}6)$$

As long as $b^2 > 4ac$, two possible solutions to this equation are given by

$$x = \frac{-b \pm [b^2 - 4ac]^{1/2}}{2a} \qquad (2\text{-}7)$$

A ball is thrown straight up into the air with an initial velocity v_o of 20 m/s. The acceleration due to gravity is downward at 9.8 m/s². How long will it take the ball to reach a height of 15 m? How fast will it then be going?
Analysis:
(1–3) Read text. Read problem. Sketch (see Figure 2-16).

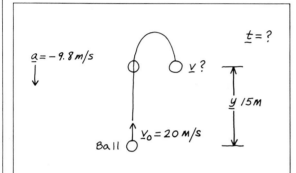

FIGURE 2-16
A sketch: throwing a ball.

(4) Given? In this problem we must establish a positive direction. This is an arbitrary choice. If we choose the upward direction as positive, the initial velocity is $v_o = 20$ m/s; however, the acceleration is negative: $a = -9.8$ m/s².
(5) Find? The time t to go a distance $y = 15$ m.
(6) Relationships.

(a) $y = v_o t + \frac{1}{2}at^2$ (?)   Yes: v_o, a, and y given; t unknown.

(b) $v = v_o + at$ (?)   No: v_o and a given; t and v unknown.

(7) Solve. We have one equation (a) and one unknown, but it is a quadratic form. This becomes clear if we rewrite (a) as

$$at^2 + 2v_o t - 2y = 0$$

If we note that $x = t$, $a = a$, $b = +2v_o$, and $c = -2y$, we can immediately write down the quadratic equation, equation 2-6, and its solution, equation 2-7. Making the proper substitutions we get

$$t = \frac{-2v_o \pm [4v_o{}^2 - 4a(-2y)]^{1/2}}{2a}$$

which reduces to

$$t = \frac{-v_o \pm [v_o{}^2 + 2ay]^{1/2}}{a}$$

Substituting values

$$t = \frac{-20 \text{ m/s} \pm [(20 \text{ m/s})^2 + 2(-9.8 \text{ m/s}^2)(15 \text{ m})]^{1/2}}{(-9.8 \text{ m/s}^2)}$$

$$t = 2.04 \text{ s} \pm 1.05 \text{ s}$$

There are two solutions:

$$t_1 = 0.99 \text{ s}$$

and

$$t_2 = 3.09 \text{ s}$$

The velocity at time t_1 is

$$v_1 = v_o + at_1$$
$$v_1 = (20 \text{ m/s}) + (-9.8 \text{ m/s}^2)(0.99 \text{ s})$$
$$v_1 = +10.3 \text{ m/s}$$

The positive sign indicates that the ball is going upward.
The velocity at time t_2 is $v_2 = -10.3$ m/s, indicating that the ball is going downward. The two roots of the quadratic equation correspond to the time required either to reach a height of 15 m or to go all the way to the top and pass the 15 m height on the way down. Often both roots of a quadratic equation have physical significance.

Missing Information

Suppose in the previous example you were asked to determine the maximum height the ball would reach. There is now some missing information, which you obtain by looking for some implied questions and answering them.

A ball is thrown straight up into the air with an initial velocity $v_0 = 20$ m/s. The acceleration due to gravity is downward $a = -9.8$ m/s². What is the maximum height reached by the ball?
Analysis:
(1–4) Read. Picture. Sketch (see Figure 2-17). Given?

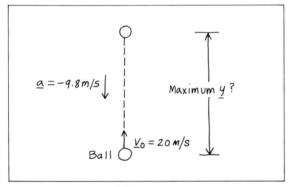

FIGURE 2-17
A sketch: the maximum height.

(5) Find? The maximum height y.
(6) Relationships.

(a) $y = v_0 t + \frac{1}{2}at^2$ (?) a and v_0 given; t and y unknown.

(b) $v = v_0 + at$ (?) a and v_0 given; t and v unknown.

(7) Solve. We could determine the maximum height y, using the relationship (a) if we knew the time t.

Solving this problem requires us to ask the implicit question, How long does it take to reach the top? We can use the second relationship (b) to determine this time. However, to solve this relationship for the time t, we need to know how fast the ball is going at the top of the trajectory (another implicit question). This was not given directly in the problem. However, a moment's thought shows that at the top of the trajectory the ball will be instantaneously at rest: $v = 0$. Solving for the time t we have

$$t = \frac{v - v_0}{a}$$

$$t = \frac{(0 \text{ m/s} - 20 \text{ m/s})}{-9.8 \text{ m/s}^2}$$

$$t = \frac{-20 \text{ m}}{-9.8 \text{ s}} \times \frac{\text{s}^2}{\text{m}}$$

$$t = 2.04 \text{ s}$$

Having solved for the time we can immediately solve for the maximum height:

$$y = v_0 t + \frac{1}{2}at^2$$

$$y = (20 \text{ m/s})(2.04 \text{ s}) + (0.50)(-9.8 \text{ m/s}^2)(2.04 \text{ s})^2$$

$$y = 40.8 \text{ m} - 20.4 \text{ m}$$

$$y = 20.4 \text{ m}$$

How can you tell if there is sufficient information to solve a problem? If we have one unknown, we need one equation with all of the other quantities known. If we have two unknowns, we need two relationships. If we have three unknowns, then, in principle at least, we can solve the problem if we have three independent relationships. If we find more unknowns than relationships, we must look for more information, either finding it implicit in the problem or finding additional equations that relate the unknown quantities.

DRAWING BY SIDNEY HARRIS

"*I tend to agree with you—especially since* $6.10^{-9} \sqrt{t_c}$
is my lucky number."

Algebra

In the preceding example we were given the initial velocity v and acceleration a and were asked to find the maximum height y. Implicit to the solution were the intermediate questions, How fast was it going at the top? And how long did it take? We solved the problem explicitly by determining the time and substituting it into the relationship for the distance traveled. Algebra makes possible a general relationship for this type of problem. The two relationships involving uniformly accelerated motion are, respectively, equations 2-4 and 2-3:

$$y = v_0 t + \tfrac{1}{2}at^2$$

$$v = v_0 + at$$

Let us, using algebra, solve for the time; substitute this expression into the distance formula; and then reduce it to the simplest form. Equation 2-3 can be rewritten as

$$t = \frac{v - v_0}{a}$$

Substituting this expression for t into the distance equation we get

$$y = v_0 t + \tfrac{1}{2}at^2$$

$$y = \frac{v_0(v - v_0)}{a} + \frac{\tfrac{1}{2}a(v - v_0)^2}{a^2}$$

Multiplying out we get

$$ay = v_0 v - v_0{}^2 + \tfrac{1}{2}v^2 - \tfrac{1}{2}(2vv_0) + \tfrac{1}{2}v_0{}^2$$

$$2ay = v^2 - v_0{}^2 \qquad (2\text{-}8)$$

This, then, is our third relationship involving velocity, acceleration, and distance. We have eliminated time as a consideration.

TRANSPORTATION PROBLEMS

We have all been stuck in freeway traffic. At one time or another the number of cars exceeds the capacity of the road, and traffic crawls to a stop. How much traffic can a freeway lane carry? From physics, and with perhaps a few assumptions about driver behavior, we can make some predictions about the behavior of a lane of freeway traffic. A mathematical **model** contains basic physical principles and a few plausible assumptions. Mathematical models play an important role in science, for they describe phenomena in simple terms by reducing them to essentials. In this section we shall apply the equations of motion for uniform acceleration (developed in the previous section) to the capacity of a freeway lane.

Obviously an eight-lane highway will carry twice as much traffic as a four-lane highway. But neglecting the effects of lane hopping, how much traffic can one lane safely carry? Cars stacked bumper to bumper can safely go only very slowly, say 5 km/hr. The total flow of traffic, the number of cars passing a particular point, will be very small. On the other hand, cars going very fast, say 200 km/hr, must maintain a large distance between them to be safe. But what is the proper spacing between automobiles? We shall consider two possibilities; the highway-patrol model and the braking model.

The Highway-Patrol Model

In the driver's manual published by your state highway department you are admonished to keep a safe distance when following other automobiles in traffic. The usual rule of thumb is one car-length of distance for every 10 mi/hr of speed. This advice is taken as one of the assumptions of the "highway-patrol" model for the capacity of a freeway lane.

Say we have a car with length l_o. At 10 mi/hr a safe distance is l_o from the car ahead; at 30 mi/hr a safe distance is $3l_o$. Generally, if $v_{10} = 10$ mi/hr and we are traveling at some speed v, we should maintain a safe distance given by

$$\left(\frac{v}{v_{10}}\right)(l_o)$$

Thus, in a freeway lane, the total length occupied by the car plus the safe distance is

$$l = l_o + \left(\frac{v}{v_{10}}\right)(l_o)$$

or

$$l = l_o\left(1 + \frac{v}{v_{10}}\right)$$

EXAMPLE

Suppose a car of length $l_o = 15$ ft is traveling at 55 mi/hr. The total distance l occupied by this car on the roadway is given by its own length plus the safe distance (see Figure 2-18).

$$l = l_o\left(1 + \frac{v}{v_{10}}\right)$$

$$l = 15\text{ ft}\left(1.000 + \frac{55\text{ mi/hr}}{10\text{ mi/hr}}\right)$$

$$l = 97.5\text{ ft}$$

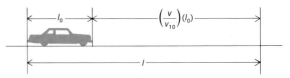

FIGURE 2-18
The highway-patrol model. The total distance occupied by a car is the length of the car l_o plus one car length for every 10 mi/hr of speed.

Suppose a string of cars of length L is traveling at some speed v. The capacity of this freeway lane is given by the total number of cars N that pass our observation point in a period of time T. We can then ask, How many cars are in this string of length L? And how long will the entire string of cars take to pass our observation point? The total number of cars that a freeway lane of length L could safely carry is given by the length L divided by the total distance l occupied by each car at a particular speed.

$$N = \frac{L}{l} = \frac{L}{l_o(1 + v/v_{10})}$$

The total time for the entire string to pass our observation point at the head of the string of cars is given by the time it takes the last car in the string to travel the distance L at speed v

$$T = \frac{L}{v}$$

EXAMPLE

If a string of cars is one mile long, $L = 5280$ ft, traveling at 55 mi/hr $= 80.7$ ft/s, and each car has a length $l_o = 15$ ft, the total number of cars in the one-mile string is

$$N = \frac{5280 \text{ ft}}{(15 \text{ ft})(1.00 + 5.50)}$$

$$N = 54.1 \text{ cars}$$

The total time for the last car to pass the observation point is just given by

$$T = \frac{L}{v} = \frac{5280 \text{ ft}}{80.7 \text{ ft/s}} = 65.4 \text{ s}$$

The total capacity, C_{hp}, of the freeway lane is just the number of cars passing a point per unit time. In this example we have

$$C_{hp} = \frac{N}{T} = \frac{54.1 \text{ cars}}{65.4 \text{ s}}$$

$$C_{hp} = 0.827 \text{ cars/s}$$

$$C_{hp} = 49.6 \text{ cars/min}$$

$$C_{hp} = 2980 \text{ cars/hr}$$

The capacity of the freeway lane is a function of velocity of the traffic flow. Generally we can write

$$C_{hp} = \frac{N}{T}$$

$$C_{hp} = \frac{\dfrac{L}{l_o(1 + v/v_{10})}}{\dfrac{L}{v}}$$

$$C_{hp} = \frac{v}{l_o(1.00 + v/v_{10})}$$

The highway-patrol model predicts that the capacity of the freeway lane increases with the speed of the traffic flow. At a speed of 70 mi/hr the model predicts a capacity of 3080 cars/hr. If highway traffic really follows the prediction of this model, freeway lane capacity would be sacrificed only 3% if the speed limit were reduced from 70 mi/hr to 55 mi/hr. Figure 2-19 shows the highway-patrol model's prediction of lane capacity as a function of traffic speed.

The Braking Model

In predicting the capacity of a freeway lane, the highway-patrol model assumes that the safe distance required for stopping increases uniformly with speed. This assumption is unrealistic. Let us examine the actual distance that an automobile requires to come to a complete stop.

EXAMPLE

For an average-size late-model station wagon, minimum stopping distance is $y_{stop} = 164$ ft when traveling at initial speed $v_o = 55$ mi/hr $= 80.7$ ft/s. The final speed $v = 0$ ft/s. What is the acceleration?

$$2 \, ay = v^2 - v_o{}^2$$

$$a = \frac{-v_o{}^2}{2 \, y_{stop}}$$

$$a = \frac{-(80.7 \text{ ft/s})^2}{2.0 \, (164 \text{ ft})}$$

$$a = -19.8 \text{ ft/s}^2$$

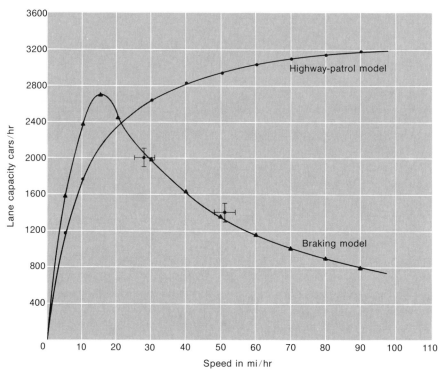

FIGURE 2-19
Freeway capacity as a function of traffic speed. ⊙ Highway-patrol model. △ Braking model.

This is a large negative acceleration (deceleration), some 60% of the acceleration due to gravity, and is typical of the optimum performance of the modern automobile. A person not wearing a seatbelt could easily be thrown forward.

Let us now consider the stopping distance at different speeds, assuming a maximum stopping deceleration $a_{stop} = 19.8 \, \text{ft/s}^2$. Considering the deceleration as a positive quantity, the stopping distance increases with the square of the velocity.

$$y_{stop} = \frac{v_o^2}{2a_{stop}}$$

Figure 2-20 shows the stopping distance as a function of speed, assuming a reaction time of 0.75 between the sighting of a danger and application of the brakes, and assuming a deceleration of about $20 \, \text{ft/s}^2$.

In the braking model for the capacity of a freeway lane, cars are spaced by the stopping distance y_{stop}. By then choosing some value of the deceleration a_{stop} we approximate our model for the capacity of a freeway lane to the observed data. If the car has a length l_o we allow an additional safe distance y_{stop}. As Figure 2-21 shows, the total distance reserved for one car is

$$l = l_o + y_{stop}$$

$$l = l_o + \frac{v^2}{2a_{stop}}$$

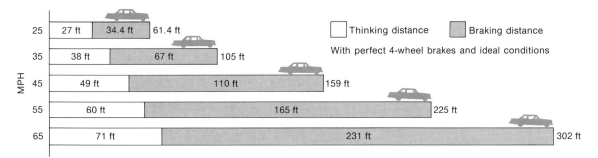

FIGURE 2-20
Stopping distance: from eye to brain to foot to wheel to road. Adapted from drivers' manual published by California State Division of Motor Vehicles.

FIGURE 2-21
The braking model. The total distance occupied by the car is the length of the car l_o plus the stopping distance.

The number of cars N on a segment of roadway of length L is

$$N = \frac{L}{l} = \frac{L}{l_o + (v^2/2a_{stop})}$$

The time T required for a string of cars to pass the observation point is just

$$T = \frac{L}{v}$$

Finally, using this braking model we can determine the capacity of a freeway lane

$$C_{brake} = \frac{N}{T} = \frac{v}{l_o + (v^2/2a_{stop})}$$

The braking model predicts freeway lane capacity quite differently. Because the stopping distance is short at low speeds, the lane capacity increases rapidly with velocity v,

until, beyond 25 mi/hr, the stopping distance required increases rapidly and the lane capacity begins to fall. This is also shown in Figure 2-20. The lane capacity is actually greater at 55 mi/hr than at 70 mi/hr.

Comparison of Models

Clearly, the highway-patrol model and the braking model give quite different predictions concerning the capacity of a freeway lane. Which model better describes the physical situation can be determined by experiment.

It is our experience that when traffic volume is light, the traffic speeds along, limited only by the conditioning encouraged by your friendly highway patrol: As rush hour approaches, the volume of traffic increases and its speed slows. Students have made observations near Boston during rush hour, when the roads are operating near capacity. One of the central arterial highways has a lane capacity of 2000 ± 100 cars/hr at an average speed of 28 ± 3 mi/hr. On Highway 128, the belt corridor around the city, the traffic load was 1400 ± 100 cars/hr per lane at an average speed of 51 ± 3 mi/hr.

These two observational data points can be plotted on our graph (Figure 2-13) of traffic-lane capacity as a function of traffic speed. These points lie nowhere near the highway-patrol model. On the other hand, we get a good agreement between our model and our data if we assume a nominal car length $l_o = 15$ ft, and a deceleration $a_{stop} = 15$ ft/s^2. This deceleration is equivalent to a panic stop, under perfect road conditions and with excellent tires and skillful application of the brakes to avoid skidding or swerving out of control. The large deceleration needed to fit our braking model to our traffic observations indicates that during rush hour most of us are pushing our luck. The frequency of rear-end collisions and multiple-car pileups on major freeways indicates that all of us are driving closer together than is safe at our customary speeds.

Another prediction of the braking model can be confirmed from experience with rush-hour driving. Suppose the lane capacity has crept upward to 2500 cars/hr and the speed has slowed to 18 mi/hr. If the traffic pressure increases to some 2550 cars/hr per lane, the speed drops to 15 mi/hr. Traffic congestion and instability brought on by lane changing, merges from on ramps, etc., will then further reduce the speed until the traffic grinds to a virtual standstill, with stop-and-go creeping along at a few miles per hour. Even when the pressure of traffic lets up the cars will continue to creep along at perhaps 7.5 mi/hr until the head of the line begins to thin out and speed up. The braking model predicts that at a capacity of 2000 cars/hr per lane, two different speeds are possible: 7.5 mi/hr and 28 mi/hr.

Review

Give the significance of each of the following terms. Where appropriate give two examples.

distance interval	magnitude	component method
time interval	direction	relative velocity
average speed	position vector	gravity
frame of reference	displacement	acceleration
point of origin	velocity	uniform acceleration
Cartesian coordinate system	scalar	instantaneous velocity
vector	head-to-tail method	model
	component	

Further Readings

Arthur Beiser, *Essential Math for the Sciences: Analytic Geometry and Calculus* (New York: McGraw-Hill, 1969). An excellent review of the basics.

Stanley Plagenhoef, *Patterns of Human Motion: a Cinematographic Analysis* (Englewood Cliffs, N.J.: Prentice-Hall, 1971). This is a fascinating study of human motion in sports, based on the analysis of motion pictures.

G. Polya, *How to Solve it: A New Aspect of Mathematical Method* (Princeton, N.J.: Princeton University Press, 1971). Problem solving discussed by a brilliant contemporary mathematician.

Problems

POSITION IN TIME

1 In Figure 2-1 Ken Rosewall is serving a tennis
ball. The drawing represents a tennis racket 67.5 cm
in length.
(a) Where is the tennis racket traveling the fastest?
(b) Where is the tennis racket traveling the slowest?
(c) How fast is the racket traveling at impact between frames 22 and 23?

2 In 1968, at the Mexico City Olympic Games,
James Hines set the world's record time for the
100 meter dash, 9.9 s. What was his average speed
in m/s, km/hr, mi/hr?

3 The land speed record for a wheeled vehicle was
achieved by Gary Gabelich at Bonneville Salt Flats,
Utah on 23 October 1970 in his automobile the
Blue Flame. He traveled a distance of 1.000 mi in
5.829 s. On a separate trial he traveled a distance of
1.000 km in 3.543 s.
(a) What was his average speed in m/s, km/hr,
mi/hr on each trial?
(b) Which trial was at a faster average speed?

4 A Boeing 747 leaves the runway with a takeoff
speed of about 165 knots and hits the runway with
a landing speed of about 125 knots. One knot is one
nautical mile per hour. One nautical mile is equal
to 1852 m. What is the plane's speed in m/s,
km/hr, and (statute) mi/hr? (a) at takeoff? (b) at
landing?

5 A child running across a playground at 2.0 m/s is
chased by a parent traveling in the same direction
at 4.0 m/s. The parent is initially 20 m from the
child. How long does it take to reach the child?

6 A parent and a child are at opposite ends of a field
40 m long. The child runs toward the parent at a
speed of 2.0 m/s, the parent toward the child at
4.0 m/s. Where do they meet?

7 A highway patrolman observes a sportscar speeding
along the freeway at 75 mi/hr. By the time the pa-

trolman is in pursuit at a speed of 90 mi/hr, the car
is one mile ahead. How long does it take the patrol
car to overtake the speeder?

8 Two race cars are traveling around a track 1.0 km
in circumference. One car is traveling at 150 km/hr.
The second car is traveling at 160 km/hr. How
many laps will they travel before the faster car
completes one lap more than the slower car?

MOTION IN THREE DIMENSIONS

9 San Francisco is about 390 km north and 350 km
west of Los Angeles. Neglecting curvature of the
earth and treating this portion as a flat plane,
(a) How far is San Francisco from Los Angeles?
(b) What is the heading on the compass from San
Francisco to Los Angeles?

10 Philadelphia is about 110 km north and 170 km east
of Washington, D.C.
(a) How far is Philadelphia from Washington,
D.C.?
(b) What is the compass heading from Washington,
D.C., to Philadelphia?

11 Baltimore is 44 km North of Washington, D.C. in a
direction 40° east of north.
(a) What is the distance between Baltimore and
Washington, D.C.?
(b) How far east is Baltimore?

12 Paris is about 175 km south of London at an angle
of 31° east of south.
(a) What is the distance between the two cities?
(b) How far south is Paris from London?

13 A sailboat travels 10 km in a direction of 37° east
of north, then 14.1 km due south. What is the final
position? Work the problem by the head-to-tail and
component methods.

14 A person travels 14.1 m northeast, then 15 m west, and finally 10 m south. What is the net displacement from the starting point? Work the problem by the head-to-tail and component methods.

15 A boat crosses a river 100 m wide, with a flow velocity of 6 km/hr. The boat travels with a velocity of 8 km/hr relative to the water in a direction perpendicular to the flow of the river.
(a) What is the boat's net velocity?
(b) Where will the boat land on the other side?
(c) How long will it take the boat to cross the river?

16 A boat crosses a river 100 m wide, with a flow velocity of 3 km/hr. The boat can travel with a speed of 8 km/hr relative to the water. The boat pilot wishes to head in such a direction that it will land exactly on the other side.
(a) In what direction must the boat head?
(b) How long will it take to cross the river?

17 An airplane flying at an altitude of 8500 meters encounters a prevailing wind from the west at 240 km/hr. The airplane is headed northwest with a speed of 900 km/hr relative to the air. What is the effective ground velocity?

18 An airline passenger has attached a key to a thread and allows it to hang in gravity. As the airplane accelerates, the key hangs at an angle of 16° to the vertical.
(a) Draw a vector diagram showing the hanging key, the acceleration due to gravity, and the acceleration of the airplane.
(b) What is the acceleration of the airplane?
(c) If the airplane starts its acceleration at a speed of 20 mi/hr, what is its speed after 20 s?

19 An airplane flying at an altitude of 8000 meters encounters a prevailing wind from the west at 200 km/hr. The pilot wishes to fly due north and has an air speed of 800 km/hr.
(a) What heading must the plane fly?
(b) What is the ground speed?

20 An automobile is traveling at 20 mi/hr a distance of 20 ft behind another car. The lead car slows to 15 mi/hr. How long before the two cars collide?

GRAVITY AND ACCELERATION

21 A sports car will accelerate from 0 to 100 km/hr in 9.5 s.
(a) What is the average acceleration?
(b) How far does it go?

22 A rapid transit train accelerates from 0 to 50 km/hr with an acceleration of 1.2 m/s².
(a) How long does it take to reach this speed?
(b) How far does it go?

23 A BART train accelerates from 15 m/s to 30 m/s at an acceleration of 1.0 m/s².
(a) How long does it take to reach this speed?
(b) How far does it go?

24 A Triumph TR-7 sports car will travel 0.40 km from a standing start in 18.5 s.
(a) What is the average acceleration?
(b) What is the final speed, assuming a constant average acceleration? (The observed final speed is 34.2 m/s. Why might your answer be different?)

25 A person throws a ball straight up into the air with a velocity of 25 m/s.
(a) How fast will the ball be going after 3 s?
(b) Where will the ball be after 3 s?
(c) What is the maximum height reached by the ball?

26 To thrill crowds, divers near Acapulco regularly dive from a ledge 36 m above the ocean.
(a) How long does it take them to reach the water?
(b) How fast are they traveling when they reach the water in m/s, km/hr?

TRANSPORTATION PROBLEMS

27 A BART train traveling at 129 km/hr will come to a complete stop in a distance of 594 m. Assuming uniform braking,
(a) What is the average acceleration?
(b) How long will it take to come to a complete stop?

28 A Ford Mustang traveling at 129 km/hr (80 mi/hr) will come to a stop in a distance of 86.3 m.

(a) What is the average deceleration?

(b) At this same deceleration how far will the car travel from the point where the brakes are applied if the car is traveling at 112 km/hr (70 mi/hr)? 88 km/hr (55 mi/hr)?

(c) If it takes 0.75 s to recognize danger and apply the brakes, what is the total distance traveled in the three above cases?

29 Four lanes of traffic in one direction are carrying 5200 cars/hr. Using the braking model shown in Figure 2-21, what is the maximum speed that the traffic will travel?

30 If the traffic in the previous problem were smoothly merged into three lanes,

(a) What would be the traffic load per lane?

(b) What would be the speed of the traffic?

(c) What would be the speed of the traffic if it merges smoothly to two lanes?

31 Traffic on the Golden Gate Bridge during the morning rush hour reaches as much as 6700 cars/hr on four lanes. What is the maximum speed predicted by the braking model? (Due to congestion at the toll plaza the actual speed is about 20–25 mi/hr.)

32 If one of the bridge lanes of the previous problem were blocked by an accident, leaving only three lanes to carry rush hour traffic, what would be the speed of the traffic flow?

ADDITIONAL PROBLEMS

33 An automobile crashes into a solid wall at 55 mi/hr. From what height would the auto have to be dropped to reach the same speed and suffer the same impact?

34 An automobile striking a concrete wall at 50 km/hr decelerates at $20\,g = 200$ m/s^2. How far does the car travel before it stops? How far does it travel if it is initially going 100 km/hr?

35 A Toyota Corolla SR-5 automobile will accelerate in first gear from 0 to 11.6 m/s in 4.0 s; in second gear to 20.6 m/s after a total elapsed time of 9.0 s;

in third gear to 30.8 m/s after a total elapsed time of 22.5 s.

(a) What is the average acceleration in each gear?

(b) Assuming uniform acceleration, what is the total distance covered in the 22.5 s?

36 Big Daddy Don Garlits, driving a rear-engine Swamp Rat dragster, set the record for the maximum velocity over a quarter mile from a standing start at the Gainesville Dragway in Florida on 19 March 1972. He achieved a top speed of 243.9 mi/hr from a standing start in just 6.175 s.

(a) Determine the average acceleration by two different methods. The two answers may not quite agree. Why?

(b) Compare your answer with the acceleration due to gravity.

37 In performance trials a Dodge Polara Custom automobile will come to a panic stop from 65 mi/hr in a distance of 230 ft from the point where the brakes are applied.

(a) What is the braking deceleration?

(b) Suppose this car has just come around a curve at 55 mi/hr and the driver sees that both lanes of traffic ahead are blocked. It takes him 0.75 s to react and apply the brakes. If the braking deceleration is the same, what is the total distance required to stop? What if the car is traveling at 80 mi/hr?

38 In a 400 m race a runner accelerates from rest at 5.20 m/s^2 for 1.78 s. He then finishes the race at constant speed.

(a) How fast is he going and how far has he traveled at the end of 1.78 s?

(b) How long does it take him to complete the remainder of the race?

(c) What is his total time and average speed? The record for the 400 m race is 43.8 s set by Lee Evans in Mexico in 1968.

39 We can make a model of the performance of a rapid transit train by assuming a constant acceleration and deceleration. We can break the journey into four parts: acceleration, travel at top speed, deceleration, and station wait. Suppose the train accelerates at a constant 4.0 km/hr-s, up to a normal op-

erating speed of 109 km/hr (68 mi/hr). It travels at top speed until it decelerates at 4.0 km/hr-s, arriving in the station at a dead stop. The total distance traveled is 3.5 km. The train then waits 20 s to take on and discharge passengers.

(a) What is the minimum time and distance to accelerate to top speed? to stop?

(b) How far and for how long will the train travel at top speed if the stations are 3.5 km apart?

(c) What is the average speed of the train including station wait time?

40 Suppose the top speed of the train in the previous problem is increased to 129 km/hr (80 mi/hr). Repeat the problem.

(a) How much time is saved by traveling at the higher speeds?

(b) How much has the average speed increased?

41 A BART train goes a distance of 0.58 km between the Embarcadero and Montgomery stations. The train accelerates and decelerates at 4.0 km/hr-s.

(a) What is the maximum speed reached by the train?

(b) What is the time required to complete the distance and the average speed, including a 20 s station wait time?

42 A modern passenger elevator travels from the main lobby to the twenty-seventh floor a total distance of 108.5 m. The elevator accelerates and decelerates at 1.2 m/s² up to a top speed of 5 m/s. How long will it take to reach the twenty-seventh floor?

43 The elevator accelerates and decelerates at 1.2 m/s². It travels between two floors a distance of 4.1 m.

(a) What is the maximum speed reached on this short run?

(b) How long will it take to complete the distance?

3

Newton's Laws of Motion

*If I have seen further than others, it is because
I have stood on the shoulders of giants.*

Sir Isaac Newton

The modern era of science begins with Isaac Newton, who, with the publication of the *Principia* in 1687, set a new style for scientific analysis. Before him such scientists as Kepler and Copernicus had published long, discursive treatises that mixed speculation with observation; and Galileo, although a brilliant observer and experimenter, had put most of his scientific arguments in the classic form of the Socratic dialogues of Plato—a mode more suited to philosophical speculation than to science. Newton's model of discourse was not the Socratic dialectic but the crystal-clear logic of Euclid's geometry.

Born Christmas Day 1642 (the same year that Galileo died) Newton was considered a bright child and had wide-ranging interests, running from sundials and kites to models of windmills and water clocks. In 1665 Newton took his bachelor's degree at Cambridge University, with no particular distinction. In 1665 the great plague of London became so severe that some 10% of its population died. In the autumn of that year the university was closed, and remained so until the spring of 1667. For those 18 months Newton, then about 23 years old, retired to his country home at Woolsthorpe, where began perhaps the most creative period of his career in mathematics and natural philosophy.

At Woolsthorpe he formulated his theory of motion and laid the basis of his mechanics of the solar system. He also developed the theory of fluxions (what we now know as differential and integral calculus), a type of analysis developed independently and simultaneously by the German mathematician and philosopher Gottfried Wilhelm Leibniz, whose notation and method were ultimately adopted.

After his return to Cambridge in 1667, Newton was primarily interested in optics. He invented the first reflecting telescope, for which he meticulously ground his own mirror. Soon embroiled in distasteful disputes over the nature of light with his senior colleagues in the Royal Society of London, Newton declined to publish the results of his investigations or to answer his critics.

Nonetheless he continued to ponder scientific problems, a major one of his times being planetary motion. In 1684 the famous astronomer Edmund Halley asked Newton for his thoughts on the subject.

He asked him what would be the orbit described by a gravitational force obeying an inverse square law. Newton immediately answered *an elipse.* "Struck with joy and amazement," to quote the record of the meeting, Halley asked him how he knew it. Why, replied he, I have calculated it; and being asked for the calculation, he could not find it, but promised to send it to him.[1]

Shortly thereafter Newton sent to Halley two different proofs. His interest in planetary motion revived, Newton recorded the principles of his mechanics in a little book and near the end of the year sent a copy to Halley. With Halley's encouragement he set to work on one of the most influential scientific books ever written, *Philosophiae naturalis principia mathematica* (*The Mathematical Principles of Natural Philosophy*). The *Principia* was published in 1687, a full 20 years after the inception of his ideas at Woolsthorpe.

NEWTON'S LAWS

The *Principia* is a simple and elegant scientific document. In a little more than a page Newton states the fundamental axioms or building blocks of his theory, traditionally identified as **Newton's laws of motion.** From his premises everything else follows with powerful logic. In the remainder of the first volume of the *Principia*, the consequences of his fundamental axioms are explored.

Newton stated his laws quite simply.[2]

Newton's First Law

Every body continues in its state of rest, or of uniform motion in a right [straight] line, unless it is compelled to change that state by forces impressed upon it.

This states the principle of inertia. The planets would fly off in a straight line never to be seen again were it not for a force from the sun compelling them to constantly change their direction of motion.

---

[1] Quoted from E. N. Da.C. Andrade, *Sir Isaac Newton: His Life and Work,* (New York: Doubleday Anchor Books, Science Study Series S42, 1958), p. 68.

[2] Sir Isaac Newton, *Principia, Vol. 1, The Motion of Bodies,* Motte's translation revised by Florian Cajori (Berkeley: University of California Press, 1966).

Newton's Second Law

The change of motion is proportional to the motive force impressed; and is made in the direction of the right [straight] line in which that force is impressed.

The change of motion is acceleration, which is a vector. The acceleration is in the direction of the applied force, so that force must also be a vector quantity. If several forces act on a body, it is the vector sum of the different forces, the **net force,** that acts to accelerate the body.

Newton's Third Law

To every action there is always opposed an equal reaction; or, the mutual actions of two bodies upon each other are always equal, and directed to contrary parts.

If you push on an object with your hand, the object pushes back with an equal and opposite force.

CONSEQUENCES OF THE SECOND LAW

The change of motion is proportional to the motive force impressed. . . .

Central to Newton's laws of motion is his idea of force. **Force** *is what acts on a body to change its velocity; that is, to produce acceleration.* The action of a force is quantitatively described by the second law, which we discuss before examining the consequences of the first and third laws of motion.

The acceleration given by Newton's second law is in the same direction as, and is proportional to, the net force applied to the body. From our experience, however, we know that the same force applied to pushing a child's tricycle, a Volkswagen, or a heavy Chevrolet gives different accelerations.

The ratio of the net force applied to a body and the acceleration of that body is the body's **inertial mass,** summarized in the equation:

$$\text{net } \mathbf{F} = m\mathbf{a} \qquad (3\text{-}1)$$

Force and Mass

The fundamental SI unit of inertial mass is the kilogram. Hence, if the kilogram, meter, and second are defined, Newton's second law then gives a definition of a unit of **force:** that which is required to accelerate an inertial mass of 1 kilogram at a rate of one meter per second squared. This unit is the **newton, N.**

$$\mathbf{F} = m\mathbf{a}$$

$$\mathbf{F} = (1 \text{ kg})(1 \text{ m/s}^2) = 1 \text{ kg-m/s}^2$$

$$\mathbf{F} = 1 \text{ newton (N)}$$

EXAMPLE

A European sports car with driver has a mass of 1540 kg. The rear wheels exert a net forward force on the road of 7 000 N. What is the car's acceleration?

$$\text{net } F = ma$$

$$a = \frac{\text{net } F}{m} = \frac{7\,000 \text{ N}}{1\,540 \text{ kg}} = 4.5 \frac{\text{N}}{\text{kg}}$$

$$a = 4.5 \text{ m/s}^2$$

EXAMPLE

A fully loaded rapid transit train has four cars with a total mass of 130 000 kg. The train accelerates uniformly at 1.2 m/s². What net force is required to accelerate the train?

$$\text{net } F = ma$$

$$\text{net } F = (1.30 \times 10^5 \text{ kg})(1.2 \text{ m/s}^2)$$

$$\text{net } F = 1.56 \times 10^5 \text{ kg} - \text{m/s}^2$$

$$\text{net } F = 1.56 \times 10^5 \text{ N}$$

Mass and Weight

An object at the earth's surface experiences constant downward acceleration because of the earth's gravitational attraction, g.[3]

$$g = 9.80 \text{ m/s}^2 = 32 \text{ ft/s}^2$$

From Newton's second law we can calculate the force required to give this acceleration:

$$W = mg \qquad (3\text{-}2)$$

At the outset we must clarify the difference between mass and weight. *The* **weight** *of an object is the downward force due to gravitational attraction.* This force depends on the local conditions of gravity. Normally we refer to an object's weight as the downward force at the earth's surface. *The* **mass** *of an object is a quantitative measure of its* **inertia,** *its resistance to acceleration.* It is related to the quantity of matter and is independent of local gravitational conditions. The weight of an object is proportional to its mass. Whether one kilogram of matter is on the surface of the earth, on the moon, on Mars, or in outer space, it possesses the same inertia; its weight, however, will be very different in those places.

EXAMPLE

What is the weight of a 1.0 kg mass at the earth's surface? (This force is sometimes referred to as a **kilogram weight,** kg-wt.)

$$W = mg = (1 \text{ kg})(9.8 \text{ m/s}^2) = 9.8 \text{ kg-m/s}^2 = 9.8 \text{ N}$$

Weight is a force!

CGS Units The system of units based on the centimeter, gram,[4] second—the **CGS system**—is still widely used in

---

[3] The "standard value" for $g = 9.806\,65$ m/s², as adopted by the General Conference in 1901. This corresponds to $g = 32.174$ ft/s².

[4] Although gram is widely abbreviated g, in this book it is abbreviated gm to avoid confusion with g, the acceleration due to gravity. For other units of mass the abbreviations kg (kilogram), mg (milligram), etc., are used where no confusion exists. The metric ton has the name tonne (t) and is equal to 10^3 kg.

biology, medicine, and some branches of physics, though these units are gradually being replaced by SI units. A unit of force can be defined using Newton's second law and CGS units. If a mass of one **gram** is accelerated at a rate of 1 cm/s^2, then the force required is one **dyne.**

$$F = ma$$

$$F = (1 \text{ gm})(1 \text{ cm/s}^2) = 1 \text{ gm-cm/s}^2$$

$$F = 1 \text{ dyne.}$$

The SI equivalent to a force of one dyne can be determined from the definition

$$1 \text{ dyne} = 1 \frac{\text{gm-cm}}{\text{s}^2} \times \frac{1 \text{ kg}}{10^3 \text{ gm}} \times \frac{1 \text{ m}}{10^2 \text{ cm}}$$

$$1 \text{ dyne} = 10^{-5} \text{ kg-m/s}^2 = 10^{-5} \text{ N}$$

EXAMPLE

What is the force of gravity acting on a 1 gm mass at the earth's surface? This is sometimes referred to as a gram weight, gm-wt.

$$W = mg$$

$$W = (1 \text{ gm})(980 \text{ cm/s}^2)$$

$$W = 980 \text{ gm-cm/s}^2 = 980 \text{ dynes}$$

Weight is a force!

Engineering Units Coexisting with the metric units over the past century have been the practical engineering units: the foot, the second, and the pound (a unit of force).

To emphasize that the pound, as a unit of weight, is a unit of force, we abbreviate pound, lbf. When we weigh ourselves, and find that our weight is 120 lbf, or 160 lbf, we are actually measuring the force of gravity acting upon us. The **pound** measure is defined in terms of the force of gravity acting on a one-kilogram mass: A 1 kg mass weighs 2.204 62 lbf; a 1 lbf weight has a mass of 0.4536 kg. (See Table 3-1.)

TABLE 3-1
Mass, gravitational acceleration, and weight in metric and engineering units.

Mass m	Gravity g	Weight mg
1 kg	9.8 m/s^2	9.8 newtons
1 gm	980 cm/s^2	980 dynes
1 slug	32.17 ft/s^2	32.17 lbf
0.0311 slug	32.17 ft/s^2	1.00 lbf

EXAMPLE

What is the weight of one pound in newtons? A 1 lb weight (lbf) has an equivalent mass of 0.4536 kg. The weight is given by mg. $W = 1.0 \text{ lbf}$. In metric units the weight is

$$W = mg = (0.4536 \text{ kg})(9.807 \text{ m/s}^2)$$

$$W = 4.448 \text{ N}$$

Since the weight is the same regardless of units,

$$1 \text{ lbf} = 4.448 \text{ N}$$

We can define a unit of mass in an engineering unit, the **slug,** using Newton's second law, and solving for the mass:

$$M = \frac{F}{a}$$

By analogy with the SI units, we can define a unit of mass such that if a force of one pound, $F = 1 \text{ lbf}$, is applied to a unit mass, there is an acceleration of $a = 1 \text{ ft/s}^2$.

$$m = \frac{1 \text{ lbf}}{1 \text{ ft/s}^2}$$

$$m = \frac{1 \text{ lbf-s}^2}{\text{ft}} = 1 \text{ slug}$$

From Newton's second law, again, the weight of an object is given by

$$W = mg$$

where the acceleration due to gravity at the earth's surface is given by $g = 32.17 \text{ ft/s}^2$.

EXAMPLE

What is the weight of a 1 slug mass?

$$W = mg$$

$$= (1 \text{ slug})(32.17 \text{ ft/s}^2)$$

$$= 32.17 \text{ slug-ft/s}^2$$

$$W = 32.17 \text{ lbf}$$

CONSEQUENCES OF THE THIRD LAW

To every action there is always opposed an equal reaction. . . .

If a person throws a ball, hand A pushes on the ball B with a force F_{AB}; the ball exerts an equal and opposite force on the hand F_{BA} (see Figure 3-1). Only one force,

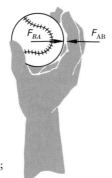

FIGURE 3-1
Action-reaction.
The hand pushes on the ball;
the ball pushes on the hand.

F_{AB}, is acting on the ball B. Only one force, F_{BA}, is acting on the hand A. These forces are equal and opposite:

$$F_{AB} = -F_{BA}$$

The negative sign indicates the force of the ball is in the opposite direction.

EXAMPLE

Block A, with a mass of 4 kg, sits atop block B, with a mass of 5 kg, which in turn sits on a table (see Figure 3-2). What are the forces on block A, block B, and the table?

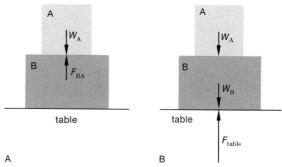

FIGURE 3-2
Left: forces acting on A. Right: forces acting on B.

The weight of block A is downward with a magnitude

$$W_A = M_A g = 39.2 \text{ N}$$

This force must be balanced by the force of block B pushing upward on block A, for block A remains stationary.

$$\text{net } F = W_A - F_{BA} = M_A a = 0.$$

The negative sign simply indicates the upward direction of the force. Since the block remains at rest, the acceleration is zero, so that the magnitude of the reaction force is

$$F_{BA} = W_A$$

The second block pushes back with a reaction force that is equal in magnitude and opposite in direction to the weight of the first block.

The forces acting on block B are its own downward weight,

$$W_B = M_B g = 49 \text{ N}$$

The downward force of block A pushing on block B, F_{AB},

$$F_{AB} = W_A$$

and the upward force of the table, F_{table}. Because block B is not accelerating,

$$\text{net } F_B = W_A + W_B - F_{table} = M_B a = 0$$

The upward force of the table is

$$F_{table} = W_A + W_B$$

Tension

Suppose an object is being pulled upward by a cable or rope. (We consider that the cable or rope itself has no mass.) The rope pulls on the object with some force T. The object reacts on the rope with an equal and opposite force T. This develops in the rope a tension T that is transmitted along the rope from the point at which the force is applied, to the point where the force acts. To produce tension in a rope we must have such action-reaction pairs of forces.

EXAMPLE

A fully loaded elevator car has a mass of 4600 kg. The elevator is supported by a set of eight cables that lift the elevator car vertically.
(a) What is the tension in the cables when the car is at rest?
(b) What is the tension in the cables when the car is accelerating upward at 1.2 m/s²?
   The forces acting on the elevator car are the weight of the car downward with a magnitude $W = Mg$, and the upward tension T in the cables (see Figure 3-3). When the car travels at uniform speed or stands at rest,

FIGURE 3-3
Forces acting on an elevator car.

its acceleration is zero. From Newton's second law we have that

$$\text{net } F = T - Mg = ma$$

where the upward direction is now considered positive. If the acceleration is zero, then

$$T = Mg$$
$$= (4600 \text{ kg})(9.8 \text{ m/s}^2)$$
$$T = 4.5 \times 10^4 \text{ N}$$

If the car is accelerating upward at 1.2 m/s², then the tension in the cables is given by

$$T = Mg + Ma$$
$$= (4600 \text{ kg})(9.8 \text{ m/s}^2) + (4600 \text{ kg})(1.2 \text{ m/s}^2)$$
$$T = 5.05 \times 10^4 \text{ N}.$$

To give a vertical acceleration, the tension must be greater than that necessary to sustain the weight of the car. What is the tension required if the car is accelerating downward at 1.2 m/s²?

Analyzing a System

A system is an orderly arrangement of objects connected or associated to form a complex unity. If we know enough about a system we can make some predictions about how it will perform. Most of the systems of concern to physi-

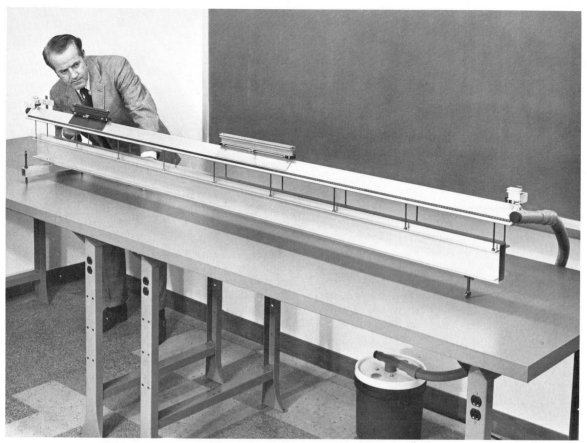

FIGURE 3-4
An air track provides an excellent demonstration of Newton's laws of motion by greatly reducing the effects of friction. Courtesy of the Ealing Corporation.

cists are generally simpler than biological systems, and it is easier to isolate those factors that influence an experiment. However, even physical systems may be sensitive to external influences. Electronic equipment and laser experiments can be extremely sensitive to mechanical vibration. Particle detectors and photographic emulsions are extremely sensitive to stray light.

To analyze a system we must isolate those elements that are within the system, or that influence it, from those that are beyond the system. Our skill at isolating relevant factors can mean the success or failure of our analysis. In this book we shall analyze simple systems, perhaps oversimplified systems. We hope to identify the major influences and describe the gross features of natural phenomena. We will often ignore more than we consider.

Air Tracks Newton's laws of motion can be applied to a complete system or to any part of a system. Consider a system often used in student laboratory work: the air track, which was invented by physicists to demonstrate Newton's first law, namely that objects tend to go in a straight line at uniform speed unless influenced by an external force. If a book is slid across a table or a wheel rolls across a floor, there is an interaction with the surface, known as friction, that tends to cause the object to stop.

The air track (as shown in Figure 3-4) avoids this difficulty by supporting the moving object, the car, on a cushion of air, so that it travels without resistance. This provides a nearly ideal system for studying Newton's laws. We ignore that a certain amount of scraping can occur to slow the car, that the track may not be perfectly level (it might dip in the middle), or that the car requires a certain amount of force (however small) to push aside the air along its path. In our analysis of the following example we ignore these imperfections. We shall consider the ideal air track. It is left to the student in the laboratory to determine whether this analysis is sufficient.

EXAMPLE

On a perfectly level, frictionless air track rides an air-track car with a mass $m_1 = 0.38$ kg. A string with negligible mass is attached to the car and extends over a frictionless pulley at the far end of the track. A second mass $m_2 = 0.10$ kg is attached to the string and hangs by the force of gravity. The string transmits a tension T from the hanging mass and from the air-track car (see Figure 3-5).

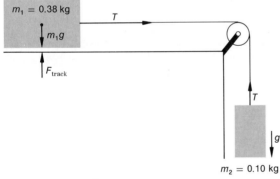

FIGURE 3-5
A falling weight pulls an air-track car.

(a) What is the acceleration of the car and hanging mass?
(b) What is the tension in the string?
   The problem can be divided into two parts to consider each mass separately. The downward weight of the car $m_1 g$ is balanced by the force of the air track pushing upward. We need consider only the net force in the horizontal direction due to the tension, T, in the string: net $F_1 = T$. If we consider the direction along the track toward the pulley as positive then we can write Newton's second law as

$$\text{net } F_1 = m_1 a$$

or

$$T = m_1 a \qquad (1)$$

The weight of mass m_2 is just

$$W_2 = m_2 g$$

and is pulling downward. The tension in the cord T is pulling upward on m_2. Net $F_2 = W_2 - T$. Assuming that the downward direction is positive, and that the cord does not stretch, so that the car and the hanging mass have the same acceleration, Newton's second law gives us

$$\text{net } F_2 = m_2 a$$
$$W_2 - T = m_2 a$$

which, substituting for W_2 and rearranging, can be written

$$m_2 g - T = m_2 a \qquad (2)$$

We have been asked to find both the acceleration a and the tension T. We have two relationships, (1) and (2), which involve these two unknowns T and a. We can therefore solve them for both the acceleration and the tension. By substituting the value for T from equation (1) into equation (2), we have

$$m_2 g - m_1 a = m_2 a$$

Solving for the acceleration we obtain our result.

$$a = \frac{m_2 g}{m_1 + m_2}$$

Substituting the values of m_1, m_2, and g into this result can easily show that the acceleration is $a = 2.04 \text{ m/s}^2$.

Limiting Values Our intuition is often useful for checking the results of analysis. Does our answer seem reasonable? What happens in the limiting cases where the hanging weight is either very large or very small? Does our analysis then give the same answers as our intuition?

If the hanging mass m_2 is much smaller than the mass of the air-track car m_1(for instance, $m_1 = 0.38$ kg and $m_2 = 0.005$ kg) we would expect the more massive car to be accelerated very slowly. In this limit we can neglect m_2 compared to m_1. The acceleration of the car becomes approximately

$$a \cong \frac{m_2 g}{m_1}$$

which approaches zero as m_2 becomes very small.

Were a very large mass (for instance, $m_2 = 30$ kg) hanging by gravity, we would expect this mass to fall nearly at the acceleration of gravity, dragging the air-track car along with it. In this limit we can neglect m_1 compared to m_2 and the acceleration becomes

$$a \cong \frac{m_2 g}{m_2} \cong g$$

This again is in agreement with our physical intuition. It gives us some confidence in our analytic results. The process of taking limits and looking at the predictions is extremely valuable in physics.

EXAMPLE

An elevator in the Transamerica building in San Francisco has a passenger car with an unloaded weight of 7180 pounds, and a designed load of 3000 pounds. The car is balanced by a counterweight of 8680 pounds, so that the lift motor does not have to lift the entire weight of the car (see Figure 3-6). What is the net force that must be applied by the lift motor to accelerate the fully loaded car upward at 4 ft/s²?

To work this problem in engineering units we must determine the mass of the loaded passenger car and the counterweight in slugs. We are given the weight. The mass is

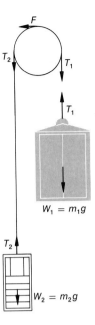

FIGURE 3-6
An elevator car is balanced by a counterweight. The motor provides a force that accelerates both masses.

$$m = \frac{W}{g}$$

where the acceleration due to gravity is 32.17 ft/s²

$$m_1 = \frac{10\ 180\ \text{lbf}}{32.17\ \text{ft/s}^2} = 316\ \text{slugs}$$

$$m_2 = \frac{8\ 680\ \text{lbf}}{32.17\ \text{ft/s}^2} = 270\ \text{slugs}$$

The magnitude of the force F applied by the lift motor gives the difference in tension

$$F = T_1 - T_2$$

By Newton's second law, the tension in rope one is

$$T_1 = m_1 a + m_1 g$$

For the counterweight

$$T_2 = m_2 g - m_2 a$$

where the minus sign indicates that the counterweight is falling when the elevator is rising.

The force F applied by the lift motor is the difference in tension of the two ropes

$$F = T_1 - T_2$$
$$F = (m_1 + m_2)a + (m_1 - m_2)g$$

The force applied by the lift motor must accelerate the entire mass of the system. It must also compensate for the difference between the car's weight and that of the counterweight. In this example the force required is

$$F = (316 + 270 \text{ slugs})(4 \text{ ft/s}^2)$$
$$+ (316 - 270 \text{ slugs})(32.17 \text{ ft/s}^2)$$

$$F = 23\underline{50} \text{ lbf} + 14\underline{50} \text{ lbf}$$

$$F = 38\underline{00} \text{ lbf}.$$

CONSEQUENCES OF THE FIRST LAW

Static Equilibrium

Newton's first law states that a body at rest remains at rest unless acted on by a net external force. As a consequence of this law, if the net force acting on a body is zero, the body remains at rest. This is the principle of **static equilibrium.** The net force acting on the body is the vector sum of all the forces acting on it

$$\text{net } \mathbf{F} = \sum_i \mathbf{F}_i = 0 \qquad (3\text{-}3)$$

If the forces acting on the body are resolved into components, then the net force in any particular direction is also zero.

$$\text{net } F_x = 0$$
$$\text{net } F_y = 0$$
$$\text{net } F_z = 0$$

Let us now consider two examples of static equilibrium.

EXAMPLE

A guy wire attached to the ground and to the top of a utility pole describes an angle of 60° to the horizontal (see Figure 3-7). The horizontal pull of the wires at the top is a tension of $F_{\text{wire}} = 750$ N.

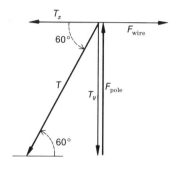

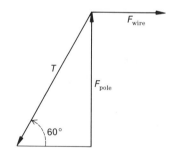

FIGURE 3-7
A utility pole. The forces acting at the top must sum to zero at equilibrium.

(a) What is the tension T in the guy wire?
(b) What upward thrust F at the top of the pole is required to balance the downward pull of the guy wire?

The horizontal pull of the wires T_x must be balanced by the horizontal component of the guy wire tension T_x.

We can resolve the guy wire tension T into horizontal and vertical components

$$T_x = T \cos 60° \qquad T_y = T \sin 60°$$

The top of the pole is in equilibrium so that in the horizontal direction,

$$\text{net } F_x = 0$$

$$F_{\text{wire}} - T_x = 0$$

We can solve for the tension in the guy wire.

$$F_{\text{wire}} = T_x = T \cos 60°$$

$$T = \frac{F_{\text{wire}}}{\cos 60°}$$

The horizontal force of the wires is $F_{\text{wire}} = 750 \text{ N}$ so that

$$T = \frac{750 \text{ N}}{\cos 60°} = \frac{750 \text{ N}}{0.50} = 1500 \text{ N}$$

The upward push at the top of the pole must be balanced by the downward pull of the guy wire. The condition for equilibrium in the vertical direction is

$$F_{\text{pole}} = T_x = T \sin 60°$$

$$= (1500 \text{ N})(\sin 60°) = (1500 \text{ N})(0.866)$$

$$F_{\text{pole}} = 1300 \text{ N}$$

However, a problem cannot always be solved by step-by-step numerical calculation. For some problems we must deal with two equations and two unknowns and solve them as simultaneous equations. The principle remains the same.

EXAMPLE

A 20 kg traffic light is suspended from two wires over the right lane of a roadway (see Figure 3-8). Wire one has tension T_1 and makes an angle $\theta_1 = 30°$ with the horizontal. The second wire has tension T_2 and makes an angle $\theta_2 = 45°$ with the horizontal, as shown in Figure 3-11. With this information let us determine the tension in each wire.

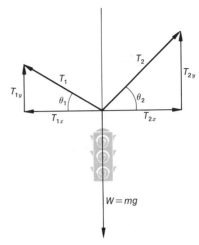

FIGURE 3-8
A 20 kg traffic light suspended over a roadway.

The tension in each wire can be resolved into x and y components.

$$T_{1x} = T_1 \cos \theta_1 \qquad T_{2x} = T_2 \cos \theta_2$$

$$T_{1y} = T_1 \sin \theta_1 \qquad T_{2y} = T_2 \sin \theta_2$$

For equilibrium—that is, for the traffic light to stand still—the forces in the horizontal and vertical directions must cancel to zero.

$$\text{net } F_x = 0$$

$$\text{net } F_y = 0$$

In the horizontal direction

$$\text{net } F_x = T_{2x} - T_{1x} = 0$$

or

$$T_2 \cos \theta_2 = T_1 \cos \theta_1 \qquad (3\text{-}4)$$

If the angles were equal, $\theta_1 = \theta_2$, then the tension in each wire would be equal. However, in this example the angles and the tensions are not equal. Since there are two unknowns, T_1 and T_2, we must find a second equation to solve the problem.

In the vertical direction the net upward force of the two balances the downward weight of the traffic light.

$$\text{net } F_y = T_{1y} + T_{2y} - mg = 0$$

$$T_1 \sin \theta_1 + T_2 \sin \theta_2 = mg \qquad (3\text{-}5)$$

This gives us two relationships, equations 3-4 and 3-5, involving the two unknowns, T_1 and T_2. Because we know the numerical values of the other quantities, in principle our problem is solved. We have done the physics, we need only solve the equations (do some algebra) and plug in some numbers (do some arithmetic).

We can solve this pair of equations by substituting the expression for T_2 from equation 3-4 into equation 3-5. We then obtain

$$T_1 \sin \theta_1 + \left(T_1 \frac{\cos \theta_1}{\cos \theta_2}\right) \sin \theta_2 = mg$$

Solving for T_1 we get

$$T_1 = \frac{mg}{\sin \theta_1 + \sin \theta_2 \left(\dfrac{\cos \theta_1}{\cos \theta_2}\right)}$$

Using numerical values, $m = 20$ kg, $g = 9.8$ m/s^2, $\theta_1 = 30°$, and $\theta_2 = 45°$, we get

$$T_1 = \frac{(20\text{ kg})(9.8\text{ m/s}^2)}{(0.50) + (0.707)\left(\dfrac{0.866}{0.707}\right)}$$

$$= \frac{196\text{ N}}{1.366}$$

$$T_1 = 143\text{ N}$$

The value for T_2 can be obtained from equation 3-4:

$$T_2 = T_1 \frac{\cos \theta_1}{\cos \theta_2}$$

$$= (143\text{ N})\left(\frac{0.866}{0.707}\right)$$

$$T_2 = 176\text{ N}$$

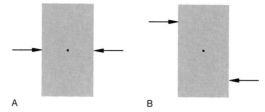

A   B

FIGURE 3-9
A. A pair of forces in translational and rotational equilibrium.
B. A pair of forces that are in translational equilibrium but will cause rotation.

The Action of a Force

An object is said to be in **translational equilibrium** when the sum of the forces in each direction is zero. Figure 3-9A shows two balanced forces acting on an object. These forces act toward a common point. If the two forces are shifted in point of application as shown in Figure 3-9 B the object is still in translational equilibrium, but will tend to rotate. We shall discuss rotation in detail in Chapter 7. For an object to be both stationary and nonrotating it is sufficient that the net horizontal and vertical force be zero, and that all of the forces act toward a common point.

The forces that act on the lower jaw of a carnivor, such as the marten, which feeds on squirrels and other small animals, are shown in Figure 3-10. When the animal tugs at some food, a force **P** is exerted. This is balanced by a

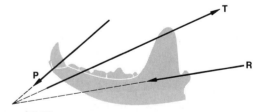

FIGURE 3-10
The forces acting on the lower jaw of the marten act toward a common point in equilibrium. The net torque is zero. From R. McNeill Alexander, *Animal Mechanics* © 1968, University of Washington Press, Seattle, and Sidgwick & Jackson, London. Used with permission.

large tension **T** of the temporalis muscle. A reaction force **R** at the pivot point holds the jaw against the cranium. These three forces act toward a common point and the jaw does not rotate. If the vector sum of the forces is also zero (see Figure 3-11) the jaw will not be accelerated.

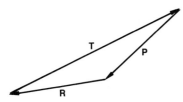

FIGURE 3-11
The net force acting on the jaw is zero in equilibrium.

Often a dog or a lion at the zoo can be seen holding a bone with its paws and gnawing on it with the back corner of the jaw. The animal is using the carnassial teeth well back in the jaw and a different muscle to provide the balancing force. Figure 3-12 shows the forces. The force of the teeth **P′** is balanced by the tension of the temporalis muscle **T** and of the masseter muscle **M**. There is little force at the pivot to the cranium. Again these three forces act toward a common point when the jaw is in rotational equilibrium.

The force of gravity acts at every point on an object that is extended in space. An object suspended from a

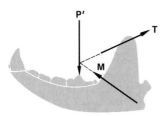

FIGURE 3-12
Gnawing employs a different set of forces to hold the jaw in equilibrium. From R. McNeill Alexander, *Animal Mechanics*. © 1968 University of Washington Press, Seattle, and Sidgwich & Jackson, London. Used with permission.

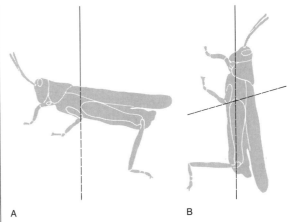

A   B
FIGURE 3-13
The locust's hind legs are attached at the center of gravity. Thus, when the locust jumps the hind legs apply their force at the center of gravity. Redrawn from R. McNeill Alexander, *Animal Mechanics* © 1968 University of Washington Press, Seattle, and Sidgwich & Jackson, London. Used with permission.

single point generally tends to rotate to some preferred position in which the **center of mass** is directly below the point of support. The force of gravity acts as if the entire mass were concentrated at this point. Thus, the center of mass is also referred to as the **center of gravity.** Figure 3-13 shows a locust suspended by a thread from two different points. The locust hangs so that the center of gravity is directly below the point of support. If the locust is suspended from two different points, its center of gravity can be located: It is at the intersection of the lines. Because the locust's large hind legs are attached very near the center of gravity, it can apply a very large jumping force at the center of mass without flipping over.

Uniform Circular Motion

Newton's first law states, in part, that an object will continue in a straight line unless acted on by a force. The moon would fly off into space were it not for the earth's gravitational attraction. The earth would fly off into space were it not for the sun's gravitational attraction. The moon's orbit about the earth and the earth's about the sun are almost circular. What force is required to sustain uniform motion in a circle?

An object in a uniform circular orbit of radius R moving at some constant speed v completes one orbit in a period of time T. The average speed is given by the total distance traveled in one orbit, the circumference $2\pi R$, divided by a given period T.

$$v = \frac{2\pi R}{T}$$

In some shorter interval of time Δt, the object will travel a distance in an arc Δs given by

$$\Delta s = v\Delta t$$

Although the speed remains constant in magnitude, the velocity is constantly changing because the object's direction is constantly changing. At one instant the velocity is $\mathbf{v}_1$; at a time Δt later it is $\mathbf{v}_2$. The change in velocity, represented as a vector, is given by

$$\mathbf{v}_2 = \mathbf{v}_1 + \mathbf{a}\Delta t$$

and is shown in Figure 3-14.

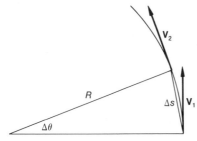

FIGURE 3-14
Uniform circular motion. The velocity is constant in magnitude and changing in direction.

If we consider the velocity at two instants separated by a time interval much smaller than the period $\Delta t \ll T$, then the angle between these two velocities is also quite small. We can draw two triangles, one representing the movement in space, the second the change in vector velocity, as shown in Figure 3-15. These two triangles are similar:

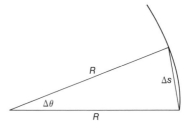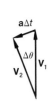

FIGURE 3-15
Uniform circular motion. The change in velocity is similar to the change in position.

Both have two equal sides, the same angle $\Delta\theta$, and the same ratio of the side opposite the angle $\Delta\theta$ to the side adjacent to the angle. The ratio of the magnitude of the displacement Δs to the radius R is the same as the ratio of the magnitude of the change in velocity Δv to the speed v.

$$\frac{\Delta s}{R} = \frac{\Delta v}{v}$$

The arc swept out in time Δt is approximately equal to the base of the triangle for small angles, and is equal to

$$\Delta s = v\Delta t$$

The magnitude of the change in velocity for small angles is given by

$$\Delta v = a\Delta t$$

Using these values we obtain

$$\frac{v\Delta t}{R} = \frac{a\Delta t}{v}$$

Solving for the magnitude of the acceleration we get

$$a = \frac{v^2}{R} \tag{3-6}$$

In the limiting case of a very small arc, this acceleration is pointing radially inward, that is, perpendicular to the direction of the motion. Such acceleration is known as **centripetal acceleration,** and, if uniform, maintains uniform circular motion. The magnitude of this acceleration is just $a = v^2/R$. Newton was able to demonstrate this result. If our object has a mass m, there must then be a **centripetal force** directed radially inward with a magnitude

$$F = mv^2/R \qquad (3\text{-}7)$$

to keep our object in its circular path.

EXAMPLE

In the roller coaster car shown in Figure 3-16, riders accelerate from 0 to 24 m/s in 4.2 seconds. The car then negotiates a 360° loop with a radius $R = 7.0$ m. (a) What is the initial acceleration of the car? (b) What is the minimum speed required to hold the car against the track at the top of the loop?

The initial acceleration is given by

$$a = \frac{\Delta v}{\Delta t} = \frac{24 \text{ m/s} - 0 \text{ m/s}}{4.2 \text{ s}} = 5.7 \text{ m/s}$$

The initial acceleration is 0.6 g, where g is the acceleration due to gravity.

The centripetal force required to keep a car of total mass M in a circular path of radius R is given by

$$F = \frac{Mv^2}{R}$$

At the minimum speed at the top of the track, this centripetal force is provided by the weight of the car.

$$F = Mg = \frac{Mv^2}{R}$$

Solving for the minimum speed we get

$$v = [gR]^{1/2}$$

which is independent of the mass of the car.

$$v = [(9.8 \text{ m/s}^2)(7.0 \text{ m})]^{1/2} = 8.3 \text{ m/s}$$

FIGURE 3-16
The "Tidal Wave" roller coaster car negotiates a 360° loop. Courtesy of Marriott's Great America®, Santa Clara, Calif.

MOMENTUM

In the *Principia* Newton defines the *quantity of motion* as "arising from the velocity and the quantity of matter conjointly." Newton's quantity of motion is what we now call **momentum:** *a vector quantity produced by the multiplication of the vector velocity* **v** *by the scalar mass, m.* The symbol **p** is often used to represent the momentum

$$\mathbf{p} = m\mathbf{v} \qquad (3\text{-}8)$$

Impulse

In Newton's second law the acceleration is proportional to the net applied force. The acceleration is the time rate of change of velocity.

$$\mathbf{a} = \frac{\Delta \mathbf{v}}{\Delta t}$$

Newton's second law can thus be related to the change in momentum

$$\text{net } \mathbf{F} = m\mathbf{a} = m\frac{\Delta \mathbf{v}}{\Delta t}$$

$$\text{net } \mathbf{F} = \frac{\Delta m\mathbf{v}}{\Delta t} = \frac{\Delta \mathbf{p}}{\Delta t}$$

If a net force is applied for a time interval Δt, the momentum changes. The product of the average net force applied multiplied by the time interval is the **impulse.** The applied impulse is equal to the change in momentum.

$$\Delta \mathbf{mv} = (\text{net } \mathbf{F})\Delta t = \text{impulse} \qquad (3\text{-}9)$$

EXAMPLE

A soccer player uses his head to hit a soccer ball of mass 0.425 kg. The ball approaches at 7.5 m/s and leaves at 12.5 m/s. High-speed motion pictures show that the duration of impact is about 23 ms. If the ball approaches and leaves along the same line of motion,
(a) What is the impulse applied to the ball?
(b) What is the average force during the time of impact?

The change in momentum of the ball along the line of motion is equal to the applied impulse

$$\Delta \mathbf{mv} = m\mathbf{v}_{\text{after}} - m\mathbf{v}_{\text{before}}$$

If we assume the positive direction is away from the player, then the magnitude of the change in momentum is

$$\Delta mv = (0.425 \text{ kg})(12.5 \text{ m/s}) - (0.425 \text{ kg})(-7.5 \text{ m/s})$$

$$\Delta mv = 5.31 \text{ kg-m/s} + 3.19 \text{ kg-m/s}$$

$$\Delta mv = 8.50 \text{ kg-m/s}$$

The magnitude of the average force is just the change in momentum divided by the time over which that change takes place.

$$F = \frac{\Delta \mathbf{mv}}{\Delta t} = \frac{8.50 \text{ kg-m/s}}{0.023 \text{ s}}$$

$$F = 3\underline{7}0 \text{ N}$$

This is a fairly large force. If the soccer ball is too heavy, say because it has gotten wet and absorbed moisture, the force against the head can be considerably greater, causing serious injury to the head or neck.
 Compare this force with other familiar forces. What would the force be if the ball were just 0.10 kg heavier?

EXAMPLE

A pitched baseball is traveling at 30 m/s toward the batter. It is struck in a line drive more or less parallel with the ground, traveling at 35 m/s. High-speed motion pictures show that the ball and bat are in contact for about 0.001 25 seconds. The ball has a mass of 0.15 kg. Assuming that the ball is traveling in approximately the same line with the bat before and after impact, what is the average force of the ball against the bat during the time of contact?

If the direction away from the batter is positive, the ball's initial velocity is $v_1 = -30$ m/s, and the final velocity $v_1' = +35$ m/s. The change in momentum is given by

$$\Delta mv = mv_1' - mv_1$$

$$= (0.15 \text{ kg})(+35 \text{ m/s}) - (0.15 \text{ kg})(-30 \text{ m/s})$$

$$= 5.25 \text{ kg-m/s} + 4.5 \text{ kg-m/s}$$

$$\Delta mv = 9.75 \text{ kg-m/s}$$

The average force during the impact is just the change in momentum divided by the time of the impact. The

average force is given by

$$F = \frac{\Delta mv}{\Delta t}$$

$$= \frac{9.75 \text{ kg-m/s}}{0.00125 \text{ s}}$$

$$F = 78\underline{0}0 \text{ N}$$

This is a large force. (No wonder bats are broken if held improperly.) Compare the magnitude of this force with other forces you know. How does this compare to your own weight in newtons?

Collisions

Newton's third law states that for every action there is an equal and opposite reaction. When two objects interact, both change their velocity: The soccer ball reverses its direction and the soccer player's head slows down; the baseball reverses direction and the bat slows down. If two objects with masses m_1 and m_2 interact, there is a force on the first because of the second, $\mathbf{F}_{12}$, and likewise a force on the second because of the first, $\mathbf{F}_{21}$. By Newton's third law, this second force is a reaction force that is equal in magnitude and opposite in direction to the force on the first. $\mathbf{F}_{12}$ and $\mathbf{F}_{21}$ form an action-reaction pair (see Figure 3-17)

$$\mathbf{F}_{21} = -\mathbf{F}_{12}$$

Interaction between two masses need not be direct physical contact; it can be due to gravitational, electrical, or nuclear forces. Interactions between two moving particles, say two billiard balls, are called collisions. During a collision the momentum of each mass changes. The

change in momentum Δmv is given by the average force F_{av} multiplied by the time Δt over which that force acts.

$$\Delta \mathbf{mv} = (\mathbf{F}_{av})\Delta t$$

If two masses interact, the initial velocities can be represented as $\mathbf{v}_1$ and $\mathbf{v}_2$. The final velocities are $\mathbf{v}_1'$ and $\mathbf{v}_2'$. The change in momentum of the first mass is given by

$$\Delta m_1\mathbf{v}_1 = m_1\mathbf{v}_1' - m_1\mathbf{v}_1 = (\mathbf{F}_{12_{av}})\Delta t$$

The change in momentum of the second mass is

$$\Delta m_2\mathbf{v}_2 = m_2\mathbf{v}_2' - m_2\mathbf{v}_2 = (\mathbf{F}_{21_{av}})\Delta t$$

The total change in momentum for the entire system is

$$\Delta m_1\mathbf{v}_1 + \Delta m_2\mathbf{v}_2 = [(\mathbf{F}_{12_{av}}) + (\mathbf{F}_{21_{av}})]\Delta t$$

However, by Newton's third law, $\mathbf{F}_{21_{av}} = -\mathbf{F}_{12_{av}}$. Thus, the total change in momentum for the entire system is zero:

$$\Delta m_1\mathbf{v}_1 + \Delta m_2\mathbf{v}_2 = 0$$

This equation states the principle of **conservation of linear momentum** (also called simply the conservation of momentum): *In any isolated system, the total change in the linear momentum of the system due to interaction within the system is zero.*

The conservation of momentum also implies that the total momentum before an interaction is equal to the total momentum after:

$$(m_1\mathbf{v}_1' - m_1\mathbf{v}_1) + (m_2\mathbf{v}_2' - m_2\mathbf{v}_2) = 0$$

This can be rearranged to give the usual statement of the conservation of momentum.

$$m_1\mathbf{v}_1 + m_2\mathbf{v}_2 = m_1\mathbf{v}_1' + m_2\mathbf{v}_2' \qquad (3\text{-}10)$$

EXAMPLE

A man kicks a soccer ball, as shown in Figure 3-18. The ball has a mass of 0.425 kg and is initially at rest

FIGURE 3-17
An action-reaction pair.

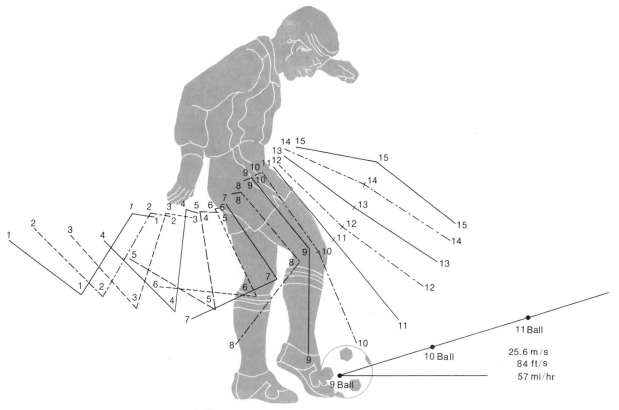

FIGURE 3-18
Kicking a soccer ball straight on. Momentum is transferred
from the foot to the ball. Courtesy of Stanley Plagenhoef.
From *Patterns of Human Motion: A Cinematographic Analysis,*
© 1971. Reprinted by permission of Prentice-Hall Inc., Englewood Cliffs, New Jersey.

$v_2 = 0$. The man's foot has an effective striking mass
of 2.2 kg, an initial velocity of $\mathbf{v}_1 = 18$ m/s, and a final
velocity $\mathbf{v}_1' = 13$ m/s. Assume that all motion near impact is along the same line. What is the ball's final velocity?

Conservation of momentum can be rewritten as

$$m_1(v_1 - v_1') = m_2(v_2' - v_2)$$

Assuming the ball is initially at rest, $v_2 = 0$, we can
solve for the final velocity of the ball v_2':

$$v_2' = \frac{m_1(v_1 - v_1')}{m_2} = \frac{(2.2\ \text{kg})(18\ \text{m/s} - 13\ \text{m/s})}{0.425\ \text{kg}}$$

$$v_2' = \frac{26\ \text{kg-m/s}}{\text{kg}} = 26\ \text{m/s}$$

The ball travels away from the foot with a velocity of
26 m/s. Its speed depends on the foot's speed before
impact and on the foot's effective striking mass. The
foot's striking mass depends on how squarely the
player addresses the ball and upon how rigid the player's body is kept during impact.

OTHER APPLICATIONS
OF NEWTON'S LAWS

The Inclined Plane

In the second corollary to his laws of motion, Newton was quick to point out that the action of a force could be resolved into two perpendicular components. Consider what happens if an object, say a toy car, is placed on an **inclined plane.** The force of gravity acts vertically downward. However, the car cannot fall straight down. It is constrained by the force of the inclined plane to move along the surface of the plane. The force of gravity can be resolved into two components (see Figure 3-19), one parallel to the plane and the other normal (perpendicular) to it.

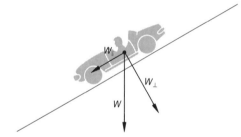

FIGURE 3-19
The force of gravity can be resolved into two components.

EXAMPLE

A toy race car, shown in Figure 3-20, has a mass of about 50 gm and sits on a track about 1 m long inclined at an angle of 37° with the horizontal.
(a) If the car rolls on its wheels without friction, what is the acceleration?
(b) With what force does the car press against the track?
(c) If a string is attached to the car to hold it at rest, what is the tension in the string if it is parallel to the track?

The weight of the car is given by

$$W = mg = (50 \text{ gm})(980 \text{ cm/s}^2) = 4.9 \times 10^4 \text{ dynes}$$

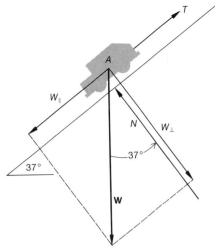

FIGURE 3-20
A toy car sits on a track inclined at an angle of 37°.

The weight of the car can be resolved into two components, one perpendicular to the surface of the track, the second parallel to the track. The components of the car's weight are given by

$$W_\perp = W \cos 37° \qquad W_\parallel = W \sin 37°$$

We can apply Newton's laws in the directions both perpendicular and parallel to the track. The acceleration in the parallel direction is given by the net parallel force.

$$\text{net } F_\parallel = W_\parallel = ma_\parallel$$

Solving for the acceleration

$$a_\parallel = \frac{W \sin 37°}{m}$$

Substituting numerical values we get

$$a_\parallel = \frac{2.94 \times 10^4 \text{ dynes}}{50 \text{ gm}}$$

$$a_\parallel = 588 \text{ cm/s}^2$$

The perpendicular component of the weight of the car is balanced by the normal force of the track push-

FIGURE 3-21
The broad jump. At takeoff the maximum vertical impulse is directed through the center of gravity to gain the maximum possible height while moving forward at top speed. From Geoffrey H.G. Dyson, *The Mechanics of Athletics*, 4th ed., © 1968, Hodder & Stoughton Limited. Used with permission.

ing back so that the acceleration perpendicular to the plane is zero, $a_\perp = 0$.

$$\text{net } F_\perp = ma_\perp$$

$$W_\perp - N = 0$$

$$N = W \cos 37°$$

$$N = 3.92 \times 10^4 \text{ dynes}$$

If a string holds the car back, the tension in the string is that necessary to let the parallel acceleration equal zero, $a_\parallel = 0$. The result is then

$$\text{net } F_\parallel = ma_\parallel$$

$$W_\parallel - T = ma_\parallel = 0$$

$$T = W \sin 37°$$

Trajectory Motion

If a ball is hurled into the air, or if a person executes a running broad jump (Figure 3-21) or dives from a diving board, the path of motion, the **trajectory motion,** will be a parabola. Galileo was the first to accurately describe this parabolic path, and Newton later studied the effect of air resistance on this motion. Using Newton's laws of motion, we can analyze such motion in two dimensions.

When a ball is thrown it has a definite speed and direction at the instant it is released. That is, the initial velocity of the ball is a vector, and can be resolved into components (see Figure 3-22). The horizontal component is given by

$$v_{x0} = v_0 \cos \theta$$

where θ is the angle with the horizontal. The vertical component is given by

$$v_{y0} = v_0 \sin \theta$$

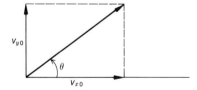

FIGURE 3-22
The initial velocity can be resolved into horizontal and vertical components.

Newton's second law for the horizontal and vertical directions can be stated

$$\text{net } F_x = ma_x$$

$$\text{net } F_y = ma_y$$

The horizontal and vertical motions can be treated separately, if we neglect the air resistance. The force of gravity acts to pull the ball downward but does not affect the horizontal motion.

In the vertical direction, the only force acting is the downward weight of the ball. If we define up to be the positive direction, then Newton's law gives us

$$\text{net } F_y = -mg = ma_y$$

The minus sign indicates downward acceleration. Using our previous results for uniformly accelerated motion, we can write down relationships for the distance traveled in the vertical direction y and the velocity in the vertical direction v_y in terms of the initial velocity in the vertical direction v_{yo}, the acceleration due to gravity g, and the time t.

$$v_y = v_{yo} - gt$$
$$y = y_o + v_{yo}t - \tfrac{1}{2}gt^2$$

In the horizontal direction no force is acting to retard the motion, so Newton's law gives us

$$\text{net } F_x = ma_x = 0$$

Since $a_x = 0$, our equations of motion in the horizontal direction are then

$$v_x = v_{xo}$$
$$x = x_o + v_{xo}t$$

The ball will move forward uniformly at constant velocity in the horizontal direction.

EXAMPLE

A baseball batter hits a pop fly at a velocity of 25 m/s at an angle of 53° to the horizontal.
(a) What are the horizontal and vertical components of the initial velocity?
(b) What is the position of the ball after 1 and 2 s respectively?

(c) What is the maximum height of the ball's trajectory?
(d) How far does the ball go?

We can find the horizontal and vertical components of the velocity by resolution of vectors:

$v_{xo} = v_o \cos\theta$	$v_{yo} = v_o \sin\theta$
$= (25 \text{ m/s})(\cos 53°)$	$= (25 \text{ m/s})(\sin 53°)$
$v_{xo} = 15 \text{ m/s}$	$v_{yo} = 20 \text{m/s}$

The ball is moving forward with a speed of 15 m/s. At the same time it initially travels upward at a speed of 20 m/s. The acceleration due to gravity is acting downward with a magnitude 9.8 m/s².

After a time $t = 1.0$ s the ball will move forward a distance of

$$x = v_{xo}t$$
$$x = (15 \text{ m/s})(1.0 \text{ s}) = 15.0 \text{ m}$$

Its distance of travel in the vertical direction equals the distance it would travel at the initial velocity v_{yo} less the distance it would travel if it were freely falling from rest.

$$y = v_{yo}t - \tfrac{1}{2}gt^2$$
$$= (20 \text{ m/s})(1.0 \text{ s}) - (0.5)(9.8 \text{ m/s}^2)(1.0 \text{ s})^2$$
$$y = 20 \text{ m} - 4.9 \text{ m} = 15.1 \text{ m}$$

It is easy to show that after a time $t = 2.0$ s the ball has moved forward a distance of 30 m and a vertical distance of 20.4 m. We can calculate the position of the ball at various times. These have been plotted in Figure 3-23, which shows the characteristic parabola for the trajectory.

At the maximum height of the ball's trajectory, the point at which the ball stops going up and starts coming down, the vertical component of velocity changes from upward to downward. At the very top the vertical component of velocity will be zero, $v_y = 0$. We can solve for the time t_m when this occurs.

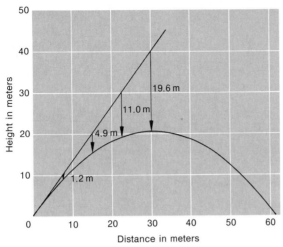

FIGURE 3-23
The trajectory of a pop-fly baseball, neglecting air resistance.

At time t_m, the vertical velocity is given by

$$v_y = 0 = v_{yo} - gt_m$$

The maximum height is then given by

$$y_m = v_{yo}t_m - \tfrac{1}{2}g(t_m)^2$$
$$y_m = v_{yo}(v_{yo}/g) - \tfrac{1}{2}g(v_{yo}/g)^2$$
$$y_m = v_{yo}^2/2g$$

This final relationship for the maximum height involves the height y_m, the acceleration, the initial vertical velocity v_{yo}, and the final vertical velocity $v_y = 0$. Can you think of another way of obtaining this same result?

What goes up must come down. If it takes a time $t_m = v_{yo}/g$ to go up to the maximum, it will take the same time to return to the ground so that the total time of flight t_t is just

$$t_t = 2v_{yo}/g$$

Meanwhile, the ball has continued to move forward in the horizontal direction at a constant speed $v_{xo} = 15$ m/s. The total distance traveled in this example is then

$$x_{total} = v_{xo}t_t$$

$$x_{total} = (15 \text{ m/s})(4.08 \text{ s})$$
$$x_{total} = 61.2 \text{ m.}$$

When the ball reaches the ground, the vertical height is again zero, $y = 0$. Does this suggest to you another way to determine the total time of flight, t_t? Also, can you determine the final vertical component of velocity? How does it compare with the initial vertical component of velocity?

Range

The total distance traveled, the **range,** of a trajectory is related to the initial velocity $\mathbf{v}_o$, the angle of the trajectory θ, the acceleration due to gravity g, and also the difference in elevation between the starting and end points. From the previous example we find that on level ground the total time of flight is

$$t_t = 2v_{yo}/g$$

The total distance traveled is just

$$x_{total} = v_{xo}t_t$$

Putting this together with the expressions for the components of the initial velocities, we get that the range is just

$$x_{total} = (v_o \cos \theta)(2v_o \sin \theta/g)$$
$$x_{total} = \frac{2v_o^2}{g} \sin \theta \cos \theta$$

From trigonometry we can use the double-angle formula:

$$2 \sin \theta \cos \theta = \sin 2\theta$$

The range is then

$$x_{total} = \frac{v_o^2}{g} \sin 2\theta \qquad (3\text{-}12)$$

The maximum range occurs when $2\theta = 90°$ or when the initial trajectory is at an angle of $\theta = 45°$.

EXAMPLE

The longest measured home run in major league baseball was hit by Mickey Mantle in 1953. The ball went a distance of 565 ft. Neglecting air resistance and assuming that the ball was hit at an optimum angle of 45°, how fast was the ball going when it left the bat?

$$x_{total} = \frac{2v_o^2}{g} \cos \theta \sin \theta$$

$$v_o^2 = \frac{x_{total} \, g}{2 \sin \theta \cos \theta}$$

$$v_o^2 = \frac{(565 \text{ ft})(32.17 \text{ ft/s}^2)}{2(0.707)(0.707)}$$

$$v_o^2 = 18\,200 \text{ ft}^2/\text{s}^2$$

$$v_o = 135 \text{ ft/s}$$

Review

Give the significance of each of the following terms. Where appropriate give two examples.

Newton's laws of motion
net force
force
inertial mass
newton
weight
mass
inertia
kilogram weight

CGS system
gram
dyne
pound
slug
static equilibrium
translational equilibrium
center of mass
center of gravity

centripetal acceleration
centripetal force
momentum
impulse
conservation of linear momentum
inclined plane
trajectory motion
range

Further Readings

E. N. Da.C. Andrade, *Sir Isaac Newton: His Life and Work* (New York: Doubleday, Anchor Books, Science Study Series S42, 1958). An interesting biography.

Isaac Asimov, *Realms of Measure* (Boston: Houghton Mifflin Co., 1961). A delightful discussion of hogsheads, barrels, firkins, fluid drams, bushels, gills, stones, long tons, and other vagaries of British and American units. The discussion of metric units is a little dated.

I. Bernard Cohen, *The Birth of a New Physics: From Copernicus to Newton* (New York: Doubleday, Anchor Books, Science Study Series S10, 1960). A fresh account of the scientific ferment following the Renaissance, by one of America's outstanding science historians.

Richard Feynman, *The Character of a Physical Law* (Cambridge, Mass.: MIT Press, 1967). A fascinating set of lectures on physical laws, by an eminent physicist.

Problems

CONSEQUENCES OF THE SECOND LAW

1 A Datsun 260-Z sports car will accelerate from 0 to 100 km/hr in 13.0 s. The car with driver has a mass of 1430 kg.
   (a) What is the average acceleration over this interval in m/s^2?
   (b) What net force is needed for this acceleration?

2 A Boeing 747 aircraft sits on the runway with a mass of 323 000 kg. The engines can develop a net thrust of 4.8×10^5 N.
   (a) What is the acceleration of the aircraft?
   (b) How fast is it going after traveling 2400 m down the runway?

3 A Pontiac Bonneville, with a total mass of 2050 kg, will accelerate from 10 m/s to 30 m/s in about 9.5 s.
   (a) What is the average acceleration?
   (b) What net force is needed for this acceleration?

4 A Chevrolet Bel-Aire with a total mass of 1800 kg is traveling at 25 m/s. The driver slams on his brakes and stops in a distance of 100 m from the point where the brakes take hold.
   (a) What is the average deceleration?
   (b) What is the average horizontal stopping force required?

5 A Boeing 707 aircraft, with a maximum takeoff mass of 150 000 kg, travels down the runway at an average acceleration of 1.6 m/s^2.
   (a) What is the net force needed to accelerate the airplane?
   (b) The takeoff speed is 165 knots (nautical miles/hour) which is about 85 m/s. How long a runway is needed to reach takeoff speed if acceleration is constant?

6 A multistage Saturn rocket with its Apollo command module has a total mass of 2.7×10^6 kg at liftoff.
   (a) What is the weight of the Saturn rocket?
   (b) At liftoff the rocket accelerates upward at 2.45 m/s^2. What net force acting on the rocket is required?
   (c) What total force from the rocket is required?

7 The lunar module ascent stage that brought the first astronauts back from the surface of the moon had a total mass of about 4550 kg and an engine that developed a thrust of 15 600 N. The acceleration due to gravity at the surface of the moon is $g_{\text{moon}} = 1.6 \text{ m/s}^2$.
   (a) What is the weight of the lunar ascent stage on the earth?
   (b) What is the weight of the lunar ascent stage on the moon?
   (c) What is the acceleration of the ascent stage at liftoff from the surface of the moon?
   (d) Could the ascent stage get off the surface of the earth? Why or why not?

CONSEQUENCES OF THE THIRD LAW

8 A block with a mass of 5 kg is attached to a string with a breaking strength of 70 newtons. What is the maximum vertical acceleration of the block that will not break the string?

9 An automobile with a mass of 1500 kg is being towed on level ground by a nylon rope with a breaking strength of 6000 N. Ignoring the effects of friction, what is the maximum acceleration that will not snap the rope?

10 A woman with a mass of 50 kg is in an elevator. The elevator accelerates from rest at 1.2 m/s^2 to a maximum speed of 5.0 m/s. The elevator then decelerates at 1.2 m/s^2 to reach the destination at rest. What is the weight of the woman in newtons—
   (a) before the elevator begins to move?
   (b) while accelerating upward?
   (c) while traveling at a uniform speed of 5.0 m/s?
   (d) while slowing down?

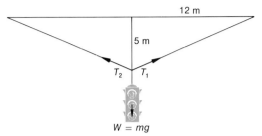

FIGURE 3-25
A traffic light hangs in the center.

11 A car with a mass of 750 gm is riding on a horizontal air track. A string attached to the car extends over a pulley to a 375 gm mass hanging under the influence of gravity.
(a) What is the acceleration of the car and the hanging mass?
(b) What is the tension in the string in dynes? in newtons?

12 A fully loaded elevator car, similar to those in the Transamerica building in San Francisco, has a passenger car with a total mass of 4600 kg, and a counterweight of 4000 kg as shown previously in Figure 3-6. The elevator accelerates upward at 1.2 m/s². The lift motor must apply what net force to
(a) give this acceleration?
(b) keep the elevator moving at constant velocity?
(c) allow the upward-moving elevator to decelerate (slow down) at 1.2 m/s²?

CONSEQUENCES OF THE FIRST LAW

13 A 15 kg light fixture is held by two wires as shown in Figure 3-24. One wire is horizontal with a tension T_1, the second is at an angle of 53° from the horizontal with a tension T_2.
(a) What is the tension T_1?
(b) What is the tension T_2?

14 A traffic light hangs in the center of an intersection between two poles 24 meters apart as shown in Figure 3-25. The center of the wire is 5.0 m below the ends. The traffic light has a mass of 15 kg. What is the tension in each wire?

15 A traffic light is suspended over a street by a bar that extends from a vertical pole. The traffic light weighs 75 lbf. The bar is supported by a wire that makes an angle of 30° with the bar as shown in Figure 3-26.

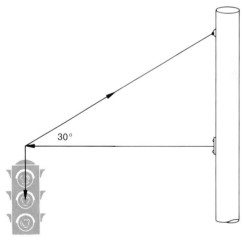

FIGURE 3-26
A traffic light suspended over a street.

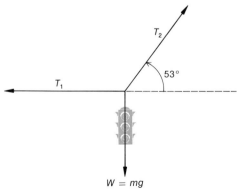

FIGURE 3-24
A light fixture is held by two wires.

(a) What is the force F that the bar is pushing outward?
(b) What is the tension in the wire?

16 A Pantera GTS sports car can handle a maximum lateral acceleration of 9.25 m/s^2. At a speed of 100 km/hr,
(a) What is its minimum turning radius?
(b) What force is required if the automobile has a mass of 1540 kg?

17 A Mazda RX-4 automobile can handle a maximum lateral acceleration of 23 ft/s^2. If the automobile is traveling at 30 mi/hr, what is the car's minimum turning radius?

18 A merry-go-round has an outer diameter of some 10 m. A child is riding a horse on it at a distance of 4 m from the center. The merry-go-round is revolving at a rate of about 4 revolutions per minute.
(a) What is the speed of the child?
(b) What is the centripetal force necessary to hold the child (whose mass = 25 kg) on the horse?

19 A 50 gm toy car travels down a plastic track at an angle. At the bottom of the track there is a loop-the-loop 30 cm in diameter (Figure 3-27). At the very top of the loop the centripetal force is supplied by gravity. What is the minimum speed that the car must have at the top of the loop to stay on the track?

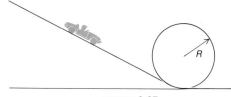

FIGURE 3-27
A loop-the-loop.

20 In the simplified Bohr theory of the hydrogen atom, an electron moves about the proton at a uniform speed of 2.2×10^6 m/s in a circular orbit of radius 5.3×10^{-11} m.
(a) What is the centripetal acceleration of the electron?
(b) The mass of the electron is 9.1×10^{-31} kg. What is the centripetal force needed to keep the electron in its supposed circular orbit? Compare this with the weight of an electron.

MOMENTUM

21 A man kicks a soccer ball with his foot. The ball has a mass of 0.425 kg, is initially at rest $v = 0$, and has a final velocity $v' = 26$ m/s. The time of impact is about 8.0 ms.
(a) What is the change in momentum of the soccer ball during impact?
(b) What is the average force against the foot during impact?

22 A man serving a 0.061 kg handball strikes it with a force of 110 N for a period of time $\Delta t = 1/80$ s. The ball is initially at rest.
(a) What is the ball's change in momentum?
(b) What is its final velocity?

23 A bullet of mass 9.0 gm is fired with a muzzle velocity of 820 m/s from a rifle of mass 3.4 kg. What is the rifle's recoil velocity?

24 A bullet of mass 9 gm with a muzzle velocity of 400 m/s penetrates 5.0 cm into a block of wood of mass 1 kg, initially at rest.
(a) What is the total momentum before the bullet strikes the block?
(b) What is the final velocity of the bullet and block?

25 An astronaut on a space walk has a mass of 90 kg. He pushes away from the space capsule with a velocity of 0.30 m/s. The capsule has a mass of 3 000 kg. What is the capsule's recoil velocity?

26 A radium-226 nucleus decays by the emission of an alpha particle with a mass of 6.64×10^{-27} kg. The alpha particle has a velocity of about 1.5×10^7 m/s. The radon-222 nucleus left behind has a mass of 3.685×10^{-25} kg. What is the recoil velocity of the radon-222 nucleus?

27 A golfer drives a 0.047 kg golf ball from the tee. The ball leaves the tee at 68 m/s. The head of the golf club is traveling at 50 m/s the instant before impact, and is traveling at 34 m/s after impact.
(a) What is the change in momentum of the golf ball?
(b) What is the change in momentum of the golf club?
(c) What is the effective striking mass of the golf club?

OTHER APPLICATIONS
OF NEWTON'S LAWS

28 A girl with a mass of 55 kg is on a rope tow on a ski slope angled 30° from the horizontal. What force must the rope apply if she is to move, without friction, uphill at a uniform speed of 4 m/s?

29 A sled and boy with a total weight of 400 N are initially at rest on an icy hill with a slope of 30° downward from the horizontal. The sled is released and starts to slide without friction.
    (a) What is the acceleration of the sled?
    (b) If the total distance to the bottom of the hill is 10 m, how fast will it be going if it started at rest at the top?

30 A 100 gm mass initially sits on a 45° frictionless inclined plane. It is attached by a rope over a pulley to an 80 gm mass hanging straight down, as shown in Figure 3-28.
    (a) When the system is released, which way will it move?
    (b) What is the acceleration of the system?

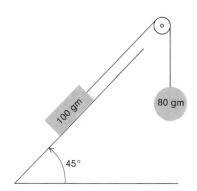

FIGURE 3-28
A 100 gm mass on a frictionless plane is attached by pulley to a hanging 80 gm mass.

31 A marble is knocked horizontally off a tabletop 75 cm above the floor. The marble strikes the floor a horizontal distance of 1.0 m from the edge of the table. How fast was it going when it left the table?

32 A ball player throws a baseball at an angle of 30° to the horizontal with a speed of 30 m/s.
    (a) What is the maximum height the ball will go?
    (b) How far will it go?
    (c) What is the velocity at the end of 3.0 s?

33 At the top of its trajectory a baseball is 48 ft higher than the bat where it started. It is traveling at 50 ft/s.
    (a) With what velocity did the ball leave the bat?
    (b) How much time has elapsed since it left the bat?
    (c) How far will the ball travel before it hits the ground?

ADDITIONAL PROBLEMS

34 A kangaroo with a mass of 65 kg leaves the ground with a velocity of 7.0 m/s on a trajectory at a 45° angle. He pushes with his legs over a distance of 60 cm, starting at rest.
    (a) What is the average acceleration?
    (b) What vertical and horizontal forces are required to give this acceleration?
    (c) How far does he leap?

35 Suppose a person can exert a force with his legs equal to twice his weight on the earth. By bending his knees he exerts this force over a distance of 0.25 m and jumps straight up.
    (a) What is his acceleration?
    (b) With what velocity does he leave the ground?
    (c) How high does he jump?
    (d) Why don't we need to know the person's weight?

36 Suppose an astronaut with a mass of 70 kg is weighted down with 30 kg of equipment, and on earth can jump vertically only a height of 11 cm. He pushes with his legs over a distance of 0.25 m.
    (a) With what average force does he jump?
    (b) If this same astronaut were on the moon where the acceleration of gravity $g' = 1.63$ m/s^2, how high could he jump?

37 A person is restrained by a seat belt across the lap. The maximum force of the seat belt against the

hip should not exceed 1400 lbf. A car is traveling at 30 m/s. A person with a mass of 60 kg is held by his seat belt.

(a) What is the maximum deceleration that would not exceed the above limit?

(b) How far would the car travel before stopping at the maximum deceleration?

38 A ride at an amusement park consists of a wheel of about 5.0 m radius. The patron stands with back against a wall at the outer edge of the wheel. The wheel begins to rotate at about 4 s per revolution. Once up to speed it is tipped vertically so that the person at the top is staring straight down.

(a) How fast is the edge of the wheel moving?

(b) What centripetal acceleration is required to make the person go in a circular orbit of 5.0 m radius at this speed?

(c) Why doesn't the person fall out of the wheel?

39 A baseball leaves the bat at a distance of 3.0 ft above the ground at an angle of 45° and a speed of 120 ft/s. The fence is 380 ft from home plate and is 35 ft high. Ignore the effect of air resistance.

(a) Is it a home run?

(b) What is the minimum speed for the ball to be a home run?

4

Forces

UNIVERSAL GRAVITATION

. . . I offer this work as the mathematical principles of philosophy, for the whole burden of philosophy seems to consist of this—from the phenomena of motions to investigate the forces of nature, and then from these forces to demonstrate the other phenomena; and to this end the general propositions in the first and second books [of the *Principia*] are directed. In the third book I give an example of this in the explication of the System of the World; for by the propositions mathematically demonstrated in the former books, in the third I derive from the celestial phenomena the forces of gravity with which bodies tend to the sun and the several planets. Then from these forces, by other propositions which are also mathematical, I deduce the motions of the planets, the comets, the moon, and the sea.

—Sir Isaac Newton[1]

Having laid the groundwork in the first two books of the *Principia* with his fundamental laws of motion, Sir Isaac Newton then set out to solve essentially every astronomical problem of his day. In Book Three of the *Principia*, Newton delineates four "rules of reasoning in philosophy" that remain cornerstones of modern science to this day.

Rule I. We are to admit no more causes of natural things than such as are both true and sufficient to explain their appearances.
Rule II. Therefore to the same natural effects we must, as far as possible, assign the same causes.

Rule III. The qualities of bodies . . . which are found to belong to all bodies within the reach of our experiments, are to be esteemed the universal qualities of all bodies whatsoever.
Rule IV. In experimental philosophy we are to look upon propositions inferred by general induction from phenomena as accurately or very nearly true, . . . till such time as other phenomena occur, by which they may either be made more accurate, or liable to exceptions.

The best scientific argument contains only as many assumptions as are necessary to explain phenomena (Rule I). Whenever the same effect is observed, the same cause should usually be assumed (Rule II). The great laws of science—whether of gravitation, evolution, genetics, or any other topic or discipline—should be assumed to apply universally, even in such places as stars, planets, and galaxies beyond the reach of direct observation (Rule III). And, finally, we must accept our explanations of nature as true until we discover phenomena that do not fit those explanations (Rule IV). As new phenomena are discovered, we modify our theories to fit the new evidence. Thus, science becomes a living, growing body of knowledge.

Newton postulated a **universal gravitation:** a force that holds the planets in their orbits around the sun, keeps the moons of Jupiter in their orbits, and in fact holds objects to the earth and keeps the moon in orbit around the earth. Between two masses m_1 and m_2 there is an attractive force that acts radially inward along the line of centers between them (see Figure 4-1). This force is proportional to the two masses and is inversely proportional to the square of the distance R between them. The constant of proportionality is a universal constant G—the

---

[1] Sir Isaac Newton, *Philosophiae Naturalis Principia Mathematica, Volume I. The Motion of Bodies,* translated by Andrew Motte (1729), revised by Florian Cajori (Berkeley University of California Press, 1966).

FIGURE 4-1
The force between two masses.

value of which was unknown to Newton. The magnitude of the force between the two masses is given by

$$F = \frac{Gm_1m_2}{R^2} \qquad (4\text{-}1)$$

Newton believed that the same force that attracts an object of mass m at the surface of the earth should also hold the moon in its orbit.

An object at the earth's surface that has a mass m experiences the gravitational acceleration of the earth:

$$g = 9.8 \text{ m/s}^2$$

This acceleration is a consequence of the gravitational attraction between the mass m and the mass of the earth M_e. The magnitude of this force of gravity is

$$F = \frac{GM_e m}{R_e^2} = mg$$

where R_e is the distance from the mass to the center of the earth. The acceleration due to gravity at the surface of the earth is then

$$g = \frac{GM_e}{R_e^2}$$

For the moon to revolve about the earth in a uniform circular orbit of radius R_m, the earth must provide a centripetal force, the magnitude of which is given by

$$F_m = \frac{GM_e M_m}{R_m^2} = M_m a_m$$

where a_m is the centripetal acceleration of the moon, which can be written as

$$a_m = \frac{GM_e}{R_m^2}$$

The gravitational acceleration at the surface of the earth can be solved on the basis of the centripetal acceleration of the moon.

$$g = \frac{R_m^2 a_m}{R_e^2}$$

EXAMPLE

The moon is at a distance of about 60 earth radii.

$$R_m = 60R_e$$

The period of the moon's orbit relative to the fixed stars—the sidereal period—is 27 days, 7 hours, and 43 minutes. The radius of the earth is about 6.38×10^6 m. From this information, determine the centripetal acceleration of the moon in its orbit around the earth and the acceleration due to gravity.

First we must calculate the velocity of the moon in its orbit. The moon goes a distance of $2\pi R_m$ in a period of time $T = 27.322$ days.

$$R_m = 60R_e = 3.83 \times 10^8 \text{ m}$$
$$T = 27.322 \text{ days} = 2.36 \times 10^6 \text{ s}$$
$$v = \frac{2\pi R_m}{T}$$
$$v = \frac{(2.0)(3.14)(3.83 \times 10^8 \text{ m})}{2.36 \times 10^6 \text{ s}}$$
$$v = 1.02 \times 10^3 \text{ m/s}$$

The centripetal acceleration of the moon is just given by

$$a_m = \frac{v^2}{R_m}$$
$$= \frac{(1.02 \times 10^3 \text{ m/s})^2}{3.83 \times 10^8 \text{ m}}$$
$$a_m = 0.002\,72 \text{ m/s}^2$$

This acceleration is, of course, much smaller than at

the surface of the earth. Using this value, the acceleration at the surface of the earth is

$$g = \frac{R_m^2}{R_e^2} a_m$$

$$= \frac{(60\,R_e)^2}{R_e^2} a_m$$

$$= (60)^2\, a_m$$

$$= (3600)(0.002\ 72\ \text{m/s}^2)$$

$$g = 9.8\ \text{m/s}^2$$

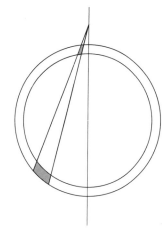

FIGURE 4-2
The earth's mass can be conceptually broken up into spherical shells. The gravitational effect on a point outside each shell acts as if the mass were concentrated at the center of the shell.

There was, however, a serious theoretical problem. In calculating the force of gravity acting on an object, Newton asserted that the gravitational attraction acted as if the entire mass of the earth were concentrated at the center of the earth. That this is true is not immediately obvious. Figure 4-2 shows an object at point P, close to the earth. The ground immediately beneath the object is so much closer and exerts a much larger force than does the far side of the earth some 12 000 km away.

The action of the earth's mass on a point above the earth's surface can be determined by treating the mass as if it were concentric spherical shells (see Figure 4-2) and treating each shell separately. For a point outside a spherical shell, the gravitational force acts as if the entire mass of the shell were concentrated at the center. This proposition of Newton's is not simple to prove. Newton broke the spherical shell up into small pieces, and added up the gravitational force due to each piece. This is the basic process of calculus.

Weighing the Sun

The principle of universal gravitation and a few simple assumptions allowed Newton, given the radius of a planetary orbit, to calculate the period of the orbit. If we consider that the moon is in a nearly circular orbit (as shown in Figure 4-3), the centripetal acceleration is provided by the force of gravity:

$$F_m = \frac{GM_e M_m}{R_m^2} = M_m a_m$$

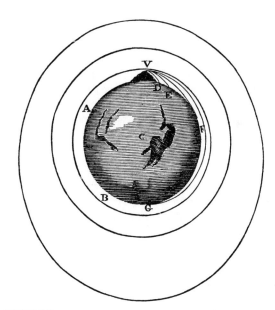

FIGURE 4-3
Newton's illustration of the action of centripetal forces. The earth's gravity provides the centripetal force for an object in circular orbit around the earth. From Sir Isaac Newton, *Principia, Volume II*, translation revised by Florian Cajori (Berkeley: University of California Press, 1966).

The centripetal acceleration of the moon is given by

$$a_m = \frac{v^2}{R_m}$$

And the orbital speed of the moon is given by

$$v = \frac{2\pi R_m}{T}$$

If we put this together, eliminating the acceleration a_m and the orbital speed, the force law becomes

$$F_m = \frac{GM_e M_m}{R_m^2} = \frac{M_m (2\pi R_m)^2}{R_m T^2}$$

This can be reduced to

$$GM_e = (2\pi)^2 \frac{R_m^3}{T^2}$$

The constant depends on the product of the gravitational constant G and the mass of the central body, in this case the earth. Similar relationships also apply to planets orbiting the sun and satellites orbiting Jupiter and Saturn. By astronomical measurement of the period and radius of the orbits, the product GM can be determined. The relative masses of the different planets can then be determined.

EXAMPLE

The moon orbits the earth with a period relative to the fixed stars of 27.322 days, at a mean radius of $R_m = 3.83 \times 10^8$ m. The earth orbits the sun with a period of 365.25 days at a distance of $R_s = 1.50 \times 10^{11}$ m. What is the relative mass of the sun compared to the mass of the earth?

For the earth the product of the gravitational constant times the mass is just

$$GM_e = (2\pi)^2 \frac{R_m^3}{T_m^2}$$

$$= (6.28)^2 \frac{(3.83 \times 10^8 \text{ m})^3}{(2.36 \times 10^6 \text{ s})^2}$$

$$GM_e = 3.98 \times 10^{14} \text{ m}^3/\text{s}^2$$

For the sun the product of the gravitational constant times the mass is just

$$GM_s = (2\pi)^2 \frac{R_{es}^3}{T_e^2}$$

$$= (6.28)^2 \frac{(1.50 \times 10^{11} \text{ m})^3}{(3.16 \times 10^7 \text{ s})^2}$$

$$GM_s = 1.33 \times 10^{20} \text{ m}^3/\text{s}^2$$

The ratio of the solar mass M_s to the earth mass M_e is just

$$\frac{M_s}{M_e} = \frac{GM_s}{GM_e} = \frac{1.33 \times 10^{20} \text{ m}^3/\text{s}^2}{3.98 \times 10^{14} \text{ m}^3/\text{s}^2}$$

$$\frac{M_s}{M_e} = 3.35 \times 10^5$$

Newton's law of universal gravitation makes it possible to determine the relative mass of any body that has an orbiting satellite. It does not, however, tell us the mass of the earth.

Weighing the Earth

In about 1798 the Englishman Henry Cavendish began a series of experiments to directly measure the gravitational force between two massive objects, and thereby to determine the universal gravitational constant G. Two lead spheres 5 cm in diameter were suspended from a slender fiber (see Figure 4-4). If a force were exerted on these spheres the fiber would twist. If the force required to twist the fiber is determined, then the force acting on the spheres can be measured. Two lead spheres about 30 cm in diameter were brought close to the smaller spheres and the maximum displacement of the fiber was observed.

FIGURE 4-4
Cavendish's apparatus. The 30 cm diameter lead spheres are marked W. The 5 cm diameter spheres are suspended at each end of rod m. The entire experiment is enclosed in a box to eliminate air currents. The lead spheres are brought close to the smaller spheres causing a small, visible displacement.

Cavendish's measurements yielded the value for the gravitational constant within 1% of the presently accepted value:

$$G = (6.674 \pm 0.012) \times 10^{-11} \text{ N-m}^2 \text{ kg}^2$$

With a measured value for the gravitational constant, the mass of the earth can immediately be determined. Consequently Cavendish is often credited with "weighing the earth."

From our previous discussion, the product of the gravitational constant and the mass of the earth is just

$$GM_\text{e} = 3.98 \times 10^{14} \text{ m}^3/\text{s}^2$$

$$M_\text{e} = \frac{3.98 \times 10^{14} \text{ m}^3/\text{s}^2}{6.674 \times 10^{-11} \text{ N-m}^2/\text{kg}^2}$$

$$M_\text{e} = 5.96 \times 10^{24} \frac{\text{kg}^2\text{-m}}{\text{N-s}^2}$$

The unit of $1 \text{ N} = 1 \text{ kg-m}/\text{s}^2$ so that

$$M_\text{e} = 5.96 \times 10^{24} \text{ kg}$$

Other Fundamental Forces

Newton had tapped a deep secret of nature in his discovery of the principle of universal gravitation. Indeed, in the preface to the *Principia* he commented:

I wish we could derive the rest of the phenomena of Nature by the same kind of reasoning from mechanical principles, for I am induced by many reasons to suspect that they may all depend upon certain forces by which the particles of bodies, by some causes hitherto unknown, are either mutually impelled towards one another, and cohere in regular figures, or are repelled and recede from one another. These forces being unknown, philosophers have hitherto attempted the search of Nature in vain; but I hope the principles here laid down will afford some light either to this or some truer method of philosophy.

It would be another 150 years before the secrets of electrical forces would be revealed, over 200 years before the nature of the atom would be understood, and 250 years before the strong and weak forces within the nucleus would be studied.

STRENGTH OF MATERIALS

The fundamental forces of nature are manifested in the behavior of materials in the world around us. We shall look at the mechanical strength of solid materials and at how forces scale from small structures to very large structures.

The material of an object placed under tension tends to stretch as the force of tension tries to pull its atoms further apart. There are two forces between atoms in a solid: repulsion and attraction. As two atoms are forced closer together, a repulsive force, caused primarily by the positive charges on their nuclei, increases. When the atoms are further apart, a force of attraction operates between the positively charged nucleus of each atom and the negatively charged electron cloud of the other. The force between two atoms depends on the structure of the atoms and is basically an electrical, or **Coulomb force** (which we discuss in detail in Chapter 13). The net force between the two atoms is the sum of the repulsive and attractive forces, which depend on the separation distance, as is

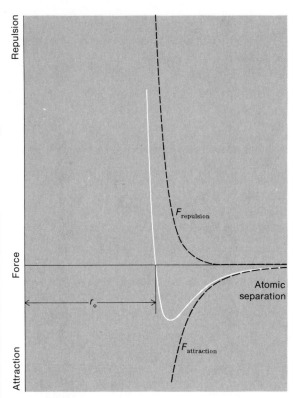

FIGURE 4-5
The force between atoms is a function of the separation between them.

shown in Figure 4-5. There is an equilibrium distance, r_0, where the net force between the two atoms is zero. For example, between iron atoms this distance is about 2.3×10^{-10}m or 0.23 nanometers (nm).

If atoms are forced closer together, as when an object is under compression, the net repulsive force provides the reaction force. If atoms are pulled apart, as when an object is under tension, the net attraction is what holds the atoms together.

Tensile Strength

When the properties of solid material are to be studied, a sample is placed in a machine that stretches or compresses it. The **stress** applied to the material is the force F per unit cross-sectional area A.

$$\text{stress} = \frac{F}{A}$$

As the material changes shape under the applied stress, it experiences strain. The **strain** is the fractional elongation or contraction of the material. If a material of length L has a change of length ΔL, the strain is

$$\text{strain} = \frac{\Delta L}{L}$$

For isotropic materials (materials of homogeneous, or uniform, composition) the strain is in the same direction as the applied stress.

The mechanical strength of a material, regardless of its size and shape, is measurable as the stress required to produce a given strain. A tensile stress-strain diagram for

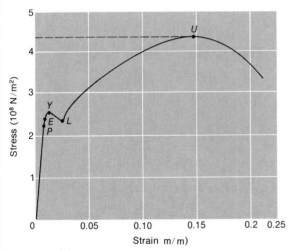

FIGURE 4-6
Stress as a function of strain for mild steel under tension.

construction (mild) steel is shown in Figure 4-6. For relatively small change in length the strain is proportional to the stress. **Young's modulus** of elasticity, Y, is defined as the ratio of the applied tensile stress to the resulting strain, or fractional elongation.

$$Y = \frac{\text{stress}}{\text{strain}} = \frac{F/A}{\Delta L/L} = \frac{FL}{A\,\Delta L}$$

The tensile force F required to produce a certain elongation ΔL is

$$F = \frac{YA\Delta L}{L}$$

Young's modulus applies up to a **proportional limit,** point P (shown in Figure 4-6) beyond which the stress is no longer proportional to strain. Up to a slightly larger stress, point E, the **elastic limit,** the material returns to its original length when the stress is removed. Beyond the elastic limit, the material begins to be deformed plastically, that is permanently: When the stress is removed the material does not return to its original dimensions. At the **yield point** Y, the material continues to become further deformed even though the applied stress has not been increased. Beyond point L the stress required to produce a given strain will again increase as the material becomes brittle. The material reaches its **ultimate tensile strength** at point U. If the load is not immediately reduced the material will stretch further, reaching the breaking point B. Young's modulus of elasticity, the proportional limit, and the ultimate tensile strength of common materials are shown in Table 4-1.

EXAMPLE

A steel wire has a diameter of 3×10^{-4} m and a length of 1.0 m. It is stretched by the weight of a 5 kg mass. How far does the wire stretch?
From equation 4-2 we can solve for the elongation of the wire:

$$\Delta L = \frac{LF}{YA} = \frac{(1.0\text{ m})(5\text{ kg})(9.8\text{ m/s}^2)}{(20 \times 10^{10}\text{N/m}^2)(7 \times 10^{-8}\text{m}^2)}$$

$$\Delta L = 0.0035\text{ m}$$

The wire will stretch by 3.5 mm.

TABLE 4-1
Properties of common metals.

Metal	Modulus of elasticity—tension 10^{10} N/m²	Proportional limit—tension 10^7 N/m²	Ultimate strength—tension 10^7 N/m²
Aluminum			
Cast 99%	6.9	6.2	9.0
Hard drawn	6.9	13.8	20
Brass			
Cast (60% Cu, 40% Zn)	9.0	13.8	31
Common rolled	9.7	17	41
Magnesium (extruded)	4.5	11.7	22
Cast iron	17	24.8	37
Steel			
Cold-rolled (0.2% C)	20.7	41	55
Hot-rolled (0.8% C)	20.7	48	83
Nickel steel, oil-quenched (3½% Ni, 0.4% C)	20.7	110	197

SOURCE: Enrico Volterra and J. H. Gaines, *Advanced Strength of Materials* (Englewood Cliffs, N.J.: Prentice-Hall, 1971), p. 15.

Compressive Strength

When two atoms in a metal are forced together, the repulsive force between them resists the motion, giving the material strength under compression. The compressive stress is just a pressure:

$$\text{stress} = \frac{F}{A}$$

The compressive strain is just the fractional contraction of the material:

$$\text{strain} = \frac{\Delta L}{L}$$

The modulus of elasticity for compression is just the ratio of compressive stress to compressive strain. Again, for limited strain, the strain is proportional to the stress.

In the human body the muscles are under tensile stress; the bones are usually under compression, which tends to hold them in their sockets. The fractional change in length under compressive and tensile loading of bone material is shown in Figure 4-7. The strength of compact bone material is quite similar for such diverse species as horse, ostrich, and human. The modulus of elasticity

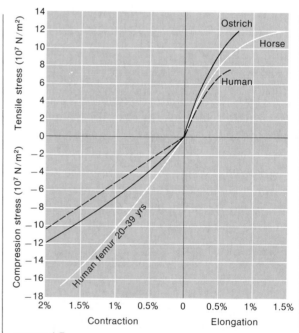

FIGURE 4-7
Stress-strain curves for wet, compact bone from the femur of humans, horse, and ostrich. Data from Hiroshi Yamada, *Strength of Biological Materials* (Baltimore, Md.: Williams and Wilkins, 1970).

under compressive stress for bone is of the order of

$$Y = \frac{\text{stress}}{\text{strain}} = \frac{10.4 \times 10^7 \, \text{N/m}^2}{(0.01)} = 10^{10} \, \text{N/m}^2$$

EXAMPLE

A person with a mass of 70 kg stands balanced on the balls of both feet (see Figure 4-8). The tibia, or shin-bone, is kept under compression by the combination of the body's (downward) weight and the tension in each Achilles tendon of about 900 newtons. The cross-section of the tibia is a hollow ring with an outer diameter of about 2.5 cm and an inner diameter of about 1.2 cm. The tibia is about 40 cm long.
(a) What is the compressional stress on the tibia?
(b) By how much will the tibia compress?
The compressional stress is the force per unit of cross-sectional area. The force on one leg is the combination of one-half the person's weight plus the tension of the Achilles tendon. The weight of the person is

$$W = mg = (70 \, \text{kg})(9.8 \, \text{m/s}^2) = 686 \, \text{N}.$$

The total compressive force acting on the tibia is approximately

$$F = \frac{686 \, \text{N}}{2} + 900 \, \text{N} = 1240 \, \text{N}$$

The cross-sectional area of the tibia is approximately

$$A = \frac{\pi}{4}[(D_0{}^2) - (D_1{}^2)]$$

$$A = \frac{\pi}{4}[(0.025 \, \text{m})^2 - (0.012 \, \text{m})^2]$$

$$A = 3.8 \times 10^{-4} \text{m}^2$$

The compressive stress is

$$\text{stress} = \frac{F}{A} = \frac{1240 \, \text{N}}{(3.8 \times 10^{-4} \text{m}^2)} = 3.3 \times 10^6 \text{N/m}^2$$

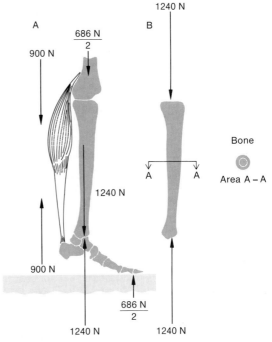

FIGURE 4-8
Compressive loading of the tibia. Adapted from Leonard H. Greenberg, *Physics for Biology and Pre-med Students,* © 1975, W.B Saunders Co. Used with permission.

The strain is

$$\text{strain} = \frac{\text{stress}}{Y} = \frac{3.3 \times 10^6 \, \text{N/m}^2}{10^{10} \, \text{N/m}^2} = 3.3 \times 10^{-4}$$

If the tibia has a length of 40 cm, then the compression ΔL is

$$\Delta L = (\text{strain})L = (0.40 \, \text{m})(3.3 \times 10^{-4}) = 1.3 \times 10^{-4} \, \text{m}$$

The tibia will compress by 0.13 mm, a small amount.

SCALING LAWS

The strength of a material is its ability to withstand a mechanical stress. The strength of an object depends on its cross-sectional area. The weight of an object depends on its volume and density. As size changes, there is a natural change in the proportions of plants, insects, and animals.

The volume of a cube of dimension D on a side is

$$V = D^3$$

If the size of the cube is increased by a linear scale factor a, so that the length of a side becomes $D' = aD$, the new volume is

$$V' = (D') = a^3 D^3$$

The volume increases by a factor of a^3. If linear size of a cube, cylinder, or sphere is increased by a factor of 2, the volume increases by a factor of 8; if size is increased by a factor of 3, volume increases by a factor of 27.

The cross-sectional area and the total surface area of a shape depend on the square of the dimension. The area of one face of a cube is

$$A = D^2$$

Increasing the linear dimensions by a scale factor a increases the area by a factor

$$A' = a^2 A$$

If the diameter of a rope is doubled, its cross-sectional area and consequently its strength are increased by a factor of 4.

Galileo was the first to comment in detail on the **laws of scaling,** those laws that describe how the size and proportions of objects are related to the strength of their materials. In his *Discourses and Demonstrations Concerning Two New Sciences* (1638) Galileo concludes:[2]

. . . you can plainly see the impossibility of increasing the size of structures to vast dimensions either in art or in nature; likewise the impossibility of building ships, palaces, or temples of enormous size in such a way that their oars, yards, beams, iron-bolts, and, in short, all their other parts

[2]Galileo Galilei, *Dialogues Concerning Two New Sciences,* translated by Henry Crew and Alfonso de Salvio (Evanston, Ill.: Northwestern University Press, 1950), pp. 125–126.

will hold together; nor can nature produce trees of extraordinary size because the branches would break down under their own weight; so also it would be impossible to build up the bony structure of men, horses, or other animals so as to hold together and perform their normal functions if these animals were to be increased enormously in height; for this increase in height can be accomplished only by employing a material which is harder and stronger than usual, or by enlarging the size of the bones, thus changing their shape until the form and appearance of the animals suggest a monstrosity.

To illustrate briefly, I have sketched a bone whose natural length has been increased three times and whose thickness has been multiplied until, for a correspondingly large animal, it would perform the same function which the small bone performs for its small animal. [See Figure 4.9] From the figures here shown you can see how out of proportion the enlarged bone appears. Clearly then if one wishes to maintain in a great giant the same proportion of limb as that found in an ordinary man he must either find a harder and stronger material for making the bones, or he must admit a diminution of strength in comparison with men of medium stature; for if his height be increased inordinately he will fall and be crushed under his own weight. Whereas, if the size of a body be diminished, the strength of that body is not diminished in the same proportion; indeed the smaller the body the greater its relative strength. Thus a small dog could probably carry on his back two or three dogs of his own size; but I believe that a horse could not carry even one of his own size.

Among animals, increase in size from small creatures like the mouse to the large species like the hippopotamus, rhinoceros, or elephant involves changes in proportion. As

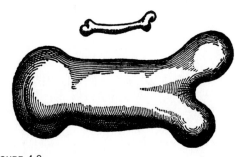

FIGURE 4-9
An illustration of scaling from Galileo's *Dialogues Concerning Two New Sciences.*

the size of the animal increases, the bones and muscles must become proportionately thicker to support the increased weight. A model of an animal can be made with the legs and torso represented as cylinders with some diameter d and length l. As the animal is scaled up in size, the diameter of the legs or torso must increase at a somewhat greater rate than the length. For our model we can assume that the diameter squared is proportional to the length l cubed:

$$d^2 = kl^3$$

The weight of a particular segment—for instance, the torso—depends on the length l times the area:

$$W = k'ld^2$$

Substituting for d^2 we get that

$$W = k'kl^4$$

or that the length or height of an animal increases as

$$l \propto W^{1/4}$$

only slowly with weight. Measurements of cattle indicate that

$$l \propto W^b$$

where $b = 0.24$, whereas measurements with primates indicate $b = 0.28$, so that this simple model, which gives $b = 0.25$, is reasonably well confirmed.

We can also solve for the relationship between diameter and weight. Substituting for l we get

$$W \propto (W^{1/4})d^2$$
$$d^2 \propto W^{3/4}$$

The diameter as a function of weight becomes

$$d \propto W^c = W^{3/8}$$

The diameter is proportional to the weight raised to a power c, where $c = 0.375$.

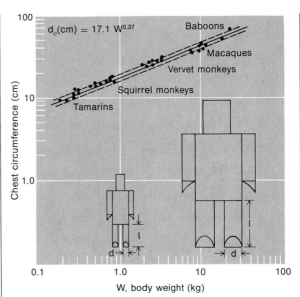

FIGURE 4-10
Chest circumference as a function of body weight for five species of primate. Courtesy of Thomas McMahon. From "Size and Shape in Biology," *Science* (23 March 1973), p. 1203. Copyright 1973 by the American Association for the Advancement of Science.

Figure 4-10 shows the chest circumference as a function of body weight for five species of primates. The solid curve indicates the diameter proportional to the weight to the 0.37 power over a wide range of species weight. This confirms our simple scaling model.

The mouse has dainty little legs to support a few grams of weight. With relatively small muscles it is able to leap several times its overall length. A dog's leg bones are more substantial. A cow has heavy leg bones to support its weight and can jump only a short distance. The elephant has very thick legs to support the weight of as much as 6000 kg of mass. The size of the great land animals of prehistoric times was ultimately limited by the strength of their bones. The great *Apatosaurus* was known to walk on dry land with a mass of 18 000 kg. Its limb bones were massive.

Surface Area

Many processes are affected by amount of **surface area.** The loss of heat by radiation or of moisture by evaporation, the absorption of the sun's energy in leaves or of oxygen from air or water, depend on the amount of surface area present. As an organism's size changes, the ratio of the surface area to the volume or weight changes.

A sphere of radius R has a surface area A and a volume V given by

$$A = 4\pi R^2 \qquad V = \frac{4}{3}\pi R^3$$

If we increase the linear size of the sphere by a factor a the radius becomes $R' = aR$. The new area and volume become

$$A' = 4\pi(aR)^2 = a^2\,4\pi R^2 = a^2 A$$

$$V' = \frac{4}{3}\pi(aR)^3 = a^3 V$$

The ratio of the surface area to volume changes as

$$\frac{A'}{V'} = \frac{a^2 A}{a^3 V} = \frac{1}{a} \times \frac{A}{V}$$

As the object's size increases, the ratio of the surface area to volume decreases.

To exchange carbon dioxide for oxygen in the blood, the lungs require an enormous surface area. The alveoli in the lungs have a diameter of only about 2×10^{-4} m. Because of this small size, the ratio of surface area to volume is very large. (The total surface area of the alveoli in the lungs approximately equals the area of a regulation tennis court!)

To a certain extent, the ratio of surface area to volume also determines the maximum size of cells. A cell needs oxygen in its energy cycle. How much is needed depends on the volume of the cell. Oxygen comes into the cell only by diffusing through the cell wall. As the cell becomes larger, the surface area per unit volume decreases. If the cell becomes too large, it becomes only marginally able to get the needed oxygen. The spherical cell can then divide into two smaller cells with the same total volume but with 26% greater area.

The laws of scaling also apply to flying creatures such as birds and insects. Whereas bees and flies can hover in place seemingly for hours, only the smallest birds, such as the hummingbird, can do so. As the weight of birds increases (from sparrows and finches to robins, mockingbirds, and the like), the relative size of the wing to the bird increases. How fast a bird can fly depends on amount of wing area. Slow-flying birds like broad-winged hawks are almost all wing. Fast-flying birds such as seagulls, barn swallows, and terns have proportionately smaller wings for their weight. As an extreme example, the albatross has proportionately so small a wing area that although graceful once aloft, it can become airborne with ease only into a head wind.

The lift of a wing depends on the shape and surface area and on the square of the air speed.

$$F_{\text{lift}} = K_{\text{lift}} A_{\text{wing}}\, v^2$$

This lift force must support the weight. As birds become larger the weight scales as a^3. Consequently the wing area must also scale roughly as a^3

$$A_{\text{wing}}' = A_{\text{wing}}\, a^3$$

If the bird were simply increased in size while maintaining the same proportions, the wing area would increase as a^2. As birds become larger they more closely resemble a flying wing. The upper limit on the mass of a bird is about 15 kg. A 70 kg person who tried to fly with wings proportioned to that of a hawk would not have sufficient wing area to get off the ground. Thus, early attempts at human-powered flying were frustrated by the laws of scaling. Only hang gliders (as in Figure 4-11) give humans enough wing area to fly.[3]

FRICTION

Friction plays an important role in the world around us. Without friction many things would fall apart—our houses when nails came loose from boards; our cars and

---

[3] In 1977 an American, Bryan Allen, won the British Kremer Prize (50 000 £) by flying the Gossamer Condor around a mile course. The craft (32 kg) and person (61 kg) had a maximum speed of 21 km/hr and flew for about 8 min.

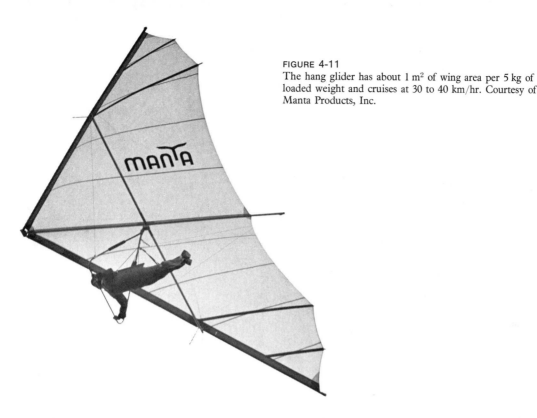

FIGURE 4-11
The hang glider has about 1 m² of wing area per 5 kg of loaded weight and cruises at 30 to 40 km/hr. Courtesy of Manta Products, Inc.

machines when bolts and nuts came loose. We would have no brakes for our cars, nor a working clutch or automatic transmission. Tires would have no traction. Yet friction is often a great nuisance, requiring extra force to overcome it. Where there is friction there is also wear.

When two surfaces slide over one another the motion always meets some resistance, the force of friction. A book sent sliding across a table quickly comes to rest, a tossed stone soon rolls to a stop. Hence the ancients believed that the natural state of objects was to be at rest. Although we shall discuss primarily the friction between two metal surfaces, the concepts developed are useful for describing other surfaces.

Static Friction

A metal surface is not really smooth. Most surfaces are visibly rough. However, even finely ground surfaces as shown in Figure 4-12 have irregularities of the order of 10^{-6} m in height, highly polished surfaces, of the order of 10^{-7} m. Rarely are surfaces really clean. Most metals

when exposed to air quickly develop an oxide coating with properties different from those of the subsurface metal. The surface is also often coated with a layer of oil, grease, or dirt that is the actual layer of contact.

Between two surfaces in apparent contact there is only a limited area of actual contact. Each surface has protrusions that slightly deform the two surfaces in contact until there is sufficient area of actual contact to support the load. The surface of construction (mild) steel will support a compressive stress or pressure of

$$P_o = 1.2 \times 10^9 \, \text{N/m}^2$$

The actual area of contact A_a depends on the **normal force** F_n, the force perpendicular to the surfaces that pushes the two surfaces together.

$$A_a = \frac{F_n}{P_o}$$

Under most conditions the actual area of contact is only a small fraction of the total surface area.

When two surfaces are pressed against one another and

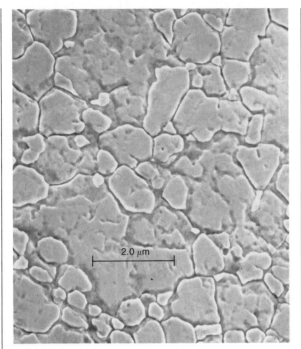

Electron micrograph (magnification 1100 x) of a smooth stainless steel surface showing surface irregularities. Courtesy of Kenneth Letsch, San Francisco State University.

are static—are not moving in relation to one another—there is an adhesive force by which the points of actual contact tend to stick together. If the two surfaces in contact are to be slid past one another, a force parallel to the surface, a shear force, is required to break this adhesive contact. The force of **static friction** F_f is the force parallel to the surface required to break the contact points. The static friction force is proportional to the area of actual contact:

$$F_f = GA_a$$

The area of actual contact depends on the force perpendicular to the surface, the **normal force** F_n, so that the force of static friction is proportional to the normal force:

$$F_f = GA_a = \frac{GF_n}{P_o}$$

The measured force of static friction between two surfaces is repeatable, but the actual value depends on many factors. Provided that the surface area of contact does not become too small, or the normal force too great, it is useful to define a **coefficient of static friction** μ_s. This coefficient relates the force required to move the surface, to the force normal to the surface F_n

$$F_f = \mu_s F_n$$

Contained in this relationship are the two laws of friction.

1. The friction force is proportional to the normal force acting against the surfaces.

2. The friction force is independent of the total area of contact.

Table 4-2 lists the coefficients of static (as well as kinetic) friction for several surfaces.

For precise standards of length machinists use gauge blocks. To give an extended measure of length these blocks can be "wrung" together side by side. They stick to one another because of the adhesion of the grease film. They can be separated with a modest force. However, if the blocks are thoroughly degreased before being "wrung" together, there will be strong metal-to-metal adhesion that can be broken only by considerable force. When the adhesive contacts are torn, the metal surface is roughened and the length standard ruined.

Glass slides on glass fairly well. A clean, dry tumbler can be dragged upside down across a clean sheet of dry glass without difficulty or particular scratching. The glass slides on the oil and dirt film that remains on the surface of even clean glass. If, however, the glass sheet is wetted with water, the tumbler slides with greater difficulty. The water tends to lift the film of oil, giving the two glass surfaces much greater contact. Upon close examination, the glass sheet shows noticeable scratches.

EXAMPLE

A book is placed on a smooth wooden board, one end of which is raised to a maximum angle—about 16°—before the book begins to slide, as shown in Figure 4-13. What is the coefficient of static friction between the book and the board?

TABLE 4-2
Static and kinetic coefficients of friction.

Material A	on Surface B	μ_s	μ_k	Condition
Hard steel	Hard steel	0.78	0.42	Dry
Hard steel	Hard steel	0.16		Light oil
Aluminum	Steel	0.61	0.47	Dry
Glass	Glass	0.94	0.40	Dry
Cast iron	Cast iron	1.1	0.15	Dry
Cast iron	Cast iron	0.2	0.07	Lubricated
Teflon	Steel	0.04	0.04	Dry
Teflon	Snow 0° C	0.02		Dry
Ice	Ice 0° C	0.1	0.02	Dry
Brake shoe	Brake drum		1.2	
Tire	Road	1.0		Dry
Tire	Road	0.2		Wet

The weight of the book is a force that can be resolved into two components: one perpendicular to the plane, the other parallel to it. The perpendicular component, the normal force, is given by the weight of the book times the cosine of the angle:

$$F_n = W \cos \theta$$

The component of the force directed parallel to the board acts to move the book down the plane:

$$F_\parallel = W \sin \theta$$

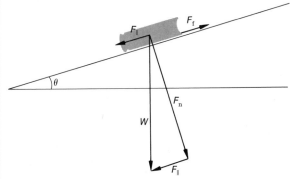

FIGURE 4-13
A book sliding on an inclined plane.

At the maximum angle of repose this force acts to overcome friction. The coefficient of static friction is the ratio of the parallel force to the normal force.

$$\mu_s = \frac{W \sin \theta}{W \cos \theta} = \tan \theta$$

$$\mu_s = \tan (16°) = 0.29$$

Kinetic and Rolling Friction

Once two surfaces are in motion relative to one another, the friction force changes. As the irregularities in the surface move over one another, the material is deformed. New adhesive junctions are established and broken. The friction force is still proportional to the normal force pushing the two surfaces together. The proportionality is given by the **coefficient of kinetic friction.** The kinetic friction is generally smaller than the static friction

$$F_f = \mu_k F_n$$

The coefficients of static and kinetic friction are shown in Table 4-2.

Friction between two moving surfaces can be reduced by lubrication. In the automobile engine, oil is pumped under pressure to lubricate the crankshaft and camshaft bearings and other moving parts. If the oil pressure is sufficient, the shaft actually rides on a film of oil rather than on the facing metal surface. The only friction is then due to the oil's viscosity, its resistance to flow.

Friction is also used in the brakes of the automobile. In drum brakes, a brake shoe with a special lining is pressed outward against the metal brake drum. The force of friction is proportional to the pressure applied at the brake cylinder. The greater this pressure, the greater the force pressing the two surfaces together, and the greater the stopping force. In disk brakes, a caliper presses a friction pad against a rotating disk attached to the wheel. As the applied force increases, so does the friction force.

Much of the engineering development in the past hundred years has involved reducing the destructive effects of metal sliding over metal. If a round object rolls over a flat surface, the point of contact between the object and the surface is instantaneously at rest. There is little time for adhesive bonds to form. The surface is deformed slightly, but if the material is elastic, the work to form the depres-sion is recovered as the object rolls on. We can define a **coefficient of rolling friction** such that the friction force is still proportional to the normal force applied between the two surfaces.

$$F_{\rm f} = \mu_{\rm r} F_{\rm n}$$

For a hard steel roller on a hard steel surface, $\mu_{\rm r}$ can be as small as 0.001. For a deformable material such as a rubber tire, the coefficient of rolling friction is still only 0.05.

In a ball bearing a series of hard steel balls, held in place by an outer collar called a race, roll around a shaft. The balls touch only at two small points of contact, where deformation can take place. Tapered cylindrical rollers have much greater areas of contact and correspondingly less deformation and friction. The hard tapered rollers roll in a tapered track as the bearing turns. Lubrication is needed only between the rollers and the metal race that keeps the rollers evenly spaced.

Review

Give the significance of each of the following terms. Where appropriate give two examples.

universal gravitation	proportional limit	static friction
Cavendish's measurements	elastic limit	normal force
Coulomb force	yield point	coefficient of satatic friction
stress	ultimate tensile strength	coefficient of kinetic friction
strain	laws of scaling	coefficient of rolling friction
Young's modulus	surface area	

Further Readings

R. McNeill Alexander, *Size and Shape* (London: Edward Arnold Publishers, 1971). An interesting short monograph on scaling laws of biological systems.

Galileo Galilei, *Dialogues Concerning Two New Sciences,* translated by Henry Crew and Alfonso de Salvio (New York: Dover Publications, 1914).

Thomas McMahon, "The Mechanical Design of Trees," *Scientific American* (July 1975). Scaling laws applied to the growth of trees.

Frank Philip Bowden and David Tabor, *Friction: An Introduction to Tribology* (New York: Doubleday, Anchor Books Science Series S71, 1973). An introduction to the nature of friction.

Problems

GRAVITATION

1 In the *Principia* Newton described the minimum velocity required for a projectile to enter an orbit at the surface of the earth, as shown in Figure 4-3. Neglecting air resistance and assuming that the centripetal acceleration is provided by the earth's gravity, what is that minimum required velocity?

2 The moon has a mass of 7.36×10^{22} kg and a radius of 1.74×10^{6} m. What is the gravitational acceleration at the surface of the moon?

3 The Apollo-11 spacecraft was in a parking orbit 185 km above the surface of the earth. The radius of the earth is 6.38×10^{6} m.
(a) What is the period of the parking orbit?
(b) How fast is the spacecraft going?

4 The Apollo-11 spacecraft was placed in a lunar orbit at a distance of 110 km from the surface of the moon. The radius of the moon is 1 740 km.
(a) What is the period of the orbit?
(b) How fast is the spacecraft going?

STRENGTH OF MATERIALS

5 A goldline twisted nylon climbing rope has a diameter of 11 mm. The rope will stretch 13% under the weight of a 100 kg mass.
(a) What is Young's modulus for the rope?
(b) If a similar rope with a diameter of 8 mm is placed under the weight of an 80 kg load, what is the percentage stretch?

6 The E_5 string on a guitar has a radius of 1.5×10^{-4} m and is under a tension of 370 N.
(a) What is the tensile stress on the string?
(b) If the string is made of nickel steel, is the string within the proportional region?
(c) Within the ultimate tensile strength?

7 The Golden Gate Bridge is suspended upon cold-rolled steel cables that are 2.54 cm in diameter.
(a) What load will a single cable support in stretching only 0.5%?
(b) What is the maximum load that a single cable can support if the ultimate strength is 55×10^{7} N/m^2?

8 A single strand of human hair will support a static weight of 170 gm and will stretch 43% before breaking. The hair has a diameter of about 1.0×10^{-4} m.
   (a) What is the ultimate tensile stress that the material will sustain?
   (b) If the hair is considered perfectly elastic, up to the breaking strength, what is Young's modulus?

9 A twisted nylon rope 11 mm in diameter is used to tow an automobile. Two strands are attached to an automobile of mass 1800 kg. The ultimate breaking stress of the rope is 1.2×10^8 N/m². What is the maximum acceleration that will not break the rope?

SCALING LAWS

10 A steel piano wire has a diameter of 8.4×10^{-4} m and an ultimate tensile strength of 2.7×10^9 N/m².
   (a) How much weight can this wire support?
   (b) If a wire 50% stronger is made from the same material, what is its diameter?

11 A steel piano wire has a diameter of 1.3×10^{-3} m and an ultimate tensile strength of 2.3×10^9 N/m².
   (a) How much weight (kg-wt) can this wire support?
   (b) If a wire with 10 times the strength is to be made from the same material, what will be its diameter?

12 Suppose that a person's mass scales as height cubed. A woman with a height of 5 ft 8 in has a weight of 130 lbf.
   (a) What would be the weight of a similarly shaped woman with a height of 4 ft 10 in?
   (b) 6 ft?
   (c) Express your answers in SI units.

13 Suppose that a person's mass scales as height cubed. A man 1.80 m tall has a mass of 70 kg.
   (a) The midget General Tom Thumb stood 35 in tall at the age of 30. What should his mass be?
   (b) The giant Robert Wadlow stood 8 ft 11.1 in at the age of 22. What should his mass be?

14 The mass of soaring birds increases with the cube of the dimension $M \propto l^3$. The wing area increases with the square of the dimension $S \propto l^2$. Consequently the mass supported is related to the wing area as $M = KS^{3/2}$. The constant $K = 10$ kg/m³.
   (a) A hang glider is designed for a loaded mass of 100 kg. What is the required wing area?
   (b) An albatross with a mass of 9.0 kg requires what wing area?

15 The lungs have a volume of the order of 3.0 liters. The lung volume is made up of thousands of small alveoli of diameter 2×10^{-4} m.
   (a) What is the surface area of a sphere of volume 3.0 liters?
   (b) How many alveoli would make up this same volume?
   (c) What is the surface area of that many alveoli?

16 A soup can is a cylinder 6.5 cm in diameter and 9.4 cm high. It contains 300 gm of soup.
   (a) What is the surface area per unit volume of the can?
   (b) What would be the dimensions of a can with the same proportions that has twice the volume?
   (c) What is the surface area per unit volume of the new can?

FRICTION

17 A skier has teflon-coated skis and is on dry snow, $\mu_s = 0.02$. What is the minimum slope necessary for him to begin sliding downhill?

18 A block of aluminum is placed on an inclined steel plane, $\mu_s = 0.61$. What is the minimum slope necessary for the block to begin sliding?

19 In football practice a lineman pushes on a blocking sled with a horizontal force of 1200 N before it begins to move. The sled has a mass of 160 kg. What is the coefficient of static friction?

20 A wooden crate with a mass of 50 kg is just pushed across a wooden floor by a horizontal force of 90 N. What is the coefficient of static friction?

21 A cast iron block rests on a lubricated cast iron surface. The coefficients of static and kinetic friction are $\mu_s = 0.20$ and $\mu_k = 0.07$ respectively. The block has a mass of 2 kg.
  (a) What is the minimum angle from the horizontal for the block to begin sliding?
  (b) Once the block begins to slide, what is the net force acting on the block down the plane?
  (c) What is the acceleration of the block?

22 An automobile with a mass of 1400 kg is resting part way up a small hill with a 1.0 m rise per 100 m of travel. The coefficient of rolling friction is 0.03.
  (a) What force is required to push the car at a uniform speed up the hill?
  (b) What force is required if the car instead has a mass of 2300 kg?

23 An automobile with a mass of 1350 kg accelerates up a 5.0% grade (a rise of 5.0 m for 100 m of travel). The coefficient of rolling friction is 0.02. The car accelerates at 2.6 m/s^2.
  (a) What force is required to accelerate the car up the hill?
  (b) What force is required to accelerate the car down the hill at the same rate?

24 An automobile is traveling on a horizontal, circular path at 140 km/hr. The friction of the tires against the road provides the centripetal force. The coefficient of static friction for the tire against the road is 0.40.

(a) What is the minimum radius of the path?
(b) Why don't we need to know the mass of the auto?

ADDITIONAL PROBLEMS

25 Jupiter's largest moon, Ganymede, has a period of 7.155 days and a mean orbit radius of 1.07×10^6 km about the planet. What is Jupiter's mass?

26 An Intelsat communications satellite was launched from Cape Kennedy, Florida, in January 1972 and placed in a geosynchronous orbit over the Gilbert Islands in the central Pacific Ocean. The satellite has a mass of 720 kg. In a geosynchronous orbit the satellite remains over the same spot on the earth, for its period is just 24 hours.
  (a) What is the radius of a geosynchronous orbit?
  (b) How far from the surface of the earth is the orbit?
  (c) How fast is the satellite going?

27 A person who jumps from a small height and lands stiff-legged with knees locked comes to an abrupt halt in 10^{-2} s. The two leg bones (the femurs), which will withstand a maximum compressive stress of about 10^8 N/m^2, must provide the force to stop the upper body. Assume that the cross section of each femur is about 1.0 cm^2 and that the body mass is about 50 kg. What is the maximum height that one can jump from, landing stiff-legged, and not break the leg bones?

5

Energy

We live in an era of much public concern and discussion about energy—about its shortage and about the need to develop new sources while conserving present ones. This is also the era of global energy politics. Those countries possessing large quantities of energy resources have become prominent in the economic and political affairs of the world.

Energy plays an important role in every aspect of life on our planet. Our technological society runs on energy, and its material affluence depends strongly on its rate of energy use. Without electricity and the energy products of crude oil, our society would quite literally grind to a halt. Moreover, all biological life depends on the use of energy from food. Our food supply is a storehouse of energy from the sun.

Yet energy is a somewhat elusive concept. On some days we have a lot of "energy." We can work tirelessly all day and into the night. Yet on other days—if we are sick, for instance—we "don't have enough energy to get out of bed." Observing a group of children at play we wonder, "Where do they get all that energy?" But subjective observations about energy, however correct, are not sufficiently precise for physics; we need to put the concept of energy in quantitative terms. In this chapter we shall define work and energy and examine the different forms of energy and the basic units of energy and power. We shall look briefly at the conversion of one form of energy to another, and at energy consumption by our society.

MECHANICAL ENERGY

Technological advances during the Industrial Revolution, in the late eighteenth and early nineteenth centuries, made it necessary to compare the efficiency of different machines and methods of doing work. If a waterwheel or a steam engine was to be designed to replace animal power, the new system's energy output must be quantitatively measured. It became accepted engineering practice to measure the output of an engine by having it lift weights. The product of the weight lifted multiplied by the distance through which the weight was lifted came to be used as a measure of mechanical work.

Work

In everyday usage "work" describes any physical or mental activity that requires exertion. We work on a car, we work on homework problems. The term also describes the condition of machines or equipment: A television set in need of repair is said to be not working. But in physics the term work has a very special meaning.

The **work** done on an object is the product of the magnitude of the applied force multiplied by the distance along which the force acts. (Conversely, **energy** is defined as the capacity of a system or object to do work.) The work done is a scalar quantity that has magnitude and units. The force

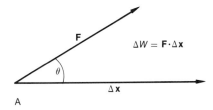

A

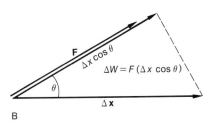

B

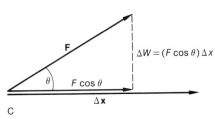

C

FIGURE 5-1
Work can be represented as the dot product of a force times a displacement: $\Delta W = \mathbf{F} \cdot \Delta \mathbf{x}$.

$\mathbf{F}$ is a vector quantity. The distance along which the force acts is a displacement $\Delta \mathbf{x}$, also a vector quantity. Suppose the applied force $\mathbf{F}$ and the displacement $\Delta \mathbf{x}$ are at an angle θ to one another, as shown in Figure 5-1 A. The displacement parallel to the direction of the applied force, as shown in Figure 5-1 B, is just

$$\Delta x_\| = \Delta x \cos \theta$$

The work done in this case is just

$$\Delta W = F \Delta x_\| = F \Delta x \cos \theta$$

Conversely, we can define **work** as the magnitude of the displacement $\Delta \mathbf{x}$, multiplied by the component of force acting

in the direction of the displacement. The component of the force parallel to $\Delta \mathbf{x}$, as shown in Figure 5-1 C, is just

$$F_\| = F \cos \theta$$

The work done is just

$$\Delta W = F_\| \Delta x = F \cos \theta \, \Delta x$$

The work done is a scalar quantity that has magnitude and units but is independent of orientation in space. The work done is the same by either of the definitions given above.

We can express the multiplication of two vectors to give a scalar result using a notation known as the **dot product.** The work done can be written as

$$\Delta W = \mathbf{F} \cdot \Delta \mathbf{x} \qquad (5\text{-}1)$$

This is a shorthand way of saying

$$\Delta W = F \Delta x \cos \theta$$

If the component of force acts in the *same direction* as the displacement, we say that the work done on the object is *positive*. If the component of force acts in the *direction opposite* to the displacement, the work done is *negative*. If the force acts at a right angle to the displacement, no work is done.

Units of Work

The SI unit for work is the **joule** (J): *the work done by a force of one newton acting through a distance of one meter:* 1 joule = 1 newton-meter

$$1 \text{ J} = 1 \text{ N-m} = 1 \text{ kg-m}^2/\text{s}^2$$

In the CGS system of units, the unit of work is the **erg:** *the work done by a force of one dyne acting through a distance of one centimeter:*

$$1 \text{ erg} = 1 \text{ dyne-cm} = 1 \text{ gm-cm}^2/\text{s}^2$$

Mechanical Energy

Mechanical work is required to lift an object against gravity. An object with a mass m has a weight mg. To lift this object at constant velocity, an upward force of $\mathbf{F} = m\mathbf{g}$ is required. If we lift the object straight up, as shown in Figure 5-2, the displacement $\Delta\mathbf{x}$ is the same direction as the force, $\theta = 0$, and positive work is done on the object. The total work required to lift an object a distance h at constant velocity is

$$\Delta W = \mathbf{F} \cdot \Delta\mathbf{x} = mgh$$

This work can be recovered if the object is lowered, because then the applied force and the displacement are in opposite directions and the work done on the object is negative. Energy can be stored by doing work in changing the position or configuration of a system. This energy of position is known as **gravitational potential energy.**

A 70 kg person is lifted a vertical distance of 5 m. What is the change in potential energy of the system?

$$\Delta W = mgh = (70 \text{ kg})(9.8 \text{ m/s}^2)(5 \text{ m})$$

$$\Delta W = 3.43 \times 10^3 \text{ kg-m}^2/\text{s}^2 = 3430 \text{ J}$$

Mechanical work is also required to accelerate an object. If a force $\mathbf{F}$ is applied to a mass m, as shown in Figure 5-3, the object accelerates as given by Newton's second law

$$\mathbf{F} = m\mathbf{a}$$

The work done in constant acceleration of the object through a distance $\mathbf{s}$ is given by

$$\Delta W = \mathbf{F} \cdot \mathbf{s} = m\mathbf{a} \cdot \mathbf{s}$$

However, from our discussion of kinematics in Chapter 2 we know that for constant acceleration

$$2\,\mathbf{a} \cdot \mathbf{s} = v_f^2 - v_i^2$$

The work done on the accelerated mass is then

$$\Delta W = m\frac{(v_f^2 - v_i^2)}{2} = \tfrac{1}{2}mv_f^2 - \tfrac{1}{2}mv_i^2$$

$$\Delta W = KE_f - KE_i$$

The quantity $\tfrac{1}{2}mv^2$ is the **kinetic energy,** the energy of

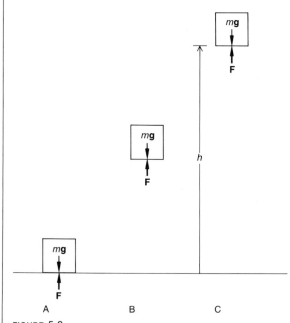

FIGURE 5-2
Mechanical work done in lifting a mass at constant velocity increases the potential energy.

FIGURE 5-3
Mechanical work done in accelerating a mass increases the kinetic energy.

FIGURE 5-4
When a mass falls, the work done by gravity is converted into kinetic energy.

motion of the mass. If an object of mass m is accelerated by gravity through a distance h (see Figure 5-4), the work done by the weight, $\Delta W = mgh$, increases the object's kinetic energy. If the applied force is in the same direction as the displacement, then positive work is done on the system. The kinetic energy of the mass increases. If the component of the force is in the direction opposite to the displacement, then negative work is done on the mass and the final kinetic energy will be smaller than the initial kinetic energy. Both energy of position, gravitational potential energy, and energy of motion, kinetic energy, are forms of **mechanical energy.**

Conservation of Mechanical Energy

In a simple system without friction or other methods of energy loss, the mechanical energy of the system is conserved. That is, total mechanical energy, the sum of the kinetic and potential energy, is constant:

$$E = KE + PE$$

In the case of gravitational potential energy, the mechanical work done to get to a certain height is independent of the path used to get there. The total energy of the system can then be written

$$E = \tfrac{1}{2}mv^2 + mgh \qquad (5\text{-}2)$$

We can often use this energy principle to determine the behavior of a system without knowing the details of the motion.

EXAMPLE

A tall roller coaster, 32 m high, has a total track length of 1150 m (see Figure 5-5). Assume that a car starts at the top of the ramp at rest, $v_o = 0$. It descends to ground level, a vertical distance of 32 m on the first ramp. Assume that the car has a mass of 1000 kg when fully loaded, and that the gravitational potential at ground level is zero.

FIGURE 5-5
The roller coaster converts gravitational potential energy into kinetic energy. Courtesy of Marriott's Great America®.

(a) What is the initial potential energy of the car at the top of the first incline?

(b) When the car has dropped a vertical distance of 16 m, what is the potential and kinetic energy (neglecting friction)?

(c) What is the maximum possible speed at the bottom of the first ramp?

The kinetic energy at the top of the ramp is zero, so that the potential energy at the top is also the total energy of the system

$$E = PE_1 = mgh_1 = (1000 \text{ kg})(9.8 \text{ m/s}^2)(32 \text{ m})$$

$$E = 4.13 \times 10^5 \text{ J}$$

Halfway down the incline some of the potential energy is converted into kinetic energy; but, neglecting friction, the total energy of the system is conserved. Because of conservation of energy, the total mechanical energy is

$$E = PE_2 + KE_2$$

The potential energy halfway down is just

$$PE_2 = mgh_2 = (1000 \text{ kg})(9.8 \text{ m/s}^2)(16 \text{ m})$$
$$PE_2 = 1.57 \times 10^5 \text{ J}$$

The kinetic energy is just

$$KE_2 = E - PE_2$$
$$= 3.14 \times 10^5 \text{ J} - 1.57 \times 10^5 \text{ J}$$
$$KE_2 = 1.57 \times 10^5 \text{ J}$$

One-half the potential energy has been converted into kinetic energy. We can use the definition of kinetic energy to calculate the velocity

$$KE_2 = \tfrac{1}{2}mv_2^2$$

which can be rewritten as

$$v_2 = \left[\frac{2KE_2}{m}\right]^{1/2}$$

Evaluating this, we obtain

$$v_2 = \left[\frac{(2.0)(1.57 \times 10^5)}{1000 \text{ kg}}\right]^{1/2}$$

$$v_2 = (314 \text{ m}^2/\text{s}^2)^{1/2} = 17.7 \text{ m/s}$$

At the base of the incline, all of the energy has been converted into kinetic energy. $KE_3 = E = 3.14 \times 10^5$ J. The final velocity is given by

$$v_3 = (2KE_3/m)^{1/2} = 25 \text{ m/s}$$

Sports

Transformation of kinetic energy into potential energy is important in many sports activities, including the high jump, the pole vault, and the diving board. To achieve maximum upward velocity in a standing vertical jump, a person bends at the knees, then simultaneously pushes upward with both legs and throws both hands upward. The legs do mechanical work to raise the center of gravity and to increase the body's kinetic energy. When the feet leave the ground the kinetic energy is converted into gravitational potential energy.

EXAMPLE

A person with a mass of 60 kg bends at the knees and pushes upward through a distance of 0.24 m in only 0.20 s (see Figure 5-6).

(a) What force must be exerted?

(b) How high can this person jump?

If the acceleration is uniform, the final takeoff speed v_1 will be twice the average speed:

$$v_1 = 2\frac{\Delta s}{\Delta t} = 2\frac{0.24 \text{ m}}{0.20 \text{ s}} = 2.4 \text{ m/s}$$

The kinetic energy at takeoff is

$$KE_1 = \tfrac{1}{2}mv_1^2 = \tfrac{1}{2}(60 \text{ kg})(2.4 \text{ m/s})^2 = 173 \text{ J}$$

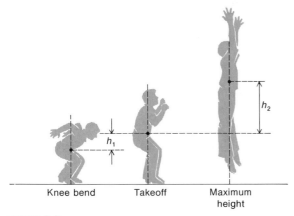

Knee bend   Takeoff   Maximum height

FIGURE 5-6
The vertical jump. Work done by the legs is converted to kinetic energy of the center of mass. As the center of mass rises, this kinetic energy is converted into potential energy.

After takeoff the kinetic energy is converted into gravitational potential energy of the center of mass. At maximum elevation the increase in potential energy equals the initial kinetic energy. The maximum elevation h_2 is given by

$$PE_2 = mgh_2 = KE_1 = \frac{1}{2}mv_1{}^2$$

$$h_2 = \frac{KE_1}{mg} = \frac{173 \text{ J}}{(60 \text{ kg})(9.8 \text{ m/s}^2)} = 0.29 \text{ m}$$

You might wish to see how high you can vertically jump from the ground and compare it with this value.

While the legs are pushing upward, mechanical work is done both to lift the person against gravity and to increase the kinetic energy. The work done in lifting the person is approximately

$$PE_1 = mgh_1 = (60 \text{ kg})(9.8 \text{ m/s}^2)(0.24 \text{ m}) = 141 \text{ J}$$

The total mechanical work done by the muscles in the jump is the sum of the potential and kinetic energy at takeoff.

$$\Delta W = KE_1 + PE_1 = 173 \text{ J} + 141 \text{ J} = 314 \text{ J}$$

The total work done is the product of the force **F** exerted by the muscles and the distance Δs through which that force acts. In the vertical jump the force and the distance are in the same direction, so the work done is their product

$$\Delta W = F \Delta s$$

The magnitude of the force is then

$$F = \frac{\Delta W}{\Delta s} = \frac{314 \text{ J}}{0.24 \text{ m}} = 13\underline{1}0 \text{ N}$$

Since the person's weight $mg = 588$ N, the force required to jump a height of 0.29 m is over twice the person's weight.

In the standing jump most persons can raise the center of gravity only between 0.3 to 0.6 m. Yet a high jumper can clear a bar at a height of 2.2 m. The center of gravity of a person 1.8 m tall is about 1.1 m above the floor and can be raised in the standing jump to perhaps 1.6 m. But this is not sufficient to clear 2.2 m.

In the high jump the athlete makes a running approach to the bar at speeds of the order of 5 m/s. In the final jumping step this forward motion is redirected vertically, driving the center of gravity upward by an average thrust of four times the body weight. A good high jumper, first, employs a forceful, fast, and long takeoff thrust, and second, has mastered the technique of curling over the bar, keeping the center of gravity as close to the bar as possible.

In the pole vault the conversion of forward kinetic energy to vertical potential energy is even greater (see Figure 5-7). The athlete approaches at the maximum speed that strength and timing allow. As he drives the end of the fiberglass pole into the box, he lunges against the pole, which bends to absorb the kinetic energy as potential energy. As the vaulter swings upward the remaining forward kinetic energy and the potential energy stored in the pole are converted into altitude. Moreover, the vaulter pulls upward against the pole, doing additional work to increase elevation as well as to improve position to achieve maximum altitude. Heights of 5.6 m have been reached in competition.

In jumping from a diving board one takes a small hop step, an upward leap, a return to the board, and a takeoff

FIGURE 5-7
In the pole vault the kinetic energy of forward motion is converted to the potential energy in the bent pole, and then to gravitational potential energy. From *The Mechanics of Athletics*, 4th ed., by Geoffrey H. G. Dyson. Copyright © 1968 by Hodder & Stoughton Limited. Used with permission.

jump from the end of the board. The work done in the small hop step is stored as potential energy in the springy board. In the upward leap the work done by the legs is added to the energy returned from the board to increase the kinetic energy. Once returned to the board this energy is again stored as potential energy. As the board begins to spring back the diver makes a final takeoff jump. The energy from the board plus the work done in the final jump increases the kinetic energy to achieve maximum height. In the ideal takeoff all of the energy from the board is absorbed as the diver points the toes at the last instant. The board should not rebound as it does in a less than perfect dive, when energy remains in the board.

GRAVITATIONAL POTENTIAL ENERGY

The acceleration due to gravity, more or less constant near the surface of the earth, becomes weaker as we move away from the earth. Newton's law of universal gravitation gives the magnitude force on a mass m at a distance R from the earth with mass M_e, and is given previously by equation 4-1:

$$F = \frac{GM_e m}{R^2}$$

To lift an object from the surface of the earth we must do work. In doing work against gravity we increase the gravitational potential energy.

We can determine the work required to lift an object of mass m from a distance R_1 from the center of the earth to a distance R_2, as shown in Figure 5-8. The work done lifting the object will be just the average force exerted times the distance traveled. If R_2 is not too different from R_1, the square of the distance is approximately

$$R^2 = R_1 R_2$$

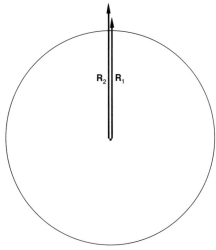

FIGURE 5-8
A displacement $\Delta\mathbf{R} = \mathbf{R}_2 - \mathbf{R}_1$ is obtained in lifting an object from position $\mathbf{R}_1$ to position $\mathbf{R}_2$.

The average force required to lift the object is then

$$F = \frac{GM_em}{R_1R_2}$$

The distance over which the force acts is

$$\Delta R = R_2 - R_1$$

The force applied is in the same direction as the displacement, so the work required in lifting the mass is

$$\Delta W = F\Delta R = \frac{GM_em}{R_1R_2}(R_2 - R_1) = GM_em\left[\frac{1}{R_1} - \frac{1}{R_2}\right]$$

The work done to go over a small increment ΔR can be repeated over many increments. The work done to get from some initial distance R_i to some final distance R_f against the pull of gravity is just

$$\Delta W = GM_em\left[\frac{1}{R_i} - \frac{1}{R_f}\right] \tag{5-3}$$

We can define the gravitational potential energy as the work required to go from infinity $R_i = \infty$ to some final radius R. The gravitational potential energy of an object with mass m in the vicinity of another mass M_e is just

$$U = -\frac{GM_em}{R} \tag{5-4}$$

We live at the bottom of a **potential energy well** produced by the earth's gravitation. For a spacecraft to go to the moon work must be done to lift it. Enormous rocket engines lift the launch vehicle against gravity. The spaceship accelerates to sufficient speed that its kinetic energy will carry it to the moon. In Figure 5-9 we see the potential energy of a 1 kg mass in the earth-moon system. The moon has a smaller mass and radius than the earth's. It is at a distance of 60 earth radii. A lunar spacecraft traveling toward the moon from earth gains potential energy while losing kinetic energy out to about 54 earth

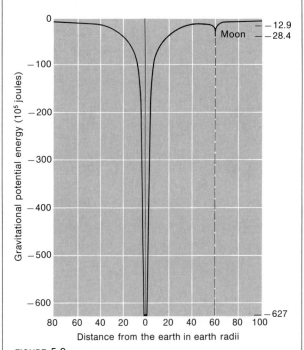

FIGURE 5-9
The gravitational potential energy well of the earth over the moon for a 1 kg mass. A space capsule traveling to the moon must increase its total energy to escape the potential well of the earth but then must decrease it to drop into orbit around the moon. It is much easier to leave the moon than to leave the earth.

radii. At this point the gravitational potential of the moon becomes important. The rocket accelerates as it gets closer to the moon. To drop into a parking orbit around the moon, the spacecraft must fire its retro-rockets to decrease its kinetic energy.

Bound Orbits

An object's total energy includes both the gravitational potential energy and the kinetic energy. In the case of a gravitational potential energy well, the total energy represents the satellite's possible range.

$$E = PE + KE$$

$$E = -\frac{GM_e m}{R} + \tfrac{1}{2}mv^2$$

If the total energy is less than zero, $E < 0$, the satellite is in a **bound orbit.** As it travels away from the earth it continues to lose velocity until it reaches zero velocity. At this point it is still within the influence of the earth's field, and it will start back.

If a satellite of mass m is in a circular orbit around the earth at a radius R, we can calculate the total energy. The earth's gravitational attraction provides the centripetal force to hold the satellite in a circular orbit.

$$F = \frac{mv^2}{R} = \frac{GM_e m}{R^2}$$

We can at once solve for the kinetic energy

$$\tfrac{1}{2}mv^2 = \tfrac{1}{2}\frac{GM_e m}{R}$$

The total energy is just

$$E = -\frac{GM_e m}{R} + \tfrac{1}{2}mv^2 = -\frac{GM_e m}{R} + \frac{GM_e m}{2R}$$

$$E = -\frac{GM_e m}{2R}$$

The uniform circular orbit is a bound orbit with total energy of less than zero.

Unbound Orbits

If the total energy of the satellite is greater than zero, $E > 0$, then the satellite is in an **unbound orbit:** The satellite is free to leave the gravitational influence of the planet. The Mariner 11 probe to Jupiter and Saturn was launched with sufficient velocity to leave the gravitational potential well of the earth entirely. As Mariner 11 approached Jupiter it sped up. But after swinging once past the planet, it headed off for a rendezvous with Saturn.

An object a distance R from a planet must have a certain velocity before it can escape from the planet's gravitational potential energy well. For the satellite to escape, total energy must be greater than or equal to zero. The **escape velocity** is just the minimum velocity required to escape the planet.

$$E = 0 = -\frac{GM_e m}{R} + \tfrac{1}{2}mv^2$$

Solving for the velocity we get that

$$v_{escape} = \left[\frac{2GM_e}{R}\right]^{1/2} \qquad (5\text{-}5)$$

EXAMPLE

What is the escape velocity from the earth with a mass of $M_e = 6.0 \times 10^{24}$ kg and a radius of 6.38×10^6 m?

$$v_{escape} = \left[\frac{(2)(6.67 \times 10^{-11}\ \text{N-m}^2)}{\text{kg}^2}\frac{(6.0 \times 10^{24}\ \text{kg})}{(6.38 \times 10^6\ \text{m})}\right]^{1/2}$$

$$v_{escape} = 1.12 \times 10^4\ \text{m/s}$$

THE MECHANICAL EQUIVALENT OF HEAT

We have sensory experience of hot and cold, properties of matter that can be measured with a thermometer (using the Fahrenheit or Celsius temperature scale). Atoms and molecules possess kinetic energy because of their random motion. If this random kinetic energy is decreased or increased, we say that **heat** has been lost or gained by the body.[1] As will be discussed in detail in Chapter 10, water is used as a standard substance to define our units of energy transfer and the Celsius temperature scale.

The **calorie** is defined as the amount of energy transfer, heat, required to raise the temperature of one gram of water from 14.5°C to 15.5°C, a difference of one degree Celsius. In the 1840s James Prescott Joule performed experiments demonstrating that mechanical work could be converted into heat and raise the temperature of liquids. In his experiments paddlewheels were immersed in a liquid and were driven by forces that were precisely measured. Calculating the work done on the system, Joule was able to determine the **mechanical equivalent of heat.** The accepted value is that one calorie (cal) is equal to 4.184 joules of energy.

$$1 \text{ calorie} = 4.184 \text{ J}$$

At one time or another almost everyone has been conscious of diet and has "counted calories." The food calorie is the kilocalorie (kcal), the amount of energy necessary to raise one kilogram of water 1°C.

$$1 \text{ kcal} = 1000 \text{ cal}$$

In engineering units, the British thermal unit (Btu) is the amount of heat required to raise the temperature of one pound mass of water from 62°F to 63°F, a difference of one degree Fahrenheit. A pound mass (lbm) is a mass with a weight on earth of one pound. The energy equivalent of the one Btu of heat is 1054.8 joules.

$$1 \text{ Btu} = 1054.8 \text{ J}$$

Conservation of Energy

The concept of energy was developed through the work of many men. In the late eighteenth century, Count Rumford had demonstrated a direct relationship between the amount of mechanical work done in boring cannons and the amount of heat liberated. In Paris in 1824, Sadi Carnot established the basic processes by which heat could be converted into useful mechanical work. At about this same time, Joseph Mayer was able to show the direct relationship between the mechanical work done in compressing a gas and the heat generated in the process. In the 1840s, Joule, having carefully measured the heat produced from various forms of both mechanical and electrical work, was the first to conclude that a quantity having the dimensions of work is conserved through all of these different conversions. (We shall discuss the conversion of thermal energy to mechanical energy in Chapter 12.)

It was Hermann von Helmholtz who made the definitive statement about energy. Helmholtz was concerned with the interrelationship of various forms of energy, including chemical energy and electrical energy. In his scientific memoirs Helmholtz states that, from investigation of known physical and chemical process,[2]

we arrive at the conclusion that nature as a whole possesses a store of [energy][3] which cannot in any way be either increased or diminished; and that, therefore, the quantity of [energy] in nature is just as eternal and unalterable as the quantity of matter. Expressed in this form, I have named the general law "the principle of the conservation of [energy]."

It is useful to define because it is conserved: Energy is a quantity having the dimensions of work. The sum of all forms of this quantity, expressed in proper units, is conserved through all interactions of nature that take place in a closed system. Energy is neither created nor destroyed; it is simply transferred from one system to another, from one form to another.

The great unfolding drama of the universe is the flow of energy. Matter, first distributed more or less uniformly throughout all space, coalesces and condenses because of

---

[1] We commonly think of heat as something that inheres in an object—the kinetic energy of its atoms and molecules. That energy is more properly called *internal energy*. **Heat** *is the transfer of energy into or out of a system.*

[2] Hermann von Helmholtz, *Scientific Memoirs, Natural Philosophy* (1853), translated by John Tyndall. Quoted in Leon N. Cooper, *An Introduction to the Meaning and Structure of Physics,* Short Edition (New York: Harper & Row, 1970), pp. 103–104.

[3] Helmholtz used the German word *Kraft,* which was originally translated as "force." His concept was identical with what we now call energy.

gravitational attraction. As gravity compresses the matter further, young stars are born and begin their own evolutionary history. These stars radiate their energy out into space, warming planets such as our own. The combination of elements present on the planet and the continuous flux of solar energy begin the process of chemical and biological evolution. This flux of solar energy sustains the life of our planet.

BIOENERGETICS

Chemical Energy

Most of the activity of our biological world is the direct consequence of chemical reactions. **Chemical energy** is stored as potential energy in configurations of different atoms and molecules. Chemical reactions redistribute this potential energy. Exothermic reactions release it in the kinetic energy of the reaction products; endothermic reactions require that energy be added, such as the energy provided by sunlight in photosynthesis.

If 2 gm of hydrogen react with 16 gm of oxygen, 1 mole or 18 gm of water is produced. The net liberation of energy Q is 2.84×10^5 joules.

$$H_2 + O \rightarrow H_2O + 2.84 \times 10^5 \text{ J/mole}$$

This is a substantial quantity of energy. If it were all converted into mechanical energy with a 30% efficiency (typical of a modern fossil fuel powerplant), it could lift a 70 kg man a distance of 124 m.

Let us look at the energy liberated in the formation of a single molecule of H_2O. One mole of a substance contains Avogadro's number of molecules

$$N_A = 6.02 \times 10^{23} \text{ molecules/mole}$$

The energy per reaction q is just

$$q = \frac{Q}{N_A} = \frac{2.84 \times 10^5 \text{ J/mole}}{6.02 \times 10^{23} \text{ molecules/mole}}$$

$$q = 4.7 \times 10^{-19} \text{ J}$$

This may seem to be a very small amount of energy, but

only because there is such a huge number of molecules in one mole of a substance.

The **electron volt** (eV) is a useful unit of energy when considering individual atoms, molecules, electrons, or elementary particles. We shall carefully define this unit in our discussion of electricity; for now we simply state that one electron volt is equal to 1.6×10^{-19} joules:

$$1 \text{ eV} = 1.6 \times 10^{-19} \text{ J}$$

Thus, in these units, the energy liberated in the formation of a single molecule of H_2O is:

$$q = \frac{4.7 \times 10^{-19} \text{ J}}{1.6 \times 10^{-19} \text{ J/eV}}$$

$$q = 2.95 \text{ eV}$$

The energies associated with chemistry are typically of the order of a few eV. For instance, the energy required to free the electron from a hydrogen atom is 13.6 eV.

Bioenergetics

An organic cell can function only if it can extract and use the chemical energy locked in organic molecules. In all living cells, from bacteria to those in the human body, the major energy-carrying molecule is adenosine tri-phosphate (**ATP**). ATP is a large organic molecule with three phosphate groups. The aromatic ring structure, *adenine*, is connected to the five-carbon sugar *D-ribose*, to which three phosphate groups are attached. In the intact cell ATP is highly charged. At pH 7.0, each of the phosphate groups is ionized, and ATP has four negative charges, ATP^{4-}. In contact with water and the proper enzymes, ATP undergoes hydrolysis and is converted into adenosine di-phosphate (ADP) while liberating considerable free energy:

$$ATP^{4-} + H_2O \rightarrow ADP^{3-} + HPO_4{}^{2-} + H^+$$

The free energy liberated in the reaction is about 2.9×10^4 J/mol or 7000 kcal/mol. This corresponds to an energy release of about 0.30 eV per molecule. Hydrolysis of ATP liberates significantly more energy than do other organic molecules. For this reason ATP is sometimes called a high-energy phosphate compound with high-energy phosphate bonds. The energy is not in fact released by breaking of the phosphate bond but rather by the transfer of charge in the chemical reaction.

As Figure 5-10 indicates, the energy stored in ATP can be used by the cell to perform mechanical work in muscular contraction, to perform osmotic work in the active transport of ions across cell membranes, and to synthesize other organic molecules. ATP is synthesized in the body by the metabolic breakdown of carbohydrates, such as glucose, and of proteins and phospholipids. Most of the ATP synthesis takes place in small bodies within the cell known as mitochondria. Chemical energy from large molecules such as glucose is used to recombine ADP and inorganic phosphate to produce ATP.

$$ADP + P + 0.3 \, eV/molecule \rightarrow ATP + H_2O$$

Glucose, $C_6H_{12}O_6$, a major body fuel, has a molecular weight of 180 gm. If one mole is oxidized it liberates 2.87×10^6 joules of energy.

$$C_6H_{12}O_6 + 6\,O_2 \rightarrow 6\,CO_2 + 6\,H_2O + 2.87 \times 10^6 \, J/mole$$

About 40% of this energy is stored in the synthesis of ATP. The remainder is dissipated as heat. The oxidation of one molecule of glucose is sufficient to synthesize about 38 molecules of ATP. The synthesis and dissociation of ATP is the primary energy cycle of terrestrial biology and has great versatility in providing the energy needs of the cell.

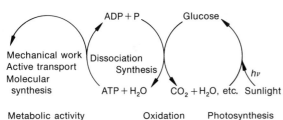

FIGURE 5-10
The synthesis and dissociation of ATP is the primary energy cycle of biology.

The ultimate source of biological energy is the sun's light, radiant energy that travels through space. Radiant energy can be treated as a wave (as we shall discuss in Chapters 18 and 20) or (as discussed in Chapter 21) also as tiny bundles of energy, energy quanta called photons ($h\nu$).

Plants absorb the energy from sunlight in **photosynthesis** of carbohydrates. Chlorophyll-a absorbs sunlight at the red and blue ends of the visible light spectrum and reflects the other colors, which is why growing plants have a characteristic green color. The energy of photons ranges from 1.8 eV in the red to about 3.0 eV in the blue.

In photosynthesis, one carbohydrate unit (CH_2O) is synthesized from carbon dioxide and water by absorbing eight photons.

$$4\,H_2O + 4\,h\nu \rightarrow O_2 + 2\,H_2O + 4\,H$$
$$CO_2 + 4\,h\nu + 4\,H \rightarrow (CH_2O) + H_2O$$

The eight absorbed photons have a total energy of about 15 eV. The energy stored in the carbohydrate unit and liberated when its carbon atom is oxidized amounts to about 5 eV. The overall efficiency of the photosynthesis in storing energy is about 33%.

Fossil Fuels

Since the introduction of the steam engine, in the nineteenth century, our civilization has run on fossil fuels. Through geological time, the energy in sunlight has been stored in organic molecules by the process of photosynthesis. These molecules, as the constituents of various kinds of plant and animal matter, were laid down, covered over, and compressed to form fossil fuels: coals, oil, natural gas, oil shale, and tar sands. Burning these fuels liberates chemical energy that is used in engines to turn the gears and wheels of our mechanized, industrial society.

Table 5-1 lists the energy released in combustion of various fuels and common light gases. The primary energy-producing elements in fossil fuels are carbon and hydrogen. The energy content of most of the fuels and gases listed is about 5×10^7 joules per kilogram, although it is much higher for hydrogen and somewhat lower for coal, which is partially oxidized.

TABLE 5-1
Energy of combustion of common fuels and light gases.

Fuel	Composition	Combustion energy content J/kg
Texas crude oil	85% C, 12% H, 1.75% S, .7% H	4.5×10^7
Gasoline	85% C, 15% H	4.6×10^7
Illinois coal, bituminous	69% C, 5% H, 1.6% N, 15% O, 1% S	2.8×10^7
Pennsylvania natural gas	67% ethane, 32% methane	5.0×10^7
West Virginia natural gas	32% ethane, 68% propane	5.0×10^7

Gas	Formula	Molecular wt gm/mole	Combustion energy content J/kg
Hydrogen	H_2	2	14.2×10^7
Methane	CH_4	16	5.5×10^7
Ethane	C_2H_6	30	4.9×10^7
Propane	C_3H_8	44	5.0×10^7
Butane	C_4H_{10}	58	4.8×10^7

EXAMPLE

The combustion of propane, C_3H_8, releases 5.0×10^7 J/kg. Propane has a molecular weight of 44 gm/mole. The reaction is

$$C_3H_8 + 7\,O_2 \rightarrow 3\,CO_2 + 4\,H_2O + Q$$

What is the energy released, in eV, per molecule of propane?

The energy released per molecule is given by

$$q = 5.0 \times 10^7 \frac{\text{joules}}{\text{kg}} \times \frac{1\,\text{kg}}{1000\,\text{gm}} \times \frac{44\,\text{gm}}{\text{mole}}$$

$$\times \frac{1\,\text{mole}}{6.02 \times 10^{23}\,\text{molecules}}$$

$$q = 3.65 \times 10^{-18} \frac{\text{joules}}{\text{molecule}} \times \frac{1\,\text{eV}}{1.6 \times 10^{-19}\,\text{joules}}$$

$$q = 22.8\,\text{eV/molecule}$$

The energy released in the combustion of one molecule of propane is 22.8 eV, liberated primarily as kinetic energy of the reaction products and appearing as heat.

POWER

Power *is the time rate at which energy is produced or used in a system.* Power has the dimensions of energy or of work per unit time.

$$P = \frac{\Delta E}{\Delta t} = \frac{\Delta W}{\Delta t} \qquad (5\text{-}6)$$

The SI unit of power is the **watt** (W), which is equal to a joule per second: 1 W = 1 J/s. The engineering unit of power is the **horsepower** (hp): One horsepower is 550 foot-pounds **(ft-lbf)** per second, equivalent to 745 watts.

$$1\,\text{hp} = 550\,\text{ft-lbf/s} = 745\,\text{watts}$$

Table 5-2 shows common units of power.

TABLE 5-2
Commonly used units of power.

Unit of power	joules/s
1 watt (W)	1.000
1 kilowatt (kW)	1.000×10^3
1 megawatt (MW)	1.000×10^6
1 Btu/hr	0.293
1 horsepower (hp)	745

Mechanical power is the time rate at which mechanical work is performed. If a force **F** acts through a distance $\Delta \mathbf{x}$ the amount of work done is given by

$$\Delta W = \mathbf{F} \cdot \Delta \mathbf{x}$$

If this work is done over a time interval Δt, the average power required is

$$P = \frac{\Delta W}{\Delta t} = \frac{\Delta(\mathbf{F} \cdot \mathbf{x})}{\Delta t} = \mathbf{F} \cdot \frac{\Delta \mathbf{x}}{\Delta t}$$

However, $\Delta \mathbf{x}/\Delta t$ is just the average velocity over the interval **v** so that the average power over the interval is

$$P = \mathbf{F} \cdot \mathbf{v} = Fv \cos \theta \qquad (5\text{-}7)$$

In a very short time interval, the average velocity approaches the instantaneous velocity. The instantaneous power is then the product of the instantaneous force times the velocity component in the direction of the force.

EXAMPLE

An elevator is traveling uniformly upward at 1.2 m/s. The net force applied is 6.7×10^3 N. What power is required to lift the elevator in watts? in horsepower?

$$P = \mathbf{F} \cdot \mathbf{v} = (6.7 \times 10^3 \text{ N})(1.2 \text{ m/s})$$

$$P = 8.0 \times 10^3 \text{ W}$$

where 1 N-m/s = 1 J/s = 1 W. It requires about 8 kilowatts of mechanical power to lift the elevator. The horsepower required is given by

$$P = (8.0 \times 10^3 \text{ W})\frac{(1 \text{ hp})}{(745 \text{ W})} = 10.8 \text{ hp}$$

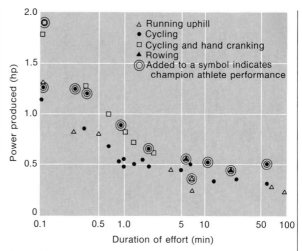

FIGURE 5-11
The power produced by a human being depends on duration of effort. A champion athlete can produce almost 2 hp briefly and 0.5 hp for extended periods. Source: D.R. Wilkie, *Journal of the Royal Aeronautical Society,* 64 (August 1960), p. 479.

The rate at which the human body can do mechanical work has been measured. Trained athletes can put out 0.5 hp or more for short periods of time, and 0.3 hp for up to an hour. An output of 0.1 hp can be sustained for a long time (see Figure 5-11).

Muscular production of mechanical power depends on oxygen consumption. As physical exertion increases, the rate of blood flow is increased by an increased rate of heart beat. This increase continues until the maximum heart rate is reached. The supply of oxygen to the muscles is then limited and the person quickly becomes exhausted. One purpose of athletic training, and the main purpose of jogging and "aerobic" exercises, is to increase the efficiency and strength of the lungs, heart, and diaphragm to increase the maximum oxygen consumption. The heart rate as a function of the power being exerted on a treadmill is shown in Figure 5-10. The oxygen supply limits the sustained power that can be exerted, and thus limits an athlete's performance.

Bicycle Riding

A bicycle rider must do mechanical work against gravity and air resistance. (Of course, much of the mechanical work that must be done in going uphill is recovered in coming back downhill.) The force required to push a rider through the air depends on the effective cross-section area of the rider A, the density of the air ρ (the Greek letter *rho*), and the square of the velocity relative to the air.

$$F = \frac{1}{2}\rho v^2 A \tag{5-8}$$

A bicycle rider's effective area is about $A = 0.33 \text{ m}^2$. The density of air is $\rho = 1.3 \text{ kg/m}^3$.

The mechanical work is done against the air resistance. Since the applied force is in the direction of motion, the power required is given by

$$P = Fv = \frac{1}{2}\rho A v^3$$

The power required increases as the cube of the velocity!

EXAMPLE

Suppose that a 70 kg bicycle rider on a 10 kg bicycle can sustain a power output of 0.30 horsepower. How fast can the rider go on level ground with no headwind? If the muscles produce mechanical energy with 25% efficiency, how much energy in kilocalories is consumed in riding the bicycle at this speed for 30 minutes?

The power in metric units is

$$P = 0.30 \text{ hp} \times \frac{745 \text{ W}}{1 \text{ hp}} = 224 \text{ W}$$

The maximum velocity is given by

$$v^3 = \frac{2P}{\rho A} = \frac{2(224 \text{ J/s})}{(1.3 \text{ kg/m}^3)(0.33 \text{ m}^2)} = 10\underline{5}0 \text{ m}^3/\text{s}^3$$

$$v = 10 \text{ m/s} = 36 \text{ km/hr}$$

At 25% efficiency the energy required to sustain this power level for 30 minutes is

$$E = (224 \text{ J/s})(1800 \text{ s})\frac{(100\%)}{(25\%)} = 1.6 \times 10^6 \text{ J}$$

Since one kilocalorie equals 4183 joules, the energy re-

quired for this activity is $3\underline{9}0$ kcal—roughly the energy content of one-third of a pound of ground beef, a cup of ice cream, or 100 gm of rice. It takes a great deal of exertion to "burn off" extra calories.

Someone riding a bicycle uphill must do work against gravity. The slope of a hill expressed as a percentage is the rise divided by the distance. A 5% slope will rise 5 m in 100 m of travel. If θ is the angle with the horizontal, $\sin \theta = 0.05$. The component of weight that is parallel to the hill is

$$F_g = mg \sin \theta \tag{5-9}$$

The force exerted by the rider to travel up the hill at a speed v is

$$F = mg \sin \theta + \frac{1}{2}\rho v^2 A$$

The power exerted by the rider is

$$P = Fv = mgv \sin \theta + \frac{1}{2}\rho A v^3$$

EXAMPLE

A 70 kg rider on a 10 kg bicycle can sustain a power output of 0.30 hp = 224 W. What is the maximum slope (Figure 5-12) that the rider can climb at 5 m/s? The rider has an area $A = 0.33 \text{ m}^2$.

FIGURE 5-12
Bicycling uphill requires work to raise the center of gravity.

We can solve the above equation for the slope

$$\sin \theta = \frac{P - \frac{1}{2}\rho A v^3}{mgv}$$

The second term in the numerator is just the power required to overcome wind resistance

$$\frac{1}{2}\rho A v^3 = \frac{1}{2}(0.33 \text{ m}^2)(1.3 \text{ kg/m}^3)(5.0 \text{ m/s})^3$$

$$\frac{1}{2}\rho A v^3 = 27 \text{ W}$$

We can then solve for the slope

$$\sin \theta = \frac{224 \text{ W} - 27 \text{ W}}{(80 \text{ kg})(9.8 \text{ m/s}^2)(5.0 \text{ m/s})}$$

$$\sin \theta = 0.050$$

The hill has a 5% slope. A rider could climb a steeper hill only at a lower speed. The maximum slope that could be climbed at this power would be a 5.7% grade.

An athlete can exert very great effort for a short time. At a given instant the body holds an amount of oxygen roughly equivalent to the rate of oxygen supply in a minute. Over short times, such as the duration of a 100-meter dash, exertion will consume much more oxygen than is supplied by respiration. During such sprints the body is functioning on stored oxygen. If exertion continues after the stored oxygen is exhausted, the athlete may collapse—as happens at or near the end of a race if an athlete has not properly paced himself or herself. The maximum work that can be done by the body anaerobically, that is, on the stored oxygen alone, is about 10 000 ft-lbf or 1.4×10^4 J. This would correspond to a power output of 0.50 hp for about 36 s.

The maximum work that a bicycle rider can do anaerobically is 1.4×10^4 J. A bicycle rider with a total mass of 80 kg sprints up a hill at a power output of 0.50 hp (Figure 5-13). What is the maximum height of the hill the rider can power over on stored oxygen?

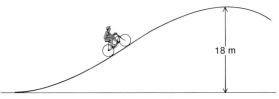

FIGURE 5-13
The amount of work the body can do on stored oxygen limits the height of a hill that can be sprinted over.

The primary effort is done in raising the rider and bicycle against gravity. We assume a low speed so that we can neglect air resistance. The maximum height h is given by

$$h = \frac{W}{mg} = \frac{1.4 \times 10^4 \text{ J}}{(80 \text{ kg})(9.8 \text{ m/s}^2)}$$

$$h = 18 \text{ m}$$

Hills higher than 18 m (about 60 ft) require that the rider gear down for a steady grind rather than trying to sprint over them.

ENERGY AND MASS

In 1905 Albert Einstein published a short paper, "Does the Inertia of a Body Depend on its Energy?" Einstein found it necessary to make a correction to Newton's second law. In Newton's mechanics a body of mass m and a velocity $\mathbf{v}$ possesses a momentum:

$$\mathbf{p} = m\mathbf{v}$$

Newton's second law can be written that the force $\mathbf{F}$ acting on a body causes a change in momentum per unit time:

$$\mathbf{F} = \frac{\Delta \mathbf{p}}{\Delta t} = \frac{\Delta m\mathbf{v}}{\Delta t}$$

Einstein found that the correct momentum at a velocity v is given by

$$\mathbf{p} = \frac{m_o \mathbf{v}}{(1 - v^2/c^2)^{1/2}}$$

This is a small correction for velocities much less than the speed of light.

In this same paper Einstein also first stated the equivalency of mass and energy. The energy associated with matter is given by

$$E = mc^2 \qquad (5\text{-}10)$$

where c is the speed of light, $c = 3.00 \times 10^8$ m/s. The mass m is the **relativistic mass** given by

$$m = \frac{m_o}{(1 - v^2/c^2)^{1/2}} \qquad (5\text{-}11)$$

m_o is the mass of the object at rest.

When the velocity is small compared to the speed of light, Einstein's equation can be related to our definition for the kinetic energy of an object. The total energy can be written

$$E = mc^2 = m_o c^2 \left(1 - v^2/c^2\right)^{-1/2}$$

If the velocity of the object is much less than the speed of light, then $v^2/c^2 \ll 1$, and we can expand the square root using the binomial theorem (see Appendix A). The total energy is then

$$E = m_o c^2 = m_o c^2 \left(1 - \tfrac{1}{2}\left(-v^2/c^2\right) + \ldots\right)$$

or

$$E = m_o c^2 + \tfrac{1}{2} m_o v^2 \qquad (5\text{-}12)$$

The total energy of the particle is the sum of the rest mass energy $m_o c^2$ plus the kinetic energy $\tfrac{1}{2} m_o v^2$. At velocities near the speed of light the total energy of a particle is

$$E = mc^2 = \frac{m_o c^2}{(1 - v^2/c^2)^{1/2}} \qquad (5\text{-}13)$$

There are two important processes by which a small amount of mass can be converted into energy: fission and fusion. **Fission** (discussed in Chapter 24) involves splitting the nucleus of uranium, thorium, or plutonium. When a large nucleus such as that of uranium-235 absorbs a neutron it becomes unstable and splits into two smaller nuclei with the release of additional neutrons. The mass of the final products is less than the mass of the initial products, with the mass difference converted into energy. A single reaction can liberate as much as 200 million electron volts (MeV) of energy, as much as 2.0×10^{13} joules per mole! The human ability to effect this great release of energy has had enormous impact on global military strategy since the 1940s. It may also be a significant source of commercial power for our society in the next quarter-century.

Fusion (discussed in Chapter 26) involves the combining of nuclei of hydrogen, deuterium, or tritium in energetic reactions. The mass of the nuclei formed is less than the mass of the initial nuclei, with the mass difference converted into energy. Fusion is the principal source of energy in stars.

ENERGY USE

Energy is conserved. The total of all forms of energy in a closed system is constant. However, by many different processes energy can be converted from one form to another. Figure 5-14 shows the six basic forms of energy and indicates some of the possible conversions from any one form to others.

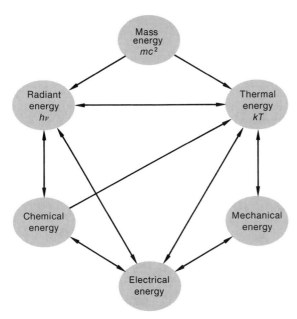

FIGURE 5-14
The fundamental forms of energy are interrelated, and often can be converted from one into another.

Some conversions are very efficient. (Efficiency is the ratio of useful work to energy consumed.) Mechanical energy can be totally converted to thermal energy by the process of friction, because work has to be done only to overcome friction, not to maintain uniform motion. Electrical energy is converted into thermal energy by the use of resistance heating. A generator can convert mechanical energy into electrical energy with efficiencies of the order of 85%. The electric motor can reverse this conversion also with high efficiency. Batteries and fuel cells can reversibly convert chemical energy into electrical energy. Absorption efficiently converts radiant energy to thermal energy, and combustion converts chemical energy into thermal energy.

The nature of thermal energy—random kinetic energy of the molecules—makes its conversion into mechanical energy relatively inefficient. The best fossil fuel power-plants have an overall efficiency of only 40%. Nuclear powerplants, which operate at a lower temperature, have only about a 33% efficiency. The automobile's internal combustion engine quite efficiently converts chemical energy to thermal energy but converts thermal energy into mechanical energy with only about 8% efficiency.

Other processes are quite inefficient. Only a portion of the spectrum of visible light is involved in photosynthesis, demonstrating an efficiency of about 30% in laboratory experiments. In usual growing conditions, the energy-storage efficiency over a year is only about 0.2%. Photoelectric conversion of sunlight directly into electricity has an efficiency of 2 to 10%. Application of this technology is at present limited by costs.

Energy is converted from one form to another as it moves through our society. Some of the radiant energy of sunlight is stored as chemical energy in plants, which over geological time are compressed to become coal. During combustion in a furnace the chemical energy of the coal is liberated to become the thermal energy in steam. The energy is then converted into mechanical energy with about 40% efficiency, which at once is converted into electrical energy for distribution. As electrical energy it can be converted into light, mechanical energy, or heat. Because of its ease of conversion, electrical energy has become the coin of the realm.

Units of Energy

In public discussions many units are used to describe quantities of energy. One author will quote the world's energy consumption in quadrillion Btus. Our consumption of domestic and foreign crude oil is measured in millions of barrels per day. A utility company reports electric energy generation in kilowatt-hours, or perhaps kilowatt-years. Our strip-mine reserves of coal in Black

TABLE 5-3
Common units of energy.

Unit of energy	joules (J)
1 ton (short) bituminous coal	2.76×10^{10}
1 ton of TNT equivalent	4.2×10^{9}
1 barrel crude oil	5.9×10^{9}
1 gallon No. 2 fuel oil	1.45×10^{8}
1 gallon regular gasoline	1.32×10^{8}
1 therm (10^5 Btu)	1.054×10^{8}
1 kilowatt-hour (kW-hr)	3.6×10^{6}
1 ft^3 natural gas	1.09×10^{6}
1 kilocalorie (kcal)	4186
1 British thermal unit (Btu)	1054
1 calorie (cal)	4.186
1 foot-pound (ft-lbf)	1.356
1 watt-second	1.000
1 million electron volts (MeV)	1.6×10^{-13}
1 electron volt (eV)	1.6×10^{-19}

Mesa, New Mexico, or in Montana are measured in tons. Our natural gas consumption is measured in cubic feet or in therms. These units are ultimately a measure of a quantity having the dimensions of work that is conserved—energy.

Even in scientific discussions there are many units, prompting the physicist Richard Feynman to state: "For those who want some proof that physicists are human, the proof is in the idiocy of all the different units which they use for energy."[4] Scientists measure energy in calories (cal), kilocalories (kcal), joules (J), watt-seconds (W-s), ergs, electron volts (eV), million electron volts (MeV), or billion electron volts (GeV), etc. Table 5-3 lists common units of energy (including some engineering units) with their SI equivalents.

Our Energy Future

During the Industrial Revolution, new energy sources were harnessed, such as fuel woods and coal as heat to run engines. At the beginning of the twentieth century, the technological development of electricity use provided an

inexpensive way to transmit energy over long distances in a readily interchangeable form. Since then, electricity has become the coin of the realm in the transmission and use of energy.

Our society consumes great quantities of energy. We dissipate energy in almost every activity of our lives. We use natural gas to cook our food, to dry our clothes, and to heat our water and our homes. Yet the supply of natural gas is dwindling. We consume 7 million barrels of oil per day driving automobiles and delivering goods in trucks, and another 9 million barrels per day heating homes, generating electricity and for other uses. Yet our domestic production of crude oil is falling. Our civilization is built of iron, steel, concrete, aluminum, copper, plastics, etc.—all of which require great quantities of energy to produce. Our society runs with electricity. When the electricity goes off, through accident or storm, the normal flow of our lives comes to an abrupt halt—often giving us a peaceful respite.

Our material well-being is directly related to our consumption of energy, and there is a great disparity in the standard of living among the prosperous industrial nations and the underdeveloped nations. A nation's material prosperity, as measured by its per capita gross national product (GNP), closely corresponds to per capita electric energy consumption, as shown in Figure 5-15.

Table 5-4 compares projected energy use in the United

---

[4]Richard Feynman, *The Character of a Physical Law* (Cambridge, Mass.: MIT Press, 1967), p. 75.

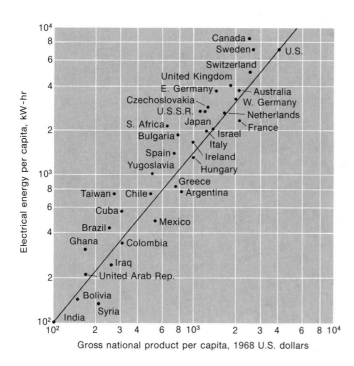

FIGURE 5-15
Per capita electrical energy consumption as a function of per capita gross national product (GNP). Our standard of living is directly related to our consumption of energy. Courtesy of University of California, Lawrence Livermore Laboratory, and the U.S. Energy Research and Development Administration.

TABLE 5-4

The energy consumption of the United States from various primary fuels, 1970 and 1985 (projected).

Source	1970	1985	Physical units	1970 10^{15} joules	1985 10^{15} joules
Oil					
Domestic	11.3	17.0	10^6 barrel/day	22 200	33 400
Imported	3.4	5.1	10^6 barrel/day	7 860	7 960
Synthetic	0	1.6	10^6 barrel/day	0	3 100
Gas					
Natural, domestic	21.8	32.1	10^9 ft^3/yr	23 600	34 700
Natural, imported	0.9	5.8	10^9 ft^3/yr	1 000	6 200
Synthetic	0	2.2	10^9 ft^3/yr	0	2 400
Coal	519	1077	10^6 tons/yr	13 800	28 600
Hydroelectric	249	309	10^9 kW-hr/yr	2 800	3 500
Nuclear electric	22	2771	10^9 kW-hr/yr	250	31 400
Geothermal electric	0.6	130	10^9 kW-hr/yr	7	1 470
Total U.S. consumption				71 500	131 800

*This projection, from the National Petroleum Council, makes the optimistic assumptions that substantial domestic oil and gas discoveries will continue and that nuclear power will be fully developed. It makes other assumptions, such as that increased production of coal will be unhampered by environmental disputes. These projections, however, were made before the oil embargo and energy crisis of 1973-74. Other assumptions would lead to even greater dependence on foreign crude oil.

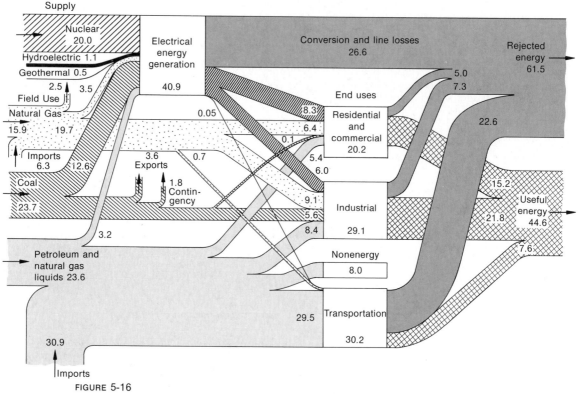

Supply

FIGURE 5-16

U.S. energy flow patterns projected to 1985. All values are in 10^{15} Btu = 1.054×10^{18} joules. The total annual energy production is estimated to be 122×10^{15} Btu. Courtesy of University of California, Lawrence Livermore Laboratory, and the U.S. Energy Research and Development Administration.

States in 1985 with actual use in 1970. Figure 5-16 shows the projected flow of energy through our society in 1985. With the exception of hydroelectric power, all of these resources are used to generate heat that is then converted into mechanical or electrical energy. All are nonrenewable energy sources. Almost 60% of the energy consumed by our society immediately enters the environment as waste heat. The remaining energy serves useful purposes before ending up as heat, or ends up as nonenergy products such as chemicals and plastics.

Review

Give the significance of each of the following terms. Where appropriate give two examples.

work
energy
dot product
joule
erg
gravitational potential energy
kinetic energy
mechanical energy
potential energy well

bound orbit
unbound orbit
escape velocity
heat
calorie
mechanical equivalent of heat
Btu
chemical energy
electron volt

ATP
photosynthesis
power
watt
horsepower
ft-lbf
relativistic mass
fission
fusion

Further Readings

Energy and Power: A Scientific American Book (San Francisco: W. H. Freeman and Company, 1971). Timely articles on many aspects of our energy use.

Farrington Daniels, *Direct Use of the Sun's Energy* (New Haven, Conn.: Yale University Press, 1964). A definitive review of the early work on solar energy.

Problems

MECHANICAL ENERGY

1 A person with a mass of 60 kg climbs a 5 m ladder leaning against a wall. The base of the ladder is 3 m from the wall. How much work must the person do to climb to the top of the ladder?

2 A person climbs a flight of stairs inclined at a 45° angle and 10 m long. How much mechanical work must a person of a mass of 60 kg do?

3 A horse is harnessed to a sled weighing 1000 lbf. The harness is attached at an angle of 30° to the horizontal. The horse pulls along the harness with a force of 600 lbf, which is sufficient to overcome friction. The horse drags the sled 500 ft. How much work did the horse do?

4 A 10 ton truck traveling at 25 mi/hr down the Grapevine Hill, on U.S. 99 outside Bakersfield, California, loses its brakes 2000 ft along the road

from the bottom of the hill. The hill is a 5% grade (it drops 5 ft for every 100 ft of travel). Neglecting friction, use the conservation of energy to find the final velocity of the truck at the bottom of the hill.

5 A teenage entrant in the annual soapbox derby race has a mass of 50 kg and sits in his racer with a mass of 70 kg. The race takes place on a hill 20 m high and at an angle of 10° from the horizontal. Assume that friction can be neglected and that the potential energy at the bottom of the hill is zero.
(a) What is the potential energy of the car and driver at the top of the hill?
(b) What is the kinetic energy of the car at the bottom of the hill? How fast is the car going?
(c) What is the net force acting to push the car down the hill? Over what distance does this force act? What is the work done by this force in pushing the car down the hill?

6 A VW with a mass of 800 kg is traveling at a speed of 72 km/hr = 20 m/s. It runs out of gas at the base of a hill.
(a) What is its kinetic energy at the base of the hill?
(b) Neglecting friction, what is the highest hill it can get over?

7 A favorite children's toy is the Mattel "Hot Wheels," small racing cars with a mass of about 50 gm having wheels with nylon bearings and thus very little friction. The cars go on a plastic track that can be elevated. A "Hot Wheel" track is elevated a distance of 75 cm above the ground and slopes downward for a distance traveled of 120 cm. If we neglect friction entirely, how fast will the car be going at the bottom of the track if it starts at rest?

8 By jumping a person can exert a force approximately twice his or her weight. A 70 kg person exerts this force vertically for a distance of 0.25 m.
(a) What is the work done?
(b) What is the change in the person's kinetic energy? Gravitational potential energy?
(c) How high can the person jump?

9 A pole vaulter with a weight of 150 lbf approaches the takeoff point at a speed of 30 ft/s.

(a) How high can the person jump if all this forward kinetic energy is directed vertically?
(b) How much additional work must the person do to clear 16 ft?

10 A flea with a mass of 0.45 mg can jump a height of 5 cm. The flea jumps by storing energy in elastic tissue *resilin*, and then releasing this energy suddenly.
(a) How much energy must be stored in the resilin for the flea to jump 5.0 cm?
(b) What is the flea's takeoff speed?

GRAVITATIONAL POTENTIAL ENERGY

11 An Apollo spacecraft headed for the moon first establishes a circular orbit 218 km above the surface of the earth. The radius of the earth is 6 371 km; the mass of the spacecraft is 53 000 kg.
(a) What is the velocity of the orbit?
(b) What is the gravitational potential energy?
(c) What is the kinetic energy?
(d) What is the total energy?

12 The first manned orbit of the moon took place with the Apollo 8 Mission to the moon. The lunar module entered into a nearly circular orbit about 112 km above the surface of the moon. The moon has a radius of 1 748 km. The spacecraft has a total mass of 16 500 kg.
(a) What is the velocity of the orbit?
(b) What is the gravitational potential energy?
(c) What is the kinetic energy?
(d) What is the total energy of the spacecraft?

13 The Apollo ascent vehicle has a mass of 5000 kg. How much energy is required to return the ascent stage from the surface of the moon to an orbit 112 km above?

BIOENERGETICS

14 One metric ton of TNT (trinitrotoluene) will liberate 4.2×10^9 joules of energy. TNT has a molecular weight of about 227 gm/mole.
(a) How much energy is released per mole?

(b) How much energy in eV is released per molecule of TNT?

15 Butane, a common fuel in cigarette lighters, is a liquid at room temperature at only modest pressures.
(a) How much energy is liberated in the combustion of one mole of butane?
(b) How much energy in eV is liberated in the combustion of one molecule of butane?

16 The intensity of the sunlight falling on the earth in October at about 43°N latitude at noon on a horizontal surface is 0.8 cal/cm²-min.
(a) What is the power falling on this surface in watts/m²?
(b) What is the power falling on this surface in Btu/hr-ft²?

POWER

17 A Boeing 707 has a loaded mass of 150 000 kg, accelerates for 54 s, and leaves the runway at a speed of about 85 m/s.
(a) What is its kinetic energy as it leaves the ground?
(b) Neglecting air resistance, what average power is required to accelerate the airplane? Express your answer in watts and horsepower. (The actual power requirements will be much greater than this to overcome the resistance of the air.)

18 A Datsun 260-Z automobile will accelerate from 0 to 34 m/s in just 20 seconds. The mass of the car and driver is 1430 kg.
(a) What is the final kinetic energy of the auto?
(b) What is the average power delivered by the engine in watts?
(c) What is the average power in horsepower? The engine is rated at 139 horsepower.

19 The world's record of 7.6 m for the rope climb was set in 1947, when Garvin Smith climbed it in just 4.7 seconds. Assume that Mr. Smith has a mass of 70 kg.
(a) How much work is done in climbing?
(b) What is the average power he must generate during his climb?

20 A person in good physical shape can pedal a bicycle with a power output of 0.1 hp all day long. The rider has an effective area of 0.33 m² and a mass of 50 kg on a 10 kg bicycle. What is the maximum speed on level ground with no wind at this power level?

21 A person and bicycle have a total mass of 80 kg and a power output of 0.1 hp.
(a) What is the maximum slope that a person can pedal up, neglecting air resistance, at a speed of 4 m/s at this power level?
(b) If the human body can generate mechanical energy with a 10% efficiency, how many kilocalories are burned up in one hour at this power level?

22 A cyclist with a total mass of 80 kg and an effective area of 0.33 m² pedals at 10 km/hr up a 10% grade.
(a) What horsepower is required?
(b) If the person has a maximum short-term energy capacity of 1.4×10^4 joules, how far up the grade can the person get before "burning out"?

23 A 50 kg woman is on a diet of 2000 kilocalories per day. Suppose the energy contained in this diet were converted into mechanical work with a 10% efficiency.
(a) How many joules of energy would be available?
(b) If this energy is used to climb stairs, how high could the person climb?

ENERGY AND MASS

24 A proton is traveling at 0.8 c.
(a) What is the momentum of the proton?
(b) What is the kinetic energy of the proton?
(c) What is the total energy of the proton?

25 An Apollo spacecraft of a mass of 18 800 kg is traveling toward the moon at a speed of 10.8 km/s.
(a) What is the spacecraft's rest mass energy?
(b) What is its kinetic energy?
(c) What is its total energy?

26 A proton in the accelerator at the Fermi National Accelerator has a kinetic energy of 400 GeV. The proton has a rest mass of 1.67×10^{-27} kg.
(a) What is the total energy of the proton?
(b) How fast is the proton going?

ENERGY USE

27 The energy from the sun falling on a collector has an intensity of about 800 watts/m² for about 6 hours in midday. Suppose the solar collector is 2 m by 3 m.
   (a) What is the power falling on the collector in watts?
   (b) If this power falls on the collector for about 6 hr, what is the total amount of energy collected, assuming 50% efficiency?
   (c) The cost of natural gas is about $1.50/10⁹ J. The cost of electricity is about $6/10⁹ J. What is the value of one day's collection of solar energy if it is used to replace natural gas? If it is used to replace electricity?

28 The price of energy is increasing as ready supplies dwindle. The price of electric power delivered to the home is about 2.4¢/kw-hr. The price of natural gas is 15¢ per therm. Gasoline is 70¢ per gallon. Crude oil runs about $12 per barrel. Determine the cost for 10⁹ joules of energy for each fuel. Which is the cheapest? Which is most expensive? (You can consult your latest utility bill or service station to update these prices.)

29 The Four Corners power plant at Farmington, New Mexico, generates 2075 megawatts of electric power. If the plant operates at a 40% thermal efficiency, it must burn coal to generate 5190 megawatts of heat. Assume that the energy content of western coal is similar to that of Illinois coal. How many metric tons of coal would be consumed in one day's operation? One year's operation?

ADDITIONAL PROBLEMS

30 A fully loaded Otis elevator car of a mass of 4600 kg travels upward at 5.0 m/s. The car is balanced by a counterweight of a mass of 4000 kg that is traveling downward at 5.0 m/s.
   (a) What is the net force applied by the motor to lift the car and counterweight at this speed?
   (b) What is the power in watts required to lift the elevator?

   (c) What is the horsepower required to lift the elevator? The maximum-rated power of the lift motor is 65 hp.

31 A moped (a power-assisted bicycle) has an engine rated at 0.85 hp. If the cyclist, bicycle, and engine have a total mass of 85 kg and an effective area of 0.33 m²,
   (a) What is the maximum speed on level ground?
   (b) If the engine has a mechanical efficiency of 10%, how many miles per gallon of gasoline would the bicycle get?

32 The top speed of a bicycle rider is limited by the drag force in moving through the air. The effective area presented by the rider is $A = 0.33$ m². Use of very hard bicycle tires can largely eliminate the friction force with the road.
   (a) If the rider can exert a maximum horizontal force of 25 N, what is the maximum speed?
   (b) If the bicycle and rider of mass 70 kg are coasting down a 5% grade, what is the maximum speed?
   (c) What is the maximum grade that the rider can go up by exerting a 25 N force parallel to the roadway?

33 The "Turn of the Century" roller coaster at Marriott's Great America, in California, has the first incline with a profile as shown in Figure 5-17. The maximum elevation at A is $h_A = 24.7$ m. It drops to a first minimum with an elevation $h_B = 3.3$ m with a radius of curvature $R_B = 24.4$ m. The rise C has an elevation of $h_C = 15.4$ m and a radius of curvature $R_C = 18.3$ m. The second dip has a minimum elevation of $h_D = 2.7$ m and a radius of curvature $R_D = 18.3$ m. Suppose that a fully loaded roller coaster car has a mass of 1000 kg and travels without friction so that mechanical energy is conserved.
   (a) Assume that the potential energy is zero at an elevation $h = 0$ m. What is the potential energy at A, B, C, and D?
   (b) If the car leaves point A with zero velocity $v_A = 0$, what is the velocity at B, C, and D?
   (c) What is the acceleration felt by a person in the car including the effect of centripetal acceleration at B, C, and D?

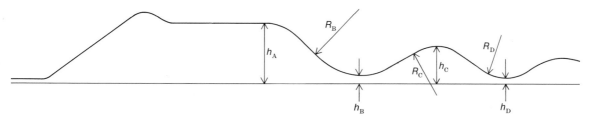

FIGURE 5-17
The first incline and hill of the "Turn of the Century" roller coaster. The first incline drops at a 45° angle from a maximum elevation of 24.7 m to a minimum elevation of 3.3 m. Information courtesy of Arrow Development Co., Mountain View, Calif.

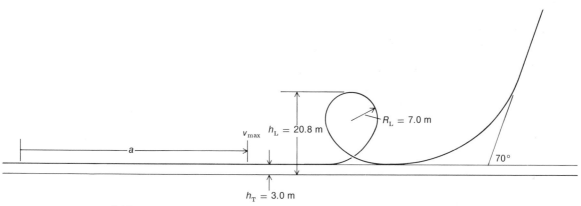

FIGURE 5-18
The "Tidal Wave" roller coaster. The giant loop reaches an elevation of 17.8 m above the track and has a radius of curvature of 7.0 m. The train reaches a top speed of 24 m/s before executing a 360° vertical loop. The entire ride forward and back takes only 50 s. Information Courtesy of Marriott's Great America®, Santa Clara, Calif.

34 The "Tidal Wave" roller coaster ride at Marriott's Great America, in California, accelerates passengers on a linear track to a top speed of 24 m/s in a distance of 70 m. The track is 3.0 m $h_T = 3.0$ m above the ground. The train then executes a vertical loop as shown in Figure 5-18. At the top of the loop the elevation is approximately $h_L = 20.8$ m. The radius of curvature of the loop is $R_L = 7.0$ m. Suppose that a fully loaded roller coaster car has a mass of 1000 kg and travels without friction so that energy is conserved.
(a) What is the car's velocity at the top of the loop?
(b) What is the acceleration felt by a rider at the top of the loop?
(c) The ride ends in a ramp that slopes upward 70°. What is the car's maximum elevation before it begins to fall backward?

6

Relativity

FRAMES OF REFERENCE

Since ancient times philosophers and scientists have been fascinated by the notion of absolute space at rest, with some eternally fixed reference point in the universe, and absolute motion through that space. Is there a point in space that has no motion? The medieval view that the earth is at rest bestowed a feeling of security, until Copernicus, Kepler, and Newton made the sun the fixed center about which the earth revolved. But there are other stars in our galaxy as well. Is the center of the galaxy at rest? Or is the point of absolute rest perhaps far distant from our own galaxy? Or does the idea of absolute rest have no meaning?

Inertial Reference Frame

Discussing in the *Principia* the motion of a particle through space, Isaac Newton stated that the motion of one body relative to another is the same whether that space is at rest or moving with uniform velocity, without rotation. Such a nonaccelerated reference frame is called an **inertial reference frame.** Because Newton's laws of motion are valid in any inertial reference frame, whether it is at rest or in motion, they avoid the issue of determining whether there is absolute motion and absolute rest.

Galileo was the first to use as an example of relative motion a ball falling from the top of a mast to the deck of a moving ship. In the *Dialogue Concerning the Two Chief World Systems* (1632), the character Simplicia held the medieval opinion, based on Aristotle: If the world were turning on its axis, a stone dropped from the top mast of a ship would land displaced by the distance the earth moved during the stone's fall. If the ship were moving, it would be displaced a distance equal to the distance the ship traveled during the fall. But Galileo argues through the character Salviati:

Now have you ever made this experiment of the ship? . . . For anyone who does will find that the experiment shows exactly the opposite of what is written; that is, it will show that the stone always falls in the same place on the ship whether the ship is standing still or moving at any speed you please [see Figure 6-1].[1]

Observer on the Ship

From the perspective of an observer on the ship, we can use a coordinate system x, y. Assuming that the ship is in uniform motion in a straight line, the acceleration of the ball in the forward direction is zero, $a_x = 0$. The initial forward velocity is also zero, $v_{xo} = 0$. The acceleration due to gravity acts downward. The ball will fall a distance s to the deck of the ship in a time given by the usual laws of motion:

$$t = (2s/g)^{1/2}$$

Since the ball possesses no forward motion relative to the ship, the ball will hit the deck directly beneath the point of release.

---

[1] Galileo, *Dialogue Concerning the Two Chief World Systems,* translated and edited by Stillman Drake (Berkeley: University of California Press, 1967), p. 144.

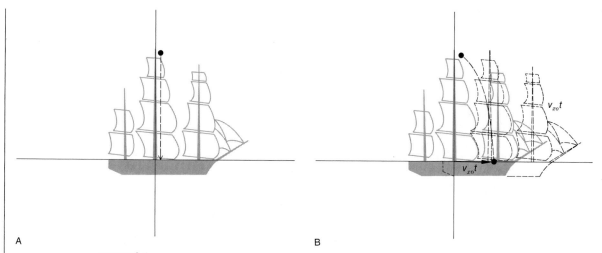

A   B

FIGURE 6-1
Galileo's experiment as viewed, A, from the moving ship, B, from the ocean, at rest. As seen from either viewpoint, a ball dropped from the top of a mast will strike the deck at the base of the mast.

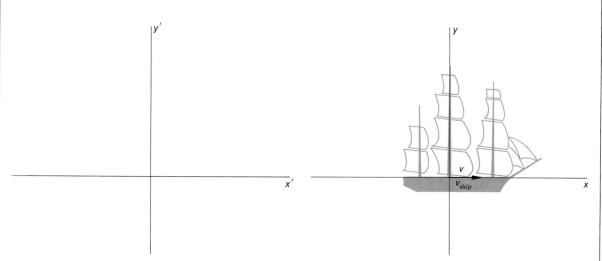

FIGURE 6-2
A Galilean transformation.

Observer at Rest

From a coordinate system x', y' at rest in the water, the boat is moving forward with a uniform velocity v_{xo}. Our new coordinate system is related to our old one by a simple coordinate transformation, a **Galilean transformation:**

$$x' = x + v_{\text{ship}}t \qquad (6\text{-}1)$$

$$y' = y$$

$$t' = t$$

In this new coordinate system (see Figure 6-2) the initial velocity of the ball is $v'_y = 0$ and $v'_{xo} = v_{\text{ship}}$. Since motion in the x'- and y'-directions can be treated separately, we find that the ball will fall a distance s' in time $t' = (2s/g)^{1/2}$, which is the same time as when the ship is at rest. Neglecting air resistance, the ball will move forward a distance $x' = v_{\text{ship}}t' = v'_{x0}t'$. The ship will move forward a distance $x' = vt'$. The ball will again land directly at the base of the mast. With Newton's and Galileo's relativity, we can analyze a system from any inertial reference system and obtain the same result.

THE SPEED OF LIGHT

Sagredo: Is light instantaneous or momentary, or does it like other motions require time?
Simplicio: Everyday experience shows that the propagation of light is instantaneous.
Salviati: I have devised a method which might accurately ascertain whether the propagation of light is really instantaneous. Let each of two persons take a light contained in a lantern, such that by the interposition of the hand, the one can be shut off or admit light to the vision of the other. . . . In fact, I have tried this experiment only at a short distance, less than a mile, from which I have not been able to ascertain with certainty whether the appearance of the opposition light was instant or not, but, if not instant, it is extraordinarily rapid.

Galileo Galilei, *Dialogue Concerning
the Two Chief World Systems*

Light appears to act instantaneously. If the two observers in Galileo's experiments were 1600 meters apart, and if the reaction time between seeing the light flash and opening the second lantern was about 0.2 seconds, we could determine a lower limit on the **speed of light.** The light would go a total distance of 3200 m in a time of less than 0.2 s. The speed of light consequently must be greater than:

$$v = \frac{\Delta x}{\Delta t} = \frac{3200 \text{ m}}{0.2 \text{ s}}$$

$$c > 1.6 \times 10^4 \text{ m/s}$$

The speed of light was first estimated in the seventeenth century, by the Danish astronomer Olaus Roemer while an assistant to the French astronomer Jean Picard. In Roemer's day it was well known that the intervals between successive reappearances of Jupiter's innermost moon, Io, from behind the planet changed during the course of the year. This moon should regularly emerge from the shadow of Jupiter every 42.5 hours. But after observations over several years, Roemer noted that when the earth is closest to Jupiter, reappearance tended to be as much as 11 minutes early, and when the earth is furthest from Jupiter, as much as 11 minutes late. This total discrepancy of about 22 minutes, Roemer concluded, is equal to the time required for the light from Jupiter's moon to cross the diameter of the earth's orbit (see Figure 6-3).

These measurements required the development of reasonably sophisticated timekeeping, so that a discrepancy of 22 minutes could be observed over a 6-month interval. New technologies often allow new discoveries in science. The diameter of the earth's orbit around the sun in Roemer's day was thought to be 2.9×10^{11} m. This gives a speed of light:

$$c \simeq \frac{\Delta s}{\Delta t} = \frac{2.9 \times 10^{11} \text{ m}}{22 \text{ min} \times 60 \text{ s/min}}$$

$$c \simeq 2.2 \times 10^8 \text{ m/s}$$

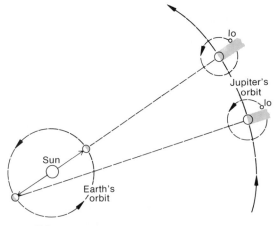

FIGURE 6-3
Roemer deduced the speed of light from the time required for light from Jupiter's moon Io to cross the diameter of the earth's orbit.

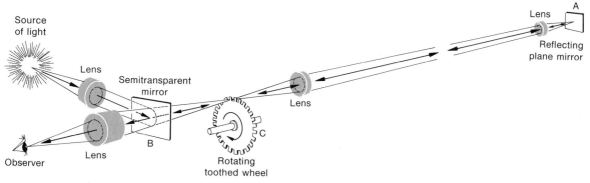

FIGURE 6-4

Fizeau used a rotating wheel with teeth, C, to chop a light beam into bursts. The light travels 8.6 km to a mirror, M, where it is reflected. The light returns to the wheel just as the next space in the wheel opens, thus allowing the duration of the light's travel to be measured. Excerpted from *Michelson and the Speed of Light* by Bernard Jaffe. Copyright © 1960 by Doubleday & Company, Inc. Reprinted by permission of the publisher.

In September 1676 Roemer predicted that in November the moon of Jupiter would reappear about 10 minutes late. His prediction was confirmed. Roemer's ideas about the finite speed of light were greeted with skepticism. However, his opinions were supported by Newton and the Dutch physicist Christian Huygens.

The first terrestrial measurement of the speed of light was not until 1849, almost 200 years later. The Frenchman Armand Fizeau reflected a beam of light past a toothed wheel rotating at high speed, as Figure 6-4 shows. The rotating wheel acted as a light chopper. A burst of light would leave the apparatus, go a distance of 8.62 km to a mirror, be reflected, and arrive just as the next notch in the rotating wheel appeared. Determining the wheel's speed of rotation allowed the speed of light to be determined. Fizeau's value was:

$$c \simeq 3.12 \times 10^8 \text{ m/s}$$

The Frenchman Jean Foucault, who had worked with Fizeau, greatly improved the accuracy of the experiment by substituting a rotating mirror for the notched wheel. His value for the speed of light in air, obtained in 1862, was

$$c = 2.977 \times 10^8 \text{ m/s}$$

(Foucault also was able to directly measure the speed of light in water, finding it less than the speed of light in air, as expected from the wave theory of light, discussed in Part V, Optics.)

Michelson's Measurements

The American physicist Albert Abraham Michelson improved significantly on the precision of Foucault's measurement of the speed of light. In 1877 Michelson used his own improvement of Foucault's rotating mirror method. A college professor, he had no money for expensive apparatus, and, like college professors today, he accomplished what he could in the few hours between and after classes. Despite the usual setbacks and delays, he established a new value of the speed of light in air:

$$c = 3.000 \ 91 \times 10^8 \text{ m/s}$$

A diligent and persistent experimental scientist who took great pride in the precision of his measurements, Michelson continually refined his optical measurements of the speed of light. Between 1924 and 1927 he made five more independent measurements. The light was reflected from a mirror on Mt. Wilson to a mirror on Mt. San Antonio (both in California) about 34 km away, as shown in Figure 6-5. He observed values for the speed of light between 299 690 000 m/s and 299 813 000 m/s and concluded that the best value was

$$c = 299 \ 796 \ 000 \text{ m/s} \pm 1 \ 000 \text{ m/s}$$

The most accurate measurement of the speed of light has been obtained, by indirect methods, since the invention of the laser. By special absorption techniques, the frequency of a particular spectrum line in methane can be

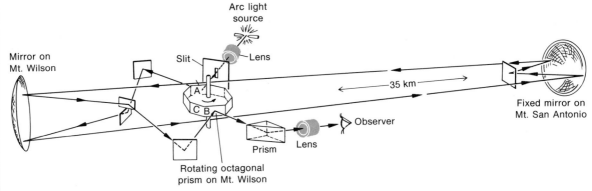

FIGURE 6-5
In Michelson's most precise measurements, in the 1920s, light was reflected from a rotating octagonal prism and a mirror on Mt. Wilson to a mirror on Mt. San Antonio, about 34 km away (as surveyed by the U.S. Coast and Geodetic Survey, to an accuracy of about 0.5 cm). From these measurements the speed of light was determined to be $c = 299\ 796\ 000$ m/s $\pm\ 1\ 000$ m/s. Excerpted from *Michelson and the Speed of Light* by Bernard Jaffe. Copyright © 1960 by Doubleday & Company, Inc. Reprinted by permission of the publisher.

compared to the frequency of the cesium transition used to define the second. By a different technique, the wavelength of the methane spectrum line can also be compared to the wavelength of the krypton transition used to define the meter. The product of the wavelength times the frequency is the speed of light. This best value[2] is

$$c = 299\ 792\ 458 \pm 1.2 \text{ m/s}$$

Michelson's value was within 0.0012% of today's accepted value.

THE LUMINIFEROUS ETHER

As early as 1678 Christian Huygens decided that light must consist of waves traveling from the source to the observer. By Huygen's time it was fairly well understood that sound consisted of waves traveling through the air, just as ripples travel over the surface of water. Huygen's theory of light, however, presented a problem: If light is a wave, it requires a medium; yet space appears to be empty. How could light reach us from the sun? Space must only appear empty, but must be filled with a luminiferous medium—that is, a medium that allows light to propagate—called the **luminiferous ether.**

In the nineteenth century Michael Faraday's and James

Clark Maxwell's studies of electricity and magnetism revived interest in the concept of ether to explain the propagation of electromagnetic waves through empty space. (We discuss electricity and magnetism in chapters 13–18.) By the end of that century it was widely accepted that light was a form of electromagnetic wave that propagated through the luminiferous ether. It was reasoned that if the ether is there, it should have some observable effect. Perhaps the motion of the earth could be measured relative to the ether.

Relative Motion

Before we examine the motion of light through a medium, we must first consider two simple examples of relative motion, both on a moving platform car: a game of catch, and sound traveling. (See Figure 6-6.) Suppose two peo-

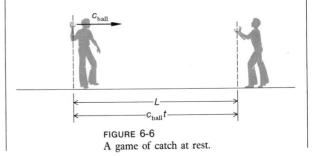

FIGURE 6-6
A game of catch at rest.

[2]E. Richard Cohen, "Fundamental Constants Today," *Research/Development* (March 1974), pp. 32–38.

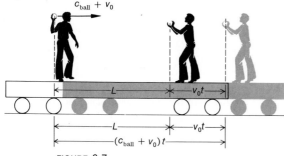

FIGURE 6-7
A game of catch on a moving platform.

ple playing catch are separated by a distance L and throw the ball with a speed c_{ball} relative to the platform, which is as yet stationary. The total time that the ball is in the air on a round trip is given by:

$$t_{total} = \frac{2L}{c_{ball}} \qquad (6\text{-}2)$$

If these two people are playing catch on a railroad car traveling with some speed v_0, and we ignore the effect of air resistance, an observer stationary on the ground sees the ball travel at a speed $c_{ball} + v_0$ for a time t_1 (see Figure 6-7). It goes a distance $L + v_0 t_1$:

$$L + v_0 t_1 = (c_{ball} + v_0) t_1$$

$$t_1 = \frac{L}{c_{ball}}$$

A similar result can be obtained for the return time. The total time to throw the ball and return it is the same whether on a stationary platform or one moving with uniform speed.

$$t_{total} = \frac{2L}{c_{ball}}$$

The moving car is an inertial reference frame. The laws of physics are unaffected by uniform motion in a straight line.

Travel through a Medium

Suppose our two people, rather than throwing a ball, throw sound. Say one fires a blank from a gun and starts a timer. When the second person hears the report from the

first gun he fires his starting gun. The first person, hearing the report from the gun of the second, stops the timer. What is the total time for the sound to travel both directions? For two persons at rest relative to the air, the answer is simple. If the speed of sound is c_s, the time to go a distance L is just given by:

$$L = c_s t$$

$$t = \frac{L}{c_s}$$

The total time to go both directions is just twice this value:

$$t_{total} = t_1 + t_2$$

$$t_{total} = \frac{2L}{c_s}$$

Suppose, however, that the two people are standing on a railroad flatcar moving at some speed v_0 (see Figure 6-8). The sound travels at speed c_s relative to the medium, the air, which is at rest. In a time t_1 the sound travels a distance $c_s t_1$. The total distance the sound must travel to reach the second man is the distance L plus the distance the train has moved $v_0 t_1$:

$$c_s t_1 = L + v_0 t_1$$

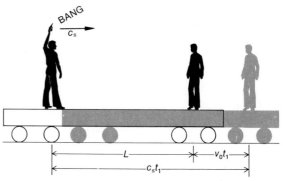

FIGURE 6-8
Sound traveling forward and heard on a moving platform.

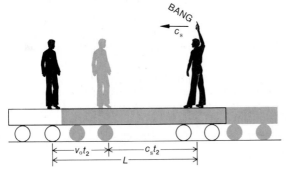

FIGURE 6-9
Sound traveling backward and heard on a moving platform.

Solving for t_1:

$$t_1 = \frac{L}{c_s - v_o}$$

On the return path (Figure 6-9) the sound goes a distance $c_s t_2$. The first man has moved forward $v_o t_2$. The total distance traveled is then

$$c_s t_2 = L - v_o t_2$$

Solving for the return time:

$$t_2 = \frac{L}{c_s + v_o}$$

The total time for the sound to travel the distance L and return is then given by

$$t_{total} = t_1 + t_2 = \frac{L}{c_s - v_o} + \frac{L}{c_s + v_o}$$

Putting everything over the least common denominator, we get

$$t_{total} = \frac{2c_s L}{c_s^{\,2} - v_o^{\,2}}$$

or

$$t_{total} = \frac{2L}{c_s[1 - (v_o/c_s)^2]} \qquad (6\text{-}3)$$

For sound waves traveling through a medium, there is a difference in time if the source and observer are moving relative to the medium.

The Michelson-Morley Experiment

While in Europe in the early 1880s Michelson developed an experiment for measuring the motion of the earth relative to the ether. If the earth is traveling through the ether, then light traveling along the direction of motion should travel at a slightly different speed than light traveling perpendicular to the motion. Michelson designed an interferometer (see Figure 6-10). In the interferometer, a partially silvered glass plate P_1 divides light from the source into two beams traveling at right angles. One beam travels to mirror A and returns through the plate P_1 to the observer. The second beam travels to a second mirror B and, being reflected off plate P_1, back to the observer. An unsilvered glass plate P_2 is placed in the second path simply so both beams go through the same thickness of glass.

Light acts as a wave. Two beams of light originating from a common source and traveling the same speed and distance will arrive in step, or in phase. The waves will reinforce one another, producing a bright region, as Fig-

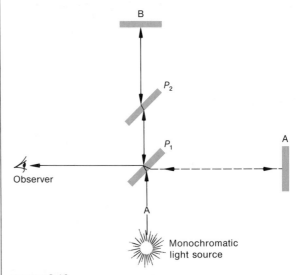

FIGURE 6-10
The basic interferometer has two light paths from the light source to the observer. The second plate P_2 is placed in the light path so that both beams will travel the same distance through glass. Excerpted from *Michelson and the Speed of Light* by Bernard Jaffe. Copyright © 1960 by Doubleday & Company, Inc. Reprinted by permission of the publisher.

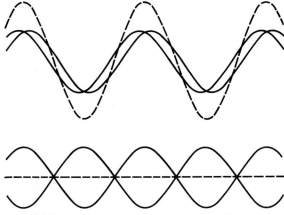

FIGURE 6-11
Interference of light waves. Top: If waves arrive in step or only slightly out of step, wave crests meet and reinforce each other, intensifying the light observed in the interferometer. Bottom: Waves exactly one-half wavelength out of step cancel each other, producing no light. From "The Michelson-Morley Experiment," by R. S. Shankland. Copyright © 1964 by Scientific American, Inc. All rights reserved.

ure 6-11 illustrates. The figure also shows that if one wave takes a longer time and arrives one-half wavelength out of step with the other, the waves cancel each other, producing no light.

If there is indeed an ether, these two light beams, once rejoined, would be out of step with each other and would produce interference (discussed in Chapter 20). Let us look at the total time required for the light to travel in each of the two directions, toward mirror A and toward mirror B in the interferometer.

Suppose that our interferometer is moving forward through the luminiferous ether in the direction of mirror A with a velocity v_0. The time the light beam takes to reach mirror A is analogous to that of the sound pitched forward on the railroad car in the earlier example. If the distance to mirror A is L_A, then total time for the light to travel to the mirror is

$$t_{1A} = \frac{L_A}{c - v_0}$$

On the return path, the time is

$$t_{2A} = \frac{L_A}{c + v_0}$$

The total time t_A for the light to go to mirror A and return is just

$$t_A = \frac{2L_A}{c[1 - (v_0/c)^2]} \qquad (6\text{-}4)$$

This is identical in form to equation 6-3, discussed earlier.

In the direction perpendicular to the forward motion we must be a little more careful (see Figure 6-12). The light will move forward from the beam-splitting plate P_1 to mirror B, for a time t_{1B}. During this time the mirror will move forward (to the right) a distance $v_0 t_{1B}$. The total distance the light must travel is determined by the triangle formed by L_B and $v_0 t_{1B}$, and is ct_{1B}, which by the Pythagorean theorem is just

$$(ct_{1B})^2 = L_B^2 + (v_0 t_{1B})^2$$

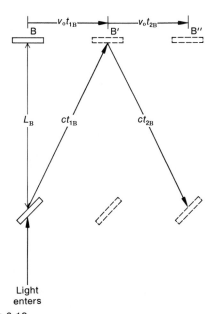

FIGURE 6-12
The light path in Michelson's interferometer perpendicular to the direction of motion.

Solving for t_{1B}, we get

$$t_{1B} = \frac{L_B}{c[1 - (v_o/c)^2]^{1/2}}$$

The same argument applies to the return path:

$$t_{2B} = \frac{L_B}{c[1 - (v_o/c)^2]^{1/2}}$$

The total time of travel for the light to mirror B and back is just:

$$t_B = \frac{2L_B}{c[1 - (v_o/c)^2]^{1/2}} \qquad (6\text{-}5)$$

If both paths of the interferometer are the same length L, it takes a slightly longer time for light to go to mirror B and back than to go to mirror A and back. The time difference between the two paths is given by

$$\Delta t = t_B - t_A$$

$$\Delta t = \frac{2L}{c} \{1 - (v_o/c)^2\}^{-1/2} - \frac{2L}{c} \{1 - (v_o/c)^2\}^{-1}$$

We can approximate this by using the fact that for small x, $x \ll 1$, $(1 + x)^n = 1 + nx$. If $x = -(v/c)^2$, and $n = -\frac{1}{2}$ and -1 respectively, then the time delay is given by

$$\Delta t = \frac{2L}{c} \{\tfrac{1}{2}(v_o/c)^2 - (v_o/c)^2\}$$

The time difference between the two branches of the interferometer is just

$$\Delta t = -\frac{L}{c} (v_o/c)^2$$

If the apparatus is rotated through $90°$, then the effect on path A and path B should be interchanged. The time difference would just be

$$\Delta t = \frac{+L}{c} (v_o/c)^2$$

The net time shift between the two observations would be

$$\Delta t_{net} = \frac{2L}{c} (v_o/c)^2 \qquad (6\text{-}6)$$

This corresponds with a path shift of $\Delta x_{net} = c\Delta t_{net}$:

$$\Delta x_{net} = 2L(v_o/o)^2$$

With funds provided in part by a grant from Alexander Graham Bell (the inventor of the telephone), Michelson set up his first experiments in a laboratory at the University of Berlin. The apparatus had to be kept at a constant temperature, and was so sensitive to vibration that vehicular traffic outside the laboratory was a disturbance, even late at night. Michelson was unable to detect any motion of the earth through the ether. He published his then-controversial results in 1881.

Back in the United States at the Case School of Applied Science in Cleveland, Ohio, and with the collaboration of chemistry professor Albert Morley, Michelson began a more sophisticated and sensitive experiment. Because the apparatus, shown in Figure 6-13, had to be isolated from mechanical vibration and yet be able to turn freely, they mounted it on a massive block of stone that itself floated in mercury. The whole experiment was mounted on a brick pier going down to bedrock.

To greatly increase the path length in their experiment, Michelson and Morley used several mirrors to reflect the beam of light back and forth many times, as shown in Figure 6-14. The effective path length for each branch of their interferometer was $L = 11$ m. The earth travels around the sun at about 30 km/s. The experiment was done using light from a sodium vapor lamp with a wavelength of about 5×10^{-7} m. If motion of the earth relative to the ether were to be detected, the net path difference should have been of the order of

$$\Delta x_{net} = 2L(v_o/c)^2$$

$$= (2.0)(11 \text{ m}) \frac{(3.0 \times 10^4 \text{ m/s})^2}{(3.0 \times 10^8 \text{ m/s})^2}$$

$$\Delta x_{net} = 2.2 \times 10^{-7} \text{ m}$$

This is a path difference of about 0.4 of the wavelength of light, or about 0.4 fringe shifts.

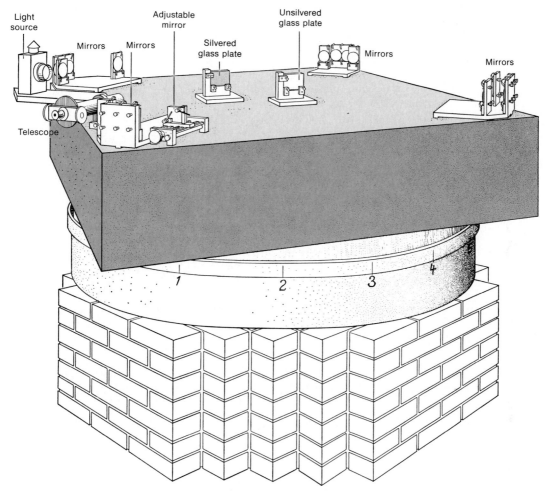

Light source

Mirrors

Mirrors

Adjustable mirror

Silvered glass plate

Unsilvered glass plate

Mirrors

Mirrors

Telescope

1 2 3 4

FIGURE 6-13

Michelson-Morley apparatus used in the decisive ether-drift experiment in 1887. The optical parts were mounted on a 5 ft square sandstone slab floating in mercury. Observations could be made in all directions by horizontally rotating the apparatus. From "The Michelson-Morley Experiment," by R. S. Shankland. Copyright © 1964 by Scientific American, Inc. All rights reserved.

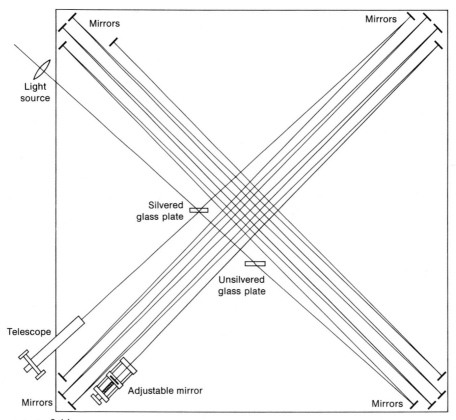

Their final observations were made on 1 July 1887 and their report, "On the Relative Motion of the Earth and the Luminiferous Ether," published that year. The greatest path difference found in repeated experiments was 0.02 fringe shifts, and the average was less than 0.01. They concluded that it was not possible to detect motion of the earth through the ether.

Lorentz-FitzGerald Contraction

Failure to measure any motion relative to the luminiferous ether cast doubt on the validity of the concept of ether itself. Twice, in 1893 and 1895, George FitzGerald at Trinity College, Dublin, proposed an idea that Hendrik Lorentz later developed independently. It was an idea that seemed outlandish but that would nevertheless save appearances: Perhaps the apparatus contracts in the direction of motion. This **length contraction** (also called **Lorentz-FitzGerald contraction**) is given by

$$L' = L_0 [1 - (v/c)^2]^{1/2} \tag{6-7}$$

where L_0 is the **proper length** measured when the apparatus is at rest relative to the ether.

If the effect of length contraction is included, then the time required for the light to reach mirror A and return in the direction of motion is

$$t_A = \frac{2L_A'}{c[1 - (v/c)^2]} = \frac{2L_A[1 - (v/c)^2]^{1/2}}{c[1 - (v/c)^2]}$$

$$t_A = \frac{2L_A}{c[1 - (v/c)^2]^{1/2}}$$

The time required for the light to reach mirror B and return in the direction perpendicular to the motion is still given by

$$t_B = \frac{2L_B}{c[1 - (v/c)^2]^{1/2}}$$

The time taken to go through either light path is the same independent of the motion through the ether. Thus, length contraction can explain the null result of the Michelson-Morley experiment while retaining the notion of the ether. Of course, the concept of length contraction is completely foreign to our experience, but the effect would become large only at speeds approaching that of light.

EINSTEIN'S SPECIAL THEORY OF RELATIVITY

Albert Einstein, born in Germany in 1879 of Jewish parents, was not thought to be too bright a child. When 16, Einstein failed the entrance exam of the University at Zurich and spent a year studying in a nearby town. His knowledge of mathematics was good but his grasp of such other subjects as biology and language was weak. In 1896 he entered the Institute of Technology to study to become a teacher. Once graduated, and now a Swiss citizen, he took a post in 1902 as a technical officer at the Swiss Patent Office in Berne, where he was to spend the next seven years. This was a germinal period in Einstein's scientific career. His work at the patent office provided him with a livelihood, while his free time was devoted to thinking about physics.

In 1905 Einstein published a paper, in the *Yearbook of Physics,* "On the Electrodynamics of Moving Bodies." Einstein examined the consequences of measuring the speed of light in two different coordinate systems. Einstein's paper, which outlined what has come to be known as the Special Theory of Relativity, contained two major assumptions.

1. The laws of physics are the same in all *inertial reference frames.* Any nonaccelerated reference frame is as good as any other.

2. Measurement of the speed of light always gives the same numerical result regardless of the relative velocity of the source or of the observer.

Einstein's paper explored the consequences of these assumptions. He explained the results of the Michelson-Morley experiment, and also came to startling conclusions about the nature of space and time.

Time

Alice sighed wearily. "I think you might do something better with the time," she said, "than wasting it in asking riddles that have no answers."

"If you knew Time as well as I do," said the Hatter, "you wouldn't talk about wasting *it.* It's *him.*"

"I don't know what you mean," said Alice.

"Of course you don't!" the Hatter said, tossing his head contemptuously. "I dare say you never even spoke to Time!"

"Perhaps not," Alice cautiously replied; "but I know I have to beat time when I learn music."

"Ah! That accounts for it," said the Hatter. "He won't stand beating. Now, if you only kept on good terms with him, he'd do almost anything you liked with the clock. . . ."

Lewis Carroll, *Alice in Wonderland*

Time is measured by clocks (as we saw in Chapter 1). Any regularly recurring phenomenon can be used as a clock: the swing of a pendulum, the back-and-forth motion of a hairspring and balance wheel, the vibration of a tuning fork or of a quartz crystal, etc. With enough care in making adjustments, all clocks could be calibrated to keep the same time and to remain in agreement. The ultimate standard of time is the atomic clock, which is driven by the vibration of a certain wavelength of light—a fundamental and reproducible standard.

To explore the consequences of Einstein's assumptions, it is useful to do a **thought experiment,** one that takes place in our imagination rather than in a laboratory. Let us create a plausible, although not particularly practical, clock using a mirror and a pulse of light. The clock, as shown in Figure 6-15, consists of a small light source S

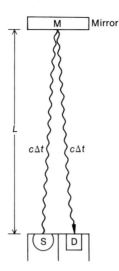

FIGURE 6-15
The light-pulse clock.

(perhaps a light-emitting diode or an electronic flash lamp) a distance L from a mirror M, and a light-sensitive detector D (say a phototransistor). The clock operates by the following sequence: A flash of light travels to the mirror and back to the detector, which triggers another pulse of light. The clock's rate of flashing is a measure of the passage of time. If properly calibrated, this clock should remain synchronized with any other accurate timekeeping instrument.

Now let us suppose that two such clocks, A and B, identical in design, are taken to the U.S. Naval Observatory Time Service laboratory and are calibrated against the atomic clocks and given the official seal of approval. Let us now take clock B into a different coordinate system—perhaps on a satellite headed toward the planet Jupiter.

Clock A, remaining on earth, is in our frame of reference. It ticks at a rate given by the time it takes light to traverse the distance L and return.

$$\Delta t_A = \frac{2L}{c}$$

The time kept by our light-pulse clock will remain synchronized with our other clocks. The time kept by a clock as observed from a coordinate system at rest with that clock is the **proper time.** Clock B, moving away from the earth at a uniform speed v, is in its own inertial reference frame. A visiting astronaut who stopped in for tea and checked the operation of the second clock would find it

ticking at a rate given by

$$\Delta t_B = \frac{2L}{c}$$

which again is in agreement with the astronaut's own quartz crystal watch. This is as it should be, by Einstein's assumptions, the laws of physics and the speed of light should be the same in any inertial reference frame.

Suppose observer A in reference frame A wishes to examine the rate at which clock B is ticking—this could be done by observing the clock flashes, or by sending a radio time signal. Let us assume that clock B is moving away from observer A at a speed v, which approaches the speed of light. Let the clock be oriented perpendicular to the direction of motion as shown in Figure 6-16.

Observer A would see that the light traveling to the mirror and returning to the detector must go over a longer path, taking a time $\Delta t'$ to go from the lamp to the mirror. In this time the mirror has moved forward a distance $v\Delta t'$. The speed of light in observer A's frame of reference is still c, so that the time for the light to traverse the distance to the mirror is just given by

$$(c\Delta t')^2 = L^2 + (v\Delta t')^2$$

Solving for the time $\Delta t'$

$$\Delta t' = \frac{L}{[c^2 - v^2]^{1/2}}$$

The return path takes the same length of time. Observer A will observe that clock B ticks at a rate $\Delta t_B'$, which is slower than the clock in reference frame A.

$$\Delta t_B' = \frac{2L}{c[1 - (v/c)^2]^{1/2}} \qquad (6\text{-}8)$$

The proper time for clock B in its own reference frame is

$$\Delta t_B = \frac{2L}{c}$$

Observer A will then find that the clock B in the satellite is running slower

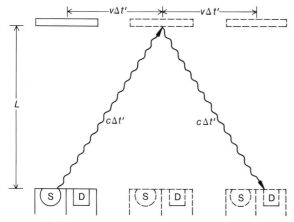

FIGURE 6-16
The light-pulse clock oriented perpendicular to the direction of motion as viewed by an observer at rest.

$$\Delta t_B' = \frac{\Delta t_B}{[1 - (v/c)^2]^{1/2}} \qquad (6\text{-}9)$$

This effect is known as **time dilation:** that is, clocks A and B are in two different coordinate systems moving in uniform motion with respect to one another. If an observer in coordinate system A looks at the clock in system B it appears that its rate of timekeeping has slowed with respect to clock A.

The relativistic factor occurs often enough that it has been given its own symbol, γ (the Greek letter *gamma*):

$$\gamma = \frac{1}{[1 - (v/c)^2]^{1/2}}$$

Observer A, at rest, observes clock B, which is moving away at speed v, as having an apparent time duration

$$\Delta t_B' = \gamma \Delta t_B$$

To the observer on the satellite in the reference frame of clock B, everything seems normal. His quartz crystal wrist watch remains synchronized with the light-pulse clock. His mental and biological experience of time also appear normal. The special theory of relativity, however, predicts that time runs at a different rate in a coordinate system traveling away at a large relative velocity. Because the speed of light is so large, the effect of time dilation is observed only in extreme cases.

EXAMPLE
How much does the rate of a clock change if a person is
(a) traveling at 2 240 km/hr in a supersonic transport plane?
(b) traveling at 36 000 km/hr in a spaceship headed for the moon?

The effective time interval is given by

$$t = \frac{t_0}{[1 - (v/c)^2]^{1/2}}$$

For small velocities, $v \ll c$, this is approximately

$$t = t_0[1 + \tfrac{1}{2}(v/c)^2]$$

The fractional change in the time interval is

$$\frac{\Delta t}{t_0} = \frac{t - t_0}{t_0} = \tfrac{1}{2}(v/c)^2$$

(a) $v = 2\,240$ km/hr $= 622$ m/s

$$\frac{\Delta t}{t_0} = \frac{\tfrac{1}{2}(622 \text{ m/s})^2}{(3.0 \times 10^8 \text{ m/s})^2}$$

$$\frac{\Delta t}{t_0} = 2.15 \times 10^{-12}$$

Observed from the earth, this would amount to about 8 nanoseconds in one hour.

(b) $v = 36\,000$ km/hr $= 10^4$ m/s

$$\frac{\Delta t}{t_0} = \frac{(0.5)(10^4 \text{ m/s})^2}{(3.0 \times 10^8 \text{ m/s})^2}$$

$$\frac{\Delta t}{t_0} = 5.56 \times 10^{-10}$$

Observed from the earth, a clock on the spaceship would appear to run slow by about 2 microseconds in one hour.

Length

In the frame of reference of clock B, all clocks, if calibrated, keep the same time. Thus, it should make no difference whether clock B is oriented perpendicular or parallel to the line of motion relative to observer A. However, it does make significant difference to observer A. Suppose clock B is aligned in the direction of motion and viewed by an observer at A (see Figure 6-17).

From A's perspective, the mirror is moving away from the source (this is analogous to the earlier illustration of sound going back and forth on the moving railroad car). To observer A, it will take a time $\Delta t_1'$ for the light to go from the source to the mirror, a distance L'. It is left to the student to show that this time is

$$\Delta t_2' = \frac{L'}{c - v}$$

The detector is approaching the light on the return path. The return time is

$$\Delta t_2' = \frac{L'}{c + v}$$

Viewed from observer A's vantage point, the total time for

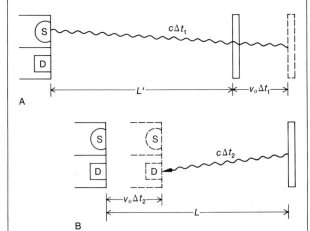

A

B

FIGURE 6-17
The light-pulse clock oriented parallel to the direction of motion as viewed by an observer at rest. A. Light travels in the direction of motion. B. Light travels opposite to the direction of motion.

the clock to tick once in this case is given by

$$\Delta t_B' = \frac{L'}{c - v} + \frac{L'}{c + v}$$

$$\Delta t_B' = \frac{2L'}{c[1 - (v/c)^2]}$$

This result is identical to equation 6-4, previously obtained for the Michelson interferometer, where the light travels parallel to the direction of motion.

If the clock is oriented perpendicular to the direction of motion, the apparent rate of time for the clock is given by equation 6-8:

$$\Delta t_B' = \frac{2L}{c\,[1 - (v/c)^2]^{1/2}}$$

If we equate the two (since the clock should run at the same rate regardless of orientation), we find the apparent length of the clock L' in the direction of motion as viewed by observer A is

$$L' = L\,[1 - (v/c)^2]^{1/2} = \frac{L}{\gamma}$$

The apparent length of our second clock in the direction of motion has decreased. This length contraction is the same result as suggested by Lorentz and FitzGerald to resolve the Michelson-Morley experiment.

Einstein's theory of relativity, first stated in 1905, makes only assumptions that the laws of physics and the speed of light are the same in any inertial coordinate system. The theory makes two predictions as to what will happen when one system is observed from another. First, a clock traveling near the speed of light past a stationary observer will appear to be running too slow: time dilation. Second, a spaceship or satellite traveling near the speed of light past a stationary observer will appear shortened in length: length contraction.

Space Travel

Suppose a spaceship sets off to explore in the vicinity of the star Altair in the constellation of Alquila the Eagle.

Altair is some 16 light years distant from our solar system. By advances in technology the spaceship is capable of traveling at 0.8 c. To an observer based on the earth it will take a total of 20 years for the spaceship to reach Altair.

$$t = L/v = 16 \text{ light-yr}/0.8\,c$$

$$t = 20 \text{ years}$$

To an observer on the earth, it will also appear that the clocks on the spaceship are running slow because of time dilation.

$$t' = \gamma t_{\text{ship}}$$

where the relativistic factor

$$\gamma = \frac{1}{[1 - (v/c)^2]^{1/2}}$$

$$\gamma = \frac{1}{[1 - (0.8)^2]^{1/2}}$$

$$\gamma = \frac{1}{0.6} = 1.67$$

the elapsed time recorded on the spaceship's clock will be just

$$t_{\text{ship}} = t'/\gamma$$

$$= 20 \text{ yr}/1.67$$

$$t_{\text{ship}} = 12 \text{ yr}$$

To the navigator aboard ship, the spaceship clock is keeping proper time. However, because of relative motion, the distance between earth and the destination near Altair has become contracted. The distance traveled is given by the Lorentz-FitzGerald contraction

$$L' = L_{\text{o}}/\gamma$$

$$L' = \frac{16 \text{ light-yr}}{1.67}$$

$$L' = 9.58 \text{ light-yr}$$

Since the ship is traveling near the speed of light, $v =$

0.8 c, it will take them a total time

$$t_{\text{ship}} = L'/v$$

$$t_{\text{ship}} = \frac{9.58 \text{ light-yr}}{0.8\,c}$$

$$t_{\text{ship}} = 12 \text{ yr}$$

An observer on the earth will say that it takes 20 years to make the journey, and that the clock aboard the spaceship is running slow. The navigator on the ship will say that his clock is running perfectly, but the distance they have traveled has contracted. The total time for the journey to Altair and back will be recorded as 40 years by an observer on earth, as only 24 years by the space travelers. If the navigator on the ship has a twin brother who remained behind on the earth, he will find his twin 40 years older while he himself has aged only 24 years.

Velocity is relative to a particular coordinate system. Each of the two twins will see the other traveling away at a certain velocity. Originally this was stated as a paradox. Which twin's clock would run more slowly? Acceleration, however, is absolute: It is possible to detect acceleration by observing the response of an inertial mass suspended from a spring. Thus, it is clear which of the two twins underwent acceleration to relativistic speeds. It is the twin who accelerated to high speed whose clock will run more slowly. Acceleration also has a relativistic effect on the rate at which clocks tick, although we can assume that this effect operates for only a short part of the total journey.

A spaceship on an express journey to a distant star will also experience length contraction. If observed as it passes a nearby space station, the ship's length in the direction of motion will appear considerably shortened. As the ship's velocity approaches the speed of light it appears more and more compressed when viewed from the side, as Figure 6-18 and Table 6-1 show.

Annual Reports

For further evidence of how time dilation affects space travel, consider again a journey to Altair. The spaceship travels at a speed of 0.8 c, a distance of 16 light-yr. Suppose that once a year during the mission the astronauts send back, by microwave-computer communication, a

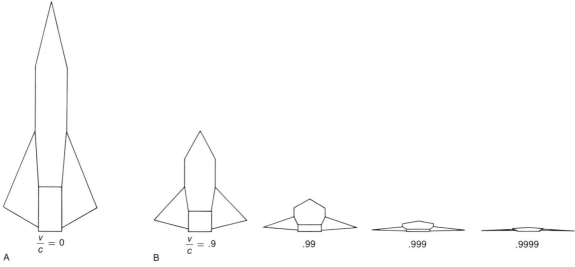

A spacecraft appears to contract as it approaches the speed of light relative to the observer.

TABLE 6-1
Length Contraction: A spaceship 20 m in length appears to contract as it approaches the speed of light.

v/c	γ	$L = L_0/\gamma$
0	1.0	20.0 m
0.9	2.3	8.7 m
0.99	7.2	2.8 m
0.999	23.0	0.9 m
0.9999	709.0	0.03 m

50-page annual report on the mission; likewise, the base station on the earth sends an annual report to the ship. The frequency of the reports is once a year, $f_o = 1.0/\text{yr}$.

The base station on earth will send out 20 reports during the time the spaceship takes to reach its destination near Altair. However, the ship will receive only 4 of these reports, for it takes 16 years for the fifth report to reach Altair. On the return trip the spaceship will receive the remaining 16 reports, which were already on the way when it reached its destination. It will also receive the 20 reports sent in subsequent years. Thus, on the return trip the spaceship will receive 36 reports. According to its own clock, the ship takes 12 years each way, so that the frequencies of receiving these reports going, f_g, and returning, f_r, are just

$$f_g = \frac{4 \text{ reports}}{12 \text{ years}} = 0.333 \text{ reports/yr}$$

$$f_r = \frac{36 \text{ reports}}{12 \text{ years}} = 3.0 \text{ reports/yr}$$

In going and coming the spaceship will receive all 40 reports sent out by the base station.

By its own clock, the spaceship will send out 12 reports on its trip to Altair. However, according to the base on the earth, it will take 16 years after the completion of the 20-year trip to Altair for the last report from the outwardbound ship to reach the earth, for a total of 12 reports in 36 years. On the return trip the spaceship will send out 12 reports, which will arrive on the earth in the remaining 4 years of the mission. The observer on the earth will find the frequency of reports going and returning as

$$f_g = \frac{12 \text{ reports}}{36 \text{ years}} = 0.333 \text{ reports/yr}$$

$$f_r = \frac{12 \text{ reports}}{4 \text{ years}} = 3.0 \text{ reports/yr}$$

On both the outbound and return trips the spaceship and the earth base will receive reports at the same rate as the other does. And each will send reports once a year according to its own time.

Every report sent is received: The 40 sent by the earth base are received by the spacecraft; the 24 sent by the spacecraft are received by the earth base. The twin on the spacecraft will be 24 years older than when he left; the twin who remained on the earth will be 40 years older.

Hafele-Keating Experiment

Our international standard of time is defined by a set of cesium atomic clocks in national observatories around the world. They are kept synchronized against a portable atomic clock that is periodically moved from one laboratory to another. In October 1971 Dr. J. C. Hafele, then at Washington University in St. Louis, and Dr. Richard Keating of the U.S. Naval Observatory undertook an experiment to observe the relativistic effects acting on atomic clocks (see Figure 6-19). They observed the behavior of 4 atomic clocks over some 600 hours to establish the stability and characteristics of each clock. During this period, all four clocks were flown around the world on commercial jet transportation, first in an easterly direction, then westerly. The clocks were in the air approximately 45 hours in each direction at altitudes of about 7 km and a speed of 280 m/s. The behavior of the

FIGURE 6-19
Four atomic clocks with portable power supplies used to test the theory of relativity. Dr. Hafele (left) and Dr. Keating (right) are greeted at an airport. Courtesy of Hewlett-Packard Co.

clocks depended on two factors; time dilation because of the speed relative to a clock at rest on the ground, and gravitational red shift because of change in altitude: If a clock is operated at a higher altitude, it tends to run a little faster than a similar clock at the earth's surface.

The aircraft flew in different directions and at different speeds and altitudes for different periods of time, but this

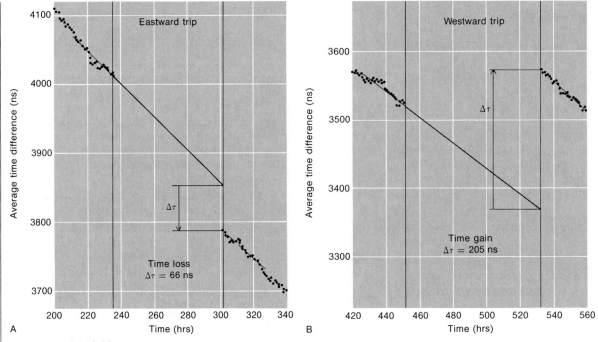

FIGURE 6-20
The average time kept by the four clocks in the Hafele-Keating experiments before and after, A, the eastward trip and, B, the westward trip plotted against the mean time kept by the U.S. Naval Observatory. The eastward trip shows a time loss of 66 ns; the westward trip shows a time gain of 205 ns. From "Around-the-World Atomic Clocks: Observed Relativistic Time Gains," by J. C. Hafele and R. E. Keating, *Science* (14 July 1972), pp. 168–170. Copyright 1972 by the American Association for the Advancement of Science.

TABLE 6-2
Predicted and observed time offsets after corrections for four cesium-beam atomic clocks traveling around the world in the Hafele-Keating experiment.

	East	West
Predicted	-40 ± 23 ns	275 ± 21 ns
Observed	-59 ± 10 ns	273 ± 7 ns

information was carefully recorded. The predicted shift in the clocks was then calculated over the actual path, and was then compared with the time shift determined by all four clocks, after considerable statistical comparison among the clocks. (Because a given atomic clock occasionally shifts its frequency of operation slightly, comparing several clocks makes it possible to determine when, and by how much, each clock changed its frequency.) Figure 6-20 clearly shows how the average time kept by all four clocks differs from the mean standard time kept by the U.S. Naval Observatory in Washington, D.C. The predicted offsets and the offset deduced from all the data are shown in Table 6-2. The Hafele-Keating experiment is a remarkable verification of the predictions of relativity. There is much other evidence in modern physics to support Einstein's special theory of relativity.

Review

Give the significance of each of the following terms. Where appropriate give two examples.

inertial reference frame
Galilean transformation
speed of light
luminiferous ether

Michelson-Morley experiment
length contraction (Lorentz-Fitz-
    Gerald contraction)
proper length

thought experiment
proper time
time dilation
relativistic factor

Further Readings

Hermann Bondi, *Relativity and Common Sense; A New Approach to Einstein* (New York: Doubleday, Anchor Science Study Series S 36, 1964).

Ronald W. Clark, *Einstein: The Life and Times* (New York: World Publishing, 1971). A comprehensive biography.

George Gamow, *Mr. Tompkins in Wonderland* (Cambridge, England: Cambridge University Press, 1965). A delightful introduction to relativity.

Bernard Jaffe, *Michelson and the Speed of Light* (New York: Doubleday, Anchor Science Study Series S 13, 1960).

R. S. Shankland, "Michelson and his Interferometer," *Scientific American* (November 1974).

Problems

SPEED OF LIGHT

1   Modern electronic instrumentation will easily record times of the order of 1 millisecond = 1 ms = 10^{-3} s; 1 microsecond = 1 μs = 10^{-6} s; or even of the order of 1 nanosecond = 1 ns = 10^{-9} s. How far will light travel in 1 ms? in 1 μs? in 1 ns?

2   A television transmitter sits atop a hill 25 km away. How long does it take the electromagnetic wave carrying the television signal, and propagating at the speed of light, to reach the antenna on your house? Express your answer in milliseconds, ms; and microseconds, μs.

3   The center of the galaxy is 30 000 light-yr from our solar system. How far is this in meters?

EINSTEIN'S SPECIAL THEORY

4   A space mission to the star Sirius some 8.7 light-yr distant from earth is undertaken at a speed of 0.6 c.
   (a) How long will it take to reach Sirius, according to earth time? According to spaceship time?

(b) What is the distance of the journey according to measurements from the moving spaceship?

(c) How long will it take to complete the mission to Sirius and return, according to earth time?

(d) One of the crew of the spaceship has a twin brother who remains behind on earth. How much older will be the twin who remained on the earth than the one who made the journey and returns?

5 A supersonic transport plane flies roundtrip from Los Angeles to New York, a distance of 4 700 km. Suppose the plane flies at a speed of 2 200 km/hr. Also suppose that the speed of light is only $c = 1 000$ m/s $= 3 600$ km/hr. Neglect the rotation of the earth.

(a) How long will it take to go to New York and return, as viewed from the control tower in Los Angeles? As viewed from a clock in the cockpit of the aircraft?

(b) How far will the distance appear to be when observed from the aircraft while flying?

(c) What is the actual time dilation and length contraction for $c = 3.0 \times 10^8$ m/s?

6 In the book *Mr. Tompkins in Wonderland*, by George Gamow, Mr. Tompkins dreams he is in a world where the speed of light is about 50 km/hr. He is riding a bicycle 1.8 m long. How long will the bicycle appear to a man at rest, if the bicycle is traveling at 25 km/hr? 40 km/hr?

7 Mr. Tompkins wishes to ride his bicycle across town to visit the zoo, which is a proper distance of 8 km away. If the speed of light is 50 km/hr, and he rides his bicycle at 25 km/hr,

(a) How long will it take as measured by his own clock?

(b) How long will it take as measured by the town hall clock?

(c) How far away will he observe the zoo to be from his starting point if he measures the distance from the moving bicycle?

8 Suppose the speed of light were 100 km/hr. A car 3 m long is traveling at 75 km/hr.

(a) How long will the car appear to the driver of the car?

(b) How long will it appear to an observer at rest?

9 A group of students is in a rocket moving at $0.8\,c$ relative to the earth. They are taking a correspondence course by television from the earth. The professor on the earth wishes to give them a one-hour exam. He begins the exam at 11:00 A.M. and finishes it promptly at noon earth time. How much time, as measured by their own watches, do the students have to take the exam?

ADDITIONAL PROBLEMS

10 An astronaut wishes to reach one of the brightest stars in the Northern sky, Arcturus, to determine if there are any habitable planets around the star. Arcturus is 40 light-yr away. He wishes to travel there in 10 years according to his own clock.

(a) Can he get there that fast?

(b) How fast must he travel?

(c) How far will the astronaut observe that he has traveled during the voyage?

(d) How long did it take as measured from the earth?

(e) If the proper mass of the ship is 5 000 kg at rest, what will be the mass energy of the ship during the flight as viewed from earth? How does this compare to the energy required to lift the ship from the surface of the earth?

11 The Concorde supersonic transport plane will travel at a speed of 2 240 km/hr at an altitude of about 7 km. The plane has a range of 6 400 km and will travel the distance in some 2.75 hours. Suppose the plane is traveling west on the equator. Because of the earth's rotation, a point on the equator travels at 460 m/s to the east. The earth has a radius of 6 400 km. For the 2.75 hr flight,

(a) If a clock on the ground is compared to a proper clock at rest with respect to the center of the earth, what is the time difference caused by the earth's motion?

(b) What is the time difference of a clock on the plane traveling west, compared to a proper clock at rest?

(c) What is the time difference of a clock on the plane traveling east, compared to a proper clock at rest?

II
FORMS OF MOTION

7

Angular Motion

TORQUE

Leverage

The Greek natural philosopher Archimedes (287–212 B.C.) first stated the principle of the **lever:** *The force the lever exerts is inversely proportional to the length of the lever arm,* the distance from the pivot point to the applied force.

There are three basic levers (see Figure 7-1). The simplest is the **first-class lever.** The pivot point, or **fulcrum,** is between the applied force and the weight (as with the teeter-totter and the beam balance). In the **second-class lever,** the fulcrum is at one end, the weight to be lifted is near it, and the applied force is at the other end (as in the wheel barrow and the pry bar). In the **third-class lever,** the applied force is between the fulcrum and the weight to be lifted. To balance the lever, the fulcrum must oppose the applied force. There are many such examples in the human body: A muscle is attached on a short lever arm to move force through a longer distance.

Consider the action of a pair of forces on a teeter-totter, shown in Figure 7-2. Of two children straddling it, the heavier child exerts a downward force F_1 and sits a distance R_1 from the pivot. The lighter child can achieve balance with a downward force F_2 by sitting slightly further from the pivot, a distance R_2. A perfectly balanced teeter-totter should move up or down a short distance with virtually no work from either child. The teeter-totter moves through a small angle $\Delta\theta$.

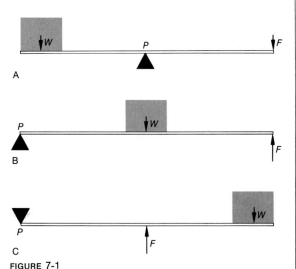

FIGURE 7-1
The three types of lever. A: first class; B: second class; C: third class. The load weight W is balanced using a pivot P by a force F.

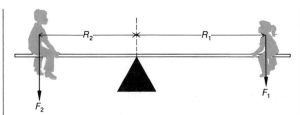

FIGURE 7-2
The teeter-totter. A lighter child can balance a heavier child.

Angles

We are all familiar with the measure of the angles of a triangle in degrees. The angles of a triangle add up to 180°. There are 360° in a circle. We can also define angular displacement as the arc swept out by the angle: The greater the angle, the greater the arc swept out. If the radius of the arc is R and the length along the arc is s, we can define an angle in radian measure as

$$\theta = s/R$$

The **radian** *is the ratio of two lengths and is dimensionless.* The unit radian (rad) is sometimes included simply to remind us that we are measuring angles. The circumference of a circle is just $2\pi R$, so that an angle of 360° is expressed in radian measure as

$$\theta = \frac{2\pi R}{R} = 2\pi \text{ radians}$$

One radian is equal to 57.3°. (Basic trigonometry is reviewed in Appendix A.)

Angular displacement is the difference between two angles:

$$\Delta\theta = \theta_1 - \theta_2$$

Each of these angles can be expressed as the ratio of an arc s to the radius of a circle R:

$$\Delta\theta = \frac{s_2}{R} - \frac{s_1}{R} = \frac{\Delta s}{R}$$

If an object with a radius R is rotated through an angular

displacement $\Delta\theta$, a point on the surface moves forward a distance $\Delta s = R\Delta\theta$, as Figure 7-3 also shows.

If the balanced teeter-totter shown in Figure 7-2 is moved through a small angle $\Delta\theta$, the first child rises a distance

$$\Delta s_1 = R_1 \Delta\theta$$

The first child's gravitational potential energy increases:

$$\Delta W_1 = F_1 \Delta s_1 = F_1 R_1 \Delta\theta$$

Because the second child goes downward, the amount of work done on that child is negative.

$$W_2 = F_2 \Delta s_2 = -F_2 R_2 \Delta\theta$$

If the teeter-totter is balanced it takes virtually zero work to cause it to move. The total energy change is zero:

$$\Delta W = \Delta W_1 + \Delta W_2 = 0$$

$$F_1 R_1 \Delta\theta - F_2 R_2 \Delta\theta = 0$$

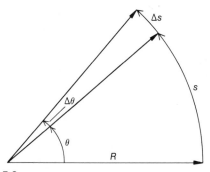

FIGURE 7-3
An angle in radian measure is defined as the ratio of the arc length s to the radius of the arc R.

Torque

The force F multiplied by the perpendicular lever arm R is called **torque,** *represented by the symbol τ (the Greek letter tau).*

$$\tau = FR_\perp$$

The work done on each child on the teeter-totter is just the applied torque times the angular displacement.

$$\Delta W = \tau \Delta \theta$$

We shall discuss chiefly forces that act in a single plane. We can then calculate the torque tending to rotate an object about an axis perpendicular to that plane. We need only determine the direction of rotation (clockwise or counterclockwise) and the magnitudes of the torques involved.

If a force F is applied on a lever arm R, the magnitude of the torque depends on the sine of the angle between them

$$\tau = RF \sin \theta$$

If the force and the lever arm are at right angles, maximum torque is produced. If the force and the lever arm are along the same line, the angle and the torque produced are zero.

Figure 7-4 shows the torque acting on a meter stick supported from one end. The meter stick's weight acts at its midpoint. The magnitude of the torque is equal to the product of the lever arm times the component of force perpendicular to the lever arm.

$$\tau = RF_\perp = R(F \sin \theta) = RF \sin \theta$$

It is also equal to the product of the force times the lever arm perpendicular to the force

$$\tau = R_\perp F = (R \sin \theta)F = RF \sin \theta$$

Equilibrium

A torque tending to produce counterclockwise rotation we consider positive; a torque tending to produce clockwise rotation, negative. If the net torque or the sum of the torques about an axis is zero, the condition is one of **rotational equilibrium:**

$$\text{net } \tau = \Sigma \tau = 0$$

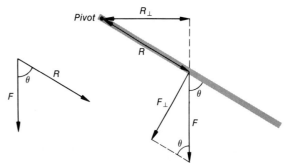

FIGURE 7-4
A meter stick suspended from one end will experience a torque. Torque is force times the perpendicular lever arm, or the lever arm times the perpendicular force.

As the axis about which to calculate the net torque, we can choose the object's natural pivot or a point where unknown reaction forces are acting.

The condition for translational equilibrium (discussed in Chapter 3, on Newton's laws of motion) is that the net force acting on an object is zero:

$$\text{net } \mathbf{F} = \Sigma \mathbf{F} = 0$$

Because force is a vector, the net force is zero in each direction, x, y, and z:

$$\Sigma F_x = 0; \qquad \Sigma F_y = 0; \qquad \Sigma F_z = 0$$

If two equal and opposite forces are applied at different points on a rigid object, the object, as shown in Figure 7-5, will be in **translational equilibrium.** However, it may tend to rotate. For an object to be at rest in space and not rotating it must satisfy the conditions of both translational and rotational equilibrium.

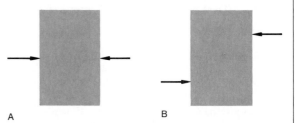

A   B

FIGURE 7-5
In both A and B there is translational equilibrium, but only in A is there rotational equilibrium.

Suppose a child with mass 25 kg is walking along a 2.0 m long plank supported at each end, as shown in Figure 7-6. The child stops 0.5 m from one end. Assume that the mass of the plank is negligible. What is the force at each end of the plank?

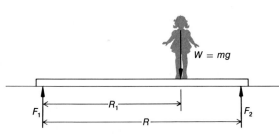

W = mg

F_1 R_1 R F_2

FIGURE 7-6
The forces supporting a child on a plank can be determined by considering torques.

The weight of the child is

$$W = mg = (25 \text{ kg})(9.8 \text{ m/s}^2) = 245 \text{ N}$$

The child's weight is balanced by the upward force at each end of the plank. The condition for translational equilibrium in the vertical direction is

$$\Sigma F_y = W - F_1 - F_2 = 0$$

where the downward direction is taken as positive.

If the child were standing at the first end, $F_1 = W$ and $F_2 = 0$. If the child were standing at the second end, $F_1 = 0$ and $F_2 = W$. If the child were in the middle, $F_1 = F_2 = W/2$. We need to find a general condition.

For rotational equilibrium, the net torque about an axis is zero. If, to compute the torque, we choose an axis through the point of application of F_1, then there are two forces that tend to make the plank rotate. F_2 tends to give counterclockwise rotation. W tends to give clockwise rotation. The net torque is

$$\Sigma \tau = RF_2 - R_1 W = 0$$

We can solve at once for F_2

$$F_2 = \frac{R_1}{R} W$$

We can then solve for F_1

$$F_1 = W - F_2 = W - \frac{R_1}{R} W = W \left[1 - \frac{R_1}{R} \right]$$

As the child goes from one end to the other the force on the first end is reduced from a maximum W to zero.

Vector Cross Product

A force $\mathbf{F}$ is a vector quantity, as is the lever arm $\mathbf{R}$. The torque τ can also be represented as a vector—the **vector cross product** of vectors $\mathbf{R}$ and $\mathbf{F}$

$$\tau = \mathbf{R} \times \mathbf{F}$$

The vector cross product defines both the magnitude and direction of the torque.

The vectors $\mathbf{R}$ and $\mathbf{F}$ are shown in Figure 7-7 A. Both lie in the plane of this page. The magnitude of the torque is indicated by the area that the two vectors enclose and depends on their magnitude and the sine of the angle between their directions as shown in Figure 7-7 B:

$$\tau = RF \sin \theta$$

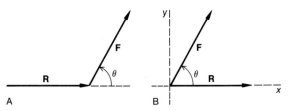

FIGURE 7-7
The intersection between two vectors defines a plane.

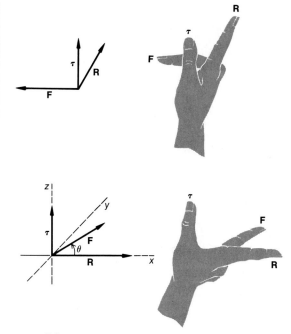

FIGURE 7-8
Torque is a vector that is perpendicular to both the lever arm *R* and the force *F*. The direction of the force is given by the right-hand rule.

Because torque is a vector, we must specify its direction. There is a unique direction in space that is perpendicular to both **R** and **F**—that is, perpendicular to the plane of this page. The direction of the torque is along this direction. We must, however, specify the positive direction for the torque. Figure 7-7 B shows the directions of the vectors **R** and **F** in the plane of the paper. If **R** rotates counterclockwise into **F**, then the positive direction of the torque is out of the paper, toward the reader. If **R** rotates clockwise into **F**, the direction of the torque is into the paper. An ordinary wood screw turned clockwise advances deeper into the wood; this is the same as the direction of the torque.

We can also specify the direction of the torque using the **right-hand rule** as shown in Figure 7-8. If the index finger of the right hand points in the direction of **R** and the middle finger points in the direction of the force **F,** the thumb points in the direction of the torque **τ,** along the axis of rotation. We shall use the right-hand rule and the vector cross product again in our discussion of magnetism in Chapter 15.

Center of Mass

If the object shown in Figure 7-9 is supported from a single pivot point and is acted on by gravity, it finds an equilibrium position so that its **center of mass** (also called the **center of gravity**) is directly below the point of support. If the object is held by another, new pivot point, the center of mass will remain below the new one. The entire object acts as if its entire mass were concentrated at the center of mass. This fact can greatly simplify the analysis of many physical problems.

Suppose we wish to calculate the center of mass of an object such as a baseball bat, shown in Figure 7-10. If the bat is allowed to pivot about the small end and is held in a horizontal position, gravity will act on each small element of mass Δm. The magnitude of torque due to each small element is just

$$\Delta \tau_i = R_i \Delta m_i g$$

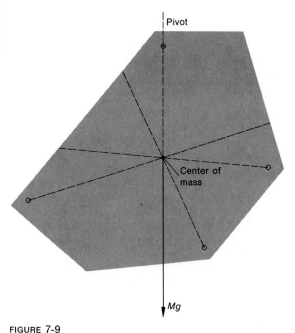

FIGURE 7-9
Center of mass. An object held freely from a single point will come to equilibrium with the center of mass directly beneath the pivot point.

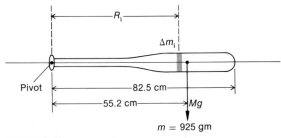

FIGURE 7-10
The baseball bat. The center of mass can be calculated by the sum of the mass elements Δm_i multiplied by the displacement from the pivot R_i. The distance to the center of mass R_c is given by $R_c = \Sigma m_i R_i / \Sigma m_i$.

The total torque acting clockwise on this object is just the sum of the contributions of each element:

$$\tau = \Sigma R_i \Delta m_i g$$

The sum of the mass times the distance from the pivot is equal to the total mass times the distance to the center of mass.

$$\Sigma R_i \Delta m_i = M R_{CM}$$

If the pivot point is moved to directly under the center of mass, the object is balanced. If the center of mass is not directly above or below the point of support, there is a net torque causing the mass to rotate.

The Biceps Muscle

Torques play an important role in the application of muscular force to the human skeleton. The forces on the forearm, for example, are applied at different points. If we are holding a ball of mass M in our hand, as shown in Figure 7-11 A, the force of gravity acts on the ball, hand, and forearm. The biceps muscle and the force at the elbow balance these forces. Because several forces act on, for example, the forearm, it is useful to accurately isolate them. A **free body diagram** represents all external forces which act on an isolated body, allowing us to determine the forces and torques required, for example, to establish rotational and translational equilibrium.

As Figure 7-11 A, a free body diagram, shows, the forearm pivots at the elbow, where an undetermined force $\mathbf{F}_{elbow}$ acts. The weight $M_f g$ of the forearm and hand acts

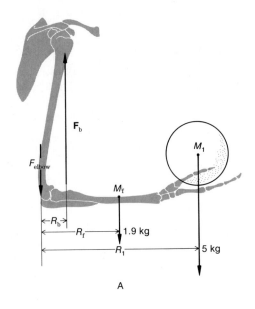

A

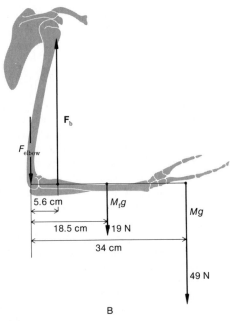

B

FIGURE 7-11
The biceps muscle. The torque produced by the biceps muscle is balanced by the torque produced by the forearm, hand, and 5 kg mass. Adapted from Alice L. O'Connell and Elizabeth B. Gordon, *Understanding the Scientific Bases of Human Motion*, © 1972, The Williams & Wilkins Co., Baltimore, Md.

at the center of mass with a lever arm R_f. The weight of the ball Mg acts with a lever arm R. These forces produce a clockwise torque about the pivot at the elbow

$$\tau_c = M_f g R_f + MgR$$

This torque must be balanced by the counterclockwise torque provided by the biceps muscle:

$$\tau_{cc} = F_b R_b$$

A hand is holding a ball of mass $M = 5$ kg a distance $R = 0.34$ m from the elbow with the forearm horizontal. The mass of the forearm and hand is $M_f = 1.90$ kg with the distance from center of mass to pivot $R_f = 0.185$ m. The biceps muscle applies a force F_b at a distance $R_b = 0.056$ m from the pivot. What is the force F_b applied by the biceps muscle? What is the reaction force at the elbow, F_{elbow}?

The weight of the forearm, hand, and ball produces a clockwise torque:

$$\tau_c = M_f g R_f + MgR$$

$$\tau_c = (1.90 \text{ kg})(9.8 \text{ m/s}^2)(0.185 \text{ m})$$

$$+ (5.0 \text{ kg})(9.8 \text{ m/s}^2)(0.34 \text{ m})$$

$$\tau_c = 18.5 \text{ N-m}$$

At equilibrium, this clockwise torque is balanced by the counterclockwise torque applied by the biceps:

$$\tau_c = \tau_{cc} = F_b R_b$$

The force applied by the biceps is just

$$F_b = \tau_c / R_b = \frac{18.5 \text{ N-m}}{0.056 \text{ m}}$$

$$F_b = 3\underline{3}0 \text{ N}$$

This force is substantially larger than the downward weight of the 5 kg mass, $W = 49$ N.

The arm is not only in rotational equilibrium (it does not rotate), it is also in translational equilibrium (it is stationary). Newton's first law tells us that the net force acting on an object at rest is zero. The upward force of the biceps is much greater than the downward force of the arm, hand, and ball. The extra force is provided by a downward force on the forearm at the elbow. We can calculate the force in the vertical direction from translational equilibrium:

$$\text{net } F = 0$$

$$F_{elbow} + M_f g + M_b g - F_b = 0$$

$$F_{elbow} = - (1.90 \text{ kg})(9.8 \text{ m/s})$$

$$- (5.0 \text{ kg})(9.8 \text{ m/s}) + 330 \text{ N}$$

$$F_{elbow} = 2\underline{6}0 \text{ N}$$

There is a downward force at the elbow of about $2\underline{6}0$ N. This force places the upper arm bone under compression.

The Leg and Hip

The conditions for translational and rotational equilibrium can be applied to the action of the hip. Three hip abductor muscles control the motion of the femur at the pelvis in the frontal plane. These three abductor muscles give a resultant force. When a person is standing on both feet the muscles exert forces at an angle to the horizontal of, typically: **A**, 89 N at 48°; **B**, 178 N at 78°; and **C**, 45 N at 86°. The resultant force **F** of about 300 N acts at an angle of about 71° to the horizontal. These muscles are attached at the greater trochanter of the femur as shown in Figure 7-12.

During walking, the entire weight of the body is alternately transferred from one foot to the other. For the body to remain balanced, its center of mass must remain directly over the point of support in the frontal plane. Figure 7-13 indicates the forces that act when the body weight is supported by a person's right leg. The weight of each leg is approximately one-seventh of the total body

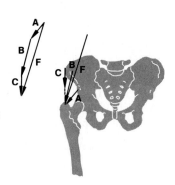

FIGURE 7-12
The resultant force of three hip abductor muscles controls the
pelvis in the frontal plane. The resultant is a single force at
an angle of about 71° to the horizontal when stance is normal.
From Marian Williams and Herbert R. Lissner, *Biomechanics
of Human Motion* (Philadelphia: W. B. Saunders Company, ©
1962).

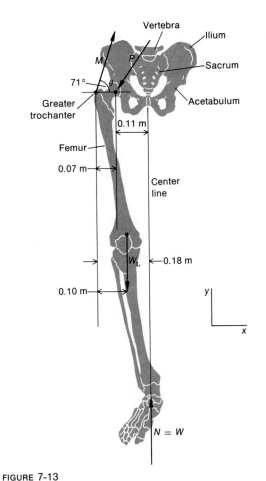

FIGURE 7-13
Standing on one foot. The sketch shows the resultant force of
the hip abductor muscles M, the reaction force P at the head
of the femur, the weight W_L of the leg, and the net weight N
supported by the leg. Adapted from Marian Williams and
Herbert R. Lissner, *Biomechanics of Human Motion* (Philadelphia: W. B. Saunders Company, © 1962).

weight, $W_L = W/7$, and acts at the center of mass of the
leg. The weight of the leg and the supported weight are
balanced primarily by the force of the hip abductor muscles M, attached at the greater trochanter, and the pressure
P where the head of the femur pivots in the hip socket
acetabulum).

We can simplify the analysis by considering a free body
diagram of the leg as shown in Figure 7-14. We treat the

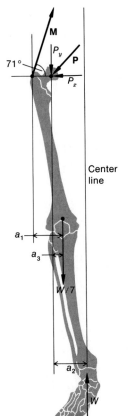

FIGURE 7-14
A free body diagram show-
ing the forces acting on the
leg and hip socket when a
person is standing on one
foot.

leg as rigid body and simply examine the forces acting on it. There are four: the reaction of the ground pressing upward supporting the total weight W; the weight of the leg itself $W/7$; the tension of the hip abductor muscles M; and the hip-socket force P. We can then apply the conditions of translational and rotational equilibrium.

EXAMPLE

A person is standing with the entire weight balanced on one foot.
(a) What is the force of the abductor muscle M?
(b) What is the magnitude and direction of the reaction force at the hip socket?
   The free body diagram for the leg is shown in Figure 7-14. The only forces acting in the horizontal plane are the components of M and P:

$$\text{net } F_x = \Sigma F_x = M \cos 71° - P_x = 0$$

In the vertical direction we have four components:

$$\text{net } F_y = \Sigma F_y = W - W_L + M \sin 71° - P_y$$

where the upward direction is considered positive. There are three unknowns: M, P_x, and P_y. We need a third equation.
   In rotational equilibrium, net torque is zero. Our choice of axis for calculating the torque is arbitrary; we choose as our axis the natural pivot of the hip socket. We can then write down the conditions:

$$\text{net } \tau = \Sigma \tau = W a_2 - M a_1 \sin 71° - W_L a_3 = 0$$

Solving for the force of the abductor muscle:

$$M = \frac{W a_2 - W_L a_3}{a_1 \sin 71°}$$

   For a 70 kg person the weight is $W = mg = 686$ N. The weight of the leg is about $W_L = 98$ N. The dimensions are shown in Figure 7-16: $a_1 = 0.07$ m;

$a_2 = 0.11$ m; $a_3 = 0.03$ m. The force of the abductor muscles is

$$M = \frac{(686 \text{ N})(0.11 \text{ m}) - (98 \text{ N})(0.03 \text{ m})}{(0.07 \text{ m})(0.946)}$$

$$M = 1100 \text{ N}$$

Having found the force M we can then solve for the two components of the reaction force at the hip socket using the condition of translational equilibrium:

$$P_x = M \cos 71° = (1100 \text{ N})(0.326)$$

$$P_x = 360 \text{ N}$$

$$P_y = W - W_L + M \sin 71°$$

$$= 686 \text{ N} - 98 \text{ N} + (1100 \text{ N})(0.946)$$

$$P_y = 1620 \text{ N}$$

   A force on the head of the femur of 1660 N [370 lbf] at an angle of 78° from the horizontal is required to support a weight of only 686 N [154 lbf] when standing on one foot.

   While a person is balancing on one foot, considerable force is applied at the pivot of the hip socket, both by the abductor muscles and the weight of the person. Persons with weak abductor muscles or painful hip joints reflexively shift their weight so that their center of mass is more nearly over the hip socket as they walk. The lever arm a_2 is much smaller and the required force of the abductor muscles and the reaction force at the joint are thus much smaller. This gives rise to a characteristic "antalgic gait," which can lead to abnormal bone growth.[1]
   Persons injured in the leg or the pelvic region are placed on crutches so that the arms and shoulders and the

---

[1] This and other conditions are discussed in detail in George B. Benedek and Felix M. H. Villars, *Physics with Illustrative Examples from Medicine and Biology, Vol. 1, Mechanics* (Reading, Mass.: Addison Wesley, 1973). The classic work on the subject is Marian Williams and Herbert R. Lissner, *Biomechanics of Human Motion* (Philadelphia, Pa.: W. B. Saunders, 1962).

good leg carry most of the body weight, taking the strain off the hip socket and abductor muscles. Later a cane is used. The cane can transfer approximately one-sixth of the body weight to the upper part of the body, reducing to five-sixths the weight placed on the foot. However, the leverage of the cane also allows the foot to be placed nearer the pivot point of the hip, and thereby greatly reduces the stress on the hip abductor muscles and hip socket.

EXAMPLE

Suppose that a person with a mass of 70 kg and a weight $W = mg = 686$ N uses a cane. The cane supports one-sixth of the weight and is placed a distance of 0.30 m to the right of the center of gravity. To balance the weight, the foot is placed a distance x to the left of the center of gravity as shown in Figure 7-15.

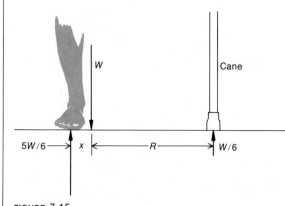

FIGURE 7-15
A cane distributes the weight between itself and the foot. The position of the foot relative to the center of gravity is determined by rotational equilibrium.

By the condition of translational equilibrium the foot must support $5W/6$. What is the distance x?

Calculating the net torque about the center of mass we find

$$\Sigma \tau = R \frac{W}{6} - x \frac{5W}{6} = 0$$

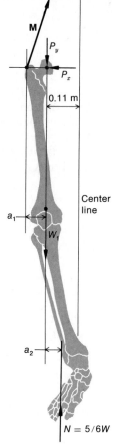

FIGURE 7-16
The free body diagram of the leg when a person is standing on one foot and using a cane.

Solving for x we get

$$x = \frac{R}{5} = 0.06 \text{ m}$$

The foot is placed 0.06 m to the left of the center of gravity. The pivot of the hip is typically 0.11 m to the left of the center of gravity so that the moment arm $a_2 = 0.05$ m, greatly reducing the torque at the hip socket.

Figure 7-16 shows the free body diagram of this case. The condition for translational equilibrium is

$$\Sigma F_x = M \cos 71° - P_x = 0$$

and

$$\Sigma F_y = \frac{5W}{6} - W_L + M \sin 71° - P_y = 0$$

The weight of the leg W_L is now almost directly below the hip socket so that its lever arm is zero. The force of the abductor muscle is applied with a lever arm $a_1 \sin 71°$ where $a_1 = 0.07$ m. The net torque about the pivot is zero.

$$\Sigma\tau = a_2 \frac{5W}{6} - a_1 M \sin 71° + 0\ W_L = 0$$

Solving for the force of the abductor muscles we get

$$M = \frac{5\ Wa_2}{6a_1 \sin 71°} = \frac{(5)(686\ \text{N})(0.05\ \text{m})}{(6)(0.07\ \text{m})(0.946)}$$

$$M = 4\underline{3}2\ \text{N}$$

The tension in the abductor muscles has been reduced from 1100 N when one foot is stood upon, to 4$\underline{3}$0 N when a cane is used. The net force at the head of the femur is 8$\underline{9}$0 N at an angle of 81°, much less than the 16$\underline{6}$0 N when no cane is used.

ANGULAR KINEMATICS

Angular Velocity

If a shaft of radius R is spinning at constant speed, a point on the surface is traveling with a tangential velocity v_t. The rate at which the angle of the shaft changes, as measured in radian measure, is the **angular velocity,** represented by ω (the lower-case Greek letter *omega*).

$$\omega = \frac{\Delta\theta}{\Delta t}$$

$\Delta\theta = \Delta s/R$, so that

$$\omega = \frac{\Delta s}{R\Delta t}$$

However, $\Delta s/\Delta t = v$, so that

$$\omega = v_t/R$$

One revolution sweeps out an angle of 2π radians. An angular velocity of 1800 rev/min = 30 rev/s = 188.5 rad/s

In a time Δt at an angular velocity ω, a shaft will sweep out an angle $\Delta\theta$:

$$\Delta\theta = \omega\Delta t$$

At constant angular velocity, if the initial angle is θ_o at time t_o, then the angle as a function of time is

$$\theta = \theta_o + \omega t$$

The acceleration required to maintain an object with a speed v in a uniform circular orbit of radius R is the **centripetal acceleration** (discussed in Chapter 3):

$$a_\perp = v_t^2/R$$

The tangential velocity is given by the angular velocity times the radius:

$$v_t = \omega R$$

The centripetal acceleration is then

$$a_\perp = \omega^2 R$$

The centripetal acceleration depends on the square of the angular velocity ω and on the radius of the object's orbit.

EXAMPLE

A Sorval vacuum ultracentrifuge spins at a top speed of 65 000 rev/min. The sample tube spins at a radius of 9 cm from the axis.

(a) What is the angular speed of the centrifuge?
(b) What is the centripetal acceleration?
(c) Express the acceleration in terms of the earth's gravitational acceleration (1 g = 9.8 m/s^2).
The angular speed is given by

$$\omega = \frac{\Delta\theta}{\Delta t}$$

$$\omega = \frac{(65\ 000\ \text{rev})(2\pi\ \text{rad/rev})}{60\ \text{s}}$$

$$\omega = 6.8 \times 10^3\ \text{rad/s}$$

The centripetal acceleration is given by

$$a = \omega^2 R$$

$$a = (6.8 \times 10^3 \text{ rad/s})^2 (9 \times 10^{-2} \text{ m})$$

$$a = 4.16 \times 10^5 \text{ m/s}^2 \times \left[\frac{1g}{9.8 \text{ m/s}^2} \right]$$

or, $a = 4.2 \times 10^4 g$

Angular Acceleration

As a spinning wheel accelerates, its angular velocity increases. The rate of change with time of the angular velocity is the **angular acceleration,** α (the lower-case Greek letter *alpha*):

$$\alpha = \frac{\omega_2 - \omega_1}{t_2 - t_1} = \frac{\Delta\omega}{\Delta t}$$

For a cylinder of radius R whose surface is rotating at a tangential velocity v_t, the angular velocity is given by $\omega = v_t/R$. The angular acceleration is then given by

$$\alpha = \frac{v_2/R - v_1/R}{t_2 - t_1} = \frac{\Delta v/R}{\Delta t}$$

$$\alpha = \frac{1}{R} \frac{\Delta v}{\Delta t}$$

$$\alpha = a_t/R$$

Where a_t is the tangential acceleration, that is, the linear acceleration of a point on the rotating surface. If the angular acceleration α is constant, the angular velocity from initial time $t_o = 0$ is given by

$$\omega = \omega_o + \alpha t$$

The average angular velocity is given by

$$\omega_{av} = \frac{\omega + \omega_o}{2}$$

The angular displacement over a time t is given by

$$\theta = \theta_o + \omega_{av} t$$

Combining these three relationships, we can obtain

$$\theta = \theta_o + \omega_o t + \tfrac{1}{2}\alpha t^2$$

The relationships for angular displacement, velocity, and constant angular acceleration are analogous to the same relationships for linear motion. Table 7-1 compares the relationships for linear and angular motion.

TABLE 7-1
Relationships of linear and angular motion.

	Linear	Angular
Displacement	Δs	$\Delta\theta = \Delta s/R$
Velocity	$v = \dfrac{\Delta s}{\Delta t}$	$\omega = \dfrac{\Delta\theta}{\Delta t} = \dfrac{v}{R}$
Acceleration	$a = \dfrac{\Delta v}{\Delta t}$	$\alpha = \dfrac{\Delta\omega}{\Delta t} = \dfrac{a}{R}$
Average velocity	$v_{av} = \dfrac{v + v_0}{2}$	$\omega_{av} = \dfrac{\omega + \omega_0}{2}$
Velocity	$v = v_0 + at$	$\omega = \omega_0 + \alpha t$
Displacement	$s = s_0 + v_0 t + \tfrac{1}{2}at^2$	$\theta = \theta_0 + \omega_0 t + \tfrac{1}{2}\alpha t^2$

NEWTON'S LAWS FOR ANGULAR MOTION

Thus far we have considered angular motion; the angular displacement, velocity, and acceleration of an object; and rotational equilibrium, the condition in which net torque is zero. Let us now consider what happens when a torque acts on an object.

Moment of Inertia

Suppose a total force F_t acts on the rim of a thin cylindrical shell of mass M, causing it to turn counterclockwise, as shown in Figure 7-17. According to Newton's second law, the applied force will cause the mass to accelerate in the tangential direction, $F_t = Ma_t$. Because it is a rotating object, there is an angular acceleration α. The tangential acceleration of a point on the rim is related to the angular acceleration α by

$$a_t = R\alpha$$

The magnitude of the torque applied to the system is

$$\tau = RF_t$$

Because a force F_t is applied to the thin shell there is a tangential acceleration of the rim, $F_t = Ma_t$. This tan-

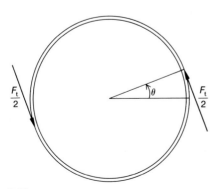

FIGURE 7-17
A thin cylindrical shell of mass M and radius R rotates under the action of a pair of forces $F/2$.

gential acceleration is related to the angular acceleration $a_t = R\alpha$. The torque acting on the thin shell is then

$$\tau = RmR\alpha = MR^2\alpha$$

The torque gives rise to an angular acceleration α, which depends on the magnitude of the torque and on the distribution of matter.

The relationship between the torque and angular acceleration can be written in the same form as Newton's laws:

$$\tau = I\alpha$$

where I is the **moment of inertia.** For a thin cylindrical shell of matter as shown in Figure 7-17, or for a single mass M traveling in a circle of radius R, the torque is

$$\tau = MR^2\alpha$$

so that the moment of inertia is

$$I = MR^2$$

If an object's matter is distributed in space, we must account for this distribution by adding the contribution of each small element of matter in the object m_i multiplied by the square of the distance from the axis R. The moment of inertia generally can be written as

$$I = \Sigma m_i R_i^2$$

Figure 7-18 shows the moment of inertia of common shapes.[2]

Objects of greater complexity, such as the rotor of a centrifuge or a flywheel, also have moments of inertia. To characterize an object's moment of inertia it is useful to determine the **radius of gyration, R_g,** the measure of the mean square distance from the rotational axis. The moment of inertia for such an object of mass M is

$$I = MR_g^2$$

---

[2]Calculating the moment of inertia generally requires the use of calculus. The moment of inertia is obtained from the integral over the mass: $I = \int R^2 dm$

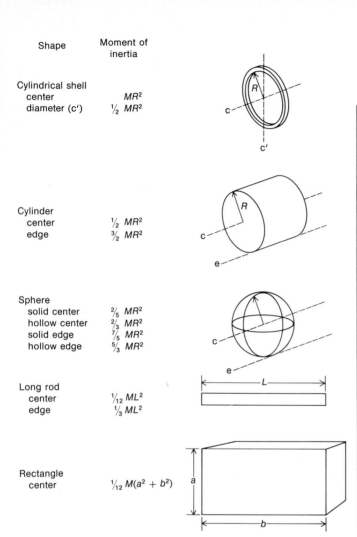

Shape	Moment of inertia
Cylindrical shell center diameter (c')	MR^2 $\frac{1}{2} MR^2$
Cylinder center edge	$\frac{1}{2} MR^2$ $\frac{3}{2} MR^2$
Sphere solid center hollow center solid edge hollow edge	$\frac{2}{5} MR^2$ $\frac{2}{3} MR^2$ $\frac{7}{5} MR^2$ $\frac{5}{3} MR^2$
Long rod center edge	$\frac{1}{12} ML^2$ $\frac{1}{3} ML^2$
Rectangle center	$\frac{1}{12} M(a^2 + b^2)$

FIGURE 7-18
The moment of inertia for different shapes about an axis through the center of mass (c) and about the edge (e).

Parallel Axis Theorem

If we know an object's moment of inertia about an axis through its center of mass, the moment of inertia about any other axis parallel to the first can easily be determined. The **parallel axis theorem** states that the resulting moment of inertia is the sum of the moment of inertia about the center of mass plus the moment of inertia of the center of mass about the new axis.

We can determine the moment of inertia of a solid sphere of mass M about an axis at the edge of the sphere. The moment of inertia about the center of mass is just

$$I_{center} = \frac{2}{5}MR^2$$

The moment of inertia of the mass about the point on the edge is just

$$I_m = MR^2$$

The total moment of inertia about a point on the edge is then

$$I_{edge} = I_{center} + I_m$$
$$I_{edge} = \frac{2}{5}MR^2 + MR^2$$
$$I_{edge} = \frac{7}{5}MR^2$$

This is shown in Figure 7-18.

EXAMPLE
A meter stick of mass M, pivoted from one end, is held in a horizontal position and then released (Figure 7-19).
(a) What is the initial angular acceleration of the stick?
(b) What is the initial linear acceleration of the end of the stick?
  The weight of the meter stick acts at the center of

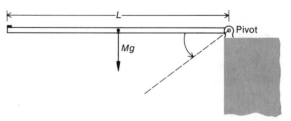

FIGURE 7-19
A falling meter stick attached to a pivot at one end. The far end of the meter stick will fall faster than a penny placed there.

TABLE 7-2

Linear and angular quantities.

Linear		Angular	
Inertia	M	Moment of inertia	$I = \Sigma M_i R_i^2$
Force	$\mathbf{F} = M\mathbf{a}$	Torque	$\tau = I\alpha = \mathbf{R} \times \mathbf{F}$
Work	$\Delta W = \mathbf{F} \cdot \Delta \mathbf{s}$	Work	$\Delta W = \tau \cdot \Delta \boldsymbol{\theta}$
Kinetic energy	$KE = \frac{1}{2}Mv^2$	Kinetic energy	$KE = \frac{1}{2}I\omega^2$
Momentum	$\mathbf{p} = M\mathbf{v}$	Angular momentum	$\mathbf{L} = I\omega = \mathbf{R} \times \mathbf{p}$
Impulse	$\Delta \mathbf{p} = \mathbf{F}\Delta t$	Angular impulse	$\Delta \mathbf{L} = \tau \Delta t$

mass at a distance $L/2$ from the end. The torque is given by

$$\tau = \mathbf{R} \times \mathbf{F} = I\alpha$$

$$\tau = L/2Mg$$

The moment of inertia of this rod is

$$I = \frac{1}{3}ML^2$$

Solving for the angular acceleration

$$\alpha = \frac{\tau}{I} = \frac{L/2Mg}{\frac{1}{3}ML^2}$$

$$\alpha = \frac{3}{2}\frac{g}{L}$$

Linear acceleration of the end of the stick is given by

$$a = L\alpha$$

$$a = \frac{3}{2}g$$

The end of the stick will accelerate somewhat faster than the normal acceleration due to gravity.

Rotational Kinetic Energy

If a force F is applied through a distance Δs, the work done is

$$\Delta W = F\Delta s$$

The displacement in angular motion is

$$\Delta s = R\Delta\theta$$

The applied torque is $\tau = RF$, so that the work done is

$$\Delta W = \tau \Delta \theta$$

If a thin hoop is spinning with an angular speed ω, then a point on the rim is traveling with a tangential velocity

$$v_t = R\omega$$

All of the mass is traveling with this velocity, so the kinetic energy of the spinning hoop is just

$$KE = \frac{1}{2}Mv_t^2$$

$$KE = \frac{1}{2}M(R\omega)^2$$

$$KE = \frac{1}{2}MR^2\omega^2$$

MR^2 is just the thin hoop's moment of inertia I. Generally the **rotational kinetic energy,** the kinetic energy of a rotating object, is given by the moment of inertia and the angular velocity.

$$KE = \frac{1}{2}Mv_t^2$$

Table 7-2 lists linear and angular quantities.

Since the invention of the potter's wheel, flywheels have been used to store energy and to deliver it smoothly. The upper limit of energy that a flywheel can store is determined by the "hoop stress," which provides the centripetal force that keeps the wheel from flying apart. The kinetic energy stored in a flywheel is determined by the moment of inertia I and the angular velocity ω.

A vacuum ultracentrifuge has a rotor with a mass 8.3 kg and an outer diameter 20 cm. The rotor spins at a top speed of 65 000 rev/min. Assume that the rotor's radius of gyration is 6 cm.

(a) What is the rotor's angular velocity?
(b) What is the rotor's moment of inertia?
(c) What is the rotor's rotational kinetic energy?
(d) What average power is required to bring the rotor up to speed in five minutes?

The angular speed is

$$\omega = \left(\frac{6.5 \times 10^4 \text{ rev}}{\text{min}}\right) \times \left(\frac{2\pi \text{ rad}}{\text{rev}}\right) \times \left(\frac{1 \text{ min}}{60 \text{ s}}\right)$$

$$\omega = 6.8 \times 10^3 \text{ s}^{-1}$$

The moment of inertia is given by

$$I = MR_g{}^2 = (8.3 \text{ kg})(0.06 \text{ m})^2$$

$$I = 0.030 \text{ kg-m}^2$$

The rotational kinetic energy is

$$KE = \tfrac{1}{2}I\omega^2 = \tfrac{1}{2}(0.030 \text{ kg-m}^2)(6.8 \times 10^3 \text{ s}^{-1})^2$$

$$KE = 6.9 \times 10^5 \text{ J}$$

The power is the work per unit time to increase the kinetic energy

$$P = \Delta W/\Delta t = \frac{6.9 \times 10^5 \text{ J}}{300 \text{ s}}$$

$$P = 2\,300 \text{ W} = 3 \text{ hp}$$

Angular Momentum

There is a parallel between linear and angular quantities. By Newton's second law, a force gives rise to a linear acceleration, and also gives the rate at which momentum changes in time:

$$\mathbf{F} = m\mathbf{a} = \frac{\Delta m\mathbf{v}}{\Delta t}$$

Likewise, a torque produces an angular acceleration, and also gives the rate at which angular momentum changes in time:

$$\tau = \mathbf{R} \times \mathbf{F} = I\alpha$$

Substituting for the force the torque:

$$\tau = \mathbf{R} \times \frac{\Delta m\mathbf{v}}{\Delta t} = \frac{\Delta \mathbf{R} \times m\mathbf{v}}{\Delta t}$$

If the torque on a system is zero, the quantity $\mathbf{R} \times m\mathbf{v}$ is conserved. This is the **angular momentum.**

$$\mathbf{L} = \mathbf{R} \times m\mathbf{v}$$

A spinning object continues to spin uniformly unless acted on by an external torque. The torque gives the rate at which angular momentum changes in time:

$$\tau = \frac{\Delta \mathbf{L}}{\Delta t}$$

We can calculate the angular momentum of a thin shell of mass M and radius R. If the thin shell is rotating at a uniform angular velocity ω, the tangential velocity of a point on the circumference is

$$v_t = R\omega$$

Since the tangential velocity v_t is perpendicular to the radius R, the magnitude of the angular momentum is

$$L = Rmv_t = RM(R\omega)$$

$$L = MR^2\omega$$

MR^2 is just the moment of inertia I of the thin shell of mass. An object with a moment of inertia I spinning about its axis with an angular velocity ω has an angular momentum

$$L = I\omega$$

Table 7-2 shows the relationship between Newton's laws for linear motion and the laws of angular motion.

In an isolated system the angular momentum is conserved and is given by the product of the moment of inertia times the angular velocity. If a system's moment of inertia is changed by redistribution of the mass, the angu-lar velocity changes, as Figure 7-20 shows. A diver doing a double somersault off the high board leaves the board with a certain angular velocity. As he tucks himself into a tight ball, his angular velocity increases. He holds the tight configuration until the right instant, when he un-folds, greatly reducing his angular velocity. If he breaks at just the right moment (this is a matter of experience, practice, and reflexes rather than calculation) he will enter the water perfectly. If he opens too late, he will land on his back; if too soon, on his face.

A ballet dancer or an ice skater will start a slow spin with the arms extended and torso erect. As the person pulls the arms in, the angular velocity increases, giving a very rapid rotation. A person wishing to release the spin simply extends the arms, increasing the moment of inertia and decreasing the angular velocity.

Modern timekeeping, accurate to a few nanoseconds, has shown that there is a small seasonal variation in the earth's rotation. This is associated with the seasonal redistribution of water, as snow, on the earth's surface.

Review

Give the significance of each of the following terms. Where appropriate give two examples.

lever	rotational equilibrium	centripetal acceleration
lever arm	translational equilibrium	angular acceleration
first-class lever	vector cross product	moment of inertia
fulcrum	right-hand rule	radius of gyration
second-class lever	center of mass	parallel axis theorem
third-class lever	center of gravity	rotational kinetic energy
radian	free body diagram	angular momentum
torque	angular velocity	

Further Readings

Richard F. Post and Stephen F. Post, "Flywheels", *Scientific American* (December 1973), pp. 17–23.

George B. Benedek and Felix M. H. Villars, *Physics with Illustrative Examples from Medicine and Biology, Vol. 1, Mechanics* (Reading, Mass.: Addison-Wesley, 1973).

Alice L. O'Connell and Elizabeth B. Gardner, *Understanding the Scientific Basis of Human Movement* (The Williams and Wilkins Co., Baltimore, Md., 1972).

Marian Williams and Herbert R. Lissner, *Biomechanics of Human Motion* (Philadelphia, Pa.: W. B. Saunders Co., 1962).

Problems

TORQUE

1 A boy with a mass of 55 kg stands on the end of a diving board 4 m long, which pivots about the far end and has a single support 1.5 m from the pivot.
(a) Draw a sketch of the forces involved.
(b) What is the torque provided by the boy?
(c) What is the upward force provided by the support?

2 A ladder 5 m long with a mass of 10 kg leans against a building, as shown in Figure 7-21. The base of the ladder is 3 m from the wall.

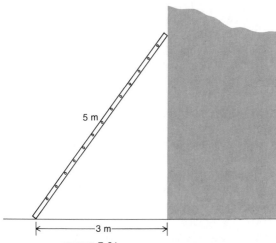

5 m

3 m

FIGURE 7-21
A ladder leaning against a building.

(a) Draw a sketch of the forces involved.
(b) What is the torque about a pivot at the base of the ladder?
(c) What is the horizontal force of the ladder against the building?

3 A man with a mass of 70 kg stands in the middle of the ladder of the previous problem.
(a) Draw a sketch of the forces involved.

(b) What is the torque, clockwise and counterclockwise, about the pivot at the base of the ladder?
(c) What is the force of the ladder against the building when the man is standing in the middle of the ladder? When the man has climbed 4 m up the ladder?

4 A person holds a 10 kg mass in his hand with his forearm extended in the horizontal position as shown previously in Figure 7-11. What is the tension in the biceps muscle and what is the reaction force at the elbow?

5 The Achilles tendon provides the tension to balance the weight when standing on the ball of the foot. Suppose a person with a mass of 80 kg is balanced evenly on the balls of both feet. The distance from the Achilles tendon to the ball of the foot is about 18 cm. The distance from the ball of the foot to the point of support of the leg bone, the tibia, is about 13 cm. The forces acting are shown in Figure 7-22. Assume that the foot acts as a rigid body.
(a) Make a free body diagram showing the forces acting on the foot.
(b) Solve for the tension in the Achilles tendon and the downward force of the tibia.

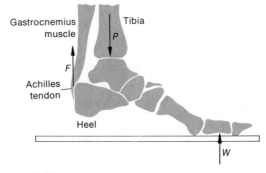

FIGURE 7-22
The Achilles tendon provides the tension for standing on the ball of the foot. From Leonard H. Greenberg, *Physics for Biology and Pre-med Students* (Philadelphia: W. B. Saunders Company, © 1975).

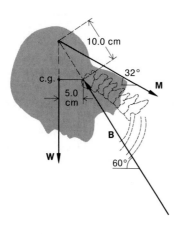

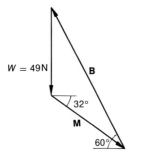

FIGURE 7-23
The weight of the head W is balanced by the force of the
neck extensor muscles M and the force at the atlanto-occipital
joint B. From Marian Williams and Herbert R. Lissner, *Biomechanics of Human Motion* (Philadelphia: W. B. Saunders
Company, © 1962).

6 The weight of the head is balanced by the tension
of the head extensor muscles M and the reaction
force B at the base of the skull (the atlanto-occipital joint). Assume that a person is leaning over as
shown in Figure 7-23. The head has a mass of
about 5.0 kg. Solve for the two forces required to
support the head.

ANGULAR KINEMATICS

7 The earth spins about its axis once in 24 hours.

The radius of the earth is 6 400 km.
(a) What is the angular velocity of the rotation of
the earth?
(b) What is the tangential velocity of the earth at a
point on the equator?
(c) What is the tangential velocity of a point at
45° North latitude?

8 A continuous-flow centrifuge operates in a speed
range from 2 000 rev/min to 55 000 rev/min. The
rotor has an effective sample radius of 4.4 cm.
(a) What is the minimum and maximum angular
velocity?
(b) What is the centripetal acceleration in each
case?
(c) Express the centripetal acceleration in
$g = 9.8 \text{ m/s}^2$.

9 A general laboratory centrifuge spins at a top speed
of 6 000 rev/min. The rotor has an effective sample
radius of 12.3 cm.
(a) What is the angular velocity of the rotor?
(b) What is the tangential velocity of a point on
the sample?
(c) What is the centripetal acceleration of the sample?
(d) Express the centripetal acceleration in g.

10 To separate protein fractions, cell fragments must
be spun at about 100 000 g for about 15 hr. If the
centrifuge has an effective sample chamber radius
of 15 cm,
(a) What is the angular velocity of the rotor?
(b) What is the tangential velocity of the sample
chamber?

NEWTON'S LAWS FOR ANGULAR MOTION

11 One end of a meter stick is allowed to pivot at the
edge of a table, as shown previously in Figure
7-19. The meter stick is horizontal with a penny
placed on top of it at the outer end. The meter
stick is then released.
(a) Which initially falls faster, the penny or the
end of the meter stick?
(b) Will the penny stay on the meter stick?

12 In the circumstances described in the previous
problem, where should the penny be placed so that

it will fall at the same rate as the meter stick on which it sits?

13 A potter's wheel consists of about 75 kg of concrete cast into a disk 0.70 m in diameter and about 0.08 m thick. The wheel spins at about 90 rev/min.
   (a) Assuming that the potter's wheel is a solid disk, what is the moment of inertia?
   (b) What is its angular velocity?
   (c) How much energy is stored in the wheel?

14 A general laboratory centrifuge spins at 6 000 rev/min and has a rotor with a mass of 2.8 kg and a diameter of 26 cm. Assume that the radius of gyration is $R_g = 9.5$ cm.
   (a) What is the moment of inertia of the rotor?
   (b) What is the angular velocity of the rotor?
   (c) How much energy is stored in the rotor?
   (d) If the rotor comes up to speed in 1 min, what is the average power required?

15 A general purpose refrigerated centrifuge spins at 5 000 rev/min and has a rotor with a mass of 14.5 kg. The rotor has an outside diameter of 39.2 cm. Assume that the radius of gyration of the rotor is 12 cm.
   (a) What is the rotor's moment of inertia?
   (b) What is its angular velocity?
   (c) How much energy is stored in the rotor at top speed?

16 The rotor in the preceding problem will accelerate from 0 to 5 000 rev/min in just 50 seconds.
   (a) What is the average angular acceleration?
   (b) What torque is required?
   (c) What is the total angular distance θ traveled while accelerating?
   (d) How much work was done in accelerating?

17 The earth has a mass of $M_e = 6.0 \times 10^{24}$ kg and a radius of 6.38×10^6 m. It rotates with a period of $T = 24$ hours.
   (a) What is the angular velocity of the earth about its axis?
   (b) If the earth is considered as a solid sphere, what is the moment of inertia?
   (c) What is the rotational kinetic energy of the earth?

18 A flywheel proposed to provide energy storage for peak-load electrical power has a mass of

2.0×10^5 kg and a radius of 2.3 m and spins with a top speed of 3 500 rev/min. Assume that it has a radius of gyration of 1.6 m.
   (a) What is the flywheel's angular velocity?
   (b) What is its moment of inertia?
   (c) What is the kinetic energy stored in the flywheel in joules? in kilowatt-hours?

ADDITIONAL PROBLEMS

19 A person with a mass of 70 kg carries a 25 kg trunk with the right hand at a distance of 25 cm to the right of his center of gravity. Walking requires that the weight be balanced for a moment on the left foot.
   (a) Where must the person place his left foot relative to the center of gravity to balance the load?
   (b) Make a free body diagram showing the forces acting on the supporting leg. Assume that the hip abductor muscles act at an angle of 70° to the horizontal.
   (c) Assuming the dimensions are similar to those shown in Figure 7-13 and that the weight of the leg is halfway between the foot and the pivot, determine the force on the hip abductor muscle and the reaction force at the hip socket.

20 A person lifting a heavy object must lift with the back relatively straight, carrying the weight with the legs. Trying to lift by bending at the waist places a great strain on the lower back. Figure 7-24, a free body diagram of the vertebral column, shows the forces acting when bending at an angle θ. The weight of the torso W_1, plus the weight of the head and arms W_2, plus the weight of any load W_L must be balanced by the force of the erector spinae muscles F_e and the reaction force R at the fifth lumbar vertebra at the base of the vertebral column. The weight of the torso is about 40% of the body weight $W_1 = 0.40 W$ and the center of gravity is at a distance of $L/2$ up the spine. The weight of the head and arms is about 20% of the body weight $W_2 = 0.20 W$ and is applied at a distance L. The gross effect of the erector spinae muscles can be approximated by a single force F_e applied at a distance of $2L/3$ from the base of the spine at an angle of 12° to the spine.

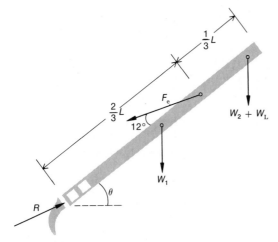

FIGURE 7-24
A free body diagram showing the forces acting on the vertebral column when a person bends over in picking up a load W_L. A very large force acts at the base of the column.

(a) Suppose a person with a total mass of 60 kg bends over at an angle of 60° to the vertical, carrying no load $W_L = 0$. What is the total force at the base of the spine?
(b) If the person bends over to an angle of 45°, what is the total force at the base of the spine?

21 A person with a total mass of 80 kg bends over to an angle of 45° to pick up a 20 kg mass.
(a) What is the force at the base of the spine? A force of 2500 N will cause the lumbar vertebral disk to compress by about 20% and bulge outward by 10%. A maximum force of 15 000 N will cause a healthy vertebral disk to rupture.
(b) If the person carries the same load with the spine at an angle of 80°, what is the force at the base of the spine?

22 A person with a mass of 60 kg sits on top of a stepladder with a vertical height of 2.0 m and two sides of equal length, as shown in Figure 7-25, at an angle of 30°. Halfway up the ladder is a horizontal brace 0.54 m long. Assume that the floor is slippery, and thus exerts only a vertical force.
(a) What is the force exerted by the floor on each leg of the ladder?
(b) What is the magnitude and direction of the force in the crossbrace?

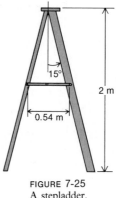

FIGURE 7-25
A stepladder.

23 A solid disk with a radius R is rolled, without slipping, down an inclined plane of angle θ as shown in Figure 7-26. The disk has a moment of inertia $I = MR_g^2 = 3MR^2/2$ about a point of contact.
(a) What is the torque acting on the disk about the point of contact?
(b) Show that the angular acceleration is $\alpha = 2g \sin \theta / 3R$.

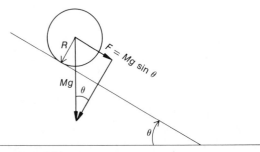

FIGURE 7-26
A disk rolls, without slipping, down an inclined plane.

(c) What is the linear acceleration of the disk down the plane?

24 A basketball, a baseball, a solid cylinder, and a hoop are rolled from rest down an inclined plane. Which will roll the fastest? Slowest? Why?

8

Harmonic Motion

Many physical systems execute a repetitive motion in time, often as a vibration about an equilibrium position. For example, displaced by a finger and released, a guitar string vibrates. A pendulum displaced from rest and released oscillates with a characteristic period. A tuning fork, if displaced, vibrates at a characteristic frequency.

SIMPLE HARMONIC MOTION

Suppose an object is hanging from a spring, as shown in Figure 8-1. By Hooke's law, the spring has a restoring force proportional to the **displacement** from the equilibrium. If we attach an object of mass m to the spring, the spring stretches so that its force balances the weight of the mass. If the mass is displaced from this **equilibrium position,** the object bounces up and down. If the object is above the equilibrium position, the spring exerts less force than the weight, and there is a net downward force. If the object is below the equilibrium position, the spring exerts a force greater than the weight, and the mass is accelerated upward. The spring provides a **restoring force,** which is proportional to the displacement from the equilibrium x. The net force acting on the mass gives its acceleration:

$$F_{\text{restoring}} = -kx = ma$$

The equation of motion for the spring and mass is just

$$a_x + \frac{k}{m} x = 0 \qquad (8\text{-}1)$$

When the restoring force is proportional to the displacement from equilibrium, the resulting motion is known as **simple harmonic motion.**

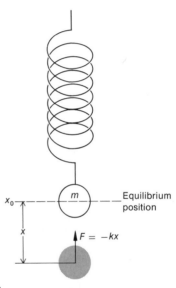

FIGURE 8-1
The spring pendulum. The pendulum has an equilibrium position, a displacement from the equilibrium, and a restoring force.

The Reference Circle

To see the form of simple harmonic motion, consider uniform motion on a **reference circle** of radius R, as shown in Figure 8-2. If the point is rotating at constant speed, the angle as a function of time is

$$\theta = \omega t$$

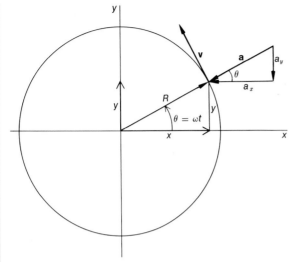

FIGURE 8-2
Uniform motion about a reference circle.

This is the solution to an equation of the form

$$a_x + \omega^2 x = 0$$

Similarly, the acceleration in the y-direction is related to the displacement in the y direction by

$$a_y + \omega^2 y = 0$$

The solution in this direction is

$$y = R \sin \omega t$$

Both the sine function and the cosine function are solutions to the equation for simple harmonic motion. The sine and cosine as a function of time are shown in Figure 8-3.

In the first illustration for the spring and mass the equation of motion (equation 8-1) was

$$a_x + \frac{k}{m} x = 0$$

A point on the reference circle travels with a speed given by $v = \omega R$. The centripetal acceleration is radially inward with a magnitude

$$a = v^2 R = \omega^2 R$$

At any particular position with angle θ we can look at the x component of the acceleration. By simple trigonometry:

$$a_x = -\omega^2 R \cos \theta$$

However, the displacement in the x-direction is just

$$x = R \cos \theta$$

The acceleration in the x-direction is proportional to and opposite in direction to the displacement in the x-direction:

$$a_x = -\omega^2 x$$

This is the condition for simple harmonic motion. Because the angle increases uniformly in time, the x displacement as a function of time is

$$x = R \cos \omega t \qquad (8\text{-}2)$$

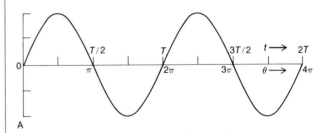

A

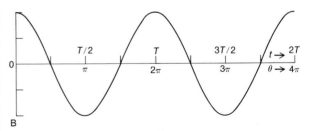

B

FIGURE 8-3
A. The sine as a function of time $y = y_0 \sin \omega t$, or angle in radians $y = y_0 \sin \theta$. B. The cosine as a function of time $y = y_0 \cos \omega t$, or angle in radians $y = y_0 \cos \theta$.

By comparison with our relationship for the reference circle, we find that

$$\omega^2 = k/m$$

The solution to this equation is then

$$x = x_0 \cos \omega t$$

$$x = x_0 \cos (k/m)^{1/2} t$$

The period of the motion is the time required for one complete revolution of the reference circle

$$\theta = \omega T = 2\pi$$

$$T = \frac{2\pi}{\omega}$$

$$T = 2\pi (m/k)^{1/2} \qquad (8\text{-}3)$$

The frequency of the motion is the recripocal of the period

$$f = \frac{1}{T} = \frac{\omega}{2\pi}$$

In the case of the spring pendulum

$$f = \frac{1}{2\pi}(k/m)^{1/2}$$

The Simple Pendulum

A **simple pendulum** (a pendulum of which the supporting thread is treated as weightless) is an example of simple harmonic motion. If the pendulum shown in Figure 8-4 is swung through an angle θ from the vertical it is displaced a horizontal distance $x = L \sin \theta$. A component of the weight tends to force the pendulum back to the equilibrium position. The magnitude of this restoring force is

$$F_x = -T \sin \theta$$

where T is the tension in the string. The y component of the tension must support the weight

$$T \cos \theta = mg$$

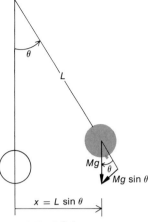

FIGURE 8-4
The simple pendulum.

Eliminating the tension T, the restoring force is

$$F_x = -mg \frac{\sin \theta}{\cos \theta} = -mg \tan \theta$$

In the limit of small displacements

$$\tan \theta = \frac{x}{L} \quad \text{and} \quad F_x = -mg\frac{x}{L}$$

The force causes the mass to accelerate. Applying Newton's second law of motion,

$$F_x = ma_x = -mg\frac{x}{L}$$

the equation of motion can be written

$$a_x + (g/L)x = 0$$

The acceleration is proportional to the displacement, resulting in simple harmonic motion. The angular frequency ω^2 is

$$\omega^2 = g/L$$

The period of the motion is given by

$$T = 2\pi/\omega$$

$$T = 2\pi(L/g)^{1/2} \qquad (8\text{-}4)$$

The period depends on the pendulum's length but is independent of its mass.

What is the period of a simple pendulum with a length of 1 m?

The period of the pendulum is given by

$$T = 2\pi (L/g)^{1/2}$$

$$T = 6.28(1.00/9.8 \text{ m/s}^2)^{1/2}$$

$$T = 2.01 \text{ s}$$

The Physical Pendulum

The simple pendulum consists of a mass M at the end of a string of length L. The mass of a **physical pendulum** is distributed along the length of the pendulum. Figure 8-5 shows a bar of mass M and length L that is swinging from one end. Gravity acting at the center of mass produces a torque that will cause the bar to rotate. The magnitude of the torque is

$$\tau = RF \sin \theta = -(L/2)Mg \sin \theta$$

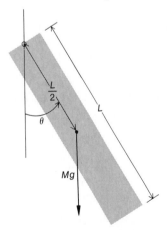

FIGURE 8-5
The physical pendulum. A bar with a mass M and length L pivots from one end.

where the minus sign indicates clockwise rotation. This torque gives rise to an angular acceleration of the bar:

$$\tau = I\alpha$$

The moment of inertia of a long rod suspended from one end is, as given earlier, in Figure 7-18,

$$I = \frac{ML^2}{3}$$

The torque can then be written

$$\tau = \frac{ML^2}{3}\alpha = -\frac{LMg}{2}\sin \theta = -\frac{LMg}{2}\theta$$

because for small angles, $\sin \theta \cong \theta$ in radian measure. Solving for the angular acceleration we obtain

$$\alpha = -\frac{3g}{2L}\theta$$

The angular displacement is proportional to the angular acceleration.

We can write this in the form of our equation of simple harmonic motion

$$\alpha + \frac{3g}{2L}\theta = 0$$

The frequency of the oscillation is given by

$$\omega^2 = \frac{3g}{2L}$$

so that the period of the motion is

$$T = \frac{2\pi}{\omega} = 2\pi \left[\frac{2L}{3g}\right]^{1/2} \qquad (8\text{-}5)$$

Because of the distribution of matter, the period of a physical pendulum, such as a solid rod, is somewhat smaller than that of a simple pendulum with the same length.

FIGURE 8-6
"The Great Wave Off Kanazawa," by Hokusai (1760–1849).
By courtesy of the Victoria and Albert Museum.

Walking

Walking legs swing back and forth much as a pendulum does. The leg is about 90 cm long. If we neglect the bending at the knee, the natural period for the leg swing is approximately

$$T = 2\pi \left[\frac{2L}{3g}\right]^{1/2} = 6.28 \left[\frac{(2)(0.90\text{ m})}{(3)(9.8\text{ m/s}^2)}\right]^{1/2}$$

$$T = 1.6\text{ s}$$

In each walking step the leg makes a half swing. For the easiest walk, the legs swing at their natural period. The time for each step is about $T/2 = 0.8$ s. To walk faster requires additional muscular effort to accelerate the legs. To walk much slower requires additional muscular effort to hold back the legs.

Short people have shorter natural periods to their walk than do taller people. The length of people's steps scales as the length of their legs:

$$L' = aL$$

The period of the step scales as the square root of the length of the legs:

$$T' = a^{1/2}T$$

The speed at which a person walks is proportional to the length of the step divided by the period

$$v' = \frac{L'}{T'} = \frac{aL}{a^{1/2}T} = a^{1/2}\frac{L}{T} = a^{1/2}v$$

The speed at which a person walks scales as $a^{1/2}$. Shorter people have a slower natural walking pace than do taller people. Smaller animals usually have slower walking gaits than larger animals.

WAVES IN MATTER

A disturbance created in a continuous medium, such as air or water, propagates through the medium as a wave. Dramatic gravity waves of the kind shown in Figure 8-6

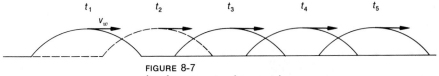

FIGURE 8-7
A pulse propagates down a string.

are generated on the ocean by the wind. Surface-tension waves can be produced on the surface of water or on a stretched string or drum. Compression waves can also travel through a medium such as air or water. For a mechanical wave to propagate, there must be a displacement from some equilibrium, a restoring force to return the system to equilibrium, and a mass to provide inertia. Although the propagation of a wave through matter moves the matter little, both momentum and energy are transmitted with waves.

Waves on a String

Waves will propagate on a stretched string. A string has a total length L and a mass M. The string is stretched with a tension T, which pulls in equal and opposite directions. If the string is displaced, a pulse propagates down it as shown in Figure 8-7. The string is displaced from the equilibrium position. The tension of the string provides the restoring force that returns the string to its equilibrium position. The string has a mass per unit length M/L. The disturbance propagates down the string with a characteristic wave velocity v_{w}. The string of course moves only side to side or up and down.

The displacement of a stretched string is perpendicular to the direction in which the disturbance propagates. The wave propagates along the string; the displacement of the string could be in any direction perpendicular to the string. A displacement can be resolved into two perpendicular components, or **polarizations.** If the displacement is perpendicular to the direction of travel, the motion is called a **transverse wave.**

Suppose a pulse in the form of an arc of radius R is traveling down a stretched string at a wave speed v. We can analyze this process by using a coordinate reference system that travels with the pulse at the wave speed. Because Newton's laws are valid in any nonaccelerating coordinate system, we will see a stationary pulse with the string moving by at a speed v_{w} as shown in Figure 8-8.

Because the string is moving in an arc of radius R there must be a centripetal force acting on a segment of the string with mass m. For a small segment defined by an angle of $\pm \frac{1}{2}\theta$ there is an arc of length $s = R\theta$. The mass m of this segment is a small fraction of the total mass M:

$$m = \frac{Ms}{L} = \frac{MR\theta}{L}$$

The centripetal force is provided by the transverse component of the tension in the string T:

$$F = 2T \sin \frac{1}{2}\theta$$

Equating the net force with the centripetal force we obtain

$$2T \sin\frac{1}{2}\theta = \frac{mv^2}{R} = \frac{MR\theta}{RL}v^2$$

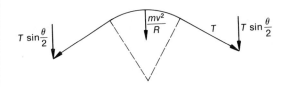

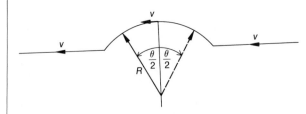

FIGURE 8-8
A traveling pulse viewed from a reference frame moving with the pulse.

For small angles $\sin \theta/2 = \theta/2$. We can then solve for the wave speed:

$$2T\theta/2 = \frac{M\theta}{L}v^2$$

or

$$v_w = \left[\frac{T}{M/L}\right]^{1/2} \tag{8-6}$$

The wave velocity for a pulse propagating on a string depends on the square root of the tension divided by the mass per unit length. The wave speed is independent of the radius of curvature of the pulse so long as our angle θ remains small, that is, less than about 30°. For extremely sharp pulses, of course, our analysis would break down.

EXAMPLE

A guitar string of mass $M = 6.0$ gm is stretched from the bridge to the nut a distance of 60 cm, under a tension of 400 N. What is the wave speed on the string?

$$v_w = \left[\frac{400\ \text{N}}{(0.006\ \text{kg})/0.60\ \text{m}}\right]^{1/2} = 200\ \text{m/s}$$

The wave speed is 200 m/s. The string is displaced from equilibrium by plucking perpendicularly to the string. The string's tension provides the restoring force to return the string to equilibrium position. The string's mass provides the inertia to resist acceleration.

If a long string is shaken at a particular frequency a sine wave is caused to propagate down the string. A wave can be characterized by a **period,** or **frequency;** an **amplitude;** a direction of travel; and a **wavelength.** If a wave has a frequency f the period of the wave is given by the reciprocal of the frequency:

$$T = 1/f$$

A wave travels through a medium with a **wave velocity** v. The wave moves forward a distance of one wavelength in a time of one period:

$$\lambda = vT = v/f$$

We can describe the propagation of a wave on a string mathematically. If a wave is propagating in the positive, x-direction and is displaced in the y-direction (see Figure 8-9), then the displacement at any point x and at any time t can be expressed as a mathematical function $y(x, t)$:

$$y(x, t) = y_0 \sin 2\pi\left(\frac{x}{\lambda} - ft\right)$$

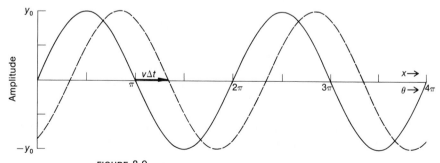

FIGURE 8-9
A wave propagating to the right can be represented by a traveling sine wave $y = y_0 \sin 2\pi(x/\lambda - ft)$.

In such a **traveling wave** a fixed pattern propagates down the string. A point where $y = 0$ travels down the string at the wave speed v. The sine function is zero when the argument has a value of $n\pi$.

$$\sin n\pi = 0$$

where n is an integer $n\pi = 0, 1, 2, \ldots$
A particular n corresponds to a particular zero point for the wave, a point of the wave's constant phase. This point then moves down the string:

$$2\pi\left(\frac{x}{\lambda} - ft\right) = n\pi$$

Solving for x we get

$$x = \frac{f}{\lambda}t + n\lambda = vt + x_0$$

The constant term simply tells us which zero we were looking at. A point of constant phase moves down the string at the wave speed, or phase velocity v:

$$v = \frac{f}{\lambda} \tag{8-7}$$

Standing Waves

A disturbance propagating along a string carries energy. If a disturbance is propagating to the left along a string fastened to a rigid support at one end (say at the origin on the x-axis as shown in Figure 8-10), the energy in the wave is either absorbed by the support or reflected. If the support is solid, a large fraction of the energy is reflected. The reflected disturbance travels to the right. If at any instant two or more disturbances act at a point on the string, the displacement of the point is the sum of the displacements made by each wave separately. This **principle of superposition** applies as long as the restoring force for the medium is proportional to the displacement of the medium. The principle of superposition applies to many types of wave motion, including waves on a string, sound waves in air and water, surface waves on water, and light waves. Because the support is rigid the sum of the incoming wave and the reflected wave at the point of

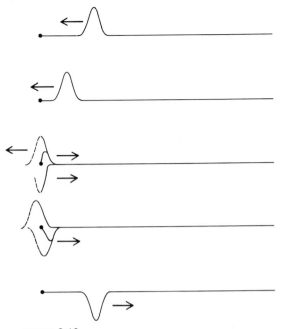

FIGURE 8-10
The reflection of a pulse from a fixed end is inverted.

support must be zero. Consequently, the reflected wave is inverted, as shown in Figure 8-10.

If a wave of frequency f and wavelength λ traveling to the left on the string is reflected by a rigid support, the reflected wave travels to the right. The resultant displacement of the string is the sum of the displacements caused by the incoming and reflected waves. If the wave is totally reflected, equal-amplitude waves travel in opposite directions along the string.

If both ends of the string are attached to rigid supports, the wave traveling to the right will also be reflected. If the string has the proper length (as determined by the frequency and speed of wave propagation), these waves alternately reinforce and cancel one another, giving rise to **standing waves.** At certain points on the string the displacements will cancel, causing points of minimum displacement, **nodes.** At other points the waves will add, giving large displacements, **antinodes.** Figure 8-11 shows simple standing wave patterns for a vibrating string of length L. For standing waves to occur, the length of the string must be an integer of half-wavelengths.

The simplest pattern, or **fundamental,** has a node (a point of zero displacement) at each end. The fundamental

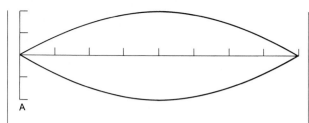

A

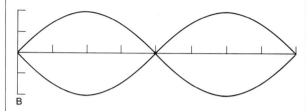

B

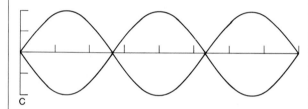

C

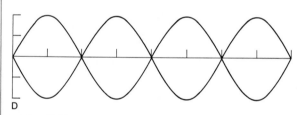

D

FIGURE 8-11
Standing wave patterns for a vibrating string of length L.
A. $L = \lambda_1/2$; $f_1 = v/\lambda_1 = v/2L$: first harmonic.
B. $L = 2\lambda_2/2$; $f_2 = v/\lambda_2 = 2v/2L$: second harmonic.
C. $L = 3\lambda_3/2$; $f_3 = v/\lambda_3 = 3v/2L$: third harmonic.
D. $L = 4\lambda_4/2$; $f_4 = v/\lambda_4 = 4v/2L$: fourth harmonic.

corresponds to $n = 1$ and is called the **first harmonic.** The length of the string is one-half wavelength. For the higher harmonics, the patterns correspond to $n = 2, 3, 4$, etc., and one or more nodes are evenly spaced along the string. The first four harmonics for a vibrating string are shown in Figure 8-10. A guitar string can be made to ring or chime at higher harmonics by plucking the string and using the finger to place a node at the proper point.

EXAMPLE

Suppose a string of mass M is stretched with some tension T between two points separated by a distance L. Each end of the string is tied down, so at $x = 0$ and at $x = L$ the amplitude must be $y = 0$. We can represent a wave traveling to the right as

$$y_A(x, t) = A \sin 2\pi\left(\frac{x}{\lambda} - ft\right)$$

where the wave's speed of propagation is of course

$$v = f\lambda$$

A wave traveling to the left is represented as

$$y_B(x, t) = B \sin 2\pi\left(\frac{x}{\lambda} + ft\right)$$

The displacement of the string at any position x at any time t is the sum of these two waves:

$$y(x, t) = y_A(x, t) + y_B(x, t)$$

At $x = 0$ the displacement of the string must be $y = 0$. Therefore

$$y(0, t) = A \sin(-2\pi ft) + B \sin(2\pi ft) = 0$$

Because $\sin(-2\pi ft) = -\sin(2\pi ft)$,

$$(-A + B) \sin(2\pi ft) = 0$$

The amplitudes of the two waves are identical in magnitude: $A = B$. Our wave is then

$$y(x, t) = A \sin 2\pi\left(\frac{x}{\lambda} - ft\right) + A \sin 2\pi\left(\frac{x}{\lambda} + ft\right)$$

We can rewrite the wave using the formula for the sine of the sum and difference of two angles:

$$\sin(a \pm b) = \sin a \cos b \pm \cos a \sin b$$

The sum of the two waves can then be written

$$y(x, t) = A \left[\sin(2\pi x/\lambda) \cos 2\pi ft - \cos(2\pi x/\lambda) \sin 2\pi ft\right]$$
$$+ A \left[\sin(2\pi x/\lambda) \cos 2\pi ft + \cos(2\pi x/\lambda) \sin 2\pi ft\right]$$

The second terms in both pairs cancel, leaving

$$y(x, t) = 2A \sin(2\pi x/\lambda) \cos 2\pi ft \qquad (8\text{-}8)$$

At the far end of the string $x = L$, the wave must go to zero:

$$y(L, t) = 0 = 2A \sin 2\pi L/\lambda \cos 2\pi ft$$

This can equal zero at all times only if

$$\sin(2\pi L/\lambda) = 0$$

The zeros of the sine function are for $\theta = n\pi$. Therefore the conditions for standing waves on a string are that

$$2\pi L/\lambda = n\pi \qquad (8\text{-}9)$$

or that

$$L = n\lambda/2 \qquad (8\text{-}10)$$

The length of the string must be an integer of half wavelengths.

Fourier's Theorem

Fourier's theorem states that every finite and continuous periodic motion can be analyzed as the sum of a series of sine waves of the appropriate frequency, amplitude, and phase. When a string of a musical instrument, such as a violin, guitar, piano, or harp, is plucked or struck, the resultant motion can be considered as the sum of possible harmonics.

If the string is plucked in the middle and released, its shape just at release is more or less as shown in Figure 8-12 A. This shape can be represented as approximately

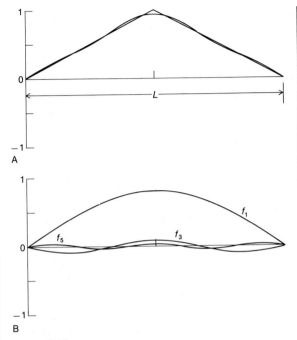

FIGURE 8-12
A string plucked in the center can be represented as a sum of odd harmonics.

the sum of the fundamental frequency plus the third and fifth harmonics at different amplitudes. This is shown in Figure 8-12 B. We include in our approximation higher harmonics so we can closely approximate the initial shape. Where the string is plucked determines what harmonics will be excited. If we pluck it exactly in the center, only the odd harmonics will be excited. The sound from a plucked string fades out as the energy in the string is transferred to the body of the instrument and then into the air. The higher harmonics tend to damp out more quickly than the fundamental; after a short time only the lowest harmonics can be heard.

Two common electrical wave forms, the sawtooth and the square wave, can be represented as the sum of many harmonics. Figure 8-13 shows the sawtooth wave and the sum of the first six harmonics that make it up. The square wave shown in Figure 8-14 can be represented as a sum of

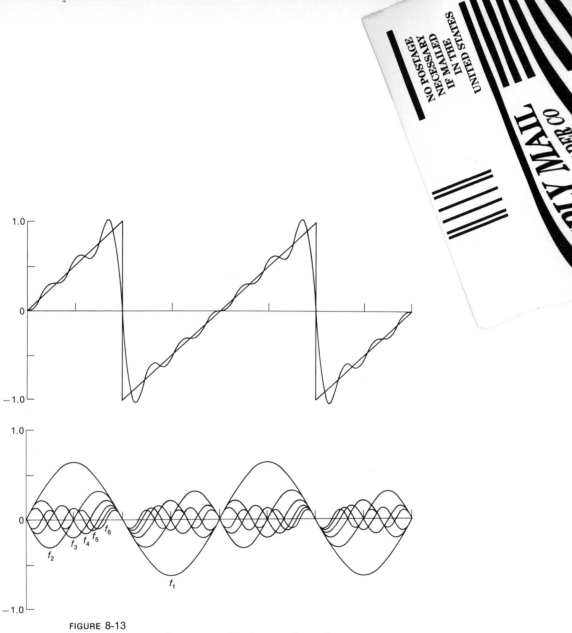

FIGURE 8-13
A sawtooth wave can be represented as the sum of many harmonics. The sum of six harmonics gives a close approximation.

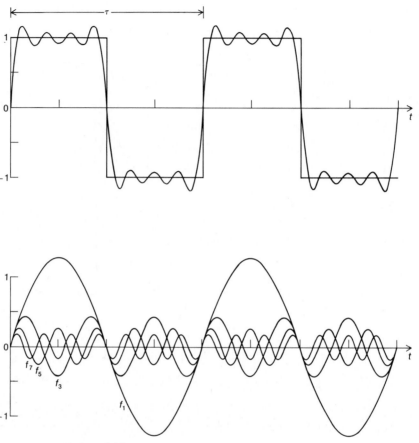

FIGURE 8-14
The square wave can be represented as the sum of odd harmonics of the fundamental.

the odd harmonics. The sum of the first four harmonics is shown. In both types of wave the high harmonics still have significant amplitude. This contributes to the quality of the sound when a loudspeaker is driven with a sawtooth or square wave. Because of the loudspeaker's frequency response, the highest harmonics are lost.

Modern electronic music synthesizers generate the fundamental tones and many harmonics. Proper mixing of harmonics can generate a wide range of tone qualities and repetitive sequences of tones, and can control the buildup and decay of tones. In the Moog synthesizer, shown in Figure 8-15, a combination of oscillators, amplifiers, gates, and sequential modular controllers gives the musician an infinite variety of tones to work with.

Compression Waves

If the end of a long aluminum rod is struck with a mallet, the molecules at the very end of the rod will be com-

FIGURE 8-15
The Moog synthesizer electronically produced waves are widely used in modern
music. Tone quality depends on the harmonics used. Courtesy of Norlin Music,
Inc.

pressed instantaneously, by an amount that depends upon
the force of the blow and the modulus of elasticity for the
material. This disturbance propagates down the rod with
a characteristic speed v_s. The material is displaced in the
same direction as the propagation of the disturbance. This
is the characteristic of a **compression,** or **longitudinal,
wave.**

Suppose there is a column of length L and a cross-
sectional area A as shown in Figure 8-16. This column
could represent a bar of steel, a column of water in a tube,
or a column of air. If the end of this column is struck by a
hammer or piston traveling at some speed u, the effect of

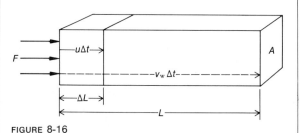

FIGURE 8-16
A compression disturbance propagates into the medium at the
wave velocity v_w. Displacement is in the same direction as the
propagation.

this blow will propagate into the column at the wave speed v_w. In a time Δt the disturbance will travel a distance

$$L = v_w \Delta t$$

Beyond this distance the material will be unaffected (as yet) by any approaching disturbance.

In the meantime the piston will have traveled a distance

$$\Delta L = u \Delta t$$

We can assume that the striking piston is so massive and the time Δt is so short that u does not change.

The modulus of elasticity E (discussed in Chapter 4) expresses the ratio of stress to strain for a material:

$$E = \frac{F/A}{\Delta L/L}$$

The maximum force F to compress the material a distance ΔL is

$$F = EA\frac{\Delta L}{L}$$

The work done on our column by this force is just the product of the average force times the distance over which the force acts:

$$W = F_{av}\Delta s$$

The average force is about one-half the maximum force $F_{av} = \frac{1}{2}F$. The distance traveled is just $\Delta s = \Delta L$. The work done, then, is

$$W = \frac{1}{2}F\Delta L = \frac{EA\Delta L}{2L}\Delta L = \frac{EA}{2L}\Delta L^2$$

This work goes into the kinetic energy of the column as the disturbance propagates into the material.

$$KE = \frac{1}{2}Mu^2$$

The mass of the column is just the product of the cross-sectional area A, the length L, and the mass density ρ: $M = AL\rho$.

$$KE = \frac{1}{2}AL\rho u^2$$

Equating the kinetic energy with the work done we get

$$\frac{1}{2}\frac{EA}{L}\Delta L^2 = \frac{1}{2}AL\rho u^2$$

$$\frac{L^2u^2}{\Delta L^2} = \frac{E}{\rho}$$

However, $L = v_w\Delta t$, and $\Delta L = u\Delta t$, so that

$$\frac{(v_w\Delta t)^2u^2}{(u\Delta t)^2} = v_w{}^2 = \frac{E}{\rho}$$

The wave velocity depends on the square root of the ratio of the modulus of elasticity E to the mass density ρ.

$$v_w = \left[\frac{E}{\rho}\right]^{1/2}$$

The wave velocity for such a compression wave in a solid medium is given by the square root of the ratio of the modulus of elasticity to the density for the particular material.

EXAMPLE

The sound of an approaching railroad train can be heard in the rails long before the train can be seen or heard by a standing person. What is the speed of compression waves in steel?

$$v_w = (E/\rho)^{1/2}$$

$$v_w = \left(\frac{2.0 \times 10^{11}\,\text{N/m}^2}{7\,800\,\text{kg/m}^3}\right)^{1/2}$$

$$v_w = 5.06 \times 10^3\,\text{m/s}$$

TABLE 8-1
The speed of sound in different materials

Material	Temperature C°	Density kg/m³	Modulus N/m²	Speed of sound m/s
Air	0	1.293	1.42×10^5	331.3
Air	20	1.205	1.42×10^5	343
Helium	0	0.179	1.69×10^5	972
Hydrogen	0	0.089	1.43×10^5	1 267
Water	15	1 000	2.1×10^9	1 450
Aluminum	20	2 700	7.0×10^{10}	5 100
Steel	20	7 800	2.0×10^{10}	5 130
Lead	20	11 300	1.7×10^{10}	1 230
Glass	20	2 600	5.8×10^{10}	470

For liquids and gases, the speed of propagation of a compression wave depends on the volume compressibility, or bulk modulus, of the material. The bulk modulus relates the fractional change in volume to the change in applied pressure.

$$F/A = B(\Delta V/V)$$

The speed of a compression wave in a gas or liquid is given by

$$v_s = (B/\rho)^{1/2}$$

Table 8-1 lists the modulus of volume compressibility or elasticity and density of various substances.

EXAMPLE

The bulk modulus for water is $B = 2.0 \times 10^9 \text{ N/m}^2$. The density of water is $\rho = 1.00 \times 10^3 \text{ kg/m}^3$. What is the speed of propagation of a compression wave in water?

$$v_s = (B/\rho)^{1/2}$$

$$v_s = \left(\frac{2.0 \times 10^9 \text{ N/m}^2}{1.0 \times 10^3 \text{ kg/m}^3}\right)^{1/2}$$

$$v_s = 1.41 \times 10^3 \text{ m/s}$$

SOUND

The Speed of Sound

Sound is a compression wave in air. A vibrating source of sound slightly displaces the air from the equilibrium position, slightly increasing air pressure and density. The increased pressure provides a force that causes the disturbance to propagate into the adjacent air. The mass of the air provides the inertia. The speed with which the disturbance propagates through the air is given by

$$v_s = (B/\rho)^{1/2}$$

The bulk modulus B for air is given by the equilibrium pressure P_o multiplied by a factor γ, known as the ratio of specific heats (which will be discussed in Chapter 12).

$$B = \gamma P_o$$

The ratio of specific heat for air is $\gamma_{air} = 1.40$. The speed of sound in air depends on the density of the gas, the pressure, and the ratio of specific heat:

$$v_s = (\gamma P_o /\rho)^{1/2}$$

EXAMPLE

For air at a temperature of 0°C, the pressure is $P_o = 1.0129 \times 10^5 \text{ N/m}^2$ and the density is

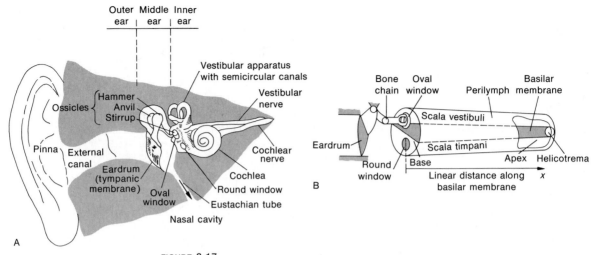

FIGURE 8-17
The ear: A. Schematic view of the ear. B. The cochlea stretched out (highly simplified). From Juan G. Roederer, *Introduction to the Physics and Psychophysics of Music,* © 1973 by Springer-Verlag, New York, Inc. Used with permission.

$\rho = 1.293 \, \text{kg/m}^3$. The ratio of specific heat is $\gamma = 1.4$. The speed of sound in air is

$$v_s = (\gamma P_o/\rho)^{1/2} = \left[\frac{(1.4)(1.0129 \times 10^5 \, \text{N/m}^2)}{(1.293 \, \text{kg/m}^3)} \right]^{1/2}$$

$$v_s = (1.097 \times 10^5 \, \text{m}^2/\text{s}^2)^{1/2}$$

$$v_s = 331 \, \text{m/s}$$

The wavelength of a sound wave depends on the frequency of the sound. At 1 000 Hz the wavelength is

$$\lambda = v/f$$

$$\lambda = \frac{331 \, \text{m/s}}{1 \, 000 \, \text{Hz}}$$

$$\lambda = 0.331 \, \text{m}$$

The human ear is sensitive to a range of frequencies roughly from 20 Hz to 20 000 Hz. This corresponds to wavelengths in the range from 16 m to 16 mm.

Hearing

The ear is divided into three sections, as shown in Figure 8-17. The outer ear collects sound and channels it to the tympanic membrane, or eardrum. The middle ear consists of the ossicles, a chain of small bones: the hammer, anvil, and stirrup. Sound pressure on the eardrum is transmitted as a force through these ossicles. The area of the eardrum is of the order of 0.55 cm². The pressure acts on this area, creating a force that the ossicles transmit to the oval window of the **cochlea.**

The oval window has an area of 0.032 cm². The actual pressure transmitted to the oval window is 15 times that exerted on the eardrum. The ossicles also protect the ear from very loud sounds, by changing their movements to reduce the effective transmission of sound energy to the oval window.

The inner ear consists of the cochlea and the semicircular canals, which sense balance and orientation. The pressure vibrations transmitted through the oval window set up high-pressure, short-wavelength sound waves in the fluid of the cochlea. These vibrations stimulate the neurons, the nerve receptors, in the **basilar membrane,**

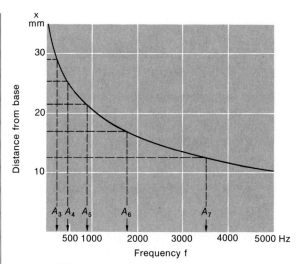

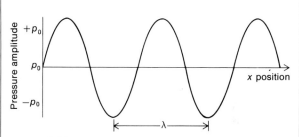

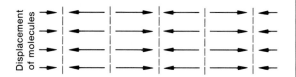

FIGURE 8-18
The position of maximum sensitivity on the basilar membrane varies with the frequency of a pure sinusoidal tone. From Juan G. Roederer, *Introduction to the Physics and Psychophysics of Music,* © 1973 by Springer-Verlag, New York, Inc. Used with permission.

shown in Figure 8-17 B. The neurological perception of sound is quite complex. However, the position at which the basilar membrane is most sensitive to a pure tone depends on the frequency, as shown in Figure 8-18.

Sound Intensity

Sound is a pressure wave. The displacement of the air molecules gives rise to a pressure disturbance that propagates through a medium, as shown in Figure 8-19. Atmospheric pressure is $1.013 \times 10^5 \, \text{N/m}^2$. A pressure disturbance with an amplitude of $p_0 = 3 \times 10^{-5} \, \text{N/m}^2$ at a frequency of 1000 Hz can be heard by a sensitive ear. This frequency is the threshold of hearing, the lowest frequency audible to the normal human ear. If the ear were more sensitive, we would be able to perceive the noise of individual molecules striking the eardrum.

When air is compressed, work is done, and is stored as potential energy. As the disturbance propagates, this energy is transmitted through space. The intensity of sound I is measured by the energy passing a unit area in space per unit of time. If p_0 is the amplitude of the sound pressure wave, and ρ the density of the medium, and v_s the speed of sound, then the average sound intensity is given by

$$I = \frac{p_0{}^2}{2\rho v_s} \qquad (8\text{-}11)$$

FIGURE 8-19
Sound is a pressure wave. The average pressure difference is 0.707 of the peak pressure difference. The pressure maximum and minimum correspond to displacement nuls.

EXAMPLE
The minimum pressure disturbance that can be heard by a sensitive ear at a frequency of 1000 Hz is approximately $p_0 = 3 \times 10^{-5} \, \text{N/m}^2$. What is the sound intensity

$$I = \frac{(3 \times 10^{-5} \, \text{N/m}^2)^2}{2(1.293 \, \text{kg/m}^3)(340 \, \text{m/s})} = 1.0 \times 10^{-12} \, \text{W/m}^2$$

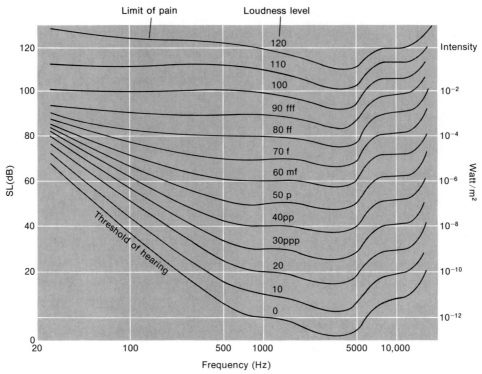

FIGURE 8-20
Curves of equal loudness compared to a pure tone at 1000 Hz
for an average group of young adults. From Juan G.
Roederer, *Introduction to the Physics and Psychophysics of
Music,* © 1973 by Springer-Verlag, New York, Inc. Used
with permission.

The loudness of a sound is a subjective impression of the observer. If **sound intensity** increases or decreases by a factor of two, the difference will be barely noticeable; if changed by a factor of ten it will be sensed as twice (or half) as loud; if changed by a hundredfold it will seem four times (or one-fourth) as loud. Sound level is customarily measured by a scale that corresponds more nearly to our perception of loudness: a logarithmic scale, the decibel scale. The sound level, *SL*, measured in the **decibel,** dB, is defined as ten times the logarithm of the ratio of the sound intensity I to the reference sound intensity I_o

$$SL = 10 \log (I/I_o)$$

The reference sound intensity is the threshold of hearing at 1000 Hz; $I_o = 10^{-12} \text{ W/m}^2$. If the sound intensity is doubled, the sound level increases by 3 dB. If the sound intensity increases by a factor of 100, the sound level increases by 20 dB.

The threshold of hearing depends on the frequency. Figure 8-20 shows the threshold intensity of sound as a function of frequency from 20 Hz to 20 000 Hz. The ear is much less sensitive at low frequencies, so we are not sensitive to noises produced by the action of our muscles or body movement. The figure also shows the sound intensity for tones of equal loudness, determined by having subjects compare the intensity of tones at different frequencies. The loudness range for musical dynamics from piano-pianissimo, *ppp,* to forte-fortissimo, *fff,* is also shown in Figure 8-15. A change of one dynamic level corresponds approximately to an intensity change of a factor of 10.

A sound louder than 120 dB is felt rather than heard, and causes pain to the ear. What sound intensity does this correspond to?

$$SL = 10 \log \frac{I}{I_o}$$

$$\log \frac{I}{I_o} = \frac{SL}{10} = \frac{120}{10} = 12$$

$$I = I_o (10^{12}) = (10^{-12} \text{W/m}^2)(10^{12}) = 1 \text{ W/m}^2$$

Beats

If two different tones with slightly different frequencies ω_1 and ω_2 are at the same amplitude, the two waves will add. The resulting pressure wave will be the sum of the two different pressure waves. The sound pressure at the observer can be represented as

$$p = p_o \sin \omega_1 t + p_o \sin \omega_2 t$$

The average frequency is

$$\omega_o = \frac{\omega_1 + \omega_2}{2}$$

The difference in frequency $\Delta\omega$ is

$$\omega = \frac{\omega_2 - \omega_1}{2}$$

With these definitions we can write the frequencies as

$$\omega_1 = \omega_o - \Delta\omega$$

and

$$\omega_2 = \omega_o + \Delta\omega$$

The amplitude of the pressure wave is then

$$p = p_o \sin (\omega - \Delta\omega) + p_o \sin (\omega + \Delta\omega)$$

The sine of the sum of two angles is

$$\sin (A + B) = \sin A \cos B \pm \cos A \sin B$$

The pressure wave can be written

$$p = p_o [\sin \omega_o t \cos \Delta\omega t - \cos \omega_o t \sin \Delta\omega t]$$
$$+ p_o [\sin \omega_o t \cos \Delta\omega t + \cos \omega_o t \sin \Delta\omega t]$$

$$p = 2p_o \sin \omega_o t \cos \Delta\omega t$$

The intensity of the sound is

$$I = \frac{(p)^2}{\rho v_s} = \frac{(2p_o)^2}{\rho v_s} \sin^2\omega_o t \cos^2\delta\omega t$$

By trigonometry

$$2 \cos^2 \Delta\omega t = 1 + \cos 2\Delta\omega t$$

The intensity of the sound is then

$$I = \frac{(2p_o)^2}{\rho v_s} \frac{\sin^2 \omega t}{2} (1 + \cos 2\Delta\omega t)$$

The sound intensity varies from a maximum down to zero and back to a maximum at a frequency of

$$2\Delta\omega = \omega_2 - \omega_1$$

The result is an audible **beat** at a frequency equal to the difference between the two original frequencies. Initially the waves may be exactly in phase and will add as shown in Figure 8-21 A. But they are soon out of phase, giving a minimum. If the frequency difference is 10 Hz, they go in and out of phase 10 times a second. If the frequency difference is 1 Hz, the sound becomes louder and softer once a second. If the frequency difference is less than about 10 Hz, the beats are clearly heard. If the difference in frequency exceeds 15 Hz, then a roughness or dissonance is heard. If the frequency difference exceeds a critical value greater than a whole musical interval, then two distinct tones are perceived. The superposition of two sounds to produce beats is represented in Figure 8-21 B.

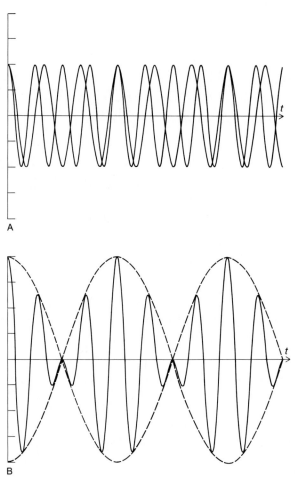

FIGURE 8-21

Beats: A. Two waves of equal amplitude and slightly different frequencies are alternately in phase and out of phase. B. Interference of the two waves produces beats. The beat frequency heard is equal to the difference in frequency.

ULTRASONIC WAVES

Ultrasonic waves are compression waves with frequencies above 20 kilohertz. Such waves have important medical applications, being used for diagnostic purposes at frequencies of 1, 5, 7, or even 10 megahertz.

Ultrasonic waves can be produced by a piezoelectric crystal, a crystal that acts as a pressure transducer, con-

verting an oscillating electric field into pressure variations. If this pressure transducer is placed in contact with the skin, the pressure variations are transmitted through the tissue beneath the skin and into the body as sound waves. The speed of sound is within a few percent of 1 540 m/s in different body components, including blood, saline muscle, fat, and kidney, liver, and brain tissue, as shown in Table 8-2.

TABLE 8-2

Ultrasonic waves. The velocity of sound and percentage reflection at an interface with muscle tissue.

Material	Velocity of sound m/s	Reflection %
Muscle	1540	0
Water	1525	0.28
Saline	1534	0.28
Fat	1540	1.04
Liver	1540	0.028
Blood	1540	0.045
Bone	3380	36.1
Air	340	99

TABLE 8-3

Note and frequency in diatonic tuning of guitar strings. The mass of a 60 cm length of each string is also shown.

Musical note	Frequency Hz	Mass gm
E_3	165	6.0
A_3	220	3.9
D_4	297	2.2
G_4	396	1.4
B_4	495	0.55
E_5	660	0.35

About 500 times per second a short burst of high-frequency ultrasonic sound energy is transmitted in a narrow beam from the transducer through the body. Some of this energy is reflected and some is absorbed in tissue. Any slight change in the speed of propagation through the medium produces reflections—weak if the change is small, strong if the change is great. Table 8-3 also shows the percentage of reflection as a wave goes from muscle tissue to another medium. In the time between energy bursts, the ultrasonic transducer is used as a receiver to detect the reflections. Since the speed of ultrasound in various tissues is approximately constant, the time required for the echo to return to the transducer is a measure of twice the distance to the tissue interface. Propagating at a speed of 1 540 m/s, the echo will take 52 microseconds to return to the transducer from an interface 4 cm away.

The resolving power for measurement of distance is approximately equal to the ultrasonic wavelength, which depends on the frequency and speed of propagation. At a frequency of one megahertz, the wavelength is about 1.5 mm; at higher frequencies, the wavelengths are shorter.

Ultrasonic diagnostic systems operate at low power levels of the order of 3 to 5 milliwatts/cm². At frequencies of 10 megahertz and below, there has been neither clinical nor animal-research evidence of damage caused by ultrasound.

Diagnostics

Today ultrasound is widely used in medical diagnosis. In a time-position scan, such as is used in echocardiography, sonic study of the heart, the narrow beam of sound waves travels along a straight path from the transducer, as shown in Figure 8-22, and is reflected at changes in the density of the heart. The returning echoes are displayed as points of light and are recorded photographically. Since the heart is changing in time, the echoes are also changing position, indicating the heart's movement. Turning the probe allows different parts of the heart to be viewed. This single record, made in only a few seconds, gives valuable information on the functioning of the heart. In the normal echocardiogram shown in Figure 8-23, the motion of the left and right ventricles, the mitral valve leaflets, and the aortic valve can be clearly seen.

Ultrasonic waves can be used to build a two-dimensional image of a plane through the body. Such ultrasonic imaging has been widely used in obstetrics, for ultrasound exposes the fetus to no known hazard and is preferred to x-rays. The uterus of a pregnant woman is an ideal subject for ultrasonic observation because it is filled with amniotic fluid, an excellent transmitter of sound. Figure 8-24

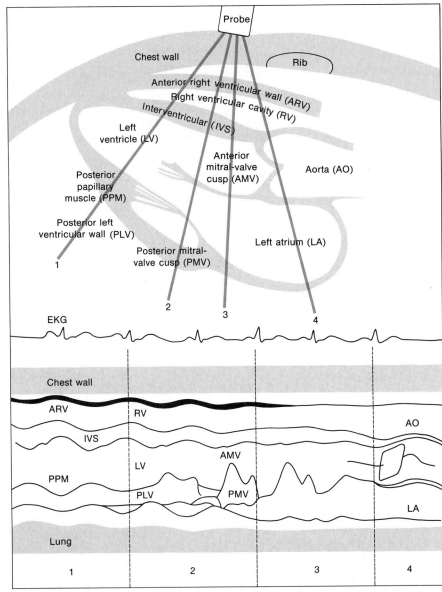

FIGURE 8-22

The cross section of the heart shows the structures through which the ultrasonic beam passes as it is aimed in four different directions: 1, the apex of the heart; 4, the base of the heart; and 2 and 3, locations in between. The time-scan record is shown diagramatically at the bottom with traces corresponding to several structures as labeled and shown in relation to a simultaneous recording from an electrocardiogram (ECG). From Gilbert B. Devey and Peter N. T. Wells, "Ultrasound in Medical Diagnosis," *Scientific American* (May 1978). Copyright © 1978 by Scientific American, Inc. All rights reserved.

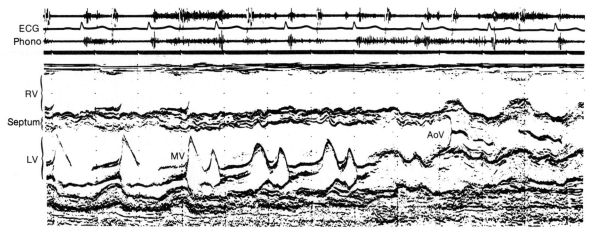

FIGURE 8-23

A normal echocardiogram. Indicated are traces corresponding to several structures: RV (right ventricle), Septum, LV (left ventricle), MV (mitral valve), AoV (aortic valve), Phono (heart sounds), and ECG (electrocardiogram). The time-scan record corresponds to the diagram in Figure 8-22. Courtesy of Nelson B. Schiller, M.D., University of California, San Francisco. From: Thomas A. Ports, Nelson B. Schiller, and Brian L. Strunk, "Echocardiography of Right Ventricular Tumors," *Circulation* (September 1977). Used by permission of the American Heart Association, Inc.

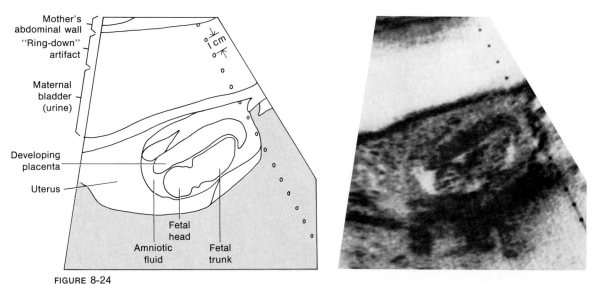

FIGURE 8-24

An ultrasonic image showing a small fetus in early pregnancy—about nine weeks after conception, according to measurements from crown to rump. The fetal head and placenta are clearly shown. The size can be estimated from the row of dots 1 cm apart. Courtesy of Roy A. Filly, M.D., University of California, San Francisco.

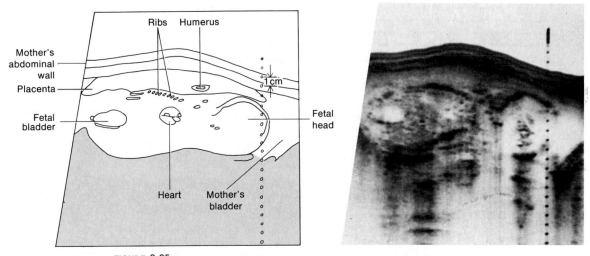

FIGURE 8-25
Ultrasonic image of a well-developed fetus—about six months after conception, according to head size. The head, heart, fetal bladder, and ribs are clearly shown. Courtesy of Roy A. Filly, M.D., University of California, San Francisco.

shows a small fetus about nine weeks after conception. The amniotic sac is only about 8 cm long. Figure 8-25 shows a well-developed fetus about six months old, with head, ribs, and internal organs visible. The ultrasound image can be used to determine the position, size, and approximate age of a fetus.

Review

Give the significance of each of the following terms. Where appropriate give two examples.

displacement	frequency	first harmonic
equilibrium position	amplitude	Fourier's theorem
restoring force	wavelength	compression wave
simple harmonic motion	wave velocity	longitudinal wave
reference circle	traveling wave	cochlea
simple pendulum	principle of superposition	basilar membrane
physical pendulum	standing wave	sound intensity
polarization	node	decibel
transverse wave	antinode	beat
period	fundamental	ultrasonic wave

Further Readings

Arthur H. Benade, "The Physics of Brasses," *Scientific American* (July 1973), and "The Physics of Woodwinds," *Scientific American* (October 1960).

Leo L. Beranek, "Noise," *Scientific American* (December 1966).

Willem A. van Bergeijk, *Waves and the Ear* (New York: Doubleday, Anchor Science Study Series S9, 1960). A good introduction to the physiology of hearing.

E. Donnell Blackham, "The Physics of the Piano," *Scientific American* (December 1965).

Gilbert B. Devey and Peter N. T. Wells, "Ultrasound in Medical Diagnosis," *Scientific American* (May 1978).

E. Eugene Helm, "The Vibrating String of the Pythagoreans," *Scientific American* (December 1967).

Problems

SIMPLE HARMONIC MOTION

1 A coil spring of length 60 cm will stretch 10 cm if a 0.2 kg mass is attached to it.
 (a) What is the force constant k for the spring?
 (b) If the mass is raised 1 cm above the equilibrium position and released, what is the period of the motion?

2 What is the length of a pendulum with a period of 1 s? 10 s?

3 A rubber band, normally 10 cm long, will be stretched to a length of 20 cm by the weight of 0.50 kg of mass. Assume that the rubber band obeys Hooke's law.
 (a) What is the force constant k?
 (b) If the 0.5 kg mass is displaced slightly from equilibrium up and down, what is the period of the resulting motion?

4 A pendulum of length L_0 has a period T_0.
 (a) If the length of the pendulum is doubled, what happens to the period?
 (b) If the period is cut in half, what must have happened to the length?

WAVES IN MATTER

5 A single piece of rubber band 9.0 cm long and with a mass of 0.2 gm is stretched to a length of 10 cm by a force of 1.0 N, and to 19 cm by a force of 3.3 N.
 (a) What is the speed of propagation of a transverse wave on the rubber band in each case?
 (b) What is the fundamental frequency of the rubber band in each case?

6 The washtub bass, a musical instrument, consists of a piece of clothesline 1.0 m long and with a mass of 0.10 kg attached at one end to a stick and at the other end to a washtub that acts as a soundbox. Pulling on the stick increases the tension in the clothesline and changes the tone.
 (a) What tension is required to produce a note with a fundamental frequency of 110 Hz (A_2)? Express the answer in N and lbf.
 (b) What is the speed of propagation of the wave along the string in this case?
 (c) What tension is required to produce a note with a fundamental frequency of 220 Hz (A_3)?

7 For the washtub bass described in the previous problem,

(a) What tension is required to produce a note with a fundamental frequency of 130 Hz (C_3)? 260 Hz (C_4)?

(b) What is the speed of propagation of the wave along the string in each case?

The musical note, frequency, and mass of a 60 cm length of each string of a guitar are shown in Table 8-3. Use this information in problems 8–10.

8 A guitar string has a total length $L = 60$ cm.
(a) What is wavelength on the string for the fundamental frequency of strings E_3, D_4, and B_4?
(b) What is the speed of propagation of the wave on each string?
(c) What is the tension in each string?

9 A guitar string has a total length $L = 60$ cm.
(a) What is the wavelength on the string for the fundamental frequency for strings A_3, G_4, and E_5?
(b) What is the speed of propagation of the wave on each string?
(c) What is the tension in each string?

10 The E_5 string is tuned too tightly with a total tension of 400 N.
(a) What is the speed of propagation of a wave on the string?
(b) What is the fundamental frequency? Is it in tune?

11 At 25°C seawater has a density $\rho = 1025 \, \text{kg/m}^3$ and a bulk modulus $B = 2.40 \times 10^9 \, \text{N/m}^2$. Distilled water has a density $\rho = 998 \, \text{kg/m}^3$ and a bulk modulus $B = 2.24 \times 10^9 \, \text{N/m}^2$. What is the speed of sound in seawater? In distilled water?

12 Pyrex glass has a density of 2320 kg/m³ and a speed of sound of 5640 m/s. What is the modulus of compression of glass?

SOUND

13 A man has a hearing test and finds that at a frequency of 1 000 Hz the minimum amplitude of tone he can hear is 30 dB. What sound intensity does this correspond to?

14 A woman who has been working near very noisy equipment has a hearing test. At a frequency of 1 000 Hz the minimum amplitude of tone she can hear is 40 dB. What is the sound intensity?

15 If the musical sound level corresponding to a musical instrument playing fortissimo, ff, is 80 dB, what is the sound intensity?

ADDITIONAL PROBLEMS

16 A hoop of mass M and radius $R = 25$ cm is suspended from a point on the rim. The hoop is displaced from the equilibrium position by some small angle θ.
(a) What is the torque on the hoop?
(b) Equate the torque with the angular acceleration and moment of inertia. Determine the period of oscillation of the hoop. How does it compare with a simple pendulum of length 25 cm?

17 Two trumpets each sound a tone at a level of forte, f, 70 dB.
(a) What is the resulting sound intensity of the two instruments together?
(b) How many instruments would be required to increase the sound level by 10 dB? 20 dB?

18 An ultrasonic wave at a frequency of 3.5 MHz propagates through tissue at a speed of 1540 m/s. At an interface between two tissue layers at a distance of 5.0 cm from the transducer there is a reflection. The transducer sends a short pulse of waves and then listens for the echo. How much time is there between the initial pulse and the echo?

9

Fluid Motion

FLUIDS AT REST

Fluids

A **fluid** *can be considered as any substance that flows when a force is applied to it.* A solid, by contrast, has internal forces between atoms that resist flow until the elastic limit of the material is reached. Both liquids and gases are fluids.

A **liquid** *is characterized by cohesive forces between adjacent molecules that give the substance a definite volume and hence a definite surface.* A liquid poured into a container adopts the shape of the container by filling it up to some level. In the absence of gravity, astronauts have observed liquids floating in containers as irregular blobs. Likewise, liquids spilled into the chambers of spacecraft float around, breaking up into small spherical droplets.

A **gas** *is characterized by a lack of cohesive forces between molecules and by lack of definite shape or volume.* A gas will become distributed throughout the volume of any container in which it is placed. Our discussion of fluids will deal mainly with liquids, but may also be applied to gases when the effect of temperature is not important.

Archimedes' Principle

If an object is suspended in water, its apparent weight is reduced by the **buoyant force,** the lifting force, of the water. **Archimedes' principle** *states that when an object is placed in a fluid, the buoyant force equals the weight of the fluid displaced by the object.* According to legend, the king

of Syracuse, wanting a crown of gold, issued the required amount of gold to his craftsmen. The crown fabricated for him was pleasing and had the proper weight, yet the craftsmen were rumored to have substituted a certain weight of silver for the gold. Archimedes was asked to find a method to determine if theft had occurred. The same mass of gold and silver has different volumes. The **density** ρ (the Greek letter *rho*) of an object is the mass per unit volume.

$$\rho = \frac{M}{V}$$

A crown made of an alloy of silver and gold would have a different volume for the same mass of one made of gold alone. But how does one measure the volume? By Archimedes' principle the weight of the fluid displaced by an object can be measured. Knowing the density of the fluid and the acceleration due to gravity, we can determine the volume by observing the buoyant force.

The SI unit for density is kg/m^3. The common unit of density is the gm/cm^3. The density of water[1] is $1.00\ gm/cm^3 = 1000\ kg/m^3$. The density of gold is

---

[1] Water has a maximum density of $0.999\,973\ gm/cm^3$ at a temperature of $3.98°$Celsius. At room temperature, $20°$C, the density of water is slightly different, $0.998\,203\ gm/cm^3$. Since 1964, when the *liter* as a measure of volume was defined as exactly $1000\ cm^3$, the cubic centimeter, cm^3, and the milliliter, ml, have been identical. Before 1964 the liter had been defined as the volume of 1 kg of water at $3.98°$C, which was $1000.028\ cm^3$. This difference in definition of the liter is insignificant except in precise analytical work.

TABLE 9-1
The density of various substances.

Solids	Density kg/m^3	Liquids	Density kg/m^3	Gases	Density kg/m^3
Cork	220 to 260	Ethyl alcohol (20°C)	789	Hydrogen (0°C)	0.090
White pine	350 to 500	Water (20°C)	998.203	Helium (0°C)	0.179
Oak	600 to 900	Water (3.98°C)	999.973	Steam (100°C)	0.597
Glass	2400 to 2800	Sea water	1025	Methane (0°C)	0.717
Cement	2700 to 3000	SAE 30 motor oil	880	Dry air (20°C)	1.205
Ice	917	Mercury (0°C)	13 600	Dry air (0°C)	1.295
Steel	7860	Mercury (20°C)	13 540	Carbon dioxide (0°C)	1.981
Copper	8920	Blood plasma	1030	Propane (0°C)	2.023
Silver	10 500	Whole blood	1050		
Gold	19 300	Sucrose solution 25%	1104		

19 300 kg/m³; the density of silver, 10 500 kg/m³. The densities of common materials are shown in Table 9-1.

An object of density ρ_1 and a volume V has a mass $M = \rho_1 V$. The weight of the object is $F_g = Mg = \rho_1 Vg$. If the object is totally submerged in a fluid with a density ρ_2, then by Archimedes' principle the fluid exerts a buoyant force (upward) equal to the weight of the fluid displaced.

$$F_b = \rho_2 Vg$$

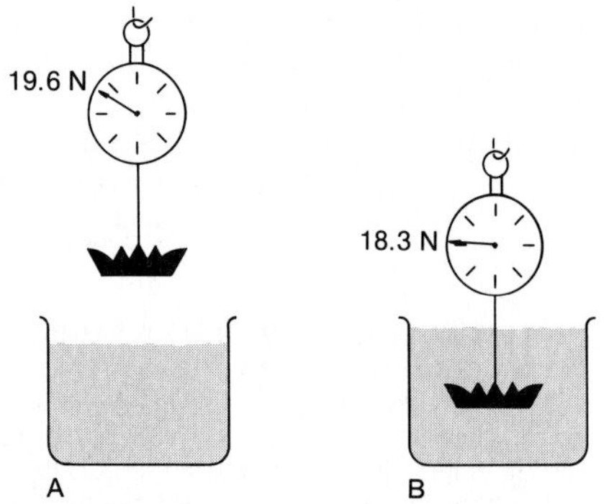

FIGURE 9-1
Archimedes' Principle. An object is buoyed up by the weight of the fluid displaced.

The net downward force on the object is

$$F_{net} = F_g - F_b$$
$$F_{net} = (\rho_1 - \rho_2) Vg \qquad (9\text{-}1)$$

If the object is denser than the surrounding fluid, the buoyant force is less than the force of gravity and the object sinks. The apparent weight of the object is reduced by an amount equal to the buoyant force (see Figure 9-1).

EXAMPLE

A crown with a mass of 2.0 kg has a weight in air of 19.6 N. The crown hanging in water has an apparent weight of 18.3 N.
(a) What is the crown's volume?
(b) What is its density?

The weight of the fluid displaced is just equal to the difference between the actual and apparent weights.

$$F_b = F_g - F_{net} = 19.6 \text{ N} - 18.3 \text{ N}$$
$$F_b = 1.3 \text{ N}$$

The volume of fluid displaced is then

$$V = \frac{F_b}{\rho_2 g} = \frac{(1.3\,\text{N})}{(1000\,\text{kg/m}^3)(9.8\,\text{m/s}^2)}$$

$$V = 1.33 \times 10^{-4}\,\text{m}^3 = 133\,\text{cm}^3$$

The density of the crown is

$$\rho_1 = \frac{F_g}{Vg} = \frac{19.6\,\text{N}}{(1.33 \times 10^{-4}\,\text{m}^3)(9.8\,\text{m/s}^2)}$$

$$\rho_1 = 15\,\underline{0}00\,\text{kg/m}^3$$

The density of the crown is somewhat less than the density of pure gold of $19\,\underline{3}00\,\text{kg/m}^3$.

EXAMPLE

Suppose the crown in the previous example were an alloy of gold with a base metal such as copper, which has a density of $8900\,\text{kg/m}^3$. How much gold and how much copper are in the crown?

The crown's volume $V = 133\,\text{cm}^3$ is made up of a volume V_1 of gold and V_2 of copper.

$$V = V_1 + V_2$$

The crown's mass $M = 2.00\,\text{kg}$ is made of a mass M_1 of gold and M_2 of copper. The mass and volume of the gold and copper are related by, respectively:

$$M_1 = \rho_1 V_1; \quad M_2 = \rho_2 V_2$$

The total mass is then

$$M = M_1 + M_2 = \rho_1 V_1 + \rho_2 V_2$$

We have two equations in two unknowns V_1 and V_2. Eliminating V_2 from the equation we get

$$M = \rho_1 V_1 + \rho_2(V - V_1) = (\rho_1 - \rho_2)\,V_1 + \rho_2 V$$

Solving for V_1 we find that

$$V_1 = \frac{M - \rho_2 V}{(\rho_1 - \rho_2)}$$

$$= \frac{2.00\,\text{kg} - (8900\,\text{kg/m}^3)(1.33 \times 10^{-4}\,\text{m}^3)}{(19\,\underline{3}00\,\text{kg/m}^3 - 8900\,\text{kg/m}^3)}$$

$$V_1 = 7.85 \times 10^{-5}\,\text{m}^3 = 78.5\,\text{cm}^3$$

The mass of gold is

$$M_1 = \rho_1 V_1 = (19\,300\,\text{kg/m}^3)(7.85 \times 10^{-5}\,\text{m}^3)$$

$$M_1 = 1.51\,\text{kg}$$

The crown is made of 75% gold by weight. Pure gold is 24 karat, so the crown is made of 18 karat gold.

Floating Objects

If the density of the object is less than that of the surrounding fluid, there is a net upward force. If the object is totally submerged there is an upward acceleration. On reaching the surface of the liquid, the object will float partially out of the water, so that the object's downward weight is just balanced by the weight of the fluid displaced. If fresh-water ice, with a density of $917\,\text{kg/m}^3$, is floating in seawater with a density of $1030\,\text{kg/m}^3$, then 11% of the volume will be above the surface. An iceberg has 89% of its volume below the surface of the sea.

The size of an oceangoing ship or oil tanker is given in terms of its displacement, the weight of the water that the ship would displace when fully loaded. Tankers currently in service have a displacement of 70 000 tons, while the new supertankers may have as much as 400 000 tons. When fully loaded, these tankers ride low in the water, displacing a maximum amount of water. When empty, their displacement is much less, and they ride high in the water.

The density of a liquid can be measured with a **hydrometer,** usually a long cylinder of a given mass and volume and weighted at one end so that it floats erect. When placed in a liquid of unknown density it will displace a volume of liquid equal to its weight. The hydrometer is calibrated to read the density directly. Quite often the hydrometer will read the liquid's specific gravity: the ratio of the density of the liquid to the density of water. Specialized hydrometers measure the degree of charge or discharge of lead–sulfuric acid batteries by measuring the density, and thus the concentration, of the battery acid. Other hydrometers measure the "gravity" of oils or the concentration of sucrose sugar solutions in water.

Pascal's Principle

In the seventeenth century the French mathematician and scientist Blaise Pascal discovered some of the principles of hydrostatics, that is, of fluids at rest. In 1648 Pascal reported his experiments with a machine that had two pistons very different in size. The machine would multiply forces such that "a man pressing the small piston will match the strength of a hundred men pressing the piston in the hundredfold greater opening." A pressure applied at a small piston will be transmitted throughout the fluid. **Pascal's principle** states that the pressure applied to a confined fluid at rest is transmitted equally to every portion of the fluid and to the walls of the confining vessel. Pascal also reported that the pressure exerted by a fluid at rest depends only on the height of the column of fluid and not on the shape of the vessel or on the volume of fluid.

Pressure *is the force per unit area that acts perpendicular to a surface*. If a force F acts perpendicular to a surface with area A, then the pressure is

$$P = \frac{F}{A}$$

The unit for pressure is the newton per square meter, N/m^2, or, in SI units, the **pascal** (Pa):

$$1 \text{ Pa} = 1 \text{ N/m}^2$$

The force acting on a surface area A by a pressure P is of magnitude $F = PA$. The larger the area, the larger the force. The force is a vector quantity directed perpendicularly toward the surface.

Automobile brakes use the principle of pressure. The driver's foot applies pressure to a master brake cylinder:

$$P = \frac{F_1}{A_1}$$

The brake fluid transmits the pressure to the brake cylin-

ders at each wheel. The fluid pushes against the pistons, which push the brake lining against the drum or disk (see Figure 9-2). The applied force is proportional to the piston area:

$$F_2 = PA_2 = F_1 \frac{A_2}{A_1} \tag{9-2}$$

Hydraulic cylinders are commonly used to move heavy equipment and to exert large forces (see Figure 9-3), as in the heavy excavator in the following example.

EXAMPLE

The hydraulic cylinder that controls the bucket has a diameter of 0.165 m, a stroke of 1.36 m, and an operating pressure of $7.0 \times 10^6 \text{ N/m}^2$.
(a) What is the maximum force exerted by this cylinder in newtons? in lbf?
(b) How much mechanical work is done in moving the length of a full stroke at this force?
(c) If this work is done in 5 seconds, how much power is required?

The area of the cylinder is

$$A = \pi d^2/4 = 2.14 \times 10^{-2} \text{ m}^2$$

The maximum force is then

$$F = PA = (7.0 \times 10^6 \text{ N/m}^2)(2.14 \times 10^{-2} \text{ m}^2)$$

$$F = \frac{1.5 \times 10^5 \text{ N}}{4.45 \text{ N/lbf}} = 3.3 \times 10^4 \text{ lbf} = 16.5 \text{ tons!}$$

The work done is $W = \mathbf{F} \cdot \mathbf{s} = (1.5 \times 10^5 \text{ N})(1.36 \text{ m}) = 2.0 \times 10^5$ J. If this work is done in 5 seconds, the power required is

$$P = \frac{\Delta W}{\Delta t} = \frac{2.0 \times 10^5 \text{ J}}{5.0 \text{ s}} = 4 \times 10^4 \text{ J/s} = 40 \text{ kW}.$$

These machines are certainly powerful.

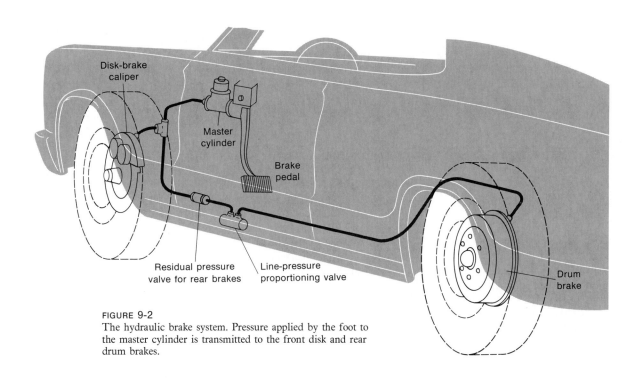

FIGURE 9-2
The hydraulic brake system. Pressure applied by the foot to
the master cylinder is transmitted to the front disk and rear
drum brakes.

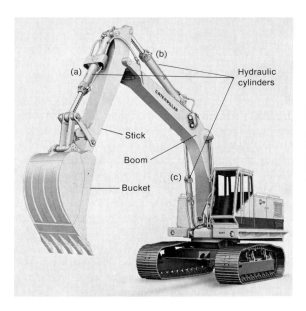

FIGURE 9-3
Hydraulic cylinders are commonly used in heavy equipment
such as this excavator. Separate cylinders control (a) the
bucket, (b) the stick, and (c) the boom (two cylinders). The
stick cylinder has a diameter of 0.20 m. The bucket and two
boom cylinders each have a diameter of 0.165 m. The cylinders operate with hydraulic fluid from a pump at a pressure
of up to 1000 lbf/in^2, or 7×10^6 N/m^2 (70 atmospheres).
Courtesy of Caterpillar Tractor Co.

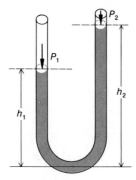

FIGURE 9-4
The U-tube manometer.

Measurement of Pressure

We can use a column of liquid to measure pressure. Both water, with a density of 1 000 kg/m³, and mercury, with a density of 13 600 kg/m³, are often used. A U-tube manometer (see Figure 9-4) consists of a glass tube with an interior diameter of about 6 mm. Downward pressure acts on each side of the manometer. If pressure P_1 is larger than P_2, the liquid in side one is forced down and the liquid in side two is raised until the pressure difference is balanced by the weight of the column of liquid.

The force acting at the bottom of the first side includes the total force at the top of the liquid column plus the force of gravity acting on the liquid column of height h_1. The weight of the liquid column, with cross-sectional area A_1, is just

$$M_1 g = (\rho_1 h_1 A_1) g$$

The total pressure acting downward on side one is

$$P_1' = P_1 + \rho g h_1$$

Similarly, the total pressure acting at the bottom of the second column is

$$P_2' = P_2 + \rho g h_2$$

The pressure at the bottom of the tube must be the same or there will be a net force pushing one way or the other.

$$P_1' = P_2'$$
$$P_1 + \rho g h_1 = P_2 + \rho g h_2$$

The pressure difference is then given by the difference in height of the two columns.

$$P_1 - P_2 = \rho g(h_2 - h_1) \qquad (9\text{-}3)$$

Because mercury is often used as the working fluid in a manometer, a common unit for measuring pressure is the millimeter of mercury, mm of Hg. This unit is so prevalent that it has been given the name **torr,** after Evangelista Torricelli, who first studied the barometer:

$$1 \text{ mm of Hg} = 1 \text{ Torr}$$

EXAMPLE

What is the pressure in N/m² associated with a pressure of 1 Torr? The pressure is

$$P = \rho g h$$

where ρ is the density of mercury and $h = 0.001$ m.

$$P = (13\ 600 \text{ kg/m}^3)(9.8 \text{ m/s}^2)(0.001 \text{ m})$$
$$P = 133 \text{ N/m}^2$$

Atmospheric Pressure

Galileo observed that the best suction pumps could lift drinking water from wells or accumulated ground water from coal mines no more than 32 ft. This limit had economic consequences in that it severely inhibited the deep mining of coal. Galileo incorrectly surmised that the limit was due to the molecular forces of water. However, two of his students, Torricelli and Vincenzo Viviani, correctly surmised that the atmosphere pressing on the water in a well or mine would force the water up a pipe from which air had been exhausted.

Torricelli studied the effect of air pressure on a column of mercury. A tube sealed at one end was filled with mercury, then inverted into a dish. Some of the mercury

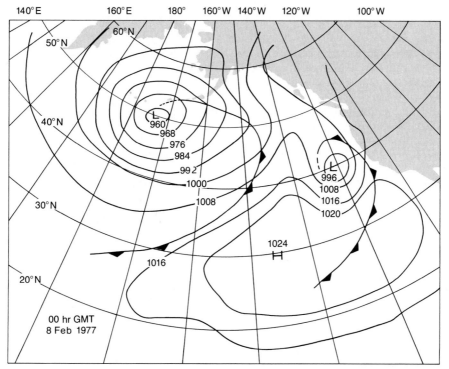

FIGURE 9-5
Surface pressure in the north Pacific Ocean in millibar on 8 February 1977. The large high-pressure center off the southern California coast pushes the winter storm track far to the north, giving the west a very dry winter. A low-pressure center has broken through, bringing a small rainstorm. A very large, deep low-pressure zone off the Aleutian Islands, west and south of Alaska, is building the next large weather system, which alas will also move inland north of California, continuing the drought. The prevailing winds are from the west. Courtesy of National Weather Service, U.S. Department of Commerce.

flowed from the tube, but some remained. Torricelli found that the pressure of the atmosphere could sustain a column of mercury only about 760 mm high. The region above the mercury was essentially a vacuum. The pressure of the air on the dish forces the mercury up the tube.

The pressure of 760 mm of Hg equals 1.0129×10^5 N/m², which can be expressed in a variety of units as shown in Table 9-2. The small variations in local atmospheric pressure are recorded daily with barometers at airports, weather stations, etc. A pressure of 10^5 N/m² correponds to a unit of pressure, the bar. A pressure of 760 mm of Hg is 1012.9 millibars. From the reported pressure measurements, contours of equal pressure are determined and plotted in millibars to show weather systems on our weather maps, as shown in Figure 9-5.

TABLE 9-2
A pressure of one atmosphere.

760 mm of Hg
760 torr
29.92 inches of Hg
33.5 ft of water
14.696 lbf/in²
2116 lbf/ft²
1.0129×10^5 N/m²
1.0129×10^5 pascal
1.0129×10^6 dyne/cm²
1.0129 bar
1012.9 millibar

Gauge Pressure

In many common pressure measurements, including blood pressure and automobile tire pressure, we measure **gauge pressure:** the pressure difference from atmospheric pressure. If the gauge pressure is zero, then the force pushing outward against the interior wall of a chamber is exactly equal to the force of the atmosphere pushing inward on the exterior wall—there is no net pressure on the wall but there is an absolute pressure pushing on the wall.

$$P_{gauge} = P_{absolute} - P_{atmosphere}$$

EXAMPLE

We inflate an automobile tire to a pressure of 29 lbf/in^2 (psi). What is the absolute pressure inside the tire?

$$P_{absolute} = P_{gauge} + P_{atmosphere}$$

$$P_{absolute} = 29 \, \text{lbf/in}^2 + 14.7 \, \text{lbf/in}^2$$

$$P_{absolute} = 43.7 \, \text{lbf/in}^2$$

In engineering units, the absolute pressure lbf/inch^2 is indicated as psia to distinguish it from gauge pressure (psi).

A blood pressure reading of 120 torr indicates that the arterial pressure is 120 mm of Hg above atmospheric pressure. A standard instrument for measuring such pressures is the U-tube manometer (shown in Figure 9-4) with one end open to the atmosphere. Another common pressure gauge uses a stiff metal tube, a Bourdon tube which is bent into an arc and connected to a dial indicator. As the difference between internal (tube) and external pressure increases or decreases, the tube straightens or curls more tightly. The mechanically connected dial indicator is calibrated to read gauge pressure in inches or millimeters of mercury, or pounds per square inch, or newtons per square meter, etc. Such gauges are often found on regulators used to control the pressure of gases delivered from high-pressure cylinders. These gauges can also read "vacuum"—that is, from zero to -760 mm of Hg, as the pressure is reduced from normal atmospheric pressure.

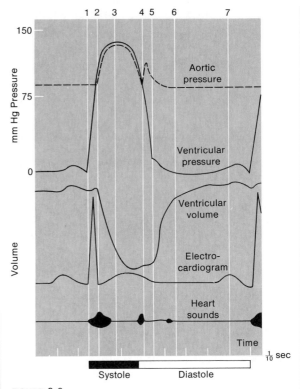

FIGURE 9-6
The heartbeat. The graph shows the pressure and volume, electrocardiogram, and heart sound relationships as a function of time for the systole and diasystole cycles. Adapted from *Biophysical Science* by Eugene Ackerman, ©1962. Reprinted by permission of Prentice-Hall, Inc., Englewood Cliffs, New Jersey.

Blood Pressure Measurement

In medicine, one of the most widely, and routinely, used physiological measurements is that of blood pressure. From medical research, the pressure and volume relationships for the ventricle and auricle contractions of the heart are well understood. A pressure-measuring tube, or catheter, inserted into the aorta can measure the pressure in this major artery as the heart beats. Figure 9-6 shows that the pressure in the aorta rises from about 85 torr to about 140 torr as the left ventricle contracts during the systole cycle. However, since installation of catheters requires surgical procedures, which are not usually expedient, consistent measurements of the patient's blood pressure are obtained indirectly by using a **sphygmomanometer.**

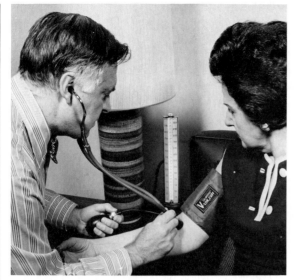

Indirect blood pressure measurement using a sphygmomanometer. Courtesy of W. A. Baum Co., Inc.

The sphygmomanometer (see Figure 9-7) consists of a pneumatic cuff, a hand pump with a release valve, and a pressure gauge (either the Bourdon tube type, or the more traditional mercury manometer shown in the figure). When inflated, the cuff applies pressure against the arm's muscles and tissues. The cuff is inflated until this pressure exceeds that of the blood flowing through the arm's large brachial artery. The artery then collapses, stopping altogether the flow of blood. The doctor or nurse then gradually reduces the pressure in the cuff and listens with a stethoscope placed over the brachial artery at the crook of the arm.

As the pressure in the cuff falls, nothing special is heard until the cuff pressure is less than the maximum pressure in the artery. At this point blood squirts through the constriction, and sounds called "Korotkoff sounds" are heard. The pressure corresponding to the onset of these sounds is the maximum pressure in the artery and is known as the systolic pressure. As the pressure in the cuff is further decreased, the sounds continue to be heard, for during part of the cycle the artery is still collapsed. Once the pressure in the cuff is lower than the minimum arterial pressure, the artery is no longer constricted and the Korotkoff sounds disappear. This marks the dyastolic pressure.

While the sphygmomanometer does not directly measure arterial pressure, it agrees with intra-arterial measurements within about 10 torr. Pressure in the upper arm, which is at the same level as the aorta, closely corresponds to the pressure in the aorta near the heart.

FLUIDS IN MOTION

Hydrodynamics is the study of fluids in motion. Pascal's and Archimedes' principles describe the behavior of a fluid at rest, and we have established an operational definition of pressure in terms of the force exerted by a liquid. Let us now consider how much work must be done to move a fluid through some pressure difference, and what are the relationships among three variables: the area of the pipe, the flow velocity, and the pressure. Finally we shall look at the viscosity that retards the smooth flow of a fluid through a pipe and limits the speed of objects falling in a liquid.

Flow Continuity

Consider an incompressible liquid flowing through a pipe. At one end the pipe has a pressure P_1 and a cross-sectional area A_1, and the fluid is flowing smoothly with a velocity v_1. At the other end of the pipe the pressure is P_2, the area A_2, and the velocity v_2. We can also consider the work required to move the fluid uphill by having the first end of the pipe be some height h_1, the other end some greater height h_2. (See Figure 9-8.)

In a time Δt the fluid in the bottom end moves forward a distance $\Delta x_1 = v_1 \Delta t$. At the top it moves forward a distance $\Delta x_2 = v_2 \Delta t$. Provided that our pipe does not bulge, and that the fluid is incompressible, the same quantity of fluid that moves forward at one end does so at the other. Quantitatively, we can write this in terms of the volumes, ΔV_1 and ΔV_2, which move forward in time Δt:

$$\Delta V_1 = A_1 \Delta x_1 = A_1 v_1 \Delta t$$

$$\Delta V_2 = A_2 \Delta x_2 = A_2 v_2 \Delta t$$

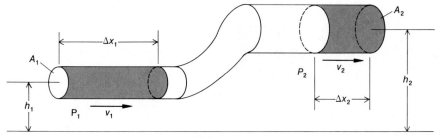

FIGURE 9-8
Fluid flow through a pipe.

These two volumes are equal:

$$\Delta V_1 = \Delta V_2$$

so that

$$A_1 v_1 = A_2 v_2$$

The volume flow rate Q is just the volume per unit time passing through the pipe. Q is the same at both ends of the pipe. **Flow continuity** expresses the volume flow rate for an incompressible fluid through a rigid pipe:

$$Q = \frac{\Delta V}{\Delta t} = A_1 v_1 = A_2 v_2 \qquad (9\text{-}4)$$

where v_1 and v_2 are velocity of flow, which is assumed to be constant across A_1 and A_2 and to be uniform in time.

EXAMPLE

Suppose we wish to inject 10 cm³ of a fluid through a hypodermic needle into a patient in 10 seconds. The hypodermic needle has an inner radius of 0.05 cm. The piston of the syringe has a diameter of 1.0 cm. What is the velocity of the fluid through the needle? Through the syringe?

The volume flow rate is

$$Q = \frac{\Delta V}{\Delta t} = \frac{10 \text{ cm}^3}{10 \text{ s}} = 1.0 \text{ cm}^3/\text{s}$$

The cross-sectional area of the syringe is $A_1 = 0.79 \text{ cm}^2$, and of the needle is $A_2 = 0.0079 \text{ cm}^2$.

The velocity of the fluid in the needle is

$$v_2 = \frac{Q}{A_2} = \frac{1.0 \text{ cm}^3/\text{s}}{0.0079 \text{ cm}^2}$$

$$v_2 = 1.27 \text{ cm/s}$$

Similarly the velocity of the fluid in the syringe is 1.27 cm/s.

Energy Conservation

The work done on or by a fluid depends on the force F and the distance through which it acts Δx. If the force pushes against the end of a cylinder of area A with a pressure P, the work done is

$$\Delta W = F\Delta x = PA\Delta x$$

Since the product of the area A and the displacement Δx is just the change in volume ΔV, the work done is

$$\Delta W = P\Delta V$$

as we discuss in Chapter 12.

The net energy change in moving a volume of fluid from one point to another depends on several factors: the work done on the system; the work done by the system; the change in potential energy because of different elevations; and the change of kinetic energy because of the change in velocity. Considering the fluid flow through the pipe shown in Figure 9-8, we can summarize how each factor contributes.

Work done on the system:

$$\Delta W_1 = F_1 \Delta x_1 = P_1 A_1 v_1 \Delta t$$

Work done by the system:

$$\Delta W_2 = F_2 \Delta x_2 = P_2 A_2 v_2 \Delta t$$

Gravitational potential energy:

$$\Delta PE_1 = \Delta Mgh_1 = (\rho A_1 v_1 \Delta t)gh_1$$

$$\Delta PE_2 = \Delta Mgh_2 = (\rho A_2 v_2 \Delta t)gh_2$$

Kinetic energy:

$$\Delta KE_1 = \tfrac{1}{2}\Delta M \langle v_1^2 \rangle = \tfrac{1}{2}(\rho A_1 v_1 \Delta t)\langle v_1^2 \rangle$$

$$\Delta KE_2 = \tfrac{1}{2}\Delta M \langle v_2^2 \rangle = \tfrac{1}{2}(\rho A_2 v_2 \Delta t)\langle v_2^2 \rangle$$

where $\langle v^2 \rangle$ is the mean square average speed. The brackets indicate the time average of the square of the velocity.

The Human Heart

With these energy relationships we can estimate the work done by the human heart in pumping blood. Venous blood enters the right auricle of the heart, as shown in Figure 9-9, from the inferior and superior vena cava at a pressure that is slightly below atmospheric pressure during the diastolic cycle. The blood is then pumped by the right ventricle through the pulmonary artery to the lungs at a pressure of about 20 torr and returns, oxygenated, to the left auricle with only a small drop in pressure. From the left auricle the blood is pumped by the left ventricle into the aorta, the body's main artery. The pressure necessary to open the heart valve to the aorta is about 80 torr. The blood pressure in the aorta then rises to about 140 torr. The average pressure at which blood is delivered to the aorta is about 100 torr for a resting person, 140 torr for an active person.

To estimate the mechanical work done by the heart we adopt a simple model, shown in Figure 9-10. The blood enters the heart at zero pressure, $P_1 = 0$, and comes to rest in the right auricle. The blood leaves the left ventricle at

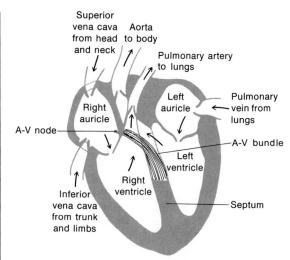

FIGURE 9-9
Diagram of the human heart. The arrows show the direction of blood flow. Adapted from *Biophysical Science* by Eugene Ackerman, ©1962. Reprinted by permission of Prentice-Hall, Inc., Englewood Cliffs, New Jersey.

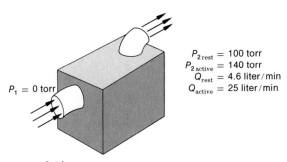

$P_{2\,\text{rest}} = 100$ torr
$P_{2\,\text{active}} = 140$ torr
$Q_{\text{rest}} = 4.6$ liter/min
$Q_{\text{active}} = 25$ liter/min

$P_1 = 0$ torr

FIGURE 9-10
A simple model used to calculate the work done by the heart.

an average pressure of 100 torr through an opening to the aorta of cross-sectional area $A_2 = 0.81$ cm². The heart rate of a resting person is about 70 beats/min and the heart is pumping about 4.6 liter/min. During strenuous exercise, the heart rate of a trained, physically conditioned person is 180 beats/min and the heart pumps about 25 liter/min. For a world-class endurance runner, the maximum blood flow rate is 35 liter/min.

We may now calculate the work done by the heart in a time Δt. Since the vena cava and the aorta are at the same level, the heart does no work against gravity; thus, we can ignore potential energy. Since the input to the heart is at zero pressure, we can ignore the work done initially on the heart by the fluid. Since it starts at rest, we can also ignore the initial kinetic energy as well. The mechanical work done by the heart in some time Δt is given by the work done at the exit pressure P_2, as well as the final kinetic energy of the blood:

$$\Delta W = \Delta KE_2 + \Delta W_2$$

$$\Delta W = \frac{1}{2}(\Delta M)<v_2{}^2> + F_2\Delta x_2$$

$$\Delta W = \frac{1}{2}(\rho A_2 v_2\Delta t)<v_2{}^2> + P_2 A_2 v_2\Delta t$$

The blood flow rate $Q = A_2 v_2$. The mechanical power exerted by the heart is the work per unit time, so that

$$\text{Power} = \frac{\Delta W}{\Delta t}$$

$$\text{Power} = \frac{1}{2}\rho A_2 v_2 <v_2{}^2> + P_2 A_2 v_2$$

$$\text{Power} = \frac{1}{2}\rho Q <v_2{}^2> + P_2 Q$$

EXAMPLE

The blood flow rate for the active heart is $Q_{active} = 25$ liter/min, and for the resting heart is $Q_{rest} = 4.6$ liter/min. What is the blood flow velocity? The blood flow rate is given by

$$Q_{rest} = 4.6 \text{ liter/min} = \frac{4.6 \times 10^{-3} \text{ m}^3}{60 \text{ s}}$$

$$Q_{rest} = 7.7 \times 10^{-5} \text{ m}^3/\text{s}$$

The average blood flow velocity is $v_{rest} = Q/A$

$$v_{rest} = \frac{7.7 \times 10^{-5} \text{ m}^3/\text{s}}{0.81 \times 10^{-4} \text{ m}^2} = 0.95 \text{ m/s}.$$

Similarly for an active person

$$Q_{active} = 25 \text{ liter/min} = 4.2 \times 10^{-4} \text{ m}^3/\text{s}$$

$$v_{active} = 5.2 \text{ m/s}$$

The heart pumps blood into the aorta only briefly during the cycle, when the pressure in the left ventricle is sufficient to open the valve into the aorta. The average blood flow velocity v_2 leaving the resting heart is 0.95 m/s averaged over the total heart cycle. The instantaneous velocity of the blood leaving the heart can be much higher. The average kinetic energy depends on the time average of the square speed $<v^2>$. The higher velocities are more important when one does the average over time. The mean square average speed of the blood has been estimated at about 3.5 times higher than the square of the average speed:

$$\langle v_2{}^2 \rangle \cong 3.5(v_2)^2$$

The density of blood is similar to that of water

$$\rho \cong 1.0 \text{ gm/cm}^3$$

EXAMPLE

The heart at rest has a blood flow rate of $Q_{rest} = 4.6$ liter/min $= 7.7 \times 10^{-5}$ m³/s; a flow velocity of 0.95 m/s; and an average pressure difference $P_{2rest} = 100$ torr $= 1.33 \times 10^4$ N/m². What is the mechanical power exerted by the heart at rest?

$$\text{Power} = \frac{1}{2}Q\rho <v_2{}^2> + P_2 Q$$

$$\text{Power} = \frac{1}{2}(7.7 \times 10^{-5} \text{ m}^3/\text{s})(1\,000 \text{ kg/m}^3)(3.5)$$
$$(0.95 \text{ m/s})^2 + (1.33 \times 10^4 \text{ N/m}^2)(7.7 \times 10^{-5} \text{ m}^3/\text{s})$$

$$\text{Power} = 0.12 \text{ N-m/s} + 1.03 \text{ N-m/s} = 1.15 \text{ watts}$$

The resting heart generates a mechanical power of about one watt. Most of this power is used to raise the pressure of the blood.

EXAMPLE

When a person is strenuously exercising, the heart moves blood at a rate of $Q_{active} = 25$ liter/min. This increased blood flow is caused partly by an increase in the heartbeat rate to a maximum of about 180 beats/

min, and partly by an increase in the volume of blood moved with each heartbeat. The average blood pressure also increases to about $P_{2\text{active}} = 140$ torr $= 1.87 \times 10^4$ N/m². What is the mechanical power generated by the heart?

Power $= \frac{1}{2}Q\rho<v_2{}^2> + P_2Q$

Power $= \frac{1}{2}(4.2 \times 10^{-4}$ m³/s$)(1000$ kg/m³$)(3.5)$
$\qquad (5.2$ m/s$)^2 + (1.87 \times 10^4$ N/m²$)(4.2 \times 10^{-4}$m³/s$)$

Power $= 19.9$ N-m/s $+ 7.9$ N-m/s

Power $= 27.8$ watts

This is considerable mechanical power, about 70% of which goes into the kinetic energy of the blood surging through the aorta, while only 30% is used to increase the blood pressure. If the heart of a highly trained athlete is pumping at 35 liters/min, the mechanical power delivered by the heart is over 60 watts, with 80% of the power going into the kinetic energy of the blood as it surges into the aorta.

Bernoulli's Equation

We can now consider the flow of a fluid through a pipe, as shown in Figure 9-8, with respect to conservation of energy. The energy into the system consists of: the work done on the system ΔW_1, the initial potential energy ΔPE_1, and the initial kinetic energy ΔKE_1. The energy out of the system is the work done by the system ΔW_2 plus the kinetic and potential energies. The energy associated with an element of fluid moving through the system is conserved:

$$\Delta W_1 + \Delta PE_1 + \Delta KE_1 = \Delta W_2 + \Delta PE_2 + \Delta KE_2$$

Writing out the different energy terms explicitly we have

$$P_1A_1v_1\Delta t + (\rho A_1v_1\Delta t)gh_1 + \frac{1}{2}(\rho A_1v_1\Delta t)<v_1{}^2> =$$
$$P_2A_2v_2\Delta t + (\rho A_2v_2\Delta t)gh_2 + \frac{1}{2}(\rho A_2v_2\Delta t)<v_2{}^2>$$

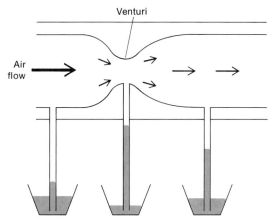

FIGURE 9-11
The venturi effect. A drop in pressure occurs when a fluid, such as air, flows past a constriction. Variation in the height of the mercury columns shows the pressure difference. All three columns indicate a pressure below atmospheric.

From the continuity equation we have that

$$Q = A_1v_1 = A_2v_2$$

Dividing both sides by $A_1v_1\Delta t$ or $A_2v_2\Delta t$, as appropriate, we derive **Bernoulli's equation** for the motion of an incompressible fluid:

$$P_1 + \rho gh_1 + \frac{1}{2}\rho<v_1{}^2> = P_2 + \rho gh_2 + \frac{1}{2}\rho<v_2{}^2>$$

The conclusion of Bernoulli's equation is that when the velocity of a fluid is greater, its pressure must be less. Figure 9-11 shows the **Venturi effect:** At a constriction of air flow, the pressure drops, as the height of liquid in the middle column shows. This principle has wide application to such diverse phenomena as the vacuum produced by an automobile carburetor, the pumping action of an aspirator, or the lift from the wings of an airplane.

EXAMPLE

A 115 horsepower two-stroke engine has a total piston displacement of 2.25 liter. The engine, running at 3600 rev/min, draws about this volume of air every two revolutions. The air horn of the carburetor has a diameter of 7.2 cm. (See Figure 9-12.)
(a) What volume of air per second is drawn into the air horn?

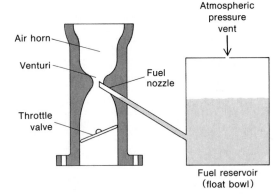

FIGURE 9-12
A vacuum is produced at the venturi of an automobile carburetor. Atmospheric pressure pushes the fuel up and out of the fuel nozzle.

(b) How fast is the air moving as it enters the air horn?

(c) Assuming a constant density of air of $\rho = 1.2 \times 10^{-3}$ gm/cm³, what is the pressure difference, in torr, from atmosphere at the air horn given by Bernoulli's equation?

Air is drawn into each cylinder only once every two revolutions. The total amount of air drawn into the carburetor is then

$$Q = \frac{\Delta V}{\Delta t} = \frac{(3600 \text{ rev})}{(\text{min})} \frac{(2.25 \text{ liter})}{(2 \text{ rev})}$$

$$Q = 67.5 \text{ liter/s} = 6.75 \times 10^{-2} \text{ m}^3/\text{s}$$

The airhorn has a diameter of 7.2 cm or an area $A = 4.1 \times 10^{-3}$ m³. The air flow velocity is $v = Q/A = 16.6$ m/s.

If we neglect the change in height $h = 0$, and assume that initially the air was at rest $v_1 = 0$, Bernoulli's equation gives us a relationship between the change in pressure and the flow velocity:

$$P_1 - P_2 = \tfrac{1}{2}\rho <v_2{}^2>$$

In this case we have

$$P_1 - P_2 = \tfrac{1}{2}(1.2 \text{ kg/m}^3)(16.6 \text{ m/s})^2$$

$$P_1 - P_2 = 165 \text{ N/m}^2 = 1.2 \text{ torr}$$

Viscosity

Whenever a fluid flows past a surface, the fluid next to the surface is in contact and is not moving. At increasing distances from the surface the velocity of the fluid increases. We can imagine layers of fluid sliding over one another. Between layers there is a certain amount of friction, which retards the flow of the faster fluid and speeds up the flow of the slower fluid. A measure of this friction is the **viscosity.** It gives rise to a net force on the surface.

Consider a liquid flowing in the x-direction adjacent to a wall with a surface area A_{wall}. The velocity v_x of the liquid in the x-direction increases from zero as distance y from the wall increases. (see Figure 9-13.)

The friction force on the wall depends on the area of the wall A_{wall} and the rate at which the velocity Δv_x changes as the distance from the wall Δy changes. This relationship of the change of velocity with distance from the wall is known as the gradient of the velocity:

$$\frac{\Delta v_x}{\Delta y}$$

The coefficient of viscosity η relates the friction force at the wall F_x to the area of the wall A_{wall} and the velocity gradient $\Delta v_x/\Delta y$:

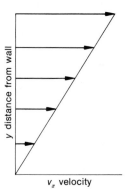

FIGURE 9-13
The velocity profile of a fluid near a wall.

$$F_x = \eta \frac{\Delta v_x}{\Delta y} A_{\text{wall}} \qquad (9.5)$$

There are laboratory instruments that measure the viscosity of liquids by measuring the friction force between two moving plates separated by a layer of fluid.

The CGS units of viscosity can be found by solving the above relationship for the viscosity η:

$$\eta = \frac{F_x}{A_{\text{wall}} \frac{\Delta v_x}{\Delta y}} = \frac{\text{dyne}}{(\text{cm}^2) \frac{(\text{cm/s})}{\text{cm}}} = \frac{\text{dyne-s}}{\text{cm}^2}$$

This unit has been given the name **poise,** in honor of the French physician Poiseuille, who first studied the flow of liquids:

$$1 \text{ poise} = 1.0 \frac{\text{dyne-s}}{\text{cm}^2} = 0.1 \frac{\text{N-s}}{\text{m}^2}$$

Poise and centipoise, commonly used for viscosity, are not SI units. Values of viscosity for liquids are listed in Table 9-3, those for gases and vapors in Table 9-4.

Lubrication is used in mechanical equipment (such as motor oil in automobiles) to minimize the friction force and wear between adjacent moving metal surfaces. The friction force between two moving surfaces separated by a film of oil is proportional to the viscosity of the lubricant.

TABLE 9-3
Viscosity of common liquids listed in centipoise (cp).
$1 \text{ cp} = 10^{-2} \text{ poise} = 10^{-2} \text{ dyne-s/cm}^2 = 10^{-3} \text{ N-s/m}^2.$

Liquid	Temperature	Viscosity
Water	20°C	1.000 cp
Water	40°C	0.653 cp
Water	100°C	0.282 cp
Acetone	15°C	0.337 cp
Carbon tetrachloride	20°C	0.969 cp
Ethyl alcohol	20°C	1.2 cp
Glycerin	20°C	1 490 cp
SAE 30 oil	16°C	60 cp
Gasoline	16°C	2.6 cp

TABLE 9-4
Viscosity of common gases in micropoise (μp).
$1 \ \mu p = 10^{-6} \text{ poise} = 10^{-6} \text{ dyne-s/cm}^2 = 10^{-7} \text{ N-s/m}^2.$

Gas	Temperature	Viscosity
Air	20°C	170.8 μp
Air	18°C	182.7 μp
Carbon dioxide	20°C	148 μp
Helium	20°C	194.1 μp
Hydrogen	20°C	87.6 μp
Propane	19°C	79.5 μp

The development of air bearings, where two moving metal surfaces are separated by a layer of air which has a very small viscosity, has allowed virtually frictionless operation of high-speed devices, such as the ultracentrifuge.

Fluid Flow through a Pipe

Let us now consider the resistance to flow of a liquid through a round pipe when the fluid has viscosity. The fluid near the wall is traveling at zero velocity. The fluid at the center is traveling at some maximum velocity v_{max}. The volume flow rate of the pipe is determined by the average velocity of the fluid averaged over the pipe's cross-sectional area A_{cross}:

$$Q = A_{\text{cross}} v_{\text{av}}$$

Figure 9-14 shows a velocity profile of fluid flow through the pipe.

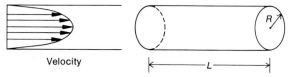

FIGURE 9-14
The velocity profile for fluid flow through a pipe of radius R.

If the fluid velocity at the center is v_{max}, the average fluid velocity is approximately

$$v_{av} = \frac{v_{max}}{4}$$

The velocity of fluid flow through the pipe is determined by balancing the force of the pressure difference against the resistance to flow caused by the viscosity. The total force due to the pressure difference for a round pipe of radius R is just

$$F_{pressure} = (P_2 - P_1)A_{cross}$$
$$F_{pressure} = (P_2 - P_1)\pi R^2$$

The friction force acts against the walls of the pipe.

$$F_{viscosity} = \eta \frac{\Delta v_x}{\Delta y}A_{wall}$$

If the pipe has a length L, the area of the wall is just

$$A_{wall} = 2\pi RL$$

The velocity gradient is approximately

$$\frac{\Delta v_x}{\Delta y} = \frac{v_{max}}{R}$$

The force due to viscosity then can be approximated as

$$F_{viscosity} = \eta \frac{v_{max}}{R}A_{wall}$$

Balancing the pressure force with the friction force we have

$$F_{pressure} = F_{viscosity}$$
$$(P_2 - P_1)A_{cross} = \eta \frac{v_{max}}{R} A_{wall}$$

Solving for the maximum speed v_{max}:

$$v_{max} = (P_2 - P_1)\left(\frac{A_{cross}}{A_{wall}}\right)\left(\frac{R}{\eta}\right)$$

$$v_{max} = (P_2 - P_1)\left(\frac{\pi R^2}{2\pi RL}\right)\left(\frac{R}{\eta}\right)$$

$$v_{max} = (P_2 - P_1)\left(\frac{R^2}{2L\eta}\right)$$

The total flow volume through the pipe is then given by

$$Q = v_{av}A_{cross}$$

$$Q = \frac{v_{max}}{4}\pi R^2$$

which can be reduced to

$$Q = \frac{(P_2 - P_1)\pi R^4}{8L\eta} \tag{9-6}$$

This same relationship, called the Poiseuille equation, was arrived at empirically by Poiseuille, when he was investigating flow pressure relationships in the small blood vessels of living animals. The above derivation is only approximate: We had to make some guesses to determine both the velocity gradient and the average flow velocity in terms of the maximum velocity. However, precise analysis using calculus gives the same result, demonstrating again how a simple model of a physical situation can give the correct functional dependence, and often the same numerical coefficients, as a more sophisticated analysis.

EXAMPLE

A hypodermic syringe is used to inject 10 cm³ of a water solution into the artery of a patient in a 10 s interval. The artery has an average pressure of 100 torr above atmospheric. The syringe is fitted with a No. 17 needle 2.0 cm long and with an inside radius of 0.05 cm.

What pressure is needed to inject the fluid into the patient? Given: The fluid flow rate is

$$Q = \frac{\Delta V}{\Delta t} = \frac{10 \text{ cm}^3}{10 \text{ s}} = 1.0 \text{ cm}^3/\text{s}$$

The radius of the needle $R = 0.05$ cm and the length $L = 2.0$ cm. We can assume the viscosity is that of water $\eta = 1.0$ cp $= 0.010$ dyne-s/cm^2.

We are asked to determine the pressure drop across the needle $P_2 - P_1$. Solving the Poiseuille equation for the pressure drop, we obtain

$$P_2 - P_1 = \frac{8QL\eta}{\pi R^4}$$

$$P_2 - P_1 = \frac{(8.0)(1.0 \text{ cm}^3/\text{s})(2.0 \text{ cm})(0.01 \text{ dyne-s/cm}^2)}{(3.14)(5.0 \times 10^{-2} \text{ cm})^4}$$

$$P_2 - P_1 = (8\,150 \text{ dynes/cm}^2)\frac{(1 \text{ torr})}{(1\,333 \text{ dynes/cm}^2)}$$

$$P_2 - P_1 = 6.1 \text{ torr}$$

It would require a pressure of 6.1 torr to inject the fluid into the air. It would take a pressure of 106 torr to inject the fluid into the patient against the average arterial pressure.

EXAMPLE

For intravenous administration of a 10% sucrose solution the supply bottle is elevated 60 cm above the patient. The fluid is fed through a flexible tube to a No. 20 needle of length $L = 3.8$ cm and an inner radius $R = 0.04$ cm, which is inserted into one of the patient's veins. The interior venous blood pressure is about 10 torr.
(a) Neglecting the viscosity of the solution in the flexible tube, what is the hydrostatic pressure at the end of the tube if the sucrose solution has a density of 1.04 gm/cm^3?
(b) What is the solution's flow rate into the patient if the flow is limited by the viscosity $\eta = 1.33$ cp of the fluid in the needle? In most clinical applications the flow is regulated by a clamp placed on the feed line or by a controller.

(a) The hydrostatic pressure P_2 is that associated with a column of fluid of height $h_2 = 60$ cm and a density $\rho = 1.04$ gm/cm^3.

$$P_2 = \rho g h_2$$

$$P_2 = (1.04 \text{ gm/cm}^3)(980 \text{ cm/s}^2)(60 \text{ cm})$$

$$P_2 = 61\,000 \text{ dyne/cm}^2 = 46 \text{ torr}$$

The venous pressure $P_1 = 10$ torr $= 13\,000$ dyne/cm^2.
(b) The volume flow rate Q is given by

$$Q = \frac{(P_2 - P_1)\pi R^4}{8L\mu}$$

$$Q = \frac{(61\,000 - 13\,000 \text{ dyne/cm}^2)(3.14)(0.04 \text{ cm})^4}{8.0 \,(3.8 \text{ cm})(0.013 \text{ dyne-s/cm}^2}$$

$$Q = 0.98 \text{ cm}^3/\text{s}$$

The Reynolds Number

Our discussion has assumed that fluid flow is smooth, or laminar. This is often not true, and when turbulence develops our simple notions of how a fluid behaves are no longer adequate. We can observe this in the flow of water from an ordinary water faucet. If the flow from the faucet is slow, then the water runs out smoothly with perhaps only a few ripples. If the flow velocity is increased above a certain rate, the flow becomes chaotic and turbulent.

The onset of turbulent motion in a room with quiet air can be observed in the plume from a cigarette. Near the cigarette the plume rises quite smoothly, as is characteristic of laminar flow, but higher up it becomes wider. At a certain point an instability begins and the motion becomes chaotic.

There is a dimensionless ratio, the **Reynolds number,** that indicates whether we can expect the motion of a wide range of different fluids to be smooth or turbulent. The Reynolds number is given by

$$N_R = \frac{D v_{av} \rho}{\eta}$$

where D is the diameter of the flow, v the average flow velocity, ρ the fluid mass density, and η the viscosity. If the Reynolds number is less than about 2000, we can expect the fluid flow to be smooth and laminar and the simple relationships developed here among pressure, flow velocity, and viscosity apply. If the Reynolds number is greater than 2000, we can expect turbulent motion, and the fluid's behavior must be determined experimentally.

EXAMPLE

Blood flows in the aorta with an average velocity of about 72 cm/s. The diameter of the aorta is about $D = 1.0$ cm; the density of whole blood is $\rho = 1.06$ gm/cm³; and the viscosity of blood is 4.7 centipoise. The Reynolds number for this flow is given by

$$N_R = \frac{Dv\rho}{\eta}$$

$$N_R = \frac{(1.0 \text{ cm})(72 \text{ cm/s})(1.06 \text{ gm/cm}^3)}{4.7 \times 10^{-2} \text{ dyne-s/cm}^2}$$

$$N_R = \frac{72 \times 1.06}{4.7} \times 10^2 \frac{\text{gm-cm/s}^2}{\text{dyne}}$$

$$N_R = 1600$$

In this example, $N_R < 2000$ and the fluid flow is smooth. At higher blood flow velocities, the flow may become turbulent, and our simple relationships among flow viscosity, flow velocity, vessel diameter, and pressure drop may no longer be applicable.

Stokes' Law

If a spherical body of radius R as shown in Figure 9-15 moves with a velocity v through a fluid with viscosity η, there is a force acting to slow the body down. This friction drag on the object is given by **Stokes' law:**

$$F_{\text{drag}} = 6\pi\eta v R$$

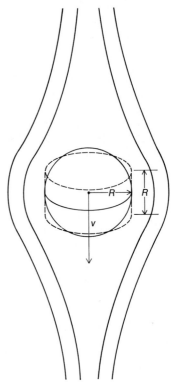

FIGURE 9-15
Stokes' law. A spherical body moving through a liquid is slowed by the force of viscosity. We can approximate the force on the sphere by considering the viscous force on a cylinder of radius R and thickness R.

Stokes' law can be derived from our definition of viscosity if we make some assumptions about the smooth flow of a fluid around the sphere. The sphere has a total surface area given by $4\pi R^2$. Of course, the moving fluid is not flowing past the entire surface. We can picture the sphere as a small cylinder of radius R and thickness R, as shown in Figure 9-15. Then the area of surface in contact with the moving liquid is approximately

$$A_{\text{wall}} = 2\pi R^2$$

The force of viscosity acting on the sphere is given by

$$F = \eta \frac{\Delta v_x}{\Delta y} A_{\text{wall}}$$

At the surface of the sphere the fluid is in contact with the sphere and traveling at velocity v. At some distance from the sphere the fluid is at rest. If the velocity of the fluid falls to zero in a distance $\Delta y = R/3$ from the edge of the sphere, then the velocity gradient is

$$\frac{\Delta v_x}{\Delta y} = \frac{v}{R/3} = \frac{3v}{R}$$

The force of viscosity on the sphere, the drag force F_{drag}, is

$$F_{\text{drag}} = \eta \frac{3v}{R} 2\pi R^2$$

or

$$F_{\text{drag}} = 6\pi \eta v R \qquad (9\text{-}7)$$

which is Stokes' law.

EXAMPLE

A steel shot of radius 1 mm and density of $\rho = 7.86$ gm/cm^3 is dropped into a cylinder of SAE 30 weight motor oil with a shot density $\rho_{\text{oil}} = 0.91$ gm/cm^3 and a viscosity of $\eta = 60$ cp. What is the terminal velocity of the shot?

The volume of the steel shot is $V = \frac{4}{3}\pi R^3 = 4.2 \times 10^{-9}$ m^3. The net downward force acting on the shot suspended in oil is

$$F_{\text{net}} = (\rho_{\text{shot}} - \rho_{\text{oil}})Vg$$

$$F_{\text{net}} = (7860 - 910 \text{ kg/m}^3)(4.2 \times 10^{-9} \text{ m}^3)(9.8 \text{ m/s}^2)$$

$$F_{\text{net}} = 2.86 \times 10^{-4} \text{ N}$$

At terminal velocity the net downward force is balanced by the drag force given by Stokes' law:

$$F_{\text{net}} = F_{\text{drag}} = 6\pi \eta v R$$

Solving for the terminal velocity

$$v = \frac{F_{\text{net}}}{6\pi \eta R}$$

$$v = \frac{2.86 \times 10^{-4} \text{ N}}{6\pi \,(0.06 \text{ N-s/m}^2)(10^{-3} \text{ m})}$$

$$v = 0.25 \text{ m/s}$$

The Centrifuge

A common clinical laboratory procedure is to take a sample of whole blood, place it in a centrifuge, and spin it down to determine the hematocrit, that is, the ratio of red blood cells to total blood volume. A more refined analysis is to take a sample of human blood serum, place it in an ultracentrifuge, and spin it down to separate the lipoproteins from the heavier proteins, globulins, and albumin. Both processes depend on the balance between the net gravitational force acting on the particle buoyed up by the liquid, and the frictional drag due to viscosity.

A particle in a liquid experiences the force of gravity pulling it downward and the weight of the fluid displaced acting as a buoyant force pushing it upward. If the particle is denser than the surrounding liquid it falls, if less dense it rises. The particle's terminal velocity depends approximately on Stokes' law. For large particles such as red blood cells, this is a relatively large velocity, and the particles separate out quickly. On the other hand, for the smaller protein molecules the terminal velocity can be quite small. The effective acceleration due to gravity g depends on the speed of the centrifuge $g = \omega^2 R$, as discussed previously.

The terminal velocity is determined on the balance between the net buoyant force and the Stokes' law viscous friction.

$$F_{\text{net}} = (\rho - \rho_{\text{serum}})\tfrac{4}{3}\pi R^3 g$$

$$F_{\text{drag}} = 6\pi \eta v R$$

Solving for the terminal velocity we have that

$$v = \tfrac{2}{9} \frac{(\rho - \rho_{\text{serum}}) R^2 g}{\eta} \qquad (9\text{-}8)$$

Ordinary centrifuges are quite adequate for separating out large particles such as red blood cells. Protein particles are much smaller, on the order of 60 nm with a density of about 1.007 gm/cm³. The serum density is typically raised to about $\rho = 1.063$ gm/cm³ by the addition of sodium chloride. For such a small molecule, assuming a viscosity of 4.7 centipoise, in a centrifuge capable of 2 260 g, separation velocity would be

$$v = \tfrac{2}{9}\,\frac{(\rho - \rho_{\text{serum}})R^2g}{\eta}$$

$$v = \tfrac{2}{9}\,\frac{(1.007 - 1.063 \text{ gm/cm}^3)(6.0 \times 10^{-6} \text{ cm})^2}{(4.7 \times 10^{-2} \text{ poise})}$$

$$\frac{(0.98 \times 10^3 \text{ cm/s})(2\,260)}{}$$

$$v = -2.1 \times 10^{-5} \text{ cm/s}$$

The separation velocity is negative, indicating that the molecule, less dense than the surrounding fluid, is buoyed upward to the top. Increasing the centripetal acceleration increases the effective weight of the fluid displaced and the buoyant force. Because this separation velocity is small, it would take days of spinning in an ordinary centrifuge to achieve significant separation.

Ultracentrifuges make effective accelerations as high as 430 000 g possible, greatly speeding up the separation process. With an effective acceleration of 220 000 g in the above example the separation velocity is

$$v = \tfrac{2}{9}\,\frac{(\rho - \rho_{\text{serum}})R^2g}{\eta}$$

$$v = \frac{2(1.007 - 1.063 \text{ gm/cm}^3)(6.0 \times 10^{-6} \text{ cm})^2}{9(4.7 \times 10^{-2})\text{poise}}$$

$$\frac{(2.2 \times 10^5)(980 \text{ cm/s}^2)}{}$$

$$v = -2.0 \times 10^{-3} \text{ cm/s}$$

The separation velocity is again negative. However, in one hour the molecule will travel about 7 cm. Different protein molecules have different radii and somewhat different densities. The smaller molecules have smaller separation velocities. When a sample is spun for a specified time, the different components (lipoproteins, albumin, cell bodies, etc.), having different separation velocities, form distinct strata. These strata can be identified by their own intrinsic colors, optical absorption characteristics, indexes of refraction, or other physical properties.

An alternate technique is to do the separation in a solution of sucrose or of saline that has a continuous variation in density from bottom to top. The particles of a sample placed in such a solution sort themselves out by density. When the density of a particle is the same as that of the surrounding solution, the net force on the particle is zero and it stops moving. The ultracentrifuge has become one of the standard instruments for analysis in cellular and molecular biology.

SURFACES

Surface Tension

A liquid has a definite volume confined by a definite surface. Although the surface of a liquid, such as water, can be easily penetrated by a heavy object, water beetles or strands of hair or bits of dust float on the surface. If carefully done, a sewing needle can be floated on the surface of water, even though it ordinarily would sink. And water placed on the surface of a waxed automobile or on water-repellent fabric forms distinct beads.

At the bounding surface of a volume of liquid is a force, called **surface tension,** that tends to contract the surface area to a minimum. Surface tension can be measured: A wire of some length l is immersed parallel to the surface and is slowly pulled out of the surface, as shown in Figure 9-16. As this is done, a film of the liquid forms on either side of the wire balancing the upward force. If

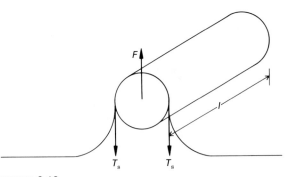

FIGURE 9-16
Surface tension can be measured by the force required to pull a wire from the surface of the liquid.

TABLE 9-5

Surface tension for various liquids.

Surface	Temperature °C	Surface tension dyne/cm
Water-air	20	72.75
Mercury-air	20	465
Ether-air	20	17
Benzene-air	20	28.85
Ethyl alcohol-air	20	22.75
Carbontetrachloride-vapor	20	26.95
Nitrogen-vapor	−203	10.53

the upward force becomes too great, the film breaks and the wire is freed from the liquid.

The surface tension, defined as the force per unit length at the edge of the surface, depends only on the substance and the temperature, and not on the shape of the surface. It is usually given for the surface of a liquid in contact with air or with its own vapor. See Table 9-5. There are surfaces on each side of a wire being withdrawn from a liquid. At the maximum force, the liquid surface pulls straight down, balancing the upward force. In this case the surface tension is given by

$$T_{\text{s}} = \frac{\text{force}}{\text{length}} = \frac{F}{2l}$$

The dimensions of the surface tension are dyne/cm or newton/m.

Work to Expand a Surface

When a liquid surface is expanded in area, molecules are brought from the interior, where they form bonds with neighboring molecules on all sides, to the surface, where they form bonds with neighboring molecules on only one side. The liquid surface contracts so that the molecules pull themselves together as close as possible. To expand the surface area, we must do work to bring more molecules to the surface. In the previous example we pulled a wire from the surface of the liquid. If we raise the wire by a distance $\Delta y = 0.1$ mm, we do work in stretching the surface. The amount of work done is

$$\Delta W = F\Delta y$$

However, at the same time we increase the surface area by an amount

$$\Delta A = 2l\Delta y$$

The work done per unit surface area to increase the surface area is just

$$\frac{\Delta W}{\Delta A} = \frac{F\Delta y}{2l\Delta y} = \frac{F}{2l} = T_{\text{s}}$$

The surface tension, then, is also a measure of the work required per unit area to increase the surface of a liquid. Consequently the units of surface tension can also be given in energy/area

$$\frac{\text{dyne}}{\text{cm}} = \frac{\text{dyne-cm}}{\text{cm-cm}} = \frac{\text{erg}}{\text{cm}^2}$$

Bubbles and Drops

A soap bubble is a thin liquid film enclosing a volume of air. The spherical shape and the pressure difference between the inside and outside of the bubble are determined by the surface tension. We can conceptually separate a soap bubble into two halves bisected by a plane through the center. The soap film has two surfaces, one on the outside and a second on the inside, each surface with its own characteristic surface tension. Since the soap film is thin, each surface has approximately the same radius. The force of the surface tension F_{surface} pulling the two halves together is balanced by the force due to the pressure difference F_{pressure} pushing the two halves apart (see Figure 9-17).

$$F_{\text{surface}} = 2T_{\text{s}}(2\pi R)$$
$$F_{\text{pressure}} = (P_{\text{in}} - P_{\text{out}})(\pi R^2)$$

Equating these two forces and solving for the difference in pressure, we get that

$$P_{\text{in}} - P_{\text{out}} = \frac{4T_{\text{s}}}{R} \qquad (9\text{-}9)$$

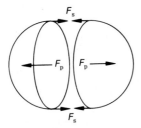

FIGURE 9-17
A soap bubble can be represented as two hemispheres. The pressure difference forces the hemispheres apart; surface tension holds them together.

EXAMPLE

A soap bubble consists of a spherical film of liquid with exterior and interior surfaces. Suppose the surface tension of the soap film is $T_s = 6$ dyne/cm. What is the pressure difference between the inside and outside of the soap bubble if the bubble has a radius $R = 2.0$ cm?

In this example the pressure difference is given by

$$P_{in} - P_{out} = \frac{(4)(6.0 \text{ dyne/cm})}{(1.0 \text{ cm})}$$

$$P_{in} - P_{out} = 2.4 \text{ dyne/cm}^2 = 0.018 \text{ torr}$$

The pressure difference for soap bubbles is small.

The surface tension on the outer surface of a freely falling droplet pulls the droplet into a spherical shape. This shape has a minimum surface area and thus a minimum surface energy. The force acting to pull the two halves of the droplet together is $F_{surface} = T_s 2\pi R$. The force from the pressure difference across the droplet is $F_p = (P_{in} - P_{out})\pi R^2$. Equating the pressure-difference force with the surface force we get that

$$P_{in} - P_{out} = \frac{2T_s}{R} \qquad (9\text{-}10)$$

Except for the numerical factor of 2 this is the result we obtained for a bubble (equation 9-9). Astronauts in space have observed these freely floating spheres of water.

A bubble within a liquid also has a single surface which is trying to compress the gas within the bubble. If a liquid is filtered to remove particles of dust and is then heated without vibration, it can be heated considerably above its boiling temperature without actually boiling. A dust-free liquid initially can form only small bubbles, which require a pressure much greater than atmospheric pressure to avoid collapsing because of the surface tension forces. The pressure difference for these bubbles is also

$$P_{in} - P_{out} = \frac{2T_s}{R}$$

Dust particles or other impurities present in a liquid provide boiling centers of modest radius. Boiling crystals (which are chemically inert) are sometimes placed in solutions to provide surfaces with large radii of curvature, so that boiling can commence near the boiling point in the absence of large pressure differences.

Respiration

A tidal volume of some 500 cm³ is moved in and out of the lungs with each breath. Because the viscosity of air is low and the diameter of the bronchial air passages is relatively large, the pressure drop during the passage of air from the nose to the lungs is small—of the order of 1 torr, even though a large volume of air is moved in a period of only about 4 s. Certain physical conditions, such as asthma or bronchitis, constrict these air passages, restricting the passage of air, and making it difficult to breathe.

The functional units in the lungs are the alveoli, tiny spherical sacks that exchange carbon dioxide for oxygen in the blood. The total surface area of these tiny alveoli is approximately equal to that of a regulation tennis court. These alveoli can respond to a difference between their interior, or intra-alveolar, pressure P_{in}, and the exterior pressure in the chest cavity, the intrapleural pressure, P_{out}. The pressure in the chest cavity is about -4 to -6 torr—that is, below atmospheric pressure. This negative pressure keeps the alveoli inflated (see Figure 9-18). If one side of the chest cavity is punctured, as by a stab wound, then the intrapleural cavity pressure increases and that lung immediately deflates.

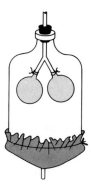

FIGURE 9-18
The expansion of the lungs can be demonstrated by two balloons sealed in a bell jar closed with a rubber diaphragm. As the diaphragm is lowered the lungs expand because of the pressure difference.

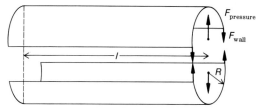

FIGURE 9-19
A blood vessel can be represented as a cylinder of radius R and length l.

Expansion and contraction of the alveoli are regulated by the motion of the diaphragm. The alveoli have an interior radius of about $R = 10^{-2}$ cm, and are lined with mucous fluid. If this fluid had the surface tension of water, the pressure difference across the alveoli needed to keep them from collapsing would be

$$P_{\text{in}} - P_{\text{out}} = \frac{4T_s}{R} = \frac{(4.0)(72 \text{ dyne/cm})}{(0.01 \text{ cm})}$$

$$P_{\text{in}} - P_{\text{out}} = 29\,000 \text{ dyne/cm}^2 = 22 \text{ torr}$$

The actual pressure difference required to expand this mucous lining is only about 1.5 torr. To reduce the surface tension of the alveolar mucous lining to $T_s = 5$ dyne/cm, the alveoli secrete a **pulmonary surfactant.** The total pressure difference across the alveoli during respiration is only 3 to 5 torr.

The Law of Laplace

Just as we calculated the surface tension required to contract a spherical surface, such as a bubble or alveoli, we can do so for a cylindrical shape, such as a blood vessel. Suppose we have a cylinder of radius R and length l as shown in Figure 9-19. Let us calculate the wall tension per unit length T_{wall} required to withstand a pressure difference $P_{\text{in}} - P_{\text{out}} = \Delta P$.

Conceptually we can separate the cylinder lengthwise into two halves. The force of the wall tension F_{wall} pulling the two halves together is balanced by the force of the pressure difference F_{pressure} pushing the two halves apart:

$$F_{\text{wall}} = T_{\text{wall}} 2l$$

$$F_{\text{pressure}} = (P_{\text{in}} - P_{\text{out}})(2Rl)$$

Equating these two forces and solving for the wall tension we get

$$T_{\text{wall}} = (P_{\text{in}} - P_{\text{out}})R = \Delta PR \qquad (9\text{-}11)$$

This last equation, known as the **law of Laplace,** relates wall tension to the radius of a cylindrical vessel and the pressure difference.

EXAMPLE
Suppose the average blood pressure is 100 torr. The lumen, the cylindrical cavity, of the aorta has a radius of 1.5 cm. A lumen of the arteriole has a radius of 1.5×10^{-2} cm. What tension per unit of wall length is required to withstand this pressure difference in each case?

The pressure difference is

$$\Delta P = 100 \text{ torr} \frac{1.01 \times 10^6 \text{ dyne/cm}^2}{760 \text{ torr}}$$

$$\Delta P = 1.33 \times 10^5 \text{ dyne/cm}^2$$

The tension in the aorta is then given by

$$T_{\text{wall}} = \Delta PR = (1.33 \times 10^5 \text{ dyne/cm}^2)(1.5 \text{ cm})$$

$$T_{\text{wall}} = 2.0 \times 10^5 \text{ dyne/cm} = 2.0 \text{ N/m}$$

The tension in the arteriole is

$$T_{\text{wall}} = \Delta PR = (1.33 \times 10^5 \text{ dyne/cm}^2)(1.5 \times 10^{-2} \text{ cm})$$

$$T_{\text{wall}} = 2.0 \times 10^3 \text{ dyne/cm} = 0.020 \text{ N/m}$$

The small-radius arterioles withstand the blood pressure with much thinner walls than do the major arteries. This is shown by the law of Laplace.

Both Laplace's law and our earlier relationship for spherical surfaces indicate a direct relationship among the surface or wall tension, the pressure difference, and the radius of curvature of the cylindrical or spherical surface: The smaller the radius of curvature, the greater the pressure difference that can be withstood for a given wall tension. Conversely, the greater the radius of curvature, the larger is the wall tension required to withstand a given pressure difference.

In the heart, the walls of the auricles have a smaller radius of curvature than the ventricles have, and thus withstand the same hydraulic pressure with a smaller wall tension and have a thinner muscle wall than do the ventricles. Both the ventricle and auricle walls are thicker than the wall of the aorta, which has a smaller radius still. In the abdominal wall, the muscle tissue withstands the pressure of the internal organs. However, if the abdominal wall is weakened, it bulges outward, reducing the radius of curvature. If there is a local separation in the abdominal wall, a hernia develops: A pocket of smaller radius of curvature bulges outward to balance the pressure with the weaker wall. Likewise, if there is a weakening in an automobile tire casing, the tire bulges outward, forming an area with a smaller radius to balance the pressure.

Review

Give the significance of each of the following terms. Where appropriate give two examples.

fluid	pressure	Venturi effect
liquid	pascal	viscosity
gas	torr	poise
buoyant force	gauge pressure	Reynolds number
Archimedes' principle	sphygmomanometer	Stokes' law
density	hydrodynamics	surface tension
hydrometer	flow continuity	pulmonary surfactant
Pascal's principle	Bernoulli's equation	law of Laplace

Further Readings

A. L. Stanford, Jr., *Foundations of Biophysics* (Academic Press, 1975).

James V. Warren, "The Physiology of the Giraffe," *Scientific American* (November 1974), pp. 96–105.

Problems

FLUIDS AT REST

1 A block of wood 20 cm by 30 cm by 50 cm with a density of 0.4 gm/cm³ is submerged 2 m below the surface of a lake.
   (a) What is the buoyant force acting on the block?
   (b) What is the initial acceleration of the block if it is released under water?
   (c) If the block floats to the surface of the lake, how much of it will stick out of the water?

2 An ice cube 3.0 cm on a side is placed in a glass of very cold water. The water is just level with the very brim of the glass.
   (a) How much of the ice cube will stick out of the water?
   (b) What will happen to the water in the glass when the ice melts? Will it overflow?

3 To determine if a crown were made of pure gold or some mixture of gold and silver, Archimedes weighed the crown and found it had a mass of 2 kg. The density of gold is 19.3 gm/cm³, the density of silver, 10.5 gm/cm³.
   (a) How much should the crown weigh in water if pure gold?
   (b) How much should the crown weigh in water if 50% gold and 50% silver by weight?

4 A hydraulic press has a cylinder 6.5 cm in diameter and a 36 cm stroke. The cylinder is driven from a hydraulic pump that operates at a maximum pressure of 7×10^7 N/m². What is the maximum force that can be exerted by the cylinder in N? in lbf?

5 An upright cylindrical tank is 10 m high and 3 m in diameter. What is the pressure in N/m² at the bottom of the tank if it is filled with
   (a) water, $\rho = 1000$ kg/m³?
   (b) gasoline, $\rho = 740$ kg/m³?
   (c) ethyl alcohol, $\rho = 790$ kg/m³?

6 A water tower for a small municipal water supply is used to maintain the water pressure by hydrostatic pressure. If the pressure in the water main 4.0 m below ground level is to be 10 atmospheres, how high does the water tower need to be?

7 An atomic submarine is cruising at a depth of 95 m off the coast of Florida. The density of sea water is $\rho = 1030$ kg/m³. The captain wishes to pump the sea water out of the ballast tanks by injecting compressed air.
   (a) What pressure must be applied to force the water out?
   (b) What is the total force exerted by the ocean on a hatch cover 0.8 m in diameter?

8 The approximate venous pressure in the hand can be observed while lying down. If the hand is at your side its veins will stand out. If you raise your hand by about 25 cm the veins will collapse, indicating a pressure difference equivalent to 25 cm of blood. What is this pressure difference in N/m², in torr? The density of blood is 1050 kg/m³.

9 The vertical distance between the brain and heart in humans is about 0.34 m. The mean arterial pressure at the brain is about 65 torr. The density of blood is 1050 kg/m³.
   (a) What is the hydrostatic pressure difference between the brain and the heart?
   (b) What is the mean arterial pressure at the heart in N/m², in torr?

10 The vertical distance between the brain and heart in a giraffe is about 4.0 m. The mean arterial pressure at the brain is about 90 torr. The density of blood is 1050 kg/m³.
   (a) What is the hydrostatic pressure difference?
   (b) What is the mean arterial pressure at the heart in N/m², in torr?

FLUIDS IN MOTION

11 A bathtub is being filled with hot water at a rate of 20 liter/min. The water flows through a standard "one-inch" pipe with an internal diameter of about 2.5 cm.
(a) What is the speed at which water flows through the pipe?
(b) If the service from the water main is through a pipe with an internal diameter of 5.0 cm, and filling the bathtub is the only present use of water, what is the flow velocity through the service pipe?

12 The area of the opening from the heart to the aorta has a diameter of about 1.0 cm. If the blood flow volume is 60 cm³/s, what is the average blood flow velocity at this point?

13 A hypodermic syringe for a 1 cm³ insulin injection has a chamber 5.7 cm long and 0.47 cm in diameter. The syringe is fitted with a No. 25 needle with a length of 1.6 cm and a lumen (internal diameter) of 0.024 cm. The contents of the syringe are injected in 2.0 s.
(a) What is the average velocity of the fluid flow in the syringe?
(b) What is the average velocity of the fluid flow in the hypodermic needle?

14 A hydraulic pump delivers 6×10^{-3} m³/s (6 liters/s) of hydraulic fluid at a pressure of 7×10^6 N/m² to the stick hydraulic cylinder shown on the excavator in Figure 9-3. The cylinder has a diameter of 0.20 m and a stroke of 1.8 m.
(a) How fast does the cylinder move?
(b) How long does it take to go the stroke distance?
(c) What power is required to move the cylinder at this speed?

15 The average flow velocity for smooth flow in a 2.5 cm diameter pipe is 7.0 cm/s. The viscosity of water is 1.0 cp.
(a) What is the volume flow in m³/s, in liter/min?
(b) What pressure drop over a distance of 30 m is required to maintain this flow in N/m², in torr?

16 Fluid with the viscosity of water flows through a No. 25 hypodermic needle with a lumen of 0.024 cm and a length of 1.6 cm. The average flow velocity is 40 m/s.
(a) What is the volume flow in m³/s, in liter/min?
(b) What pressure drop is required to maintain this flow in N/m², in torr?

17 A golf ball with a mass of 150 gm and a diameter of 4.4 cm is dropped into a freshwater pond. Its sinking is retarded only by the water's viscous friction and buoyant force. The viscosity of water is 1.0 cp. What is the ball's maximum speed of fall?

18 A column of cigarette smoke 0.5 cm in diameter rises vertically upward. Assume the viscosity of smoke is 170 μp. If the diameter of the column expands to 1.5 cm as it rises, what is the maximum flow velocity at which the flow remains smooth? Hint: Use the Reynolds number.

19 The left ventricle of the heart pumps blood into the aorta. Assume that this is analogous to a simple piston and chamber. On each beat of the heart a volume of 70 cm³ of blood is pumped into the aorta at an average pressure of 105 torr.
(a) Calculate the amount of work done by the heart on the fluid in a single contraction.
(b) If the heart rate is 70 beats/min, what is the mechanical power applied by the heart? (Note this neglects the kinetic energy of the blood.)

20 Under strenuous activity, the left ventricle of the heart pumps a volume of 120 cm³ of blood at an average pressure of 145 torr with each beat.
(a) Calculate the amount of work done by the heart on the fluid in a single contraction.
(b) If the heart rate is 185 beats/min, what is the mechanical power applied by the heart? (Note this neglects the kinetic energy of the blood.)

21 During strenuous exercise the lung pressure difference can be as large as 40 torr and the volume of air exchanged as much as 3.0 liter with each breath.
(a) How much mechanical work is done with one such large breath?
(b) How much power is required to breathe 40 breaths/min?

22 The water pressure at a faucet is 2.7 atmospheres above atmospheric pressure. The faucet has a diameter of 1.3 cm. Water has a density of 1.00 gm/cm^3.
   (a) From Bernoulli's equation, determine the maximum speed at which the water can be ejected from the faucet at this pressure.
   (b) What is the volume flow rate at this maximum speed?

SURFACES

23 A bicycle tire with a tube radius of 2.25 cm is inflated to a gauge pressure of 3×10^5 N/m^2.
   (a) What is the wall tension along the small radius of the tube?
   (b) If the tube has a major radius of 35 cm, what is the wall tension in the long direction?

24 A basketball has a radius of 12 cm and is inflated to a gauge pressure of 8.0×10^4 N/m^2. What is the wall tension?

25 The pressure difference between the inside and outside of a spherical balloon of diameter 20 cm is 2.4 torr.
   (a) What is the pressure difference in N/m^2?
   (b) What surface tension is required to hold this pressure difference?

26 A garden hose with an inner radius of 0.25 in withstands a pressure of 30 lbf/in^2.
   (a) What is the wall tension?
   (b) What wall tension would be required if the hose had a radius of 1.0 in?

ADDITIONAL PROBLEMS

27 A jet of water issues from a fountain nozzle 5.0 cm in diameter at a speed of 10 m/s.
   (a) What is the volume flow rate in liter/min?
   (b) How high does the water jet go?
   (c) What power is required?

28 An automobile engine draws about 70 liters of air per second. The venturi has a diameter of 2.5 cm. Assuming that the density of air is constant at 1.2 kg/m^3,
   (a) What is the average air flow velocity through the venturi? (For comparison the speed of sound is about 340 m/s.)
   (b) What is the pressure drop at the venturi in N/m^2, in torr?

29 In the hematocrit determination of the volume of closely packed red blood cells, a sample of whole blood with a suitable anticoagulate is centrifuged at about $2260\,g$ for about 30 min. A red blood cell has a radius of about 5×10^{-6} m and a density of 1098 kg/m^3. The blood serum has a density of 1030 kg/m^3 and a viscosity of 4.7 cp. What is the maximum separation velocity of the red blood cells if,
   (a) the blood experiences normal gravity of one g?
   (b) the blood experiences $2260\,g$?
   (c) How long would it take the red blood cell to fall a distance of 1.0 cm in each case?

30 A soda straw is 20 cm long with an inside diameter of 4.0 mm. A person sips 0.5 liter of cola in 25 s. Assume the viscosity and density of the cola are the same as water. What pressure difference is required to sip the soda in torr?

31 When water leaves a faucet it is accelerated as it falls under gravity. As the speed increases, the cross-sectional area required to maintain a given volume flow rate decreases. Assume that water leaves a faucet with a circular opening 1.0 cm in diameter at a speed of 5 cm/s, and flows smoothly.
   (a) What is the volume flow rate?
   (b) What is the cross-sectional area of the stream at a distance of 2 cm, 4 cm, and 10 cm from the orifice?

III

THERMAL PHYSICS

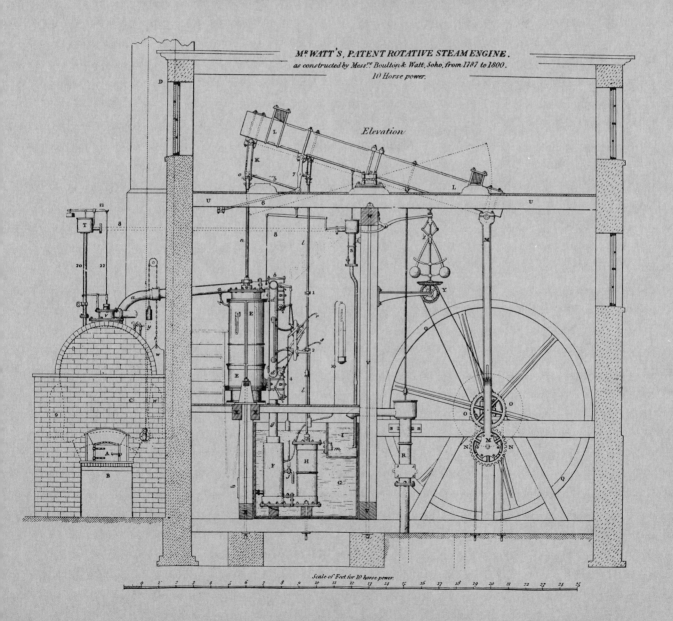

Mʳ WATT'S, PATENT ROTATIVE STEAM ENGINE.
as constructed by Messʳˢ. Boulton & Watt, Soho, from 1787 to 1800.
10 Horse power.

10

Thermal Energy

TEMPERATURE

We have physiological sensations of heat and cold, but to make quantitative statements of an object's temperature we need means of measurement. A **thermometric property** of a material—any property that changes when the material is heated or cooled—provides us with a measure of temperature, in effect a **thermometer.** Our common scales of temperature, the **Celsius** (°C) and **Fahrenheit** (°F), divide the interval between the melting point of ice (0°C, 32°F) and the boiling point of water (100°C, 212°F) into equal units.

Thermal Expansion

When a substance is heated its molecules move further apart, increasing all linear dimensions, as Figure 10-1 shows. (Even a hole drilled in a metal plate expands when the metal is heated—contrary to the common-sense expectation that the hole should shrink.) Thermal expansion is one of the simplest thermometric properties. For many substances, the fractional change in length is proportional to the change in the temperature ΔT:

$$\frac{\Delta L}{L} = \alpha \Delta T$$

where α is the **coefficient of thermal expansion** and has the units of °C^{-1} or °F^{-1}. Table 10-1 shows the coefficient of thermal expansion for various materials.

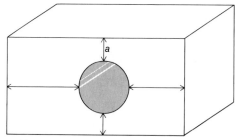

FIGURE 10-1
Thermal expansion. All linear dimensions of a heated solid expand uniformly.

The effects of thermal expansion must be accommodated in the design of concrete roadways, bridges, and other structures by the use of expansion joints. In winter, roadways and steel bridges contract, leaving spaces between sections. In summer, the structures expand.

EXAMPLE

In industry two cylindrical parts often must have a very tight fit. Suppose a brass sleeve of inner diameter of 1.995 cm at 20°C must be slipped over a steel shaft of diameter 2.000 cm. How hot must the brass sleeve be made so that it slips over the shaft? (Once in place, both the sleeve and the shaft reach the same temperature and the sleeve cannot be removed.)

$$\Delta T = \frac{1}{\alpha} \frac{\Delta L}{L}$$

$$\Delta T = \frac{(0.005 \text{ cm})}{(19 \times 10^{-6} \, ^{\circ}\text{C}^{-1})(2.000 \text{ cm})}$$

$$\Delta T = 132 \, ^{\circ}\text{C}$$

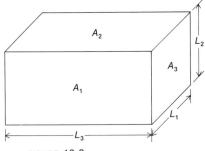

FIGURE 10-2
The volume of a rectangular solid.

TABLE 10-1
Coefficients of thermal expansion of common materials.

Material	Thermal expansion coefficient $10^{-6} \, ^{\circ}\text{C}^{-1}$
Aluminum	24
Brass	19
Copper	14
Steel 1.2% carbon	10.5
Brick	9.5
Concrete	10-14
Glass, pyrex	3.6
Glass, crown	9
Pine wood, across grain	34.1
Pine, parallel to grain	5.4

Volume Expansion

Since the linear dimensions of a solid expand when heated, the volume also expands. Imagine a rectangular solid of dimensions L_1, L_2, L_3. The respective faces of the solid have areas A_1, A_2, A_3 as shown in Figure 10-2. The volume of the solid is

$$V = L_1 L_2 L_3 = A_1 L_1 = A_2 L_2 = A_3 L_3$$

Each linear dimension of the solid expands:

$$\Delta V = A_1 \Delta L_1 + A_2 \Delta L_2 + A_3 \Delta L_3$$

$$\Delta V = A_1 L_1 \frac{\Delta L_1}{L_1} + A_2 L_2 \frac{\Delta L_2}{L_2} + A_3 L_3 \frac{\Delta L_3}{L_3} = 3V \frac{\Delta L}{L}$$

The expansion of each linear dimension is given by the coefficient of thermal expansion

$$\frac{\Delta L}{L} = \alpha \Delta T$$

The coefficient of volume expansion β is about three times the coefficient of linear expansion

$$\frac{\Delta V}{V} = 3\alpha \Delta T = \beta \Delta T$$

EXAMPLE
The volumetric expansion of mercury is used in the clinical thermometer. The coefficient of expansion of mercury is $\beta_{\text{Hg}} = 182 \times 10^{-6} \, ^{\circ}\text{C}^{-1}$ and that of glass is $30 \times 10^{-6} \, ^{\circ}\text{C}^{-1}$. A volume of mercury, say 0.5 cm³, is enclosed in a glass bulb ending in a small capillary tube of a radius 0.01 cm. When the mercury expands it rises in the capillary tube. However, the glass also expands. How much does the mercury rise for a 1°C change in temperature?
The change in the volume of mercury is

$$\Delta V_{\text{Hg}} = V \beta_{\text{Hg}} \Delta T = (0.5 \text{ cm}^3)(182 \times 10^{-6} \, ^{\circ}\text{C}^{-1})(1 \, ^{\circ}\text{C})$$

$$\Delta V_{\text{Hg}} = 91 \times 10^{-6} \text{ cm}^3$$

The glass cavity also expands slightly:

$$\Delta V_{\text{glass}} = V \beta_{\text{glass}} \Delta T$$

$$\Delta V_{\text{glass}} = (0.5 \text{ cm}^3)(30 \times 10^{-6} \, ^{\circ}\text{C})(1 \, ^{\circ}\text{C})$$

$$\Delta V_{\text{glass}} = 15 \times 10^{-6} \text{ cm}^3$$

The net change in volume of the liquid

$$\Delta V_{\text{net}} = \Delta V_{\text{Hg}} - \Delta V_{\text{glass}}$$

rises in the capillary tube a distance Δh:

$$\Delta V_{\text{net}} = \pi R^2 \Delta h$$

The rise of the liquid for a 1°C change in the temperature is

$$\Delta h = \frac{\Delta V_{\text{net}}}{\pi R^2}$$

$$\Delta h = \frac{76 \times 10^{-6} \text{ cm}^3}{(3.14)(0.01 \text{ cm})^2} = 0.24 \text{ cm}$$

Celsius and Fahrenheit

The mercury thermometer is calibrated at two reference points: At a pressure of one atmosphere, the boiling point of water is 100°C or 212°F, the melting point of ice is 0°C or 32°F. The expansion of mercury is proportional to temperature. If the lumen (the cavity of the tube) of the mercury thermometer is uniform, the temperature at an intermediate point should simply be proportional to the length of the column of mercury.

Figure 10-3 shows two identical mercury thermome-

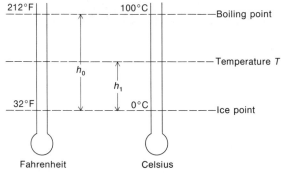

Fahrenheit   Celsius

FIGURE 10-3
A mercury thermometer can be calibrated in the Fahrenheit or Celsius temperature scale.

ters. Since the Fahrenheit and Celsius scales depend on the length of the column of mercury, we can establish the relationship between the two scales:

$$\frac{T_{\text{F}} - 32}{212 - 32} = \frac{h_1}{h_{\text{o}}} = \frac{T_{\text{C}} - 0}{100 - 0}$$

where T_{F} is the temperature in °F and T_{C} is the temperature in °C. Solving for each temperature we obtain

$$T_{\text{C}} = \frac{100°\text{C}}{180°\text{F}} (T_{\text{F}} - 32°\text{F}) = \frac{5}{9} (T_{\text{F}} - 32°\text{F})$$

or

$$T_{\text{F}} = \frac{9}{5} (T_{\text{C}}) + 32°\text{F} \qquad (10\text{-}1)$$

The temperature interval in Celsius is larger than the temperature interval in Fahrenheit.

$$5°\text{C} = 9°\text{F}$$

The Behavior of Gases

The behavior of gases has been studied since the seventeenth century, when Robert Boyle, assisted by Robert Hooke, experimented with vacuums and air pressure. In 1662 Boyle stated what is known as **Boyle's law of gases:** *If temperature remains constant, the volume of a gas varies inversely with the pressure:*

$$P_1 V_1 = P_2 V_2$$

Over a hundred years later, in 1787, Jacques Charles communicated to Joseph Gay-Lussac his observation that the volume of all gases expands equally with the same increase in temperature.

Charles did many experiments aloft in a hydrogen-filled balloon (reporting after one ascent to a height of 2.7 km that the air at that altitude is thin and cold). Further investigations, in 1802, demonstrated what is now called variously the Gay-Lussac law or the Charles law:

The volume of many gases increases linearly with temperature:

$$V = A(T - T_o)$$

where the slope A depends on the quantity of gas present (see Figure 10-4).

A similar relationship (also shown in Figure 10-4) can be found between pressure and temperature at constant volume:

$$P = B(T - T_o)$$

If we extrapolate the volume occupied by a gas at constant pressure to zero volume, or if we extrapolate the pressure of a gas at constant volume to zero pressure, we find that there is a lowest temperature, an **absolute zero temperature:**

$$T_o = -273.15°C = -459.67°F$$

Any substance with a thermometric property can be used to establish a temperature scale. However, because scales using different substances will not agree exactly with one another, it is desirable to establish a temperature scale that is independent of the particular substance used.

Such a thermometer can be constructed from a constant volume of a gas by measuring the pressure as a function of temperature. Different gases will not agree exactly. For example, a thermometer using carbon dioxide does not agree with one using air or helium. However, as a gas is made dilute by decrease in its density and pressure, its molecules become spread far enough apart to be virtually unaffected by one another: They are in the condition of an **ideal gas.** When sufficiently diluted, all gases approach the same linear relationship between pressure or volume and temperature; they act as an ideal gas and satisfy Charles Law exactly. (For example, helium has only weak forces between atoms, and acts in effect as an ideal gas.) The properties of such an ideal gas can be used to establish a **thermodynamic temperature scale** that is independent of the particular thermometric substances used in the thermometer, and that varies linearly with the temperature from absolute zero.

The **Kelvin temperature scale** is such a scale. Two temperature points, absolute zero and the triple point of water, are needed to determine a straight line. A **triple point cell,** shown in Figure 10-5, is a sealed chamber in which water and water vapor are in equilibrium. If this cell is cooled, ice begins forming. At the triple point the three phases (solid ice, liquid water, and water vapor) coexist at a single temperature of 0.01°C or 273.16 kelvin (abbreviated K, not °K, and expressed simply as "Kelvin," not "degrees Kelvin"). The Kelvin temperature scale is related to the Celsius scale by

$$T_K = T_{°C} + 273.15$$

where absolute zero corresponds to $-273.15°C$. A temperature interval of one Kelvin is the same as one degree Celsius.

The engineering thermodynamic temperature scale is the Rankine scale (°R) and is related to the Fahrenheit scale by

$$T_{°R} = T_{°F} + 459.67$$

where absolute zero corresponds to $-459.67°F$. A temperature interval of one degree Rankine is equal to one degree Fahrenheit.

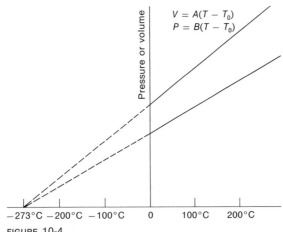

FIGURE 10-4
Volume and pressure of a gas are linear functions of temperature.

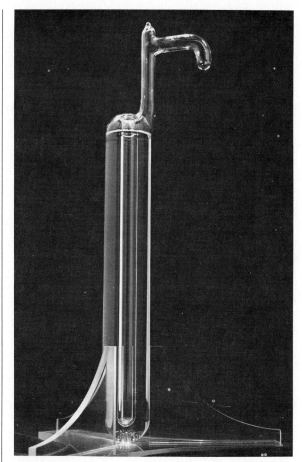

FIGURE 10-5
The triple point cell is a cylindrical pyrex container surrounded by thermal insulation nearly filled with highest purity water and permanently sealed. Partial freezing of the sealed water with subsequent melting establishes the triple point temperature of 0.0100°C, or 273.1600 K. Different triple point cells agree within 0.0002 K. Repeated measurements with a single cell agree to within 0.0001 K. Courtesy of National Bureau of Standards.

The Practical Temperature Scale

Although the ideal-gas thermometer can be used to define the thermodynamic temperature scale, it is not a particularly practical method. In Paris in 1968 the International Committee of Weights and Measures adopted the International Practical Temperature Scale: a set of 11 reproducible equilibrium temperature references (shown in Table 10-2) that correspond as accurately as possible with

TABLE 10-2

The International Practical Temperature scale: a set of 11 reproducible temperature standards. The boiling point measurements are at a pressure of one atmosphere.

Assigned temperature	Equilibrium state
1337.58 K	Freezing point of gold
1235.08 K	Freezing point of silver
692.73 K	Freezing point of zinc
373.15 K	Boiling point of water
273.16 K	Triple point of water
90.188 K	Boiling point of oxygen
54.361 K	Triple point of oxygen
27.102 K	Boiling point of neon
20.28 K	Boiling point of hydrogen
17.042 K	Boiling point of hydrogen at 33 330.6 N/m² pressure
13.81 K	Triple point of hydrogen

the Kelvin scale and thus serve as calibration points for working thermometers.

Because mercury freezes at $-38.87°C$ and boils at $356.58°C$, a mercury thermometer is useful only over a limited range of temperatures, so other thermometric properties must be used. The electrical resistance of a pure platinum wire changes with temperature in a very reproducible manner. The **platinum resistance thermometer** is a coiled length of pure platinum wire with a diameter of between 0.05 mm and 0.5 mm and is of sufficient length to have a resistance of about 25 ohms at 0°C (units and methods of measuring electrical resistance are discussed in Chapter 13). The resistance changes by about 0.3% for a change of 1 K. The platinum resistance thermometer is the standard instrument used for measuring temperatures between 13.81 K and 903.89 K, and can be calibrated to within about ± 0.0003 K at the triple point of water (273.1600 K).

For temperatures between 903.89 K and 1337.58 K, the respective freezing points of antimony and gold, the standard instrument of temperature measurement is the **thermocouple:** a junction between two dissimilar metals, such as pure platinum and a platinum alloy with 10% rhodium. At such a junction there is a small electrical potential difference, or voltage, the magnitude of which is directly related to temperature. Figure 10-6 shows a typical thermocouple. An ice-water bath 0°C is often used as the reference junction.

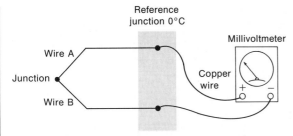

FIGURE 10-6
The thermocouple, a junction of two dissimilar metals, generates a voltage that depends on the temperature.

Different thermocouples have different ranges of use. A copper-constantan (a nickel-copper alloy) junction is accurate to $\pm0.5°C$ in the range from $-185°C$ to $400°C$. The platinum-10% rhodium thermocouple is useful from $0°C$ to about $1750°C$. Figure 10-7 shows typical millivolt outputs of various thermocouples with the reference junction at $0°C$.

The Thermistor

A **thermistor** is a solid-state device the electrical resistance of which changes rapidly with temperature—resistance typically being some 10 000 ohms at $0°C$, dropping

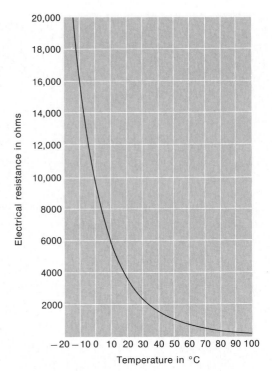

FIGURE 10-8
The resistance of a thermistor changes rapidly with temperature.

to 3000 ohms at $25°C$ and 200 ohms at $100°C$ (see Figure 10-8). Thermistors are used as the temperature-sensing element in many electronic circuits operating in the $-75°C$ to $100°C$ range. Digital clinical thermometers (see Figure 10-9), now widely used in hospitals, house a thermistor at the tip of the probe, covered by a disposable, sterile plastic and aluminum sheath. The electrical resistance is converted to the corresponding temperature, in either $°F$ or $°C$, within about 15 s, signaling audibly when the sensor reaches the patient's temperature. The temperature is accurate to about $\pm0.2°F$.

HEAT

Thermal Equilibrium

Two objects placed next to one another exchange energy: The hotter object cools, the cooler heats up, until by thermal interaction they reach **thermal equilibrium,** the

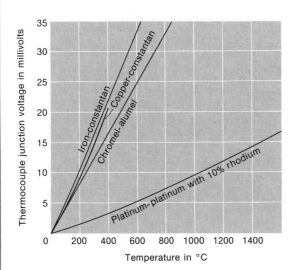

FIGURE 10-7
The thermocouple output in millivolts as a function of temperature depends on the type of junction.

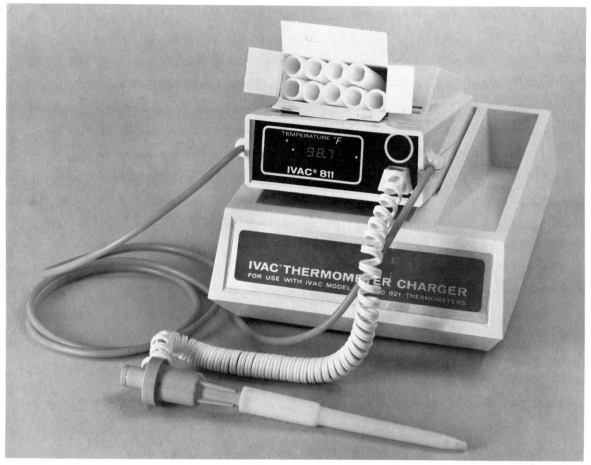

FIGURE 10-9
The clinical digital thermometer uses a thermistor to sense the
temperature. Courtesy of IVAC Corporation.

point at which neither gains or loses net energy. (If object
A, say a thermometer reading 20°C, is in thermal equilibrium with object B, they are at the same temperature. If
object C, also at 20°C, is placed next to object B it is also
in thermal equilibrium.) The net energy transferred from
one system to another as the result of thermal interaction
is known as **heat.**

The increase in temperature with the transfer of energy
into water is often used as a measure of heat. The **specific
heat** C_s of a substance is the amount of heat required to
raise 1 gm of it a temperature interval of 1°C. The heat
ΔQ required to increase the temperature of a mass m of a
substance by a temperature difference ΔT is

$$\Delta Q = mC_s\Delta T \qquad (10\text{-}2)$$

The specific heat is

$$C_s = \frac{1}{m}\frac{\Delta Q}{\Delta T}$$

One **calorie** (cal) of heat is required to raise one gram of
water from 14.5°C to 15.5°C. Thus, the specific heat of
water is $C_s = 1.00 \text{ cal/gm-°C}$. Most substances have a
specific heat somewhat smaller than that of water. Table
10-3 shows the specific heat of various substances as well

TABLE 10-3
Specific heat of various materials. Also shown is temperature change when 1 cal of heat is added to 1 gm of the substance.

Material	Specific heat C_s cal/gm-°C	Temperature rise ΔT °C
Water	1.00	1.0
Body tissue	0.83	1.2
Wood (pine)	0.67	1.5
Methyl Alcohol	0.60	1.7
Water vapor (37°C)	0.40	2.5
Air (37°C)	0.25	4.0
Glass	0.20	5.0
Steel	0.11	9.1
Copper	0.093	10.7
Lead	0.0306	32.7

as temperature change when 1 cal of heat is added to 1 gm of a material.

The engineering unit of heat is the **British thermal unit** (Btu): the amount of heat required to raise the temperature of a pound mass (lbm)[1] of water by 1°F. The specific heat of water in engineering units is

$$C_s = 1.00 \text{ Btu/lbm-}°F$$

EXAMPLE

What is the energy of one Btu expressed in calories?

$$\Delta Q = mC_s\Delta T$$

One lb of water has a mass of 454 gm. The specific heat of water is $C_s = 1.00$ cal/gm-°C. 9°F is equal to 5°C. The Btu is equal to

$$1 \text{ Btu} = (454 \text{ gm})(1.00 \text{ cal/gm-}°C)(5/9°C)$$

$$1 \text{ Btu} = 252 \text{ cal}$$

---

[1] A pound mass (lbm) is the amount of a substance that weighs one pound on earth. 1.0 lbm = 1/32.17 slug. We are not here interested in inertial properties but only in how much material is present.

In the eighteenth century there was much speculation about the phenomenon of heat. Some believed that one object heated another by physically transferring a fluid, caloric, to it. In about 1798 Count Rumford observed the generation of heat when cannons are bored. He forcefully stated that heating involves microscopic motions. Friction can generate an endless quantity of heat without any change in the nature of the material.

In 1840, James Prescott Joule, to establish the quantitative relationship between mechanical work and heat, began a long series of experiments. He used three independent methods: having a falling weight run an electric generator that heated a resistor, giving an observable amount of heat; compressing a gas by doing mechanical work; rotating a crank driving a set of paddle wheels immersed in a liquid and measuring the increase in temperature for the work expended. In 1849, Joule published a paper clearly stating that the quantity of heat produced by friction is always proportional to the energy expended, and that the energy required to heat 1 lbm of water 1°F (1 Btu) is 772 lbf-ft (the accepted value is now 778 lbf-ft). The **mechanical equivalent of heat** is the energy required to heat 1 gm of water 1°C and is equal to 4.184 joules.

$$1 \text{ cal} = 4.184 \text{ J}$$

or, in engineering units,

$$1 \text{ Btu} = 1054 \text{ J}$$

The First Law of Thermodynamics

The **first law of thermodynamics** restates the law of conservation of energy: *The heat ΔQ into an isolated system is equal to the sum of the change of internal energy of the system ΔU plus the work done by the system ΔW:*

$$\Delta Q = \Delta U + \Delta W \qquad (10\text{-}3)$$

The change in the system's internal energy is indicated by the change in temperature:

$$\Delta U = mC_s\Delta T$$

If mechanical work is done *on* the system, such as by friction, ΔW is negative. Even though no heat flows into the system, $\Delta Q = 0$, the internal energy of the system will increase,

$$\Delta U = -\Delta W$$

increasing the temperature of the system. The system can also convert internal energy to mechanical work without the flow of heat. Mechanical work is done at the expense of internal energy of the system

$$\Delta W = -\Delta U = -mC_s\Delta T$$

and the temperature of the system must fall. The amount of work that can be extracted from a thermal system is discussed further in Chapter 12. In any case, energy must be conserved.

EXAMPLE

A 52 gallon electric water heater contains about 200 kg of water. Its electric heating element puts out energy at the rate of 5 000 watts. The water is initially at a temperature of 20°C. What is the water temperature after one hour if all of the heat is absorbed by the water? The heat generated by the heating element is

$$\Delta Q = (5\,000 \text{ J/s})\frac{(1 \text{ cal})}{(4.18 \text{ J})}(3600 \text{ s})$$

$$\Delta Q = 4.3 \times 10^6 \text{ cal}$$

The first law states

$$\Delta Q = \Delta U = mC_s\Delta T$$

The change in temperature is

$$\Delta T = \frac{\Delta Q}{mC_s} = \frac{4.3 \times 10^6 \text{ cal}}{(2.0 \times 10^5 \text{ gm})(1.0 \text{ cal/gm-°C})}$$

$$\Delta T = 21.5 \text{ °C}$$

The final temperature is then

$$T_f = T_i + \Delta T$$

$$T_f = 20°C + 21.5°C$$

$$T_f = 41.5°C$$

How long will it take to reach an operating temperature of 65°C?

Calorimetry

The first law of thermodynamics and the concept of specific heat make possible **calorimetry:** measurement of the energy transfer between objects. A calorimeter is a thermally isolated system consisting of a metal can that contains two or more substances of known masses. Heat transfer can be determined by measuring temperature changes. It is thus possible to determine the specific heats of materials or the energy absorbed or liberated in chemical reactions or in combustion of fuels or foods.

A simple water calorimeter consists of a metal can of mass m_A and specific heat C_A at temperature T_A, filled with water of mass m_B and specific heat C_B at temperature T_B. The water and the can are in thermal equilibrium $T_A = T_B$. A third object, perhaps a copper mass m_C, is at some initial temperature T_C and some unknown specific heat C_C. The experiment consists of measuring the initial masses and temperatures, dropping the copper into the water, and measuring the final temperature T_F. The total energy change for the system is given by

$$\Delta Q = m_A C_A(T_A - T_F) + m_B C_B(T_B - T_F) + m_C C_C(T_C - T_F)$$

If no heat is lost or gained to the surrounding environment, then $\Delta Q = 0$. One can then solve for the unknown specific heat C_C of the copper.

With an oxygen bomb calorimeter it is possible to measure the energy of combustion of fuels or foods (see Figure 10-10). About one gram of a sample to be burned is placed in a strong chamber, the bomb, with a small fuse

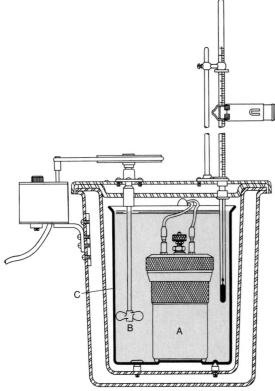

The oxygen bomb calorimeter consists of a stainless steel bomb (A) surrounded by water (B) in a bucket (C). A paddle stirs the water. Simple models have only a thermally insulating jacket surrounding the bucket. Courtesy of Parr Instrument Company.

wire. The chamber is sealed and filled with oxygen to a pressure of 30 to 55 atmospheres. The bomb is placed in the calorimeter bucket containing about 2000 gm of water. The fuse ignites the contents of the bomb. The energy released in combustion heats the metal of the bomb, the water in the bucket, and the bucket itself. To minimize heat loss to the surroundings, the bucket is insulated and surrounded by an airspace.

In sophisticated calorimeters, the bucket is surrounded by a water jacket that automatically remains at the same temperature as the water in the bucket. This effectively reduces to zero the heat loss to the surroundings. The heat liberated in reaction is related to the change in temperature:

$$\Delta Q = m_A C_A \Delta T + m_B C_B \Delta T + m_C C_C \Delta T$$

where m_A and C_A; m_B and C_B; and m_C and C_C are the mass and specific heat of the calorimeter bomb, the water, and the bucket respectively. In practice, the calorimeter is calibrated with a precise quantity of benzoic acid that has an accurately known energy of combustion.

True accuracy also requires small corrections for the energy liberated in burning the fuse wire and the heat of formation of the oxides of nitrogen and sulfur in the oxygen-rich atmosphere. A small measured amount of combustible material is sometimes added to the sample to achieve ignition. For some food samples, the material is partially dried so that ignition and complete combustion can be achieved.

EXAMPLE

The oxygen bomb has a mass of 2.9 kg and is made of steel with a specific heat of 0.11 cal/gm-°C. The steel bucket has a mass of about 0.7 kg and is filled with 2 kg of water. 1.0 gm of gasoline is burned in the calorimeter giving rise to a temperature difference of 4.6°C. What is the energy of combustion of 1 gm of gasoline?

$$\Delta Q = (m_A C_A + m_B C_B + m_C C_C)\Delta T$$

$$\Delta Q = \left[(2900 \text{ gm})\frac{(0.11 \text{ cal})}{\text{gm-°C}} + (2000 \text{ gm})\frac{(1.0 \text{ cal})}{\text{gm-°C}} \right.$$
$$\left. + (700 \text{ gm})\frac{0.11 \text{ cal}}{\text{gm-°C}} \right](4.6°C)$$

$$\Delta Q = (2396 \text{ cal/°C})(4.6°C) = 11\,000 \text{ cal}$$

Phase Transitions

Matter exists in three phases: solid, liquid, and gas. Adding heat to a liquid increases its internal energy, and thus its temperature until it begins to change phase into a vapor. This temperature at a pressure of one atmosphere is the **boiling point** of the liquid. It requires energy to remove molecules from the liquid. The amount of energy is the **latent heat of vaporization,** L_v, usually given in calories per gram. During a phase transition, the internal energy of the liquid changes without changing the temperature. If an amount of mass Δm is converted from a liquid to a vapor, the change in internal energy is given by

$$\Delta U = \Delta m L_{\text{v}} \qquad (10\text{-}4)$$

Converting a solid into a liquid, which occurs at a temperature, the **freezing point,** requires energy to remove molecules from the crystal structure of the solid. The change in internal energy as a solid becomes a liquid is given by

$$\Delta U = \Delta m L_{\text{f}} \qquad (10\text{-}5)$$

where L_{f} is the **latent heat of fusion.**

Suppose that to 1 gm of ice at a temperature of $-50°\text{C}$ and one atmosphere pressure, heat is added and the in-

crease in temperature is observed. The specific heat of ice is about

$$C_{\text{ice}} = 0.50 \text{ cal/gm-}°\text{C}$$

so that for every 0.50 calories of heat added the temperature increases by $1°\text{C}$. The ice increases in temperature until the melting point is reached, as shown in Figure 10-11. To melt the ice, energy must be added equal to the latent heat of fusion of the ice:

$$L_{\text{f}} = 80 \text{ cal/gm}$$

While the ice is melting, the temperature does not change. As heat is added, ice changes to water. A mixture of ice in equilibrium with water is a useful temperature reference for $0°\text{C}$.

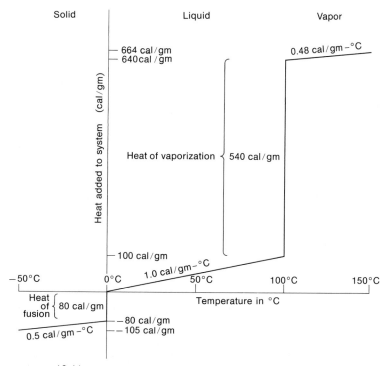

FIGURE 10-11
Heat added to the system as a function of temperature for water at a pressure of one atmosphere. The heat of fusion and the heat of vaporization are shown.

Once the ice is melted, the temperature begins to increase. The specific heat of water is

$$C_{water} = 1.0 \text{ cal/gm-}°C$$

The temperature increases by 1°C for each calorie of heat added until the boiling point is reached at 100°C. At the boiling point any additional heat converts water to water vapor without increasing the temperature. The latent heat of vaporization of water at 100°C is approximately

$$L_v = 540 \text{ cal/gm}$$

Once the water has been converted to steam the addition of about 0.48 calories of heat increases the temperature by 1°C. Steam at one atmosphere pressure has a specific heat

$$C_{steam} = 0.48 \text{ cal/gm-}°C$$

EXAMPLE

A stainless steel bowl with a mass $m_A = 100$ gm and a specific heat of 0.11 cal/gm $-$ °C contains a mass $m_B = 400$ gm of water. The bowl and the water at $T_1 = 5°C$ are placed in an insulated box to minimize heat transfer to the surroundings. A mass $m_C = 50$ gm of ice at $T_2 = -20°C$ is then placed in the bowl and allowed to come to equilibrium. What is the final temperature of the ice water?

If all the ice melts, its change in internal energy is the energy required to heat the ice to 0°C, the energy required to melt the ice, and the energy to heat the melted ice to the final temperature T_f:

$$\Delta U_{ice} = m_C C_{ice}(T_2 - 0°C) + m_C L_f + m C(0°C - T_f)$$

The change in internal energy of the water is

$$\Delta U_{water} = m_B C_B(T_1 - T_f)$$

The change in internal energy of the bowl is

$$\Delta U_{bowl} = m_A C_A(T_1 - T_f)$$

The first law of thermodynamics requires that energy be conserved for our system. No mechanical work is done on or by the system: $\Delta W = 0$. Because our system is isolated from the surroundings there should be no heat transfer $\Delta Q = 0$. Instead there is a transfer of internal energy from the bowl and water to the ice. Our condition for equilibrium is given by

$$\Delta Q = 0 = \Delta U_{ice} + \Delta U_{water} + \Delta U_{bowl}$$

This gives us one equation. There is only one unknown, the final temperature:

$$0 = m_C C_C(0°C - T_2) + m_C L_f + m_C C_C(T_f - 0°C) + (m_A C_A + m_B C_B)(T_f - T_1)$$

$$0 = (50 \text{ gm})\left(0.5 \frac{\text{cal}}{\text{gm-}°C}\right)(0°C - (-20°C))$$

$$+ (50 \text{ gm})(80 \text{ cal/gm}) +$$

$$(50 \text{ gm})\left(1.0\frac{\text{cal}}{\text{gm-}°C}\right)(T_f - 0°C)$$

$$+ \left[(100 \text{ gm})\left(0.11\frac{\text{cal}}{\text{gm-}°C}\right) + (400 \text{ gm})\left(1.0\frac{\text{cal}}{\text{gm-}°C}\right)\right]$$
$$(T_f - 15°C)$$

$$0 = +500 \text{ cal} + 4000 \text{ cal}$$
$$+ (50 \text{ cal/}°C)(T_f - 0°C)$$
$$+ (411 \text{ cal/}°C)(T_f - 15°C)$$

$$T_f = 3.6°C$$

KINETIC THEORY OF GASES

The Atomic Hypothesis

From our vantage point late in the twentieth century, the concept of the atom is accepted as a simple truth. The quantum discoveries earlier in the century, which explained the functioning of the atom in minute detail, allowed a new understanding of solids, which led in turn to the age of the transistor and the computer.

Yet the idea of the atom is ancient. The Greek philosopher Democritus (born about 460 B.C.) proposed that

1. Atoms are infinite in number and absolutely identical in all respects except shape and size.

2. Atoms are ceaselessly in motion. This motion is, like the atoms themselves, eternal.

Much of the early work in modern chemistry was done with gases. For example, in the eighteenth century Joseph Priestly and Antoine Lavoisier first isolated and studied the properties of oxygen.

The modern notion of the atom was first stated by John Dalton, in 1803: ". . . the ultimate particles of all homogeneous bodies are perfectly alike in weight, figure, etc., . . . every particle of hydrogen is like every other particle of hydrogen, etc."[2]

In 1809 the French chemist Joseph Louis Gay-Lussac published a paper on the combining of volumes of gases. When two volumes of gas react to form a third volume, the initial and final volumes are in simple ratios. Two years later the Italian physicist Amedeo Avogadro published an essay in which he stated that

> . . . the quantitative proportions of substances in compounds seem only to depend on the relative number of molecules which combine, and on the number of composite molecules which result. It must then be admitted that very simple relations also exist between the volumes of gaseous substances and the numbers of simple or compound molecules which form them. The first hypothesis to present itself in this connection, and apparently even the only admissible one, is the supposition that *the number of integral molecules in all gases is always the same for equal volumes.*[3]

Avogadro's hypothesis was largely ignored until about 1860, when the Italian chemist Stanislao Cannizzaro saw

---

[2] Quoted in Andrew G. Van Melsen, *From Atomos to Atom* (Pittsburgh, Pa.: Duquesne University Press, 1952), p. 136.

[3] Emphasis added. Quoted in Peter Wolff, *Breakthroughs in Chemistry* (New York: The New American Library, 1967), p. 126. Chapter 5 of Wolff's book contains a section of Avogadro's essay followed by a discussion of the atomic hypothesis.

how it could be used to distinguish between the atomic weight and the molecular weight of important gaseous elements. The **molecular weight** of a substance is the weight of the molecule compared to the atomic weight of the isotope carbon-12 equal to 12 000 000.

The gram molecular weight, the **mole** (mol), is the amount of a chemical substance equal to one gram multiplied by the molecular weight of the molecule. One mole of carbon-12 atoms has a mass of 12.000 000 gm. One mole of hydrogen H_2 molecules contains the same number of molecules as one mole of oxygen O_2, which contains the same number of molecules as one mole of water H_2O. **Avogadro's number** gives the number of molecules in a gram molecular weight or mole. The current accepted value for Avogadro's number is

$$N_A = (6.022\ 045 \pm 0.000\ 031) \times 10^{23}/\text{mol}$$

To three significant figures Avogadro's number is

$$N_A = 6.02 \times 10^{23}/\text{mol}$$

EXAMPLE

At one atmosphere pressure and 0°C, nitrogen has a density of 0.001 2506 gm/cm³. Liquid nitrogen at a temperature of −195.8°C (77 K) has a density of 0.8081 gm/cm³. Nitrogen has a molecular weight of 28.0134 gm/mol.
(a) What is the volume occupied by one mole?
(b) What is the average distance between molecules?
The volume occupied by one mole of nitrogen gas is

$$V = \frac{28.0134\ \text{gm/mol}}{0.001\ 2506\ \text{gm/cm}^3} = 22\ 400\ \text{cm}^3/\text{mol}$$

$$V = 22.4 \times 10^{-3}\ \text{m}^3/\text{mol}$$

The volume occupied by one mole of nitrogen gas in these conditions is 22.4 liters. By Avogadro's law this should be the same volume for one mole of most gases.

The mean spacing for nitrogen gas, or any gas, at 0°C and one atmosphere pressure can be found by first determining the volume v occupied by one molecule.

The volume occupied by one molecule is given by the molar volume divided by the Avogadro's number.

$$v = \frac{V}{N_A} = \frac{22.4 \times 10^{-3} \text{ m}^3/\text{mol}}{6.02 \times 10^{23}/\text{mol}} = 3.72 \times 10^{-26} \text{ m}^3$$

If we consider this volume to be a cube of dimension d, its volume is $v = d^3$. The mean distance between molecules is then

$$d = (v)^{1/3} = (3.72 \times 10^{-26} \text{ m}^3)^{1/3}$$

$$d = 3.34 \times 10^{-9} \text{ m} = 3.3 \text{ nm}$$

The gas molecules are separated typically by 3.3 nm.

In the liquid state of nitrogen the molecules are closely packed and the volume is greatly reduced. The molar volume of liquid nitrogen at $-195.8°C$ (77 K) is

$$V = \frac{28.0134 \text{ gm/mol}}{0.8081 \text{ gm/cm}^3} = 34.67 \text{ cm}^3/\text{mol}$$

$$V = 34.67 \times 10^{-6} \text{ m}^3/\text{mol}$$

By a calculation similar to that for the gas, the volume v occupied by one liquid molecule is

$$v = 5.76 \times 10^{-29} \text{ m}^3$$

and the mean distance between liquid molecules is

$$d = 3.86 \times 10^{-10} \text{ m} = 0.39 \text{ nm}$$

The mean distance between molecules of the liquid is 0.39 nm (compared to the mean distance between gas molecules of 3.3 nm). Most of the volume occupied by the gas is empty space.

An Ideal Gas

If the density of a gas is sufficiently low, its molecules are relatively far apart and it behaves as an ideal gas: Its pressure or its volume depends linearly on the absolute temperature. Boyle's law states that at constant tempera-

ture and quantity of gas the volume is inversely proportional to the pressure:

$$V \propto \frac{1}{P}$$

The Charles or Gay-Lussac law states that at constant pressure the volume of a gas is proportional to the absolute temperature in Kelvin:

$$V \propto T$$

Avogadro's law states that the same number of moles of a gas occupy the same volume of space. The volume is proportional to the number of moles, n:

$$V \propto n$$

The volume of a gas is then proportional to the number of moles n and the absolute temperature T, and is inversely proportional to the pressure P

$$V \propto \frac{nT}{P}$$

We can use a constant R, the **ideal gas constant,** to express this proportionality:

$$V = \frac{nRT}{P}$$

We can write this in the more familiar form of the **ideal gas law:**

$$PV = nRT \qquad (10\text{-}6)$$

The current accepted value for the ideal gas constant is

$$R = 8.134\,41 \text{ joules/mol-K}$$

$$R = 1.987 \text{ cal/mol-K}$$

$$R = 0.082 \text{ liter-atmosphere/mol-K}$$

All dilute gases behave as an ideal gas. Many gases behave as an ideal gas at room temperature, if room temperature is not too close to the temperature (and pressure) at which the gas condenses to a liquid.

Pressure of a Gas

A gas consists of a great number of widely separated particles in constant motion, usually moving freely through space in agreement with Newton's laws of motion. A nitrogen molecule at 0°C has a mean speed of about 500 m/s and travels a distance of about 100 nm before colliding elastically with another nitrogen molecule. Because of the great speed and small distances, a nitrogen molecule experiences some 10^9 collisions a second. At one instant a molecule may be at rest, at another it is struck and recoils with large velocity. The velocity of a given molecule is continually changing in both magnitude and direction.

By considering the motion of a collection of molecules with different velocities, we can develop a **kinetic theory** to determine the pressure of an ideal gas. Let us consider a gas consisting of N molecules at a temperature T, occupying a volume V. The molecules are traveling with a mean speed v in random directions. Molecules near the wall and headed toward it will strike the wall. We can calculate the average force acting on a small section of wall with area A.

From our discussion of Newton's laws of motion we know that a force can be represented as the time rate of change of momentum:

$$\mathbf{F} = \frac{\Delta \mathbf{p}}{\Delta t}$$

Suppose that a molecule is traveling toward the wall with velocity v_x as shown in Figure 10-12. The molecule with a mass m will have a momentum in the x direction of

$$p_x = mv_x$$

If this molecule bounces elastically off the wall it will possess a final momentum in the x direction of

$$p_x{}' = -mv_x$$

The change in the particle's momentum in the collision is

$$\Delta p_x = p_x - p_x{}'$$

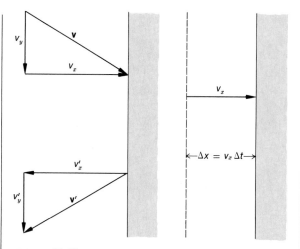

FIGURE 10-12
The pressure of an ideal gas can be calculated from a simple model.

$$\Delta p_x = mv_x - (-mv_x)$$

$$\Delta p_x = 2mv_x$$

Each molecule striking the wall transfers to it a momentum of $2mv_x$. The wall, being massive and held in place by the forces among its atoms, does not move. The force acting on the wall is given by the momentum transfer per unit time, which depends on the number of molecules ΔN striking an area of the wall A in a time Δt, multiplied by the momentum transfer of each molecule Δp_x. The force on the wall will be

$$F_x = \frac{\Delta N}{\Delta t} \Delta p_x$$

In a time Δt those molecules within a distance

$$\Delta x = v_x \Delta t$$

will strike the wall. The molecules within this distance are within a volume $A\Delta x$. However, only about one-third of those molecules will be headed in the $+x$ or $-x$ directions, and of that third only one-half will be headed in the $+x$ direction, toward the wall. The number that will strike the wall in time Δt is just one-sixth the density $\rho = N/V$ times the volume swept out:

$$\Delta N = \frac{1}{6}\frac{N}{V}Av_x\,\Delta t$$

The total force exerted on the wall is then

$$F = \frac{\Delta N}{\Delta t}\Delta p_x$$

$$F = \left(\frac{1}{6}\frac{N}{V}Av_x\right)(2mv_x)$$

The pressure is just the force per unit area

$$P = \frac{F}{A} = \left(\frac{1}{3}mv_x{}^2\right)\left(\frac{N}{V}\right)$$

The mean kinetic energy of a molecule is just

$$KE = \tfrac{1}{2}mv_x{}^2$$

The total number of particles of the gas is the number of moles n times Avogadro's number N_A

$$N = nN_A$$

We can then write the theoretical pressure of an ideal gas as

$$PV = n\frac{2}{3}N_A(KE)$$

This is similar in form to the experimentally determined ideal gas law

$$PV = nRT$$

By equating the two we can solve for the mean kinetic energy of a molecule

$$KE = \tfrac{1}{2}mv_x{}^2 = \frac{3}{2}\frac{RT}{N_A} = \frac{3}{2}kT$$

The constant k is the Boltzmann constant:

$$k = \frac{R}{N_A} = 1.38 \times 10^{-23}\ \text{J/K}$$

The velocity of a molecule with the mean kinetic energy (the root mean square velocity) is given by

$$v_{\text{rms}} = \left[\frac{3kT}{m}\right]^{1/2} = \left[\frac{3RT}{N_A m}\right]^{1/2} = \left[\frac{3RT}{M_A}\right]^{1/2} \quad (10\text{-}7)$$

where M_A is the molecular weight $M_A = N_A m$. The pressure of an ideal gas is proportional to the mean kinetic energy of the molecules, which in turn is proportional to the temperature in Kelvin.

Very few gas molecules have small speeds or large speeds. The most probable speed v_p for a gas with molecular weight M_A is

$$v_p = \left[\frac{2RT}{M_A}\right]^{1/2}$$

and is somewhat less than the root mean square speed. The distribution of speeds is shown in Figure 10-13. Table 10-4 shows the root mean square speeds for different gases at 0°C.

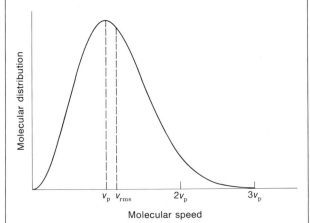

FIGURE 10-13
The distribution of speeds of gas molecules.

TABLE 10-4
Root means square speed of molecules of various gases at 0°C.

Gas	Molecular weight gm/mol	v_{rms} m/s
H_2	2.0	1845
He	4.0	1305
N_2	28.0	488
O_2	32.0	461
Ne	20.2	584
Ar	39.9	413
Xe	131.3	228
Rn	222.0	175

EXAMPLE

What is the root mean square speed of a nitrogen molecule at 0°C? Nitrogen has a molecular weight of 28.0134 gm/mol.

$$v_{\text{rms}} = \left[\frac{3RT}{M_A}\right]^{1/2} = \left[\frac{(3)(8.13 \text{ J/mol-K})(273 \text{ K})}{0.028 \text{ kg/mol}}\right]^{1/2}$$

$$v_{\text{rms}} = 488 \text{ m/s}$$

Vapor Pressure and Evaporation

Because of their motion, molecules tend to leave the surface of a liquid. To do so a molecule must have an energy higher than the average. The net effect is a cooling of the liquid. To remove 1 gm of water vapor from the surface at 100°C requires 538.7 cal—and slightly larger energy at lower temperatures. The heat of vaporization of water at different temperatures is shown in Table 10-5.

If a liquid such as water is placed in a sealed evacuated chamber, the molecules leave the liquid surface, forming a vapor. This process continues until **saturated vapor pressure** obtains: That is, an equilibrium is established between molecules leaving and entering the liquid surface. The saturated vapor pressure depends on the tem-

TABLE 10-5
Saturated vapor pressure, heat of vaporization, and vapor density for water as a function of temperature.

Temperature °C	Vapor pressure torr	Heat of vaporization cal/gm	Vapor density gm/m³
0	4.58	595	4.86
10	9.2	590.2	9.36
20	17.5	584.9	17.2
30	31.8	579.6	30.2
40	55.3	574.2	50.9
50	92.5	568.4	93
60	149.4	562.8	129
70	233.7	556.9	196
80	355.1	551.1	240
90	525.8	544.9	417
100	760.0	538.7	587
110	1074.6	532.3	808
120	1489.1	525.6	1091
130	2026.2	518.6	1448

perature. For example, that of water at a temperature of 100°C is one atmosphere, and of course the water boils. The vapor pressure decreases rapidly with temperature, as is also shown in Table 10-5.

If at any temperature the atmosphere is saturated with water vapor, the humidity is 100%. As temperature decreases, the atmosphere can no longer hold this quantity of moisture, and the vapor condenses—on surfaces as dew or frost, or in the air as fog, clouds, mist, rain, or snowflakes. The temperature at which the air becomes saturated with moisture and condensation begins is the dewpoint.

On hot days our personal comfort depends on our ability to perspire and on evaporation of that perspiration. On a hot, muggy day, when the temperature is, say, 32°C (90°F) and the humidity ranges near 100%, we can perspire profusely and all that happens is that we get wet. If, on the other hand, the temperature is 32°C but the humidity is low, we can still feel comfortable. As the saying goes, "It's not the heat, it's the humidity."

Evaporation requires that the saturated vapor pressure at the temperature of the liquid differ from the vapor pressure of water in the air. From the two million sweat glands on its surface the human body is capable of sweating at a rate of 4 liter/hr for short periods. Under normal conditions the body evaporates moisture through the skin even before sweating begins. This accounts for about 0.7 liter/day. In addition there is a loss due to the expiration of water vapor from the lungs of about 0.3 liter/day. These two forms of evaporation dissipate 25% of the basal resting body heat.

EXAMPLE

A person walking in direct sunlight in a dry desert environment where the ambient air temperature is about 100°F will sweat about 1 liter/hr. What heat loss rate does this correspond to if all the water evaporates?

$$\frac{\Delta Q}{\Delta t} = \frac{\Delta m}{\Delta t} L$$

The latent heat of evaporation is about 575 cal/gm at this temperature. The heat loss is

$$\frac{\Delta Q}{\Delta t} = \frac{(10^3 \text{ cm}^3/\text{hr})(1.0 \text{ gm/cm}^3)}{(3600 \text{ s/hr})}(575 \text{ cal/gm})(4.18 \text{ J/cal})$$

$$\frac{\Delta Q}{\Delta t} = 670 \text{ watts of cooling!}$$

Review

Give the significance of each of the following terms. Where appropriate give two examples.

thermometric property
thermometer
Celsius
Fahrenheit
coefficient of thermal expansion
Boyle's law of gases
absolute zero temperature
ideal gas
thermodynamic temperature scale
Kelvin temperature scale
triple point cell

platinum resistance thermometer
thermocouple
thermistor
thermal equilibrium
heat
specific heat
calorie
British thermal unit
mechanical equivalent of heat
first law of thermodynamics
calorimetry

boiling point
latent heat of vaporization
freezing point
latent heat of fusion
molecular weight
mole
Avogadro's number
ideal gas constant
ideal gas law
kinetic theory
saturated vapor pressure

Further Readings

Sanborn C. Brown, *Count Rumford: Physicist Extraordinary* (New York: Doubleday, Anchor Science Study Series S 28, 1962). A delightful account of the American scientist Benjamin Thompson, who rose to fame as a Bavarian nobleman and a count of the Holy Roman Empire.

Problems

TEMPERATURE

1 The normal temperature of the human body is 98.6°F. What is this temperature in °C? in Kelvin?

2 Dry ice evaporates at a temperature of −87.5°C. What is this temperature in °F? in K?

3 At what temperature are the values of the Celsius and Fahrenheit temperature scales equal?

4 The first expeditionary force to the planet Mars finds that the resident Martians have developed a temperature scale based on the freezing point of ethyl alcohol (0°M = −120°C) and on the boiling point of alcohol (100°M = 80°C).
   (a) Write a relationship between the Martian and Celsius temperature scales.
   (b) At what point are the two temperature scales equal?
   (c) The normal temperature of a human is 37.0°C. What is this temperature in °M?

5 Sections of a concrete roadway are 30 m long at 20°C. Spaces are left between sections to allow for expansion. If the maximum design temperature is 60°C and the coefficient of thermal expansion is $12 \times 10^{-6} °C^{-1}$, how much space must be left between sections at 20°C?

6 In an aluminum plate 10 cm square and 0.5 cm thick is a drilled hole with a 2.0000 cm diameter at 20°C. If the plate is heated to 100°C, and the coefficient of expansion is $11 \times 10^{-6} °C^{-1}$, how big is the hole?

7 A clinical glass thermometer consists of 0.8 cm³ of mercury in a glass bulb ending in a capillary of diameter 0.02 cm. How much will the mercury rise if the temperature goes from 34°C to 39°C? The volume expansion of mercury is $182 \times 10^{-6} °C^{-1}$ and of glass is $30 \times 10^{-6} °C^{-1}$.

HEAT

8 An electric space heater puts out 1350 watts of power as heat.
   (a) How many calories of heat does it put out per second?
   (b) How many Btu per hour does it put out?

9 A gas wall furnace puts out 50 000 Btu/hr as heat.
   (a) How many calories of heat does it put out per second?
   (b) How many watts of power does it put out?

10 A gas water heater has a power output of 35 000 Btu/hr.
   (a) How many calories of heat does it put out per second?
   (b) How many watts of power does it put out?
   (c) The heater contains 30 gallons of water, about 240 lbm. How long should it take to heat a tank of water from 50°F to 150°F?

11 An electric stove has a heating element with a heat output of 1400 watts. A pan containing 1 kg of water at 15°C is placed on this heating element on the stove.
   (a) If the heating element transfers 80% of its heat to the water, how many cal/s are added to the water?
   (b) How much heat is required to raise the water to the boiling point?
   (c) How long will it take to heat the water?

12 One system for storing heat from solar energy consists of blowing warmed air past about 12 tons of fist-size rocks. If the rocks have a mass of 12 metric tons and a specific heat of 0.2 cal/gm-°C, and are raised to a temperature 30°C above room temperature, how much energy can be stored in the rocks in cal? in joules?

13 Heat collected from solar energy is stored in a 2000 gallon water tank. The tank contains about 7.5 metric tons of water. If the water is raised 30°C above the working temperature, how much energy can be stored in the tank in cal? in joules?

14 In a calorimetry experiment, a 500 gm brass ball with a specific heat of 0.093 cal/gm-°C is heated to 100°C. The ball is dropped into an aluminum calorimeter with a mass of 200 gm and a specific heat of 0.0217 cal/gm-°C that contains 300 gm of water. Both the calorimeter and the water are initially at 20°C. What is the final temperature of the water?

15 Boiling water with a mass 450 gm is poured into a cup with a mass of 250 gm and a specific heat of 0.26 cal/gm-°C. A silver spoon with a mass 25 gm and specific heat of 0.056 cal/gm-°C sits in the cup. The cup and the spoon are initially at 20°C. What is the final temperature?

16 A glass of ice tea contains 100 gm of ice in equilibrium with 300 gm of tea. Assume that the tea has the same specific heat as water. The glass has a mass of 200 gm and a specific heat of 0.20 cal/gm-°C. How much heat is required to heat the glass of ice tea to room temperature of 20°C?

17 The capacity of air conditioners and refrigeration units is sometimes measured in "tons of refrigeration." One ton of refrigeration corresponds to the amount of cooling given by melting one ton of ice in one day. What is the capacity of a "one ton" air conditioner
(a) in cal/day?
(b) in Btu/hr?
(c) in watts?

KINETIC THEORY

18 A full No. 1A cylinder of helium gas has a volume of 42 liters and is at a pressure of 170 atmospheres at a temperature of 20°C. The gas is released to atmospheric pressure and is allowed to come to room temperature of 20°C. Assume that helium acts as an ideal gas.
(a) What is the final volume of gas in the cylinder?
(b) How many moles of gas are in the cylinder?

19 At a temperature of 20°C and one atmosphere pressure, the contents of a No. 2 cylinder of dry nitrogen have a volume of 2.30 m³. This volume of gas is placed in the cylinder with a volume of 16.8 liters.
(a) What is the pressure in the cylinder at 20°C?
(b) How many moles of gas are in the cylinder?

20 Argon is commonly used as an inert atmosphere. A cylinder of argon has a pressure of 6000 psia at a temperature of 20°C. If the cylinder is left out in the sun and is warmed to a temperature of 60°C, what is the pressure in the cylinder?

21 A classroom 10 m long, 3 m wide, and 2.5 m high is full of air at a temperature of 20°C with an average molecular weight of 29 gm/mol.
(a) How many moles of air are in the room?
(b) How many grams of air are in the room?
(c) If the specific heat of air is 0.25 cal/gm-°C, how much heat is required to raise the air temperature by 5°C?

22 A box 40 cm long, 30 cm wide, and 20 cm high is full of air at one atmosphere pressure and 20°C.
(a) How many moles of air are in the box?
(b) How many grams of air are in the box?

23 Air consists of 20% oxygen with a molecular weight of 32 gm/mol, and about 80% nitrogen with a molecular weight of 28 gm/mol. At a temperature of 27°C,
(a) What is the rms velocity of each molecule?
(b) What is the mean kinetic energy of each molecule in joules?
(c) What is the mean kinetic energy of one mole of each gas?

ADDITIONAL PROBLEMS

24 A major river is 10 m deep and 100 m across and flows at 5 km/hr. A 1000 MW electric power plant discharges 2000 MW of thermal power as waste

heat into the river. If the cooling water is thoroughly mixed with the river, how much does the temperature of the river increase?

25 A bathtub is filled with about 80 kg of water at 40°C.
(a) How much energy is required to heat this water from 10°C to 40°C in calories? in joules? in Btu?
(b) If this water is heated electrically with 100% efficiency at a cost of 4¢/kw-hr, how much would it cost? One kw-hr = 3.6×10^6 J.
(c) If this water is heated with natural gas with

70% efficiency at a cost of 20¢/therm, how much would it cost? One therm = 10^5 Btu.

26 A 52 gallon water heater contains about 200 kg of water. This water is heated from 10°C to 65°C.
(a) How much energy is required to heat this water in cal, J, and Btu?
(b) If this water is heated electrically with 100% efficiency and with electricity costing 4¢/kw-hr, how much would it cost?
(c) If this water were heated with natural gas with 70% efficiency at a cost of 20¢/therm, what would be the cost?

11

Thermal Energy Transfer

Temperature is related to the internal energy of a substance. Heat is the transfer of energy from one substance to another. There are three mechanisms by which energy is transmitted from one system to another: conduction, convection, and radiation.

CONDUCTION

The end of a silver spoon placed in a cup of hot coffee or hot soup itself gets hot, and may burn our fingers. A stainless steel spoon does not get as hot. The atoms of the spoon in the hot liquid vibrate more energetically than the atoms at the cold end of the spoon. By collisions the energetic atoms transfer energy to their neighbors. The newly energized atoms in turn transfer energy to their neighbors. Each atom remains vibrating around its original position. This transfer of thermal energy by collisions of atoms is thermal **conduction.** Heat passes from the hot liquid through the silver spoon more quickly than through the steel spoon.

Figure 11-1 shows the temperature as a function of distance across a piece of material with a thickness d, separating two regions with temperatures T_1 and T_2. There will be the conduction of energy through this material. The rate of energy transfer is in units of energy per second—cal/s, Btu/hr, joules/s, or watts. The magnitude of the energy transfer is directly proportional to the area of contact A and to the temperature difference $T_2 - T_1$, and is inversely proportional to the thickness of the surface d.

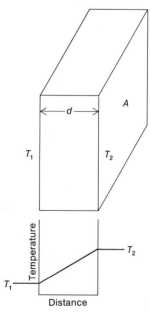

FIGURE 11-1
Thermal conduction is directly proportional to the temperature difference $T_2 - T_1$ and the cross-sectional area A and is inversely proportional to the thickness d.

Thermal Conductivity

The heat transfer characteristics of many materials can be expressed by a simple coefficient of **thermal conductivity** K:

244

TABLE 11-1
Thermal conductivity of various materials.

Material	Thermal conductivity W-m/m²-K
Metals	
Lead	35
Steel	46
Brass	110
Aluminum	210
Copper	380
Silver	420
Building materials	
Urethane foam	0.020
Glass wool insulation	0.036
Masonite board	0.048
Celotex board	0.049
White pine, across grain	0.11
Cardboard	0.21
Paper	0.12
Red brick	0.62
Concrete	0.92
Window glass	1.0
Gypsum	1.3
Other materials	
Air 20°C	0.023
Water 20°C	0.59

$$\frac{\Delta Q}{\Delta t} = KA \frac{T_2 - T_1}{d} \qquad (11\text{-}1)$$

The thermal conductivity gives the power in watts per square meter transmitted by a one Kelvin temperature difference across a thickness of one meter. The SI units of thermal conductivity are W-m/m²-K. Other common units are

$$\frac{1.0 \text{ cal}}{\text{s-cm}^2} \frac{\text{cm}}{°C} = 418.4 \frac{\text{W-m}}{\text{m}^2\text{-K}}$$

and

$$\frac{1.0 \text{ Btu}}{\text{hr-ft}^2} \frac{\text{in}}{°F} = 0.144 \frac{\text{W-m}}{\text{m}^2\text{-K}}$$

Table 11-1 shows the values of thermal conductivity for various substances. By far the best conductors of heat are metals, which transfer energy primarily by the conduction electrons. (Thus, good electrical conductors are good thermal conductors.) Metals tend to feel cold to the touch, for heat is rapidly conducted away from the hand. Silver, with a thermal conductivity of 420 W-m/m²-K, conducts heat more readily than steel. Cooking utensils are often made of aluminum or copper-clad stainless steel for rapid transfer of heat.

Materials such as brick, window glass, concrete, and water have lower thermal conductivities, of the order of 1 W-m/m²-K. A large surface area, however, can still conduct significant quantities of heat; a concrete floor can feel cold, as can a window glass on a cold day. Common solid insulating materials, having a thermal conductivity of less than 0.1 W-m/m²-K, significantly reduce heat loss by conduction. The emphasis on energy conservation has focused much attention on the use of such materials in homes and commercial buildings to reduce heat loss in winter and heat gain (and air conditioning loads) in summer.

A very good insulating material is still, dry air. Dry air trapped in the material is the source of the insulating properties of down and polyester fiber filled sleeping bags, urethane foam, glass wool, and cotton and wool clothing. A thin layer of still air about 6 mm thick next to a window surface limits heat loss through the pane.

EXAMPLE

A person with a skin temperature of 33.5°C is lying atop a thin blanket of thickness 3 mm on a waterbed at a temperature of 26.5°C (80°F). Assume that the thermal conductivity of the blanket is 0.05 W-m/m²-K and that the contact area between the person and the waterbed is 0.75 m². What is the energy transfer from the person to the waterbed? The normal metabolic heat generation rate is 80 watts. The heat loss is

$$\frac{\Delta Q}{\Delta t} = KA \frac{T_2 - T_1}{d}$$

$$\frac{\Delta Q}{\Delta t} = \frac{(0.05 \text{ W-m})(0.75 \text{ m}^2)(33.5°C - 26.5°C)}{\text{m}^2\text{-K} \qquad 0.003 \text{ m}}$$

$$\frac{\Delta Q}{\Delta t} = 87.5 \text{ watts!}$$

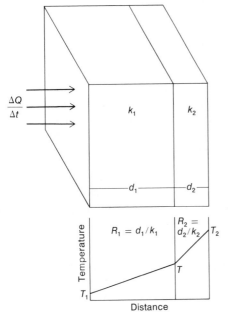

FIGURE 11-2
The heat flow by conduction through two layers can be represented by the sum of R-values for each layer. $R = d/K$. The total heat flow is given by $Q_9 = \dfrac{A(T_2 - T_1)}{R_1 + R_2}$.

Even though the waterbed is not "cold," a thin blanket will not prevent rapid heat loss that could lower the body temperature enough to promote a chill. Either a sufficiently thick foam pad must be used or the waterbed must be heated to close to body temperature. Campers sleeping outdoors should sleep with insulating material underneath to minimize heat loss to the ground.

Heat Conduction Across Two Layers

If there are two layers of material, as shown in Figure 11-2, with thickness d_1 and d_2 and thermal conductivities K_1 and K_2, the interface between the two layers will be at some temperature T, between T_1 and T_2. The heat flow through each material is the same:

$$\frac{\Delta Q}{\Delta t} = KA \frac{T_1 - T}{d_1} = K_2 A \frac{T - T_2}{d_2}$$

TABLE 11-2
R-values for typical construction materials, in engineering units. Heat loss is given by the product of the surface area times the temperature difference divided by the R-value: $Q = A(T_2 - T_1)/R$.

Material	R-value $\dfrac{ft^2 \text{-} {}^\circ F}{Btu/hr}$
Plasterboard, $\frac{1}{2}$ in.	0.45
Plywood, $\frac{3}{4}$ in.	0.93
Wood shingles	0.94
Pine, $1\frac{1}{2}$ in.	1.89
Typical wall (plasterboard, $3\frac{1}{2}$ in. air space, stucco)	2.55
Fiberglass, $2\frac{1}{4}$ in.	7.0
Fiberglass, $3\frac{1}{2}$ in.	11.0
Fiberglass, 6.0 in.	19.0

The intermediate temperature T can be eliminated, giving

$$\frac{\Delta Q}{\Delta t} = A \frac{(T_2 - T_1)}{d_1/K_1 + d_2/K_2} = \frac{A(T_2 - T_1)}{R_1 + R_2}$$

The engineering unit for rating insulation is the **R-value:** $R = d/K$. When several layers of insulation material are placed next to one another the R-values add. Table 11-2 shows typical R-values for common building materials. Recommended ceiling insulation has $R = 19$; wall insulation of $R = 11$ is required of new construction in California.

EXAMPLE
Conduction of heat from the core of the human body to the surface is relatively slow. Let us represent the body's surface tissue as a layer of thickness $d_1 = 5$ cm, which has a conductivity similar to water $K_1 = 0.59$ W-m/m²-K. This is surrounded by a layer of tissue fat of thickness $d_2 = 2$ cm, with a thermal conductivity $K_2 = 0.21$ W-m/m²-K. The body's interior is at a temperature $T_1 = 37^\circ$C, the skin $T_2 = 33.5^\circ$C. What is the heat loss per m² area? The heat loss per unit area is just

$$\frac{1}{A}\frac{\Delta Q}{\Delta t} = \frac{T_2 - T_1}{d_1/K_1 + d_2/K_2}$$

$$\frac{1}{A}\frac{\Delta Q}{\Delta t} = \frac{33.5 - 37.5°C}{[0.05/.59 + 0.02/0.21]\,[\text{m}/(\text{W-m-m}^2\text{-K})]}$$

$$\frac{1}{A}\frac{\Delta Q}{\Delta t} = -19.4\ \text{W/m}^2$$

The minus sign indicates heat loss. The basal heat produced in the body is about 80 W, which must be dissipated over approximately 2 m² of body surface. The heat loss calculated in the model above is much less than actual heat loss.

Heat Exchangers By distributing blood the arteries bring the core temperature of 37°C much closer to the surface. The temperature gradient drop occurs almost entirely across the layer of fat. Thin persons, with only a thin layer of body fat, tend to feel colder than heftier persons. Blood leaving the heart is quite close to core temperature 37°C, considerably warmer than the hands and feet. (The fingertip is as low as 20°C.) Pumping hot blood directly to the extremities would give rise to a large heat loss. But our circulatory system is sophisticated, resembling a counterflow heat exchanger. The warm, outward-flowing arterial blood loses heat by conduction to the cooler returning venous blood, as is shown in Figure 11-3. Counterflow heat exchangers have many applications in industry to transfer the heat from one fluid to another with high efficiency.

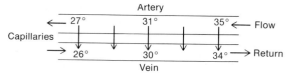

FIGURE 11-3
Heat exchangers. The exchange of heat cools warm arterial blood and warms cool venous blood.

CONVECTION

Fluids such as air or water are easily transported. In **convection,** thermal energy is transferred by physical motion of a fluid. Warmed fluid displaces cooler fluid,

giving a net transfer of thermal energy. The energy transfer ΔQ depends on the mass of fluid moved Δm, the specific heat of the fluid C_s, and the temperature difference

$$\Delta Q = \Delta m C_s (T_2 - T_1) \qquad (11\text{-}2)$$

The tendency for warm fluid to rise and cooler fluid to fall produces circulation, or **natural convection,** as can be seen in smoke rising from a cigarette or a chimney, afternoon breezes over sun-heated land, cold drafts near large windows in winter, or a rolling boil when water is heated in a pan. A fan or pump circulating a fluid is an instance of **forced convection.** Many homes are heated or cooled by the circulation of room air through a furnace or air conditioner. An electric hair dryer forces air past heating coils to produce a stream of warm air.

A significant convective energy-loss mechanism of buildings is **infiltration** of cold outside air into the interior, where it displaces warm inside air. For a typical house, infiltration produces about one complete exchange of air per hour. This rate can vary by a factor of two either way, according to wind conditions and the use of storm windows and weather stripping. Infiltration of cold air can easily amount to one-third of the heat loss from a residence on cold winter days.

Mass and Volume Flow

A gas or liquid has a specific heat at constant pressure C_s. To raise a certain mass Δm of the fluid through a given temperature difference $T_2 - T_1$ requires an amount of heat ΔQ:

$$\Delta Q = \Delta m C_s (T_2 - T_1)$$

If this mass is moved in time Δt, there is a flow of heat due to convection

$$\frac{\Delta Q}{\Delta t} = \frac{\Delta m}{\Delta t} C_s (T_2 - T_1)$$

The mass flow rate $\Delta m/\Delta t$ depends on the density of the fluid ρ and the volume flow rate $\Delta V/\Delta t$.

$$\frac{\Delta m}{\Delta t} = \rho \frac{\Delta V}{\Delta t} = \rho v A$$

where v is the speed of the flow and A is the cross section.

EXAMPLE

A gas furnace heats air with a density of 1 200 gm/m^3 and specific heat 0.25 cal/gm-°C from 20°C to 80°C. The heater's fan blows 9 m^3/min of room-temperature air through the heater. What is the output heat in cal/s, watts, and Btu/hr?
The heat output is given by

$$\frac{\Delta Q}{\Delta t} = \rho \frac{\Delta V}{\Delta t} C_s (T_2 - T_1)$$

The volume flow rate is

$$\frac{\Delta V}{\Delta t} = 9 \frac{m^3}{min} \times \frac{1\ min}{60\ s}$$

$$\frac{\Delta V}{\Delta t} = 0.15\ m^3/s$$

The heat delivered is

$$\frac{\Delta Q}{\Delta t} = \left(1200 \frac{gm}{m^3}\right)\left(0.15 \frac{m^3}{s}\right)\left(0.25 \frac{cal}{gm\text{-}°C}\right)(60°C)$$

$$\frac{\Delta Q}{\Delta t} = 2.70 \times 10^3\ cal/s \times \frac{(4.18\ J)}{(1\ cal)}$$

$$\frac{\Delta Q}{\Delta t} = 11.3 \times 10^3\ J/s \times \frac{(1\ Btu)}{(1054\ J)} \times \frac{(3600\ s)}{(1\ hr)}$$

$$\frac{\Delta Q}{\Delta t} = 38.5 \times 10^3\ Btu/hr$$

The heat output is 2700 cal/s or 11 kilowatts or 38 500 Btu/hr. Gas heaters deliver about 80% of their combustion energy as heat. Wall furnaces typically have an energy output of 30 000 Btu/hr. Heat is delivered to the room by convection of room air past the heat exchanger. Central furnaces for homes have an energy input of 40 000 Btu/hr to 150 000 Btu/hr. The heat is delivered to the rooms of the house by convection through ducts under the floor or in the attic.

Wind Chill   A windy fall day feels colder than a calm day at the same temperature. Even if the temperature is near freezing with little wind, we can feel quite comfortable with a few normal precautions. If the wind speed increases much above 10 km/hr, we can feel cold. If the temperature drops to $-10°C(14°F)$, only a light breeze will make us feel very cold. On a $-10°C$ winter evening in Chicago with an 18 km/hr wind blowing off Lake Michigan it feels bitterly cold. If the wind increases to 45 km/hr there is a risk of immediate frostbite unless sufficient protection is used. Our comfort is related to the rate of heat loss. At a heat-loss rate of 350 W/m^2 we feel quite cool, at 1 600 W/m^2 exposed flesh freezes. Figure 11-4 shows the heat-loss rate and danger level for various temperatures and wind speeds. Wind greatly reduces the layer of insulating air next to the surface and greatly increases heat loss.

The ever-present danger for people exposed to the elements (say by being caught in a sudden storm in the mountains) is **hypothermia,** a sudden drop in body temperature below its normal 37°C. A drop of the core body temperature to 31°C, brings unconsciousness, a drop to 26°C brings death. The body temperature drops when heat loss exceeds the metabolic heat production. But even if the core temperature is maintained, there is serious risk to fingers and toes. Chilblains, caused by restricted circulation, produce swelling by accumulation of fluids in the fingers, toes, ears, and cheeks. Frostbitten tissue can become frozen and cells be killed. The frozen part may become gangrenous and require amputation.

RADIATION

The blacksmith's art depended on knowing how to judge the temperature of a piece of hot iron by its color: If very hot it glows a dull red; heated further it glows a bright red; heated still further it glows a bright yellow-white. Similarly the embers of a fire and the element of an electric stove or toaster glow a bright cherry red. An incandescent light bulb glows with a white light. The color of a hot object is related to its temperature.

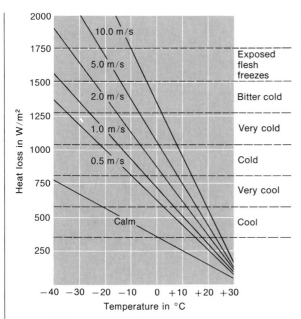

FIGURE 11-4
The wind chill factor is related to heat loss, outdoor temperature, and wind speed. At any given temperature the heat loss depends on the wind speed. At +10°C and a wind speed of 1.0 m/s (3.6 km/hr), the heat loss is the same as on a calm day with a temperature of −16°C.

A hand placed near glowing coals feels warmth. Even if the hand is too far from the fire for thermal conduction and convection of energy to be significant, there is a direct transfer of energy from the fire to the hand, by **radiation.** Thermal energy emitted from the surface of a body by radiation is virtually identical to light, except that instead of being seen it is felt as heat. The energy emitted is **radiant energy.** Radiation requires no medium for transmission: Radiant energy can travel through a vacuum. The sun's radiant energy, for example, travels through the vacuum of interplanetary space to warm the earth.

Black-Body Radiation

An object that is white reflects all sunlight falling on it. An object that is gray absorbs half the light that falls on it and reflects the other half. An object that appears black absorbs all light falling on it and reflects none. An object that absorbs all radiant energy that falls on it is called a **black body.** A good absorber of radiant energy is also a good emitter of radiant energy. The radiant energy from a black body depends not on the object's material but on its temperature only. The radiation from a piece of iron is the same as from a piece of charcoal at the same temperature.

In 1879, Josef Stefan demonstrated experimentally that the total energy radiated by a black body is proportional to the absolute temperature raised to the fourth power. In 1884, this proportionality was justified theoretically by Ludwig Boltzmann. The **Stefan-Boltzmann law** for the power radiated from a surface can be written:

$$W = A\sigma T^4 \qquad (11\text{-}3)$$

where T is the temperature in Kelvin, A is the surface area, and σ (the Greek letter *sigma*) is the Stefan-Boltzmann constant:

$$\sigma = 5.670 \times 10^{-8} \text{ W/m}^2\text{-K}^4$$

$$\sigma = 1.741 \times 10^{-9} \text{ Btu/hr-ft}^2\text{-}°\text{R}^4$$

The total power radiated by the surface increases rapidly with temperature.

Figure 11-5 shows the spectrum of black-body radiation for different temperatures. The wavelength of visible light is in the range 0.40 to 0.7 μm (1 μm = 1 microm-

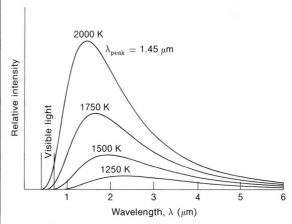

FIGURE 11-5
Black-body radiation. As the temperature of a black body increases, the wavelength of maximum radiation moves toward the visible. The total power radiated increases as the temperature to the fourth power.

eter $= 10^{-6}$ m). Almost all of the energy at the temperatures shown lies in the infrared region of the spectrum, where the wavelengths are longer than visible light. The wavelength of peak (greatest) emission λ_{peak} decreases as temperature increases, and is inversely proportional to the temperature in Kelvin:

$$\lambda_{peak} = \frac{2.898 \times 10^{-3} \text{ m-K}}{T} \qquad (11\text{-}4)$$

This is the **Wien displacement law.** We discuss blackbody radiation further in Chapter 21.

If an object is heated to 2 000 K, the tail of the blackbody radiation spectrum lies in the visible region and is seen as a dull red glow, although the wavelength of maximum intensity remains in the infrared. If the temperature increases to 4 000 K the glow is a bright orange. At 6 000 K (comparable to the temperature of the sun's surface atmosphere) the maximum peak is in the visible spectrum and the object glows white hot. As the temperature increases, the total radiation increases as the temperature in Kelvin to the fourth power.

At temperatures above the freezing point of gold (1 337.58 K), the international practical temperature scale is defined by the Stefan-Boltzmann law of radiation. The optical pyrometer is the standard instrument for measuring high temperatures for industry. Regulating the current through the tungsten lamp of an optical pyrometer controls the temperature of the filament. By comparing the intensity of light emitted by this filament with the intensity of light emitted from a source—the interior of a steel furnace, for instance—the temperature can be measured within $\pm 6°$C.

Solar Radiation

The solar photosphere, the atmosphere at the sun's surface, radiates very much like a black body at a temperature of about 5800 K. The maximum radiation wavelength is given by the Wien displacement law

$$\lambda_{peak} = \frac{2.898 \times 10^{-3} \text{ m-K}}{5.8 \times 10^3 \text{ K}}$$

$$\lambda_{peak} = 0.500 \times 10^{-6} \text{ m} = 0.5 \ \mu\text{m}$$

The maximum wavelength of 0.5 μm is in the visible. The power radiated at the surface of the sun is given by the Stefan-Boltzmann law:

$$W/A = \sigma T^4$$

$$W/A = (5.67 \times 10^{-8} \text{ W/m}^2\text{-K}^4)(5.8 \times 10^3 \text{ K})^4$$

$$W/A = 6.42 \times 10^7 \text{ W/m}^2$$

The sun has a radius $R = 6.96 \times 10^8$ m and a surface area $4\pi R^2 = 6.08 \times 10^{18}$ m^2. The total output power of the sun is

$$W = (6.42 \times 10^7 \text{ W/m}^2)(6.08 \times 10^{18} \text{ m}^2)$$

$$W = 3.90 \times 10^{26} \text{ W}!$$

The sun radiates energy into space in all directions. The intensity of solar radiation at the top of the earth's atmosphere is given by the power radiated by the sun spread over a sphere of the radius of the earth's orbit $A' = 4\pi R'^2$, where $R' = 1.5 \times 10^{11}$ m. The area is

$$A' = 2.83 \times 10^{23} \text{ m}^2$$

The intensity at the top of the atmosphere is

$$W/A' = \frac{3.90 \times 10^{26} \text{ W}}{2.83 \times 10^{23} \text{ m}^2}$$

$$W/A' = 1.38 \times 10^3 \text{ W/m}^2$$

The intensity of the light from the sun as a function of wavelength is shown in Figure 11-6. There is significant absorption of the incoming solar radiation in the atmosphere. The short-wavelength, ultraviolet rays from the sun would cause serious biological damage were they not strongly absorbed by ozone, O_3, which is present in the stratosphere above 20 km. There is at present serious concern that chlorine and fluorine ions released in freon aerosol propellants may react to deplete this vital ozone layer.[1] Hence, these propellants have been banned for consumer use.

---

[1] See Gloria B. Lubkin, "Fluorocarbons and the Stratosphere," *Physics Today* (October 1975), pp. 34–39.

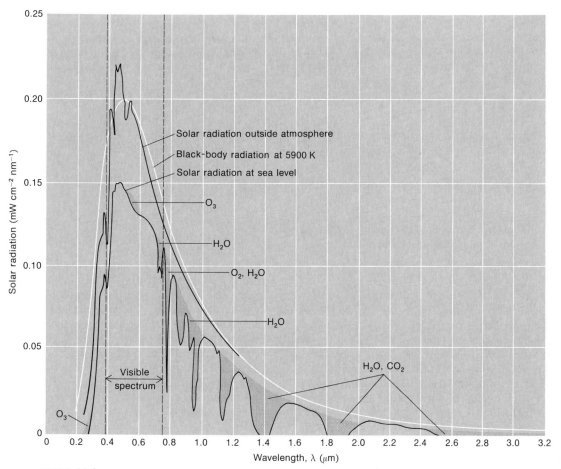

FIGURE 11-6

Spectral radiation curves. The solar radiation outside the atmosphere is similar to the curve for black-body radiation at 5900 K. Solar radiation is absorbed as it penetrates the atmosphere. There is strong absorption by water vapor and carbon dioxide at longer wavelengths in the infrared. There is also strong absorption by ozone, O_3, of the ultraviolet at short wavelengths.

In the visible solar spectrum there is scattering and absorption of light by dust in the atmosphere, reflection of light by clouds, and some absorption by oxygen and water vapor. At infrared wavelengths, there is strong absorption of the incoming solar energy by molecular energy bands of oxygen, O_2, ozone, O_3, and water vapor, H_2O. In the far infrared there is strong absorption by carbon dioxide, CO_2. The absorption of infrared radiation significantly reduces heat loss from the earth's surface. Depending on local conditions, only about 50% of the solar radiation incident at the top of the atmosphere reaches the earth's surface. Under clear skies and little dust this can be as much as 70%.

The sun is hot. If your car has been standing out in the sun on a clear summer day, it becomes hot enough to give you a minor burn. The steering wheel can be too hot to handle. A metal sheet painted dull black absorbs almost all radiation falling on it. A good radiation absorber is a good emitter. The power absorbed is given by

$$W = IA$$

where I is the incident solar energy flux and A the surface area. The power radiated by blackbody radiation is

$$W/A = \sigma T^4$$

A metal plate sitting in the hot sun, where the total solar energy flux is, for instance, 700 W/m^2, heats up until the power lost by conduction, convection, and radiation equals the power absorbed. The total heat into a solar collector is balanced by the heat withdrawn by the heat-exchange fluid and the heat lost from the surface. The simplest collector is a sheet of metal painted black, with a channel for water or air flow. Often, to minimize heat loss, glass is placed over the collector surface and heavy insulation under the surface.

Ground Radiation

The earth's surface is typically at a temperature of about $288 \text{ K} (15°C)$. The earth itself radiates as a black body. The wavelength of maximum radiation is in the far infrared portion of the spectrum. The amount of power radiated per unit area by such a surface is just given by

$$W/A = \sigma T^4$$

$$W/A = (5.67 \times 10^{-8} \text{ W/m}^2\text{-K}^4)(288 \text{ K})^4$$

$$W/A = 390 \text{ W/m}^2$$

This **ground radiation** is comparable to the solar radiation reaching the earth's surface.

If the earth had a very thin atmosphere with little water vapor or carbon dioxide, day-to-night temperature variation would be extreme. Absorption of infrared radiation in the atmosphere reduces the effect of radiation cooling. Figure 11-7 shows the percentage transmission of the atmosphere in the infrared spectra. Water vapor and carbon dioxide cause strong absorption bands. Ozone causes a small absorption band. A good absorber is also a good emitter of radiation. Although the night sky is transparent in the visible spectrum, it is opaque from 5 to 8 μm and at far-infrared wavelengths. The ground radiation is a maximum at about 10 μm. The atmosphere radiates as a black body with a temperature of $263 \text{ K} (-10°C)$, except at wavelengths near 10 μm, where its emission character is that of the higher and colder ozone layer. Figure 11-7 also shows the estimated atmospheric thermal emission.

The amount of heat that the sky radiates back to the earth is about that of a black body at a temperature of 263 K.

$$W/A = \sigma T^4$$

$$W/A = (5.67 \times 10^{-8} \text{ W/m}^2\text{-K}^4)(2.63 \times 10^2 \text{ K})^4$$

$$W/A = 271 \text{ W/m}^2$$

The net heat loss from the earth's surface is the difference between the heat radiated away from the surface less the heat radiated back from the atmosphere.

$$W/A_{\text{net}} = \sigma(T_2^4 - T_1^4)$$

$$W/A_{\text{net}} = 390 - 271 \text{ W/m}^2 = 119 \text{ W/m}^2$$

The heat-loss rate from the earth's surface depends critically on the temperature of the first few kilometers of the atmosphere. If the atmosphere is clear and cold, then

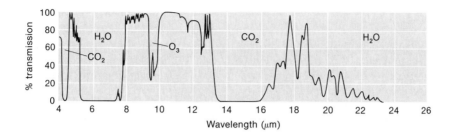

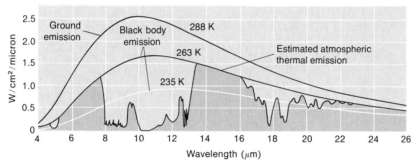

FIGURE 11-7
Infrared transmission through the atmosphere and estimated
emission of radiation from the atmosphere. From David M.
Gates, *Energy Exchange in the Biosphere* (New York: See 11-11
Harper and Row), 1962.

immediately after sunset the ground cools rapidly by radiation. Leaves, grass, and the ground cool much more rapidly than the surrounding air, and frost or dew forms. If the air near the ground is cooled, a ground fog may form. In the summer, if air warmer than ground level temperature is trapped at an altitude of 1 km after the sun goes down, the evening will remain warm. Downward radiation from the sky may exceed that from the ground and it will stay warm late into the evening.

Black-body radiation is one of the human body's primary heat-loss mechanisms. A room with an air temperature of 15°C can feel warm or cold depending on the wall or floor temperature. If the walls are cold there is a net heat loss from radiation and we feel chilly. If one wall is warm or if there is a hot radiant heater in the room, we can feel quite comfortable even though the air in the room may be a bit chilly.

EXAMPLE

The skin temperature is about 33°C. Suppose we are standing next to a cold wall at a temperature 10°C. What is the net heat loss per unit area from radiation loss?

$$W/A_{net} = \sigma(T_2^4 - T_1^4)$$
$$= (5.67 \times 10^{-8} \text{ W/m}^2\text{-K}^4)[(306 \text{ K})^4 - (283 \text{ K})^4]$$
$$W/A_{net} = 138 \text{ W/m}^2$$

If we are fully clothed, this heat loss is acceptable, but the exposed areas of our skin will feel a definite heat loss.

Radiation Pictures

An object's effective surface temperature can be directly measured by measuring the intensity of its infrared radiation. An infrared camera measures the power radiated from a surface in the infrared spectra from 2 to 5.6 μm. An indium antimonide (InSb) crystal kept at the temperature of liquid nitrogen (77 K) produces an electrical signal that is proportional to the incident infrared radiation. Infrared-transmitting germanium lenses and horizontal and vertical scanning prisms scan 16 pictures per second with 100 lines per picture. Figure 11-8 shows the electro-optical parts of an AGA System 680 scanning infrared camera.

The thermal camera can measure the absolute tempera-ture levels of objects from −30°C to 2000°C and display them on a special television screen, in 10 shades between black and white to represent different temperature intervals. The temperature intervals can be made as small as 1°C or large enough to cover a 2000°C range. Thermal photographs now have such diverse applications as detecting heat leakage from buildings, routinely inspecting electrical power-distribution systems to detect hot spots before equipment failure occurs, product testing of manufactured items, and quality control of all types of semiconductor devices.

Thermal pictures have become important in medical diagnosis. Skin temperature varies considerably over the body. If a blockage or other condition restricts blood flow, surface temperature decreases. Other abnormal conditions

FIGURE 11-8
The System 680 infrared camera unit showing the internal arrangement of the electro-optical components. The lenses and prisms are made of germanium, which transmits infrared radiation. The detector is an indium antimonide (InSb) crystal, which is sensitive to infrared radiation with a wavelength of 2 to 5.6 μm. The crystal is kept at liquid nitrogen temperature (77 K). Courtesy of AGA Corporation.

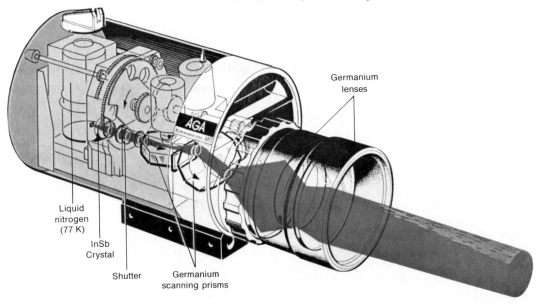

Germanium
lenses

Liquid
nitrogen
(77 K)

InSb
Crystal

Shutter

Germanium
scanning prisms

show as warm spots. Although each person's thermal patterns are unique, the human body is symmetrical. Asymmetries in thermal patterns can provide valuable clues to diagnosis of rheumatic diseases, problems of peripheral circulation, and accute appendicitis in children.

Perhaps thermal pictures have been used most significantly to screen women for breast cancer. Figure 11-9 shows a typical gray-tone breast thermogram. The white triangular area above the patient's right breast is a hot spot requiring further x-ray study.

Although thermal scanning equipment is expensive, operating costs are negligible and the procedure exposes the patients to no radiation and requires only a few minutes. It can detect breast cancer in the preclinical stage, before it shows up in x-rays. Thermography identifies those at high risk for breast cancer, because a change in the thermal pattern is often the earliest sign of breast cancer.

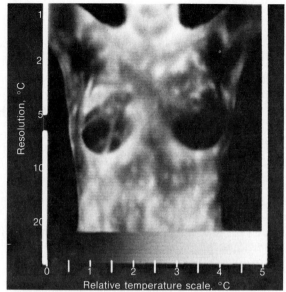

FIGURE 11-9
A typical gray-tone breast thermogram. There is a suspicious area (hot) evidenced by the white triangle in the upper quadrant above the patient's right breast. The thermogram covers a 5°C range of temperatures. Courtesy of AGA Corporation.

Review

Give the significance of each of the following terms. Where appropriate give two examples.

conduction
thermal conductivity
R-value
convection
natural convection

forced convection
infiltration
hypothermia
radiation
radiant energy

black body
Stefan-Boltzmann Law
Wien displacement law
ground radiation

Further Readings

James G. Edinger, *Watching the Wind: The Seen and Unseen Influences on Local Weather* (New York: Doubleday, Anchor Science Study Series S49, 1967).

David M. Gates, *Energy Exchange in the Biosphere* (New York: Harper & Row, Publishers, Inc., 1962). A classic study.

H. E. Landsberg, *Weather and Health: An Introduction to Biometeorology,* (New York: Doubleday, Anchor Science Study Series S59, 1969).

Problems

CONDUCTION

1 An ice chest with an interior 0.40 m long, 0.24 m wide, and 0.30 m deep is made of styrofoam 2 cm thick. The thermal conductivity (K) of styrofoam is about $K = 0.020$ W-m/m²-K. The ice chest is left out on a hot day $T = 35°C$. The interior of the chest is kept near 10°C by melting ice and cold beer.
   (a) What is the interior surface area of the chest?
   (b) What is the heat loss from the chest?
   (c) How much ice melts in 1 hr to compensate for the heat loss?

2 A person with a skin temperature of 33.5°C is lying on a urethane foam pad with a conductivity of 0.020 W-m/m²-K. Assume that the water-bed underneath the pad is at a temperature of 26.5°C (80°F) and that the area of contact is 0.75 m². How thick must the foam pad be to reduce the heat loss to 100 W?

3 A layer of fat tissue 2 cm thick with a thermal conductivity of 0.21 W-m/m²-K separates the interior body temperature of 37°C from the skin temperature of 33.5°C.
   (a) What is the heat loss per m² of area?
   (b) What is the heat loss if the layer is 1 cm thick?

4 A 30 gallon water-heater tank has interior dimensions of about 32.5 cm in diameter and 140 cm high, and is surrounded by about 3 cm of glass wool insulation. The tank's insides are maintained at a temperature of 60°C (140°F), its outside is 20°C.
   (a) What is the heat-loss rate?
   (b) If the tank contains about 110 kg of water, how much would this water cool in 1 hr if no water were drawn off or no heat were added?

5 A 3 m by 4 m ceiling consists of 1.2 cm of gypsum wallboard mounted to the studs. The attic above the ceiling reaches a temperature of 55°C when the room temperature is 25°C.

(a) What heat is transferred by conduction through the ceiling, in watts?
(b) If 8 cm (about 3 in) of glass wool insulation is installed in the attic, what is the heat transfer?

6 The ceiling of the previous problem is insulated with 16 cm (about 6 in) of glass wool insulation. What is the heat transfer through the ceiling?

7 The recommended level for insulation of a house has an R-value of R-11; that is,

$$R = \frac{d}{K} = 11 \, \frac{ft^2\text{-}°F}{Btu/hr}$$

A ceiling is insulated with R-11 material. The attic above has a temperature $T_2 = 37°F$. The bottom side of the insulation has a temperature $T_1 = 70°F$. The ceiling has an area of 250 ft². What is the heat-loss rate through the insulation, in Btu/hr? in watts?

CONVECTION

8 A solar energy collector is 1.2 m by 2.4 m (about 4×8 ft). If at noon the collector directly faces the sun, the influx of energy from the sun is about 800 W/m². The solar collector heats water from 15°C to 30°C with about 70% efficiency.
   (a) What is the mass flow rate required to carry off this heat in kg/s?
   (b) What is the volume flow rate in liter/min?

9 A solar energy collector has dimensions 1.2 m by 2.4 m and collects solar energy with about 70% efficiency. The average solar influx is 800 W/m². Water enters the collector at 30°C and flows at 2 liter/min. What is the water's outlet temperature?

10 An air conditioner cools 360 ft³/min of air from 35°C to 15°C. How much heat is removed from the room in watts? In Btu/hr?

11 A one-story house is about 7.6 m wide and 12 m long and has ceilings 2.4 m high. On a winter day, cold outside air at 5°C replaces warm interior air at 20°C once each hour. What is the infiltration heat loss in cal/hr? in watts?

12 An oil furnace with 70% efficiency has a power input of 100 000 Btu/hr and is equipped with a blower that moves 20 m³/min of air.
   (a) What is the heat delivered to the air in watts?
   (b) If the air enters at a temperature of 15°C, what is the outlet temperature?
   (c) If fuel oil has an energy content of 4.5×10^7 J/kg, how much fuel is consumed in 1 hr?

13 A 2 000 Megawatt thermal fossil-fuel powerplant generates electricity with 40% overall efficiency. Cooling water dissipates 60% of its power. The cooling water enters at 12°C and leaves at 20°C.
   (a) How much waste heat is released by the powerplant per second?
   (b) How much water must flow through the condensers to carry off the waste heat in kg/s? in liter/hr?

BLACK-BODY RADIATION

14 The solar flux is typically 750 W/m². A large mirror of diameter 2 m reflects 90% of the incident sunlight and focuses it onto the window of a furnace 0.1 m in diameter. Assume that the furnace cavity acts as a blackbody. What is the maximum temperature reached by the furnace?

15 A solar furnace with a window 0.1 m in diameter is to be constructed. Assume that the furnace cave acts as a blackbody. A large mirror reflects 90% of the incident sunlight into the furnace. The solar flux is 750 W/m². What diameter mirror is needed to reach an equilibrium temperature of 2 000 K?

16 Sunlight of an intensity 700 W/m² falls on a sunbather's backside, which has an area of about 0.3 m². Caucasian skin absorbs 65% of the light energy. For the darker skin of other races the absorption can be 80%.
   (a) What is the energy absorbed by the person?
   (b) How does this compare with the basal heat production of 80 W?

17 The palm of a hand, about 8 cm by 10 cm, at a temperature of about 25°C is placed several inches from a hot radiator at a temperature of 100°C. Assume that both the hand and the radiator act as blackbodies. What is the net heat transfer to the hand?

18 The palm of a hand, 8 cm by 10 cm, at a temperature of 25°C is placed several inches from an ice cube tray at a temperature of 0°C. Assume that both the hand and the ice cube tray act as blackbodies. What is the net heat transfer from the hand?

12

Thermodynamics

THE HEAT ENGINE

Everyone knows that heat can produce motion. That it possesses vast motive-power no one can doubt, in these days when the steam-engine is everywhere so well known.

To heat also are due the vast movements which take place on the earth. It causes the agitations of the atmosphere, the ascension of clouds, the fall of rain . . . , the currents of water which channel the surface of the globe, and of which man has thus far employed but a small portion. Even earthquakes and volcanic eruptions are the result of heat.

. . . Nature, in providing us with combustibles on all sides, has given us the power to produce, at all times and in all places, heat and the impelling power which is the result of it. To develop this power, to appropriate it to our uses, is the object of heat-engines.

The study of these engines is of the greatest interest, their importance is enormous, their use is continually increasing, and they seem destined to produce a great revolution in the civilized world.[1]

These words, written by a 23-year-old Frenchman, Sadi Carnot, were published in 1824 in a book that was meant to be a nontechnical treatise on **heat engines:** devices, such as the steam engine, that convert heat into useful work. Carnot's work was remarkable in that it preceded

---

[1] Sadi Carnot, *Reflections on the Motive Power of Fire, and on Machines Fitted to Develop that Power,* edited by E. Mendoza (New York: Dover, 1960), p. 3. (First published in 1824.)

by some 20 years Joule's demonstration of the equivalence of heat and mechanical energy. The book was favorably reviewed but was ignored by scientists and engineers. It quickly passed into obscurity and Carnot himself died, at the age of 36, in a cholera epidemic in 1832.

Émile Clapeyron rediscovered Carnot's book, recognized its significance, and in 1834 translated its argument into mathematical language—only to meet the same professional neglect that Carnot had earlier suffered, until Clapeyron's essay finally was translated into German and published in 1843 in the *Annals of Physics and Chemistry.* Thus, through Clapeyron's advocacy, Sadi Carnot, ignored in his own lifetime, laid the groundwork for thermodynamics.

Carnot had published his book at the height of the Industrial Revolution, that period beginning in the mid-eighteenth century when mechanical power (most notably from steam engines) was adapted to perform useful work. The name most closely associated with the steam engine is that of James Watt. (See Figure 12-1 for an early depiction of a Watt engine.) But Watt did not invent the steam engine. Rather, he introduced major refinements to the Newcomen engine. That engine, invented by Thomas Newcomen, and the Savery engine, invented by Thomas Savery, were steam-powered devices that had been in use since the late seventeenth century. The Newcomen engine was used primarily for lifting ground water from mines and wells; the Savery engine, designed for the same use but less effective at it, was used for lifting water limited distances and, beginning around 1750, for turning ma-

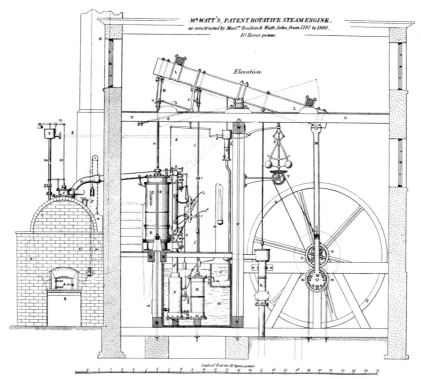

FIGURE 12-1

The Watt engine as depicted in an early illustration. The engine transformed the vertical action of the piston rod n into a rotary motion through a flywheel Q. Watt's special contribution included the steam condenser F, surrounded by water G. Courtesy of Eugene S. Ferguson, the Hagley Museum. From John Farey, *Treatise on the Steam Engine* (London, 1827).

chinery. Besides his refinements to the Newcomen engine, Watt's other major accomplishment is the one that probably linked his name so closely with the device: In 1769 he obtained a patent on a more efficient version of the steam engine that gave him an effective monopoly on it until the end of the century.

THE FIRST LAW OF THERMODYNAMICS

In 1845 a young Scottish mathematician, William Thompson (1824–1907), who later became Lord Kelvin, was working as a laboratory assistant at the *Collège de France* in Paris, helping to precisely measure the properties of gases. After reading Clapeyron's essay and being unable to locate a copy of Carnot's book in all of Paris, he pursued his own theoretical studies. In 1848 he published

the first in a series of papers entitled "On an Absolute Thermometric Scale Founded on Carnot's Theory. . . ," which began the research that ultimately led to the thermodynamic temperature scale bearing his name.

Kelvin's work was quickly followed by that of Rudolf Clausius, who in 1850 published a paper in which was first stated the **first law of thermodynamics:**

In all cases in which work is produced by the agency of heat, a quantity of heat is consumed which is proportional to the work done; and conversely by the expenditure of an equal quantity of work an equal quantity of heat is produced.[2]

---

[2]Rudolf Clausius, "On the Motive Power of Heat," in the Dover edition, above, of Sadi Carnot, *Reflections on the Motive Power of Fire*, p. 112.

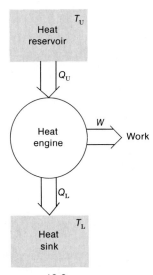

FIGURE 12-2
A heat engine does useful work.

(This is also a statement of conservation of energy, discussed in Chapter 10.)

The first law is illustrated by the schematic drawing of a heat engine in Figure 12-2. The engine draws a quantity of heat Q_U from the upper reservoir, or heat source, at a temperature T_U. During a cycle, a quantity of work W is done. A quantity of heat Q_L is then discharged to the heat sink at a temperature T_L. Conservation of energy tells us that

$$Q_U = Q_L + W$$

The equivalence between the mechanical energy of work and the thermal energy of heat is given by the mechanical equivalent of heat, first accurately determined by James Prescott Joule:

$$1 \text{ calorie} = 4.184 \text{ joules}$$

Heat Engines and Work

A heat engine has a working fluid, which is usually steam but can also be freon, mercury vapor, heated air, or the hot combustion products of gasoline. A heat engine works through a sequence of operations that do useful work and return the system to its initial configuration. We now examine the change in internal energy and work done in four cases: constant volume, constant temperature, constant pressure, and adiabatic (not accompanied by gain or loss of heat) expansion.

Energy is conserved. Useful work done by an ideal gas can come from one of two sources: from the working fluid's **internal energy** or from **heat** that flows into the system. This can be succinctly stated as

$$\Delta Q = \Delta U + \Delta W \qquad (12\text{-}1)$$

If a gas expands in a cylinder, it exerts a pressure on an area A. The area moves through a distance Δx. The work done by the gas is

$$\Delta W = \mathbf{F} \cdot \Delta \mathbf{x}$$
$$\Delta W = PA\Delta x = P\Delta V \qquad (12\text{-}2)$$

The pressure of an ideal gas is given by

$$PV = nRT$$

The change in internal energy of a gas is related to the **specific heat at constant volume** C_v.

$$\Delta U = mC_v\Delta T = nMC_v\Delta T \qquad (12\text{-}3)$$

where the mass of the gas m is equal to the number of moles present n times the molecular weight M of the gas. The product MC_v is the molar heat capacity at constant volume.

Constant Volume If a gas is heated under conditions of **constant volume** (isovolumetric heating) $\Delta V = 0$, the work done by the gas is zero:

$$\Delta W = P\Delta V = 0$$

Heat added to the gas increases its internal energy, raising the temperature from T_1 to T_2

$$\Delta Q = \Delta U = nMC_v\Delta T = nMC_v(T_2 - T_1)$$

Constant Temperature If a constant mass of gas is allowed to expand in contact with a heat reservoir and the process is done slowly enough, heat will flow from the reservoir into the gas. The gas and the reservoir remain at the same temperature. In this expansion at **constant temperature** (isothermal expansion), the change in internal energy of the gas is zero:

$$\Delta U = nMC_v\Delta T = 0$$

The heat flow ΔQ into the system is equal to the work done ΔW by the system:

$$\Delta Q = \Delta W = P\Delta V = nRT\frac{\Delta V}{V}$$

If the gas expands from a volume V_1 to a volume V_2, the pressure, of course, drops. From the ideal gas law we have that

$$P_1V_1 = P_2V_2 = nRT$$

where the temperature is assumed to be constant. The work done ΔW in changing the volume a small increment, ΔV, is proportional to nRT and times the fractional change in volume. The resulting work done in expanding from V_1 to V_2 is just[3]

$$Q = W = nRT \ln(V_2/V_1) \qquad (12\text{-}4)$$

Constant Pressure If a gas is allowed to expand in volume from V_1 to V_2 at **constant pressure** P (isobaric expansion), the work done is

$$\Delta W = P\Delta V \qquad (12\text{-}5)$$

$$\Delta W = P(V_2 - V_1)$$

for expansion V_2 is greater than V_1 and the work is

---

[3] The natural logarithm of x is $\ln x = y$. Then $x = e^y$ where $e = 2.718\,28 \ldots$ Using the notation of calculus:

$$W = \int_{V_1}^{V_2} nRT\frac{dV}{V} = nRT \ln\left(\frac{V_2}{V_1}\right)$$

positive. From the ideal gas law we have that

$$V_1 = \frac{nRT_1}{P}; \qquad V_2 = \frac{nRT_2}{P}$$

$$\Delta W = P\left(\frac{nRT_2}{P} - \frac{nRT_1}{P}\right) = nR(T_2 - T_1)$$

If the volume expands at constant pressure, the temperature must also increase. The work done can be written as

$$\Delta W = nR\Delta T$$

If the temperature increases, internal energy increases:

$$\Delta U = nMC_v\Delta T$$

To raise the temperature of the gas, and to do the mechanical work in expansion, heat must flow into the system:

$$\Delta Q = \Delta U + \Delta W$$

$$\Delta Q = nMC_v\Delta T + nR\Delta T$$

$$\Delta Q = n(MC_v + R)\Delta T = nMC_p\Delta T \qquad (12\text{-}6)$$

C_p is the **specific heat at constant pressure,** MC_p is the molar heat capacity at constant pressure; R is the gas constant. It has long been recognized that the difference between the molar heat capacity at constant pressure and at constant volume of many gases is constant:

$$MC_p - MC_v = R$$

The **ratio of specific heats** is

$$\gamma = \frac{MC_p}{MC_v} = \frac{MC_v + R}{MC_v} = 1 + \frac{R}{MC_v} \qquad (12\text{-}7)$$

Adiabatic Expansion A gas can be isolated and allowed to expand in the absence of heat flow, $Q = 0$. This is called **adiabatic expansion.** The energy to do the work in expanding must come from the internal energy of the gas. The gas will cool.

$$\Delta Q = \Delta U + \Delta W = 0$$

$$nMC_v\Delta T + P\Delta V = 0$$

From the ideal gas law this becomes

$$nMC_v\Delta T = -\frac{nRT}{V}\Delta V$$

As the gas expands, the temperature of the gas drops. The fractional change in absolute temperature is proportional to the fractional change in volume:

$$\frac{\Delta T}{T} = -\frac{R}{MC_v}\left(\frac{\Delta V}{V}\right)$$

The initial condition V_1, T_1 and final condition V_2, T_2 are related by

$$\ln\frac{T_1}{T_2} = -\frac{R}{MC_v}\ln\frac{V_1}{V_2}$$

This can be written as

$$\frac{T_1}{T_2} = \left(\frac{V_1}{V_2}\right)^{-R/MC_v}$$

The ratio of the ideal gas law for the initial and final states is

$$\frac{P_1 V_1}{P_2 V_2} = \frac{nRT_1}{nRT_2} = \left(\frac{V_1}{V_2}\right)^{-R/MC_v}$$

This can be rewritten as

$$P_1 V_1^\gamma = P_2 V_2^\gamma \qquad (12\text{-}8)$$

where

$$\gamma = 1 + \frac{R}{MC_v}$$

is the ratio of specific heats. One can show that the work done in adiabatic expansion from V_1 to V_2 is given by the change in internal energy of the gas:

$$W = \frac{P_1 V_1}{\gamma - 1} - \frac{P_2 V_2}{\gamma - 1} = \frac{nR(T_1 - T_2)}{\gamma - 1} \qquad (12\text{-}9)$$
$$= nMC_v(T_1 - T_2)$$

THE CARNOT CYCLE

Sadi Carnot saw the heat engine as a cycle—a series of operations that when completed returns the engine to its original configuration so that the cycle can be repeated. He also correctly concluded that any useful work done by an engine, and any heat rejected by it, must come from the original heat source. Mechanical work is done by "letting down" thermal energy from a high-temperature source to a lower-temperature sink. Efficiency increases as the temperature difference between the source and sink increases. Using the best data available, Carnot determined the maximum efficiency of a steam-powered engine, and compared the efficiencies of eight well-known steam engines of his day.

Carnot stated that to be efficient, the operations of a heat engine should take place in a **reversible manner**—that is, sudden changes in state are to be avoided, for a swift and irreversible process causes energy to be dissipated without useful work being done. For maximum efficiency, each operation of the cycle should be reversible. Consequently, in principle a heat engine can be run backward by putting mechanical work into it, causing heat to go from a lower-temperature to a higher-temperature reservoir. (This is in fact the principle of the refrigerator, which we discuss later.)

If a gas expands slowly in contact with a heat reservoir at a constant temperature, heat flows into the gas from the reservoir, keeping the gas in turn at a constant temperature. If, on the other hand, a gas is slowly compressed in contact with a reservoir at constant temperature, heat flows back into the reservoir. When done slowly, isothermal expansion and contraction are reversible processes.

If a gas undergoes adiabatic expansion (is isolated from a source of heat and allowed to expand), the work is done at the expense of the internal energy of the gas and its temperature drops. If a gas undergoes adiabatic compression, the work done in compressing the gas goes into the internal energy of the gas and the temperature rises. Adiabatic expansion and compression are also reversible processes.

We can compute the work gained from a heat engine operating between a heat source at temperature T_U and a heat sink at temperature T_L, using gas at an initial pressure P_o, volume V_o, and temperature T_L. A reversible cycle is represented on a pressure-volume diagram, Figure

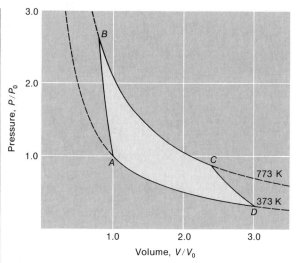

FIGURE 12-3
The Carnot cycle for an ideal gas operating between 737 K and 373 K isotherms.

12-3. The dotted lines show isothermal contours for an ideal gas at 373 K and 773 K. A **Carnot cycle** consists of the following sequence of operations.

1. Adiabatic compression of the initial gas from condition A at temperature T_L to point B at temperature T_U: Work must be done on the system.

2. Isothermal expansion of the gas from B to C: Work is done by the system and heat flows into it at temperature T_U.

3. Adiabatic expansion of the gas from C to D: Work is done by the system and the temperature drops to T_L.

4. Isothermal compression of the gas from D to A: Work must be done on the system and heat flows into the heat sink at temperature T_L.

Table 12-1 shows the work done and heat gain for each step of the Carnot cycle. We have already calculated the work done in isothermal and adiabatic expansion.

The total heat that flows into the heat engine from the source during the isothermal expansion B to C is given by equation 12-4:

$$Q_U = nRT_U \ln \frac{V_C}{V_B}$$

The heat flow out of the engine during D to A is given by

TABLE 12-1

The Carnot cycle, with heat flow into the system and the work done by the system shown for each step.

Step		ΔQ	ΔW
1	Adiabatic compression	0	$nMC_v(T_L - T_U)$
2	Isothermal expansion	$nRT_U \ln \dfrac{V_C}{V_B}$	$nRT_U \ln \dfrac{V_C}{V_B}$
3	Adiabatic expansion	0	$nMC_v(T_U - T_L)$
4	Isothermal compression	$-nRT_L \ln \dfrac{V_A}{V_D}$	$-nRT_L \ln \dfrac{V_A}{V_D}$

$$-Q_L = -nRT_L \ln \frac{V_A}{V_D}$$

The work done in adiabatic compression A to B is canceled out by the work done in adiabatic expansion C to D. Thus, the net work done is the difference between the work done by the system in isothermal expansion at the upper temperature and the work done on the system at a lower temperature T_L. It can be shown that the ratio of the volumes $V_A/V_D = V_B/V_C$. The net work done by the system is then just

$$W = nR(T_U - T_L) \ln \frac{V_C}{V_B}$$

Carnot Efficiency

The efficiency of the Carnot cycle, the **Carnot efficiency,** operating between temperatures T_U and T_L is given by the ratio of the work done to the heat into the system:

$$\text{Eff} = \frac{W}{Q_U} = \frac{Q_U - Q_L}{Q_U} \tag{12-10}$$

$$\text{Eff} = \frac{nR(T_U - T_L) \ln V_C/V_B}{nRT_U \ln V_C/V_B}$$

$$\text{Eff} = \frac{T_U - T_L}{T_U}$$

The maximum efficiency of a heat engine is its Carnot efficiency. It depends only on the temperature difference between the heat source and heat sink divided by the

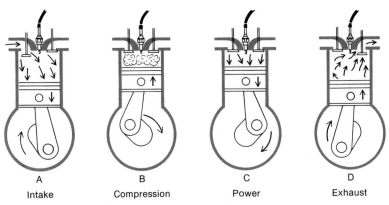

FIGURE 12-4
The Otto cycle as it occurs in the four-stroke spark-ignition engine in most automobiles.

absolute temperature of the heat source. The heat sink is usually at the temperature of our local environment, or of a nearby river, lake, or ocean. Efficient operation of a heat engine requires the highest possible upper temperature. The strength of the materials used is the limiting factor on that temperature. The combustion chambers of coal- and oil-fired powerplants can be made of materials that allow operation at relatively high temperatures (500°C), at a 40% overall efficiency. The reactors of nuclear powerplants must be made of material that requires operation at somewhat lower temperatures (300°C), at an overall efficiency of only about 33%. Hot water of geothermal brines and solar heated water, both low-grade forms of heat, cannot be very efficiently converted to mechanical work in a heat engine because of the limitation given by the Carnot efficiency.

EXAMPLE

A heat engine is to be operated using a heat source of water solar heated to 65°C (150°F) and a heat sink of groundwater at 15°C. What is the Carnot efficiency of such an engine?

$$\text{Eff} = \frac{T_U - T_L}{T_U} = \frac{338\,\text{K} - 288\,\text{K}}{338\,\text{K}}$$

$$\text{Eff} = \frac{50\,\text{K}}{338\,\text{K}} = 13\%$$

This is the maximum theoretical efficiency. If a heat engine works at 90% of the Carnot efficiency, from a solar collector operating at 75% thermal efficiency, and

the solar energy flux is 700 W/m², what actual mechanical work per second is derived from such an engine?

$$Q_1 = (0.75)(700\,\text{W/m}^2) = 525\,\text{J/m}^2$$

$$W = (\text{Eff})\,Q_1$$

$$W_{\max} = (0.13)(525\,\text{J/m}^2)$$

$$W_{\max} = 68\,\text{J/m}^2$$

$$W = (0.9)(68\,\text{J/m}^2) = 61\,\text{J/m}^2$$

With modest technology, about 60 watts of power per square meter of collector could be extracted while the sun is shining.

The Otto Cycle

The steam engine is an external-combustion engine (the steam is created outside the cylinder containing the piston). It was eventually realized—and Carnot himself had proposed in his book—that an internal-combustion engine (one in which some flammable gas-air mixture would be exploded in the cylinder) would be more compact and efficient. One such engine was designed and built by Jean Joseph Lenoir and, in 1860, hooked up to a small carriage. In 1876 Nikolaus Otto built a more efficient engine, the prototype of the engine used in most automobiles. The operation of the four-stroke internal-combustion gasoline engine (see Figure 12-4) approximates the Otto cycle, of which there are four stages: intake, compression, power,

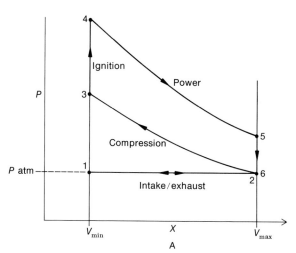

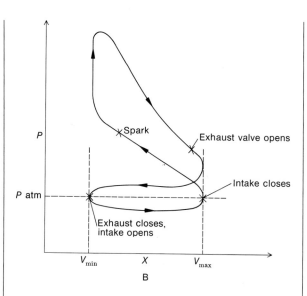

FIGURE 12-5

The Otto cycle: A. An idealized pressure-volume diagram. During the intake stroke the piston expands the volume at constant pressure, drawing in the gasoline-air mixture. Ideally, no work is done during the intake stage. The mixture is then quickly compressed adiabatically, increasing temperature and pressure. The spark plug fires near the point of maximum compression, liberating a quantity of heat Q_U, which increases the temperature and pressure at constant volume. During the power stroke the gases expand adiabatically, doing mechanical work. Finally, during the exhaust stroke, the spent gases are cleared at constant pressure. The Otto cycle is not as efficient as a Carnot cycle.

B. A pressure-volume diagram for a real spark-ignition cycle, which is not as efficient as the idealized cycle. The engine's maximum compression is limited by the gasoline-air mixture's ignition temperature. Compression increases the temperature of the mixture. If compression is too great, the ignition temperature is reached prematurely, causing preignition, which costs the engine efficiency and power. Because combustion takes time, often the spark is advanced to begin ignition slightly before the piston reaches top dead center. In the real engine the piston must do work to get air in and out of the cylinder. From William C. Reynolds, *Thermodynamics,* 2nd ed. (New York: McGraw-Hill Book Company, 1968), pp. 271, 274.

and exhaust. Figure 12-5 A shows a pressure-volume diagram for an ideal Otto cycle. The real spark-ignition engine does not perform as well as the idealized Otto cycle as Figure 12-5 B shows.

Heat Pumps

Reversible heat-engine cycles can be run backward by putting mechanical work into the engine to move heat from a lower reservoir to an upper reservoir. This is the principle of the **heat pump,** as shown in Figure 12-6. The expansion refrigerator and the air conditioner are both heat pumps.

An electric motor drives a compressor that compresses the working fluid, usually a member of the freon family,

such as Freon-12 (dichloro-difluoro methane, $Cl_2F_2C_2$). Because the gas has been compressed it is above the temperature of the environment. In a set of radiator coils, usually of copper or aluminum tubing, the gas loses heat Q_U to the environment, at upper temperature T_U. These coils can be seen on the back of a refrigerator or attached to an air conditioner's cooling fins. Under pressure of 10 atmospheres Freon-12 liquifies at a temperature of about $40°C(140°F)$. The liquid freon passes through an expansion valve to the evaporator, where it vaporizes, expands rapidly, and cools. Falling below the temperature of the local surroundings, it absorbs heat Q_L from them at a temperature T_L. The low-pressure gas is then compressed again, completing the cycle.

The **coefficient of performance** measures the amount of heat delivered to the upper reservoir Q_U compared to

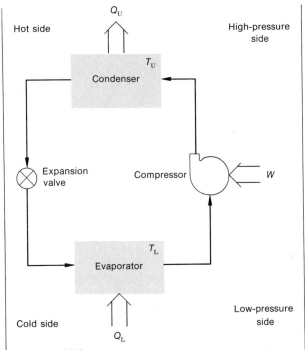

FIGURE 12-6
The heat pump. A heat engine can move heat from a cold region to a warm region by doing mechanical work.

the amount of mechanical work W the system does to move this heat. It is the reciprocal of the efficiency:

$$\text{COP} = \frac{Q_U}{W} = \frac{Q_U}{Q_U - Q_L} \qquad (12\text{-}11)$$

For a reversible or Carnot refrigerator the coefficient of performance is given by

$$\text{COP} = \frac{Q_U}{Q_U - Q_L} = \frac{T_U}{T_U - T_L}$$

The smaller the temperature difference, the greater the coefficient of performance. If the condenser is at a temperature of $T_U = 40°C = 313$ K and the evaporator is at $T_L = 10°C = 283$ K, the theoretical maximum coefficient of performance is

$$\text{COP} = \frac{T_U}{T_U - T_L} = \frac{313\text{ K}}{313\text{ K} - 283\text{ K}}$$

$$\text{COP} = 10.3$$

EXAMPLE

An air conditioning unit has a rated cooling capacity of evaporating 6000 Btu/hr, using 750 watts of power to do so. American air conditioners are rated with an energy efficiency ratio, EER, which is the ratio of the cooling power (you guessed it—in Btu/hr) to the electrical power required to drive the unit in watts. (How's that for mixed units!)

$$\text{EER} = \frac{6000\text{ Btu/hr}}{750\text{ W}} = 8\text{ Btu/W-hr}$$

The utility company recommends an EER of greater than seven for efficient room air conditioners.
(a) How much heat is moved in one hour in joules?
(b) How much work is done in one hour in joules?
(c) What is this air conditioner's coefficient of performance?

The heat removed at the evaporator in one hour is

$$Q_L = 6000\text{ Btu}(1054\text{ J/Btu}) = 6.32 \times 10^6\text{ J}$$

The electrical work done in one hour is

$$W = (750\text{ J/s})(3600\text{ s})$$

$$W = 2.70 \times 10^6\text{ J}$$

The heat discharged at the condenser is

$$Q_U = Q_L + W$$

$$Q_U = 9.02 \times 10^6\text{ J}$$

The coefficient of performance is

$$\text{COP} = \frac{Q_U}{W} = \frac{Q_L + W}{W} = \frac{Q_L}{W} + 1$$

$$\text{COP} = \frac{9.02 \times 10^6\text{ J}}{2.7 \times 10^6\text{ J}} = 3.34$$

The air conditioner with an ERR rating of 8 Btu/W-hr removes 2.34 joules of heat for every 1.00 joule of mechanical work done and discharges 3.34

joules of heat to the environment. The EER rating makes an air conditioner look better than it really is.

The air conditioner is not performing near maximum theoretical efficiency. The expansion of the gas at the expansion volume is not a reversible process. Theoretically, mechanical work could be extracted from the expansion and fed back to the compressor. Modifying the unit to do so would add considerable complication and cost but would improve energy efficiency. Energy is also required to operate the fans that move air through the evaporation coils.

THE SECOND LAW OF THERMODYNAMICS

The **second law of thermodynamics** has been stated many different ways, and there is no one best way. The law was strongly implied by Carnot: "Wherever there exists a difference of temperature, motive power can be produced." Carnot argued that if a cycle could be found that is more efficient than a reversible steam cycle, then a second heat engine could move more heat back to the source than was removed by the first engine, to restore the original conditions:

This would not only be perpetual motion, but an unlimited creation of motive power without consumption of either caloric or of any other agents whatever. Such a creation is entirely contrary to ideas now accepted, to the laws of mechanics and of sound physics. It is inadmissible.[4]

Clausius also argued, in 1850, that there is a second principle of thermodynamics: "It is impossible for a self acting machine, unaided by any external agency to convey heat from one body to another at a higher temperature."[5]

Kelvin quickly followed with his own publication in 1851, giving credit to Clausius but claiming independent discovery of his own second law of thermodynamics:

It is impossible, by means of inanimate material agency to derive mechanical effect from any portion of matter by cooling it below the temperature of the coldest of the surrounding objects.[6]

The Second Law of Thermodynamics is a fundamental postulate about the direction in which processes go. It asserts the impossibility of creating perpetual motion machines. With the combined authority of Clausius and Kelvin, the laws of thermodynamics were firmly established and the caloric theory was abandoned. This set the stage for the real understanding of heat engines, and for tremendous industrial development.

Entropy

The first law of thermodynamics tells us what is possible to do while conserving energy. The second law places serious restrictions on what is possible. Clausius extended the analysis of the second law to define a quantity, the **entropy,** which is a mathematical function that depends only on the state of a system (the temperature, pressure, and type and quantity of matter). If a system goes through a reversible process—if, for instance, heat ΔQ is transferred from part A to part B of a system—then the entropy of part B increases by an amount

$$\Delta S_B = \frac{\Delta Q}{T}$$

where T is the thermodynamic temperature in Kelvin. Of course, this same amount of heat is lost by part A. Its entropy change is

$$\Delta S_A = -\frac{\Delta Q}{T}$$

The system's total change in entropy is zero.

If heat is transferred suddenly, violently, or irreversibly, the total change in the system's entropy is greater

---

[4]Sadi Carnot, *Reflections on the Motive Power of Fire*, p. 12.

[5]Rudolf Clausius, *ibid.*, p. 134.

[6]Quoted in John F. Sandfort, *Heat Engines* (New York: Doubleday, Anchor Science Study Series S 27, 1962), pp. 86–87.

than zero. The total entropy S of a closed system never decreases; it either remains the same or increases:

$$\Delta S \geq 0$$

The entropy of part of a system can decrease, but only if the entropy of its environment increases. For a heat engine going through a reversible cycle, the total entropy change of the system including the upper heat source, the heat engine, and the lower heat sink is zero.

Reversible Cycles

Earlier we discussed the heat flow and work done by a Carnot cycle. In a heat engine, whenever there is a heat flow Q_U into an engine from a heat source at temperature T_U, the entropy of the engine increases:

$$\Delta S_1 = \Delta Q / T = Q_U / T_U$$

If an amount of heat Q_L flows out of the system at a temperature T_L, the entropy of the system decreases by

$$\Delta S_2 = \Delta Q / T = -Q_L / T_L$$

The net change in entropy of the engine is

$$\Delta S = \Delta S_1 + \Delta S_2$$
$$\Delta S = Q_U / T_U - Q_L / T_L$$

From Table 12-1 we can see that $Q_U / T_U = nR \ln(V_C / V_B)$ and that $-Q_L / T_L = nR \ln(V_A / V_D)$. Also, for the Carnot cycle $V_A / V_D = V_C / V_B$. For this reversible cycle,

$$Q_U / T_U - Q_L / T_L = 0$$

the net entropy change is zero. The maximum theoretical efficiency, or Carnot efficiency, depends on the temperatures of the heat source and heat sink. If the system is not completely reversible, the entropy change is positive:

$$\Delta S = Q_U / T_U - Q_L / T_L \geq 0$$
$$T_L / T_U \geq Q_L / Q_U$$

The actual efficiency is less than the Carnot efficiency:

$$\text{Eff} = 1 - Q_L / Q_U \leq 1 - T_L / T_U$$

Isothermal Expansion

Suppose a system consisting of n moles of a gas in a chamber of volume V in contact with a heat reservoir at a temperature T. Allowing the gas to expand yields useful work given by equation 12-4. If the gas expands isothermally to a volume $2V$, heat flows from the reservoir to the system in an amount equal to the work done:

$$\Delta W = \Delta Q_s = nRT \ln \frac{2V}{V}$$

The entropy of the system increases by an amount

$$\Delta S_s = \Delta Q / T = nR \ln 2$$

The entropy of the heat reservoir decreases by the same amount. The net entropy change of the universe in this reversible expansion is zero.

$$\Delta S = \Delta S_r + \Delta S_s = 0$$

The system can be restored to its original volume by doing mechanical work. Heat will flow out of the system into the reservoir. The same work gained in the expansion can be used to compress the gas and the change in the system's entropy is zero. The processes of isothermal expansion and compression are reversible.

Free Expansion

Gas contained in a volume V in contact with a heat reservoir at a temperature T is allowed to expand to a volume $2V$, freely, without doing work. This **free expansion** is an irreversible process. The molecules simply wander out to fill a larger volume of space; the volume doubles and the pressure halves. Since no work is done by the gas, no internal energy is lost, and the temperature of an ideal gas does not change, and no heat flows from the heat reservoir. The change in entropy of the heat reservoir is zero. There is, however, an increase in the entropy of the system, because it has undergone an irreversible process: To restore the system to its original volume, mechanical work must be done. As the gas is compressed at constant temperature T, the work done on the gas is

$$\Delta W = -nRT \ln \frac{V}{2V} = nRT \ln 2$$

Since the gas is maintained at constant temperature and constant internal energy, this same amount of heat must flow out of the gas into the heat reservoir:

$$\Delta Q = nRT \ln 2$$

The volume of the system has returned to its original state, and the entropy of the reservoir has been increased by an amount

$$\Delta S = nR \ln 2 > 0$$

In the cycle of free expansion followed by isothermal compression to an original state, mechanical work has been converted to heat and the entropy of the universe has increased.

Entropy and Information

Kinetic theories of matter developed during the latter part of the nineteenth century brought further knowledge of entropy. Maxwell proposed a method by which it would be possible in principle to violate the second law of thermodynamics. This was **Maxwell's demon,** which caused much philosophical debate. Suppose a chamber of gas is divided into compartments separated by a little door, as shown in Figure 12-7. The keeper of the door is a little demon with very specific instructions. If a fast molecule approaches the door he is to open it for an instant and let it pass. A slow molecule is to be refused entry. The mean kinetic energy and thus the temperature of the molecules in one chamber will gradually increase at the expense of the other. Once a temperature difference has been developed, a heat engine could be run, in violation of the second law.

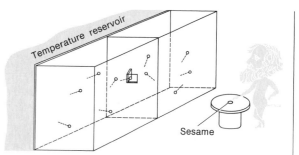

FIGURE 12-7
The second law of thermodynamics can apparently be violated by Maxwell's demon controlling a door between two chambers. The demon allows only fast molecules through the door. From Harold J. Morowitz, *Entropy for Biologists,* Copyright © 1970, Academic Press, Inc., New York. Used with permission.

Some entropy is associated with knowledge of a system's state. Suppose two chambers are separated by a door. Suppose there is one molecule in the system and in equilibrium with a heat reservoir at temperature T. If the particle's location is known, it is in principle possible to rig up a piston of some sort and extract useful work from the system. On the average, the useful work in allowing the system to expand to twice its volume is

$$\Delta W = kT \ln 2$$

Having established that the molecule is on the other side, we could reverse our machine and gain additional work. With such a trap-door machine, we could, in principle, keep extracting work from the system indefinitely. We would have a machine whose sole result is to extract thermal energy and convert it into work in violation of the second law of thermodynamics. It requires information to determine which side the particle is on. But to obtain information about the state of the system we must either increase the entropy somewhere within it or we must do work.

Energy Flow

An isolated system left to itself tends to an equilibrium condition of maximum disorder or randomness, maximum entropy. Yet life around us, however chaotic, possesses a high degree of orderliness. All biological activity is involved with creating organized structure and patterns of behavior—sometimes referred to as negative entropy.

Biological systems are not equilibrium systems. They are dynamic systems the viability of which depends on the continuous flow of energy. Energy flow through a system, in fact, acts to organize that system.

The ultimate source of all energy in the biosphere is the sun's radiation. Some of this radiation is absorbed, for example, by photosynthesis in plants and is converted to chemical energy (as discussed in Chapter 5). This ordered energy flowing through the biosphere is used for creating structures or doing work. The available energy is continually degraded, with some portion converted at each step to random thermal energy. Eventually all of this energy is converted to heat and leaves the earth as infrared radiation. It is this flow of energy that gives rise to activity. Plants and animals continually do chemical and mechanical work in creating complex molecules and in transporting fluids across membranes. Animals also do mechanical work in moving about. This flow of energy allows the organism to maintain a highly organized state without violating the second law of thermodynamics. If the flow of energy is shut off, the organism begins to decay.

That a continuous flow of energy is required to maintain a state of orderliness can be illustrated by examples from everyday life. A box of blocks or a bag of marbles neatly sitting on a shelf in a child's room represents an ordered state. The most probable configuration for the blocks or marbles is to be scattered all over the child's room and probably throughout the house: a state of increased randomness and entropy. To restore a state of orderliness, children must be taught to "pick-up-after-themselves," that is, to direct some of their energy flow to the task of ordering their toys. Without this energy flow, everything quickly degenerates to a state of randomness and confusion.

The decreased entropy or the increased orderliness in a system is at the expense of an increase in entropy of the environment. Suppose a chair is to be made. The necessary supplies—wood, a box of screws, some glue, etc.— are obtained from the local lumberyard and taken into a neat, clean workshop (a place for everything and everything in its place). The next afternoon, the chair has been fabricated. It is a masterpiece of form and function, executed with skill and precision. The shop is a mess— tools scattered, sawdust, scraps of wood, the box of screws spilled, etc. The entropy decrease in creating the well-ordered chair is more than compensated for by the dust and confusion and entropy increase of the shop. We must put out the energy to organize our environment. We must work just to keep entropy from winning.

In the context of our complex, highly organized industrial society, this fact has been raised to the status of a law of nature—Murphy's law: "If something can go wrong it will." Great care, attention, and expenditure of energy are necessary to keep all of the machines and systems of our society functioning. The most probably eventuality is that something will go wrong.

Review

Give the significance of each of the following terms. Where appropriate give two examples.

heat engine
first law of thermodynamics
internal energy
heat
specific heat at constant volume
constant volume
constant temperature

constant pressure
specific heat at constant pressure
ratio of specific heats
adiabatic expansion
reversible manner
Carnot cycle
Carnot efficiency

heat pump
coefficient of performance
second law of thermodynamics
entropy
free expansion
Maxwell's demon

Further Readings

Eugene S. Ferguson, "The Origins of the Steam Engine," *Scientific American* (January 1964).

D. K. C. MacDonald, *Faraday, Maxwell, and Kelvin* (New York: Doubleday, Anchor Science Study Series S 33, 1964).

Harold J. Morowitz, *Energy Flow in Biology: Biological Organization as a Problem in Thermal Physics* (New York: Academic Press, 1968). An excellent, sophisticated discussion of energy flow and entropy. For more advanced treatment see, by the same author, *Entropy for Biologists: An Introduction to Thermodynamics* (New York: Academic Press, 1973).

John F. Sandfort, *Heat Engines: Thermodynamics in Theory and Practice* (New York: Doubleday, Anchor Science Study Series S 27, 1962).

Problems

THE FIRST LAW OF THERMODYNAMICS

1 Using data given in Watt's patent claim, Carnot did a calculation of the work done by isothermal expansion. Steam enters the cylinder of diameter 0.4 m at a pressure of 4 atmospheres at a temperature of 148°C. The steam expands at constant temperature as the piston moves from 0.3 m to 1.2 m. 1 atmosphere $= 1.013 \times 10^6$ N/m^3.
   (a) What is the final pressure of the gas?
   (b) How much work is done by the gas?

2 Steam at a temperature of 200°C and a pressure of 15.3 atmospheres enters a cylinder of diameter 0.5 m. At constant temperature and in contact with a heat source, the steam expands from a displacement of 0.4 m to 1.6 m.
   (a) What is the steam's final pressure?
   (b) How much work is done by the steam?

3 Steam at a temperature of 330°C and a pressure of 6.0 atmospheres enters a cylinder with a volume of 0.1 m^3. The steam then expands adiabatically to a final volume of 0.4 m^3. The ratio of specific heat for steam is $\gamma = 1.3$.
   (a) What is the steam's final temperature?
   (b) What is the steam's final pressure?
   (c) How much work is done in the expansion?

THE CARNOT CYCLE

4 A nuclear powerplant provides steam at an upper temperature of 300°C and a lower temperature of 40°C.
   (a) What is the powerplant's maximum thermodynamic efficiency?
   (b) If the plant produces 1000 megawatts of electric power at the Carnot efficiency, how much heat per second is required to run the generators?
   (c) How much heat per second is discharged to the environment?

5 A coal-fired powerplant provides steam at an upper temperature of 540°C (1000°F) and has a condenser at a lower temperature of 40°C.
   (a) What is the powerplant's maximum thermodynamic efficiency?

(b) If the plant produces 1000 megawatts of electric power with this maximum efficiency, how much heat per second is needed to run the generators?

(c) How much heat per second is discharged to the environment?

6   A room air conditioning unit has a capacity to remove 6000 Btu/hr of heat from the air on the cooling side at a temperature of 10°C. The unit requires 850 watts of power. The condenser side operates at a temperature of 45°C.

(a) How much energy per second is absorbed on the cooling side?

(b) How much energy per second is discharged in the condenser?

(c) What is the actual coefficient of performance?

(d) What is the maximum theoretical coefficient of performance?

IV
ELECTRICITY AND MAGNETISM

13

Electric Charges and Fields

THE NATURE OF ELECTRICITY

Electricity is a great mystery to many. Only rarely do we directly perceive the effects of electricity—for example, static electricity in our clothes or in freshly brushed hair, or the occasional shock on a dry winter day when we walk across a carpet and touch a person or an object. If we poke around in electrical equipment or try to remove a spark plug wire from a running engine, we may get a nasty shock. The tingling sensation is dramatic evidence that the nervous system is itself electrical.

For most students this is the first introduction to electricity and magnetism. A number of new, interrelated concepts must be grasped separately and only later coherently integrated. At first this may be a bit confusing, but patient, careful reading and perseverence, and applying the ideas in the problems and doing the laboratory work that is customary with this course, should make the central ideas clear.

In this and the next five chapters we discuss fundamental electrical phenomena and examine modern concepts of electricity and magnetism: In this chapter we examine *electrostatics*, electric charge at rest; in Chapters 14 and 15, electric *currents* and *magnetic fields*, charge in uniform motion. The ideas of charge, current, electric field, electrical potential difference (voltage), magnetic field, electromotive force, electrical work, and electric power are interrelated throughout these three chapters and are applied in Chapters 16 and 17 to electric signals, electric power, and electric safety. In chapter 18 we conclude our discussion of electrical phenomena by examining the consequences of accelerated electric charge, *electromagnetic radiation*.

Electric Charge

The ancient Greeks observed that a piece of amber, when rubbed, attracted dust and chaff. Electron, the Greek word for amber, came to be applied to such phenomena. In 1734, Charles du Fay declared electricity is composed of two distinct fluids, or electricities: vitreous, associated with rubbed glass, rock crystal, wool, and fur; and resinous, associated with rubbed amber, opal, silk, thread, paper, and many other substances. He reported:

The characteristic of these two electricities is that a body of, say, the vitreous electricity repels all such as are of the same electricity; and on the contrary, attracts all those of the resinous electricity.[1]

In other words, like charges repel, unlike charges attract. A **charge** is any quantity of electricity.

Electrical attraction and repulsion can be observed in an **electroscope,** typically a thin gold leaf suspended from a metal plate, both hung from an insulated support

---

[1] Charles du Fay, quoted in Leon N. Cooper, *An Introduction to the Meaning and Structure of Physics, Short Edition* (New York: Harper and Row, 1970), p. 185.

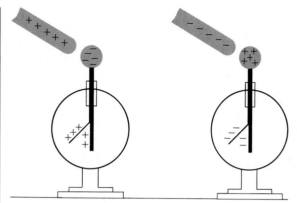

FIGURE 13-1
An electroscope can detect both positive and negative charges.

inside a metal box that removes the influence of stray charge. Rubbing a rubber rod with cat fur induces a charge on the rod and the opposite charge on the fur. Bringing the charged rod near the electroscope attracts the opposite charge to the top of the electroscope and repels the like charge, as shown in Figure 13-1. Whether the rod is initially charged positive or negative, the result is the same: The gold leaf stands away from its support by repulsion of like charge on the leaf. An uncharged electroscope cannot distinguish between charges of different signs. Each charge produces the same effect.

We can charge an electroscope by a process of induction, which involves the following sequence: A positive rod brought near the electroscope attracts negative charge to the top, leaving positive charge on the leaf. The leaf deflects, as shown in Figure 13-2 A. If the top of the electroscope is touched, the charge on the leaf escapes (Figure 13-2 B). The negative charge at the top is held in place by attraction of the rod's positive charge. Withdrawing the rod redistributes the remaining negative charge and the leaf indicates the presence of charge (Figure 13-2 C). When an electroscope is being charged by induction, the rod never touches the electroscope and the opposite charge is left behind.

A charged electroscope can distinguish the sign of a charge brought near it: A positive charge makes the leaf collapse as the negative charge is attracted to the top of the electroscope; a negative charge pushes more negative charge toward the leaf, repelling it further as shown in Figure 13-3.

The researches of Benjamin Franklin, conducted between about 1744 and 1751, significantly advanced understanding of electricity. His book *Experiments and Observations on Electricity Made in Philadelphia* was widely read in Europe and translated into several languages. Whereas du Fay had thought of electricity as made up of two fluids, Franklin thought of it as a single fluid. If a body gains an excess of this fluid, it is electrified, or charged. Franklin believed that this electric fluid was not created, but was redistributed by the act of rubbing. A

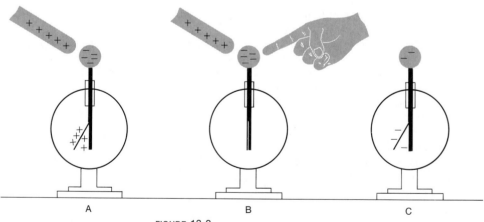

A   B   C

FIGURE 13-2
Charging an electroscope by induction.

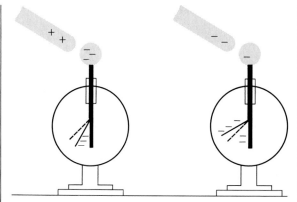

FIGURE 13-3
A charged electroscope responds differently to positive and negative charges.

glass rod rubbed with a silk cloth gains electric fluid. Franklin called this excess of electric fluid or charge *positive* electricity. The silk cloth, on the other hand, must lose the same amount of fluid, becoming *negative*. Franklin's choice of positive and negative as the signs of charge was arbitrary, but they remain in use to this day.

Coulomb's Law

To determine how electric forces vary with the distance between charges, Charles Augustin de Coulomb, in about 1785, constructed a torsion balance that could accurately measure small forces between charges. His experiment was similar to that of Cavendish, who measured (as described in Chapter 4) the gravitational force between masses. Two spheres were balanced on a rod and suspended from the torsion balance, a fine fiber or wire. A charge was placed on one sphere. When another charge was brought near that sphere, the electrical attraction or repulsion exerted a force on the sphere. The torsion balance rotated by an amount proportional to the force applied. By varying the distances between the charged spheres, Coulomb verified, within the limits of experimental accuracy, what is known as **Coulomb's law:** *The force between two electric charges is inversely proportional to the square of the distance between the centers of charge, and the force acts along the line of centers:*

$$\mathbf{F} = k_e \frac{Q_1 Q_2}{R^2} \mathbf{i}_R$$

where $\mathbf{i}_R$ is a *unit vector*, a dimensionless vector pointing radially outward along the line joining the two charges. This vector indicates the direction of the force.

The SI unit of charge is the **coulomb,** C, defined as one ampere-second, A-s. The ampere is defined and discussed in Chapter 14.) The unit of force is the newton, the unit of distance the meter. The charge, force, and distance, in SI units, are related by a constant factor k_e,[2] with a value of

$$k_e = 8.9876 \times 10^9 \text{ N-m}^2/\text{C}^2$$

$$k_e \cong 9.0 \times 10^9 \text{ N-m}^2/\text{C}^2$$

Electric Field

Coulomb's law describing interaction between two charged particles implies that there is action at a distance. How can two charges that are not in contact with one another, or not exchanging particles, interact over a great distance? They interact through an **electric field.**

The electric field is analogous to the gravitational field, by which also masses interact at a distance (though gravitational forces are always attractive, whereas electric forces may attract or repel). The presence of a mass distorts or creates a condition in the space around it: This gravitational field $\mathbf{g}$ exists in the space around some mass M whether or not another mass is present to interact with it. The presence of other masses does not affect the field produced by the first mass. Because of the gravitational field, another mass m experiences a force, the magnitude of which depends on the strength of the gravitational field $\mathbf{g}$ and the size of the mass m:

$$\mathbf{F} = m\mathbf{g}$$

Thus, two masses interact not directly but through the medium of a gravitational field. The net gravitational field at a point in space is simply the vector sum of the gravitational fields produced by all of the other masses.

---

[2]Many texts use the constant $K_e = 1/4\pi\epsilon_o$, where $\epsilon_o = 8.854 \times 10^{-12} \text{ m}^2/\text{N-C}^2$.

Likewise, according to the concept of the electric field, developed through the work of Michael Faraday in the early nineteenth century, the presence of an electric charge distorts or creates a condition in the space around the charge. This condition, known as an electric field **E,** exists around a charge whether or not another charge is present. The presence of other charges does not affect the electric field produced by the first charge. The net electric field at a point in space is simply the vector sum of the electric fields produced by all of the charges. A small charge in an electric field, a **test charge,** experiences a force **F** that depends on the electric field strength **E** and the size of the charge q:

$$\mathbf{F} = q\mathbf{E}$$

Thus, two charges interact not directly but through the medium of the electric field.

Suppose a positive test charge q is in the vicinity of a second charge Q. The force between them is

$$\mathbf{F} = k_e \frac{qQ}{R^2}\mathbf{i}_R$$

The force is proportional to the charge on the test particle. In the vicinity of a charged particle Q, we can define an electric field that represents the force per unit charge acting on a small positive test charge q:

$$\mathbf{E} = \frac{\mathbf{F}}{q}$$

The electric field has the units of newtons per coulomb, N/C.

The electric field in the vicinity of a positive charge Q is

$$\mathbf{E} = \frac{\mathbf{F}}{q} = k_e \frac{qQ}{qR^2}\mathbf{i}_R$$

$$\mathbf{E} = k_e \frac{Q}{R^2}\mathbf{i}_R$$

The electric field around a positive charge points radially outward, the direction in which a positive test charge

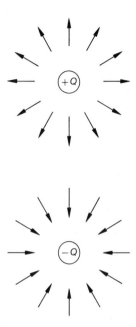

FIGURE 13-4
The electric field around a positive charge points radially outward. The electric field around a negative charge points radially inward.

would move. The electric field around a negative charge points radially inward. Figure 13-4 depicts the electric field around positive and negative charges.

Voltage

To move a charge against an electric field requires work. A test particle with a charge q in an electric field **E** experiences a force:

$$\mathbf{F} = q\mathbf{E}$$

To move the particle we must apply a force:

$$\mathbf{F}_{applied} = -q\mathbf{E}$$

To move the particle a distance $\Delta\mathbf{x}$ at a constant velocity we do an amount of work:

$$\Delta W = \mathbf{F}_{applied} \cdot \Delta\mathbf{x}$$

The work done is equal to the change in potential energy. The electrical **potential difference,** or change in **voltage,** is the *change in electrical potential energy per unit test charge:*

$$\Delta V = \frac{\Delta W}{q} = -\mathbf{E} \cdot \Delta \boldsymbol{x}$$

Conversely, the component of the electric field E_x in the direction of Δx is related to the change in electrical potential in the direction of Δx:

$$E_x = -\frac{\Delta V}{\Delta x}$$

The unit for the electrical potential is the **volt,** V, which is *work per unit charge or joule per coulomb:*

$$1\,V = 1\,\text{N-m/C} = 1\,\text{J/C}$$

To move a charge through an electric potential difference requires electrical work. Our reference potential, what we call zero volts, can be arbitrary. Quite often the earth is used as a zero reference, or so-called ground, because it absorbs considerable charge without significantly changing its potential. On the other hand, we may consider zero potential to be at a position so far from the charge that the charge has an insignificant effect. The electrical potential or voltage is defined only with respect to a reference voltage.

Work is done only in moving a charge in the direction of the electric field. If a charge is displaced a distance Δx perpendicular to the direction of the electric field $\mathbf{E}$ the work done in moving the charge is zero and the electrical potential is constant. This defines an **equipotential surface:** *The lines of constant electrical potential are everywhere perpendicular to the electric field.*

If a particle with a charge q is allowed to move through an electrical potential difference V, the work done on the charged particle is

$$W = qV$$

The electrical work is converted into an increase in the particle's kinetic energy. If the particle is accelerated from rest, the final kinetic energy is

$$KE = \tfrac{1}{2}mv_f^2 = qV$$

The work done on a charged particle depends only on the difference in electrical potential between the starting and end points; it is independent of the size of the electric field. Particle energies are often expressed in terms of electron volts, eV: One **electron volt** is *the amount of work required to accelerate an electron through a potential difference of one volt:*

$$1\,\text{eV} = (1.6 \times 10^{-19}\,\text{C})(1\,\text{V}) = 1.6 \times 10^{-19}\,\text{J}$$

EXAMPLE

An electron with a charge $-e = -1.6 \times 10^{-19}\,\text{C}$ is accelerated through a potential drop of $-32\,\text{kV}$. What is the work done on the electron? What is the final velocity?

$$W = qV = (-1.6 \times 10^{-19}\,\text{C})(-3.2 \times 10^4\,\text{V})$$

$$W = 5.12 \times 10^{-15}\,\text{J}$$

The electron energy in this example is 32 keV.

The final velocity can be determined from the kinetic energy.

$$v_f{}^2 = \frac{2KE}{m} = \frac{2(5.12 \times 10^{-5}\,\text{J})}{(9.1 \times 10^{-31}\,\text{kg})}$$

$$v_f = 1.1 \times 10^8\,\text{m/s}$$

Voltage on a Charged Sphere

If a positive charge Q is distributed on a sphere of radius R_0, a positive test charge near the sphere experiences a repulsive force. To overcome this force and move the test charge closer to the sphere, work must be done.

The force acting on the test charge at a distance R from the center of the sphere is radially outward

$$\mathbf{F} = k_e \frac{qQ}{R^2} \mathbf{i}_R$$

To move the test charge a small distance $\Delta\mathbf{x}$ toward the charged sphere we must do an amount of work:

$$\Delta W = -\mathbf{F} \cdot \Delta\mathbf{x}$$

The test charge will move from $\mathbf{R}_1$ to $\mathbf{R}_2$, along a line through the center, a distance

$$\Delta\mathbf{x} = \mathbf{R}_2 - \mathbf{R}_1 = (R_2 - R_1)\mathbf{i}_R$$

The average position of the particle is approximately

$$R^2 = R_1 R_2$$

The average work done in moving from R_1 to R_2 is given by

$$\Delta W = -\mathbf{F} \cdot \Delta\mathbf{x} = k_e \frac{qQ}{R_1 R_2}(R_1 - R_2)$$

$$= k_e qQ \left(\frac{1}{R_2} - \frac{1}{R_1} \right)$$

This argument is identical to that for gravitational potential energy in Chapter 4 and arrives at essentially the same result. Although the discussion above applies for a small step Δx, the result is valid between any initial and final positions. If we start at infinity $R_1 = \infty$, with zero potential energy, and move to a distance $R_2 = R$, the work done on test charge q is

$$\Delta W = k_e \frac{qQ}{R}$$

This is just the increase in potential energy of the test charge q. The electrical potential is the work per unit test charge. The potential at position R is

$$V = \frac{\Delta W}{q} = k_e \frac{Q}{R}$$

The potential goes to zero at infinity.

At all points at a given radius R from the charged sphere, the potential is constant. The equipotential surface around a point charge is a sphere. The equipotential surfaces around a single charge are shown in Figure 13-5.

EXAMPLE

In the Bohr model of the hydrogen atom, the electron moves around the proton in a circular orbit of radius 0.528×10^{-10} m. The charge on the proton is $Q = 1.6 \times 10^{-19}$ C. What is the voltage at the position of the electron orbit? The voltage is given by

$$V = k_e \frac{Q}{R}$$

whether the test charge has a positive or negative sign. That an electron is circulating at a radius of 0.528×10^{-10} m is irrelevant to determining the voltage at that distance:

$$V = \frac{(9.0 \times 10^9 \text{ N-m}^2/\text{C}^2)(1.6 \times 10^{-19} \text{ C})}{0.528 \times 10^{-10} \text{ m}}$$

$$V = 27.273 \text{ N-m/C}$$

$$V = 27.3 \text{ V}$$

The potential energy of the electron at this position is

$$W = qV$$

$$W = (-1 \text{ e})(27.3 \text{ V})$$

$$W = -27.3 \text{ eV}$$

The potential energy of the electron is negative, indicating that removing the electron from the atom requires work. The electron is in a potential energy well. The electron's potential energy in joules is

$$W = (-27.3 \text{ eV})(1.6 \times 10^{-19} \text{ J/eV})$$

$$W = -4.37 \times 10^{-18} \text{ joules}$$

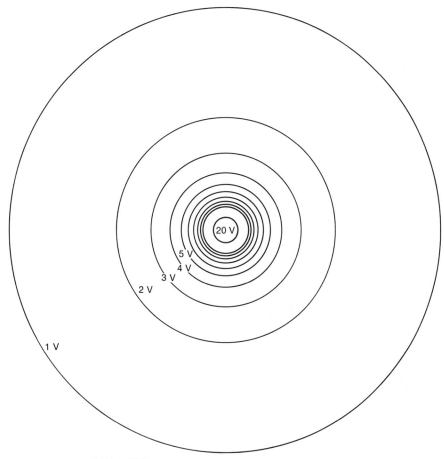

FIGURE 13-5
The equipotential lines around a point charge are concentric spheres.

Charge on a Sphere

Benjamin Franklin was first to recognize that because like charges repell, a charge placed on a hollow metal conductor tends to distribute itself on the outside surface, spreading out to the widest extent possible. Because metal is a conductor, a potential difference between two points on the metal surface would produce a flow of charge that would eliminate the potential difference. Consequently, a metal is an equipotential surface. The electric field is always perpendicular to that surface.

If a charge Q is distributed on a sphere of radius R, the magnitude of the electric field is given by

$$E = k_e \frac{Q}{R^2}$$

The area of the sphere is given by

$$A = 4\pi R^2$$

If the charge is distributed uniformly over a surface, the

charge per unit surface area, or **surface charge density,** σ (lower-case Greek letter *sigma*) can be defined:

$$\sigma = \frac{Q}{A}$$

$$Q = \sigma A = 4\pi R^2 \sigma$$

The magnitude of an electric field in the vicinity of a surface is then given by

$$E = k_e \frac{Q}{R^2} = k_e \frac{4\pi R^2 \sigma}{R^2} = 4\pi k_e \sigma$$

The electric field in the immediate vicinity of a surface is proportional simply to the surface charge density σ.

Franklin recognized that on a curved metal surface the largest charge distribution, and consequently the greatest electric field near the surface, would be where the curvature is greatest (see Figure 13-6). Franklin found that a sharp point would leak electricity to the atmosphere much more quickly than a large round surface. He put this idea to good practical and commerical advantage in the lightning rod. A sharply pointed rod (conductor) placed on a rooftop and connected to the ground by a heavy wire leaks charge to the atmosphere. Thus, the electric potential near a building (the tallest are in most danger) is prevented from becoming so large that a damaging lightning stroke would hit the building. Should lightning strike, its charge is safely conducted to ground.

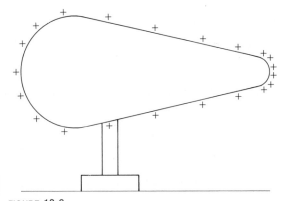

FIGURE 13-6
Electric charge remains on the outside surface of a hollow conductor and is concentrated where radius of curvature is smallest.

SUPERPOSITION OF ELECTRIC FIELDS

The electric field around a charge has both magnitude and direction; it is a vector quantity. If more than one charge is present, by the principle of **superposition** the total electric field produced is the vector sum of the fields due to each charge. The presence of a second point charge does not change the effect of the first point charge; it only adds to it. Figure 13-7 shows the electric field at a point P produced by two charges of magnitude $+Q$ separated by a distance D. The field's magnitude and direction at any point in space can be calculated. Figure 13-8 shows the electric field lines surrounding the charges. At the midpoint between the two equal charges, the electric field is zero.

The electric field at a point in space can also be determined when the charges differ in magnitude or in sign. Assume that a positive charge Q_1 is a distance D from a negative charge Q_2, which has the larger magnitude. Let us determine the point, along a line through the two

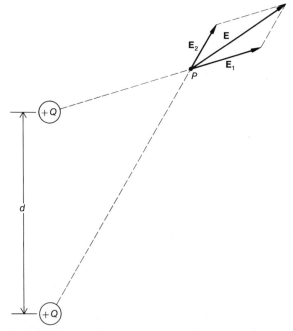

FIGURE 13-7
The total electric field $\mathbf{E}$ at point P is the vector sum of the fields produced by all charges.

FIGURE 13-8
The electric field around two equal charges.

electric fields $\mathbf{E}_1$ and $\mathbf{E}_2$ point in opposite directions. $\mathbf{E}_2$ is produced by the stronger charge Q_2; $\mathbf{E}_1$ is produced by the weaker but closer charge Q_1. If x is the distance to the left of Q_1 where the two fields cancel, we can then write that the magnitude of the electric field at x is

$$E = E_1 + E_2 = k_e \frac{Q_1}{R_1^2} + k_e \frac{Q_2}{R_2^2} = 0$$

$$E = k_e \left[\frac{Q_1}{x^2} + \frac{Q_2}{(x + D)^2} \right] = 0$$

We need only solve for the value of x.

EXAMPLE

Suppose charges $Q_1 = 10^{-9}\,\text{C}$ and $Q_2 = -2 \times 10^{-9}\,\text{C}$ are separated by a distance $D = 0.10\,\text{m}$, as shown in Figure 13-9. At what point on the line through the charges is the electric field zero? Setting the above result equal to zero and solving, we get

$$Q_1(x^2 + 2Dx + D^2) = -Q_2 x^2$$

$$\frac{(Q_1 + Q_2)}{Q_1} x^2 + 2Dx + D^2 = 0$$

This is in the form

$$ax^2 + bx + c = 0$$

The solutions to this equation are given by

$$x = \frac{-b \pm (b^2 - 4ac)^{1/2}}{2a}$$

charges, where the electric field goes to zero. Figure 13-9 shows the electric fields that each of two charges produce to their right, between them, and to their left. Between the two charges, the electric field $\mathbf{E}_1$ points away from Q_1 to the right. The electric field $\mathbf{E}_2$ points toward Q_2, also to the right. The two electric fields point in the same direction and always add. To the right of Q_2, $\mathbf{E}_2$ points to the left. $\mathbf{E}_1$, which points to the right in this region, is produced by the smaller charge Q_1 so that $\mathbf{E}_1$ is always smaller than $\mathbf{E}_2$. In the region to the left of Q_2, the

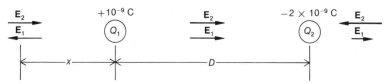

FIGURE 13-9
Electric fields from different charges are superimposed. Sometimes the fields add and sometimes they cancel.

For this problem:

$$a = \frac{(Q_1 + Q_2)}{Q_1} = -1$$

$$b = 2D = 0.2\,\text{m}$$

$$c = D^2 = 0.01\,\text{m}^2$$

There are two solutions to the equation:

$$x = +0.24\,\text{m}$$

and

$$x = -0.04\,\text{m}$$

The latter solution implies that the zero field point would be to the right of Q_1, for which the equation is invalid. This solution must be thrown out. The electric field is zero 0.24 m to the left of Q_1.

Electric Dipole Fields

An **electric dipole** consists of a charge $+Q$ and a charge $-Q$ separated by some distance d (see Figure 13-10). Let us examine the electric field at a point P on the line halfway between the two charges. Since the magnitude of the charges and the distance to them are the same, the magnitude of the electric field produced by each charge is the same:

$$E_A = E_B = E = k_e \frac{Q}{R^2}$$

The electric fields $\mathbf{E}_A$ and $\mathbf{E}_B$ point in quite different directions. We can determine the resultant electric field by the addition of components. In the y-direction the resultant electric field is

$$E_y = E_{Ay} + E_{By}$$

The two components are

$$E_{Ay} = -E \cos \theta$$

$$E_{By} = -E \cos \theta$$

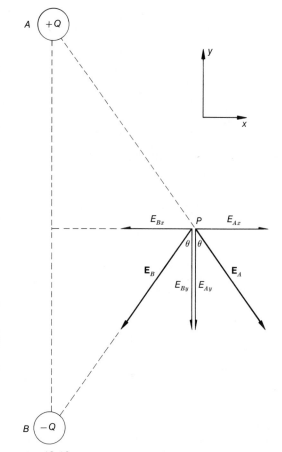

FIGURE 13-10
The electric field produced by two unlike charges is the vector sum of the electric field due to each charge.

The resultant is

$$E_y = -2E \cos \theta$$

In the x-direction the resultant electric field is

$$E_x = E_{Ax} + E_{Bx}$$

The components are

$$E_{Ax} = E \sin \theta$$

$$E_{Bx} = -E \sin \theta$$

The resultant is

$$E_x = 0$$

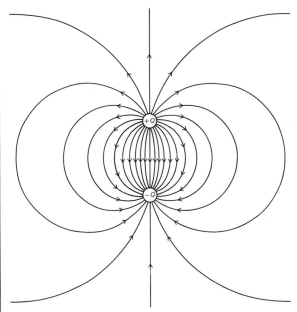

FIGURE 13-11
The electric field lines in the vicinity of equal and opposite charges.

represented by a vector that shows the direction in which a positive charge would move. Figure 13-11 shows the electric field in space in the vicinity of equal positive and negative charges.

Two equal charges $+Q$ and $-Q$ separated by a small distance d form an electric dipole. If a distance R is much greater than d, $R \gg d$, the electric field caused by the positive charge is almost exactly the electric field caused by the negative charge. A small field remains that is proportional to the **electric dipole moment,** the product of the charge times the separation distance:

$$\mathbf{p} = Q\mathbf{d}$$

where $\mathbf{d}$ is a displacement vector that points from the negative to positive charge. The electric field in the mid-plane of the dipole is given by

$$\mathbf{E} = -k_e \frac{\mathbf{p}}{R^3}$$

The electric field of the dipole falls off faster with distance than does the electric field around a charge. Figure 13-12 shows the electric field pattern around a small dipole.

The resultant electric field at the midplane between the two charges points in the negative y-direction. The electric field can be defined at every point in space and be

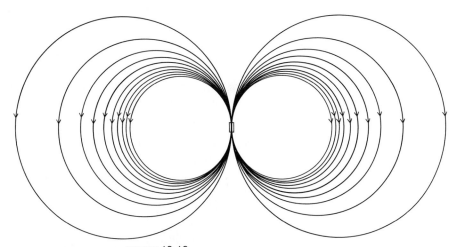

FIGURE 13-12
The electric field around a small electric dipole.

EXAMPLE

A water molecule consists of an oxygen and two hydrogen atoms. Because the electrons spend more time in the vicinity of the oxygen atom, it has a net negative charge of about $Q = -0.66$ e, and each hydrogen atom has a net positive charge of about $Q_2 = +0.33$ e. Figure 13-13 shows the location of the charge in a water molecule. Because of the charge separation, the water molecule possesses an electric dipole moment. The net charge of 0.66 e is separated by a net distance of about 0.057 nm in the y-direction. The magnitude of the dipole moment is

$$\mathbf{p} = Qd\,\mathbf{i}_y$$

$$\mathbf{p} = (0.66)(1.6 \times 10^{-19}\text{ C})(0.57 \times 10^{-10}\text{ m})\mathbf{i}_y$$

$$\mathbf{p} = 6.0 \times 10^{-30}\text{ C-m}\,\mathbf{i}_y$$

The magnitude of the electric field at a distance of 0.5 nm in the midplane of the dipole is

$$E = -k_e\frac{\mathbf{p}}{R^3} = -\frac{(9 \times 10^9\text{ N-m}^2/\text{C}^2)(6.0 \times 10^{-30}\text{ C-m})}{(5.0 \times 10^{-10}\text{ m})^3}\mathbf{i}_y$$

$$\mathbf{E} = -4.3 \times 10^8\text{ N/C}\,\mathbf{i}_y$$

Because of the electric dipole moment, there is a relatively large electric field around many molecules. The

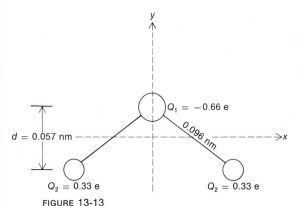

FIGURE 13-13
A water molecule has an electric dipole moment.

TABLE 13-1
The electric dipole moment of various molecules.

Molecule	Dipole moment 10^{-30} C-m
Carbon dioxide CO_2	0.00
Carbon monixide CO	0.4
Nitrous oxide N_2O	0.83
Hydrogen bromide HBr	2.6
Hydrogen chloride HCl	3.4
Ethyl alcohol C_2H_5OH	3.7
Ammonia NH_3	5.0
Water H_2O	6.0
Sulfur dioxide SO_2	5.5
Glycene	50
Egg albumin	830

molecules of many substances are strongly attracted to one another because their electric dipole moments produce strong electric fields. Table 13-1 shows the electric dipole moment of different molecules.

CAPACITORS

A **capacitor,** *a device for storing electrical charge, consists of two metal surfaces separated by a space that does not conduct charge.* The earliest capacitor was the Leyden jar (named after the Dutch city where it was invented in the early eighteenth century): Metal foil covers the inside and outside surfaces of a glass jar as shown in Figure 13-14. Protruding slightly from the jar is a rod topped with a knob and attached at its bottom to the inside foil. A charge Q transferred to the knob is immediately distributed over the entire surface of the inside foil. Only a small portion of the original charge remains on the knob. Repeated chargings can cause considerable electric charge to be stored in the jar. In the eighteenth century, as we saw, electricity was thought to be a fluid. The Leyden jar was thought to store charge by condensing this fluid, and hence was called a condenser. To this day capacitors in

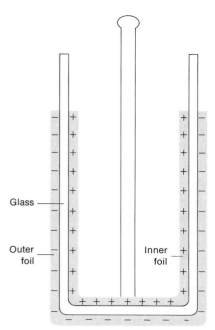

Glass

Outer foil

Inner foil

FIGURE 13-14
A Leyden jar "condenses" electricity, that is, stores electric charge.

some applications—as, for example, in automobile ignition systems—are called condensers. The Leyden jar, now only a scientific curiosity, could store relatively large quantities of charge, and thus allowed larger electrical effects to be studied.

A simple capacitor consists of two parallel conducting metal plates of area A separated by a small distance d, as shown in Figure 13-15. A charge $+Q$ placed on the upper plate tends to attract a charge $-Q$ on the lower plate. The electric field between the plates points from the positive to the negative plate. If the width of the plates is much larger than the separation distance d, then the electric field between the plates is uniform and is proportional to the surface charge density $\sigma = Q/A$. The magnitude of the electric field is given by

$$E = 4\pi k_e \sigma = 4\pi k_e \frac{Q}{A}$$

The larger the charge on the capacitor, the larger is the electric field between the plates.

Inserting a dielectric material between the capacitor plates modifies the capacitor's behavior. A dielectric ma-

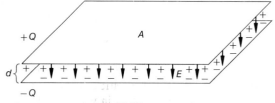

FIGURE 13-15
The parallel-plate capacitor.

terial, or simply a **dielectric,** is an insulator, a material in which electrons are bound to a particular atom or molecule and are not free to move through the material. The electrons in a dielectric respond to an electric field by being displaced slightly from the positive charge of the atom or molecule. This produces an induced electric dipole moment, which weakens the electric field. If the dielectric consists of molecules with electric dipole moments (as pure water does), these dipole moments also weaken the electric field. Figure 13-16 shows the effect of the atomic or molecular dipole moments within the dielectric. The charges on the dipoles cancel everywhere except at the edge, where a bound charge forms. The **dielectric constant** K is a measure of the ability of a material to weaken an electric field. If **E** is the electric field in the absence of a dielectric, the electric field within

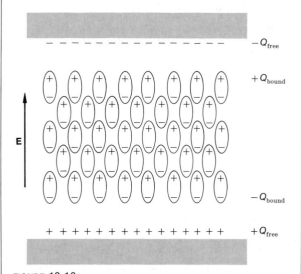

FIGURE 13-16
Bound electric charge on the surface of a dielectric material reduces a capacitor's electric field.

a dielectric is $\mathbf{E}/K$, as shown in Figure 13-17. Table 13-2 shows typical values of the dielectric constant for various materials. The dielectric constant in vacuum is 1.000. If the space between the plates of a capacitor is filled with a material with dielectric constant K, the magnitude of the electric field between the plates is given by

$$E = 4\pi k_e \frac{Q}{KA}$$

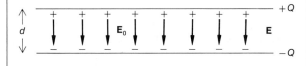

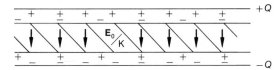

FIGURE 13-17
The dielectric material reduces the strength of the electric field.

TABLE 13-2
The dielectric constants for various materials.

Material	K
Vacuum	1.0000
Air	1.0006
Carbon tetrachloride	2.24
Glass	4 to 7
Wax	2.25
Mica	2.5 to 7
Polyethylene	2.3
Epoxy resin	3.4 to 5.0
Polystyrene	2.5
Nylon	3.5
Methyl alcohol	33.6
Pure water	80.4

Voltage across a Capacitor

The voltage difference between the two plates of a capacitor is the magnitude of the electric field times the distance between the plates:

$$V = Ed$$

If the entire region between capacitor plates separated by a distance d is filled with dielectric material with constant K, the voltage between the plates is given by

$$V = 4\pi k_e \frac{Q}{KA} d$$

The charge Q on the capacitor is directly proportional to the voltage across the capacitor:

$$Q = \frac{KA}{4\pi k_e d} V = CV$$

where the **capacitance** C gives the proportionality between charge and voltage. The unit of capacitance is the **farad,** F, which is coulomb per volt:

$$1\,F = 1\,C/V$$

The capacitance depends only on the dielectric constant and the capacitor's geometry (its surface area and the distance between plates). For a parallel-plate capacitor the capacitance is given by

$$C = \frac{KA}{4\pi k_e d}$$

EXAMPLE
A compact capacitor used in electronic circuits consists of two long strips of metal foil separated by dielectric paper and rolled into a small cylinder. If each foil strip is 6 cm wide and 30 m long and the dielectric paper is 0.010 cm thick with a dielectric constant of $K = 2.2$, what is the capacitance of this device?

The area of the plates (0.06 m × 30 m) is 1.8 m². The capacitance is

$$C = \frac{(2.2)(1.8 \text{ m}^2)}{(4\pi)(9.0 \times 10^9 \text{ N-m}^2/\text{C}^2)(10^{-4} \text{ m})}$$

$$C = 0.35 \times 10^{-6} \text{ C}^2/\text{N-m}$$

But

$$1 \text{ V} = 1 \text{ N-m/C}$$

The capacitance is

$$C = 0.35 \times 10^{-6} \text{ C/V}$$

or

$$C = 0.35 \ \mu\text{F (microfarad)}$$

Capacitors in Parallel and in Series

If a group of capacitors are connected in parallel, as shown in Figure 13-18, the voltage across each capacitor is the same

$$V = V_1 = V_2 = V_3$$

The total charge stored by the three capacitors is

$$Q = Q_1 + Q_2 + Q_3$$

$$Q = C_1 V + C_2 V + C_3 V = (C_1 + C_2 + C_3)V$$

$$Q = (C_1 + C_2 + C_3)V = CV$$

The *total capacitance* for three capacitors in *parallel* is just the sum of the capacitances.[3]

$$C = C_1 + C_2 + C_3$$

If a group of capacitors are connected in series, as

<hr />

[3]This is the opposite of how resistors behave, as we see in Chapter 14. Resistors in series add: $R = R_1 + R_2 + R_3$, while resistors in parallel add as reciprocals: $1/R = 1/R_1 + 1/R_2 + 1/R_3$.

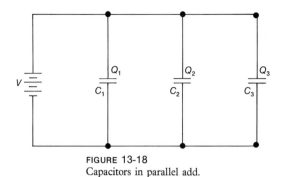

FIGURE 13-18
Capacitors in parallel add.

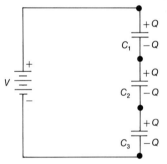

FIGURE 13-19
Capacitors in series add as reciprocals.

shown in Figure 13-19, the charge on each capacitor is the same:

$$Q = Q_1 = Q_2 = Q_3$$

The combined effect of several capacitances in *series* can be represented by a single capacitance, the *effective capacitance*, C_{eff}. The total voltage across the three capacitors is given by the sum

$$V = V_1 + V_2 + V_3$$

$$V = \frac{Q}{C_1} + \frac{Q}{C_2} + \frac{Q}{C_3} = \frac{Q}{C_{\text{eff}}}$$

TABLE 13-3
Properties of commercial capacitors.

Type	Dielectric material	Capacitance range	Maximum voltage
Variable	air	5 to 500 pF	500 V
Ceramic	barium titanate	10 pF to 1 μF	2 000 V
Oil	paper in oil	0.01 to 1 μF	10 000 V
Mica	mica	1 to 5000 pF	10 000 V
Film	Mylar, Teflon, polystyrene polycarbonate	0.001 to 50 μF	1 000 V
Electrolytic	tantalum oxide	0.01 to 3000 μF	*
	aluminum oxide	0.1 to 100 000 μF	*

*The maximum voltage of electrolytic capacitors is dependent on the capacitance value.

For capacitors in series, the reciprocal of the effective capacitance is the sum of the reciprocals of the individual capacitances:

$$\frac{1}{C_{\text{eff}}} = \frac{1}{C_1} + \frac{1}{C_2} + \frac{1}{C_3}$$

Energy Storage in a Capacitor

If a voltage V is placed on a capacitor there will be a charge Q such that

$$Q = CV$$

Adding an increment of charge ΔQ to the capacitor requires an amount of work:

$$\Delta W = \Delta Q V$$

If the capacitor is first discharged, $V = 0$, placing the first increment of charge on it takes virtually no work. As the voltage builds up in the capacitor, each increment of charge takes more work to add than did the previous increment. The average work required to charge a capacitor up to some value of charge Q and voltage V is just the product of the charge Q times the average voltage, $\frac{1}{2}V$:

$$W = \frac{1}{2}QV = \frac{1}{2}CV^2 = \frac{1}{2}\frac{Q^2}{C}$$

Thus electrical energy can be stored in a capacitor.

Types of Capacitors

There are many types of capacitor, each with its own properties (see Table 13-3). Ceramic, oil, mica, and film capacitors are designed to withstand large voltages. The voltage rating reflects the size of the electric field that the dielectric between the capacitor plates can withstand. A typical ceramic capacitor is shown in Figure 13-20 A. Electrolytic capacitors, shown in Figure 13-20 B, C, and D, are used to store large amounts of charge at relatively low voltages. An electrolytic capacitor consists of a metal surface in contact with an electrolyte, a solution that conducts charge by chemical ions. A voltage placed across the metal and the electrolyte causes a very thin layer of metal oxide to form that serves as the dielectric for the

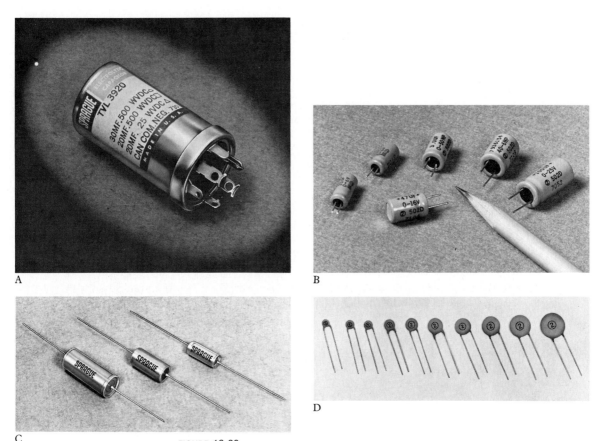

FIGURE 13-20
Typical capacitors: A. ceramic; B. aluminum electrolytic;
C. miniature aluminum electrolytic; D. sintered tantalum.
Courtesy of Sprague Electric Company.

capacitor. Because the dielectric is so thin, large capacitance values can be obtained with relatively small size, but the maximum voltage across the capacitor is limited. When electrolytic capacitors are hooked up, in a circuit, plus and minus signs must be carefully observed. In capacitors hooked up backward, the oxide layer disappears and the capacitor, instead of storing charge, conducts electricity. Aluminum electrolytic capacitors, shown in Figure 13-20 B, are commonly used in televisions, radios, stereos, etc. Miniature and tantalum electrolytic capacitors, shown in Figure 13-20 C and D, are widely used in industrial and military electronics.

Review

Give the significance of each of the following terms. Where appropriate give two examples.

charge
electroscope
Coulomb's law
coulomb
electric field
test charge
potential difference

voltage
volt
equipotential surface
electron volt
surface charge density
superposition
electric dipole

electric dipole moment
capacitor
dielectric
dielectric constant
capacitance
farad

Further Readings

Jerry B. Marion, *A Universe of Physics: A Book of Readings* (New York: John Wiley and Sons, 1970). Articles on electricity by Benjamin Franklin and Charles Augustin de Coulomb and others that are of historical interest.

Problems

THE NATURE OF ELECTRICITY

1 A helium nucleus contains two neutrons and protons. The protons have a charge $+e$ and are separated by a distance of the order of 2.4×10^{-15} m. ($e = 1.6 \times 10^{-19}$ C.) What is the force between the two protons?

2 A gold nucleus has a charge $Z = +79\,e$. An alpha particle with a charge $Z = +2\,e$ approaches within a distance of 10^{-13} m. What is the force between the charges?

3 An electron has a charge $-e$ and a mass $m_e = 9.1 \times 10^{-31}$ kg. What electric field is required to support the electron's weight?

4 An electron is accelerated in an oscilloscope through a potential drop of 4.0 kV. What is the electron's speed?

5 Between a pair of plates 1.0 cm apart there is a voltage of 300 V. In each plate there is a small hole, located opposite the other. An electron with a speed of 5.0×10^6 m/s, traveling to the right, enters the hole in the left plate. The electric field points to the right. What is the electron's final velocity?

6 A proton has a charge $+e$ and a diameter of the order of 2.4×10^{-15} m. The reference potential is $V = 0$ at infinity.
  (a) What is the potential at a distance of 2.4×10^{-15} m from the center of the proton?

(b) What is the electric field at this distance?

(c) If one proton is to interact with another, it must be brought within this distance. What is the energy required to overcome the force of coulomb repulsion in joules? in eV?

7 A copper nucleus has a charge of $+29$ e. The radius of the inner shell electron's orbit is about 1.8×10^{-12} m. The reference voltage is $V = 0$ at infinity.

(a) What is the voltage at a distance of 1.8×10^{-12} m from the center of the nucleus?

(b) What is the electric field at this distance?

(c) How much work is required to bring an electron from infinity to this distance in joules? in eV?

8 In the Bohr model of the hydrogen atom, the electron is considered to be in orbit around the proton at a radius of 5.29×10^{-11} m.

(a) What is the coulomb force on the electron?

(b) What is the electric field acting on the electron?

(c) What is the voltage at this distance if the reference is $V = 0$ at infinity?

9 The largest electric field that can be sustained in dry air without ionization and sparking is about 10^6 V/m. A charge is placed on a conducting sphere of radius 0.3 m located far from the earth.

(a) What is the maximum charge that can be placed on the sphere?

(b) What is the voltage on the sphere with reference to $V = 0$ at infinity?

10 The largest electric field that can be sustained in dry air without ionization and arcing is about 10^6 V/m. How large a sphere would be required to hold a charge of 0.1 C?

11 A polished aluminum sphere has a radius of 10 cm and is charged to a potential of 20 000 V, with respect to ground.

(a) What charge is on the sphere?

(b) What is the electric field in the vicinity of the sphere?

12 A hollow conduction sphere as shown in Figure 13-21 has a radius of 6 cm and a charge of $Q = 2 \times 10^{-9}$ C distributed uniformly on the surface.

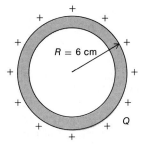

FIGURE 13-21
A hollow conducting sphere.

(a) What is the electric field just outside the surface?

(b) What is the electric potential at the surface?

(c) What are the electric field and potential at the center of the sphere?

13 A charge of $Q = +10^{-9}$ C is placed on a small sphere and lowered by a nonconducting thread into the center of an uncharged hollow conducting sphere without touching the sphere, as shown in Figure 13-22. The large sphere has a radius of 5 cm and is electrically isolated from its surroundings.

(a) What are the charges on the inner and outer surfaces of the hollow sphere?

(b) What is the electric field just inside and outside the surface of the hollow sphere?

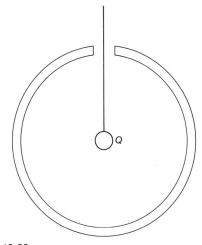

FIGURE 13-22
A charged sphere is suspended at the center of a hollow neutral conducting sphere.

14 The outside of the hollow conducting sphere in the previous problem is momentarily touched to ground.
   (a) What are the charge and electric field at the inner and outer surfaces of the hollow sphere?
   (b) If the small sphere of charge is withdrawn from the isolated hollow sphere, what are the electric field and charge on the inner and outer surfaces of the hollow sphere?

SUPERPOSITION OF ELECTRIC FIELDS

15 Two small charged bodies with a charge $Q_1 = 10^{-9}$ C and a charge of $Q_2 = -4 \times 10^{-9}$ C are placed 0.5 m apart.
   (a) What is the electric field halfway between the charges?
   (b) Where is the electric field zero?

16 Two small metal spheres of radius 2 cm are placed so that their centers are 0.4 m apart. Sphere one is placed at a voltage of -100 V; sphere two at $+300$ V.
   (a) What is the charge on each sphere?
   (b) What are the electric field and voltage halfway between the spheres?
   (c) Where is the electric field zero?

17 Charges Q_A and Q_B are $L = 10$ cm apart. Point P is perpendicular to the line joining the charges, $H = 8.66$ cm from the midpoint between them, as shown in Figure 13-23. $Q_A = Q_B = +10^{-8}$ C. What are the magnitude and direction of the electric field at P?

18 Charges Q_A and Q_B are of $L = 10$ cm apart. $Q_A = +10^{-8}$ C; $Q_B = -10^{-8}$ C. Point P is perpendicular to the line joining the charges, H = 10 cm from the midpoint between them, as shown in Figure 13-23. What are the magnitude and direction of the electric field at P?

19 The water molecule is shown in Figure 13-13. What are the magnitude and direction of the electric field at a point
   (a) 0.2 nm above the oxygen atom?
   (b) 0.2 nm below the oxygen atom?
   (c) 0.2 nm to the right of the hydrogen atom?

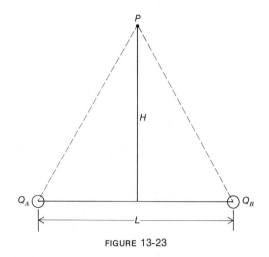

FIGURE 13-23

CAPACITORS

20 A parallel-plate capacitor has two round plates 0.2 m in diameter separated by a sheet of polystyrene 0.25 mm thick. What is the capacitance?

21 A parallel-plate capacitor has two round plates 0.2 m in diameter separated by small spacers making an air gap 0.1 mm thick. The capacitor is charged to a voltage of 300 V.
   (a) What is the capacitance?
   (b) What is the charge on the capacitor?
   (c) What is the electric field strength?

22 A capacitor formed by two parallel plates 0.1 m square and 0.25 mm apart has a charge of $+Q$ and $-Q$ placed on the plates. $Q = 1.5 \times 10^{-8}$ C.
   (a) What is the electric field between the plates?
   (b) What is the voltage difference between the plates?
   (c) If the region between the plates is filled with polyethylene with a dielectric constant $K = 2.3$, what is the voltage difference between the plates?

23 A capacitor formed by two parallel plates 0.1 m square and 0.25 mm apart, with air between, has a potential difference of 300 V.
   (a) What is the charge on each plate?
   (b) If the region between the plates is filled with polystyrene with a dielectric constant $K = 2.5$, what is the charge on each plate?

24 A displacement transducer can be made using a parallel-plate capacitor. Two circular plates 5 cm in diameter and 0.5 cm apart are charged to a voltage of 100 V.
 (a) What are the capacitance and charge on each plate?
 (b) The plates are moved to 0.4 cm apart. What are the new capacitance and charge on each plate? This change in capacitance can be detected and used to record displacement.

25 Three capacitors have values 6 μF, 3 μF, and 2 μF. What is the largest capacitance that can be achieved using these capacitors? Draw the circuit.

26 Three capacitors have values 6 μF, 3 μF, and 2 μF. What is the smallest capacitance which can be achieved using these capacitors? Draw the circuit.

27 A 30 μF capacitor and a 20 μF capacitor are connected in parallel with a 60 V battery.
 (a) Calculate the voltage and charge on each capacitor.
 (b) What is the effective capacitance?
 (c) What is the energy stored in each capacitor?

28 A 30 μF capacitor and a 20 μF capacitor are connected in series with a 60 V battery.
 (a) Calculate the voltage and charge on each capacitor.
 (b) What is the effective capacitance?
 (c) What is the energy stored in each capacitor?

29 In a 16 μF capacitor in a defibrillation unit, used to resuscitate heart attack victims, 400 watt-seconds of energy are stored.
 (a) What voltage is the capacitor charged to?
 (b) How much charge is stored on the capacitor?

14

Electric Currents

CURRENT

Electric current *is the flow of electric charge,* usually in a wire or other conductor. The SI unit of current, the **ampere,** A, is defined in terms of the force between two parallel wires carrying current, as we discuss in Chapter 15. As we discuss in this chapter, however, there are other practical ways of measuring current in amperes. An electric current I is the rate at which charge flows through a conductor:

$$I = \frac{\Delta Q}{\Delta t}$$

If a current I flows for a period of time t, an electric charge Q flows past a point on a wire:

$$Q = It$$

The SI unit of charge is the **coulomb,** C. If a current of one ampere flows for one second, the charge flowing past a point is one coulomb:

$$Q = It = (1 \text{ A})(1 \text{ s}) = 1 \text{ A-s} = 1 \text{ C}$$

The Speed of an Electron

The principal **charge carriers** in an ordinary copper wire are electrons, which, by a convention that Benjamin Franklin established, are said to carry negative charge.

The current I, the **conventional current,** *is the flow of positive charge.* Electrons travel in the opposite direction of conventional current. In a metal such as copper, for every atom there is about one conduction electron, an electron free to move through the metal. Thus the number density of conduction electrons in the metal is about equal to the number density of atoms. Copper, with a mass density of 8.96 gm/cm³, and an atomic weight of 63.54 gm/mol, has an electron number density of about 8.4×10^{28} electrons/m³. Let us consider how fast the electrons are traveling in the wire at a given current.

Figure 14-1 shows a portion of a wire. The charge carriers, here electrons, drift through the conductor with a velocity v_d, the drift velocity. In time Δt they will move forward a distance

$$\Delta x = v_d \, \Delta t$$

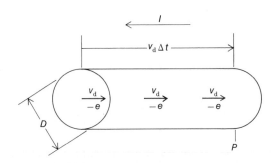

FIGURE 14-1
Conduction electrons move slowly through a wire.

As the charge carriers move forward, all of the charge in a volume,

$$\Delta Vol = A \Delta x = Av_d \Delta t$$

will move past an observation point P. If the charge carriers have a charge q and a density n, the total charge passing point P in time Δt is

$$\Delta Q = qn \Delta Vol = qnAv_d \Delta t$$

The current through the wire is then

$$I = \frac{\Delta Q}{\Delta t} = qnv_d A$$

Because each electron has a charge $q = -e$, the total current through the wire is

$$I = -en_e v_d A \qquad (14\text{-}1)$$

The minus sign indicates that the conventional current (the flow of positive charge) is opposite to the drift of electrons.

EXAMPLE

A No. 10 gauge copper wire 2.95 mm in diameter is 16 m long and carries a current of 20 A. What is the drift velocity of the electrons? How long would it take to get from one end of the wire to the other if the current remains steady?

$$v_d = \frac{-I}{en_e A}$$

The cross-sectional area of the wire is

$$A = \frac{\pi D^2}{4} = \frac{\pi(2.95 \times 10^{-3}\,\text{m})^2}{4} = 6.83 \times 10^{-6}\,\text{m}^2$$

$$v_d = \frac{-20\,\text{C/s}}{(1.6 \times 10^{-19}\,\text{C})(8.4 \times 10^{28}/\text{m}^3)(6.83 \times 10^{-6}\text{m}^2)}$$

$$v_d = -2.2 \times 10^{-4}\,\text{m/s}$$

The electrons move very slowly through the wire, and, as the minus sign indicates, move in the direction opposite to the conventional current. At this speed it would take a particular electron almost 20 hours to travel the length of the wire.

Although the drift velocity of the electrons in a wire is very small, the effects of a change in electric potential acting at one end of a wire are transmitted through the wire at the speed of light. The electrons in a metal act as an incompressible fluid. An amount of charge ΔQ_1 moving forward a distance Δx_1 in the wire pushes ahead the electrons before it, as shown in Figure 14-2. The electrons at the far end of the wire are also pushed forward by the same amount. At every point in the wire the current will be the same. If the wire is wider at some point, there the charge carriers go more slowly (as does a fluid flowing through a pipe), but the total current remains the same.

FIGURE 14-2
Electrons in a metal act as an incompressible fluid.

VOLTAGE AND ELECTROMOTIVE FORCE

The **battery** was developed by the Italian physicist Alessandro Volta, who presented his concept to the Royal Society of London in 1800. Pairs of copper and silver metal disks were separated by fiber spacers moistened with a slightly acidic solution. An electric voltage was generated, causing a current to flow. (You can make a small battery with a penny and any true silver coin—not a nickel-copper hybrid—separated by a piece of paper moistened with saliva.)

Volta had discovered electrochemical phenomena the significance of which others at once seized upon. Within a few weeks Anthony Carlisle and William Nicholson of

England had achieved the electrolysis of water. Sir Humphry Davy soon developed a battery with 2000 pairs of metal plates separated by an acid solution to supply a powerful carbon arc lamp, which he demonstrated to the amazement of everyone. The experiments of Michael Faraday, André Ampère and others were made possible by the development of batteries that could move electric currents for extended periods of time.

A battery is a source of electric potential difference or **electromotive force** (emf), *the ability to do electrical work*. A battery converts chemical energy into electrical work. The measure of electromotive force is the volt. (We are all familiar with the $1\frac{1}{2}$ volt flashlight battery and the 6 and 12 volt automobile batteries.) One coulomb of charge moving through an electric potential difference of one volt will do one joule of work.

Electrical potential is analagous to gravitational potential. Increasing the gravitational potential energy of a mass of water by pumping it to a greater height increases its ability to do mechanical work (see Figure 14-3). Likewise, raising an electric charge to a higher electric potential by a battery increases the ability of the charge to do work.

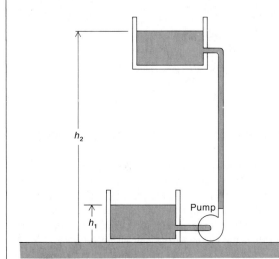

FIGURE 14-3
A chemical battery raises the electrical potential energy of electrical charge just as a pump raises the gravitational potential energy of water.

Electrochemistry

In 1834, Michael Faraday published the conclusions of his studies on the passage of electricity through solutions. A salt such as copper sulfate ($CuSO_4$), when dissolved in water, dissociates into a positive copper ion (Cu^{++}) and a negative sulfate ion (SO_4^{--}). An ion is an atom that has gained or lost one or more electrons. A singly charged metal ion is missing one electron and has a charge of $+e$; a doubly charged ion is missing two electrons and has a charge of $+2e$. If electric current is passed between two electrodes, or conductors, placed in a copper sulfate solution, copper metal will be deposited on the negative electrode. Faraday showed that the mass of metal deposited is proportional both to the total electric charge Q passing through the solution and to the atomic weight divided by the charge on each ion. For a singly charged metal ion, such as silver (Ag^+), one electron will deposit one atom. For a doubly charged metal ion, such as copper (Cu^{++}), two electrons will deposit one atom. The quantity of electric charge that will deposit one mole, one gram atomic weight, of a singly ionized metal ion is given the name **Faraday, $\mathfrak{F}$**:

$$1\ \mathfrak{F} = 96\ 487\ \text{coulombs}$$

The amount of metal that will be deposited at an electrode is given by

$$M = \frac{M_A Q}{z \mathfrak{F}} \qquad (14\text{-}2)$$

where M_A is the atomic weight, Q is the quantity of charge passing through the solution, and z is each ion's degree of ionization. A Faraday is equal to the charge on one mole, Avogadro's number, of electrons. An electron has a negative charge and a value

$$q = -\frac{\mathfrak{F}}{N_A} = -\frac{96\ 487\ \text{C/mol}}{6.02 \times 10^{23}/\text{mol}}$$

$$q = -e = -1.60 \times 10^{-19}\ \text{C}$$

The magnitude of the electron charge is represented by the symbol e. The charge on an electron is negative, $-e$.

EXAMPLE

A current of 0.5 A flows for 10 min through a solution of copper sulfate (Cu^{++})(SO_4^{--}). How much copper is deposited? The copper ion has a charge of $z = 2$. The atomic weight of copper is 63.54 gm/mol. The charge flowing in 10 min is

$$Q = It = (0.5 \text{ C/s})(600 \text{ s})$$

$$Q = 300 \text{ C}$$

The mass of copper deposited is

$$M = \frac{M_A Q}{z \mathcal{F}} = \frac{(63.54 \text{ gm/mol})(300 \text{ C})}{(2)(96\,490 \text{ C/mol})}$$

$$M = 0.10 \text{ gm of copper}$$

Batteries

A battery converts chemical energy to electrical energy. Chemical reactions within the battery move electrons from a low potential energy state to a high potential energy state. Figure 14-4 shows the electrical symbol for a battery. The electron flow through the external circuit is from the electron-rich terminal ($-$) to the electron-poor terminal ($+$). The conventional current flows through the external circuit from the battery's positive terminal to its negative terminal.

Batteries can be made from many different combinations of materials. Alkaline batteries, shown in Figure 14-5, use zinc and a potassium hydroxide electrolyte.

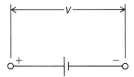

FIGURE 14-4
The battery symbol. The longer line indicates the positive, or electron-poor, terminal; the shorter line, the negative, or electron-rich, terminal.

Nickel-cadmium batteries are rechargeable. Mercury-oxide batteries have a large energy storage capacity for a given volume and have a long service life. The commonly used zinc-carbon "flashlight" battery depends on the reaction of the zinc metal can with the ammonium chloride electrolyte.

A primary laboratory standard for the calibration of voltage measurements is a **Weston cell,** a saturated cadmium sulfate solution in contact with two electrodes: a mercury sulfate and mercury electrode ($+$) and a cadmium sulfate and cadmium-mercury amalgam electrode ($-$). At a temperature of 20.00°C the cell has an electromotive force of 1.0183 V.

The lead-acid battery is a common device for storing substantial quantities of electrical energy. These batteries can have as many as 90 plates and can give "cold cranking power"—currents as high as 500 amperes for periods of 30 seconds to turn over an automobile engine on a cold morning. In the battery, a lead Pb (negative) plate and a lead-oxide PbO_2 (positive) plate are paired and suspended in a solution of sulfuric acid. In discharging, the lead becomes lead sulfate, giving two electrons to the negative plate,

$$Pb + H_2SO_4 \rightarrow 2 e^- + (Pb^{++})(SO_4^{--}) + 2 H^+$$

and leaving two hydrogen ions in solution. At the positive plate, lead oxide is reduced to lead sulfate,

$$(Pb^{++++})(O^{--})_2 + 2e^- + H_2SO_4 \rightarrow (Pb^{++})(SO_4^{--}) + H_2O + O^{--}$$

absorbing two electrons. The net reaction in the transfer of two electrons from the negative to the positive plate is

$$Pb + PbO_2 + 2 H_2SO_4 \rightleftharpoons 2 PbSO_4 + 2 H_2O$$

Each unit of charge gains electrical potential of about 2 V. As the battery is discharged, the reaction converts sulfuric acid to lead sulfate and water. The sulfuric acid becomes less concentrated, and consequently the density of the solution between the plates decreases, from 1.27 gm/cm³ to 1.15 gm/cm³. A hydrometer can measure the specific gravity of the battery acid, to determine the state of charge. If the lead-acid battery has not been so discharged

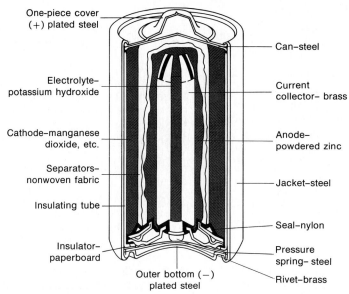

FIGURE 14-5
Cross section of an alkaline battery. Zinc powder reacts with potassium hydroxide to provide electron transfer. Courtesy of Union Carbide Corporation.

that the plates are damaged, it can be recharged. Doing electrical work to transfer electrons returns the lead sulfate to lead and lead oxide.

The electrical power delivered by a battery is the product of the electromotive force E and the current I.

$$P = EI$$

The total energy stored in a battery is the product of the power multiplied by the time the power can be delivered:

$$W = Pt = EIt = EQ$$

The current multiplied by the time is just the total charge produced by the battery, $Q = It$. Storage batteries are often rated in ampere-hours.

EXAMPLE

A heavy-duty 12 V storage battery will deliver a current of 390 A for 30 s and of 25 A for 100 min, and has a rating of 62 A-hr.

(a) What power and how much energy does it deliver in 30 s?
(b) What power and how much energy will it deliver at 25 A?
(c) What is the total energy storage of the battery?

The electrical power is given by

$$P = EI = (12\,V)(390\,A) = 4.68 \text{ kilowatts}$$

The energy delivered is

$$W = Pt = (4.68 \times 10^3\,W)(30\,s)$$
$$W = 1.4 \times 10^5 \text{ joules}$$

At 25 amperes the power level is

$$P = EI = (12\,V)(25\,A) = 300 \text{ watts}$$

The energy delivered is

$$W = Pt = (300\,W)(100\,min)(60\,s/min)$$
$$W = 1.8 \times 10^6 \text{ joules}$$

The total charge delivered is

$$Q = It = (62 \text{ A-hr})(3\,600 \text{ s/hr})$$

$$Q = 2.23 \times 10^5 \text{ coulomb}$$

The total energy stored in the battery is

$$W = QV = (2.23 \times 10^5 \text{ C})(12 \text{ V})$$

$$W = 2.68 \times 10^6 \text{ joules!}$$

Batteries in Parallel and in Series

To draw a large current at a given voltage, several batteries can be connected in parallel. To accomplish this effect in a wet cell battery, many pairs of plates are used; the positive plates are connected to one another and the negative plates to one another. The cell's voltage depends on each electron's electrochemical change in potential energy as it passes from one plate to another. The current capacity depends on the plates' effective surface area. A modern lead-acid battery has 15 plates in each cell.

Batteries can be connected directly in parallel only if each generates the same emf. Otherwise one battery may be charged at the expense of the others. Figure 14-6 shows the symbol for batteries arranged in parallel. The voltage is the same as the voltage across each cell. The currents from each cell add:

$$I_{\text{total}} = I_1 + I_2 + I_3 + \cdots$$

If batteries are connected in series, the emfs add. In an automobile battery 3 wet cells are connected in series to produce 6 volts, or 6 cells for 12 volts. In a transistor radio, 6 dry cell batteries with an emf of 1.5 V are connected in series to produce a voltage of 9 volts. The current through each cell is the same. Figure 14-7 shows batteries connected in series.

$$V_{\text{total}} = V_1 + V_2 + V_3 + \cdots$$

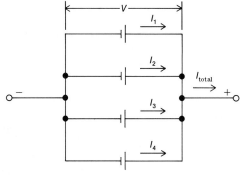

FIGURE 14-6
Batteries in parallel. Batteries may be directly connected in parallel only if each cell has the same emf. The current from the cells adds.

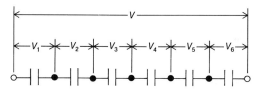

FIGURE 14-7
Batteries connected in series. The same current flows through each battery. The voltages add.

EXAMPLE
A lead-acid battery, in moving two electrons from the negative to the positive plate, converts one atom of lead Pb to an ion Pb^{+2}. The battery is rated at 50 A-hr at 12 V.
(a) What is the total charge produced in discharging the battery?
(b) How much lead oxide is converted to lead sulfate in each cell?
(c) How much lead is converted to lead sulfate in the battery?
The atomic weight of lead is $M_{\text{Pb}} = 207.21$ gm/mol and the molecular weight of lead oxide is $M_{\text{PbO2}} = 239.21$ gm/mol. The total charge delivered is given by

$$Q = It = 50 \text{ A-hr}$$

$$Q = (50 \text{ A})(3600 \text{ s}) = 1.8 \times 10^5 \text{ coulomb}$$

Both the lead and the lead oxide that are converted to

lead sulfate are given by equation 14-2. In each case, two units of charge, $z = 2$, are transferred for each atom of lead.

$$M = \frac{M_{Pb}Q}{z\mathcal{F}} = \frac{(0.207 \text{ kg/mol})(1.8 \times 10^5 \text{ C})}{(2)(9.649 \times 10^4 \text{ C/mol})} = 0.19 \text{ kg}$$

In each cell 0.19 kg of lead are converted to lead sulfate in the negative plates discharging the battery. Similarly 0.22 kg of lead oxide is converted to lead sulfate at the positive plates in discharging the battery.

RESISTANCE

Ohm's Law

A simple electrical circuit consists of a source of electromotive force, such as a battery, and a current path through some material. For many common materials the current flow is proportional to the electrical potential difference, or **voltage drop,** across the material. The ratio of voltage drop V to the current I is the **resistance** R:

$$R = \frac{V}{I}$$

The flow of electric current through a material resembles the flow of a fluid through a pipe, as shown in Figure 14-8. The volume flow through the pipe (discussed in Chapter 9) depends on the pressure difference across the pipe's length. The greater the pressure difference (that is, the voltage drop), the greater the volume flow, or current. The "resistance" of the pipe to fluid flow equals the ratio of pressure difference to volume flow, and depends both on the pipe's size and the liquid's viscosity.

The unit of electrical resistance is the **ohm,** Ω (the Greek capital letter *omega*): the resistance between two points in a circuit is equal to the potential difference, in volts, between these two points divided by the current, in amperes, flowing between them. Common units of resist-

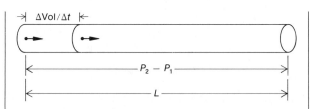

$$\text{Flow} = \frac{\Delta \text{Vol}}{\Delta t} = \frac{\pi r^4}{8\eta L}(P_2 - P_1) = \frac{P_2 - P_1}{\text{"}R\text{"}}$$

FIGURE 14-8
Electric current and fluid flow are analogous. The volume flow rate, or "current," through a pipe is proportional to the pressure difference, or "potential drop," across the pipe. The "resistance" to flow depends on the pipe's, or circuit's, length and cross-sectional area.

ance are the kilohm (kΩ) and the megohm (MΩ). In a circuit diagram the symbol for an electrical resistance, or simply a **resistor,** is a jagged line, as shown in Figure 14-9. The relationship between voltage and current is expressed by **Ohm's law:**

$$V = IR$$

We can describe the behavior of a simple circuit, consisting of a battery and a resistor shown in Figure 14-10, in terms either of conventional current or of electron current. Conventional current I flows in the direction of positive charge carriers, from the battery's positive pole, through the resistance, and back to the battery's negative pole. Conventional current always flows through the external circuit from positive voltage to negative voltage, and always enters the positive end of a resistance and leaves at the negative end.

The magnitude of the voltage drop across the resistance, given by Ohm's law, is the same for conventional current and electron current. Many engineering discussions use the language of electron current. The funda-

FIGURE 14-9
The symbol for a resistor.

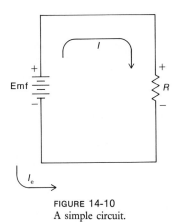

FIGURE 14-10
A simple circuit.

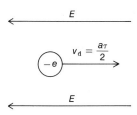

FIGURE 14-11
The electric field accelerates the electrons for a time τ, until they collide with the crystal lattice.

mental definitions of electric and magnetic fields are based on the assumption of positive charge and the flow of positive charge carriers. As is customary in physics textbooks, we shall discuss circuits from the perspective of conventional current—even though the electrons flow the other way!

Resistivity

The flow of current through a wire depends on the electron density n_e, the drift velocity of the electrons v_d, and the cross-sectional area of the wire A. The conventional current is given by equation 14-1:

$$I = -en_e v_d A$$

If an electric field is applied to the material, the electrons accelerate (see Figure 14-11)—but only briefly, for they quickly collide with atoms of the metal. (The electrons can be thought of as bouncing through a lattice of metal atoms.)

The force on the electron in one dimension gives an acceleration

$$F = qE = ma$$

$$a = \frac{qE}{m}$$

An electron will accelerate for a short time τ before colliding and coming to rest. The electron accelerates to a maximum speed from rest of

$$v_{max} = a\tau$$

The electron's average drift speed is just half the maximum speed:

$$v_d = \frac{v_{max}}{2} = \frac{a\tau}{2} = -\frac{eE\tau}{2m_e}$$

where the minus sign indicates that negative charge is carrying the current. This drift velocity is much smaller than the speed of random thermal motion of conduction electrons in a metal.

If a wire has a cross section A and an electron density n_e, the current through the wire is

$$I = -en_e v_d A = \left[(-e)n_e \frac{(-e)}{2m_e} A \right] E$$

If the wire of length L has a potential difference V across it, the magnitude of the electric field is

$$E = \frac{V}{L}$$

The current through the wire is then

$$I = \left[\frac{(-e)^2 n_e \tau A}{2 m_e L} \right] V = \frac{V}{R}$$

TABLE 14-1
Resistivity of common materials.

Substance	Resistivity Ω-m
Conductors	
Silver	1.59×10^{-8}
Copper	1.724×10^{-8}
Gold	2.44×10^{-8}
Aluminum	2.82×10^{-8}
Tungsten	5.6×10^{-8}
Brass	7.8×10^{-8}
Iron	1.1×10^{-7}
Platinum	1.0×10^{-7}
Constantan	4.9×10^{-7}
Mercury	9.8×10^{-7}
Nichrome	1.0×10^{-6}
Carbon	4×10^{-5}
Semiconductors	
Germanium	2.0×10^{0}
Silicon	3.0×10^{4}
Insulators	
Dry maple wood	3×10^{8}
Glass	10^{11}–10^{13}
Mica	2×10^{15}
Quartz	5×10^{17}

The current is proportional to the voltage. This is just a statement of Ohm's law, where the proportionality is given by the resistance R:

$$V = IR$$

The resistance of the wire is given by

$$R = \left[\frac{2m_e}{e^2 n_e \tau} \right] \frac{L}{A}$$

The resistance of a material depends on its cross-sectional area A and its length L. A thin wire has more resistance than a thick wire; a long wire, more than a short wire of the same diameter. The **resistivity** of a material gives the relationship between the length and area of a material and its electrical resistance to the flow of current. The symbol for resistivity is ρ (the lowercase Greek letter *rho*):

$$R = \rho \frac{L}{A} \qquad (14\text{-}3)$$

ρ has the units of ohm-meters (Ω-m). Table 14-1 lists the resistivity of various substances.

EXAMPLE

A No. 10 gauge copper wire 16 m long has a diameter of 2.59×10^{-3} m and carries a current of 20 A. Copper wire has a resistivity of 1.72×10^{-8} Ω-m.
(a) What is the resistance of the wire?
(b) What is the voltage drop across the wire?
The resistance of the wire is given by

$$R = \frac{\rho L}{A} = \frac{(1.72 \times 10^{-8} \ \Omega\text{-m})(16 \ \text{m})}{\frac{\pi}{4}(2.59 \times 10^{-3} \ \text{m})^2}$$

$$R = 0.0522 \ \Omega$$

The voltage drop is just

$$V = IR = (20 \ \text{A})(0.0522) \ \Omega$$

$$V = 1.04 \ \text{V}$$

TABLE 14-2
Diameter and resistance per meter of American Wire Gauge (AWG) annealed copper wire.

Gauge AWG No.	Diameter mm	Resistance/meter $10^{-3}\Omega$/m
18	1.02	21.0
16	1.20	13.2
14	1.63	8.30
12	2.05	5.22
10	2.59	3.28
8	3.26	2.06
6	4.11	1.30
4	5.19	0.817
2	6.54	0.522
0	8.25	0.323
00	9.27	0.256

If aluminum wire of the same gauge were used, the resistance would be 0.084 Ω and the voltage drop 1.7 V. Table 14-2 shows the resistance of copper wires of various diameters.

Metals have a low resistivity, carrying large currents with relatively little voltage drop, and thus are good conductors of electricity. Ceramic material, glass, and quartz are insulators, having a very high resistivity and being very poor conductors. Silicon and germanium crystals are neither good conductors nor good insulators; they are **semiconductors.**

Resistivity is determined by the number of charge carriers that are free to move in the material and the rate at which they collide with the atoms in the material. A metal with about one free conduction electron per atom has a veritable sea of electrons free to carry current. Because of random thermal motion, the rate at which conduction electrons collide with the atoms in the metal increases with temperature.

The resistance of many materials changes linearly with temperature, over a wide range. Figure 14-12 shows the resistivity of pure copper. The linear region extends from $-200°C$ to $+300°C$. In a metal known as a **superconductor,** the resistance drops to zero at a temperature a few degrees above absolute zero. To reach the superconducting state the metals must be cooled below a critical temperature T_c. Superconductivity of mercury was dis-

covered in 1911 by the Dutch physicist Kamerlingh Onnes. Figure 14-13 records Onnes's observations in cooling mercury below the critical temperature $T_c = 4.2 \text{ K}$. Table 14-3 lists several superconducting materials and their critical temperatures. Superconductivity

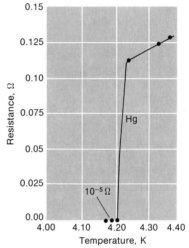

FIGURE 14-13
Superconductivity. The resistance of mercury suddenly decreases at the critical temperature $T_c = 4.2 \text{ K}$. By permission from C. Kittel, *Introduction to Solid State Physics,* 4th ed. (New York: John Wiley & Sons, Inc., 1971).

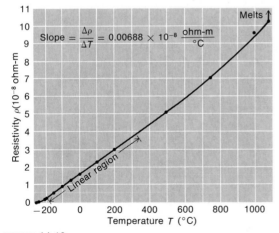

FIGURE 14-12
The resistivity of copper increases uniformly from $-200°C$ to $+300°C$. From George Shortley and Dudley Williams, *Principles of College Physcis,* 2nd ed. © 1967. Reprinted by permission of Prentice-Hall, Inc., Englewood Cliffs, New Jersey.

TABLE 14-3
Superconducting materials.

Material	Critical temperature (K)
Aluminum (Al)	1.2
Indium (In)	3.4
Mercury (Hg)	4.2
Vanadium (V)	5.0
Lead (Pb)	7.2
Niobium (Nb)	9.2
GaNb$_3$	14.5
Nb$_3$Sn	18.0

cannot be explained by a simple picture of thermal agitation of electrons. Its explanation involves modern quantum ideas.

Insulators and Semiconductors

In a simple model of a crystal, the outer-shell electrons, the *valence electrons,* form shared electron chemical bonds with neighboring atoms. Insulators have a high resistivity because the valence electrons are tightly bound in the crystal structure and very few electrons are free to move through the crystal lattice. Table 14-1 lists the resistivities of various insulators.

In a pure crystal of silicon or germanium, the valence electrons are not so tightly bound. By thermal motion a few of these electrons are knocked loose from their crystal bonds, becoming *conduction electrons,* free to move through the crystal. A material in which this happens is a semiconductor. Table 14-1 also shows the resistivity of pure silicon or germanium, which depends on temperature.

Introducing foreign or *impurity atoms* into the crystal lattice of a semiconducting material changes its resistivity. An impurity atom that has five valence electrons (as do phosphorus, arsenic, and antimony) is a *donor:* Four of the valence electrons are used in forming crystal bonds, while the remaining electron is only weakly bound and is donated, becoming a conduction electron. There are exactly as many conduction electrons as there are impurity donor atoms. In such a semiconductor, called an *n-type* semiconductor (*n* for negative), negative charges, electrons, carry the current.

An impurity atom that has three valence electrons (as do boron, aluminum, gallium, and indium) is an *acceptor:* All three valence electrons form bonds with the crystal, while the one unshared bond forms a "hole." An electron from an adjacent atom can move into this hole, creating in turn a new hole with a net positive charge between adjacent atoms. As electrons leave one atom to fill an adjacent hole, the holes can "migrate" through the crystal, giving a positive current of holes. There are exactly as many holes as there are impurity acceptor atoms. In such a semiconductor, called a *p-type* semiconductor (*p* for positive), positive charges, holes, carry the current.

Joule's Law

The electrical power P produced by a source of emf E with a current of magnitude I is given by

$$P = EI$$

The electrical power dissipated as heat in a resistor R with a current I and a voltage drop $V = IR$ is given by

$$P = VI = (IR)(I) = I^2R$$
$$P = I^2R$$

This latter relationship is known as **Joule's law.** In the 1840s, by precise calorimetry experiments, James Prescott Joule established the relationship between electrical energy and thermal energy. The dissipation of electrical energy as heat, sometimes known as **Joule heating,** has practical applications in everything from electric toasters and frying pans to electric heaters and incandescent light bulbs.

EXAMPLE

A 100 watt light bulb operates from a voltage of 120 volts. (For now we ignore the fact that it is alternating current.)
(a) What average current flows through the bulb?
(b) What is the resistance of the bulb?
The current through the bulb is given by

$$I = \frac{P}{V} = \frac{100 \text{ J/s}}{120 \text{ V}} = 0.833 \text{ A}$$

The resistance is the ratio of the voltage to the current:

$$R = V/I = \frac{120 \text{ V}}{0.833 \text{ A}}$$

$$R = 144 \ \Omega$$

The power dissipated in the resistance is

$$P = I^2R = (0.833 \text{ A})^2(144 \ \Omega) = 100 \text{ W}$$

ELECTRIC CIRCUITS

Resistors in Series

When current, supplied by a source of electromotive force (such as a battery), flows through a resistor, electrical power is dissipated as heat. The sum of the power dissipated in the resistor equals the power supplied by the battery.

For resistors in series, Ohm's law gives the voltage drop across each resistor: $V = IR$. Figure 14-14 shows three resistors in series with a battery. The sum of the voltage drops across the three resistors is equal to the emf of the battery:

$$V_0 = V_1 + V_2 + V_3$$
$$V_0 = I_1R_1 + I_2R_2 + I_3R_3$$

Because there is only a single current path, the same current must flow through each resistor

$$I_1 = I_2 = I_3 = I$$

The voltage drop across the three resistors is

$$V_0 = I(R_1 + R_2 + R_3) = IR_{series}$$

Resistances in series add:

$$R_{series} = R_1 + R_2 + R_3$$

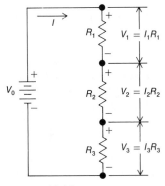

FIGURE 14-14
Resistors in series with a battery.

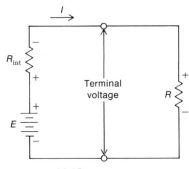

FIGURE 14-15
The internal resistance of a battery.

Figure 14-15 shows a resistance R in series with a battery with electromotive force (emf) E. Because moving charge through the battery requires work, the battery has an **internal resistance** R_{int}, which is also shown. The battery's E is equal to the sum of the voltage drops across the internal resistance plus the external resistance:

$$E = IR_{int} + IR$$

As current is drawn from the battery, the **terminal voltage,** the voltage at the terminals, is less than the emf E, having been reduced by the voltage drop across the battery's internal resistance:

$$V_T = E - IR_{int}$$

The maximum current that can be drawn from a battery is limited by the battery's internal resistance, which depends on the battery's size, construction, and state of discharge.

EXAMPLE

A 12 V automobile battery will maintain a current of 500 A for 30 seconds before the terminal voltage falls below the minimum of 7.2 volts needed to start the car. If the terminal voltage is 7.2 V, what is the battery's internal resistance?

$$R_{int} = \frac{E - V_T}{I} = \frac{12.0\,V - 7.2\,V}{500\,A}$$

$$R_{int} = 0.0096\,\Omega$$

Resistors in Parallel

Resistors connected in parallel are shown in Figure 14-16. The voltage drop across each resistor in parallel is the same:

$$V_0 = V_1 = V_2 = V_3$$
$$V_0 = I_1 R_1 = I_2 R_2 = I_3 R_3$$

The combined effect of several resistors in parallel can be represented as a single resistance, an **effective resistance.** The current through each resistance is given by

$$I_1 = \frac{V_0}{R_1}; I_2 = \frac{V_0}{R_2}; I_3 = \frac{V_0}{R_3}$$

The total current drawn from the battery is simply the sum of the currents through each resistor:

$$I = I_1 + I_2 + I_3$$
$$I = \frac{V_0}{R_1} + \frac{V_0}{R_2} + \frac{V_0}{R_3} = \frac{V_0}{R_{eff}}$$

The effective resistance of three resistors in parallel is given by

$$\frac{1}{R_{eff}} = \frac{1}{R_1} + \frac{1}{R_2} + \frac{1}{R_3}$$

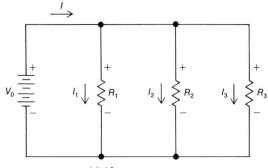

FIGURE 14-16
Resistors in parallel add as reciprocals:
$$\frac{1}{R_{eff}} = \frac{1}{R_1} + \frac{1}{R_2} + \frac{1}{R_3}.$$

Three light bulbs with resistances of 240 ohms, 192 ohms, and 144 ohms respectively are connected in parallel with a voltage of 120 volts dc.
(a) What current is drawn by each bulb?
(b) What is the effective resistance?
(c) What is the total current drawn by the three bulbs?
The voltage across each bulb is 120 V, so the current through the first bulb is

$$I_1 = \frac{V}{R_1} = \frac{120 \text{ V}}{240 \text{ } \Omega} = 0.50 \text{ A}$$

Similarly the current through the other two bulbs is 0.63 A and 0.83 A respectively, for a total of 1.96 A. The effective resistance is given by

$$\frac{1}{R_{eff}} = \frac{1}{R_1} + \frac{1}{R_2} + \frac{1}{R_3}$$

$$\frac{1}{R_{eff}} = \frac{1}{240 \text{ } \Omega} + \frac{1}{192 \text{ } \Omega} + \frac{1}{144 \text{ } \Omega}$$

$$R_{eff} = 61.3 \text{ } \Omega$$

The total current is given by

$$I = \frac{V}{R_{eff}} = \frac{120 \text{ V}}{61.3 \text{ } \Omega}$$

Series-Parallel Circuits

The analysis of many electrical circuits can be simplified by replacing resistances in parallel with an effective resistance, which may itself be in turn combined in series with other resistors. Figure 14-17 shows a simple **series-parallel circuit.** Resistors R_1 and R_2 are in parallel with one another and are in series with R_3. The effective resistance of R_1 and R_2 is

$$\frac{1}{R_{eff}} = \frac{1}{R_1} + \frac{1}{R_2} = \frac{1}{600 \text{ } \Omega} + \frac{1}{1200 \text{ } \Omega}$$

$$R_{eff} = 400 \text{ } \Omega$$

This effective resistance R_{eff} is in series with resistor R_3,

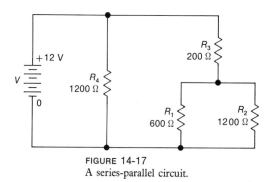

FIGURE 14-17
A series-parallel circuit.

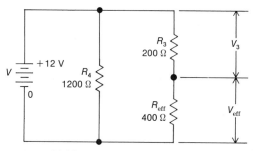

FIGURE 14-18
A series-parallel circuit can be simplified by stages.

as shown in Figure 14-18. The pair is hooked across the battery voltage V in parallel with R_4. The current I_3 through R_3 and R_{eff} can be calculated from Ohm's law:

$$I = \frac{V}{R_{\text{total}}} = \frac{V}{R_3 + R_{\text{eff}}} = \frac{12\text{ V}}{200\ \Omega + 400\ \Omega}$$

$$I_3 = 20\text{ mA (milliamperes)}$$

The voltage drop across the effective resistance can also be calculated. Since the same current flows through both R_3 and R_{eff},

$$V_{\text{eff}} = I_{\text{eff}}R_{\text{eff}}$$

$$V_{\text{eff}} = (20\text{ mA})(400\ \Omega) = 8\text{ V}$$

There is an 8 volt drop across R_{eff} and a four volt drop across R_3: $V_3 = I_3R_3 = 4$ V. We can now backstep and calculate the current through R_1 and R_2, since the voltage drop V_{eff} is across both R_1 and R_2:

$$I_1 = \frac{V_{\text{eff}}}{R_1} = \frac{8\text{ V}}{600\ \Omega} = 13.3\text{ mA}$$

Similarly the current through R_2 is

$$I_2 = \frac{V_{\text{eff}}}{R_2} = \frac{8\text{ V}}{1200\ \Omega} = 6.66\text{ mA}$$

The total current through R_1 and R_2 is just

$$I_3 = I_1 + I_2 = 13.33\text{ mA} + 6.67\text{ mA}$$

$$I_3 = 20\text{ mA}$$

as was previously shown.

The total current from the battery could also be found by further simplifying the circuit, as shown in Figure 14-19. The effective resistance R' of all the resistors is 400 Ω. The total current is then

$$I_{\text{total}} = \frac{V}{R'} = \frac{12\text{ V}}{400\ \Omega} = 30\text{ mA}$$

The total current is the sum of currents I_3 and I_4.

$$I_4 = \frac{V}{R_4} = \frac{12\text{ V}}{1200\ \Omega} = 10\text{ mA}$$

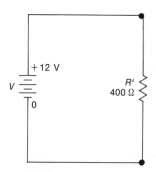

FIGURE 14-19
The simplified circuit for total current.

If I_3 as calculated above is added, the total current is

$$I_{total} = I_3 + I_4 = 20\,mA + 10\,mA = 30\,mA$$

in agreement with the value calculated using the total effective resistance.

Systematically reducing parallel combinations of resistances to an effective resistance value, and combining with resistances in series, allows series-parallel circuits to be analyzed. This approach is of limited utility: It cannot be used if there are emfs in more than one branch of a circuit, or if the circuit is complex, as we see next.

Kirchhoff's Laws of Current and Voltage

Not all electrical circuits are simple series, parallel, or series-parallel circuits. Some circuits are complex. Two such circuits are shown in Figures 14-20, the potentiometer, and 14-21, the Wheatstone bridge. These circuits require a more fundamental, sophisticated analysis, based on Kirchhoff's laws of current and voltage. Charge is not created or destroyed; it simply flows around. **Kirchhoff's current law** is a statement of conservation of electrical charge. In any circuit at any junction between two or more wires, *the sum of the currents going into the junction is equal to the sum of the currents leaving the junction.* What goes in must come out.

Using Kirchhoff's current law requires that the direction of the current in each branch of the circuit be specified.[1] This can be done by using reasonable intuition about how the currents will flow in the circuit. If we happen to choose the wrong direction for a particular current, the value comes out negative. Once we have chosen a direction for each current, we must retain it throughout our analysis.

In the **potentiometer** circuit shown in Figure 14-20 there are three junctions labeled (1), (2), and (3). For junction (1), the current in is equal to the current out.

---

[1] Kirchhoff's laws of current and voltage can be applied by any one of several equivalent methods. We shall use only the branch-current method.

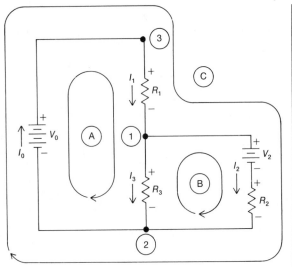

FIGURE 14-20
The potentiometer circuit.

There is one relationship for each of the three junctions:

(1) $I_1 = I_2 + I_3$

(2) $I_3 + I_2 = I_0$

(3) $I_0 = I_1$

Not all of these relationships are independent, for (2) and (3) could be combined to give (1). Quite generally, if there are N junctions there will be $N - 1$ independent relationships or equations.

In any circuit around any closed loop, the energy gain for the charge carrier from the sources of electromotive force is equal to the energy loss in the resistances. Assigning a direction to the current in each branch of the circuit allows the positive and negative ends of each resistance to be specified. The conventional current leaves the negative ($-$) end of the resistance and flows into the positive ($+$) end. **Kirchhoff's voltage law** is a statement of conservation of energy: *Around any closed loop in a circuit, the sum of the voltage gains is equal to the sum of the voltage drops.* Traveling around the loop from $-$ to $+$ across a resistor or battery is a voltage gain; traveling from $+$ to $-$ is a voltage drop.

In the potentiometer shown in Figure 14-20, there are three possible closed loops labeled (A), (B), and (C), giving us three possible relationships:

$$\text{A} \quad V_0 = I_1 R_1 + I_3 R_3$$
$$\text{B} \quad I_3 R_3 = I_2 R_2 + V_2$$
$$\text{C} \quad V_0 = I_1 R_1 + I_2 R_2 + V_2$$

Loop Ⓒ is equivalent to the sum of loops Ⓐ and Ⓑ. Quite generally, if there are M possible loops for a circuit, there will be $M - 1$ independent relationships. It is worth noting that we are using conventional current flow to keep our signs straight. We have also made initial guesses in assigning the direction of the currents in each branch of the circuit. If we guess the wrong direction, the current turns out negative.

There are four unknowns in the potentiometer: I_0, I_1, I_2, and I_3. We can choose any two of the current relationships such as ② and ③ and any two of the voltage relationships such as Ⓐ and Ⓑ to give four equations in four unknowns. At this point we have stated the problem analytically. All that the solution requires is patience and some perseverance with algebra. These four relationships can be quickly reduced to two simultaneous equations:

$$V_0 - V_2 = I_1 R_1 + I_2 R_2$$

and

$$V_0 = I_1 (R_1 + R_3) - I_2 R_3$$

These can be straightforwardly solved to give

$$I_1 = \frac{(R_3 + R_2) V_0 - R_3 V_2}{R_1 R_3 + R_1 R_2 + R_2 R_3}$$

and

$$I_2 = \frac{R_3 V_0 - (R_1 + R_2) V_2}{R_1 R_3 + R_1 R_2 + R_2 R_3}$$

These results can be obtained by direct substitution and reduction, or by matrix methods.

The Null Potentiometer

If a potentiometer circuit, as previously shown in Figure 14-20, is set up and adjusted so that no current flows in the second branch of the circuit, $I_2 = 0$, it is called a **null**

potentiometer. For instance, in a slide-wire potentiometer commonly used in laboratory measurement, the ratio of the resistance R_3 to the total resistance $R_1 + R_3$ can be changed by a sliding contact. The electrical potential of a source of emf can then be measured without drawing current and consequently without regard to the internal resistance. A Weston cell can be used as a calibration reference to give accurate measurements of electrical potential.

If no current is drawn from the battery, I_2 is zero and the potential of the battery is the same as the potential drop across R_3. The same current flows through both R_1 and R_3:

$$I_0 = I_1 = I_3$$

We can solve for I_1:

$$I_1 = \frac{V_0}{(R_1 + R_3)}$$

The potential V_2 is given by

$$V_2 = \frac{R_3}{(R_1 + R_3)} V_0$$

The Wheatstone Bridge

The **Wheatstone bridge** circuit shown in Figure 14-21 is a complex circuit commonly used in electronic instrumentation to detect changes in the resistance of such transducers as thermistors or strain gauges. To measure resistance values, a balanced Wheatstone bridge circuit is used. If the bridge is balanced, there is no current through the meter M. An unknown resistance value is placed in R_4. The other resistance values are adjusted so that the meter current is zero. The voltage drop $V_1 = I_1 R_1$ is equal to $V_3 = I_3 R_3$. Similarly, the voltage drop $V_2 = I_2 R_2$ is equal to $V_4 = I_4 R_4$. This implies that

$$\frac{R_1}{R_1 + R_2} = \frac{R_3}{R_3 + R_4} \quad \text{and} \quad \frac{R_2}{R_1 + R_2} = \frac{R_4}{R_3 + R_4}$$

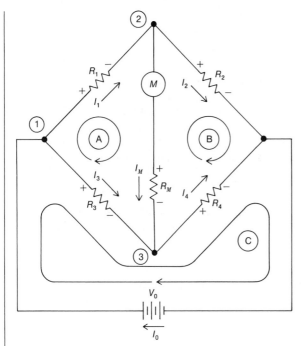

FIGURE 14-21
The Wheatstone bridge circuit.

Kirchhoff's current law applied to the circuit gives us three additional relationships:

① $I_0 = I_1 + I_3$

② $I_1 = I_2 + I_m$

③ $I_m + I_3 = I_4$

We have six unknowns in the circuit: I_0, I_1, I_2, I_3, I_4, I_m. We have six linearly independent equations that we can solve straightforwardly, if tediously, for the current through the meter I_m.

We can use relationships ② and ③ to eliminate I_1 and I_4. Our equations then become

①' $I_0 = I_2 + I_3 + I_m$

Ⓐ' $0 = I_2 R_1 - I_3 R_3 + I_m(R_1 + R_m)$

Ⓑ' $0 = I_2 R_2 - I_3 R_4 - I_m(R_m + R_4)$

Ⓒ' $V_0 = I_3(R_3 + R_4) + I_m R_4$

The last three equations, being only functions of I_2, I_3, and I_m, can be solved as a set of simultaneous equations. The general case with algebra is messy. With specific numerical values the equations rapidly simplify. When solved, the current through the meter is given by

$$I_m = \frac{R_1(R_3 + R_4) - R_3(R_1 + R_2)}{R_1 R_2(R_3 + R_4) + R_3 R_4(R_1 + R_2)} V_0$$
$$\qquad\qquad\qquad\qquad + R_m(R_1 + R_2)(R_3 + R_4)$$

The condition for the balanced bridge can then be written

$$\frac{R_4}{R_2} = \frac{R_3}{R_1}$$

If the Wheatstone bridge circuit is unbalanced, a current flows through the meter resistance R_m. We can solve for the currents in the circuit by straightforward application of Kirchhoff's laws, using the branch-current method.

Kirchhoff's voltage law applied to the circuit gives three independent relationships for the loops Ⓐ, Ⓑ, and Ⓒ, as shown in Figure 14-21:

Ⓐ $I_3 R_3 = I_1 R_1 + I_m R_m$

Ⓑ $I_4 R_4 + I_m R_m = I_2 R_2$

Ⓒ $V_0 = I_3 R_3 + I_4 R_4$

where the presumed direction of the current in each branch of the circuit is also shown on the circuit.

EXAMPLE

Wheatstone bridge circuits often use a thermistor to indicate temperature (as was described in Chapter 10). The thermistor is inserted for R_3 and typically has a resistance $R_3 = R_T = 16.3$ kΩ at a temperature of 0°C. Typical bridge resistance values of $R_2 = R_4 = 5.0$ kΩ and $R_1 = 16.3$ kΩ are chosen so that the bridge is balanced at 0°C. A mercury battery provides a voltage $V_0 = 1.35$ V. The thermistor resistance at 40°C is $R_T = 2.66$ kΩ, as is shown in Table 14-4. What is the current through the meter at 40°C?

At 0°C the bridge is balanced, so the current through the meter is zero. At 40°C the current is given

TABLE 14-4
*The characteristics of a thermistor and the response of a Wheatstone bridge circuit with temperature.**

Temperature °C	Thermistor resistance R_T kΩ	Meter current μA
0	16.3	0
25	5.0	29
37	3.0	45
40	2.66	49

*$R_1 = 16.3$ kΩ. $R_2 = R_4 = 5$ kΩ. $R_3 = R_T$. The meter resistance is $R_m = 6$ kΩ. $V_0 = 1.35$ V.

by the previous equation. Substituting values in kΩ we get

$$I_m = \frac{[16.3(2.66 + 5) - 2.66(16.3 + 5)] (1.35 \text{ V}/\text{k}\Omega)}{2.66(5)(16.3 + 5) + 16.3(5)(2.66 + 5) + 6(2.66 + 5)(16.3 + 5)}$$

$$I_m = 49 \ \mu\text{A}$$

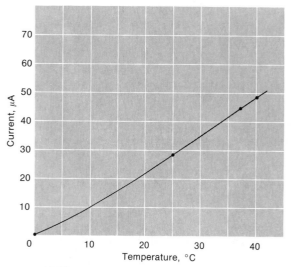

FIGURE 14-22
Thermistor bridge circuit. The meter current I_m is shown as a function of temperature. Over a narrow temperature range the circuit response is linear.

The meter current as a function of temperature is shown in Table 14-4 and is plotted in Figure 14-22. Although the magnitude of the current increases uniformly with temperature, the thermistor resistance does not, so the variation of the meter current depends on R_T in a complicated way. The meter current response is also not linear. Over a much narrower range of temperature, such as from 35°C to 40°C (95°F to 104°F), the thermistor circuit is approximately linear.

THE MULTIMETER

The Meter Movement

To measure voltage and electric current we use meters—the **voltmeter** and **ammeter** respectively. At the heart of each such meter is the **meter movement.** Most meter movements are of the D'Arsonval type: a small wire coil suspended in the field of a permanent magnet, as shown in Figure 14-23. When a current flows through the coil, a twisting torque is produced, causing the meter to deflect (how this occurs is discussed in detail in Chapter 15). An indicator needle attached to the coil points to a calibrated

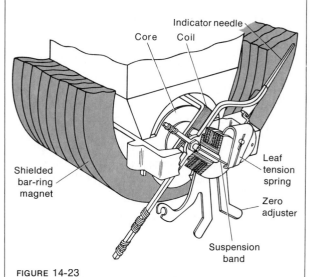

FIGURE 14-23
The D'Arsonval meter movement with a taut suspension band. The coil has 1344 turns of fine wire suspended in a magnetic field. Courtesy of Triplett Corporation.

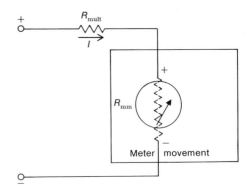

FIGURE 14-24
FIGURE 14-24
A multiplier, a series resistance, converts a meter movement into a voltmeter.

scale. The coil is sometimes delicately mounted on a jewel pivot with a suspension spring to balance the torque produced by the current through the coil. In some meters the coil is held by a taut band of metal (marked "leaf tension spring" in the figure) that serves as both support and return spring. The needle is displaced in proportion to the current through the coil. The meter movement is calibrated so that a known current will give a full-scale deflection. In a typical meter, a Triplett model 60, the meter movement gives a full-scale deflection with a current of 50 μA and has an internal resistance of 2200 ohms.

If such a meter movement were connected directly across a 1.5 V flashlight battery, the current would be

$$I = V/R = 1.5 \text{ V}/2200 \, \Omega = 680 \, \mu\text{A}$$

A current of 680 μA being much larger than the meter's full-scale deflection, the indicator needle would point off the scale (stopped only by a peg that keeps the needle from going too far). The voltage that gives a full-scale deflection of this meter movement is only

$$(50 \, \mu\text{A})(2200 \, \Omega) = 110 \text{ mV}$$

The Voltmeter

A large-value resistor called a **multiplier** is often placed in series with a meter movement to make a voltmeter (shown in Figure 14-24). The multiplier is chosen so that some voltage, such as 10 V, 30 V, 100 V, or 300 V, will register as a full-scale deflection. The scale on the meter face is calibrated to be read directly in volts. Different multipliers allow one meter to read different ranges of voltage. The multiplier resistance R_{mult} is chosen so that the maximum voltage V_{max} will give full-scale current I_{fs} through the meter movement:

$$V_{\text{max}} = I_{\text{fs}}(R_{\text{mult}} + R_{\text{mm}})$$

Figure 14-25 shows a multipurpose meter with several different voltage ranges. The meter switch position indicates the full-scale voltage range. A voltmeter typically has a large resistance so that it draws relatively small current in measuring the voltage. It is used to measure the

FIGURE 14-25
The volt-ohm-milliammeter. The same meter movement can have several uses. Courtesy of Triplett Corporation.

difference in electrical potential between two points in a circuit. That is, the voltmeter measures the voltage *across* two points in a circuit.

EXAMPLE

Figure 14-26 shows the voltmeter circuit. The meter movement has a full scale deflection with 50 μA and an internal resistance of 2.0 kΩ. A resistor chain is attached to a selector switch in such a way that different ranges can be selected. The switch shown indicates the full-scale reading. With the selector switch on 10 V, the voltage required to give a full-scale deflection of the meter is given by

$$V = I(R_{mm} + R_0 + R_1 + R_2 + R_3)$$

$$V = (50\ \mu A)(2.0\ k\Omega + 4.3\ k\Omega + 13.7\ k\Omega \\ + 43.2\ k\Omega + 137\ k\Omega)$$

$$V = (50\ \mu A)(200.2\ k\Omega)$$

$$V = 10.0\ \text{volts full scale}$$

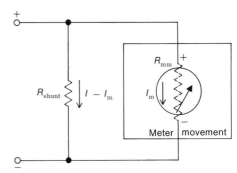

FIGURE 14-27
The shunt resistance of an ammeter carries most of the current.

The Ammeter

A typical meter movement requires a current of 50 μA to give a full-scale deflection. This typically gives a voltage drop across the meter of 100 mV. If we wish to measure a large current such as 100 mA or 1000 A, we must provide a low-resistance path, or **shunt,** that allows most of the current to bypass the meter, as shown in Figure 14-27.

The ammeter circuit for a common multimeter is shown in Figure 14-28. To give some protection to the meter an additional resistance $R_0 = 4.32$ kΩ is placed in series with the meter movement so that the effective

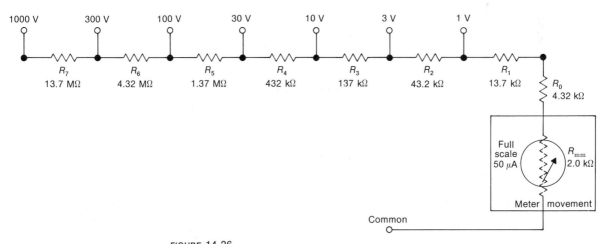

FIGURE 14-26
A voltmeter circuit for multiple ranges. The resistance R_0 is adjusted so that the total resistance of the meter movement plus R_0 is 6.32 kΩ. The multiplier is the sum of the resistances between terminals.

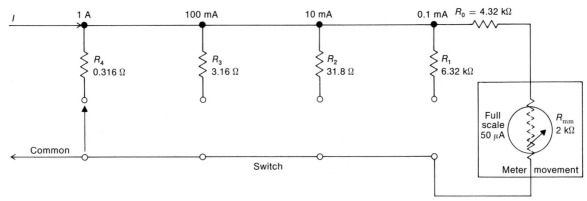

FIGURE 14-28
The ammeter circuit with multiple ranges.

resistance of the meter is about $R_{\mathrm{m}} = 6.32 \ \mathrm{k\Omega}$. A current of 50 μA through this resistance requires a voltage of 316 mV. To make an ammeter with a full-scale measurement of 1.00 ampere, the shunt resistance is chosen such that the voltage across the shunt is 316 mV when the current through the shunt is 1.00 A minus the meter current of 50 μA:

$$R_{\mathrm{shunt}} = \frac{V_{\mathrm{shunt}}}{I_{\mathrm{shunt}}} = \frac{316 \ \mathrm{mV}}{(1.00 \ \mathrm{A} - 50 \ \mu\mathrm{A})}$$

$$R_{\mathrm{shunt}} = 0.316 \ \Omega$$

The ammeter has a very low resistance. Do not connect an ammeter across a battery, for large currents may damage the shunts and the meter movement. An ammeter is connected in series with the current being measured. The current must flow *through* the ammeter. To hook up an ammeter to measure the current flowing in a wire in the circuit one must disconnect the wire, connect the wire to the ammeter, and then reconnect the ammeter to the circuit.

Review

Give the significance of each of the following terms. Where appropriate give two examples of each.

electric current	ohm	series-parallel circuit
ampere	resistor	Kirchhoff's current law
coulomb	Ohm's law	potentiometer
charge carrier	resistivity	Kirchhoff's voltage law
conventional current	semiconductor	null potentiometer
battery	superconductor	Wheatstone bridge
electromotive force	Joule's law	voltmeter
Faraday	Joule heating	ammeter
Weston cell	internal resistance	meter movement
voltage drop	terminal voltage	multiplier
resistance	effective resistance	shunt

Further Readings

D.K.C. MacDonald, *Faraday, Maxwell, and Kelvin* (New York: Doubleday, Anchor Science Study Series S 33, 1964). An interesting study of early researches in electricity.

Problems

CURRENT

1 A battery rated at 120 ampere-hours is discharged.
   (a) How many coulombs of charge flow?
   (b) How many electrons flow past a point?

2 An automobile battery delivers a starting current of 500 A for 30 s.
   (a) How many coulombs of charge flow?
   (b) How many electrons?

3 Aluminum metal is commercially produced from bauxite ore (Al_2O_3) by electrolysis. Aluminum is triply ionized Al^{+++}. It has an atomic weight of 27 gm/mol and a density of 2.7 gm/cm^3.
   (a) How much charge must flow to produce a plate of aluminum 1.0 cm thick and 1 m square?
   (b) If the electrolysis current is 1000 A, how long will it take to deposit enough metal for the plate?

4 A standard flashlight battery will give a current of 0.5 A for about 150 min. At the negative electrode, the zinc gives up two electrons: $Zn \rightarrow Zn^{++}$. Zinc has a molecular atomic weight of 65.37 gm/mol. In discharging the battery,
   (a) How much charge flows?
   (b) How much zinc is consumed?

5 A No. 12 copper wire has a diameter of 2.05 mm and carries a current of 20 A. The density of conduction electrons in the copper is about $n = 8.4 \times 10^{22}$ electrons/cm^3.
   (a) What is the speed of the electrons in the metal?
   (b) If the wire is 20 m long, how much time will an electron take to go the distance?

6 A vibrating reed electrometer will measure a current as low as 10^{-17} A. How many electrons per second is it registering?

VOLTAGE AND ELECTROMOTIVE FORCE

7 An automobile battery has 6 cells with a total emf of 12 volts. The battery is rated at 96 A-hr if discharged slowly over a 20 hr period.
   (a) What total charge is delivered by the battery?
   (b) What is the battery's total energy storage?

8 A 12 V battery has a cold cranking power of 500 A for 30 s.
   (a) What charge is delivered?
   (b) What energy and instantaneous power are delivered?

9 A fresh 12 V battery will sustain a current of 25 A for 140 min.
   (a) What charge is delivered?
   (b) What energy and instantaneous power are delivered?

10 A fresh 1.5 V flashlight battery (D-cell) will discharge at 0.5 A for 200 min. An alkaline battery will last about 1000 min at a current of 0.5 A.
   (a) What charge is delivered by each battery?
   (b) What energy and instantaneous power are delivered by each battery?

11 A color television picture tube has an acceleration voltage of 32 kV and an electron beam current of about 100 μA. What is the electrical power in the electron beam?

12 A single lightning flash can transfer a charge of 3 C to the ground across a potential of the order of 10^7 V. The flash can have a current of as high as 10 kA.
(a) How much energy is released in the lightning flash?
(b) At a current of 10 kA, how long would it take to transfer the charge?
(c) What is the instantaneous power released?

RESISTANCE

13 Winding 30 m of 0.8 mm-diameter constantan wire on a spool produces a resistor. The restivity of constantan is 49×10^{-8} Ω-m. What is the resistance of the wire?

14 No. 18 wire, commonly used in extension cords, has a diameter of 1.0 mm. Copper has a resistivity of 1.72×10^{-8} Ω-m. What is the resistance of both wires of a 5 m-long extension cord?

15 No. 12 wire, commonly used in house wiring, has a diameter of 2.05 mm. Copper has a resistivity of 1.72×10^{-8} Ω-m. Aluminum has a resistivity of 2.63×10^{-8} Ω-m.
(a) What is the resistance of a 20 m length of copper wire?
(b) Of the same length of aluminum wire?

16 An electric heater is rated at 1320 watts at an operating voltage of 120 V.
(a) What current is drawn by the heater?
(b) What is the resistance of the heater?

17 A 200 watt light bulb is operated from a voltage of 120 V.
(a) What current is drawn by the light bulb?
(b) What is the resistance of the light bulb?

ELECTRICAL CIRCUITS

18 A battery has an emf of 1.50 V and an internal resistance of 0.50 Ω. If the positive terminal is accidently connected to the negative terminal, making a short circuit, what is the maximum current drawn from the battery?

19 A mercury cell battery has an emf of 1.35 V. It is connected in series with a resistance $R_1 = 50$ Ω, at which time the terminal voltage is 1.20 V. What is the internal resistance of the cell?

20 An automobile storage battery has an emf of 6.0 V and an internal resistance of 0.005 Ω. What is the terminal voltage when the starting motor is drawing a current of 150 A?

21 A string of Christmas tree lights consists of eight bulbs connected in series with a 120 V power source.
(a) What is the voltage drop across each bulb?
(b) If one bulb is removed from its socket, what is the voltage drop across that socket? What is the voltage drop across all the other bulbs? Why?

22 A 12 V battery has a terminal voltage of 8 V when the current drawn from the battery is 100 A.
(a) What is the internal resistance of the battery?
(b) What is the load resistance?
(c) How much power is delivered by the battery?
(d) How much power is delivered into the load resistance? Why are they different?

23 Given the circuit shown in Figure 14-29,
(a) What is the effective resistance?
(b) How much current is drawn from the battery?
(c) What is the current in the 6 Ω resistor?

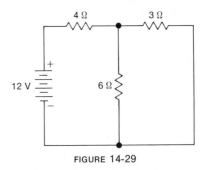

FIGURE 14-29

24 Given the circuit shown in Figure 14-30,
(a) What is the effective resistance between points a and b?
(b) What is the voltage across c and d?

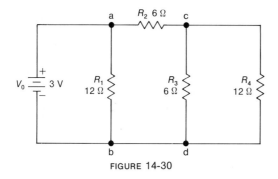

FIGURE 14-30

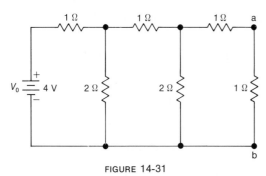

FIGURE 14-31

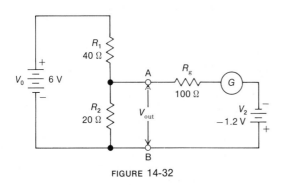

FIGURE 14-32

25 Given the circuit shown in Figure 14-31,
  (a) What is the total current drawn from the battery?
  (b) What is the voltage across a to b?

26 Figure 14-32 shows an unbalanced potentiometer circuit.
  (a) Write down Kirchhoff's current and voltage laws for the circuit.

  (b) Solve to determine the current through the galvanometer resistance R_g.
  (c) Solve to determine the current through R_2 and thus V_{out}.

THE MULTIPLIER

27 A biophysicist needs a voltmeter that reads 20 V full scale and has available a meter movement with a resistance of 100 Ω that requires 50 mA for full-scale deflection. Draw the circuit and give the values of resistance required to make the voltmeter.

28 A sensitive meter movement gives a full-scale deflection with a current of 50 μA and has an internal resistance of 3000 Ω. Draw the circuit and give values of resistance required to make a voltmeter that reads 1.0 V full scale.

29 A sensitive meter movement with an internal resistance of 500 Ω gives a full-scale deflection with a current of 100 μA. Draw the circuit and give values of resistance required to make an ammeter that will read 2 A full scale.

30 A meter movement with a resistance of 100 Ω requires a current of 0.5 mA for full-scale deflection. Draw the circuit and give values of resistance required to make an ammeter that will read 100 A full scale.

31 A bare meter movement has a resistance of 2180 Ω and gives a full-scale deflection with a current of 45 μA. We wish to construct a series-parallel circuit including the meter so that a current of 50 μA through the circuit gives full-scale deflection of the meter and the total resistance of the circuit is 5000 Ω.
  (a) Draw the circuit.
  (b) What resistance values are required?

ADDITIONAL PROBLEMS

32 The density of standard annealed copper is 8.89 gm/cm^3, that of aluminum, 2.70 gm/cm^3. A copper wire 100 m long is drawn to a diameter of 2.6 mm.
  (a) What is the resistance of the wire?

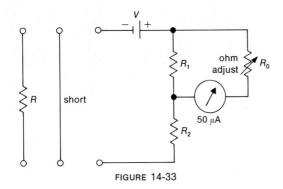

FIGURE 14-33

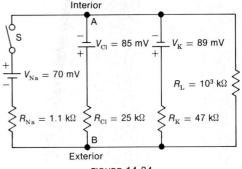

FIGURE 14-34

(b) For an aluminum wire to have the same resistance, what must be its diameter?
(c) What is the weight of each wire?
(d) Which wire has the least resistance per unit weight?

33 An immersion heater is used to heat a cup of water for tea. To heat 200 gm of water from 20°C to 100°C in 4 minutes,
(a) How much energy is required?
(b) How much power?
(c) If the heater is run from 120 V (dc), how much current is drawn?
(d) What is the resistance of the heater?

34 A voltmeter circuit similar to that shown in Figure 14-26 uses a meter movement with a resistance $R_{mm} = 2.2$ kΩ and a full-scale current of 50 μA. The other resistance values are $R_0 = 960$ kΩ, $R_1 = 55$ kΩ, $R_2 = 180$ kΩ, $R_3 = 960$ kΩ, $R_4 = 4.8$ MΩ.
(a) Draw a sketch of the circuit showing resistance values.
(b) What voltage ranges is the meter used for?

35 An ammeter circuit, similar to that shown in Figure 14-28 uses a meter movement with a resistance of $R_{mm} = 2.8$ kΩ and a full-scale current of 50 μA. The other resistance values are $R_0 = 2.8$ kΩ, $R_1 = 455$ Ω, $R_2 = 42.0$ Ω, $R_3 = 4.17$ Ω, $R_4 = 0.415$ Ω.
(a) Draw a sketch of the circuit showing resistance values.
(b) What current ranges is the meter used for?

36 An ohmmeter circuit, shown in Figure 14-33, uses a meter movement with a full-scale deflection of 50 μA and a 15 V battery. The other resistance val-

ues are $R_1 = 55$ kΩ and $R_2 = 180$ kΩ. R_0, which includes the meter resistance, is adjusted so that when the terminals are shorted the meter movement reads full scale.
(a) Draw a sketch of the circuit showing resistance values.
(b) What is the value of R_0? What is the current through and voltage drop across each resistor?
(c) If a resistor R is placed across the terminals, what value of resistance gives half-scale deflection? In this case, what is the current through and voltage drop across each resistor?

37 The electrical potential difference across a resting axon membrane can be represented by an electrical circuit as shown in Figure 14-34. The sodium, potassium, and chlorine ion differences act as batteries. There is an effective resistance to the flow of each ion current across the membrane. There is also a large leakage resistance R_L across the membrane, which allows a small current to flow. The resting membrane is pictured with the switch S open. The sodium battery is disconnected from the circuit, which means sodium ions cannot move across the membrane.
(a) What is the voltage across the membrane V_{AB}, as given by this circuit?
(b) What is the current in each branch of the circuit?

38 When the axon membrane is depolarized, as would happen if the switch S in Figure 14-34 were closed, sodium ions can move across the membrane.
(a) What is the voltage across the membrane as given by this circuit?
(b) What is the current in each branch of the circuit?

15

Magnetic Fields

MAGNETISM

Some 2000 years ago the Chinese were already aware that lodestones attract small pieces of iron and other lodestones. Early Western traders brought that knowledge back with them from China. William Gilbert, physician to Queen Elizabeth I, published, in 1600, *De Magnete*. In this, the definitive Renaissance study on magnets, he detailed his researches into the properties of lodestones. He concluded, among other things, that the earth acted as a giant lodestone, and that the force of attraction between the earth and the moon was magnetic. He produced a small model of the earth made from lodestone to demonstrate the orientation of a compass at the surface. Gilbert also observed that when a magnet or lodestone is broken, the pieces are also magnets, each having two poles that attract or repel a pole of the other piece. Opposite poles attract, and like poles repel. Although a useful tool of navigation, magnetism remained a curiosity and a subject of philosophical speculation until the nineteenth century.

In 1820, the Danish physicist Hans Christian Oersted announced his discovery that a steady electric current deflects a magnetic compass needle. This first experimental evidence of a connection between magnetic and electrical phenomena had not been possible until there were chemical batteries that could generate large, sustained electric currents. A current-carrying wire exerts a force on

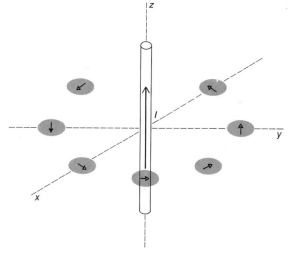

FIGURE 15-1
A current-carrying wire deflects a compass needle.

a nearby magnet to orient it perpendicular to the wire, as the compass needle in Figure 15-1 shows. If the current exerts a force on the magnet, then by Newton's third law, the magnet exerts an equal and opposite force on the current.

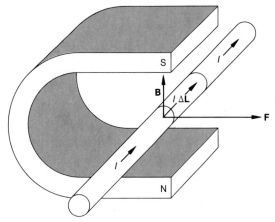

A magnetic field exerts a force on a current element. The force is perpendicular to the directions of both the magnetic field and the current element.

The Magnetic Field

In the vicinity of any lodestone or permanent magnet is a magnetic field, or what Gilbert called a sphere of influence. By 1821, Michael Faraday demonstrated that if a current is flowing in a wire placed in the gap of a permanent magnet (as shown in Figure 15-2), there will be a force on that wire. Measuring the force on a current in a conductor gives an *operational definition* for the strength of a **magnetic field.**[1]

A short segment of wire of length ΔL carrying a current I is a **current element,** $I\Delta L$, which is a vector quantity. I represents the conventional current. (With all due respect to the electrons, the laws of magnetism are defined by assuming that positive charges carry the current.) The magnitude of the magnetic field B is equal to the ratio of the maximum force acting on the current element F to the magnitude of the current element $I\Delta L$:

$$B = \frac{F}{I\Delta L}$$

---

[1] If a physical quantity is defined in terms of operations or measurements by which its value can be obtained, that quantity has been given an *operational definition*. Such a definition is contrasted to a conceptual definition, which is based on theoretical principles. Operational definitions play an important role in defining many physical quantities, as, for instance, mass, temperature, electric current, and, here, magnetic fields.

TABLE 15-1
Units of magnetic field strength B (also called magnetic flux density or magnetic induction).

Unit	SI equivalent
tesla	$1\ \text{T} = 1\ \text{N/A-m}$
weber/meter^2	$1\ \text{Wb/m}^2 = 1\ \text{N/A-m}$
gauss	$1\ \text{gauss} = 10^{-4}\ \text{N/A-m}$
gamma	$1\ \text{gamma} = 10^{-9}\ \text{N/A-m}$

The SI unit for magnetic field strength, N/A-m, is given the name, **tesla** T (after Nicholas Tesla, who studied the properties of magnetic fields, at the end of the nineteenth century). Table 15-1 lists other common units for magnetic field strength.

A magnetic field is a vector quantity; it has magnitude and direction in space. The force on a current element is perpendicular to both the current element and the direction of the magnetic field. Figure 15-2 shows a simple situation for a current element $I\Delta L$ in a magnetic field **B**. We can represent the vector relationship for the force acting on a current element in a magnetic field by the **vector cross product:**

$$\mathbf{F} = I\Delta \mathbf{L} \times \mathbf{B} \qquad (15\text{-}1)$$

The vector cross product was discussed in Chapter 7 and is briefly reviewed in Appendix A. In general, the magnitude of the force **F** is given by the product of the magnitudes of the two vectors $I\Delta \mathbf{L}$ and **B** multiplied by the sine of the angle θ between them, as shown in Figure 15-3:

$$F = I\Delta LB \sin \theta$$

If $I\Delta \mathbf{L}$ and **B** are parallel, the vector cross product is zero; if they are at right angles $\theta = 90°$, the force is a maximum.

The direction of the force is perpendicular to both the current element $I\Delta \mathbf{L}$ and the magnetic field **B**. For example, if both vectors lie in the plane of this page, the vector **F** points either out of the paper or into the paper as given by the **right-hand rule:** If the index finger of the right

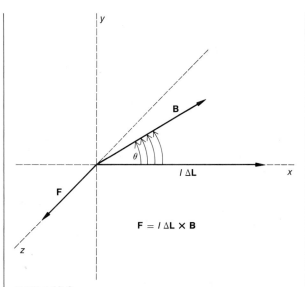

FIGURE 15-3
The cross product vector points in a direction perpendicular to the plane formed by the vectors $I\Delta\mathbf{L}$ and $\mathbf{B}$.

$$\mathbf{F} = I\,\Delta\mathbf{L} \times \mathbf{B}$$

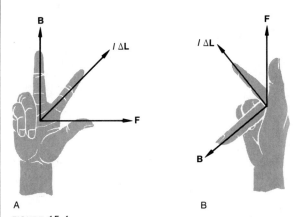

A   B

FIGURE 15-4
The right-hand rule. A. Viewed palm up. B. Viewed with the palm vertical and to the left.

hand points in the direction of the first vector $I\Delta\mathbf{L}$, and the second finger points in the direction of the second vector $\mathbf{B}$, then the thumb points in the direction of the resultant vector $\mathbf{F}$, as shown in Figure 15-4.

EXAMPLE

A wire segment 10 cm long carries a current of 10 A. It is placed in a magnet with a field strength of 0.2 tesla, as shown in Figure 15-2. What is the force acting on the wire if the direction of current is into the paper?

The force will act perpendicular to both the magnetic field and the current. By the right-hand rule, the force will be to the right. The magnitude of the force is given by

$$F = I\Delta LB \sin\theta$$

$$F = (10\text{ A})(0.1\text{ m})(0.2\text{ N/A-m})(1.0)$$

$$F = 0.2\text{ N}$$

This is not a particularly small force.

The Galvanometer Movement

The force between a current-carrying wire and a magnetic field is the basis for the action of a **galvanometer.** A rectangular coil of wire with N turns, a width $2R$, and a length L is placed in the magnetic field of a permanent magnet, as shown in Figure 15-5. Often a soft iron core is placed within the coil so that the magnetic field is uniform and perpendicular to the wire. When a current I flows in the coil of wire, the force due to the magnetic field is

$$F = NILB$$

This force produces a torque that tends to rotate the coil. Both ends of the coil are attached to a torsion spring or band that applies a restoring torque to balance that produced by the magnetic field:

$$\tau = 2RF = 2RNILB = k\theta$$

The angular displacement,

$$\theta = \frac{2NRLB}{k}I$$

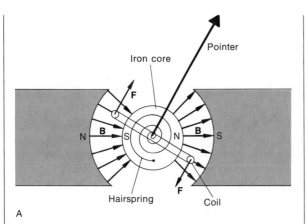

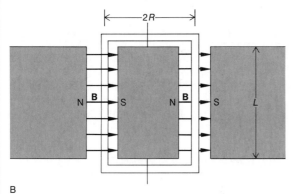

B

FIGURE 15-5
The galvanometer movement has an angular displacement proportional to the current through the coil. A. End view. B. Side view.

is proportional to the current through the coil. Attached to the galvanometer coil and calibrated to give a measure of current are a scale and indicator needle. The resistance of the galvanometer is simply the resistance of the coil of N turns of wire. Such a galvanometer is commonly used as a meter movement, as was discussed in Chapter 14.

Force on a Charged Particle

For a single particle with charge q and velocity v, the current element $I\Delta\mathbf{L}$ is given by

$$I\Delta\mathbf{L} = \frac{\Delta Q}{\Delta t}\Delta\mathbf{L} = \Delta Q\frac{\Delta\mathbf{L}}{\Delta t} = q\mathbf{v}$$

$q\mathbf{v}$ points in the direction in which a positive charge would travel. The force acting on a charged particle in a magnetic field is

$$\mathbf{F} = q\mathbf{v} \times \mathbf{B} \qquad (15\text{-}2)$$

If an ion or electron with charge q and mass m is accelerated to some velocity $\mathbf{v}$, and passes perpendicularly through a magnetic field of strength $\mathbf{B}$, it will experience a force perpendicular both to the field and to its direction of motion. This provides a centripetal force to move the particle in a circular path of some radius R, as shown in Figure 15-6. As the particle moves through the magnetic field, its direction changes, but the force continues to be perpendicular to the direction of motion. The centripetal force acting on the particle must have a magnitude:

$$F = \frac{mv^2}{R} = qvB$$

The radius of curvature of the particle in the magnetic field, the **cyclotron radius,** is just

$$R = \frac{mv}{qB} \qquad (15\text{-}3)$$

A charged particle in a uniform magnetic field will travel in a circular orbit. The time required to do so is

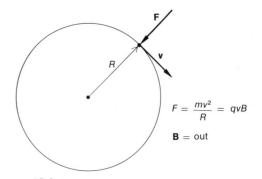

FIGURE 15-6
A charged particle moving perpendicular to a magnetic field travels in a circle.

given by the circumference of that orbit $2\pi R$ divided by the speed v:

$$T = \frac{2\pi R}{v} = \frac{2\pi m v}{vqB} = \frac{2\pi m}{qB} \qquad (15\text{-}4)$$

The time required for a single orbit is independent of both the size of the orbit and the orbital speed, so long as the particle's velocity is much less than the speed of light. Figure 15-7 shows an electron beam in a magnetic field. The electrons are traveling counterclockwise. Which way does the magnetic field point? Figure 15-8 shows the track of an energetic electron in the magnetic field of a liquid-hydrogen bubble chamber. Why is it spiraling inward?

The Mass Spectrometer

At the Cavendish Laboratory in Cambridge, England, after World War I, F. W. Aston developed the **mass spectrometer** to carefully measure the masses of atoms. He determined the precise mass of the majority of the

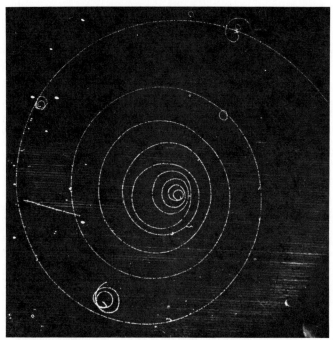

FIGURE 15-8
An energetic electron track in a liquid-hydrogen bubble chamber. Charged particles in a magnetic field travel in a circular path. As a particle loses energy the radius of curvature decreases. The magnetic field strength is 1.8 T. The maximum radius of the orbit is about 5 cm and the initial energy is about 27 MeV. Courtesy of Lawrence Berkeley Laboratory, University of California.

FIGURE 15-7
An electron beam (the bright circle) travels counterclockwise in an evacuated tube in a magnetic field produced by a pair of Helmholtz coils. Electrons striking a small quantity of background gas (krypton) produce the bright glow. In which direction does the magnetic field point? Courtesy of Ealing Corporation.

elements and showed that most elements had isotopes—atoms with the same chemical properties but slightly different masses. The most common isotope of carbon, carbon-12, has an atomic weight of 12.0000 gm/mol. The **atomic mass unit** (amu) is defined as equal to $\frac{1}{12}$ the mass of a carbon-12 atom. One mole of carbon-12 has a mass of 0.012 kg and comprises Avogadro's number, 6.02×10^{23} of atoms. Consequently:

$$1 \text{ amu} = \frac{1}{12} \frac{0.012 \text{ kg}}{6.02 \times 10^{23}} = 1.66 \times 10^{-27} \text{ kg}$$

Aston found that neon, for instance, is composed of three isotopes: (90.5%) Ne-20, with a mass 19.99 amu; (0.3%) Ne-21, with a mass 20.99 amu; and (9.2%) Ne-22, with a mass 21.99 amu. The isotope masses of all elements are very nearly equal to an integer number of atomic mass

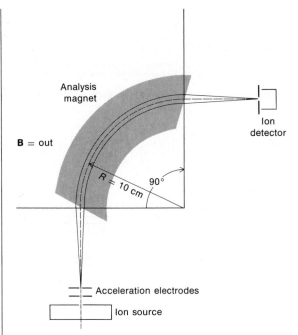

Analysis magnet

B = out

$R = 10$ cm

$90°$

Ion detector

Acceleration electrodes

Ion source

FIGURE 15-9
A high-resolution mass spectrometer. Sweeping the magnetic field from 0.4 T to 16 T allows molecular masses from 2 to 2400 amu to be observed. The magnet is shaped to optimally focus the ion beam.

units. (Aston's work played an important role in the study of the nucleus, as we shall discuss in Chapter 24.)

A modern mass spectrometer is shown in Figure 15-9. An ion source produces ionized atoms by electron bombardment. The ions are accelerated through some constant voltage difference V_{acc} and pass into a uniform magnetic field where they are deflected into a circular path. Those ions with the proper radius of curvature, typically $R = 0.10$ m, strike a detector and are recorded as an electrical signal. The magnet is shaped to focus the diverging ion beam onto the detector.

The velocity of the ion beam is given by the kinetic energy:

$$KE = \tfrac{1}{2}mv^2 = qV_{acc}$$

The radius of curvature of the particles striking the de-

tector was given previously by equation 15-3:

$$R = \frac{mv}{qB}$$

Knowing the charge on the ions, typically $q = +e$, allows solving for the mass of the ion. A well-designed magnetic analyzer can resolve the mass to about 1 part in 1000. By sweeping the magnetic field from 0.04 T to 1.6 T with an accelerating voltage of approximately 500 V, the mass spectrometer measures molecular and atomic ion masses from 2 to 2400 amu.

EXAMPLE
A duPont Type 21-490B mass spectrometer has an analysis magnet with a maximum field strength of 1.6 N/A-m. At this field strength it will bend a singly charged positive ion with a mass of 2400 amu in a radius of $R = 10$ cm.
(a) What is the velocity of the ions?
(b) What is the kinetic energy in eV?
   The velocity of the ions is given by

$$v = \frac{qBR}{m}$$

$$v = \frac{(1.6 \times 10^{-19}\ \text{C})(1.6\ \text{N/A-m})(10^{-1}\ \text{m})}{(1.67 \times 10^{-27}\ \text{kg/amu})(2400\ \text{amu})}$$

$$v = 6.4 \times 10^3\ \text{m/s}$$

The kinetic energy of the ions is

$$KE = \tfrac{1}{2}mv^2$$

$$KE = (0.5)(1.67 \times 10^{-27}\ \text{kg/amu})(2400\ \text{amu}) \times (6.4 \times 10^3\ \text{m/s})^2$$

$$KE = \frac{(8.2 \times 10^{-17}\ \text{J})}{(1.6 \times 10^{-19}\ \text{J/eV})} = 511\ \text{eV}$$

The Cyclotron

As was discussed earlier, a charged particle traveling in a uniform magnetic field takes the same time to complete an orbit regardless of the size of the orbit (see equation 15-4). In the 1930s, Ernest O. Lawrence, at Berkeley, California, used this fact to accelerate charged particles to high energies in his *cyclotron*. Figure 15-10 shows the basic plan of the cyclotron. Hydrogen ions, protons, are introduced at the center of the cyclotron, and an alternating voltage is applied across two hollow electrodes in the vacuum chamber. The ions are accelerated in the electric field as they pass between the electrodes. Their energy and their radius of curvature increase. By the time they have gone half-way around the cyclotron, the electric field has reversed direction, so the particles are again accelerated. This process continues until they reach the extractor, where they leave as an energetic beam of particles.

Under Lawrence's direction, the 184 in (4.7 m) cyclotron was constructed at Berkeley in the early 1940s. Figure 15-11 shows the electromagnet's iron core during construction. Finished after World War II, the cyclotron, with a maximum magnetic field of 1.5 T, could accelerate protons to energies of 350 MeV. Its advent began a new era of high-energy particle physics.

ORIGIN OF MAGNETIC FIELDS

Hans Christian Oersted's announcement, in 1820, of the magnetic effects of electric current created much excitement. Immediately others began to experiment, such as André Marie Ampère, Jean Baptiste Biot, and Felix Savart in France. Ampère reasoned that if a current exerts a force on a magnet and a magnet exerts a force on a current, then perhaps a current exerts a force on a current. Ampère was able to measure the force between two current-carrying wires. If the currents in the wires flow in the same direction, the wires attract. If the currents flow in opposite directions, the wires repel. Figure 15-12 shows two parallel currents. Current I_1 flows in a long wire. A second length of wire ΔL_2 is parallel to the first at a

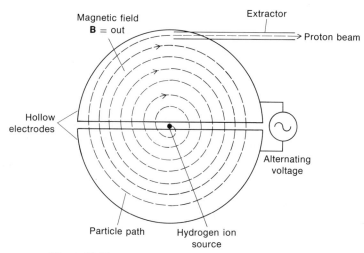

FIGURE 15-10
The basic plan of a cyclotron. As the particle's speed and energy increase, the radius of the orbit increases. The time required to complete one revolution remains constant until the particle approaches the speed of light.

FIGURE 15-11
The iron core of the electromagnet of the 184 inch cyclotron,
under construction in Berkeley in the early 1940s. The magnet
weighs some 4000 tonnes. Courtesy of Lawrence Berkeley
Laboratory, University of California.

distance R, and has a current I_2 flowing in the same direction as I_1. Because of the current in wire 1, F_{21}, there will be an attractive force on wire 2:

$$F_{21} = 2k_m \frac{I_1 I_2 \Delta L_2}{R} \qquad (15\text{-}5)$$

The constant[2] k_m is *defined* to be

$$k_m = 10^{-7} \, \text{N/A}^2$$

---

[2] Many texts use the constant $k_m = \mu_o/4\pi$, where $\mu_o = 4\pi k_m = 4\pi \times 10^{-7} \, \text{N/A}^2$.

This is the defining relationship for the SI unit of current, the **ampere,** A. For two straight parallel wires separated by a distance $R = 1.0$ m, each carrying a current $I_1 = I_2 = 1.0$ ampere (A) in the same direction, the force on a section $\Delta L_2 = 1.0$ m of the second wire is

$$F_{21} = \frac{2k_m I_1 I_2 \Delta L_2}{R} = \frac{2(10^{-7} \, \text{N/A}^2)(1.0 \, \text{A})^2(1 \, \text{m})}{1 \, \text{m}}$$

$$F_{21} = 2 \times 10^{-7} \, \text{N}$$

If in the same direction, the currents attract; if in opposite directions, they repel. The magnitude of the force is proportional to the current in wires I_1 and I_2.

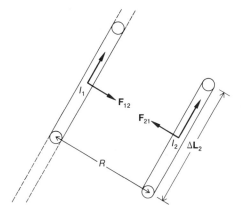

Michael Faraday

Michael Faraday in the early nineteenth century made significant discoveries in electricity, magnetism, and chemistry. Born the son of a blacksmith near London in 1791, Faraday received little formal education and was apprenticed to a bookbinder. It was here that he was attracted to scientific books. A client of the bookbinder gave Faraday tickets to a series of lectures by the illustrious young chemist Sir Humphry Davy. Faraday attended the lectures and took careful notes, which he wrote up, illustrated, bound, and sent to Sir Humphry, also expressing a wish to become his assistant. In 1813, at the age of 21, Faraday thus became a laboratory assistant in the Royal Institution. He spent the next eight years studying primarily chemistry.

In 1821, by request of the British journal *Annals of Philosophy*, Faraday undertook a review and survey of the developments, experiments, and theories in the field of electromagnetism. He quickly moved from reviewing previous work to doing his own experiments. He was able to demonstrate that if a single magnetic pole were free to move near a current-carrying wire, it would travel around the wire in a circle. Figure 15-13 shows two versions of Faraday's electromagnetic rotator. Each cup was filled with mercury to complete the electrical circuit. A steady current would give rise to continuous rotation of the wire around the pole or the magnetic pole around the wire.

Around any current-carrying wire a magnetic-field circulates, as shown in Figure 15-14. This magnetic field, which caused Oersted's compass needles to deflect, is also the source of the force between two current-carrying wires. The magnitude of the magnetic field around a long, straight wire is given by

$$B = 2k_{\mathrm{m}} \frac{I}{R} \qquad (15\text{-}6)$$

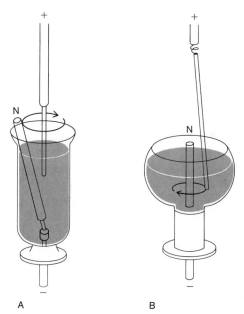

A   B

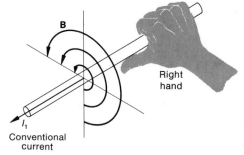

FIGURE 15-14
The "rule of thumb" for magnetic fields. If the thumb of the right hand points in the direction of conventional current flow, the fingers of the right hand curve around the wire in the direction of the magnetic field.

where R is the distance from the wire. To determine the direction of such a magnetic field, we need another convention, which is also shown in Figure 15-14. This is right-hand rule number two, or the "rule of thumb". If the extended thumb of the right hand pointed in the direction of the conventional current flow, and if one (mentally) grasped the wire with the right hand, the fingers would curl around the wire in the direction of the magnetic field.

Force between Parallel Currents

A current I_1 in a wire generates a magnetic field $\mathbf{B}_1$, which interacts with a current element $I_2\Delta\mathbf{L}_2$ of a second wire. There is a force $\mathbf{F}_{21}$ on wire 2 because of the field $\mathbf{B}_1$ produced by wire 1. That force is given by

$$\mathbf{F}_{21} = I_2\Delta\mathbf{L}_2 \times \mathbf{B}_1$$

If $I_2\Delta\mathbf{L}_2$ and $\mathbf{B}_1$ are perpendicular, then the magnitude of the force is given by

$$F_{21} = I_2\Delta L_2 B_1 = I_2\Delta L_2 2k_m\frac{I_1}{R}$$

$$F_{21} = 2k_m\frac{I_1 I_2\Delta L_2}{R}$$

This is the same result as obtained by Ampère for the force between two currents.

A long, straight wire carries a current $I_1 = 20$ A. The direction of the current I, is into the paper, as shown in Figure 15-15. In a second, parallel wire 1.0 cm from the first, a current $I_2 = 20$ A flows in the opposite direction to the first current.
(a) What are the strength and direction of the magnetic field around the first wire at a distance of 1.0 cm?
(b) What is the direction of the force on the second wire?
(c) What is the magnitude of the force on a 1.0 m section of the second wire?

The magnitude of the field is given by equation 15-6:

$$B_1 = k_m\frac{2I_1}{R} = \frac{2(10^{-7}\,\text{N/A}^2)(20\,\text{A})}{(10^{-2}\,\text{m})}$$

$$B_1 = 4.0 \times 10^{-4}\,\text{N/A-m}$$

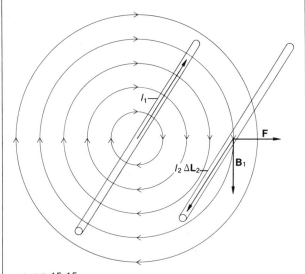

FIGURE 15-15
The interaction between two wires. The current element $I_2\Delta\mathbf{L}_2$ of wire two interacts with the magnetic field $\mathbf{B}_1$ produced by wire one. If the currents are in opposite directions, the wires repel.

As the conventional current I_1 points inward, the direction of the magnetic field around the conductor will be clockwise, as shown in the figure.

The direction of the force acting on the current element $I_2\Delta L_2$ is given by

$$\mathbf{F} = I_2\Delta\mathbf{L}_2 \times \mathbf{B}_1$$

As Figure 15-15 also shows, the direction of the magnetic field at the second wire is downward. The direction of the current element $I_2\Delta\mathbf{L}$ is out of the paper. The direction of the cross product is to the right. The second wire is repelled. In this case, the field and the current are at right angles ($\sin \theta = 1$) so that the force is a maximum. The magnitude of the force is given by

$$F = I_2\Delta L_2 B_1 \sin \theta = I_2\Delta L_2 B_1$$

$$F = (20\ \mathrm{A})(1.0\ \mathrm{m})(4.0 \times 10^{-4}\ \mathrm{N/A\text{-}m})$$

$$F = 8.0 \times 10^{-3}\ \mathrm{N}$$

For opposite currents, the force tends to move the wires apart.

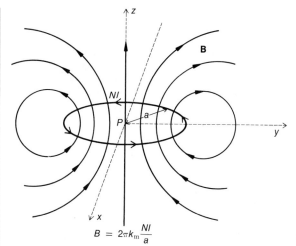

$$B = 2\pi k_\mathrm{m}\frac{NI}{a}$$

FIGURE 15-16
A magentic field at the center of a ring of radius a.

Current Ring and Solenoid

In a **current ring** of radius a consisting of N turns of wire with a current I, each short section of wire contributes to the magnetic field. The magnetic field at the center of the ring is the sum of the contributions of the sections:

$$B = 2\pi k_\mathrm{m}\frac{NI}{a} \qquad (15\text{-}7)$$

If the current in the ring shown in Figure 15-16 is counterclockwise when viewed from the top, the direction of the magnetic field, given by the rule of thumb, is upward, in the $+z$-direction.

Two thin coils of radius a with N turns and a current I separated by a distance a, as shown in Figure 15-17, form

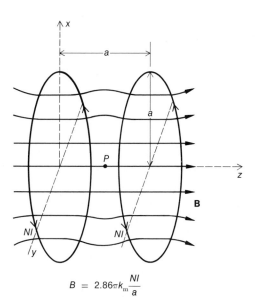

$$B = 2.86\pi k_\mathrm{m}\frac{NI}{a}$$

FIGURE 15-17
A magnetic field at midpoint P of Helmholtz coils. Two coils of radius a are separated by a distance a.

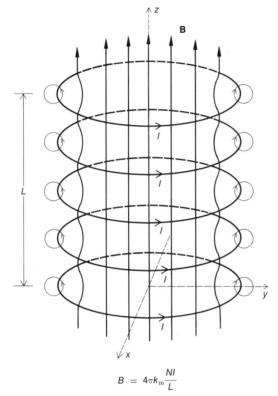

$$B = 4\pi k_{\mathrm{m}} \frac{NI}{L}$$

FIGURE 15-18
A magentic field at the center of a long solenoid of length L with N turns, each with current I.

a pair of *Helmholtz coils.* The magnetic field at the midpoint between the coils is given by

$$B = 2.86\pi k_{\mathrm{m}} \frac{NI}{a} \qquad (15\text{-}8)$$

The magnetic field over a large volume between the two coils is nearly uniform.

   If a number of coils of radius a are placed side by side forming a **solenoid,** as shown in Figure 15-18, each of the N coils contributes to the magnetic field. If the length of the solenoid is much greater than the radius $L \gg a$, the magnetic field inside the solenoid becomes nearly uniform and depends only on the number of coils per unit length N/L:

$$B = 4\pi k_{\mathrm{m}} \frac{NI}{L} \qquad (15\text{-}9)$$

Magnetic Dipole Field

A wire ring of radius a has an area $A = \pi a^2$. If a current I flows in the ring, the magnetic field can be calculated at any point in space surrounding the current ring. In a plane perpendicular to the ring and at a large distance—that is, a distance of several radii from the ring—the magnetic field describes a dipole pattern, as shown in Figure 15-19. The strength of the **magnetic dipole m** is given by a product of the current I multiplied by the area A of the ring. The vector **m** points in the direction of the magnetic field at the center of the ring, the z-direction, $\mathbf{i}_z$.

$$\mathbf{m} = IA\mathbf{i}_z$$

Along the z-axis, the magnetic field points away from the ring and is

$$\mathbf{B} = 2k_{\mathrm{m}} \frac{\mathbf{m}}{R^3}$$

In the plane of the current ring, the magnetic field is downward $(-\mathbf{i}_z)$. The magnetic field is

$$\mathbf{B} = -k_{\mathrm{m}} \frac{\mathbf{m}}{R^3} \qquad (15\text{-}10)$$

where R is the distance from the center of the ring.

The Earth's Magnetic Field

The earth possesses a magnetic field quite similar to that produced by a magnetic dipole. The axis of the earth's magnetic dipole is inclined at an angle of about $11\frac{1}{2}$ degrees from the earth's rotational axis, as shown in Figure 15-20, so that the direction of magnetic north (compass north) is not identical with geographic north (true north). The difference between compass north and true north is given by the *magnetic declination.* In 1970 in London, England, compass north was about 7° west of true north, and in Washington state, compass north was about 22° east of true north. Nor is the earth's magnetic field horizontal (except in a few places): If a compass needle is suspended in a vertical plane oriented in the direction of magnetic north in the northern hemisphere, it will point downward at some dip angle, because the

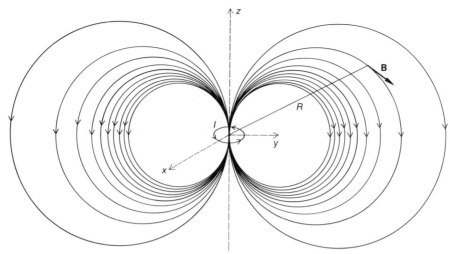

FIGURE 15-19
A magnetic dipole field around a current ring.

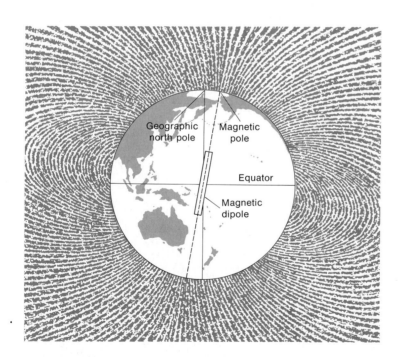

FIGURE 15-20
The earth's magnetic field is much like the field produced by a magnetic dipole at the earth's center. From Frank Press and Raymond Siever, *Earth*, 2nd ed. (San Francisco: W. H. Freeman and Company). Copyright © 1978.

magnetic field lines are converging toward the north magnetic pole. As Figure 15-21 shows, in North America the earth's magnetic field is about 0.6×10^{-4} T, pointing north and downward at an angle of about $70°$.

Measurements of the magnitude and direction of the earth's magnetic field in space began in 1958, with the Pioneer 1 space probe. They were pursued in great detail by the first Interplanetary Monitoring Platform, the IMP

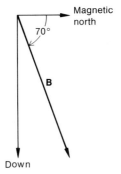

FIGURE 15-21
Magnetic dip. The earth's magnetic field in the Northern Hemisphere points downward.

satellite, launched into an elliptical orbit in 1963 with an apogee, or maximum distance from the earth, of some 30 earth radii. As Figure 15-22 shows, the dipole field of the earth is considerably distorted by the solar wind, the flow of charged particles from the sun. Toward the sun at a distance of some 10 earth radii from the earth's center is the *magnetopause:* a transition between two regions, one dominated by the earth's magnetic field and the other dominated by the interplanetary magnetic field and solar wind.

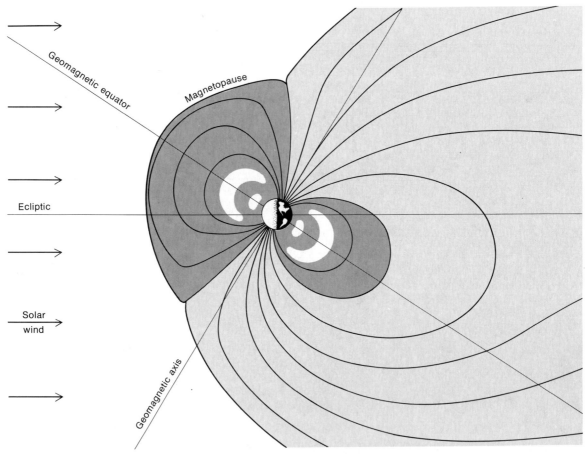

FIGURE 15-22
Solar wind distorts the earth's magnetic field. The white regions close to the earth indicate the Van Allen radiation belts, which contain energetic particles trapped in the earth's magnetic field. From Laurence J. Cahill, Jr., "The Magnetosphere." Copyright © 1965 by Scientific American, Inc. All rights reserved.

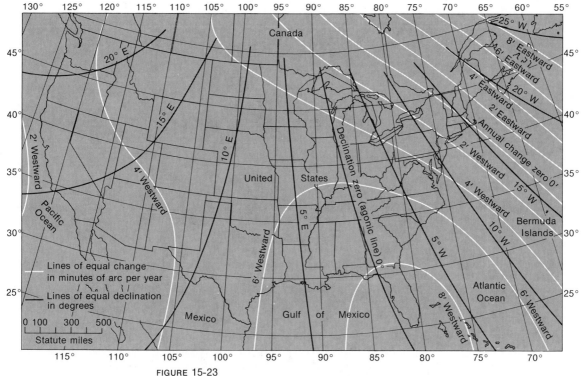

FIGURE 15-23
Magnetic declination. The direction of magnetic north is usually not the same as the direction of the geographic North Pole, and it changes slightly from year to year. U.S. Geological Survey in cooperation with National Oceanic and Atmospheric Administration.

The earth's magnetic field changes orientation with time. In 1580, the magnetic declination of London was 11° east of north; in 1760, due north; in 1820, 24° west of north; and today, 7° west of north. Figure 15-23 shows the present lines of equal magnetic declination and of annual changes in declination for North America. In addition to these relatively recent recorded changes, geological evidence indicates that the apparent position of the earth's magnetic pole has wandered over the past 2.6 billion years as the earth's outer layers, or plates, drifted, causing continents to collide and to drift apart. Moreover, there is strong evidence that direction (the polarity) of the earth's magnetic field has reversed several times in the last 4 million years.

Generation of the earth's magnetic field is not yet understood. Undoubtedly, a large electric current is induced in the earth's molten core. The earth is spinning in the path of the solar wind. Apparently, there is some dynamo effect that induces the current that produces the earth's magnetic field. Solar prominences, flares, and sunspot cycles are known to induce small fluctuations in the earth's magnetic field.

EXAMPLE

The earth's solid core has a radius of about 1000 km. The magnetic field at the earth's surface at a radius of 6400 km has a value of about 0.6×10^{-4} T.
(a) What is the strength of the magnetic moment at the center of the earth, as indicated by the magnetic field at the earth's surface?

(b) If it consists of a current ring of radius 1000 km, what is the magnitude of the current?

The magnitude of the magnetic field given by a current ring in the plane of the ring is given by equation 15-10:

$$B = \frac{k_{\mathrm{m}} m}{R^3}$$

The magnitude of the magnetic moment is

$$m = \frac{BR^3}{k_{\mathrm{m}}} = \frac{(0.6 \times 10^{-4}\,\mathrm{N/A\text{-}m})(6.4 \times 10^6\,\mathrm{m})^3}{(10^{-7}\,\mathrm{N/A^2})}$$

$$m = 157 \times 10^{21}$$

$$m = 1.6 \times 10^{23}\,\mathrm{A\text{-}m^2}$$

The ring current is given by

$$I = \frac{m}{A} = \frac{m}{\pi R^2}$$

$$I = \frac{(1.6 \times 10^{23}\,\mathrm{A\text{-}m^2})}{(3.14)(10^6\,\mathrm{m})^2}$$

$$I = 5.0 \times 10^{10}\,\mathrm{A}$$

The current required is about 50 billion amperes.

Permanent Magnets

A magnet acts as if it has **magnetic poles.** A magnet suspended to freely rotate aligns itself with the earth's magnetic field. One end of the magnet is a north-seeking (N) pole, the other end a south-seeking (S) pole. Like magnetic poles repel, unlike poles attract. If we treat the earth as a magnet, at the geographic north pole we must place an S (south-seeking) magnetic pole to attract the N (north-seeking) pole of the compass. By convention, the magnetic field lines point from an N (north-seeking) pole to an S (south-seeking) pole. Because the earth's magnetic field has a large dip angle, a vertical piece of iron, such as a door frame in your school building, will become mag-

netized. In the northern hemisphere the bottom end becomes a north-seeking pole, the upper end a south-seeking pole. The steel frameworks of many buildings produce stray magnetic fields strong enough to completely obscure the direction of north to a compass inside the building.

The atoms of *ferromagnetic* materials such as iron, cobalt, and nickel have strong magnetic dipoles because of unpaired electron spins. The magnetic dipoles of adjacent atoms tend to align in the same direction, forming small crystal regions, or **domains.** If such a magnetic material is placed in an external magnetic field, the atoms tend to align along the direction of the applied field. The domains of an unmagnetized piece of iron ore aligned in alternate directions, which cancel, give zero net magnetic field, as shown in Figure 15-24. As the material becomes magnetized, the domains aligned with the field grow in size at the expense of those aligned in other directions. The domains of a permanent magnet are aligned primarily in one direction. If a permanent magnet is broken in half, each piece is a magnet with north and south poles.

ELECTROMAGNETIC INDUCTION

Faraday's Law

In 1831, after years of persistent experimentation, Michael Faraday demonstrated that a changing magnetic field induces an electric current. As Figure 15-25 shows, around two different sections of an iron ring (some 22 mm thick and 15 cm in diameter) Faraday wound two separate coils of copper wire, each made up of many turns. Coil A was attached to a battery, coil B to a galvanometer. When the switch was closed in the circuit of coil A, the galvanometer attached to coil B was deflected for an instant. When the switch was opened, the galvanometer was deflected in the opposite direction.

Once Faraday was on the right track, his experiments proceeded rapidly. Later in the same year he demonstrated that moving a bar magnet through a coil produces a current, as Figure 15-26 shows. Faraday, however, was not the first to discover the effect of electromagnetic induction. A year earlier, the American scientist Joseph Henry had done so. But unfortunately, his teaching and administrative duties gave him little time for research and

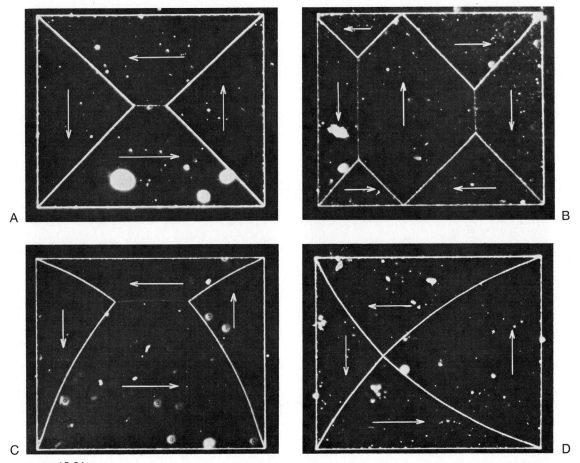

FIGURE 15-24
Magnetic domains of a Ni-Co whisker 165 μm wide and about 0.3 μm thick. A colloidal suspension is attracted to the domain boundaries. A. A complicated zero magnetic field pattern. B. The simplest possible zero magnetic field pattern. C. Reversible response of domains to an applied magnetic field of 3.6×10^{-4} tesla to the right. D. Reversible response to an applied magnetic field of 3.9×10^{-4} tesla upward. Courtesy of R. W. DeBlois, General Electric Company, Research and Development Center.

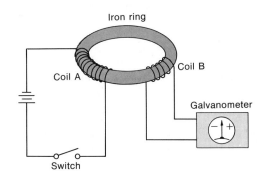

FIGURE 15-25
Faraday's experiment. The galvanometer deflects momentarily when the switch to coil A is closed.

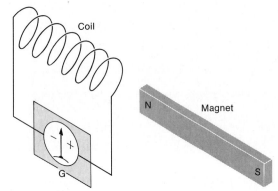

FIGURE 15-26
A moving magnet induces an electric current in a wire coil.

of the magnetic field and the area perpendicular to that magnetic field:

$$\Phi = \mathbf{B} \cdot \mathbf{A} = BA \cos \theta$$

The vector area $\mathbf{A}$ has a magnitude equal to the area of the coil and a direction along the axis of the coil. The SI unit of flux is the weber, Wb. Table 15-2 shows the units of magnetic flux.

TABLE 15-2
Units of magnetic flux, $\Phi = \mathbf{B} \cdot \mathbf{A}$.

Unit	SI equivalent
weber (Wb)	$1 \text{ Wb} = 1 \text{ T-m}^2 = 1 \text{ N-m/A} = 1 \text{ V-s}$
maxwell	$1 \text{ maxwell} = 10^{-8} \text{ Wb}$
gauss-cm^2	$1 \text{ gauss-cm}^2 = 1 \text{ maxwell} = 10^{-8} \text{ Wb}$

by the time he reported his experiments Faraday had already published the results of his independent discoveries.

Faraday and Henry discovered that a changing magnetic field generates an **electromotive force** that in turn produces a current flow. The changing magnetic field can be produced either by changing a current or by moving a magnet and a coil toward or away from one another.

The emf induced in the coil is proportional to the rate at which the magnetic flux through the coil changes. As Figure 15-27 shows, the **magnetic flux** Φ is the product

The magnetic flux can change because the magnetic field changes,

$$\Delta\Phi = \Delta\mathbf{B} \cdot \mathbf{A}$$

or because the area changes,

$$\Delta\Phi = \mathbf{B} \cdot \Delta\mathbf{A}$$

or because the angle changes,

$$\Delta\Phi = BA(\Delta \cos \theta)$$

Faraday's law states that the emf generated is proportional to the rate at which the flux changes in time:

$$\text{emf} = -\frac{\Delta\Phi}{\Delta t} \qquad (15.11)$$

Lenz's Law

Whenever the magnetic flux through a coil changes, an emf is generated in the coil. In each situation, we must determine the direction of the emf and of the induced

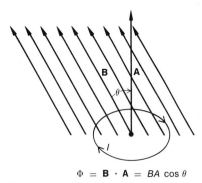

$$\Phi = \mathbf{B} \cdot \mathbf{A} = BA \cos \theta$$

FIGURE 15-27
The magnetic flux is the product of the magnetic field times the normal surface area.

current flow. **Lenz's law** states that the direction of the induced current will produce a magnetic field to oppose the change in magnetic flux that causes it. Thus, if the magnetic flux through a loop of wire is increasing, the induced current will produce a field that tends to cancel the increase in magnetic flux. Figure 15-27 shows a single loop in an increasing magnetic field. As viewed from the top, the current will flow clockwise, producing a downward magnetic field in the center of the coil that tends to cancel the increasing magnetic flux.

In a wire moving perpendicular to a magnetic field, emf is induced. Figure 15-28 shows a wire moving to the right in a magnetic field **B,** which points out of the paper. Assume that this wire is sliding to the right at some velocity **v** on contacts that maintain a complete circuit. The wire, of length L, makes some angle θ with the velocity vector in a plane perpendicular to the magnetic field. In a time Δt the area surrounded by the circuit will increase by

$$\Delta A = v\Delta t L \sin \theta$$

as indicated in the figure. The emf generated in the moving wire is

$$\text{emf} = -\frac{\Delta \Phi}{\Delta t} = \mathbf{B} \cdot \frac{\Delta \mathbf{A}}{\Delta t} = BLv \sin \theta$$

The emf generated can also be written as

$$\text{emf} = \mathbf{L} \cdot (\mathbf{v} \times \mathbf{B}) \tag{15-12}$$

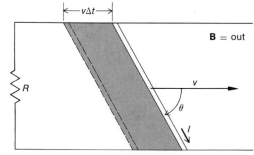

FIGURE 15-28
A conductor moving through a magnetic field generates an electromotive force by changing magnetic flux.

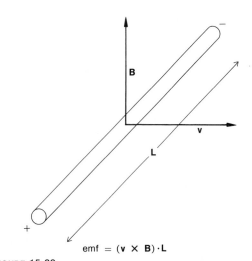

$$\text{emf} = (\mathbf{v} \times \mathbf{B}) \cdot \mathbf{L}$$

FIGURE 15-29
A conductor moving perpendicular to a magentic field generates an electromotive force because of the force acting on the charge carriers.

If, in Figure 15-29, the magnetic field points out of the paper, then an induced current will tend to reduce the magnetic flux to the left of the wire. The current direction is downward. The vector product $(\mathbf{v} \times \mathbf{B})$ also points downward. One consequence of the behavior expressed by Faraday's law is that a conductor moving through a magnetic field generates an emf. This fact makes possible the electric generator, as we shall discuss in Chapter 16.

Moving Charge Carriers

If a wire moves perpendicular to a magnetic field with some velocity **v,** force is exerted on the charge carriers in the wire, which are also moving with velocity **v.** The force on the charge carriers is

$$\mathbf{F} = q\mathbf{v} \times \mathbf{B}$$

In the frame of reference moving along with the wire, the charge carriers show no apparent motion. Nevertheless, there is a force acting on them, which, in the moving reference frame, appears as an induced electric field. The magnitude of this induced electric field is given by

$$E = \frac{F}{q} = v \times B$$

Summing this induced electric field over the length of the wire gives an induced emf:

$$\text{emf} = E \cdot L = L \cdot (v \times B)$$

This same emf was obtained previously by considering the emf produced by a changing magnetic flux. The emf acting on charge carriers moving through a magnetic field is the basis of the *Hall effect device*, which is used to measure the strength of magnetic fields. As Figure 15-30 shows, an external voltage source establishes a current of charge carriers I_H in a thin strip of semiconducting material. The charge carriers move at a drift velocity v_d perpendicular to the magnetic field. The current I_H through the device is proportional to the drift velocity, the charge on the carrier $+e$, the density of the carriers n, and the cross-sectional area of the strip A. The total current is given by

$$I_H = env_d A$$

There will be an induced emf, the Hall voltage V_H, perpendicular to both the drift velocity and the magnetic field. The magnitude of the Hall voltage is proportional to

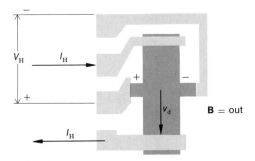

FIGURE 15-30
The Hall effect. When a current I_H flows in a magnetic field B, a Hall voltage is generated. Hall effect devices are very small.

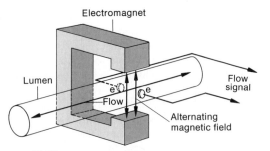

FIGURE 15-31
A blood flow rate transducer uses the Hall effect to generate a voltage between the electrodes, e. Courtesy of Gould, Inc., Measurement Systems Division.

the width of the Hall device L_H, the drift velocity v_d, and the strength of the magnetic field:

$$V_H = L_H \cdot (v_d \times B) = L_H v_d B = \frac{I_H L_H B}{enA}$$

The output voltage across the Hall device is directly proportional to the normal component of magnetic field **B**.

The Hall effect is used in clinical and research medicine to measure blood flow rate. Figure 15-31 shows a schematic of a blood flow rate transducer. An electromagnet produces a magnetic field perpendicular to the flow of blood through the lumen of an artery or vein. Small electrodes make contact with the outer surface of the blood vessel. Because blood is a conducting fluid that contains ions, a Hall voltage is induced. The platinum electrodes can become polarized from electrochemical reaction, producing dc (direct current) potentials from several millivolts to several volts. To prevent this, the magnetic field is periodically reversed, typically at a frequency of 400 Hz to 1000 Hz. The emf is measured by an ac (alternating current) amplifier, which rejects the dc component of induced voltage. Hall effect probes are available in diameters from 1 mm to 40 mm and are inserted around the vein or artery. The electrode contacts do not need to pierce the artery wall. Figure 15-32 shows a 5 mm and a 24 mm Hall effect probe suitable for aorta and coronary application. A 5 mm lumen probe has a flow range from 100 milliliter/minute to 1000 milliliter/minute.

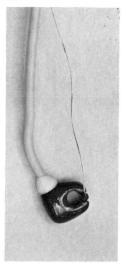

FIGURE 15-32
Cuff-type flow probes designed for medical applications. The flow probes are slipped around major blood vessels during surgery to monitor blood flow rate. Courtesy of Gould, Inc., Measurement Systems Division.

EXAMPLE

A blood flow rate transducer is designed to fit over an artery of lumen size 5 mm. Blood flows through the vessel at a speed of 300 cm³/min. A magnetic field of 4.0×10^{-3} T is perpendicular to the blood flow.
(a) What is the velocity of the blood flow through the vessel?
(b) What is the emf generated across the lumen?
   The blood flow velocity is a function of the volume flow rate and the cross section of the lumen:

$$\frac{\Delta Q}{\Delta t} = Av$$

The flow velocity is then given by

$$v = \frac{1}{A}\frac{\Delta Q}{\Delta t} = \frac{(300 \text{ cm}^3/\text{min})(10^{-6} \text{ m}^3/\text{cm}^3)}{\pi(2.5 \times 10^{-3} \text{ m})^2(60 \text{ s/min})}$$

$$v = 0.255 \text{ m/s}$$

The Hall voltage generated across the lumen is

$$V_{\text{H}} = L_{\text{H}}v_{\text{d}}B = (5 \times 10^{-3} \text{ m})(0.255 \text{ m/s})$$
$$\times (4.0 \times 10^{-3} \text{ N/A-m})$$

$$V_{\text{H}} = 5.1 \times 10^{-6} \text{ V} = 5.1 \ \mu\text{V}$$

This is a small voltage but large enough to be measured.

Inductance

A current flowing in a coil of wire produces a magnetic field that is proportional to the current I_1. Intercepting this field with a second coil, as shown in Figure 15-33, produces in this second coil a magnetic flux proportional to the current in the first coil. Any change in the current I_1 induces in the second coil an emf E_2 that resists the change in magnetic field. This gives rise to **mutual inductance** M:

$$E_2 = -N_2 \frac{\Delta \Phi}{\Delta t} = -M \frac{\Delta I_1}{\Delta t}$$

The mutual inductance relates the emf produced in coil 2 by the change in the magnetic flux through coil 2 that is produced by the current in coil 1. The magnitude of the mutual inductance depends on the geometry of the coil and the number of turns of wire. The SI unit for inductance is the **henry**, H:

$$1 \text{ H} = 1 \text{ V-s/A}$$

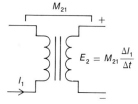

FIGURE 15-33
Mutual inductance. A changing current in coil 1 induces a voltage in coil 2.

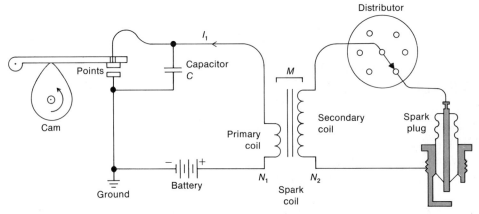

FIGURE 15-34
Because of mutual inductance M of the spark coil, interrupting the current in the primary coil I_1, by opening the points, produces a large voltage in the secondary coil.

The automobile spark coil uses the principle of mutual inductance. Figure 15-34 shows an automobile ignition system. A relatively large current flows in the spark coil's primary circuit, setting up a large magnetic flux in the second coil, which consists of very many turns. When the points in the distributor open, the current in the primary circuit abruptly stops, collapsing the magnetic field and giving a large time rate of change in the magnetic flux. This generates in the secondary coil a large emf that the distributor conveys to the appropriate spark plug. A capacitor (the condensor) is placed across the points so that any emf generated in the primary coil by the collapsing magnetic field appears across the capacitor, and does not arc or pit the points.

When a current flows in a coil of wire, a magnetic field is set up proportional to the current. The coil itself intersects the magnetic flux produced by this current. A change of the current in the coil changes the magnetic flux, setting up an emf that in turn resists the change in the current. This gives rise to a **self-inductance.**

$$\text{emf} = -N\frac{\Delta\Phi}{\Delta t} = -L\frac{\Delta I}{\Delta t}$$

The self-inductance relates the emf produced in the coil because of the change in the magnetic flux through the coil to the change in current in that coil.

The magnitude of the self-inductance depends on the geometry of the coil and the number of turns of wire. Figure 15-35 shows the symbol for an inductor. The SI unit for self-inductance is also the henry.

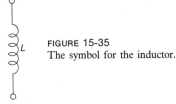

FIGURE 15-35
The symbol for the inductor.

Figure 15-36 shows typical inductors used in electronic circuits. Both have inductance values of a few millihenrys(mH). Winding the coil on a core of magnetic material greatly increases the inductance. Placing a small, movable magnetic slug in an inductor makes it possible to adjust the value of inductance. Large inductors wound on iron cores may have inductances from 0.1 to several henrys.

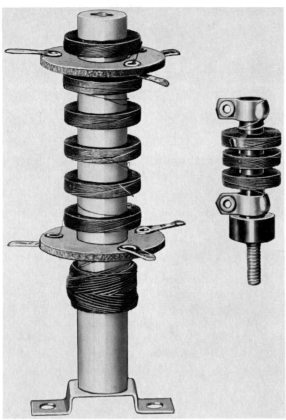

FIGURE 15-36
Typical inductors used in electronic circuits. Courtesy of Bell
Industries, J. W. Miller Division.

Review

Give the significance of each of the following terms. Where appropriate give two examples.

magnetic field
current element
tesla
vector cross product
right-hand rule
galvanometer
cyclotron radius
mass spectrometer

atomic mass unit
ampere
current ring
solenoid
magnetic dipole
magnetic poles
domains
electromotive force

magnetic flux
Faraday's law
Lenz's law
mutual inductance
henry
self-inductance

Further Readings

Laurence J. Cahill, Jr., "The Magnetosphere," *Scientific American* (March 1965).
James A. Van Allen, "Radiation Belts around the Earth," *Scientific American* (March 1959).
Robert R. Wilson and Raphael Littauer, *Accelerators: Machines of Nuclear Physics* (New York: Doubleday, Anchor Science Study Series S 17, 1960).

Problems

MAGNETISM

1 Two long, straight wires separated by 4.0 mm carry a current of 15 A 60 Hz. What is the maximum force on a 1 m length of one of the wires?

2 A current of 20 A is flowing in a wire from east to west. A second current of 20 A is flowing from south to north. Where the two wires cross they are 1.0 cm apart. What is the force on a 10 cm segment of the first wire?

3 A meter movement has a magnetic field of 0.2 T. The coil has 1344 turns of #50 wire, a width of about 1 cm, and a length, perpendicular to the field, of about 1.5 cm. The return spring has a spring constant $k = 1.67 \times 10^{-8}$ N-m/degree. The meter gives a full-scale deflection of 95°. What current is required to give a full-scale deflection?

4 The meter described in problem 3 has a current through the coil of 10 μA. What is the deflection of the meter?

5 A wire carrying a current of 5.0 A is perpendicular to the magnetic field of a laboratory magnet with a strength of 0.30 T. The wire is 10 cm long in the field. What is the force on the wire?

6 A straight wire has a current of $I_1 = 20$ A flowing from east to west. One cm away is a second wire with a current $I_2 = 20$ A flowing from west to east.

(a) What is the magnetic field caused by the first wire at a distance of 1.0 cm? Make a sketch showing the direction of the field.

(b) What is the force acting on a 1.0 m length of the second wire? Show on your sketch the direction of this force.

7 A wire 20 cm long is suspended horizontally east to west in one arm of a sensitive balance. The wire is perpendicular to a horizontal magnetic field of 1.2 T pointing north to south.

(a) If the segment of wire has a mass of 6 gm, what current is required to support the weight of the wire?

(b) The balance is adjusted so that with no current flowing it is in equilibrium. If a current of 20 A flows through the wire, how much weight must be added to restore the equilibrium?

8 A mass spectrometer accelerates ions to 500 eV into a trajectory with a radius of curvature of 0.10 m.

(a) For a singly charged positive ion with a mass of 200 amu (1 amu $= 1.66 \times 10^{-27}$ kg), what magnetic field is required?

(b) The minimum mass for a singly charged ion is 4 amu. What magnetic field is required?

(c) The maximum field strength is 1.6 T (16 kilogauss). What is the maximum observable mass of a singly charged ion?

9 A mass spectrometer accelerates ions to 1000 V into a mass spectrometer with a radius of curvature of 0.10 m.

(a) The maximum mass for a singly charged ion that can be observed is 1200 amu. What is the maximum field strength?

(b) What magnetic field is required to detect a singly charged mass of 600 amu?

10 The electron beam in a color television tube is accelerated to 28 keV. The beam is deflected by a magnetic field produced in the deflection yoke through a maximum angle of 55 degrees, with a radius of curvature of about 2.0 cm.

(a) What is the electron velocity?

(b) What magnetic field is required?

11 Some color television sets have an acceleration voltage of 34 kV.

(a) If the radius of curvature is 2.0 cm, what magnetic field is required?

(b) To a person facing the front of the picture tube, what direction is required of the magnetic field to deflect the beam to the left? To deflect the beam upwards?

12 In 1932, E. O. Lawrence produced an early-model cyclotron that produced a beam of 1.2 MeV protons. The magnet had a diameter of about 28 cm. Assume that the maximum radius of the orbit is 14 cm and that the protons are nonrelativistic.

(a) What is the magnetic field strength?

(b) What is the frequency of the proton orbit?

ORIGIN OF MAGNETIC FIELD

13 A current of 20 A dc flows in a single copper wire. The wire has a diameter of 2.0 mm.

(a) What is the magnetic field strength at the surface of the wire?

(b) What is the magnetic field strength at a distance of 1.0 cm from the center of the wire?

(c) If the conventional current in the wire is pointing out of the paper, what is the direction of the magnetic field?

14 A coil of 50 turns of wire and a radius of 5 cm has a current of 1.0 A.

(a) What is the magnetic field at the center of the coil?

(b) What current is needed to produce a magnetic field equal in magnitude to the earth's magnetic field?

15 A nuclear magnetic resonance spectrometer has a pair of Helmholtz coils placed inside the gap of a larger electromagnet with a field strength of 1.4 T. The Helmholtz coils change the magnetic field slightly and thereby adjust the resonance frequency. The coils have 50 turns of wire each and have a radius and separation of 5 cm. What current is required to produce a magnetic field change of 5.6×10^{-6} T?

16 The earth's magnetic field at the surface points toward magnetic north, and in northern latitudes points downward at an angle of about 70°. The magnitude of the earth's field is about 0.6×10^{-4} T. A horizontal wire is strung from east to west between two battery terminals; the east end attached to the negative, the west end attached to the positive.

(a) In what direction will the wire be deflected? Make a sketch.

(b) If the current in the wire is 5 A, what is the magnitude of the force on a 50 cm segment of the wire?

17 If a horizontal wire is strung from north to south between two battery terminals, the north end connected to the positive, the south end to the negative, the magnetic field is as described in the previous problem.

(a) In what direction will the wire be deflected?

(b) If the current in the wire is 5 A, what is the force on a 50 cm segment of the wire?

18 A flat, circular coil is suspended in the earth's magnetic field by a thread tied at a point on the edge of the coil. The coil is free to rotate about its vertical diameter. When a steady current is established, the coil rotates to become aligned with the earth's field. When the coil comes to rest, a student looks through the coil toward the north. In what direction around the coil do the electrons move?

19 The alignment of color television sets requires that the electron beam be properly shielded from the earth's magnetic field. If the magnetic field has a magnitude of 0.6×10^{-4} T pointing downward at $70°$, what are the radius of curvature of the 28 keV electron beam and the direction of its deflection if it is initially pointed east?

ELECTROMAGNETIC INDUCTION

20 Two circular coils of radius 10 cm are 10 cm apart, one above the other. The lower coil has a conventional current flowing clockwise when viewed from above.
   (a) What is the direction of the magnetic field produced by the lower coils?
   (b) If the current in the lower coil suddenly goes to zero, what will happen in the upper coil while the current is changing?

21 A rectangular coil of wire 2 m long and 20 cm wide has 25 turns and is connected to a voltmeter with an internal resistance of 100 ohms. The coil is pulled out lengthwise perpendicular to the field of a magnet with a strength of 0.6 T, at a uniform speed of 0.3 m/s.
   (a) What is the emf generated by the coil?
   (b) What current flows?
   (c) What force must be exerted to keep the coil moving at a constant speed?

22 A Hall effect blood flow transducer fits over the aorta with a lumen diameter of 16 mm. A magnetic field of 8.0×10^{-3} T is perpendicular to the blood flow. An average Hall voltage of 4.0μV is observed.
   (a) What is the average blood flow velocity?
   (b) What is the average volume flow rate in cm^3/min?

16

Electrical Signals

SIMPLE ELECTRICAL CIRCUITS

Sine Waves

The screen of an oscilloscope connected to a sine wave generator will display a voltage **sine wave,** a sinusoidally varying voltage signal, as shown in Figure 16-1. The voltage sine wave is a function of time:

$$V = V_p \sin \omega t$$

where V_p is the voltage amplitude, or **peak voltage** and ω (the lowercase Greek letter *omega*) is the **angular frequency** measured in units of radians per second (rad/s). The sine function will repeat itself in a period T where

$$\omega T = 2\pi$$

The **frequency** f, measured in units of cycles per second, or hertz (Hz), is the reciprocal of the period:

$$f = \frac{1}{T} = \frac{\omega}{2\pi}$$

Consequently, the voltage sine wave may also be written:

$$V = V_p \sin 2\pi f t$$

If a sinusoidal voltage source is connected across a resistor R in the simple circuit shown in Figure 16-2,

there will be a current proportional to the applied voltage, as given by Ohm's law:

$$I = \frac{V}{R} = \frac{V_p \sin \omega t}{R}$$

The **peak current** I_p is equal to the peak voltage V_p divided by the resistance:

$$I_p = V_p/R$$

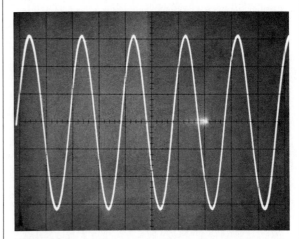

FIGURE 16-1
A sine wave voltage signal displayed on an oscilloscope. Vertical, 2 V/division; horizontal, 0.5 ms/division.

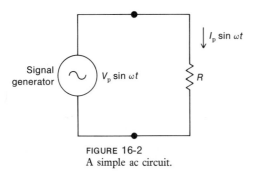

FIGURE 16-2
A simple ac circuit.

Thus, the current is also a function of time:

$$I = I_p \sin \omega t$$

Because the current reverses direction every half cycle, it is called **alternating current,** ac.

The **instantaneous power** dissipated in the resistor is given by Joule's law:

$$P = IV = (I_p \sin \omega t)(V_p \sin \omega t)$$

Thus, the instantaneous power is

$$P = I_p V_p \sin^2 \omega t = I_p^2 R \sin^2 \omega t = \frac{V_p^2}{R} \sin^2 \omega t$$

as shown in Figure 16-3. The average power over one

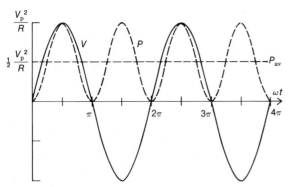

FIGURE 16-3

The instantaneous voltage and power. $P = \dfrac{V_p^2}{R} \sin^2 \omega t$.

The average power $P_{av} = \dfrac{1}{2} \dfrac{V_p^2}{R}$.

cycle requires the average value of $\sin^2 \omega t$, which can be written as

$$\sin^2 \omega t = \frac{1}{2} - \frac{1}{2} \cos 2\omega t$$

In averaging over one cycle, we get

$$\langle \sin^2 \omega t \rangle = \frac{1}{2} - \frac{1}{2} \langle \cos 2\omega t \rangle$$

The average of $\cos 2\omega t$ is zero, because during any one cycle it is positive and negative for equal durations. Thus

$$\langle \sin^2 \omega t \rangle = \frac{1}{2}$$

The **average power** over one cycle is

$$P_{av} = \frac{1}{2} I_p V_p = \frac{1}{2} I_p^2 R = \frac{1}{2} \frac{V_p^2}{R}$$

The **root-mean-square voltage,** V_{rms} is the square root of the mean value of the square of the voltage:

$$V_{rms} = \langle V_p^2 \sin^2 \omega t \rangle^{1/2}$$

$$V_{rms} = \left[\frac{V_p^2}{2} \right]^{1/2} = \frac{V_p}{\sqrt{2}} = 0.707\, V_p \quad (16\text{-}1)$$

Similarly, the root-mean-square current is

$$I_{rms} = \frac{I_p}{\sqrt{2}} = 0.707\, I_p$$

Figure 16-4 shows the peak voltage V_p and rms voltage V_{rms}. The average power dissipated in a resistance is just the product of the rms current I_{rms} and rms voltage V_{rms}:

$$P_{av} = \frac{I_p V_p}{2} = I_{rms} V_{rms} \quad (16\text{-}2)$$

Consequently, the rms values of alternating current and voltage are given when specifying ac power requirements.

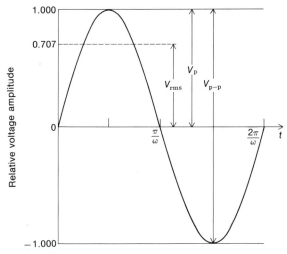

A sine wave voltage signal, showing the peak voltage V_p, the peak-to-peak voltage V_{p-p}, and the root-mean-square voltage V_{rms}.

Square Waves and Pulses

A common electrical signal is a short **pulse.** Initially at zero, the voltage rises rapidly to a maximum peak voltage V_p and remains at that voltage for some time. After some duration T_D, the voltage returns to the zero or base line. Such signals are often used in timing experiments or in counting events. The ideal square wave has sharp corners rising neatly to a maximum, having a flat top and returning quickly to zero. If the pulse is very short, such as, for instance, a counting pulse in a particle detector or a digital pulse in a computer, the duration of the pulse becomes comparable to the rise and fall time, giving a somewhat rounded shape. An ideal square pulse and a real pulse are represented in Figure 16-5.

The **square wave** is a series of square pulses. Starting at zero, the voltage rises rapidly to a maximum V_m, where it remains for a duration T_1. The voltage then rapidly

EXAMPLE

A lamp is rated at 300 watts at $V_{rms} = 120$ V at 60 Hz. What is the peak voltage? What is the rms current?

The peak voltage is related to the rms voltage by

$$V_p = V_{rms}/0.707 = 120 \text{ V}/0.707 = 170 \text{V}$$

120 V rms is 170 V peak. At an average power of 300 W, the rms current is

$$I_{rms} = P_{av}/V_{rms} = 300 \text{ W}/120 \text{ V}_{rms} = 2.5 \text{ A}$$

The peak current is

$$I_p = I_{rms}/0.707 = 2.5 \text{ A}/0.707 = 3.54 \text{ A}$$

The resistance can be found from the ratio of either the peak or the rms voltage to its respective current:

$$R = V_p/I_p = V_{rms}/I_{rms} = 120 \text{ V}/2.5 \text{ A} = 48 \text{ }\Omega$$

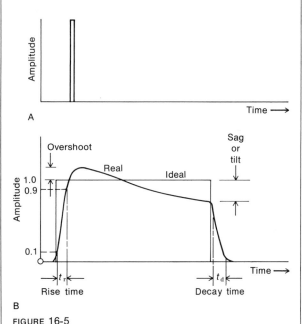

A single square pulse: A. normal time scale; B. expanded time scale showing finite rise and fall times. From A. James Diefenderfer, *Principles of Electronic Instrumentation,* © 1972 by W. B. Saunders Co.

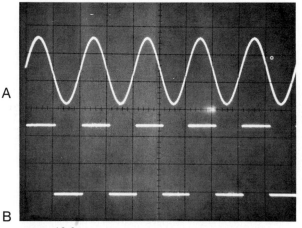

A

B

FIGURE 16-6
Common wave forms displayed on an oscilloscope. A. Sine wave. B. Square wave. Vertical, 5 V/division; horizontal, 2 ms/division.

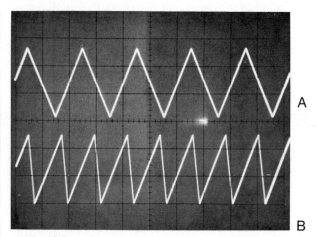

A

B

FIGURE 16-7
Common wave forms displayed on an oscilloscope. A. Triangle wave. B. Sawtooth-wave. Vertical, 5 V/division; horizontal, 2 ms/division.

returns to zero for a duration T_2, then returns to maximum. This pattern repeats itself with a period for a complete cycle of

$$T = T_1 + T_2$$

Although the ideal square wave has a very fast rise time, there is in practice a limit to the rate at which the voltage slope can increase, just as there is for the single pulse. The square wave generator often serves the important function of clocking the operations of digital computers and calculators. Figure 16-6 shows sine wave and square wave signals.

Other common wave forms are the triangle and the sawtooth, both shown in Figure 16-7. The triangle wave consists of a voltage ramp that increases uniformly to a maximum value and then decreases uniformly to a minimum value. The sawtooth wave is similar to the voltage ramp from the sweep generator in an oscilloscope.

RC CIRCUITS

Response of a Capacitor to a Sine Wave

If a voltage is applied to a capacitor, as shown in Figure 16-8, charge flows into the capacitor:

$$Q = CV$$

If the applied voltage is sinusoidal,

$$V = V_0 \sin \omega t$$

then the charge on the capacitor is also sinusoidal in time:

$$Q = CV_0 \sin \omega t = Q_0 \sin \omega t$$

The charge on the capacitor is proportional to the voltage across the capacitor.

The current flow into or out of the capacitor depends on the rate at which the charge changes:

$$I = \Delta Q / \Delta t$$

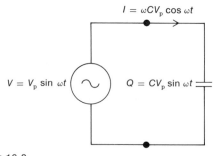

$$I = \omega C V_p \cos \omega t$$

$$V = V_p \sin \omega t \qquad Q = C V_p \sin \omega t$$

FIGURE 16-8
The charge on a capacitor has the same time dependence as the voltage.

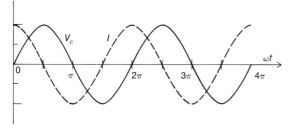

FIGURE 16-9
ICE: The current I in a capacitor C leads the voltage $E = V_c$.

When the charge is zero, its rate of change is large, so the current flow is large. When the charge reaches a maximum, the current flow goes to zero. As the capacitor discharges, the current flow reverses direction, reaching a maximum as the charge goes to zero. Figure 16-9 shows the charge on the capacitor and the current into the capacitor as a function of time. The current is 90° out of phase with the charge and reaches a maximum a quarter of a cycle *earlier* than the voltage. The current *leads* the voltage. The current is related to the charge by

$$I = \Delta Q/\Delta t = Q_0\, \omega \cos \omega t = \omega C V_0 \cos \omega t$$

The lower the frequency ω, the smaller the current flow. At high frequencies, the current flow can become very large unless limited by other elements in the circuit. The current does not flow through the capacitor—the current simply passes back and forth into and out of it. As the capacitor's upper plate becomes positive, it attracts electrons to the negative plate. As the upper plate becomes negative, it repels electrons. Charge on the other side of the capacitor is pushed back and forth. As the frequency approaches zero (in dc) the current flow in and out of capacitor goes to zero.

The ratio of the peak voltage to the peak current through a capacitor is a measure of the capacitor's impedance (which has the same units as resistance). The capacitor's impedance or **capacitive reactance,** is inversely proportional to the frequency and the capacitance:

$$X_c = \frac{1}{\omega C} = \frac{1}{2\pi f C}$$

$$V_0 = I_0 X_c$$

The current and voltage, however, are out of phase.

EXAMPLE

A 0.01 μF capacitor is attached to a 60 Hz 120 V source.
(a) What is the impedance of the capacitor at this frequency?
(b) What current would be drawn?
The impedance is

$$X_c = \frac{1}{\omega C} = \frac{1}{2\pi f C} = \frac{1}{(6.28)(60/s)(0.01 \times 10^{-6}\,\text{F})}$$

$$X_c = 2.65 \times 10^5\ \Omega = 265\ \text{k}\Omega$$

The peak current is

$$I_0 = V_0/X_c = (120\ \text{V})(1.414)/(265\ \text{k}\Omega) = 0.64\ \text{mA}$$

When an alternating voltage is placed across a capacitor, the instantaneous power is the product of the current times the voltage:

$$P = IV = (I_0 \cos \omega t)(V_0 \sin \omega t)$$

$$P = I_0 V_0 \cos \omega t \sin \omega t$$

By trigonometry, $2 \cos \omega t \sin \omega t = \sin 2\omega t$, so that

$$P = I_0 V_0 \frac{\sin 2\omega t}{2}$$

Energy alternately goes into and out of the capacitor. When $\sin 2\omega t$ crosses zero, the net energy stored in the capacitor is zero. The capacitor is a means of storing charge. Although it takes electrical work to do so, that work returns to the system when the capacitor is discharged. No energy is lost in the capacitor. Averaged over a cycle, the power into a capacitor is zero:

$$P_{av} = I_0 V_0 \left\langle \frac{\sin 2\omega t}{2} \right\rangle = 0$$

The Feed-through Capacitor

An electrical circuit containing both resistors R and capacitors C is sometimes called an **RC circuit.** Many electrical circuits use a **feed-through capacitor** to transmit an electrical signal to the next stage of the circuit while blocking dc voltage. Figure 16-10 shows a simple series RC circuit. If the current in the circuit is

$$I = I_0 \cos \omega t$$

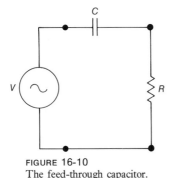

FIGURE 16-10
The feed-through capacitor.

there will be a voltage drop across both the capacitor and the resistor:

$$V_c = \frac{1}{\omega C} I_0 \sin \omega t = X_c I_0 \sin \omega t \qquad (16\text{-}3)$$

$$V_R = R I_0 \cos \omega t \qquad (16\text{-}4)$$

Through each element of the series circuit, the current will be the same, but the voltage across the capacitor will be 90° out of phase with the current. The current and voltage of the resistor are in phase. The total voltage across the circuit is

$$V = V_R + V_C = R I_0 \cos \omega t + X_C I_0 \sin \omega t$$

As Figure 16-11 shows, the effective **impedance** of a

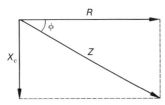

FIGURE 16-11
The impedance can be represented as a vector sum of the capacitive reactance and resistance.

resistor in series with a capacitor can be represented by a vector diagram, and is given by the Pythagorean theorem:

$$Z = (R^2 + X_c^2)^{1/2}$$

The angle ϕ represents the **phase angle** between the applied voltage and the current:

$$\tan \phi = X_c/R$$

The peak current through the circuit is given by the peak voltage divided by the impedance:

$$I_0 = V_0/Z$$

The voltage across the circuit is

$$V = V_0 \cos (\omega t - \phi) = I_0 Z \cos (\omega t - \phi)$$

The voltage across the resistor is

$$V_R = I_0 R \cos \omega t = \frac{R}{(R^2 + X_c^2)^{1/2}} V_0 \cos \omega t$$

If the impedance of the capacitor is small compared to the resistance, then almost all of the voltage is across the resistor and the phase shift will be negligible.

At a frequency at which $X_c = R$,

$$\tan \phi = X_c/R = 1$$

the phase shift will be 45°. The output voltage will be

$$V_R = \frac{V_0 \cos \omega t}{\sqrt{2}}$$

The frequency at which $X_c = R$, sometimes called the **cutoff frequency,** f_c, is given by

$$R = X_C = \frac{1}{2\pi f_c C} \text{ or } f_c = \frac{1}{2\pi RC}$$

The voltage across the resistor is then given by

$$V_R = \frac{1}{[1 + (f_c/f)^2]^{1/2}} V_0 \cos \omega t$$

The phase shift ϕ is given by

$$\tan \phi = f_c/f$$

At frequencies higher than f_c the phase shift will approach zero, and the voltage output will be almost the full amplitude. Figure 16-12 shows the voltage across the resistor as a function of frequency. At the cutoff frequency, the voltage has fallen off to $V = V_0/\sqrt{2}$; tan $\phi = 1$; and the phase shift is 45°. The figure shows also the voltage across the capacitor as a function of frequency.

EXAMPLE

A 0.5 μF capacitor feeds a signal to a 560 Ω resistor. What is the cutoff frequency?

$$f_c = \frac{1}{2\pi RC} = \frac{1}{(6.28)(560\ \Omega)(0.5 \times 10^{-6}\ \text{F})}$$

$$f_c = 280\ \text{Hz}$$

A feed-through capacitor may limit the low-frequency response of a circuit. This frequency response would be unsatisfactory for an audio circuit designed to reproduce sound recordings for the frequency range (20 Hz to 20 000 Hz) of human hearing.

Response to a Voltage Step

If a capacitor is charged by battery to a voltage V and then discharged through some resistance R, the charge takes a finite time to leak off. Figure 16-13 shows such a circuit.

A capacitor fully charged to voltage V_0 has a charge $Q_0 = CV_0$. If a switch S is thrown so that the capacitor begins to discharge through a resistor R there will initially be a current I_0 such that

$$V_0 = I_0 R$$

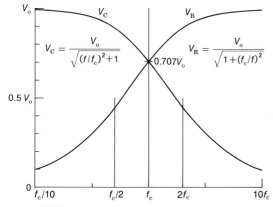

FIGURE 16-12
The voltage across the resistor V_R and the capacitor V_C as a function of frequency f. At the cutoff frequency f_c, the voltage across each is $0.707\ V_0$ and the voltage and current are out of phase by 45°.

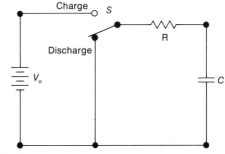

FIGURE 16-13
Discharging and charging a capacitor through a resistor.

As the capacitor discharges, the voltage and the discharge current decrease.

At an intermediate time t, the charge on the capacitor is $Q = CV$ and the current is given by $V = IR$. If we make a Kirchhoff voltage loop with the resistor and capacitor, then

$$IR + \frac{Q}{C} = 0$$

The current is the rate at which charge changes.

$$I = \frac{\Delta Q}{\Delta t}$$

These two relationships can be combined to give

$$\frac{\Delta Q}{\Delta t} = -\frac{Q}{RC} = -\frac{Q}{\tau}$$

Where τ is the **RC time constant:**

$$\tau = RC$$

Resistance has the dimensions of ohms (V/A = V-s/C), capacitance of farads (C/V). The RC time constant $\tau = RC$ has the dimensions of second

$$(1 \text{ farad})(1 \text{ ohm}) = (\text{V-s/C})(\text{C/V}) = \text{s}$$

The rate of change of charge is proportional to the magnitude of the charge. This gives exponential behavior:

$$Q = Q_0 e^{-t/\tau} = Q_0 e^{-t/RC}$$

The voltage across the capacitor is given by

$$V_c = \frac{Q}{C} = \frac{Q_0}{C} e^{-t/\tau} = V_0 e^{-t/\tau} \qquad (16\text{-}5)$$

as shown in Figure 16-14.

In many processes in biology, chemistry, physics, engineering, and even economics, the rate of change in some

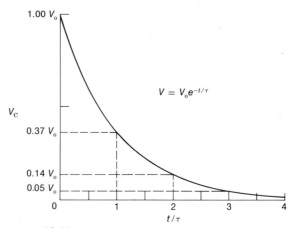

FIGURE 16-14
The discharge of a capacitor through a resistor. The voltage across the capacitor shows exponential decay with a characteristic time constant $\tau = RC$.

quantity y is proportional to that quantity. If b is the constant of proportionality, then

$$\frac{\Delta y}{\Delta t} = by$$

This can be written that the fractional change in y is proportional to Δt:

$$\frac{\Delta y}{y} = b\Delta t$$

This can be shown to imply exponential behavior:

$$y = y_0 e^{bt}$$

where y_0 is the value at $t_0 = 0$. The constant of proportionality b can be positive or negative. The number e is the base of the (Napierian) **natural logarithms:**

$$e = 2.718\,281\ldots.$$

If

$$y = \ln x = \log_e x$$

then

$$x = e^y$$

The discharge of a capacitor obeys this exponential behavior where the constant of proportionality $b = -1/RC = -1/\tau$. The time constant $\tau = RC$ gives the time for the charge on the capacitor to fall off to $1/e = 0.368$ of its initial value.

$$t = 0 \qquad e^{-0} = 1.000$$

$$t = \tau \qquad e^{-1} = 0.368$$

$$t = 2\tau \qquad e^{-2} = 0.135$$

$$t = 3\tau \qquad e^{-3} = 0.050$$

Within three time constants, $t = 3RC$, the charge on the capacitor falls off to 5% of its initial value, as shown in Figure 16-14.

Another useful concept is the **half-life,** $T_{1/2}$, the time required for a decaying exponential to fall to one-half its initial value. The voltage will equal $V = V_0/2$ in a time $T_{1/2}$. This can be written

$$V = V_0/2 = V_0 e^{-T_{1/2}/\tau}$$

$$\tfrac{1}{2} = e^{-T_{1/2}/\tau}$$

Taking the logarithm of each side we get

$$-\ln 2 = -T_{1/2}/\tau$$

Or the half life is given by

$$T_{1/2} = (\ln 2)\tau = 0.693\,RC$$

If we charge a capacitor through a resistance, the rate of charging depends on the same RC time constant. When the circuit shown in Figure 16-12 is being charged, the voltage around the closed circuit can be written as

$$IR + Q/C = V_0$$

Initially the charge and voltage on the capacitor are zero and the initial current $I_0 = V_0/R$. The final charge on the capacitor is given by

$$Q = CV_0$$

As Figure 16-15 shows, the voltage across the capacitor as a function of time is simply

$$V = V_0(1 - e^{-t/\tau}) \qquad (16\text{-}6)$$

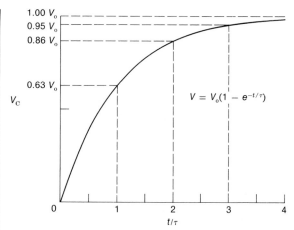

FIGURE 16-15
Charging a capacitor through a resistor. The voltage across the capacitor shows an exponential curve $V_C = V_0(1 - e^{-t/\tau})$.

Response to a Square Wave

A square wave generator produces a voltage signal V_s that alternates between values V_A and V_B. V_A and V_B may be equally spaced about zero, or one of them may be zero. If this voltage signal drives an RC circuit, as shown in Figure 16-16, the response of the circuit depends on the period of the square wave $2T$, compared to the RC time constant for the circuit.

Figure 16-17 shows the voltage across the capacitor in an RC circuit driven by a square wave. Figure 16-17 A shows the applied square wave voltage. If the period of the square wave $2T$ is long compared to the RC time constant, then the voltage across the capacitor shows (see Figure 16-17 B and C) typical exponential charging and

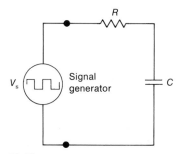

FIGURE 16-16
A square wave voltage signal V_s drives an RC circuit.

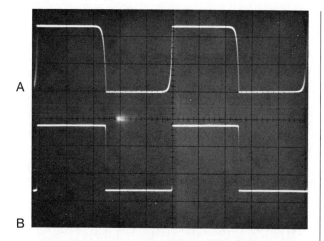

A

B

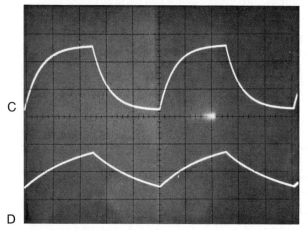

C

D

FIGURE 16-17
Response of an RC circuit to a square wave. The oscilloscope trace shows the voltage across the capacitor. Horizontal, 0.2 ms/division; vertical, 5 V/division.
A. Applied square wave $T = 0.5$ ms;
B. $RC = 0.01$ ms;
C. $RC = 0.10$ ms;
D. $RC = 3.0$ ms.

decay as discussed previously. If the period of the square wave is short compared to the time constant (see Figure 16-17 D), then the capacitor will become only partially charged and discharged.

If the period $2T$ of the voltage signal V_s is much shorter than the time constant, $T \ll RC$, then there is not suffi-

cient time for significant charge to flow into the capacitor. The capacitor becomes only partially charged and the voltage drop across the resistance remains approximately constant. The current is then approximately

$$I \cong \frac{V_s}{R}$$

The voltage across the capacitor is proportional to the charge that accumulates in the capacitor in time t:

$$V_C = \frac{Q}{C} = \frac{It}{C} = \frac{V_s}{RC} t$$

The voltage increases uniformly with time and represents a summation of the input voltage with time.[1]

Figure 16-18 shows the voltage across the resistor in the RC circuit. If the period of the square wave $2T$ is short compared to the time constant $2T \ll RC$, the voltage across the capacitor remains small and the signal voltage appears across the resistor,

$$V_R \cong V_s$$

as shown in Figure 16-18 A. If the period is about the same as the time constant $2T \approx RC$, then current flows in the resistor as the capacitor charges and discharges (see Figure 16-18 B). If the period of the input signal $2T$ is much larger than the time constant, $T \gg RC$, the capacitor charges rapidly and voltage drop appears across the resistor for a short time only, as shown in Figure 16-18 D. The charge on the capacitor is approximately that given by the input voltage:

$$Q = CV_s$$

The voltage across the resistor depends on the current:

$$V_R = IR = R\frac{\Delta Q}{\Delta t} = RC\frac{\Delta V_s}{\Delta t}$$

---

[1] In the language of calculus, the voltage across the capacitor is

$$V_C = \frac{Q}{C} = \frac{1}{C} \int I dt = \frac{1}{RC} \int V_s \, dt \text{ for } T \ll RC$$

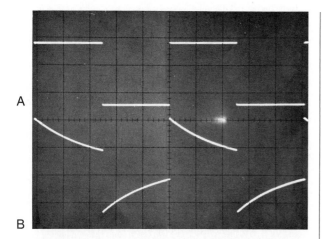

A

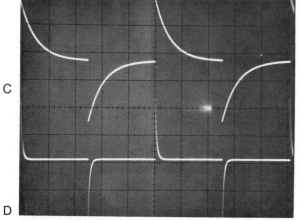

B

C

D

FIGURE 16-18
Response of an RC circuit to a square wave. The oscilloscope
trace shows the voltage across the resistor. Horizontal,
0.2 ms/division; vertical, 5 V/division.
A. Applied square wave $T = 0.5$ ms;
B. $RC = 3.0$ ms;
C. $RC = 1.0$ ms;
D. $RC = 0.01$ ms.

The voltage across the resistor is proportional to the rate
at which the source voltage changes.[2]

―――――――――

[2] In the language of calculus, the voltage across the resistor is

$$V_R = IR = R\frac{dQ}{dt} = R\frac{d(CV_s)}{dt} = RC\frac{dV_s}{dt} \text{ for } T \gg RC$$

A perfect square wave would have an infinite rate of
voltage change, which would correspond to a sudden
spike. Introducing a square wave signal to an RC circuit
with a short time constant produces a series of positive
and negative pulses across the resistor (see Figure
16-18 D).

LCR CIRCUITS

An electrical circuit containing both resistors R and inductors L is sometimes called an **LR circuit.**

LR Circuits

Figure 16-19 shows an LR circuit attached to a voltage
source that is a function of time. The inductor acts as an
emf in the circuits. By Kirchhoff's voltage law, the sum of

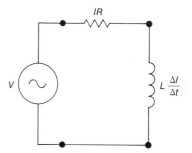

FIGURE 16-19
An LR circuit driven from a sine wave.

the emfs around the circuit is equal to the sum of the
voltage drops:

$$V - L\frac{\Delta I}{\Delta t} = IR$$

This can be written as

$$V = L\frac{\Delta I}{\Delta t} + IR$$

The voltage drop across the inductance is

$$V_L = L \frac{\Delta I}{\Delta t}$$

The voltage across the inductor depends on the rate at which the current changes. If the current in the circuit is

$$I = I_0 \cos \omega t$$

its rate of change is given by

$$\lim_{\Delta t \to 0} \frac{\Delta I}{\Delta t}$$

At $\omega t = 0$, the cosine is a maximum and the slope is zero. At $\omega t = \pi$, the cosine is zero and the slope is $-\omega$.

$$\frac{\Delta I}{\Delta t} = -\omega I_0 \sin \omega t$$

The voltage across the inductor is then given by

$$V_L = L \frac{\Delta I}{\Delta t} = -\omega L I_0 \sin \omega t$$

Although the voltage is proportional to the peak current, it is out of phase with the current. The voltage across the inductor V_L reaches a maximum value earlier than the current, as shown in Figure 16-20. The voltage E across an inductor leads the current I; E in L leads I (**ELI**). In a capacitor, the current I leads the voltage E: I in C leads E

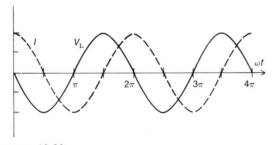

FIGURE 16-20
ELI: The voltage $E = V_L$ across an inductor L leads the current I.

(**ICE**). This has been captured in the little saying "ELI the ICE man".

The proportionality between the current through and voltage across an inductor is given by the **inductive reactance:**

$$V_L = -X_L I_0 \sin \omega t \qquad (16\text{-}7)$$

where $X_L = \omega L$. The inductive reactance increases linearly with frequency. For example, whereas the inductive reactance of a one henry inductor at 60 Hz is

$$X_L = 2\pi f L = 2\pi(60/\text{s})(1.0 \text{ V-s/A}) = 377 \text{ ohms}$$

at 100 kHz the reactance has increased to 6.28×10^5 ohms. Because an inductor in a circuit effectively stops high-frequency currents, inductors are often used in circuits as chokes, used to block unwanted high-frequency signals. Chokes are commonly used to isolate citizen band radios from electrical noise produced by automobile ignitions and generators.

For the LR circuit shown in Figure 16-19, the total voltage across the circuit is

$$V = -X_L I_0 \sin \omega t + R I_0 \cos \omega t$$

The impedance of the circuit is

$$Z = [X_L{}^2 + R^2]^{1/2}$$

The phase shift ϕ is given by

$$\tan \phi = \frac{-X_L}{R}$$

The total voltage across the circuit is

$$V = I_0 Z \cos(\omega t - \phi)$$

Series LC Resonance

Figure 16-21 shows an LC series circuit driven by a sinusoidal voltage source. If the current in the circuit is

$$I = I_0 \cos \omega t$$

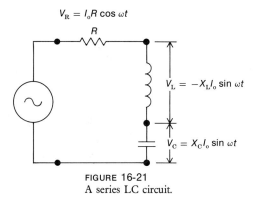

$$V_R = I_0 R \cos \omega t$$

$$V_L = -X_L I_0 \sin \omega t$$

$$V_C = X_C I_0 \sin \omega t$$

FIGURE 16-21
A series LC circuit.

the voltage across the capacitor and the inductor will be out of phase with it. The voltage across the capacitor is

$$V_C = \frac{Q}{C} = X_C I_0 \sin \omega t$$

The voltage across the inductor is

$$V_L = L \frac{\Delta I}{\Delta t} = -X_L I_0 \sin \omega t$$

where $X_C = 1/\omega C$ and $X_L = \omega L$.

The total voltage across the entire circuit including the resistance is

$$V = I_0 R \cos \omega t + (X_C - X_L) I_0 \sin \omega t$$

The total voltage can be written

$$V = I_0 Z \cos(\omega t - \phi)$$

where the impedance of the series circuit is

$$Z = [R^2 + (X_C - X_L)^2]^{1/2} \qquad (16\text{-}8)$$

At a **resonant frequency** $\omega_0 = 2\pi f_0$, the inductive reactance is equal to the capacitive reactance. Thus, $Z = R$ and the impedance of the circuit is purely resistive and at a minimum. The resonant frequency is given by

$$X_C - X_L = \frac{1}{\omega_0 C} - \omega_0 L = 0$$

or

$$\omega_0 = 2\pi f_0 = \frac{1}{[LC]^{1/2}}$$

At the resonant frequency for a series LC circuit, the current through the circuit is at a maximum, while the voltage across the circuit is at a minimum. The voltages across both the inductor and the capacitor are large, but they are out of phase by 180°. So the net voltage across the reactance is zero.

ELECTRICAL SIGNALS IN BIOLOGY

Biological activity generates many important electrical signals. These signals originate in the disturbance of the electrical potential established across nerve and muscle cell membranes that can propagate along a cell and from cell to cell. This electrical activity is associated with the propagation of nerve impulses and with muscle contraction. These electrical signals are routinely measured and recorded, as, for instance, the *electrocardiogram* (ECG) records the electrical activity of the heart and the *electroencephalogram* (EEG) records the electrical activity of the brain. We shall discuss briefly the generation of electrical potentials and signals by nerve cells, and examine the signals produced by the heart and brain.

Extending outward from the body of a nerve cell is a long slender nerve fiber, the **axon:** a long cylindrical conductor filled with an electrolytic gel, the *axoplasm,* surrounded by a thin (typically 7 to 10 nm) resistive membrane. The radius of the axon can be smaller than 1 μm or as large as 500 μm, as in the giant axon of the squid. In humans, the axon radius is limited to about 20 μm. Many axons are surrounded by a sheath of fatty material, *myelin,* as shown in Figure 16-22. Between adjacent myelin sections there are openings, or *Nodes of Ranvier,* where the thin axon membrane is exposed.

The axon may be as much as a meter in length, as are those extending from the brain to low in the spinal cord or from the spinal cord to a fingertip. The nerves that run through our bodies are bundles of many axons. Far from the cell body, the axon branches into fine nerve endings that are separated from the next cell by a small gap, the *synapse.* Nerve endings transfer signals across the synapse to the next nerve or muscle cell by a chemical neurotransmitter, *acetylocoline.* An electrical potential established across the cell membrane transmits the nerve impulses along the axon.

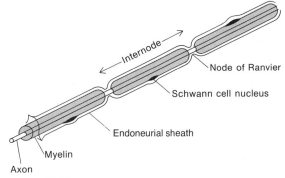

The myelinated nerve fiber. A sheath of myelin encloses the nerve axon. At intervals nodes of Ranvier interrupt the sheath. Membrane depolarization and repolarization occur at the nodes. Because of the myelin sheath, impulses propagate rapidly along the axon. From D. J. Aidley, *The Physiology of Excitable Cells* (Cambridge: Cambridge University Press), © 1971. Used with permission.

Membrane Potential

Inserting a microelectrode into a nerve or muscle fiber and measuring the voltage relative to an electrode outside the cell reveals that the voltage inside the cell is electrically negative compared to the voltage outside. This electric potential difference across the cell membrane is known as the **resting potential** and is typically -70 mV. While we focus our discussion on nerve fibers, muscle fibers show similar electrical phenomena. Nerve and muscle cells are surrounded by a lipid-protein membrane with a high electrical resistivity. Because sodium, potassium, and chlorine ions are present, the fluid inside and outside the cell is a good conductor of electric current. And because the cell maintains a difference in concentration of some ions across the membrane, an electric potential difference develops across the membrane.

If a membrane separates a concentrated salt solution (Na^+Cl^-) from a dilute salt solution, as shown in Figure 16-23, the concentrated sodium (Na^+) and chlorine (Cl^-) ions tend to migrate through the membrane. If movement of the chlorine ions through the membrane is greater than that of the sodium ions, then, for a time, more chlorine ions migrate through the membrane, giving a net negative charge to the dilute side and leaving a positive charge on the concentrated side. The membrane acts as the dielectric of a capacitor separating the two charge layers. An electric field is established that slows the migration of chlorine

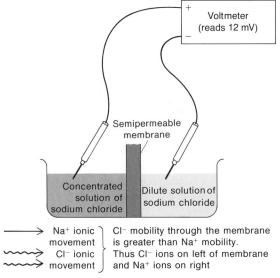

The Nernst potential. A difference in ionic concentration across a semipermeable membrane can generate electrical potentials. From Peter Strong, *Biophysical Measurements*, © 1970 by Tektronix, Inc. All rights reserved. Reproduced by permission of Tektronix, Inc.

ions and speeds the migration of sodium ions through the membrane. The electrochemical potential established by a concentration difference across a membrane is given by the **Nernst potential.** This potential establishes an electric field in the direction that equalizes the ion concentration on both sides of the membrane. The concentration c is the number of ions n (expressed in moles) per unit volume V:

$$c = \frac{n}{V}$$

If a single species of ion with ion charge Ze has a concentration $c_1 = n_1/V$ on one side of the membrane and concentration $c_2 = n_2/V$ on the other side, then the Nernst potential across the membrane is given by

$$V = \frac{kT}{Ze} \ln \frac{c_1}{c_2} \qquad (16\text{-}9)$$

where the Boltzmann constant $k = 1.38 \times 10^{-23}$ J/mol-K and T is the absolute temperature in Kelvin.

A singly charged positive ion species such as sodium is in solution at a temperature of 310 K (37°C). The concentration on the two sides of a membrane differs by a factor of 10: $c_1/c_2 = 10$. What is the Nernst potential across the membrane?

$$V = \frac{(1.38 \times 10^{-23}\ \text{J/K})(310\ \text{K})}{1.60 \times 10^{-19}\ \text{C}} \ln 10$$

$$V = (26.7\ \text{mV})(2.3) = 62\ \text{mV}$$

The Nernst potential across the membrane is 62 mV.

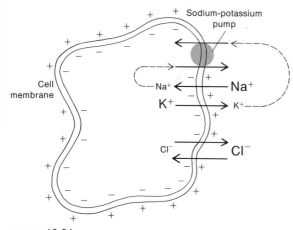

FIGURE 16-24
The active transport of sodium and potassium ions across the membrane, to balance the ion diffusion through the membrane, maintains the cell's resting potential.

The active transport of sodium and potassium ions throughout the cell maintains a concentration difference across the membrane, as shown in Figure 16-24. Table 16-1 displays the nominal concentrations of the major ions in the cell at the resting potential, and the Nernst potential for each ion species. The resting potential across the membrane is typically −70 mV. Because the mem-

TABLE 16-1
Ionic concentrations inside and outside a nerve cell, and the Nernst potentials that a single ion species would develop.

Ion	Outside cell mol/m³	Inside cell mol/m³	Nernst potential mV
K⁺	5	150	−90
Na⁺	150	15	+60
Cl⁻	125	10	−70

brane thickness is typically 5 nm, a large electric field appears across the membrane. The magnitude of this field is

$$E = -\frac{\Delta V}{\Delta x} = -\frac{-70\ \text{mV}}{5\ \text{nm}} = 1.4 \times 10^7\ \text{V/m!}$$

The electric field in a parallel-plate capacitor with dielectric constant K is as given in Chapter 13

$$E = 4\pi k_e \frac{Q}{KA}$$

so the charge per unit area on the membrane surface is

$$\frac{Q}{A} = \frac{KE}{4\pi k_e} = \frac{(5.7)(1.4 \times 10^7\ \text{V/m})}{4\pi(9 \times 10^9\ \text{N-m}^2/\text{C}^2)}$$

$$\frac{Q}{A} = 7.06 \times 10^{-4}\ \text{C/m}^2$$

There is considerable charge on each side of the nerve membrane. The membrane separating the interior of the cell from the exterior acts as a capacitor, and its ability to store charge per unit surface charge is given by the *membrane capacitance*.

Figure 16-25 shows an electric circuit that represents the electrical potential across the axon membrane. Each difference in ion concentration generates a Nernst potential—in the figure, a battery—to drive that ion species across the membrane. There is also, for each ion species, a resistance to flow across the membrane, represented as a

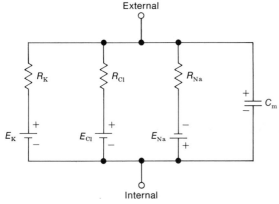

FIGURE 16-25
An equivalent electrical circuit for a section of a nerve or muscle membrane. Typically $E_K = -90$ mV, $E_{Cl} = -70$ mV, and $E_{Na} = +60$ mV.

leakage resistance. For the resting membrane, the leakage resistances for sodium and potassium ions are very large; for chlorine ions, small. The chlorine ions are relatively free to pass through the membrane.

Action Potential

The potential across a cell membrane was first measured in the 1940s, when it was discovered that certain species of squid had giant nerve cells easily penetrable by electrodes then available. In the 1950s, refinements in glass microelectrodes allowed the first direct study of mammalian nerve and muscle cells. Glass microelectrodes have a fine tip, about 1 μm in diameter, filled with a conducting solution. Inserted through the cell wall, the tip establishes electrical contact with the cell's interior, as shown schematically in Figure 16-26. The potential can be measured relative to a common electrode outside the cell. A small exterior electrode allows the potential immediately exterior to the membrane also to be measured.

The resting potential across a nerve cell membrane is about -70 to -90 mV relative to the exterior common electrode. The observed potential difference is close to the Nernst potential for the distribution of potassium ions.

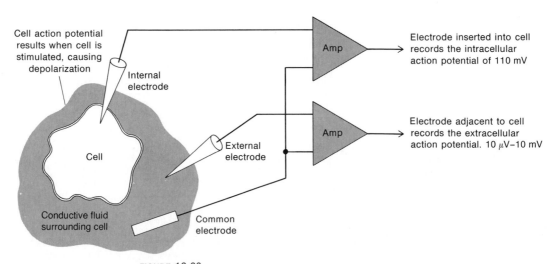

FIGURE 16-26
The interior and exterior cell potentials can be measured through microelectrodes. From Peter Strong, *Biophysical Measurements*, © 1970 by Tektronix, Inc. All rights reserved. Reproduced by permission of Tektronix, Inc.

Passing freely through the membrane, the chlorine ions adjust their ion concentration to be compatible with the observed potential difference.

If a current stimulus to a nerve cell locally raises the potential inside the membrane to −50 to −60 mV, the response of the cell membrane changes markedly. The potential in the cell increases within about 0.5 ms to a maximum value of about +20 mV, as shown in Figure 16-27. The membrane abruptly depolarizes. It also becomes quite permeable to sodium ions, which rapidly flow into the cell, and to potassium ions, which flow out of the cell. Figure 16-28 shows calculated sodium and potassium conductances for a giant squid axon as a function of time. Conductance is the reciprocal of resistance. Almost as fast as the membrane depolarizes, it begins to repolarize. The sodium conductance rapidly falls, as does the potassium conductance shortly thereafter. The resting concentration of sodium and potassium ions is restored by the active transport mechanism, which selectively pumps ions through the membrane—sodium ions out, potassium ions in. The mechanism for this *sodium-potassium pump* is still an active area of research. The energy that drives the sodium-potassium pump is from ATP. The potential across the membrane finally falls to its resting potential in about 2 milliseconds.

The cycle of depolarization and repolarization of the nerve membrane generates a rapidly changing potential difference, the **action potential.** As observed by an elec-

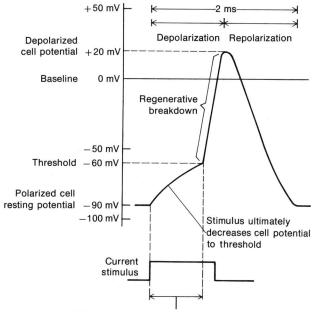

FIGURE 16-27
Cell action potential internally recorded with a microelectrode. The cell is stimulated by a current pulse that decreases the cell potential to a threshold value. From Peter Strong, *Biophysical Measurements,* © 1970 by Tektronix, Inc. All rights reserved. Reproduced by permission of Tektronix, Inc.

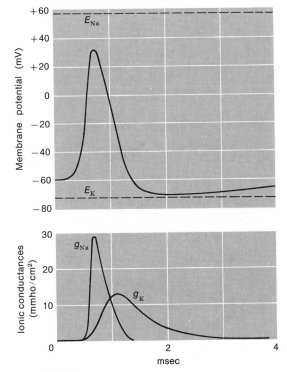

FIGURE 16-28
Sodium and potassium ion conductances during the propagation of the action potential in the giant axon of a squid. The conductance g_{Na} is the reciprocal of the membrane resistance r_{Na}. These curves were calculated from a theoretical model that predicts the membrane potential as a function of time. From D. J. Aidley, *The Physiology of Excitable Cells* (Cambridge: Cambridge University Press), © 1971. Used with permission.

trode placed inside the cell body, the voltage gives the peak shown in Figure 16-28.

The first theoretical analysis of membrane depolarization and repolarization, based on the electrical properties of the axon and ion conductance through the membrane, was published by A. L. Hodgkin and A. F. Huxley in 1952. This theory of the action potential models the behavior of the active membrane in great detail. Here we have only sketched the starting points for the theory on which modern work on active membrane phenomena is based.

If stimulated at some point, a nerve membrane will at first depolarize only over a local region. However, the effect of this potential will be transmitted along the axon, but will be attenuated. That is, the magnitude of the potential will diminish with distance, falling off to about one-half value in 0.1 to 1 mm depending on the nerve being studied. The potential disturbance, however, travels sufficiently far to trigger depolarization of the membrane some distance away. The depolarization then propagates down the axon or muscle fiber.

The electric potential difference and charge flow across the membrane provide the energy to propagate the signal. The action potential propagates at speeds from 1.0 m/s to 120 m/s. The larger the nerve fibers, the more quickly the impulse propagates, traveling from head to foot in a time of 16 ms.

Recall that between the myelin sheaths in many nerve axons, at intervals of about 1 mm, there are openings, the Nodes of Ranvier. Ions can flow into or out of the axon only at these nodes. The nodes are sufficiently close that the action potential at one node triggers a depolarization at the next node. The impulse jumps from node to node. The action potential propagates much faster on myelinated axons.

The Electroencephalogram

The transmission of neutral impulses gives rise to small electrical signals along the axons. The human brain continuously generates these impulses. Electrodes placed on the scalp observe this electrical activity. The signals are small, having an amplitude of only 50 μV peak to peak, in a frequency range from 0.5 Hz to 30 Hz. For convenience, the normal frequency range has been divided into five bands.

delta	(δ)	0.5 Hz–4 Hz
theta	(θ)	4 Hz–8 Hz
alpha	(α)	8 Hz–13 Hz
beta	(β)	13 Hz–22 Hz
gamma	(γ)	22 Hz–30 Hz

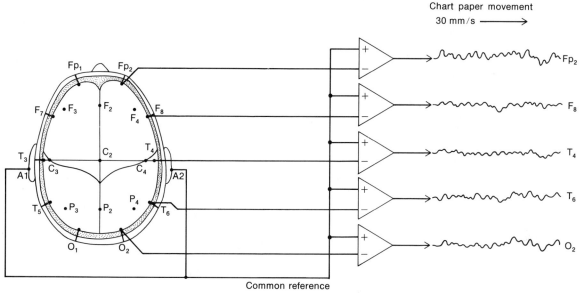

In a relaxed person with closed eyes, the alpha frequency is often observed but disappears with concentrated thought.

For the last 40-odd years, brain activity has been recorded. Electrodes are attached to the scalp at as many as 20 locations and the electrical signals are recorded on 8, 16, or as many as 32 channels. Often pens record the observations directly on moving paper. The voltage signal at each electrode is relative to a reference potential, often the common potential of the ears, as shown in Figure 16-29.

The **electroencephalogram (EEG)** is widely used in clinical medicine for monitoring the depth of anesthesia in patients during surgery and for monitoring the state of alertness, the dividing line between sleep and wakefulness. The EEG is also valuable in diagnosing epilepsy and brain injuries. Observation of electrical signals at the surface of the scalp gives only crude information about the brain's intricate, subtle electrical activity. In a way it is like affixing a sensitive microphone to the exterior wall of a concert hall to hear a symphony orchestra play Beethoven or Bartok. While the detail is lost, the basic pattern of activity is still observable.

The Electrocardiogram

When a nerve impulse reaches a synapse, the muscle cell membrane depolarizes, giving rise to a mechanical contraction of the muscle cell. The resulting potential observed at the surface of the muscle tissue will be the collective exterior potential of all muscle cells as they first polarize and then depolarize, as shown in Figure 16-30. The different phases of contraction of the heart (the most important muscle in the body) generate significant electrical potentials.

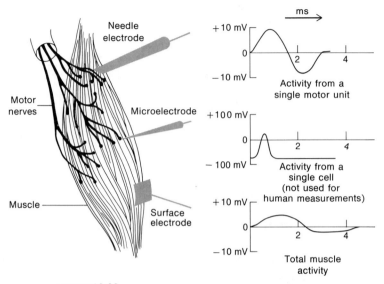

FIGURE 16-30
Electromyogram: the recording of the electrical potential produced by a muscle because of depolarization and repolarization of muscle cells. An electromyogram can be made by using either needle electrodes or surface electrodes. From Peter Strong, *Biophysical Measurements,* © 1970 by Tektronix, Inc. All rights reserved. Reproduced by permission of Tektronix, Inc.

A unique, specialized group of nerve cells, the *sinoatrial* (SA) node, regulates the normal heartbeat, by spontaneously depolarizing at a rate to provide the heart's own pacemaker. As Figure 16-31 shows, the SA node initiates the depolarization and contraction of the atrial muscle cells (see Figure 16-31 A, B, C). Near the top of the right ventricle, a second specialized group of cells, the *atrioventricular* (AV) node, depolarizes (see Figure 16-31 D) when the atrial depolarization reaches it. The action potential from this node is delayed by about 0.1 s. When the AV node fires (depolarizes), it propagates a membrane depolarization along specialized fibers to initiate, starting at the bottom of the heart, depolarization and contraction of the ventricle heart muscle (see Figure 16-31 E, F, G). The electrical activity associated with the polarization and depolarization of the heart muscle gives rise to electrical activity observable at the surface of the skin.

Early in this century, the Dutch physiologist Willem Einthoven pioneered many electrocardiograph techniques. The right arm (RA), the left arm (LA) and the left leg (LL), as shown in Figure 16-32, are three common electrodes for sensing electrical signals produced by the heart. Einthoven demonstrated that, as consequence of Kirchhoff's voltage law, the sum of the potentials of this "**Einthoven triangle**" should equal zero.

The **electrocardiogram** shows the electrical activity along an axis between two electrodes. Figure 16-33 shows a common ECG wave form and related heart action. Atrial depolarization produces the P wave; ventricular depolarization, the QRS spike. The T wave from ventricular repolarization soon follows. The entire cycle takes about 1 s depending on the heart rate. The electrocardiogram voltage signal is typically about 1 mV in amplitude and can reach 5 mV.

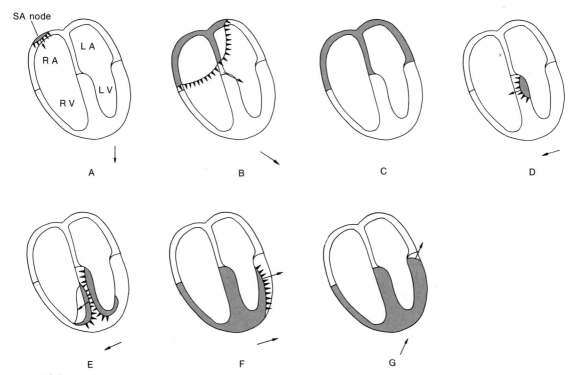

A wave of depolarization sweeps over the heart. A. Depolarization beginning at the SA node. B. Atria nearly depolarized. C. Atria completely depolarized. There is no net dipole moment. This state persists while the AV node conducts. D. Beginning of depolarization of left ventricle. E and F. Continuing ventricular depolarization. G. Ventricular depolarization nearly complete. Courtesy of Russell K. Hobbie. From "The Electrocardiogram as an example of Electrostatics," *American Journal of Physics* (June 1973). Copyright 1973. Used with permission.

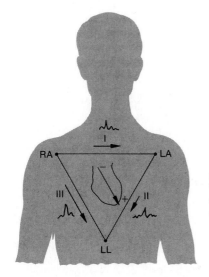

FIGURE 16-32
The Einthoven triangle represents a simple view of the heart's electrical activity. The electric field generated by the electric dipole of the heart activity appears as voltage signals.

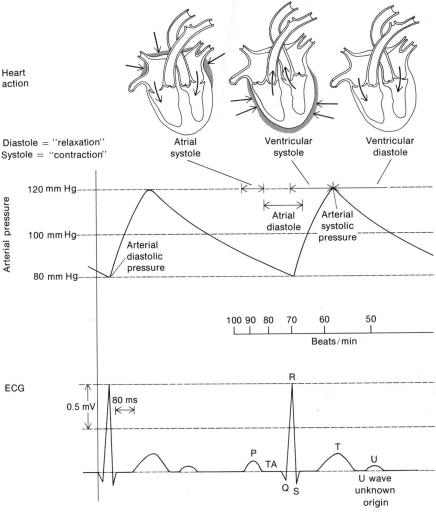

Heart
action

Diastole = "relaxation"
Systole = "contraction"

Atrial
systole

Ventricular
systole

Ventricular
diastole

120 mm Hg

Atrial
diastole

Arterial
systolic
pressure

100 mm Hg

Arterial
diastolic
pressure

80 mm Hg

Arterial pressure

100 90 80   70   60   50

Beats/min

ECG

R

0.5 mV

80 ms

P

T

TA

U

Q   S

U wave
unknown
origin

FIGURE 16-33
The ECG wave form, related heart action, and arterial blood
pressure are a function of time. From Peter Strong, *Biophysical Measurements*, © 1970 by Tektronix, Inc. All rights reserved. Reproduced by permission of Tektronix, Inc.

Across the membrane of each heart muscle fiber there is an electrical potential difference. Indeed, the membrane acts as a capacitor with equal positive and negative charge. When the heart muscle is polarized and in the resting state, polarization of the muscle fibers produces no net dipole electric field.

When the SA node fires, a depolarization wave begins to spread over the heart muscle at about 0.3 m/s. As the membrane depolarizes and charges rush in, the potential across the membrane goes to zero or even becomes positive, as shown in Figure 16-34 A. The electrical effect of the depolarization wave can be represented as a single

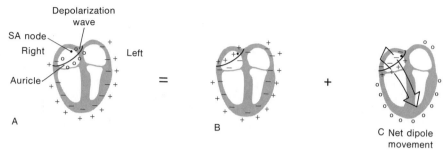

FIGURE 16-34
The depolarizing wave across the heart muscle, A, can be represented as, B, a closed dipole surface that has no net external field and as, C, a single dipole layer that has a net electric dipole moment. This dipole moment changes in amplitude and direction as the heart beats.

electric dipole layer across the depolarization wave front, as shown in Figure 16-34 C. The single dipole layer is equivalent to an electric dipole moment **m** that changes with time as the heart beats.

As a simple model of the heart, the heartbeat produces an electric dipole of changing magnitude and direction. In the medical literature, this total dipole moment is often called the "electric force vector" or "activity" of the heart. The Einthoven triangle in Figure 16-32 shows the projections of this dipole in the frontal plane.

Review

Give the significance of each of the following terms. Where appropriate give two examples.

sine wave
peak voltage
angular frequency
frequency
peak current
alternating current
instantaneous power
average power
root-mean-square voltage
pulse
square wave

capacitive reactance
RC circuit
feed-through capacitor
impedance
phase angle
cutoff frequency
RC time constant
natural logarithm
half-life
LR circuit
ELI

ICE
inductive reactance
resonant frequency
axon
resting potential
Nernst potential
action potential
electroencephalogram
Einthoven triangle
electrocardiogram

Further Readings

Robert Galambos, *Nerves and Muscles: An Introduction to Biophysics* (New York: Doubleday, Anchor Science Study Series S 25, 1962). A clear introductory discussion of the electrical behavior of nerves and muscles.

Bernard Katz, *Nerve, Muscle and Synapse* (New York: McGraw-Hill, 1966). A thorough discussion of the propagation of impulses along nerves to muscles.

Problems

SIMPLE ELECTRICAL SIGNALS

1 A 120 V rms, 60 Hz line voltage is displayed on an oscilloscope.
(a) What is the period of the sine wave?
(b) What is the peak voltage?

2 A light bulb has an average power dissipation of 100 W when connected to the 120 V rms, 60 Hz line voltage.
(a) What is the rms current?
(b) What is the resistance?
(c) What is the peak current through the bulb?

3 An electric heater has an average power dissipation of 1320 W when connected to the 120 V rms, 60 Hz line voltage.
(a) What is the resistance?
(b) What are the rms and peak currents?
(c) If the frequency of the line voltage is 50 Hz, what are the rms and peak currents?

4 A sine wave audio oscillator is tuned at concert A, 440 Hz.
(a) What is the period of the wave?
(b) If an oscilloscope display is 10 cm across, and the sweep rate is 2 ms/cm, how many full cycles will be displayed?

5 A sine wave pattern is displayed on an oscilloscope screen as shown in Figure 16-1. If the sweep rate is 100 μs/cm and the vertical deflection 2 V/cm,

(a) What are the period and frequency of the sine wave?
(b) What is the amplitude?

6 A sine wave signal displayed on an oscilloscope has a peak amplitude of 3.0 cm and a period of 4.2 cm. The vertical and time-base displays are set at 100 mV/cm and 1 μs/cm.
(a) What are the period and frequency of the sine wave?
(b) What is the rms voltage of the signal?

RC Circuits

7 A 2200 pF feed-through capacitor connects 120 V rms 60 Hz line voltage to a 1000 ohm load resistance.
(a) What is the capacitive reactance?
(b) What is the impedance of the circuit?
(c) Is the circuit more resistive or capacitive?
(d) What are the rms current and phase shift?

8 A 20 pF stray capacitance connects the 120 V rms, 60 Hz line voltage to a 500 ohm load resistance.
(a) What is the capacitive reactance?
(b) What are the impedance and rms current?

9 A 0.002 μF capacitor connects the 120 V rms, 60 Hz line voltage to a 10^5 ohm load resistance.

(a) What are the capacitive reactance and impedance?

(b) Is the circuit more resistive or capacitive?

(c) What are the peak current and phase shift?

10 A sine wave voltage signal with peak voltage of 2.0 V and frequency of 1 kHz is observed on an oscilloscope. The signal is applied to a 0.001 μF feed-through capacitor to the 1 Megohm input resistance of the oscilloscope.

(a) What is the peak voltage across the capacitor?

(b) What is the observed amplitude on the oscilloscope?

(c) What is the phase shift between the applied signal and the voltage at the scope?

11 A 10 μF capacitor is charged to 10 V and then discharged through a digital voltmeter that acts as a resistor. The voltage falls to 5.0 V in 95 s.

(a) What is the RC time constant?

(b) What is the resistance of the meter?

12 A 10 μF capacitor is charged through a 3300 Ω resistor from a 6.0 V battery.

(a) What is the RC time constant?

(b) If the voltage starts at zero, how long will it take to charge to 90% of the battery voltage?

13 A 500 Hz square wave with a peak-to-peak amplitude of 2.0 V is displayed on an oscilloscope with a 1 Megohm input resistance through a 0.001 μF feed-through capacitor.

(a) What is the time constant for the RC circuit?

(b) Sketch the wave form as it would appear on the slope. Indicate time and voltage axes.

14 A voltage generator produces a square wave of amplitude 2 V peak-to-peak at a frequency of 1 kHz. The generator is connected to a 560 Ω resistor through a 0.01 μF capacitor.

(a) What is the time constant for this circuit?

(b) Sketch the voltage across the resistor as a function of time. Show both voltage and time axes.

15 A physician is observing the encephalograph signal on an oscilloscope that has a resistance of 10^6 Ω and an input feed-through capacitor of 0.01 μF. The oscilloscope records the voltage across the 1 MΩ input resistance. The physician observes the alpha

rhythm signal at 10 Hz with an amplitude of 27 μV on the scope. What is the magnitude of the actual alpha rhythm signal?

LCR CIRCUITS

16 In a defibrillator circuit, the charge on a 16 μF capacitor at 7 kV is discharged in series through a 0.1 henry inductor to shape the pulse. This gives a natural ringing of the current. Because of the resistance, 50 Ω, in the circuit, the current damps out after $1\frac{1}{2}$ cycles. Most of the energy is dissipated in the first half cycle.

(a) What is the natural ringing frequency of the LC circuit if the resistance is neglected?

(b) What is the initial charge on the capacitor?

(c) If the charge is discharged in one half period, how long does it take? What is the average current?

ELECTRICAL SIGNALS IN BIOLOGY

17 The concentrations of the principal ions across a frog muscle membrane are: external, Na^+ (120 mol/m³), K^+(2.5 mol/m³), Cl^-(120 mol/m³); internal, Na^+(9.2 mol/m³), K^+(140 mol/m³), Cl^-(4 mol/m³). Assume a temperature of 37°C.

(a) What is the Nernst potential because of each ion species?

(b) If the resting potential is equal to the chlorine Nernst potential, what is the resting potential across the membrane? Is the inside positive or negative compared to the outside?

18 The concentrations of ions across the resting squid axon membrane are: external, Na^+(460 mol/m³), K^+(10 mol/m³), Cl^-(540 mol/m³); internal, Na^+(50 mol/m³), K^+(400 mol/m³). Assume a temperature of 37°C.

(a) What is the Nernst potential due to the sodium and potassium ions?

(b) If the inside membrane potential is -60 mV compared to the external potential, what is the internal concentration of chlorine ions?

17

Electric Power and Safety

ELECTRIC POWER

Edison and Steinmetz

Two persons, Thomas Alva Edison and Charles Proteus Steinmetz (see Figure 17-1) dominated the development of electric power in the United States at the end of the nineteenth century. Edison, a chemist and inventor, and best known for developing the electric light, in 1879, believed that a direct current (dc) was the only useful way to distribute electric power and was firmly committed to its development. In 1882 the Edison Electric Light Company began installing dc lighting systems.

The first alternating current (ac) system was demonstrated in Paris in 1883. George Westinghouse purchased the American patent rights and set up a small company, the Westinghouse Electric Company, to distribute alternating current for electric lighting in Buffalo, New York. Electric streetcars, elevators, and machines soon followed.

Steinmetz, German-born and an ardent socialist during his student days, fled the political conditions of Germany and arrived in New York City in about 1890. Although penniless and knowing little English, he brought knowledge of the latest European developments in electric power. He soon became deeply involved in theoretical problems concerning the design of streetcar motors and transformers. Dozens of companies were designing electrical equipment, but success was hit or miss. In 1892, before the American Institute of Electrical Engineers, Steinmetz presented a paper on hysteresis that gave

FIGURE 17-1
Thomas Alva Edison and Charles Proteus Steinmetz examining the damage done to insulators by high voltage "artificial lightning." Courtesy of General Electric Company.

sound physical principles for the design of electrical equipment. The same year Steinmetz joined the General Electric Company, formed by Edison and others, to become the chief theoretician for the electrification of America. His published lectures on electricity educated two generations of engineers.

The dominance of ac power was assured in 1893, when an ac generating system was incorporated into a hydroe-

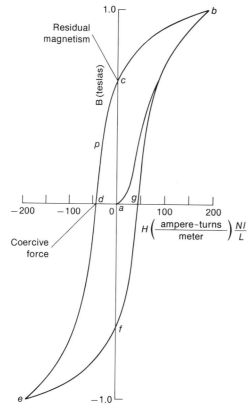

FIGURE 17-2
Hysteresis. The magnetization of a piece of iron placed in a magnetic solenoid. The magnetic induction B depends on the magnetic material and the current in the solenoid. From Paul Lorrain and Dale R. Corson, *Electromagnetic Fields and Waves*, 2nd ed., W. H. Freeman and Company. Copyright © 1970.

lectric plant being built at Niagara Falls, in New York. With cheap hydroelectric power available, electric-furnace industries, such as those producing abrasives, silicon, and graphite, and electrochemical industries such as aluminum, soon provided an ever-increasing demand for electrical power.

Hysteresis

If a piece of iron is placed in a magnetic solenoid with N turns of wire in a length L, with a current I, the iron becomes magnetized, as shown in the curve (a–b) in Figure 17-2. The magnetic field in the iron becomes saturated as the magnetic domains of the iron become aligned. If the current is removed, the magnetic field in the iron does not return to zero (b–c in the figure). Thus, *residual magnetism* remains: The iron is still magnetized and to become demagnetized requires a current in the opposite direction (c–d), a *coercive force*. The effect of the magnetization of a piece of iron lagging behind the magnetizing current is known as **hysteresis.** The iron shown in Figure 17-2 requires a coercive force of only 40 A/m. However, permanent magnets of aluminum, nickel, and cobalt alloy (Alnico) require coercive forces as large as 40 000 A/m. Obviously they are quite difficult to demagnetize.

If the reverse current is made even larger (d–e), the magnetic field in the iron will become saturated in the opposite direction. A transformer core, a motor coil, or an electromagnet consists of a piece of iron wound with a wire coil. If an alternating current is applied to the coil, the iron will become magnetized first in one direction and then in the other. The magnetic field–current curve for iron shown in Figure 17-2 is known as a hysteresis loop. In each cycle, b–d–e–g–b, the direction of the magnetization changes, and a certain amount of electric energy is dissipated in the iron as heat. This hysteresis loss can cause an improperly designed magnet, motor coil, or transformer core to become very hot.

Alternating Current

The basic electric generator is a coil of wire rotating in a magnetic field (see Figure 17-3). The coil moves with a velocity given by the generator's angular rotation speed. If the radius of the coil is R and the angular rotation speed is ω, the rotational speed is

$$v = \omega R$$

If a wire of length L moves with a speed v in a magnetic field, then an *electromotive force*, emf, is induced, as given previously by equation 15-12:

$$\text{emf} = \mathbf{L} \cdot (\mathbf{v} \times \mathbf{B}) = LvB \sin \theta$$

where θ is the angle between the velocity and the magnetic field. For a rotating coil, the angle θ is a function of

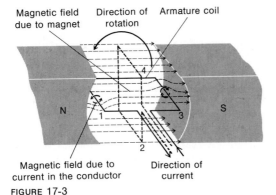

Magnetic field due to magnet   Direction of rotation   Armature coil

N   S

Magnetic field due to current in the conductor   Direction of current

FIGURE 17-3
A simple ac generator. A single coil rotates in a magnetic field.

time $\theta = \omega t$. The coil has two wires of length L moving through the field. The velocity depends on the angular velocity $v = \omega R$. The emf generated by the rotating coil is sinusoidal:

$$\text{emf} = 2L\omega RB \sin \omega t$$

Figure 17-4 shows the emf generated by a rotating coil. The total cross-sectional area of the coil is $A = 2LR$. If the coil has N turns of wire, the **generator emf** is

$$\text{emf} = NAB\omega \sin \omega t \qquad (17\text{-}1)$$

We can also determine the emf for a rotating coil by Faraday's law. The magnetic flux through the coil is given by

$$\Phi = \mathbf{B} \cdot \mathbf{A} = BA \cos \theta = BA \cos \omega t$$

For a coil with N turns the emf is then[1]

---

[1] The sine function gives the rate at which the cosine function changes with time. When the cosine is a maximum or minimum, its rate of change is zero and the sine is zero. In the limit that $\Delta t \to 0$

$$\lim_{\Delta t \to 0} \frac{\Delta(\cos \omega t)}{\Delta t} = \frac{d(\cos \omega t)}{dt} = \omega \sin \omega t$$

The rate of change approaches the derivative of calculus.

$$\text{emf} = -N\frac{\Delta \Phi}{\Delta t} = -NAB\frac{\Delta(\cos \omega t)}{\Delta t}$$

$$\text{emf} = -NAB(-\omega \sin \omega t) = NAB\omega \sin \omega t$$

which is the same result as equation 17-1 above. For a generator spinning at 60 Hz, the angular speed is

$$\omega = 2\pi f = 2\pi(60 \text{ Hz}) = 377 \text{ rad/s}$$

If the generator with a coil of a fixed number of turns N and an area A is spinning at constant speed, the emf can be adjusted by changing the magnetic field B.

If current is being drawn from the generator, there will be a force on that current that tends to slow the generator and produce a **back torque.** For a generator to maintain constant speed, work must be done against this back torque. The magnitude of the force on one wire of length L perpendicular to the field is

$$F = I\,|\mathbf{L} \times \mathbf{B}| = ILB$$

as shown in Figure 17-5. The torque produced by this force is

$$\tau = |\mathbf{R} \times \mathbf{F}| = RF \sin \theta = RF \sin \omega t$$

If the coil is rectangular with N turns, the total torque for two wires of length L in each turn is

$$\tau = 2NRILB \sin \omega t$$

or, using that the area of a single coil is $A = 2RL$, the back torque is

$$\tau = NABI \sin \omega t \qquad (17\text{-}2)$$

The magnitude of the torque for a single coil of N turns varies with time. If the generator is to maintain constant speed, mechanical work must be done on it. The magnitude of the power into the generator is given by the torque τ times the angular velocity ω:

$$P = \tau\omega = NIAB\omega \sin \omega t$$

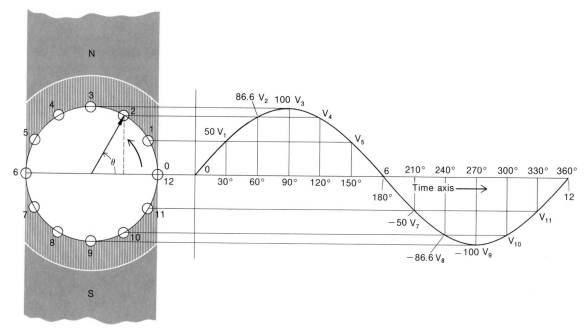

FIGURE 17-4
A sine wave emf is generated by a single coil rotating in a
magnetic field.

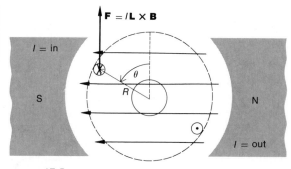

FIGURE 17-5
Back torque is produced by the current flowing in the generator coil.

The electric power delivered by the generator is just

$$P = \text{emf}I = NIAB\,\omega \sin \omega t$$

The mechanical work done against the back torque of the

generator is equal to the electrical work supplied by the
generator to the circuit. If more current flows, generating
greater power, the back torque increases.

Electric generators can convert mechanical power to
electric power with great efficiency, often greater than
95%. The only energy losses are the joule heating in the
coils and mechanical friction of the bearings. Generators
are often wound with several sets of magnetic field coils,
so that the magnetic field will reverse directions several
times in one revolution. Figure 17-6 shows a four-pole
generator that generates 60 Hz of alternating voltage
while rotating at only 30 times a second.

Commercial electric power in the United States is
generated at 60 Hz with three phases. Figure 17-7 shows a
three-phase generator: The magnetic field, supplied by a
direct current, rotates while the emf is generated in stationary coils. As the magnetic field rotates, the magnetic
flux through three stationary coils changes with time. The
three windings produce oscillating voltages out of phase
by 120°. Figure 17-8 shows the three-phase voltage output as a function of time.

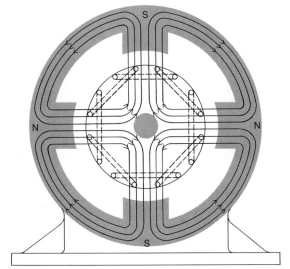

FIGURE 17-6
A four-pole ac generator showing magnetic field windings.

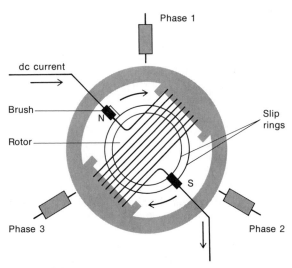

FIGURE 17-7
A three-phase generator can be represented as a north-south magnetic field rotating past three coils. Each coil produces one phase. From Hans Glavitsch, "Computer Control of Electric-Power Systems." Copyright © 1974 by Scientific American, Inc. All rights reserved.

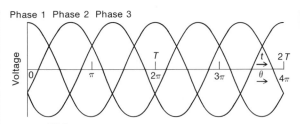

FIGURE 17-8
Three-phase voltage. Three sinusoidally varying voltages are 120° out of phase with one another. Most long-distance power transmission uses three-phase power.
Phase 1, $V_1 = V_o \cos(\omega t)$;
phase 2, $V_2 = V_o \cos(\omega t + 2\pi/3)$;
phase 3, $V_3 = V_o \cos(\omega t + 4\pi/3)$.

Direct Current Generators and Motors

Figure 17-9 shows a simple direct current generator. Steel pole pieces and magnetic field windings produce a constant magnetic field. An *armature* of two coils wound on a slotted iron core (not shown) rotates in the magnetic field. The free ends of the armature coils are connected to a split-ring *commutator,* several segments of hard copper assembled in a cylinder and insulated from the supporting armature shaft. A pair of carbon *brushes* rest on the commutator and maintain current flow in one direction (dc) from the coil to the external circuit. As the coils rotate in the magnetic field, an emf is generated and transmitted to the external circuit through the brushes. Using several coils smooths out the voltage, as shown in Figure 17-9.

Figure 17-10 shows a simplified dc generator with multiple coils. Each coil consists of two wires of length L at a radius R. As the armature rotates with angular velocity ω, each wire moves through the magnetic field with a speed $v = \omega R$. The emf generated by a single wire is

$$\text{emf} = \mathbf{L} \cdot (\mathbf{v} \times \mathbf{B}) = LB\omega R \sin \theta$$

where θ is the angle between $\mathbf{R}$ and the midplane of the field, as shown. The total emf produced depends on the number of coils N and the average moment arm R':

$$R' = \langle R \sin \theta \rangle_{\text{av}} = \frac{2}{\pi} R = 0.64R$$

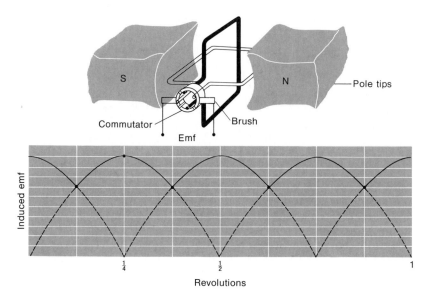

FIGURE 17-9
A two-coil generator showing magnetic poles. The commutator and brushes switch the coils to smooth the voltage to the load. This is a simple dc generator.

Induced emf

Revolutions

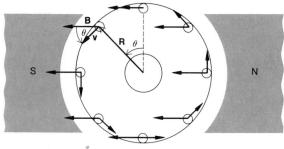

FIGURE 17-10
An emf is generated in each coil as it moves through the magnetic field: emf = 0.64 $NAB\omega$.

The emf of a dc generator with multiple coils is approximately

$$\text{emf} = 0.64\, NAB\omega \qquad (17\text{-}3)$$

where $A = 2RL$ is the area of a coil. The split-ring commutator keeps the emf and current flow in the same direction. The emf depends on the angular speed of rotation ω and on the magnetic field strength.

If the generator is connected to an external circuit, some current I flows in the armature coils. As Figure 17-11 shows, a force is exerted on a current in a magnetic field. The magnitude of the force on a single wire of L is

$$F = |I\mathbf{L} \times \mathbf{B}| = ILB$$

The magnitude of the torque produced by a single-turn

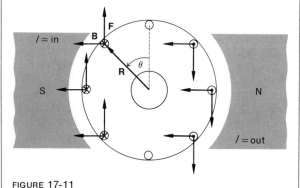

FIGURE 17-11
The magnetic field exerts a force on the generator current to produce a back torque: $\tau = 0.64\, NABI$.

coil (with two wires, one on each side of the coil) is

$$\tau = 2 \, |\mathbf{R} \times \mathbf{F}| = 2RLIB \sin \theta$$

where the angle θ depends on the position of the coil. The total torque depends on the number of coils N and on the average moment arm $R' = 0.64R$. This back torque, which tends to slow the generator down, is

$$\tau = 2NR'LIB = 0.64NABI \qquad (17\text{-}4)$$

The generator will slow down unless an external torque is applied to counteract the back torque.

The electric power provided by the generator emf is

$$P = \text{emf} I = 0.64NAB\omega \, I$$

The mechanical power required to overcome the back torque and to keep the generator turning at constant speed is given by

$$P = \tau\omega = 0.64NABI\omega$$

and is equal to the electric power. The generator converts mechanical power to electric power.

The operation of a dc motor is similar to that of a dc generator. Current from an external power source flows in the armature coils and produces a **motor torque,** causing the motor to turn. A split-ring commutator directs the current through the coils so that the torque is always in the same direction. The torque produced by a motor is the same as the back torque of a generator

$$\tau = 0.64NABI$$

Once rotation begins, the coils are moving in a magnetic field with a speed $v = \omega R$, and the motor produces a electromotive force, or **back emf,** that opposes the external power source. The back emf E_b, which opposes the supply voltage, depends on the rotation speed of the motor:

$$E_b = 0.64NAB\omega$$

The flow of current from the supply voltage E_s depends on the resistance of the armature coils R_c and the back

emf E_b. The total current through the armature coil is given by

$$E_s - E_b = IR_c$$

or

$$I = \frac{E_s - E_b}{R_c}$$

If the motor is spinning freely, doing no work, it will reach a rotation speed such that the back emf is nearly equal to the source voltage; very little current is drawn from the voltage source and little electric power is consumed. But if the motor is doing significant mechanical work, it must draw significant current from the source. In a dc motor the back emf decreases as the motor slows in response to the load, and the current increases. When a motor is started from rest, the maximum current is limited only by the resistance of the armature coil:

$$I = \frac{E_s}{R_c}$$

because the back emf is zero. If the motor is stalled, or if it is not designed to handle high starting currents, it can overheat. It is the large current drawn by a motor when starting that momentarily dims household lights when, say, a refrigerator goes on.

The electric power drawn by a motor from a source is

$$P = E_s I$$

The source voltage is the sum of the back emf and the voltage drop of the coils

$$E_s = E_b + IR_c$$

The electric power drawn from the source is

$$P = E_b I + I^2 R_c$$

which is equal to the mechanical power delivered by the motor plus the joule heating of the armature coil. A small amount of power is lost because eddy currents are induced in the armature core as the magnetic field changes with time. Small amounts are also lost energizing the magnetic field coils and overcoming friction. Electric motors efficiently convert electric energy to mechanical work.

Transformers

A **transformer** consists of two coils, the primary coil with N_1 turns and the secondary coil with N_2 turns, wound on the same iron core. Both coils share the magnetic flux they produce. As the magnetic flux changes, a back emf E_1 is produced in the first coil that resists the flow of current in the primary circuit.

$$E_1 = -N_1 \frac{\Delta \Phi}{\Delta t}$$

In the second coil, a forward emf E_2 is generated.

$$E_2 = -N_2 \frac{\Delta \Phi}{\Delta t}$$

Eliminating the changing flux between the two equations, we get that the emf E_2 in the second coil is simply related to the electromotive force applied to the first by the ratio of the number of turns in each coil:

$$E_2 = \frac{N_2}{N_1} E_1 \qquad (17\text{-}5)$$

Figure 17-12 gives a schematic representation of a transformer. If N_2 is greater than N_1, we have a step-up transformer. At a power station, a 12 000 V ac generation voltage can be stepped up to 60 kV or 230 kV ac for long-distance transmission; a power supply transformer

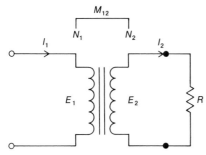

FIGURE 17-12
A transformer steps up or steps down ac voltage with high efficiency.

steps up the voltage from 120 V ac to 300 V ac. If N_1 is greater than N_2, we have a step-down transformer, which, for instance, reduces the 12 kV ac distribution voltage to 240 V ac for service to houses.

Transformers are very efficient. For an ideal transformer, the power delivered to the primary coil is equal to the electrical power delivered by the secondary:

$$P = I_1 E_1 = I_2 E_2$$

The same amount of power can be transmitted at higher voltages with a smaller current. Joule heating from the resistance of power lines causes the primary energy loss in long-distance transmission of electric power. The power loss is proportional to $I^2 R$. Reducing the current minimizes transmission losses.

Electric Power Distribution

Electric power (the coin of the energy realm) can be efficiently transmitted to the point of need and there transformed into other forms of energy. It can be turned on or off at the flick of a switch. We take it for granted except for those unusual times when "the lights go out." The great Northeast power blackout in November 1965 and the blackout in New York City in July 1977 gave us traumatic realization of our dependence on electric power.

Electric power cannot be stored; it must be generated as it is needed. If the demand increases, powerplant output must increase. If an electric power system loses some generating capacity by breakdown or accident, the demand on the system may exceed the supply, causing the system to become overloaded. This may cause other plants to automatically shut down, which increases the overload on the remaining powerplants and can lead to the collapse of the entire generating system.

Hydroelectric dams, fossil-fuel (coal, oil, or natural gas) and nuclear plants, and, in California, geothermal steam-powered plants all generate electric power. Steam or water under pressure provides mechanical power to drive a turbine that turns the generator. The electric power is transformed to a common transmission voltage of from 60 to 230 kilovolts. The electric power from many different plants is tied into a common power grid. As the demand

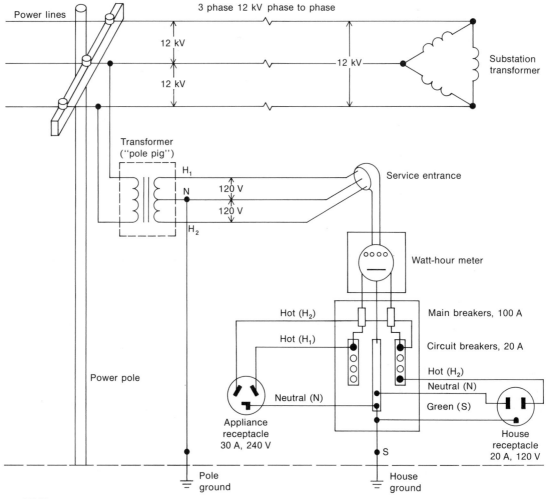

Power lines

3 phase 12 kV phase to phase

12 kV

12 kV

12 kV

Substation transformer

Transformer ("pole pig")

H₁

120 V

N

120 V

H₂

Service entrance

Watt-hour meter

Power pole

Hot (H₂)

Hot (H₁)

Neutral (N)

Appliance receptacle 30 A, 240 V

Main breakers, 100 A

Circuit breakers, 20 A

Hot (H₂)

Neutral (N)

Green (S)

House receptacle 20 A, 120 V

S

Pole ground

House ground

FIGURE 17-13
Local electric power distribution. Three-phase 12 kVrms power is delivered from the substation transformer to the local neighborhood. A transformer, or "pole pig," taps one phase of the 12 kV line and lowers the voltage to 240 Vrms with a center tap (N). Three lines are run to the service entrance, through the watt-hour meter, to the circuit breakers for distribution.

for electric power changes during the day, different plants are switched on or off. The frequency of the alternating current is kept at exactly 60 Hz (within 0.0001%). This stability not only keeps our electric clocks accurate, but, more important, allows widely separated power generators to contribute to the single electric power grid.

Electric power is transported long distances generally in three phases. The wires carry three different voltages,

separated in phase by 120°. The voltage rating is the voltage between any two phases. At the substation, a three-phase step-down transformer reduces this voltage from, typically, 230 kV to 12 kV for distribution to local neighborhoods.

At the local power pole, a transformer, affectionately called a "pole pig," further steps down the voltage to 240 V rms with a center tap N, shown in Figure 17-13.

This produces two phases 180° apart. The voltage between the center tap N and either end of the transformer H_1 or H_2 is 120 V rms ±5%.

In normal household service, the center tap of the transformer, the neutral wire N, is electrically grounded by a wire that runs down the pole into the earth. This neutral wire is also electrically grounded at the service panel, giving solid electrical contact with the earth. Electrical code requirements are that, in new installations, only the *neutral wire* (N) can be *white*. The "*hot*" *wires* (H_1 and H_2) may be *red* or *black* or any other color except green. *Green* is reserved for the *safety ground S*.

The power lines enter a house through the service entrance, running to the *watt-hour meter* and the main breaker circuit. Normal house service provides 100 A at 240 V, which gives a maximum power of 24 kilowatts. A 240 V appliance receptacle is connected to the two hot wires H_1 and H_2 with a circuit breaker in each line. A 120 V house receptacle is wired from one of the hot wires (H_2) to the neutral return N. The electric power is distributed from the main fuses to the different house circuits. It is imperative that each circuit in the house have a fuse on the hot side of the line. And all switches must be placed on the hot side. Without a circuit breaker, a short circuit could draw a current of 100 A before the main breakers trip, producing a serious hazard of electrical fire.

ELECTRICAL SAFETY

Electric Shocks

At one time or another, we have all received a mild shock from electricity. High-voltage wires present obvious dangers and are kept at a safe distance. But the 120 V ac wiring in out homes, schools, and offices also presents real hazards.

Voltage is the potential to do electrical work, which can be done only if a current flows. It is electric current that produces the effect on the human body—70 V can be as lethal as 700 V. The current depends on both the voltage of the source and the resistance of the current path.

Electrical resistance within the human body is relatively low, about 500 ohms head to foot or about 100 ohms ear to ear. The path of least resistance usually follows the nervous system, which is a good conductor of electricity. To a certain extent we are protected by our skin. Skin may have a contact resistance of more than 10 000 ohms if it is dry or oily, smaller if it is wet, and particularly if it is in contact with conducting solutions, as, for instance, when immersed in a full bathtub. The larger the contact area, the smaller is the contact resistance. Table 17-1 shows the body resistance for various conditions of contact.

TABLE 17-1
Resistance between the body and a conductor for various skin contact conditions.

Condition	Resistance	
	Dry	Wet
Finger touch	40 kΩ–1 MΩ	4–15 kΩ
Hand holding wire	15–50 kΩ	3–6 kΩ
Finger-thumb grasp	10–30 kΩ	2–5 kΩ
Hand holding pliers	5–10 kΩ	1–3 kΩ
Palm touch	3–8 kΩ	1–2 kΩ
Hand around drill handle	1–3 kΩ	0.5–1.5 kΩ
Two hands around 1½″ pipe	0.5–1.5 kΩ	250–750 Ω
Hand immersed in water		200–500 Ω
Foot immersed in water		100–300 Ω

SOURCE: Adapted from Ralph M. Lee, "Electrical safety in industrial plants," *IEEE Spectrum* (June 1971), p. 51–55.

If the surface of the skin is broken by a scrape or blister, the electrical contact resistance is small. In contact with voltage sources greater than about 600 V the protective barrier of the skin is electrically punctured and only the internal body resistance impedes the current flow. This is also true if the skin barrier is bypassed by medical insertion of a catheter. Usually, 500 ohms is the minimum body resistance used for estimating electric shock hazards.

Electric current produces the physiological effect of **electric shock.** Figure 17-14 shows the range of effects of ac currents at 60 Hz. A rms current through the body of only 1 mA is perceptible as a tickle. A current of only 10 mA is painful. A person who grips a wire and thereby conducts a current of 20 mA may experience involuntary muscular contraction and not be able to let go.

An electric shock in the current range of 50 mA to 3 A rms may often be fatal—either from ventricular fibrillation or from respiratory arrest because the current interfered with the nerve center controlling respiration. At 50 mA breathing can become labored. A current path through the heart may induce **ventricular fibrillation,** in which the heart muscle is desynchronized, losing its normal rhythm. The ventricle wall ripples continuously, pumping almost no blood. Unless venticular fibrillation is stopped quickly, death can result. Within 60 to 90 seconds, the heart itself, cut off from its own supply of fresh oxygen, will quickly degenerate. Without oxygen, the brain dies. Ventricular fibrillation is also a major cause of death in heart attacks.

Above a current of 3 A rms, a forceful contraction of the heart muscle prevents fibrillation. Breathing stops and the victim may have serious burns. There is a good chance of survival if the victim can be given immediate treatment, including artificial respiration. Once the current stops, the heart usually resumes its normal beating.

It is paradoxical that a large electric current is less often lethal than a small. A very small current applied near the heart can trigger ventricular fibrillation, but a large current can stop it. Most hospital emergency rooms and most emergency medical treatment teams have a defibrillator, which consists of two large metal pads with a power supply. The pads are placed in good electrical contact on the chest on two sides of the heart. A current as large as 6 A and an energy up to 400 J are applied as a very short pulse through the heart muscle, causing it to contract suddenly. When the current stops, the heart relaxes, and quite often will resume its normal beat. Figure 17-15 shows a portable defibrillator unit in a simulated medical emergency.

The chief damage from the passage of large currents is electrical burns and physiological shock. Most of the resistance is at the points of contact. A current entering or leaving the body over a small surface area gives a large current density and considerable joule heating. The electrical power dissipated in a resistance increases as the square of the current:

$$P = I^2R$$

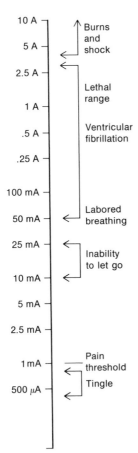

FIGURE 17-14
The physiological effect of rms electric current at 60 Hz.

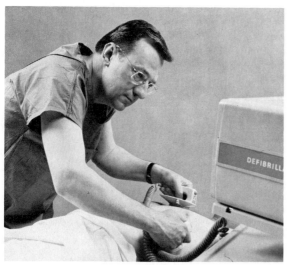

FIGURE 17-15
A portable defibrillator in a simulated cardiac emergency. A current of up to 6 A and up to 400 J of electrical energy can be discharged through the heart muscle in a single pulse. Courtesy of Hewlett-Packard, Medical Products Group.

Even one watt of power deposited over a small area feels hot. Ten watts of power can produce a burn. Larger currents can produce severe burns.

In estimating electric shock hazards, 500 ohms is commonly taken to be the minimum resistance of the human body. At 12 V across 500 ohms, the current flow is only 24 mA. The shock, which would be unpleasant but not particularly dangerous, would be produced by solid electrical contact with a 12 V automobile battery. With 120 V rms across 500 ohms, the current would be 240 mA rms. Household electricity can be dangerous! If the current path were through the heart, it could be fatal. An electric current of 240 mA at 120 A rms produces an average power of almost 30 watts, so an electrical burn could also result.

The Safety Ground

Much electrical wiring and all industrial wiring is equipped with three-prong sockets. In addition to the normal hot wire (black) and neutral return (white), there is a third prong, the **safety ground,** which provides an electrical path to earth separate from that of the neutral return. Figure 17-16 shows a safety ground receptacle.

Electric motors in particular require use of a safety ground. Because of the wear and tear of such equipment from vibration, a power cord can often become frayed, particularly where it enters the equipment. Or the motor coils can become electrically shorted against the case. The case of, say, an electric drill or power saw motor can become electrically hot unless it is specially "double insulated." We often use power tools in cramped locations near metal pipes or in damp areas where we can come in solid electrical contact with the earth. An electric drill, power saw, floor polisher, or other power tool with a hot case can become a lethal hazard. Figure 17-17 shows the proper use of the safety ground. Other household appliances such as a refrigerator, vacuum cleaner, or television set with defective wiring can become *dangerous*.

The safety ground pin of a three-prong plug is connected to the case or frame of the tool, appliance, or machine. If there is an electrical short in the equipment,

FIGURE 17-16
The safety ground. An electrical outlet with a third prong or safety ground. Courtesy of Daniel Woodhead Company.

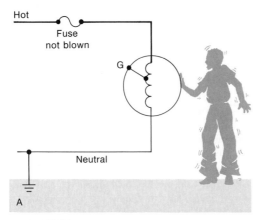
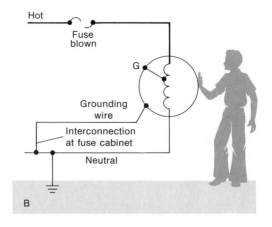

FIGURE 17-17
Third-wire grounding. Unless a motor is "double insulated" it is important that the third-wire safety ground be used. Normally this wire carries no current, but if there is an electrical short circuit from the motor coils to the motor case, the safety ground provides an immediate current path to ground, blowing the fuse. This eliminates the potentially lethal situation of an appliance with a hot case. A. A defective motor without safety ground: dangerous. B. A defective motor with safety ground: safe. From H. P. Richter, *Wiring Simplified*, 31st ed. Copyright 1974 by Park Publishing, Inc. Used with permission.

the case will safely conduct the electric current to ground, and blow a fuse in the process. Note: It will safely conduct the current to ground *provided that we actually use the ground!* In many people's tool boxes can be found a "cheater" plug for adapting a three-prong plug to a two-prong wall socket. The cheater plug has a short green safety ground wire. This should be connected to a solid electrical ground, but rarely is. Often we use power tools in locations that lack three-prong outlets, and may be tempted to get out the diagonal pliers and cut off the third prong so that it will not be a nuisance. If we *do not use* the safety ground, it *does no good.* If the ground wire is not connected or if the ground is defective, as it often is with old equipment (and improperly assembled new equipment), it offers no protection.

Hospital Electrical Safety

Electrical accidents in hospital environments range from outright electrocution to subtle microshocks. In electrocution, a person comes in contact with faulty equipment that is electrically energized, provides a current path to ground, and dies. Most such accidents can be prevented by scrupulous maintenance of equipment.

The more subtle hazard is that of microshocks. A current as small as 10 μA passing through the heart may bring on ventricular fibrillation. If the body's protective skin is bypassed by insertion of a blood pressure catheter near the heart or by a pacemaker, or if electrocardiogram electrodes are attached to the chest, then a dangerous situation may result.

Figure 17-18 A shows the electrical hazards that may result from connecting a patient to a blood pressure catheter to monitor heart action. The patient is firmly grounded through the catheter to the safety ground on the instrument. If the patient is also grounded at a second point—for instance, by the leg lead of an electrocardiogram electrode—then there will be a 500 Ω resistance path to ground through the patient's body.

Between the transformer wires and the catheter instrument case there is usually some stray capacitance. If this capacitance is $C = 2200$ pF, at 60 Hz, there is a capacitive reactance of

$$X_c = \frac{1}{2\pi fC} = \frac{1}{(6.28)(60 \text{ Hz})(2200 \text{ pF})} = 1.2 \times 10^6 \ \Omega$$

This gives a leakage current of about 100 μA rms because of the line voltage. If the safety ground resistance is 1 ohm, the leakage current divides between the 1 Ω

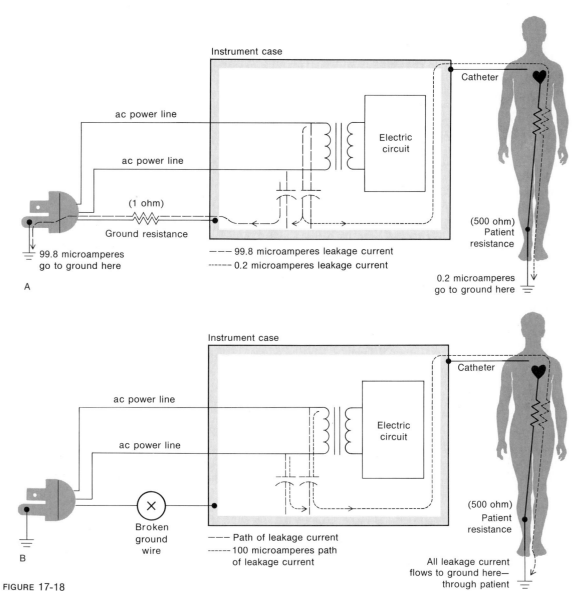

Instrument case

ac power line

ac power line

(1 ohm)

Ground resistance

Electric circuit

Catheter

(500 ohm) Patient resistance

99.8 microamperes go to ground here

– – – 99.8 microamperes leakage current
----- 0.2 microamperes leakage current

0.2 microamperes go to ground here

A

Instrument case

ac power line

ac power line

Electric circuit

Catheter

(500 ohm) Patient resistance

Broken ground wire

– – Path of leakage current
----- 100 microamperes path of leakage current

All leakage current flows to ground here— through patient

B

FIGURE 17-18

Leakage current. A. With a solid safety ground, the leakage current passes harmlessly to ground. B. With a defective safety ground, leakage current flows through the patient to ground. From "Patient Safety," Application Note AN718. Copyright 1971 by Hewlett-Packard Co. Used with permission.

ground resistance and the 500 Ω patient resistance, which are in parallel. If a safety ground is solid, most of the leakage current flows through it and only about 0.2 μA rms flows through the patient, producing no hazard.

If the safety ground path of the catheter instrument is broken because of a faulty ground plug or because an attendant used an ungrounded two-prong cheater plug, a potentially serious hazard results. If a patient is grounded to the leg lead of an electrocardiogram instrument, or if the patient touches the grounded bed frame, the entire leakage current from the catheter instrument, some 100 μA rms, may pass through the patient's heart to ground, as shown in Figure 17-18 B. This current may be sufficient to produce ventricular fibrillation. Pressure transducers are normally used only on patients with diagnosed heart conditions—that is, those patients who are particularly vulnerable. Even if a patient becomes aware that something is wrong, he or she may be too seriously ill to be able to convey distress. It is extremely important that electrical equipment be properly maintained and grounded. Figure 17-19 shows a special test instrument that can detect very small leakage currents and is used to insure that hospital equipment is properly grounded.

FIGURE 17-19
A test instrument will detect leakage currents as small as 5 μA, caused by improper ground conditions. Courtesy of Daniel Woodhead Company.

Review

Give the significance of each of the following terms. Where appropriate give two examples.

hysteresis

generator emf

back torque

motor torque

back emf

transformer

electric shock

ventricular fibrillation

safety ground

Further Readings

Jonathan Norton Leonard, *Loki: the Life of Charles Proteus Steinmetz* (Garden City, N.Y.: Doubleday, 1929).

Peter Strong, *Biophysical Measurements* (Tektronix, Inc., Beaverton Oregon 97005, 1970). An excellent introduction to biomedical instrumentation. Chapter 17, "Grounding-Safety," is quite useful.

Hans Glavitsch, "Computer Control of Electric-Power Systems," *Scientific American* (November 1974).

Problems

ELECTRIC POWER

1 An ac generator consists of a single coil of 600 turns 10 cm long and 8 cm across. The coil rotates at 1200 revolutions per minute (rpm) in a magnetic field of 0.05 N/A-m. What is the peak emf?

2 An ac generator consists of a coil with 50 turns of wire 20 cm long and 8 cm across. The coil rotates at 3600 rpm.
   (a) What is the speed of the coil in the magnetic field?
   (b) If the magnetic field is 0.3 T, what is the maximum instantaneous emf?
   (c) An instantaneous current of 5 A flows when the emf is a maximum. What is the back torque?

3 A dc generator consists of 440 turns wound on an armature that spins at 50 rev/s in a magnetic field of 0.10 tesla. The coils are 20 cm long and 10 cm across.
   (a) What is the emf generated?
   (b) If a current of 10 A flows in the generator what is the back torque?
   (c) How much mechanical power must be applied to keep the generator turning?

4 A dc motor consists of 100 turns wound on an armature. The coils are 25 cm long and 8 cm across. The magnetic field is 0.09 tesla, and a current of 30 A flows in the coil.

   (a) What is the torque produced by the motor?
   (b) The motor rotates at 50 rev/s. What is the back emf?
   (c) The coils have a total resistance of 0.4 Ω. What is the supply voltage?
   (d) How much mechanical power is delivered, and how much energy is lost in heating the armature coil?

5 An iron-core transformer steps down voltage from 12 kV to 240 V. The secondary has 340 turns.
   (a) How many turns has the primary coil?
   (b) If the secondary delivers 100 A rms at 240 V rms, what is the current in the primary circuit?

6 An electric train transformer steps down the voltage from 120 V rms ac to 5 V rms and is rated at 25 watts. The secondary has 100 turns.
   (a) What is the maximum current in the primary circuit?
   (b) How many turns are in the primary circuit?

7 Utility companies send each customer a bill for the total electrical energy supplied during the previous month. A customer is charged for the number of kilowatt-hours (kW-hr). The present rate is about 3.6 ¢/kW-hr and is rising. (You can check your local utility bill for the current rate.) What is the cost of running
   (a) Ten 100 watt light bulbs for 8 hours a day for 30 days?

(b) A 300 watt color television for 120 hours?

(c) A 500 watt washing machine for 8 hours (excluding the cost of heating the water)?

8 A 4.5 meter extension cord is plugged into a voltage source of 120 V rms and used to hook up an electric frying pan drawing a current of 12 A rms. The cord, of No. 18 wire, is rated at 10 A. Each wire has a resistance per meter of $21.0 \times 10^{-3} \, \Omega/m$.

(a) What are the total resistance and the voltage drop across the extension cord?

(b) What is the electrical power dissipated in the extension cord? (The extension cord can get warm.)

(c) What is the electrical power delivered to the frying pan?

9 The normal electrical service to a residence is a maximum of 100 A rms at 240 V rms. A sufficiently large wire must be used so that the voltage drop between the transformer at the power pole and the distribution panel at the house is less than 3% at full rated current.

(a) What is the maximum resistance between the pole and the house?

(b) The resistances of wires are as shown in Table 14-2 (Chapter 14). If the distance between the house and the pole is 50 m, what gauge of copper wire must be used?

(c) If aluminum wire is used, a larger diameter must be used. Why?

(d) What is the maximum electrical power delivered from the pole. If the current flows only in the two hot wires (H_1 and H_2 as shown in Figure 17-13), how much of this power is dissipated in the wires?

10 An electric 5 burner stove has a maximum power output of 9 kW and is powered with 240 volts.

(a) What is the maximum current drawn by the stove if all burners are on high?

(b) The wire resistances are shown in Table 14-2. If the stove is 15 m from the distribution panel and the maximum voltage drop across the wires should be less than 3%, what gauge of wire must be used in hooking up the electric stove?

ELECTRICAL SAFETY

11 A person accidently establishes good electrical contact with a 6 V automobile battery to ground. The current path is from the left arm to the right hand with a resistance of about 1000 ohms.

(a) What is the current flow?

(b) Is it dangerous?

12 A person accidently establishes good electrical contact with a 12 V automobile battery to ground. The current path is from the right hand to the right forearm with a resistance of 500 ohms.

(a) What is the current flow?

(b) How dangerous is it?

13 A person accidently establishes good electrical contact with 120 V of household electricity. The current path is from the right hand to the right leg with a resistance path of about 1000 ohms.

(a) What is the current flow?

(b) Is it dangerous?

14 A person accidently establishes good electrical contact with 240 V of electricity. The current path is from right hand to right forearm with a resistance of about 500 ohms.

(a) What is the current flow?

(b) Is it dangerous?

15 A current of 20 μA through the heart muscle can cause ventricular fibrillation. Assume that a patient attached to a pacemaker provides a 500 ohm path to ground and that no more than 10 μA of current is allowed to pass through the patient. What is the maximum voltage difference from the pacemaker to ground that can be permitted?

16 Because of a capacitor failure in the power supply of an oscilloscope, 300 V is placed on its metal case. A hospital resident preparing an encephalograph touches a metal switch on the oscilloscope with his right hand. His left hand is touching the grounded metal handle of a stainless steel cabinet. The attendant provides a resistance path of 1000 Ω. What happens to the resident?

17 A defibrillator is a high-voltage current source used to shock the heart into proper rhythm when it is undergoing ventricular fibrillation. Two electrodes with a large surface area are applied to the chest with suitable conducting gel to establish good electrical contact. About 400 joules of electrical energy are discharged through the patient in about 10 ms.
(a) What is the average power delivered to the patient?
(b) If the body resistance between the electrodes is 50 Ω, what is the average voltage?
(c) What is the average current?

18 The heating element of an electric toaster shorts against the case giving a resistive path of 20 ohms between the hot 120 V rms cord and the case. A person places his hand against the toaster while lightly touching the metal grounded sink, making a resistance path of about 5000 ohms. What is the current through the person? Is it dangerous?

19 Hospitals commonly use stainless steel refrigerators that stand on rubber feet to give electrical isolation from ground. Because of bad wiring, a 600 ohm resistance path is established between the hot 120 V rms cord and the stainless steel case. A person who is electrically grounded grasps the handle of the refrigerator, making a resistance path of 1000 ohms. What is the shock current? Is it dangerous?

20 A stainless steel refrigerator has a capacitive coupling of 0.3 μF between the hot 120 V rms cord and the metal case.
(a) What is the capacitive reactance at 60 Hz?
(b) If a person touches the handle, making a resist-ance path of 1000 ohms, what is the overall impedance?
(c) What is the leakage current?

21 A current of 1 mA rms at 60 Hz is the threshold of perceptible electric shock. What approximate capacitance is required to couple a current of 1 mA from a 120 V rms power source through a 1000 ohm resistance path to ground?

22 An ungrounded monitoring instrument in a hospital has a capacitive coupling of 0.001 μF between the 120 V rms 60 Hz power supply voltage and the case of the instrument. If a patient who is grounded to an electrocardiograph instrument touches the monitoring instrument, making a resistance path of 10 000 ohms, what leakage current will flow?

23 A 60 W light bulb can be somewhat useful for checking the quality of electrical grounds. Care must be taken to avoid contact with conductors while making the tests.
(a) The bulb is connected between the hot 120 V rms line and the green wire safety ground. The bulb does not glow with full brightness. What does this imply? What if the bulb does not glow at all?
(b) The bulb is connected between the neutral return and the safety ground and glows with full brightness. What does this indicate?
(c) The bulb is connected between the metal frame of an old refrigerator and a metal sink. The bulb glows faintly. If the two-prong plug on the refrigerator is reversed, the bulb goes out. What does this imply? Is the refrigerator hazardous?

18

Electromagnetic Waves

ELECTROMAGNETIC WAVES

James Clerk Maxwell

Faraday's discoveries of induction indicated that electric fields and magnetic fields are somehow connected. The electromotive force that a changing magnetic field produces in a wire is not unlike an electric field induced in the wire; a conductor moving through a magnetic field experiences an apparent electric field. Lacking the formal mathematical training necessary to appreciate Coulomb's, Ampère's, and others' theoretical explanations of the effects of electric currents and magnets, Faraday visualized the interaction of currents and magnets more simply, as magnetic lines of force. These lines of force could be clearly seen in the behavior of iron filings.

At the time of Faraday's death in 1867, James Clerk Maxwell had begun to lay the foundations of a unified theory of electricity and magnetism based on the concepts of magnetic and electric fields. Maxwell, born in Edinburgh, Scotland, in 1831, published his first scientific paper at the age of 14. When only 24 years old, he became a professor of physics at Aberdeen College. His scientific interests were wide ranging: In 1856 he won a prize for an essay on the structure of the rings of Saturn; in 1859 he published his first paper on the kinetic theory of gases; and, with the Austrian physicist Ludwig Boltzmann, he laid much of the foundation for statistical mechanics. Maxwell was also the first to clearly state the three-color theory of vision.

Maxwell thought of electric and magnetic fields as real physical quantities and attempted to understand their behavior as such. Henry and Faraday had demonstrated what is known as Faraday's law of induction: A changing magnetic field generates an electromotive force in a wire. Maxwell further generalized on this law: A changing magnetic field generates an electric field even in free space away from any conductors. Oersted and Ampère had shown that a constant current generates a constant magnetic field. Maxwell, noting symmetry between magnetic effects and electric effects, postulated that if a changing magnetic field can produce an electric field, then a changing electric field should produce a comparable magnetic field.

This postulate, together with Coulomb's law and the fact that magnetic poles are always created in pairs, completed the laws of electricity and magnetism. Maxwell was able to summarize these laws of electricity and magnetism in mathematical terms to comprehensively, accurately, and precisely describe all known electromagnetic phenomena of his day. His equations also described a new phenomenon, the **electromagnetic wave,** as shown in Figure 18-1. A changing magnetic field gives rise to an electric field, a changing electric field gives rise to a magnetic field. The coupling of the electric and magnetic fields produces a wave that propagates through a vacuum without energy loss. Maxwell's equations further predict that the speed of propagation is given by

$$c^2 = \frac{k_e}{k_m} = \frac{9 \times 10^9 \text{ N-m}^2/\text{C}^2}{10^{-7} \text{ N/A}^2}$$

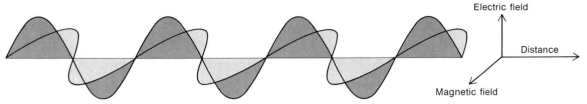

The speed of propagation of these waves is just

$$c = 3 \times 10^8 \, \text{m/s}$$

which corresponds, by no mere coincidence, with the speed of light. These waves can propagate through empty space, and can be reflected and refracted, just as light can. These similarities of behavior led Maxwell to the conclusion that light must be an electromagnetic wave.

A stationary charge Q gives rise to an electric field **E.** If the charge is moving with velocity **v,** it gives rise to a magnetic field **B.** If the charge is accelerating, it gives rise to an electromagnetic wave.

Heinrich Hertz

Not until 1886, 10 years after Maxwell's death, was the physical propagation of these electromagnetic waves experimentally demonstrated. The first to do so, Heinrich Hertz, began by studying sparks produced by an induction coil, which is similar to an automobile ignition spark coil. An induction coil consists of a transformer with a large number of turns in the secondary coil compared to the primary coil, as was discussed in Chapter 15. The primary coil is hooked to a battery through a switch. When the switch abruptly interrupts the current flow in the primary coil, a large emf develops across the secondary induction coil and the spark gap. Figure 18-2 shows one of Hertz's experimental setups.

The secondary coil is attached to a metal rod about 0.6 m long with a small gap in the middle. Square metal plates 0.4 m on a side are attached to the ends of the rod. The plates adjust the capacitance of the secondary circuit. The inductance of the secondary coil and the capacitance of the wire form an **LC resonant circuit,** which "rings" to produce high-frequency waves.

Figure 18-3 shows a schematic of the secondary circuit. A current flowing in the inductor produces a magnetic field. At A, the instant when the charge on the capacitor is zero, the magnetic field begins to collapse, causing a large

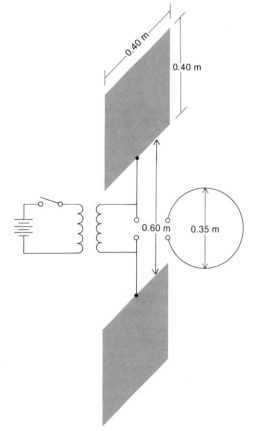

FIGURE 18-2
Hertz's experimental spark gap for studying wave propagation.

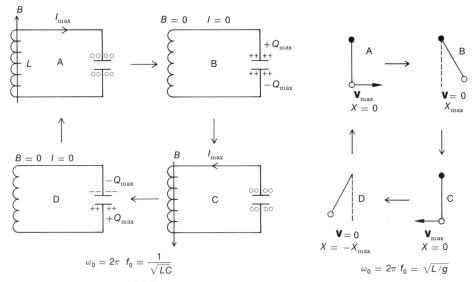

$$\omega_0 = 2\pi \ f_0 = \frac{1}{\sqrt{LC}}$$

$$\omega_0 = 2\pi \ f_0 = \sqrt{L/g}$$

FIGURE 18-3

An LC resonant circuit is analogous to a pendulum. When charge is stored on the capacitor, energy is stored in the electric field, just as gravitational potential energy is stored in a pendulum. When current flows through the coil, energy is stored in the magnetic field, just as the moving pendulum has kinetic energy. In the absence of resistance, energy is conserved. The energy swings back and forth between the electric field and the magnetic field, just as the energy in a pendulum swings back and forth between gravitational potential energy and kinetic energy.

current to flow that will charge the capacitor. The current will flow until the magnetic field completely collapses, B. All of the energy initially stored in the magnetic field is now stored in the electric field of the capacitor. The capacitor will begin to discharge, reversing the current and building up a magnetic field in the reverse direction until, C, the charge on the capacitor is zero. The reverse magnetic field will then collapse, charging the capacitor in the reverse direction, D. The capacitor will discharge, building up the original magnetic field, A. The oscillation of the circuit can be excited either by inducing a magnetic field, perhaps from a primary coil, or by placing charge on the capacitor. The current swings back and forth just as a pendulum does. With each swing it will lose a small part of its energy in resistive heating. The ringing **frequency** of such a resonant circuit is given by

$$\omega_0 = 2\pi f_0 = \frac{1}{[LC]^{1/2}}$$

as was discussed with reference to series LC resonance circuits, in Chapter 16. From the inductance and capacitance of his circuit, Hertz found that the oscillation frequency was about $f = 10^8$ Hz.

To observe the propagation of electromagnetic waves he used a detector coil, a wire loop, either circular (diameter 0.35 m) or square (0.6 m on a side), with a spark gap on one side. When the detector coil was held some distance from the radiating coil, the changing magnetic field of the electromagnetic wave would induce in the detector coil an emf that was observed as a spark. He was able to observe such sparking at a distance from the source of radiation that was too large for the effect to be explained by mutual inductance.

In a series of experiments, by observing standing waves, he was able to determine the wavelength.[1] (He

[1] Heinrich Hertz, *Electric Waves,* translated by D. E. James (New York: Dover, 1962), originally published in 1893.

encountered considerable difficulty with his observations, because the waves unexpectedly reflected from metal objects in the room.) He finally determined that the **wavelength** was about 2.8 m, which gave a propagation velocity of

$$v = \lambda f \cong (2.8 \text{ m})(10^8 \text{ Hz}) = 2.8 \times 10^8 \text{ m/s}$$

which is very close to the speed of light, $c = 3 \times 10^8$ m/s.

In a later series of experiments, he constructed a much shorter radiating rod, which he placed at the focus of a parabolic reflector.[2] He obtained higher-frequency waves with a wavelength of 0.33 m, and was able to detect the waves at a distance of 16 m from the apparatus. In these refined experiments he studied the reflection, refraction, and polarization of the waves. By stretching conducting wires on a frame parallel to one another at a separation of 3 cm, he made a polarizing screen. When the screen wires were parallel to the radiating rod, the waves did not pass; when the screen wires were perpendicular to the radiating rod, the waves passed undiminished. This demonstrated that the electric field of the wave was parallel to the radiating rod. The electric field would produce currents parallel to the wires that would reflect the wave. Hertz even studied the refraction of the waves by passing them through a large (1.2 m by 1.5 m) prism cast from pitch. He found the index of refraction of the waves was 1.69 compared to 1.5–1.6 for light in similar materials. He found that a sheet of tinfoil completely reflected the waves. The waves did not cast a well-defined, sharp-edged shadow.

Hertz's experiments conclusively verified Maxwell's theory of electricity and magnetism. Furthermore, his observed waves exhibited all of the properties normally associated with visible light. We discuss the propagation of light in detail in Chapters 19 and 20.

ELECTROMAGNETIC RADIATION

Electromagnetic waves can be generated by an oscillating current in a wire or **antenna.** The waves propagate outward into space. Around a wire suspended vertically

carrying a current I, a magnetic field will be set up. When the current is turned on, it takes a finite time for an observer at some distance from the wire to detect the effect of the current. The magnetic field produced by the current propagates out from the wire at the speed of light. The work required to build up the magnetic field can be considered as energy stored in the magnetic field. If the current is slowly reduced to zero, the magnetic field collapses, giving its energy back to the wire. This induces the electromotive force in a wire with changing current, as was described by Faraday's law (Chapter 16). Suppose, however, we suddenly reverse the current in the wire. At a large distance from the wire, the magnetic field is pointing in one direction, and the effect of the reverse current has not yet been felt. Close to the wire the field has already reversed direction, and propagates outward at the speed of light. If we periodically reverse the direction of the current, and thus the direction of the magnetic field near the wire, a wave will propagate away from the conductor, much as ripples propagate outward from a pebble thrown into a still pond. As the magnetic field propagates outward, it carries with it energy in the form of electromagnetic radiation. Because of the changing magnetic field, there will also be an oscillating electric field perpendicular to both the magnetic field and the direction of the propagation, as shown in Figure 18-1. Figure 18-4 shows the electric field produced by an oscillating antenna. A verti-

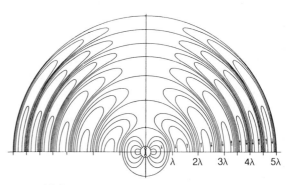

FIGURE 18-4
The electric field lines produced by an oscillating vertical antenna current. From Francis Sears and Mark Zemansky, *University Physics*, 4th ed., © 1970 by Addison-Wesley, Reading, Mass. Reprinted with permission.

[2] This experiment is described in William F. Magie, *A Source Book in Physics* (New York: McGraw-Hill, 1935), pp. 549–561.

cal antenna radiates waves with the electric field polarized in the vertical plane.

Today we watch television images that have been re-layed halfway around the world by satellite, we listen to stereophonic radio in our automobiles, and we engage in direct two-way communications on CB radios. However primitive Hertz's discoveries of a little less than a hundred years ago now seem, they launched the communications revolution that we enjoy (or, sometimes, endure). By 1901, the Italian inventor Marconi had successfully trans-mitted telegraph signals from England to Newfoundland, Canada, a distance of 2900 km. The development of the vacuum tube by the British physicist John Fleming, in 1904, and the American inventor Lee DeForest, in 1907, made possible voice communication, and in 1922 the first commercial radio broadcasting began in the United States. We shall now briefly discuss the propagation of electromagnetic waves in AM, CB, FM radio and tele-vision.

AM Radio

The AM radio signal is broadcast from a tall antenna usually located on flat terrain. The information signal to be broadcast modulates the amplitude of the antenna current and thus also the intensity of the radiated wave. At the radio station, an electrical signal is generated that is proportional to the sound pressure fluctuations of speech, music, etc. This electrical signal then modulates the am-plitude of a high-frequency **carrier wave.** The informa-tion is transmitted by **amplitude modulation** of the radio waves, thus the term AM radio. Figure 18-5 shows the original signal and the modulated carrier wave. AM radio is broadcast in a frequency range from 550 KHz to 1600 KHz with a wavelength of 550 m to 190 m.

The height of the antenna is approximately one-fourth the wavelength, $\lambda = c/f$, of the electromagnetic wave. An amplifier drives a large oscillating current up and down the antenna, producing an oscillating magnetic and elec-tric field that propagates away from the antenna. The electric power into the antenna can be as much as 50 000 watts. The electric field vector defines the direction of polarization of the electromagnetic wave. Driving a cur-rent up the antenna also puts electrical charge on the

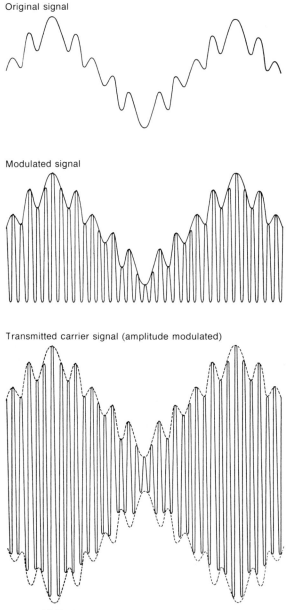

Original signal

Modulated signal

Transmitted carrier signal (amplitude modulated)

FIGURE 18-5
Amplitude modulation of a broadcast radio signal. From Henri Busignies, "Communication Channels." Copyright © 1972 by Scientific American, Inc. All rights reserved.

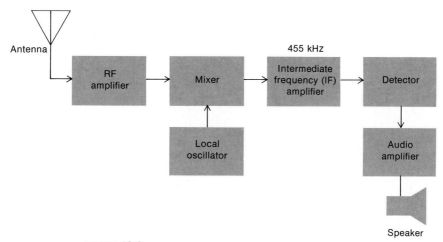

FIGURE 18-6
A superheterodyne radio receiver uses a local oscillator to produce beats with the incoming radio signal, thereby producing an intermediate frequency signal that is amplified and detected.

antenna. Both the AM radio antenna and its electric field are vertical.

Many kilometers from the radio transmitter, the radiated power has fallen off in intensity and is quite weak. But the oscillating electric field is sufficient to produce a small oscillating current in a wire such as a car's radio antenna. Figure 18-6 shows a functional schematic for a superheterodyne AM radio receiver. The small current signal from the antenna is converted into a voltage signal and amplified several hundred times in a radio frequency (RF) amplifier. This amplifies all signals over a fairly wide range of the radio spectrum and is not very selective.

However, to hear only one radio station, we must select a particular radio frequency and exclude the others. Although a highly selective tuned amplifier with several stages can be built, to change stations on it might take a skilled technician 20 minutes. This problem is avoided in the **superheterodyne receiver.** A mixer and local oscillator convert all incoming signals to the same **intermediate frequency** (IF). The frequency of the local oscillator can be controlled by an LC resonant circuit with an adjustable capacitor. (The finned tuning capacitor can be seen inside many radios, connected to the tuning knob. Automobile radios often have variable inductors with

fixed capacitors, but the effect is the same.) The oscillating voltage produced by the local oscillator is mixed with the signal from the RF amplifier. The resulting output is at the beat frequency, which is the difference in frequency between the RF signal f_{RF} and the local oscillator f_L:

$$f_{\text{beat}} = f_{RF} - f_L$$

A carefully tuned, frequency-selective amplifier known as the intermediate frequency amplifier (the IF strip) is tuned to a fixed frequency—for instance, $f_{\text{beat}} = 455$ kHz. Adjusting the local oscillator frequency selects a particular radio station from the broadcast band. This one signal is then amplified another thousand times. The envelope of the radio signal is picked off in the detector, usually a diode, eliminating the intermediate frequency and reproducing the original audio signal from the radio station. This audio signal is amplified and converted to sound by the loudspeaker.

Most household AM radios rely on the magnetic wave for reception. In the back of the radio is a loop stick, a coil of fine wire wrapped around a core of ferrite magnetic material about 1 cm in diameter and 15 cm long. The magnetic intensity of the wave induces a relatively large

FIGURE 18-7
From *Big Dummy's Guide to C.B. Radio,* © 1976 The Book
Publishing Co., Summertown, Tenn. 38483.

magnetic field oscillating at the carrier frequency. The rapidly changing magnetic flux in a wire coil produces a small emf proportional to the radio signal. This emf is then amplified and detected.

Citizens Band Radio

Citizens band radio, officially called the Citizens Radio Service, was established in 1947 by the Federal Communications Commission (FCC) as a low-cost short-distance service for people who need to contact their homes or businesses while driving automobiles. By 1976 there were 5 million licensed operators; there are expected to be 15 million by 1979. One need only be 18 years of age and file an application to receive a CB license.

The citizens radio band now consists of 40 channels in the frequency range from 26.965 MHz to 27.43 MHz separated by 0.01 MHz or 0.02 MHz, and assigned to specific frequencies. Quartz crystals control the channel frequencies. The transmitter-receiver, or transceiver, broadcasts an amplitude modulated radio signal. The power output of a citizen band transmitter is limited to 4 watts. To increase distance, stationary transmitters often broadcast from directed-beam antennas, which are horizontally polarized. Mobile CB units broadcast vertically polarized waves from vertical antenna mounted on the automobile roof or trunk. Most interstate trucks are now equipped with CB radios (see Figure 18-7).

FM Radio

In FM radio, the information signal modulates the frequency of the current in the transmitting antenna. **Frequency modulation** (FM) was introduced in 1930, to

Original signal

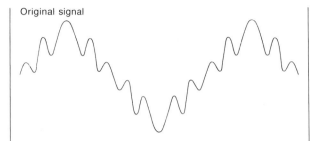

Transmitted carrier signal (frequency modulated)

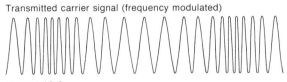

FIGURE 18-8
Frequency modulation of a broadcast radio signal. From
Henri Busignies, "Communication Channels." Copyright ©
1972 by Scientific American, Inc. All rights reserved.

improve the quality of radio transmission. The amplitude
of the audio signal increases or decreases the frequency of
the carrier as shown in Figure 18-8. Frequency modulation is relatively insensitive to noise produced by lightning discharges, sparking commutator brushes on old
motors, and faults in automobile ignition systems and in
other electrical equipment. The information transmission
depends not on the amplitude of the signal but only on the
shift in frequency.

FM radio is broadcast in a frequency range from
88 MHz to 108 MHz with a wavelength from 3.4 m to
2.8 m. The electric vector of the FM wave is typically
polarized in the horizontal plane. FM requires the use of
an external antenna. FM frequencies lie between channel
6 and channel 7 of the television band. (Thus, a TV
antenna makes an excellent FM antenna.) A superheterodyne radio receiver is also used for FM reception.

Television

The television broadcast band consists of three regions:
the VHF channels 2 to 6 (54 MHz to 88 MHz) and
channels 7 to 13 (174 MHz to 216 MHz), and the UHF
channels 14 to 83 (470 MHz to 890 MHz). The audio

signal of television is frequency modulated, while the
video and color signal is amplitude modulated. In the
United States, every television picture consists of 525
lines, and 30 pictures per second are broadcast. If each
line were broken up into 200 pieces of information, this
would require

$$30 \times 525 \times 200 = 3.15 \times 10^6$$

—over three million bits of information per second. Because the modulated television signal is broadcast with the
electric field polarized horizontally, the receiving antenna
is also horizontal. A total band width of 6 MHz is reserved for each television channel. Channel 2 has a wavelength of 5.5 m, channel 13 one of 1.4 m. The VHF
television antenna shown in Figure 18-9 uses a pair of
long elements that work best on the longer-wavelength
channels and a pair of shorter elements that work best at
higher frequencies. The UHF channels have much
shorter wavelengths (0.64 m to 0.34 m) and require a
separate antenna. In local reception areas, a pair of "rabbit
ears," which are not exactly horizontal, are sufficient to
pick up the signal. Rabbit ears must be adjusted to opti-
mize reception for each channel.

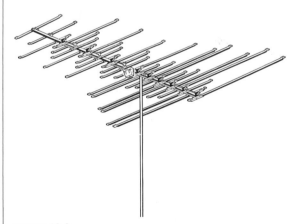

FIGURE 18-9
VHF television antenna. Several elements are used to increase
antenna performance. Short elements are effective at higher
frequencies, long elements at lower frequencies. The antenna
is designed to give maximum signal strength over the VHF
band from channel 2 to 13. Courtesy of Winegard Company.

In a television set, which also is a superheterodyne receiver, the signal from the antenna or cable is amplified and then mixed with a local oscillator signal to convert it to an intermediate frequency. The picture carrier has an intermediate frequency of 45.75 MHz. The video, color, and sound information is then decoded and processed to produce the television picture and sound.

Television transmitters are placed at the greatest possible height (on the highest hill, tower, or building) to obtain the greatest possible distance to the horizon. Television and FM signals are not reflected by the ionosphere and consequently require "line of sight" transmission. Television signals are reflected by conducting materials such as buildings and hills. If a signal reaches an antenna from two paths of different lengths—one direct from the transmitter and a second reflected signal—ghost images are produced.

EXAMPLE

A ghost image appears shifted by 10% across the line of the television picture. What is the path difference between the direct picture signal and the ghost picture signal?

A television picture consists of 15 750 lines/second. Each line takes 63 μs. A 10% shift across the line corresponds to a time delay of 6.3 μs. This corresponds to a path difference of

$$\Delta L = c\Delta t = (3 \times 10^8 \text{ m/s})(6.3 \times 10^{-6} \text{ s}) = 1.9 \text{ km}$$

The Electromagnetic Spectrum

The spectrum of electromagnetic waves extends from the low-frequency, extremely long-wavelength radiation, such as that associated with 60 Hz power lines, to the extremely high-frequency, short-wavelength radiation, such as gamma radiation. The **electromagnetic spectrum** is shown in Figure 18-10. The wavelength and the frequency are related (in a vacuum) by the speed of propagation of light c:

$$c = f\lambda$$

By international agreement the FCC has allocated different bands, or ranges of the frequency spectrum, for different purposes. The low-frequency band (LF), from 30 kHz (kilohertz) to 300 kHz has been allocated primarily for marine radio navigation, including the Loran C beacons. The medium-frequency band (MF), from 300 kHz to 3000 kHz, contains the AM radio band, 550 kHz to 1600 kHz. The high-frequency band (HF), from 3 MHz to 30 MHz, contains the long-distance shortwave radio bands as well as aeronautical beacons and citizen band radio. Very specific frequencies have been allocated for specific purposes. The very-high-frequency band (VHF), from 30 MHz to 300 MHz, contains land mobile units, aircraft instrument-landing systems, satellite telemetry systems, as well as broadcast television channels 2 to 6 (54 MHz to 88 MHz), broadcast FM (88 MHz to 174 MHz), and television channels 7 to 13 (174 MHz to 216 MHz). The UHF band from 300 MHz to 3000 MHz contains television broadcast channels 14 to 83 in a band from 470 MHz to 890 MHz. Above this is the superhigh-frequency band (SHF), from 3 GHz (gigahertz) to 30 GHz, used for radio navigation, point-to-point microwave communication links, and satellite-to-earth communications.

The low-frequency (LF), medium-frequency (MF), and high-frequency (HF) bands are reflected by the upper layers of the ionosphere, as shown in Figure 18-11. The shortwave radio frequencies in the HF band are reflected off the E and F layers of the ionosphere at considerable altitude and can travel great distances. They are affected by variation of the ionosphere during the transition from day to night. Large solar flares and intense sunspot activity can disturb the ionosphere sufficiently to interrupt shortwave transmission for several days.

The ionosphere is essentially transparent to VHF, UHF, and SHF portions of the electromagnetic spectrum.[3] Because transmission from the radiating antenna to the receiver must essentially be line-of-sight in these bands, television and FM radio signals have a limited range.

---

[3] Because of this transparency, several astronomers have speculated on how the earth must appear to an alien observer in outer space. See, for example, W. T. Sullivan, S. Brown, and C. Wetherill, "Eavesdropping: The Radio Signature of the Earth," *Science* (27 January 1978), pp. 377–388.

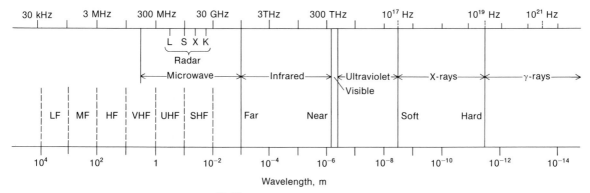

FIGURE 18-10
The major divisions of the electromagnetic spectrum.

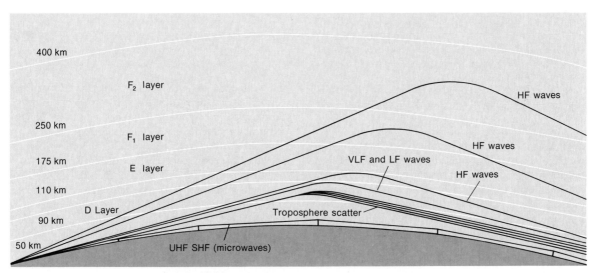

FIGURE 18-11
The ionosphere reflects low-frequency, LF, medium-fre-
quency, MF, and high-frequency, HF, electromagnetic waves.
Higher-frequency waves, in the VHF, UHF, etc., are not re-
flected and must follow line-of-sight paths. From Henri
Busignies, "Communication Channels." Copyright © 1972 by
Scientific American, Inc. All rights reserved.

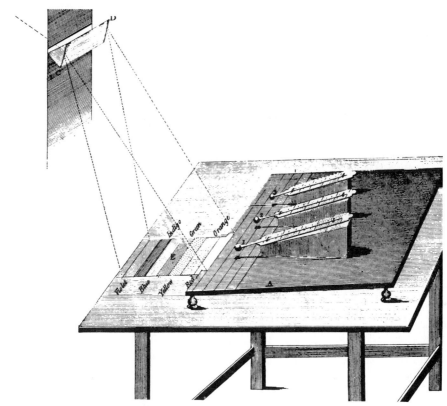

FIGURE 18-12
Infrared radiation was discovered by F. W. Herschel. Courtesy, Burndy Library.

Frequencies above 890 MHz are in the microwave region of the electromagnetic spectrum: s-band radar from 1.55 GHz to 6.2 GHz; x-band radar from 5.2 GHz to 10.9 GHz. At these frequencies, the radiation is often carried in hollow copper or aluminum waveguides of rectangular cross section. The radiation can also be transmitted through the air, and microwave horns often feed radiation to a parabolic reflector for point-to-point transmission. Microwave links with parabolic reflectors are used to connect remote television crews with the main studios, and for ground-to-satellite communication. Two narrow microwave bands have been reserved for astronomical observations of deep space. A 0.18 m wavelength, corresponding to a transition in the O–H radical, can signal the presence of oxygen in space. A 0.21 m wavelength, corresponding to a transition between different spin states of the hydrogen atom, is used to observe the density of hydrogen in space.

The microwave region ends at about 300 GHz, where the wavelength becomes so short (less than 1 mm) that very different equipment must be used. There is no clear demarcation line between the microwave and the higher-frequency infrared regions of the spectrum.

In about 1800, the astronomer F. W. Herschel, having produced a spectrum of sunlight with a prism, discovered that a thermometer placed in the red portion of the visible spectrum registered a higher temperature than when placed in the blue end. Moving the thermometer to a region below the red, infrared, produced a still higher temperature as shown in Figure 18-12. In the years that followed, this infrared radiation was studied, and by 1850 it was established that it is a form of invisible light.

In 1917, two scientists, E. F. Nichols and J. D. Tear, detected and spectroscopically examined heat radiation in the extreme infrared (with a wavelength of 4.2×10^{-4} m). Being also able to study shorter-wavelength (2.2×10^{-4} m), electrically produced radiation, they overlapped the microwave spectrum with the infrared spectrum and showed that the two forms of radiation were identical. The infrared band extends from about 3×10^{11} Hz (300 GHz) up to about 4×10^{14} Hz, or 400 terahertz (THz).

Visible light (the part of the electromagnetic spectrum visible to our eyes) extends from a wavelength of approximately 7×10^{-7} m to 4×10^{-7} m. It is probably no accident of evolution that this spectrum band corresponds to the maximum intensity of the light reaching the surface of the earth from the sun. In the whole spectrum of electromagnetic frequencies, the visible band is a small slice, from 4 to 7×10^{14} Hz. For life on earth, it is a most important slice.

Wavelengths shorter than 4×10^{-7} m are in the ultraviolet region of the spectrum. As the wavelengths become shorter, the ultraviolet spectrum becomes the x-ray spectrum, at a frequency of about 10^{17} Hz. X-rays are generated from inner-shell electron transitions of atoms, as we shall discuss in Chapter 22. At a frequency of about 10^{19} Hz the gamma-ray region of the spectrum begins. Gamma rays originate from the nucleus of the atom and extend to frequencies of 10^{22} Hz and beyond.

The entire electromagnetic spectrum, from low-frequency (60 Hz) power-line radiation through high-frequency (10^{22} Hz) gamma radiation, is essentially the same type of propagating oscillating electric and magnetic field. Frequencies and wavelengths differ, but over some 20 orders of magnitude of frequency the speed of propagation in vacuum is the same. Although electromagnetic radiation affects matter differently as frequency changes, the wave itself remains essentially unchanged.

BIOLOGICAL HAZARDS OF ELECTROMAGNETIC RADIATION

The known biological hazards of electromagnetic radiation for frequencies below that of visible light are associated with local body heating from the absorption of energy. Since D'Arsonval, in 1890, first demonstrated the heating effect on humans, electromagnetic radiation has been used therapeutically in diathermy to heat tissues and increase the blood flow.

Possible electromagnetic-radiation damage to body tissue differs with the frequency and intensity, the duration of exposure, and the nature of the tissue. At frequencies less than infrared (400 THz) electromagnetic waves act almost entirely as an oscillating electric field; at greater frequencies they can be absorbed as quanta of energy, photons. Ultraviolet, x-ray, gamma-ray, and cosmic-ray photons can ionize single atoms or molecules in tissue, by the photoelectric effect. (We discuss the photoelectric effect and the effects of ionizing radiation in Chapters 21 and 25.)

At frequencies less than 150 MHz, the body is transparent to radiation and absorbs no appreciable energy. At frequencies of 150 MHz to 1.2 GHz (wavelengths from 2.00 m to 0.25 m), approximately 40% of the energy incident on the body is absorbed. Electromagnetic radiation can penetrate to a depth of about one-tenth of the wavelength varying according to tissue characteristics. At these frequencies, radiation can penetrate 0.20 m to 0.025 m into the body. If a person were exposed to an intense beam of this radiation, the whole body would be heated with little or no heating of the skin's thermal sensors. This, called whole-body heating, would be very hazardous, because a person would feel no heat while undergoing substantial body heating—in effect being roasted without knowing it until the overall body temperature increased, and the person broke out in a sweat.

At frequencies between 1 GHz and 3 GHz, in the UHF region (a wavelength between 0.1 m and 0.3 m), from 20% to 100% of the incident radiation is absorbed, depending on tissue water content. At frequencies from 3 GHz to 10 GHz, in the microwave region, the top layers of the skin absorb most of the energy, giving an immediate sensation of warmth. At about 3 GHz, the lens of the eye is the most sensitive body part. Above 10 GHz (wavelengths of 0.03 m or less), the skin surface acts as a reflector and absorber with some heating, because the waves can penetrate only a short distance into the skin. Table 18-1 summarizes these biological effects.

Whole-Body Heating

At frequencies between 150 MHz and about 1200 GHz, deeper tissues absorb most of the energy and significant whole-body heating can result. If the heat is not dissipated as fast as it is produced, body temperature rises. The body can normally dissipate 10 mW/cm^2 from its surface; considerably more under favorable circum-

TABLE 18-1
Summary of biological effects of microwaves

Frequency	Wavelength m	Site of major tissue effects	Major biological effects
Less than 150 MHz	Above 2.00		Body is transparent
150 MHz–1.2 GHz	2.00–0.25	Internal body organs	Damage to internal organs from overheating
1 GHz–3 GHz	0.30–0.10	Lens of the eye	Lens of the eye particularly susceptible; tissue heating
3 GHz–10 GHz	0.10–0.03	Top layers of the skin, lens of the eye	Skin heating with sensation of warmth
Above 10 GHz	Less than 0.03	Skin	Skin surface reflects or absorbs with heating effects

SOURCE: Wellington Moore, Jr., "Biological Aspects of Microwave Radiation; A Review of Hazards," USHEW Bureau of Radiological Health (July 1968).

stances. If the heat is sufficient to cause a rise in body temperature, the symptoms of fever appear. An excessive rise in temperature from microwave radiation produces tissue exactly as would a fever of any origin, and death can result. It has been estimated that maintaining a human body temperature of 2°C above normal requires a continuous exposure to a microwave flux of 100 mW/cm².

With the rapid development and widespread use of microwave equipment since World War II, military and civilian authorities have set limits for permissible exposure of personnel. A review of the experimental data in 1953 showed that a level of about 100 mW/cm² was necessary to produce any biologically significant effect. To allow for uncertainties, a safety factor of 10 was incorporated and the **maximum safe exposure level** was set at 10 mW/cm².

The Eye

The rate at which a body organ can dissipate heat determines its susceptibility to microwave radiation. Blood flow determines the rate of heat removal. The lens of the eye is particularly sensitive, for it has neither a blood supply nor the cell-repair mechanisms that most other organs have. Continued radiation to the lens of the eye can raise the temperature by as much as 10 to 20°C. Damage to cells of the lens is generally irreversible; the damaged cells slowly lose their transparency and a cataract forms. A temperature increase of 10 to 14°C can produce opacity.

In 1952, the first case of cataract formation from microwave exposure was reported. For about one year, a technician operating a microwave generator at frequencies between 1.5 GHz and 3 GHz was exposed in his daily work to a power density of 100 mW/cm²—10 times the maximum safe exposure level, 10 mW/cm²—which produces significant heating of the eye.

Injury from electromagnetic radiation has been confined to a few isolated cases of accidental exposure. Concerned for environmental safety, the Bureau of Radiological Health has conducted survey of microwave power levels in the vicinity of the Washington, D.C., airport and found that all measurements were less than 0.01 mW/cm², three orders of magnitude smaller than the maximum safe exposure level. Most of the observed radiation could be ascribed to commercial broadcasting or to the high-power radar at the airport.

The bureau has also conducted extensive testing of microwave ovens (an estimated 17 million will be in use in 1980). Microwave ovens operate at 2.45 GHz or 0.9 GHz and could pose a health hazard. However, they are carefully designed and checked so that the leakage with the door closed is less than 1 mW/cm^2. Electromagnetic radiation in our home and urban environment does not pose any known health hazard.

Review

Give the significance of each of the following terms. Where appropriate give two examples.

electromagnetic wave
LC resonant circuit
frequency
wavelength

antenna
carrier wave
amplitude modulation
superheterodyne receiver
intermediate frequency

citizen band radio
frequency modulation
electromagnetic spectrum
maximum safe exposure level

Further Readings

Henri Busignies, "Communication Channels," *Scientific American* (September 1972).
Donald G. Fink and David M. Lutyens, *The Physics of Television* (New York: Doubleday, Anchor Science Study Series S 8, 1960).
The Big Dummy's Guide to C.B. Radio (Summertown, Tenn. 38483: The Book Publishing Co., 1976). A humorous yet informative introduction.

Problems

ELECTROMAGNETIC WAVES

1 An oscillator circuit from an electrosurgical knife produces high-power electrical current at a frequency of 2 MHz. A resonant circuit with a $6.3 \, \mu\text{F}$ capacitor determines the frequency. What value of inductance is required?

2 A resonance circuit similar to that used by Heinrich Hertz consists of a 5 millihenry (mH) inductor in series with a 50 pF capacitor. What is the resonant frequency?

ELECTROMAGNETIC RADIATION

3 An FM radio station broadcasts at 103.9 MHz on the dial. What is the free-space wavelength of the FM radio waves?

4 The UHF television band from channel 14 through 83 extends from 470 MHz to 890 MHz. What are the free-space wavelengths at each end of the UHF television band?

5 A television receiving antenna is designed to be of optimal length for channel 6 with a frequency of 88 MHz. The antenna should be $\frac{1}{2}$ wavelength long. How long is that?

6 A CB radio broadcasts at a frequency of 27 MHz. A stationary antenna is designed to be $\frac{5}{8}$ of a wavelength long. How long is that?

7 A microwave antenna operates at a frequency of 915 MHz. If the microwave energy penetrates to a depth of $\frac{1}{5}$ the wavelength, how thick can a roast be and still be efficiently cooked?

8 A spectrum line used by astronomers to observe hydrogen in space has a wavelength of 21 cm. What is the frequency of the radiation? What spectrum band does it correspond to?

V

OPTICS

19

Geometrical Optics

Fortunate Newton, happy childhood of science! . . . Nature to him was an open book, whose letters he could read without effort. The conceptions which he used to reduce the material of experience to order seemed to flow spontaneously from experience itself, from the beautiful experiments which he ranged in order like playthings and describes with an affectionate wealth of detail. . . .
Reflexion, refraction, the formation of images by lenses, the mode of operation of the eye, the spectral decomposition and the recomposition of the different kinds of light, the invention of the reflecting telescope, the first foundations of colour theory, the elementary theory of the rainbow pass by us in procession. . . .

Albert Einstein[1]

HUYGENS' PRINCIPLE

Our understanding of the nature of light has emerged over the past three centuries. Light has a dual nature, behaving in some circumstances as if it were a particle, in others as if a wave. This dual nature of light became evident with the work of Christian Huygens and Isaac Newton at the end of the seventeenth century. In his *Treatise on Light* (1690), Huygens presented what is known as **Huygens' principle:** *Light propagates as a wave, and each point on the wave acts as a new source of waves.* Huygens saw an analogy between the propagation of light and the propagation of sound:

We know that by means of the air which is an invisible body, sound spreads around the spot where it is produced by a movement which is passed on successively from one part of the air to another; . . . Now there is no doubt at all that light also comes from the luminous body to our eyes by some movement impressed on the matter which is between the two; . . . it will follow that this movement, impressed on the intervening matter, is successive; and con-

sequently it spreads, as sound does, by spherical surfaces and waves.[2]

The wave fronts propagate outward from a light source just as the ripples on the surface of water propagate outward from the splash of a falling pebble. Figure 19-1 shows Huygens' method of construction of waves. The wave progresses because each point on it produces a new wave. The new waves combine to form the next wave front.

Newton, who published *Opticks*, his study of light, in 1704, thought that light acted as small **particles,** or "corpuscles." In a uniform medium such as air, glass, or water, these particles traveled in straight lines, known as **rays,** which defined the shortest distance between two points. In passing from one medium to another, a light ray

---

[1] Foreword to Sir Isaac Newton, *Opticks* (New York: Dover Publications, 1952), p. lix.

[2] Quoted in Cecil J. Schneer, *The Evolution of Physical Science* (New York: Grove Press, 1960), p. 235.

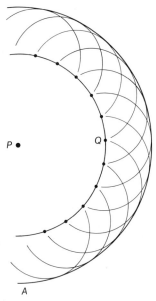

FIGURE 19-1
Huygens' method of wave front construction. Each point on the wave becomes a new source of waves.

is bent; that is, the angle of the ray to the surface changes. Newton describes the behavior of light in terms of straight lines and angles using geometrical constructions, that is **geometrical optics.**

Huygens' *Treatise on Light* and Newton's *Opticks* presented two contrary views of the nature of light: light as a wave, light as a particle. Because of Newton's prestige, his ideas dominated eighteenth-century scientific thinking. But Newton did not have the last word on the subject. The view of light as a wave (discussed in Chapter 20) dominated the nineteenth century; the wave-particle view of light emerged in the twentieth century, as we discuss in Chapters 21 and 22.

GEOMETRICAL OPTICS

Using five axioms, **Newton's axioms,** that clearly define the basic principles of geometrical optics, and using simple geometrical constructions, Newton analyzed the passage of light through prisms and simple lenses, reflection

of light by flat and curved mirrors, and the image formation of the eye:

> Axiom I. The Angles of Reflexion and Refraction lie in one and the same Plane with the Angle of Incidence.
> Axiom II. The Angle of Reflexion is equal to the Angle of Incidence.
> Axiom III. If the refracted Ray be returned directly back to the Point of Incidence, it shall be refracted into the Line before described by the incident Ray.
> Axiom IV. Refraction out of the rarer Medium into the denser is made towards the Perpendicular; that is, so that the Angle of Refraction be less than the Angle of Incidence.
> Axiom V. The Sine of Incidence is either accurately or very nearly in a given ratio to the Sine of Refraction.[3]

Reflection

In a uniform medium, a beam or ray of light follows a straight line. A ray incident on (that is, falling on or striking) a plane mirror is reflected. Figure 19-2 shows how Newton's axioms can be used to construct the image formed by a plane mirror. Axiom I states that the angles of **incidence** θ_i and of **reflection** θ_r, as measured from the **normal,** the perpendicular to the surface, lie in the same plane. Thus, we can draw the incident and reflected rays in two dimensions. Axiom II states that the angle of reflection is equal to the angle of incidence:

$$\theta_r = \theta_i$$

A ray ① from object point A, incident at an angle θ_i to the mirror normal, is reflected at an angle θ_r. A second ray ②, incident at, say, a 30° angle to the mirror normal, is reflected at 30°. The two rays appear to be coming from a position A' behind the mirror surface. In fact, many rays converge to form the image A'. A third ray ③, normally incident to the mirror, confirms the location of the image A'. Similarly, we can construct the image B' of some other point B, as shown in the figure.

---

[3] Isaac Newton, *Opticks*, p. 5. Spelling and capitalization as in original.

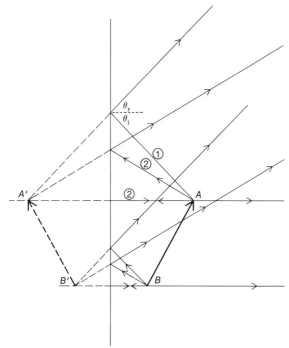

FIGURE 19-2
The plane mirror. Ray tracing reveals the position and orientation of an image in a plane mirror.

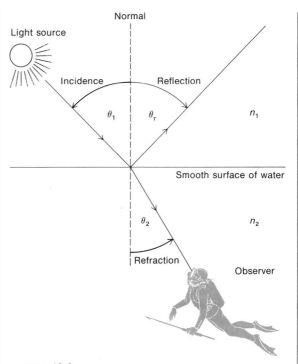

FIGURE 19-3
Reflection and refraction of a ray incident on a smooth surface.

Refraction

A ray of light incident on a smooth, transparent surface, such as a pool of still water as shown in Figure 19-3, is partially reflected and partially refracted as it enters the water. By Axiom I, the angles of incidence, reflection, and refraction lie in the same plane. Axiom IV states qualitatively and Axiom V states quantitatively **Snell's law** of refraction:

$$n_1 \sin \theta_1 = n_2 \sin \theta_2 \qquad (19\text{-}1)$$

The respective optical densities of the materials n_1 and n_2 are given by the **index of refraction,** a measure of the refracting power of a material compared to vacuum. The index of refraction of air is very close to unity. Table 19-1 lists the index of refraction of various materials. A ray of light going from an optically less dense material to an optically denser material (say from air to water) is refracted *toward* the normal, that is, toward the perpendicular to the surface. A ray of light going from an optically denser material to an optically less dense material (water to air) is refracted *away* from the normal.

Apparent Depth

The bottom of a pool of quiet water appears nearer to the surface than it is because of refraction at the surface. In Figure 19-4, two observers are looking at a stone at the bottom of a 3 m deep diving pool. The observer on the diving board is looking normal to the surface. A ray of light coming from the stone along his line of sight is undeviated. The second observer is looking at an angle to

TABLE 19-1
Index of refraction of various materials in sodium yellow (589 nm) light.

Air 15°C, 760 torr	1.000 293
Water 20°C	1.333
Acetone	1.357
Ethyl alcohol	1.360
30% sucrose in water	1.381
Tolulene	1.494
Safflower oil	1.477
80% sucrose in water	1.490
Fused quartz	1.458
High-dispersion crown glass	1.520
Canadian balsam	1.530
Styrene	1.545
Rock salt	1.544
Amber	1.546
Light flint glass	1.575
Heavy flint glass	1.650
Heaviest flint glass	1.890
Diamond	2.44

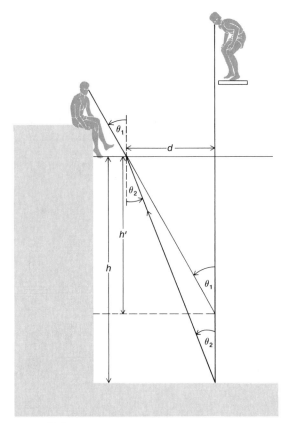

FIGURE 19-4
Apparent depth. For small angles of incidence the pool has an apparent depth of $h' = h/n$.

the surface. A ray of light from the stone to this observer is refracted at the surface. If, at the surface of the pool, the distance between the lines of sight of the two observers is d, the apparent depth and actual depth of the pool are given by

$$\tan \theta_1 = d/h' \qquad \tan \theta_2 = d/h$$

$$h' = h\frac{\tan \theta_2}{\tan \theta_1}$$

For angles less than 15 degrees

$$\tan \theta = \frac{\sin \theta}{\cos \theta} \cong \sin \theta$$

The apparent depth is

$$h' \cong h\frac{\sin \theta_2}{\sin \theta_1} = \frac{h}{n} \qquad (19\text{-}2)$$

In our example there are two observers. To a single observer the bottom of the pool will appear closer than it actually is. Our two eyes give the perception of depth, for each eye has a slightly different view. The optical medium has the effect of reducing apparent depth.

The Prism

Newton was much concerned with improving telescopes. As the magnification of the telescope was increased, the image was fringed with color and unclear. To study this effect, Newton undertook a systematic study of the refraction of light through a prism. He observed that light of different colors would be refracted by differing amounts, and that sunlight could thereby be broken up into colors.

TABLE 19-2
The index of refraction of glass is a function of wavelength and color of light.

Color	Wavelength (nm)	Index of refraction High-dispersion crown	Heavy flint
Ultraviolet	361	1.546	1.705
Violet	434	1.533	1.675
Blue	486	1.527	1.664
Yellow	589	1.520	1.650
Red	656	1.517	1.644
Deep red	768	1.514	1.638
Infrared	1200	1.507	1.628

The index of refraction of a piece of glass depends on the color or wavelength of light. Table 19-2 shows the index of refraction for high-dispersion crown and heavy flint glass for different wavelengths of light. Passing through a prism, a ray of light at the blue end of the spectrum is refracted through a larger angle than light at the red end of the spectrum.

Figure 19-5 shows an equilateral prism of high-dispersion crown glass. The apex angle A is 60°; the dotted lines are perpendicular to each surface, normal. The ray of light is incident at an angle θ_1 of about 50°. The prism will have maximum dispersion if the angle of first refraction is about equal to 30° so that the ray travels through the

prism parallel to its base. The first angle of refraction θ_2 is given by Snell's law:

$$\sin \theta_2 = \frac{n_1 \sin \theta_1}{n_2}$$

For an incident angle of 50°, the first angle of refraction varies from 30.0° in the violet to 30.4° in the deep red.

The triangle formed by the two dotted lines and the ray through the prism gives the angle of incidence at the second surface θ_2. The angle formed by the intersection of the dotted lines is $120° = 180° - A$. The sum of the interior angles of a triangle is 180°.

$$180° = \theta_2 + \theta_2' + 180° - A$$

$$\theta_2' = A - \theta_2$$

where A is the apex angle. The second angle of refraction of the ray, θ_1' going to a less dense medium is given by

$$\sin \theta_1' = \frac{n_2 \sin \theta_2'}{n_1}$$

The final angle of refraction varies from 50.1° in the violet to 48.4° in the red. Thus, the prism can separate colors. Separation of light of different wavelengths is known as **dispersion.** The variation of the index of refraction with the wavelength of light determines the dispersion of a prism.

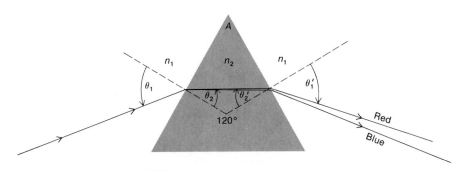

FIGURE 19-5
Refraction of light through a high-dispersion crown glass prism.

Internal Reflection

A ray of light going from an optically dense medium to one less dense is refracted away from the normal. The angle of refraction is given by

$$\sin \theta_2 = \frac{n_1 \sin \theta_1}{n_2}$$

Figure 19-6 shows the angle of refraction for various angles of incidence at the water-air interface as the angle of incidence θ_2 increases from 20° to 50°. At the **critical angle** of incidence θ_2, the angle of refraction is 90° and there is total **internal reflection.** No light is refracted through the surface; all of the light is reflected. The critical angle for internal reflection is given by

$$\sin \theta_c = \frac{n_1}{n_2} \sin 90° = \frac{n_1}{n_2} \qquad (19\text{-}3)$$

For water, with an index of refraction $n = 1.33$, the critical angle is $\theta_c = 48.75°$. For heavy flint glass, with an index of refraction of $n = 1.65$, the critical angle is 37.3°. If the angle of incidence from the dense medium is greater than the critical angle, the internal reflection is total. Total internal reflection is used in optical instruments to reverse or invert images. Figure 19-7 shows its use in the prisms used in binoculars.

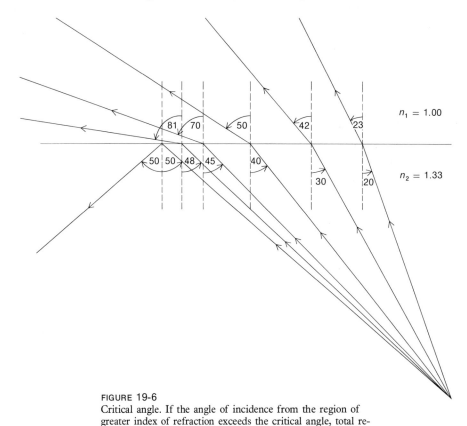

FIGURE 19-6
Critical angle. If the angle of incidence from the region of greater index of refraction exceeds the critical angle, total reflection results; $\sin \theta_c = n_1/n_2$. For air $n_1 = 1.0$; for water $n_2 = 1.33$. The critical angle is $\theta_c = 48.75°$.

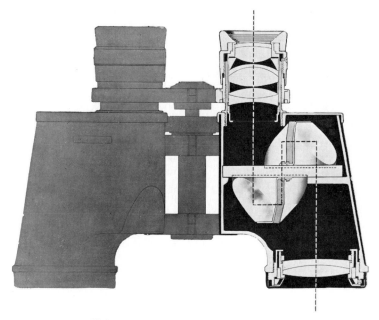

A **light pipe,** or light guide, consists of a bundle of tiny glass fibers held together by a bonding material that keeps the fibers separated. The narrow fibers usually consist of two parts: A highly polished cylindrical glass fiber with a high index of refraction is surrounded by a thin coating with a low refractive index. The critical angle for internal reflection can be as low as 50°. Light entering the pipe is continually reflected internally and follows the path of the fiber. In a properly made light pipe, almost all light loss is caused by absorption in the glass, not by internal reflections. Figure 19-8 shows that the exit angle for light leaving the light pipe is the same as the entry angle. Figure 19-9 shows the fiber optics light guide now being tested for voice, video, and data transmission by Bell Laboratories.

The Rainbow

Newton was able to explain the colors of the rainbow. A rainbow forms through the combined effects of refraction, dispersion, and internal reflection of sunlight by rain-

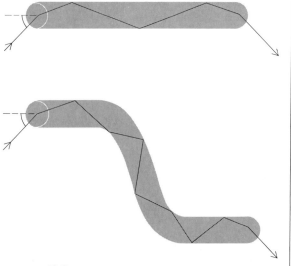

FIGURE 19-8
The light pipe. The light's angle of exit from a fiber is the same as the angle of entry even if the fiber is bent. From Narinder S. Kapany, "Fiber Optics." Copyright © 1960 by Scientific American, Inc. All rights reserved.

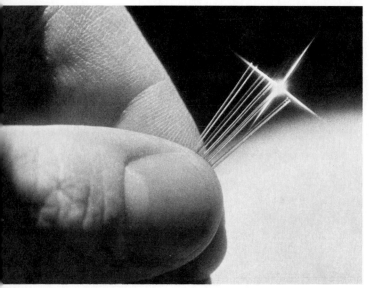

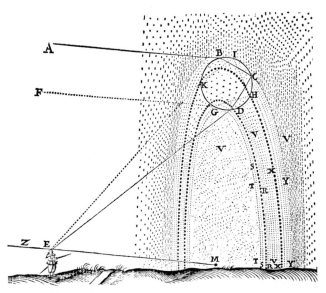

FIGURE 19-10
The rainbow. René Descartes published the correct explanation of the rainbow in 1637. The primary and secondary
rainbows are formed by refraction and reflection in spherical
water droplets.

drops. Figure 19-10 shows an illustration, published by
René Descartes in 1637, of refraction of a ray of light in
a rainbow. A ray of light AB incident on the spherical
drop is refracted at the first surface. At the second surface
BCD, it undergoes partial internal reflection. At the final
surface, as it passes out to the observer DE, it is refracted.
Each ray is strongly reflected in the direction of maximum
deviation. The angle of maximum deviation for red light
is 138°, as shown in Figure 19-11. The red primary
rainbow is observed at an angle of $\theta = 42°$. Since the
index of refraction of water changes with color, the corresponding angle for violet light is 40°. The red edge of the
primary rainbow is to the outside.

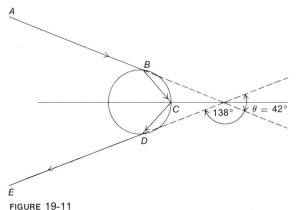

FIGURE 19-11
The angle of maximum deviation for the primary rainbow in
a spherical water drop. Because of the dispersion of water, the
colors appear at slightly different angles.

As Figure 19-10 shows, the secondary rainbow is produced by two internal reflections FGHIE. The secondary
rainbow is much fainter and the order of colors is reversed. The red edge of the secondary rainbow occurs at
an angle of $\theta = 50.5°$ and is to the inside of the violet
rainbow at an angle of 54°.

The presence of the rainbow in the sky depends on the
position of the sun relative to the observer. If the sun is at
an angle of greater than 42° above the horizon, the rainbow cannot be seen from the ground. The rainbow stands
highest in the sky in the late afternoon, when the sun's
rays are almost horizontal.

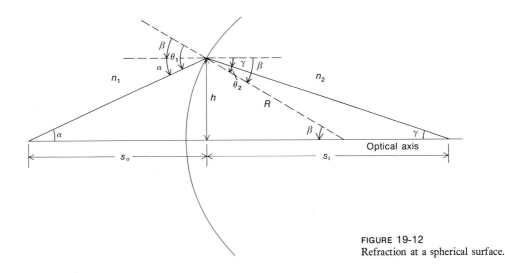

FIGURE 19-12
Refraction at a spherical surface.

THIN LENSES

Refraction from a Spherical Surface

Newton applied his principles of geometrical optics to determine the path of rays through a spherical refracting surface. A lens surface can be formed by grinding glass or another optical material to a spherical contour. Figure 19-12 shows such a spherical surface, with radius R. A ray from a point at a distance s_0 from the surface makes an angle α with the optical axis. This ray is incident at the surface at some angle θ_1. A little trigonometry gives the angle of incidence:

$$\theta_1 = \alpha + \beta$$

This ray will be refracted through some angle θ_2 given by Snell's law:

$$n_1 \sin \theta_1 = n_2 \sin \theta_2$$

A little thought (see Figure 19-12) shows that the angle of refraction is

$$\theta_2 = \beta - \gamma$$

We make the approximation that the angles are small. For angles less than about 15°, the sine and the tangent of an angle are approximately equal to the angle in radian measure, as Table 19-3 shows. In this small-angle approximation, Snell's law becomes

$$n_1 \theta_1 = n_2 \theta_2$$
$$n_1(\alpha + \beta) = n_2(\beta - \gamma)$$

which can be written

$$n_1\alpha + n_2\gamma = (n_2 - n_1)\beta$$

The ray contacts the spherical surface at some distance h from the optical axis. The distance from the object to

---

TABLE 19-3
Small angles. For small angles, the sine of the angle is approximately equal to the angle in radian measure, as is the tangent of the angle. For angles of 15° or less they are equal to within 1% or better.

15° = 0.2618 radians	sin 15° = 0.2588	tan 15° = 0.268
5° = 0.08727 radians	sin 5° = 0.08716	tan 5° = 0.0875
1° = 0.01745 radians	sin 1° = 0.01745	tan 1° = 0.01745

the surface is s_o, from the surface to the image point, s_i. The angles can be related to these distances:

$$\alpha \simeq \tan \alpha = \frac{h}{s_o}; \quad \beta \simeq \sin \beta = \frac{h}{R}; \quad \gamma \simeq \tan \gamma = \frac{h}{s_i}$$

Snell's law then gives us

$$\frac{n_1 h}{s_o} + \frac{n_2 h}{s_i} = \frac{(n_2 - n_1)h}{R}$$

or simply

$$\frac{n_1}{s_o} + \frac{n_2}{s_i} = \frac{(n_2 - n_1)}{R} \qquad (19\text{-}4)$$

Rays from an object at a distance s_o in a region with an index of refraction n_1 will be brought to a focus at a distance s_i in the second region with index of refraction n_2.

Lenses

A **thin lens** consists of a thin piece of glass with two spherical surfaces close together, in effect, back to back. The rays refracted at one surface are refracted again at the other surface. The refraction at each surface can be analyzed to determine the path of the ray. If an object, a point source of light at an infinite distance, is on the *optical axis*, the axis of symmetry through the center of the lens, the incoming rays are all parallel to the optical axis. The **converging lens** shown in Figure 19-13 refracts these

parallel rays to a single **focal point.** The **focal distance** from the center of the lens to the focal point is determined by the difference in index of refraction between the lens material (glass, $n_2 = n$) and the surrounding medium (air, $n_1 = 1.00$); and by the radii of curvature of the refracting surfaces. For thin lenses, the focal point on each side of the lens is at the same focal distance. If a thin lens is turned around, its focusing properties are the same. The focal distance f is given by the **lensmaker's equation:**

$$\frac{1}{f} = (n_2 - n_1)\left[\frac{1}{R_1} - \frac{1}{R_2}\right]$$

For glass $n_2 = n$ in air, $n_1 = 1.00$. The focal distance is given by

$$\frac{1}{f} = (n - 1.00)\left[\frac{1}{R_1} - \frac{1}{R_2}\right] \qquad (19\text{-}5)$$

R_1 and R_2 refer are the radii of curvature of the first and second surfaces encountered, respectively, in passing through the lens. The signs of the radii of curvature depend on whether the surface is concave or convex as a ray passes through the lens along the optical axis.

1. If a ray is traced through the lens (as shown in Figure 19-13) and if the surface encountered is *convex*, bowed outward, its radius of curvature is considered *positive*. In the figure, a ray is traced from the left. The first surface encountered is convex; therefore R_1 is *positive*.

2. If a ray is traced through the lens and if the surface encountered is *concave*, bowed inward, its radius of curvature is considered *negative*. In the figure, a ray is traced from the left. The second surface encountered is concave; therefore R_2 is negative.

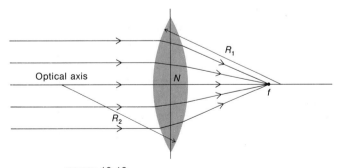

FIGURE 19-13
A converging lens focuses parallel rays of light to the focal point f.

EXAMPLE

A converging lens has an index of refraction $n = 1.65$, and radii $R_1 = +7.6$ cm and $R_2 = -7.6$ cm. What is the focal length?

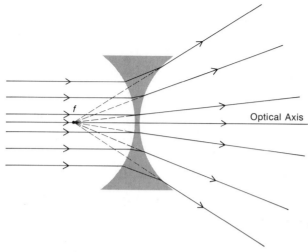

FIGURE 19-14
A diverging lens. Parallel rays will emerge as if from the focal point f behind the lens.

The focal length is given by

$$\frac{1}{f} = (1.65 - 1.00)\left[\frac{1}{0.076 \text{ m}} - \frac{1}{-0.076 \text{ m}}\right] = 17.1/\text{m}$$

$$f = 0.058 \text{ m} = 5.8 \text{ cm}$$

Figure 19-14 shows a **diverging lens.** Parallel incident rays not on the optical axis are refracted away, forming divergent rays. The light will appear to be coming from the focal point f *in front of the lens*. The focal length of a diverging lens is negative. The focal length can be calculated by the lensmaker's equation.

EXAMPLE

A diverging lens (see Figure 19-14) is made of glass with an index of refraction of 1.52. Approached from the left, the first surface is concave, $R_1 = -5.0$ cm, the second is convex, $R_2 = +5.0$ cm. The lensmaker's equation gives us

$$\frac{1}{f} = (n - 1)\left[\frac{1}{R_1} - \frac{1}{R_2}\right]$$

$$= (1.52 - 1.00)\left[\frac{1}{-0.050 \text{ m}} - \frac{1}{+0.050 \text{ m}}\right]$$

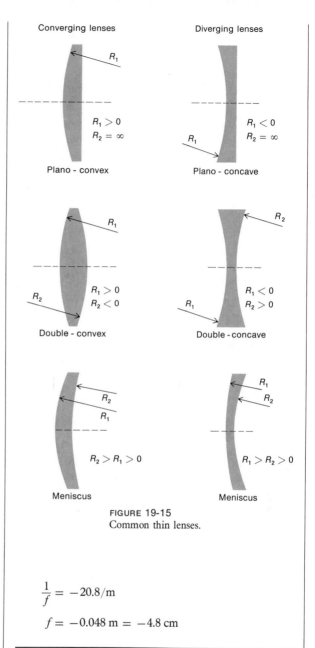

FIGURE 19-15
Common thin lenses.

$$\frac{1}{f} = -20.8/\text{m}$$

$$f = -0.048 \text{ m} = -4.8 \text{ cm}$$

Figure 19-15 shows common converging and diverging thin lenses. The shape and the index of refraction of the glass determines the focal length of each lens. A plano-lens has one flat surface (infinite radius of curvature). A meniscus lens, commonly used in eyeglasses, has two surfaces which are bowed in the same direction, but with differing radii of curvature.

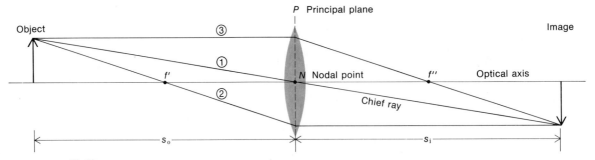

FIGURE 19-16
Ray tracing. The image formed by a lens can be determined by tracing two or more rays: (1) The chief ray through the nodal point N. (2) A ray through the front focal point f' to the principal plane P, which emerges parallel to the optical axis. (3) A ray parallel to the optical axis to the principal plane P, which emerges passing through the second focal point f''.

Ray Tracing

The position and size of an image formed by a thin lens can be determined by the method of **ray tracing.** From each point on an object, light rays travel outward in all directions in straight lines. Some of these rays can be gathered by a lens to form an image. The image can be located by tracing the path of two or three different rays from a point off the optical axis: (1) a chief ray through the nodal point, (2) a ray through the front focal point, and (3) a ray parallel to the optical axis (see Figure 19-16).

The **chief ray** (1) is the ray that goes through the **nodal point** N of the lens and is undeviated in angle. For thin lenses there is a single nodal point, the optical center of the lens. For thicker lenses there are two nodal points N and N' displaced by some distance, and the chief ray is displaced because of refraction.

A ray (2) through the front focal point f' continues to the **principal plane** P of the lens and leaves the lens parallel to the optical axis, and will intersect with the chief ray at the focal point.

For a thin lens, the principal planes for the front and rear focus are identical, and the principal plane P is at the physical center of the lens. In a thick lens, separate principal planes may be associated with each focus.

A ray (3) entering the lens parallel to the optical axis and continuing to the principal plane will be refracted through the rear focal point (f''). For now, we concern ourselves only with the images produced by thin lenses.

If the object distance s_o, the distance from the object to the lens, is greater than the focal distance of the converging lens, it will produce a **real image,** an image that is inverted and can be focused on a screen. The image distance s_i is positive and is on the opposite side of the lens from the object. The image distance equals or exceeds the focal distance.

A diverging lens cannot by itself produce a real image. Figure 19-17 shows image formation by a concave lens. A ray (1) from the object traveling parallel to the optical axis will diverge and appear to be coming from the focal point on the object side. A ray (2) headed for the far focal point will leave the lens in a direction parallel to the optical axis. The chief ray (3) is undeviated. The image will appear on the same side of the lens as the object, so that the image distance is negative. It forms a **virtual image.** It is erect (that is, not inverted), and cannot be focused on a screen without the use of another lens.

The Thin-Lens Equation

Image formation by a thin lens is determined by focal length. Figure 19-18 shows a converging lens used to form a real image. An object of height h_o is at a distance s_o in front of the lens. An image of height h_i is at a distance s_i beyond the lens. A ray of light (1) traveling from the object parallel to the optical axis AB is refracted through the far focal point F to form the image. This ray, BFC, crosses the optical axis at an angle β. This forms two similar triangles, FCD and FBN. The tangent of angle β can be written for each triangle:

$$\tan \beta = \frac{BN}{NF} = \frac{CD}{DF} = \frac{h_o}{f} = \frac{h_i}{s_i - f}$$

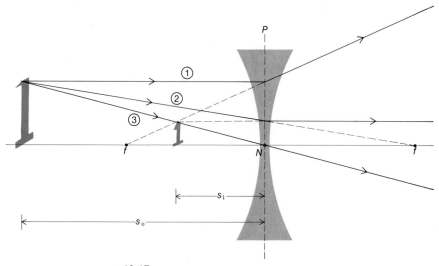

FIGURE 19-17
A diverging lens produces a virtual image, which lies behind
the lens.

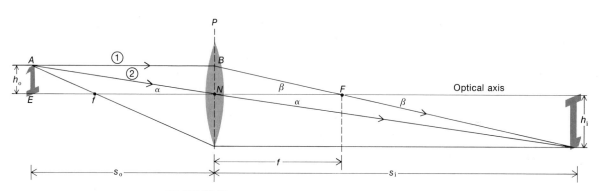

FIGURE 19-18
Linear magnification is proportional to the ratio of the image
distance to the object distance: $M = -s_i/s_o$.

where $BN = h_o$, $CD = h_i$, $NF = f$, and $DF = s_i - f$.
This can be rewritten as

$$\frac{h_i}{h_o} = \frac{s_i - f}{f}$$

The chief ray ② through the nodal point N crosses the
optical axis at an angle α. This forms two similar triangles, NAE and NCD. The tangent of angle α can be

written

$$\tan \alpha = \frac{AE}{EN} = \frac{CD}{DN} = \frac{h_o}{s_o} = \frac{h_i}{s_i}$$

where $AE = h_o$, $CD = h_i$, $EN = s_o$, and $DN = s_i$. This
can be rewritten as

$$\frac{h_i}{h_o} = \frac{s_i}{s_o}$$

Combining this with the previous result, we get

$$\frac{h_i}{h_o} = \frac{s_i}{s_o} = \frac{s_i - f}{f}$$

Dividing by s_i, we get

$$\frac{1}{s_o} = \frac{s_i - f}{s_i f} = \frac{1}{f} - \frac{1}{s_i}$$

The object distance s_o and the image distance s_i are related to the focal length f by the **thin-lens equation:**

$$\frac{1}{s_o} + \frac{1}{s_i} = \frac{1}{f} \qquad (19\text{-}6)$$

The sign conventions for the object distance, image distance, and focal length for the thin-lens equation are:

1. The *object distance* s_o is *positive* if the object is on the same side of the lens as the light source. The object distance is *negative* if the object is on the side to which the light goes, as is often the case when the image formed by a first lens serves as the object for a second lens.

2. The *image distance* s_i is *positive* if the image is on the same side of the lens to which the light goes. The image distance is *negative* if the image is on the side from which the light comes.

3. The *focal length* is *positive* if the lens is converging, *negative* if the lens is diverging.

EXAMPLE

A diverging lens has a focal length of $f = -5.8$ cm. An object is placed 10 cm from the lens, as shown in Figure 19-17. Where is the image?

The thin-lens equation gives the image distance.

$$\frac{1}{s_i} = \frac{1}{f} - \frac{1}{s_o} = \frac{1}{-5.8 \text{ cm}} - \frac{1}{10 \text{ cm}}$$

$$s_i = -3.7 \text{ cm}$$

The image distance is negative, indicating that the image is on the same side as the object. The image is erect and virtual.

Linear Magnification

If the image distance is larger than the object distance, the image undergoes **linear magnification** as shown in Figure 19-18. The chief ray ② forms some angle α with the optical axis. The tangent of the angle α is proportional to two ratios: the ratio of the image height h_i to the image distance s_i; and the ratio of the object height h_o to object distance s_o.

$$\tan \alpha = \frac{h_i}{s_i} = \frac{h_o}{s_o}$$

The linear magnification of the image is just the ratio of image size to object size

$$M = -\frac{h_i}{h_o} = -\frac{s_i}{s_o}$$

where the minus sign indicates that the image is inverted. The magnification is simply related to the ratio of image distance to object distance. In a microscope objective or in a photographic enlarger, the object distance is much smaller than the image distance and the magnification by the objective is greater than unity. Conversely, in a camera, the object distance and the magnification are less than unity. A photographic image is normally much smaller than the object.

Solving the thin-lens equation for the image or object distance, we obtain

$$s_i = \frac{f s_o}{s_o - f} \qquad s_o = \frac{f s_i}{s_i - f}$$

The linear magnification can then be written

$$M = -\frac{s_i}{s_o} = \frac{f}{f - s_o} = \frac{f - s_o}{f} \qquad (19\text{-}7)$$

A 35 mm camera (using a film with a frame size 24 mm × 36 mm) has a lens with a focal length of 50 mm. A person 1.75 m tall is photographed from a distance of 6 m.

(a) What are the image distance and height? What is the magnification?

(b) A telephoto lens with a focal length of 135 mm is used to make the same photograph. What are the image distance and height?

The 50 mm lens has a focal length of 0.050 m. At an object distance of 6.0 m the image distance is given by

$$\frac{1}{s_i} = \frac{1}{f} - \frac{1}{s_o} = \frac{1}{0.050 \text{ m}} - \frac{1}{6.0 \text{ m}}$$

$$s_i = 0.050\ 42 \text{ m}$$

The image distance very nearly equals the focal length. The linear magnification is

$$M = -\frac{s_i}{s_o} = -\frac{0.050\ 42 \text{ m}}{6.0 \text{ m}} = -0.0084$$

The image size is just

$$h_i = Mh_o = 15 \text{ mm}$$

The student can show that when the telephoto lens is used, the image distance is 0.138 m and the image height is 40 mm—greater than the width of the film.

Two Thin Lenses

The refracting power of a lens is measured in **diopters,** the reciprocal of the focal length measured in meters:

$$P = \frac{1}{f}$$

A 50 mm (0.050 m) focal length lens has a power of $+20$ diopters. A converging lens has a positive focal length and refracting power. A diverging lens has a negative refracting power.

If two thin lenses separated by a small distance are used in combination, their power as measured in diopters adds.

In Figure 19-19, two converging lenses are separated by a short distance c. An object placed a distance s_o from the first lens produces an image at a distance s_i, given by the thin-lens equation:

$$\frac{1}{s_i} = \frac{1}{f_1} - \frac{1}{s_o}$$

This image becomes the object for the second lens. Because of c, the separation between the lens, the new object distance s_o' is

$$s_o' = -(s_i - c)$$

The object distance is considered *negative* because the light is coming from the other side of the lens. The final image position is

$$\frac{1}{s_i'} = \frac{1}{f_2} - \frac{1}{s_o'} = \frac{1}{f_2} + \frac{1}{s_i - c}$$

$$\frac{1}{s_i'} = \frac{1}{f_2} + \frac{1}{s_i(1 - c/s_i)}$$

Substituting $1/s_i$ from the first equation gives

$$\frac{1}{s_i'} = \frac{1}{f_2} + \left[\frac{1}{f_1} - \frac{1}{s_o}\right]\left[\frac{1}{1 - c/f_i + c/s_o}\right]$$

If we assume that the distance between the lenses is much smaller than the object distance $c \ll s_o$ and than the focal length $c \ll f_1$, the result becomes simply

$$P_1 + P_2 = \frac{1}{f_1} + \frac{1}{f_2} = \frac{1}{s_i'} + \frac{1}{s_o}$$

For thin lenses close together, the powers of the lenses P_1 and P_2 add.

Lens Aberrations

Because the index of refraction of glass depends on the wavelength of light, the focal length of light at the red end of the spectrum is slightly different from that at the blue end of the spectrum. These differences cause color fring-

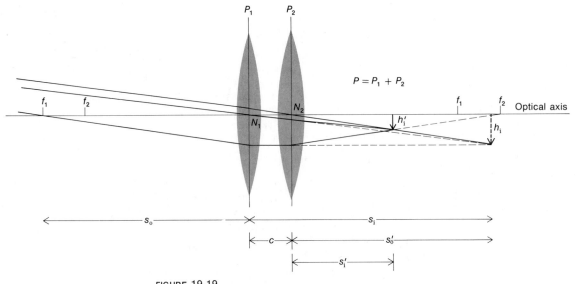

FIGURE 19-19
Two thin lenses separated by a small distance c. The image formed by the first lens h_i becomes a virtual object for the second lens. If c is small, the powers of the two lenses add.

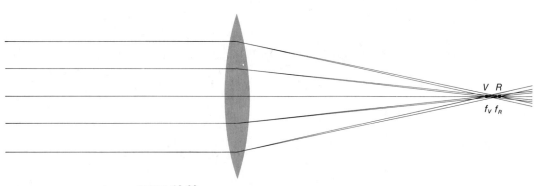

FIGURE 19-20
Chromatic aberration. Because of the dispersion of optical glass, a lens has different focal lengths at the red and violet ends of the spectrum.

ing of the image, which is particularly bothersome in astronomical observation of stars or in microscopic observation of fine detail. This flaw in focusing is known as **chromatic aberration,** shown in Figure 19-20.

For high-dispersion crown glass the index of refraction shown in Table 19-2 can vary from 1.533 in the violet to 1.514 in the deep red, giving, for a double convex converging lens with a 10 cm radius of curvature, a focal length of 9.38 cm in the violet to 9.67 cm in the deep red.

Spherical aberration is a lens defect whereby rays traveling parallel to the optical axis but at different distances from it form their images at different points. The

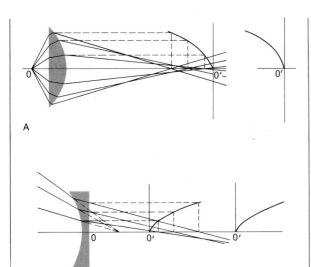

A

B

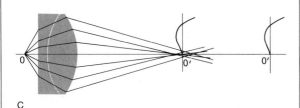

C

FIGURE 19-21
Spherical aberration. The focal position of rays passing at different distances from the optical axis is shown, A, for a convex lens, B, for a concave lens, C, for a combination of a convex and concave lens. Courtesy of Nikon Instrument Division, Ehrenreich Photo-Optical Industries, Inc.

spherical aberration of a lens can be measured by plotting the point of focus for different incident parallel rays as a function of the distance from the axis. Figure 19-21 shows the spherical aberration of a convex lens, a concave lens, and two lenses placed in combination to minimize this defect.

Spherical and chromatic aberration can both be minimized by an appropriate combination of converging and diverging lenses. A converging lens of crown glass and a diverging lens of flint glass can be cemented together to form an **achromatic doublet.** It focuses properly at both the red and violet ends of the spectrum; however, the intermediate colors, yellow and green, are not sharply focused. Similarly, spherical aberration can be corrected by a combination of convex and concave lenses. A lens

system corrected for both spherical aberration and chromatic aberration is called an *apochromatic lens.* These are commonly used in superior microscope objectives where accurate focusing of color images is desired.

A lens corrected for spherical aberration produces a clear image of a point near the optical axis. However, points off the optical axis are not focused well. Light passing through different portions of the lens is focused at different points, giving rise to a cometlike blur appearing outside the focused image. Figure 19-22 A shows this lens aberration, called **coma.**

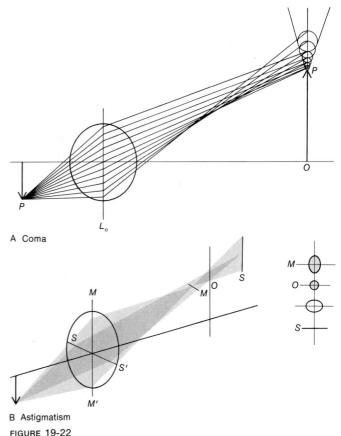

A Coma

B Astigmatism

FIGURE 19-22
Aberrations produced in forming the image of a point off the optical axis. A. Coma. B. Astigmatism. Courtesy of Nikon Instrument Division, Ehrenreich Photo-Optical Industries, Inc.

Light from a point off the optical axis passes through two different perpendicular planes of the lens. Each plane focuses a sharp image at a different point, *IM* or *IS*, as shown in Figure 19-22 B. This lens aberration is known as **astigmatism.** The sharpest focus occurs at a position between *IM* and *IS*, where the image of the point is the smallest. Near the optical axis of a spherical lens, astigmatism is not a problem, but image quality is degraded as the edge of the field of view is approached. Astigmatism is particularly troublesome in photomicroscopy, for it is desirable that the entire field of view be in sharp focus.

In the eye astigmatism can occur if the cornea has different radii of curvature along different perpendicular planes. Then, even rays near the optical axis are brought to focus at two different points (determined by the plane the rays travel through).

OPTICAL INSTRUMENTS

The Magnifying Glass

A magnifying glass is a converging lens. As Figure 19-23 shows, the object to be viewed is placed at a distance s_o less than the focal length f of the lens. The image, at a distance, s_i, formed on the same side of the lens as the object, is virtual and erect and cannot be focused on a screen. The distance from the observer to the virtual image is typically taken to be $s_i = -25$ cm, a distance for comfortable close viewing. The linear magnification is just

$$M = \frac{s_i}{s_o} = \frac{s_i(s_i - f)}{s_i f} = \frac{-s_i + f}{f}$$

$$M = \frac{25 \text{ cm}}{f} + 1 \qquad (19\text{-}8)$$

where f is the focal length in centimeters. The magnification is positive, indicating an erect final image.

The Light Microscope

In the light microscope, a short-focal-length lens, the **objective lens,** produces a real image that is magnified by the **eyepiece lens** to form a final virtual image. Figure 19-24 shows the optical path for a light microscope. The condenser lens on the light source focuses the illumination into a parallel beam. To uniformly illuminate the microscope's entire field of view, the field diaphragm and condenser field lens limit and control the light reaching the final condenser lens. The objective lens forms an image that the doubly reflecting prism in the microscope head reflects to a comfortable viewing angle. The image formed by the objective lens is viewed through the eyepiece. The eye then focuses the light to form an image on the retina.

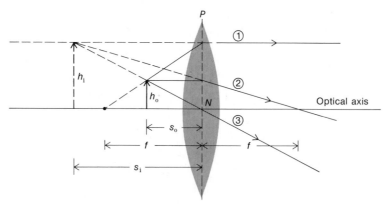

FIGURE 19-23
A magnifying glass forms a virtual image at 25 cm.

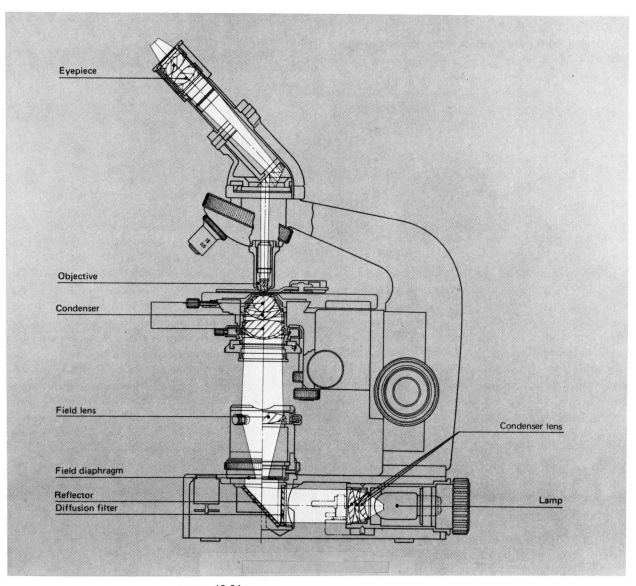

Eyepiece

Objective

Condenser

Field lens

Field diaphragm

Reflector
Diffusion filter

Condenser lens

Lamp

FIGURE 19-24
The optical path of a common microscope. Courtesy of Nikon
Instrument Division, Ehrenreich Photo-Optical Industries, Inc.

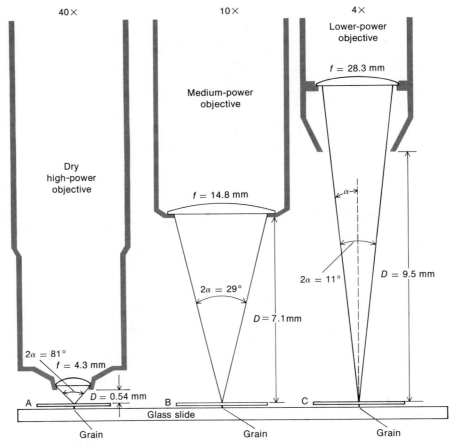

FIGURE 19-25
Microscope objectives. The angular aperture 2α and free working distance D are shown.

The Objective The objective lens forms a real inverted image near the eyepiece focus. Figure 19-25 and Table 19-4 give the geometries of several such lenses. The objective's position is adjustable so that lenses of different focal length can form the image at the focal plane of the eyepiece. A shorter-focal-length objective lens placed close to the object being viewed will form a larger image. An objective lens with a focal length f_o produces a real image with a negative magnification:

$$M_o = -\frac{s_i}{s_o} = -\frac{s_o f_o}{s_o(s_o - f_o)} = \frac{f_o}{f_o - s_o}$$

For a 40x objective lens $M_o = -40$ and a focal length of 4.3 mm, we can solve for the object distance s_o:

$$s_o = f_o - \frac{f_o}{M_o} = 4.3 - \frac{4.3}{-40} = 4.41 \text{ mm}$$

The object distance is very close to the focal length: $s_o \cong f_o$. The image distance is

$$s_i = -M_o s_o = 176.3 \text{ mm}$$

The total distance from the object to the eyepiece focal

TABLE 19-4
Microscope objective lenses.

Magnification	Focal Length (mm)	Image distance (mm)	Free working distance (mm)	Angular aperture (°)	Numerical aperture
4x	28.3	141	9.5	11	0.10
10x	14.8	163	7.1	29	0.25
40x	4.3	176	0.54	81	0.65
100x (oil immersion)	1.8	182	0.16	106 (n = 1.556)	1.25

plane is about $L = 175$ mm. The objective magnification is approximately

$$M_o = -\frac{L}{f_o}$$

Using an objective lens with a shorter focal length increases the microscope's magnifying power. Usually light, focused by the condenser lens system, is transmitted from below through the sample into the microscope objective. The *angular aperture* 2α, defined by the light cone entering the objective lens, determines the ability of the objective to collect light from the sample. The angular aperature of low-power objectives may be as small as 11°, as shown in Figure 19-25; however, short-focal-length lenses have large apertures. Table 19-4 lists the characteristics of typical microscope objective lenses.

Filling the region between the objective lens and the cover glass with an oil with index of refraction n decreases the apparent object distance,

$$s_o' = s_o/n$$

which further increases both the magnification and the angular aperture of the lens. Figure 19-26 compares the oil-immersion lens and the dry-system lens. Having the same index of refraction as the cover glass, the oil eliminates the refraction both at the cover glass and at the face of the objective lens.

Short-focal-length objective lenses must be corrected for spherical aberration. An aplanatic lens, which has a large angular aperture with no spherical aberration, is

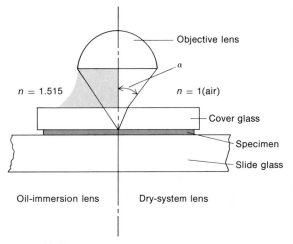

FIGURE 19-26
The microscope objective lens, comparing the oil-immersion lens with the dry-system lens. Courtesy of Nikon Instrument Division, Ehrenreich Photo-Optical Industries, Inc.

commonly used in oil-immersion objectives. The aplanatic lens, shown in Figure 19-27, is hemispheric in shape. The flat surface is immersed in oil so that the entire volume below the spherical refracting surface has a uniform index of refraction. A thin, 0.18 mm, cover glass is placed over the sample to be viewed. Light is refracted only at the spherical surface. If the lens has a radius of curvature R and an index of refraction n, then the object position A, the aplanatic point, is at a distance $d_1 = R/n$ from the center of curvature. Rays even at large angles

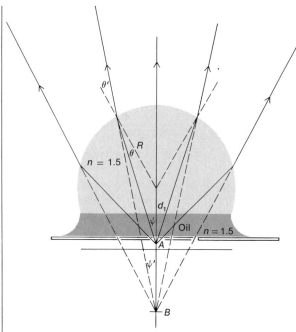

FIGURE 19-27
An aplanatic lens gathers light over a large angular aperture. An object at A will form an apparent image at B with no spherical aberration. This principle is used in the oil-immersion objective lens.

from this point will appear to originate from a second aplanatic point B at a distance $d_2 = Rn$ from the center of curvature. Such a lens is free of spherical aberration and can have a very large angular aperture.

One or more meniscus lenses with the proper curvature can be placed above the spherical surface to further focus the rays without spherical aberration. Only chromatic aberration ultimately limits magnification by this process. Figure 19-28 shows a 12-element aplanatic oil-immersion microscope objective lens with a spherical lens, two meniscus lenses, 4 achromatic doublets, and a final thick lens to form the final image.

The Eyepiece The image formed by the objective lens is viewed through the eyepiece. The simplest eyepiece is a magnifying lens. The eyepiece magnification for a single lens with a focal length f_e and an image distance $s_i = -25$ cm is

$$M_e = -\frac{s_i}{s_o} = \frac{-s_i + f_e}{f_e} = \frac{25 \text{ cm}}{f_e} + 1$$

The overall magnification of the microscope is the product of the magnification of the objective lens M_o times the magnification of the eyepiece M_e. Total magnification M is

$$M = M_o M_e \cong -\frac{L}{f_o}\left[\frac{25 \text{ cm}}{f_e} + 1\right]$$

For instance, a 40x objective and a 10x eyepiece give a 400x overall magnification.

A single-lens eyepiece presents two difficulties: limited field of view (only rays refracted into the pupil of the eye are collected and focused on the retina); and chromatic aberration (producing significant color fringing of the final image unless achromatic lenses are used). These difficulties can be corrected by using an eyepiece consisting of two lenses. The first lens, called the *field lens* is placed slightly before the image formed by the objective lens. The second lens, called the *eye lens* (it is closest to

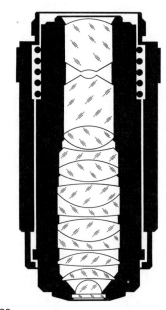

FIGURE 19-28
An apochromatic oil-immersion objective lens with 12 elements has a magnification of $100\times$ and a numerical aperture of 1.32. The multiple lenses are in pairs to reduce chromatic aberration and astigmatism. Courtesy of E. Leitz, Inc.

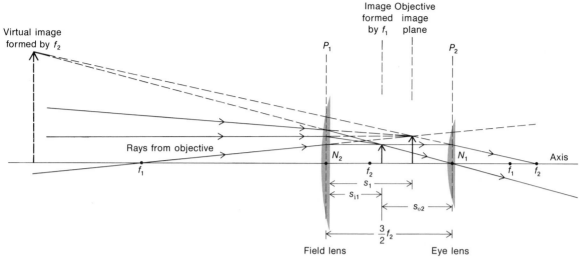

Image Objective
formed image
by f_1 plane

Virtual image
formed by f_2

Rays from objective

P_1 P_2

N_2 N_1 Axis

f_1

f_2 f_1 f_2

s_1

s_{i1}

s_{o2}

$\frac{3}{2}f_2$

Field lens Eye lens

FIGURE 19-29

The Huygens eyepiece. The eye lens has a focal length f_2. The field lens has a focal length $2f_2$. The two lenses are separated by a distance of $\frac{3}{2} f_2$. The image formed by the objective lens falls between the two lenses. This combination of lenses greatly reduces chromatic aberration and increases the field of view.

the eye), then magnifies the image produced by the field lens. Figure 19-29 shows the *Huygens eyepiece*. The proper choice of focal lengths for the field and eye lenses, and proper separation between them, can both reduce chromatic aberration and increase the field of view of the eyepiece.

The Refracting Telescope

Telescopes make distant objects appear larger. The moon with an angular size of about 0.5°, has an angular size of about 3.5° viewed through a pair of 7x binoculars (in effect, a short telescope). The refracting astronomical telescope consists of an objective lens and an eyepiece. The objective lens forms a real inverted image of the distant object (see Figure 19-30). The eyepiece magnifies this image, which remains inverted. The **angular magnification** is the ratio of the apparent angular size β to the angular size viewed by the naked eye, α:

$$m = -\beta/\alpha$$

The minus sign indicates that the final image is inverted. If an object of height h_o is at some large distance s_o, the first image h_1 will be at the focal point of the first lens f_o. The angular size of the object in radian measure is approximately

$$\alpha \cong \tan \alpha = \frac{h_o}{s_o} = \frac{h_1}{f_o}$$

The apparent angular size of the first image viewed through the eyepiece lens will be about

$$\beta \cong \tan \beta = \frac{h_1}{f_e}$$

Whether the final virtual image is at 25 cm or at infinity, the angular size will be almost the same.

The angular magnification for the astronomical telescope is given by

$$m = -\frac{\beta}{\alpha} = -\frac{f_o}{f_e} \qquad (19\text{-}10)$$

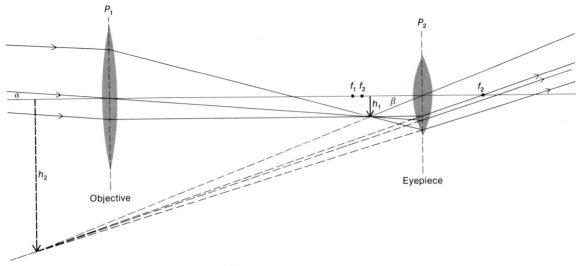

FIGURE 19-30
The astronomical telescope gives an inverted image.

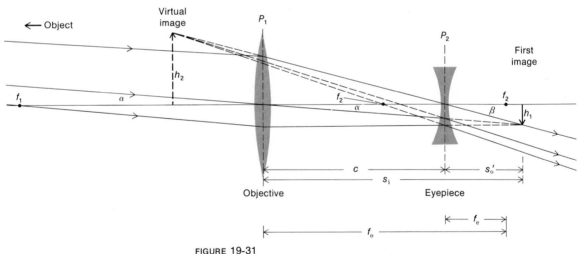

FIGURE 19-31
The Galilean telescope gives an erect image.

The astronomical telescope gives an inverted final image.

The Galilean telescope, or opera glass, has a converging objective lens and a diverging eye lens, and gives an erect image. However, it has a small field of view. If the distance between the lenses is just $c = f_o - f_e$, their rear focal points coincide. If s_i is the image distance produced by the first lens, the object distance for the second lens is $s_o' = c - s_i$. The object distance is negative. Figure 19-31 shows the ray diagram for the Galilean telescope. The objective lens forms near its focal point a real image h_1 of a distant object. Because the eye lens has a negative focal length f_e, the rays diverge and form a virtual image h_2, which is erect. The angular size of the original object in radian measure is

$$\alpha \cong \tan \alpha = \frac{h_o}{s_o} = \frac{h_1}{f_o}$$

The angular size of the final image β is

$$\beta \cong \tan \beta = \frac{h_1}{f_e}$$

The angular magnification of the Galilean telescope is

$$m = -\frac{\beta}{\alpha} = -\frac{f_o}{f_e}$$

where f_e is negative.

The Reflecting Telescope

As an improvement to astronomical telescopes of his day, Newton invented the reflecting telescope, using a concave metal mirror that he ground to a radius of curvature of about 0.32 m.

In Figure 19-32, an illustration from Newton's *Opticks*, the concave mirror takes the place of the objective lens. Newton used an internally reflecting prism so that the focus position could be viewed directly with the eye. Although the image was faint because of absorption on the metal mirror surface, Newton's early telescope had a magnification of about 35x. Modern reflecting, or *Newtonian telescopes*, use this same principle; the viewing image is formed by the objective with an achromatic eyepiece.

Many amateur astronomers have spent hours with glass, pitch, and grinding compound, patiently repeating Newton's work in grinding their own astronomical reflectors. The grinding of mirrors for reflecting telescopes has reached its perfection with the Russian 6 m (236 in) and the Palomar 5 m (200 in) telescopes. Figure 19-33 shows the optical path of a small modern reflecting telescope.

A spherical concave mirror of radius R is shown in Figure 19-34. A ray parallel to the optical axis, incident at an angle ϕ to the perpendicular to the surface at P, will be reflected through an angle ϕ. It arrives at the focal point f reflected through an angle 2ϕ. This forms an obtuse triangle RfP, where the angle $\theta = 180° - 2\phi$. The side of the triangle L is related to the radius of the circle R by the law of sines:

$$\frac{\sin \phi}{L} = \frac{\sin \theta}{R} = \frac{\sin (180° - 2\phi)}{R} = \frac{\sin 2\phi}{R}$$

The side of the triangle L is equal to

$$L = R \frac{\sin \phi}{\sin}$$

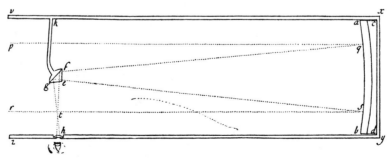

FIGURE 19-32
Newton's illustration, in his *Opticks*, of a reflecting telescope.

FIGURE 19-33
The optics of a modern reflecting telescope include a concave primary reflecting mirror and a convex secondary mirror to project the image through the hole in the primary mirror, for either direct photographic exposure or optical viewing. The front plate is specially ground to remove astigmatism, spherical aberration, and coma. Courtesy of Questar Corporation.

At the focal distance f, the rays converge:

$$f = R - L = R\left[1 - \frac{\sin \phi}{\sin 2\phi}\right]$$

where R is the radius of curvature of the mirror. If the angle 2ϕ is small, then

$$\frac{\sin \phi}{\sin 2\phi} \cong \frac{\phi}{2\phi} = \frac{1}{2}$$

If the angle ϕ is small, then $f = R/2$. If the diameter of the lens mirror is small compared to the radius of curvature, then the spherical aberration of the mirror is kept to a minimum. For a spherical mirror the image distance s_i is related to the object distance s_o by the focal distance f:

$$\frac{1}{f} = \frac{2}{R} = \frac{1}{s_o} + \frac{1}{s_i} \tag{19-11}$$

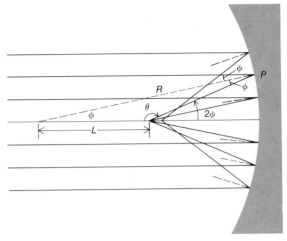

FIGURE 19-34
A concave mirror. The angle of reflection equals the angle of incidence to the mirror surface. A spherical mirror has no chromatic aberration but exhibits spherical aberration, reflecting parallel rays off-axis to a slightly different focal point.

As Figure 19-35 shows, the image formation of a spherical mirror can be determined by tracing rays. A ray ① parallel to the optical axis leaves through the focal point f. A ray ② through the focal point f leaves parallel to the optical axis. Finally, a ray ③ through the center of curvature is reflected back along the line of incidence.

A spherical convex mirror of radius R, such as a hubcap or Christmas tree ornament, will produce a virtual image behind the mirror surface with a negative image distance. The focal length for the convex mirror is $f = -R/2$. Figure 19-36 shows the image formation of a convex mirror.

THE OPTICS OF HUMAN VISION

The Structure of the Eye

Figure 19-37 shows the basic structure of the human eye and Table 19-5 lists the dimensions of its major optical components. The **cornea** is the eye's primary refracting surface. The outer surface of the cornea has a radius of curvature of about 7.7 mm. The cornea is about 0.5 mm thick and its material has an index of refraction of 1.376. Between the cornea and the lens and iris is a fluid, the *aqueous humor*, which has an index of refraction of 1.336.

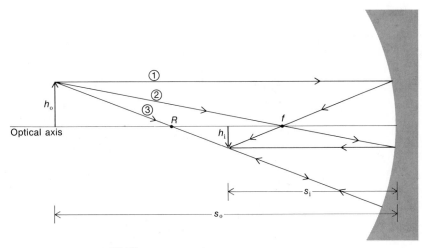

FIGURE 19-35
Image formation by a concave mirror. The focal point is in
front of the mirror. If the object distance s_o is greater than the
focal distance f, a real image is formed.

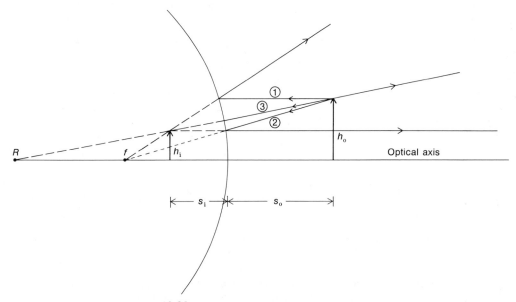

FIGURE 19-36
Image formation by a convex mirror. The focal point lies behind the mirror surface. The final image is virtual.

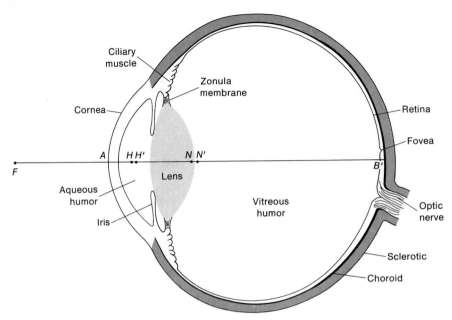

FIGURE 19-37
The principal features of the human eye.

TABLE 19-5
Optical dimensions of the eye in mm
(n = index of refraction).

Cornea (*n* = 1.376)	
Outer radius	+ 7.7
Inner radius	− 6.8
Thickness	0.5
Aqueous humor (*n* = 1.336)	
Thickness	3.1
Lens (*n* = 1.41)	
Outer radius	+ 10.0
Inner radius	+ 6.0
Thickness	3.6
Vitreous humor (*n* = 1.336)	
Thickness	17.1

Because of this small difference in indexes of refraction, the interior surface of the cornea has little refracting power.

The *iris* controls the size of the lens aperture (and gives us our "eye color"). The iris can vary in diameter from between 1 mm to 5 mm. The light collecting area of the pupil can change by a factor of about 25.

The *crystalline lens* of the eye has an index of refraction of about 1.41; it is a double convex lens with radii of curvature of about 10 mm and 6.0 mm. The lens of the human eye serves only the function of **accommodation,** changing the focal distance of the eye slightly. The lens is held under tension of the *zonula membrane,* attached to the *ciliary muscle.* For near vision, the ciliary muscle contracts, releasing the tension on the zonula, allowing the lens to become more spherical, with a radius of curvature of about 5.5 mm on each surface. (In the eye of a fish, the cornea is in contact with water and the lens is the primary refracting surface.) In the eyes of all species, thin layers of transparent cells make up the lens, developing first in the center and accumulating outward throughout life, although growth slows with age. The innermost cells receive the oxygen and nutrients only indirectly, through the aqueous humor. With age, the lens grows in size and becomes stiff. A young eye has an accommodation power of about 14 diopters; by age 35 it is reduced to about 8 diopters, by age 50 to about 1 to 3 diopters (see Figure 19-38).

The retina, about 17 mm behind the lens, has a radius of curvature of about 12 mm. Between the lens and retina is a fluid, the *vitreous humor,* which has an index of

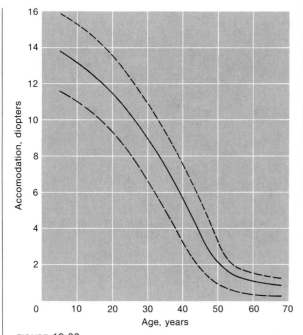

FIGURE 19-38
The typical accommodation range of the eye lens, in diopters, as a function of age. Past the age of 40 accommodation decreases rapidly.

refraction of 1.336. It had long been believed, beginning with such early natural philosophers as Galen in the second century A.D., that the source of vision was the lens. It was Johann Kepler who, in 1604, first realized that the retina is in fact a screen that receives the image produced by the lens and cornea. The light-sensitive cells of the retina were not discovered until about 1835, after the light microscope had come into systematic use.

Image Formation by the Eye

The cornea, the eye's primary refracting surface, has an index of refraction of $n_2 = 1.376$ and an outer radius of curvature of about 7.7 mm in air with $n_1 = 1.000$. The refraction of a spherical surface was given by equation 19-4:

$$\frac{n_2 - n_1}{R_1} = \frac{n_1}{s_1} + \frac{n_2}{s_2}$$

The cornea's effective focusing power depends on the change of index of refraction $n_2 - n_1$ and the radius of curvature R_1:

$$P_1 = \frac{1}{f_1} = \frac{n_2 - n_1}{R_1}$$

$$P_1 = \frac{1}{f_1} = \frac{(1.376 - 1.000)}{0.0077 \text{ m}} = 48.8 \text{ diopters}$$

The second surface of the cornea has much smaller refracting power:

$$P_2 = \frac{1}{f_2} = \frac{n_3 - n_2}{R_2}$$

$$P_2 = \frac{1.336 - 1.376}{.0068 \text{ m}} = -5.9 \text{ diopters}$$

The combined power of the two refracting surfaces gives an effective focal power of

$$P_1 + P_2 = 48.8 - 5.9 \text{ diopters} = 42.9 \text{ diopters}$$

The lens of the eye, with an index of refraction of $n_4 = 1.41$ with radii of curvature of 10.0 mm and -6.0 mm, is surrounded by fluid with an index of refraction of $n_3 = 1.336$. The effective power of the lens is given by

$$P' = (n_4 - n_3)\left[\frac{1}{R_3} - \frac{1}{R_4}\right]$$

$$= (1.41 - 1.336)\left[\frac{1}{0.010 \text{ m}} - \frac{1}{-0.0060 \text{ m}}\right]$$

$$P' = 19.7 \text{ diopters}$$

If both surfaces of the lens have a radius of curvature of about 5.5 mm, the refracting power of the lens is about 28 diopters. The refracting power of this eye can change by about 8 diopters.

The cornea and the lens are about 3.1 mm apart and together form a thick-lens system, the image formation of which is shown in Figure 19-39. The two principal planes P_1 and P_2 do not quite coincide. The principal points H and H' are 1.3 mm and 1.6 mm behind the front apex of the cornea. The front focal point F is 17.06 mm from the

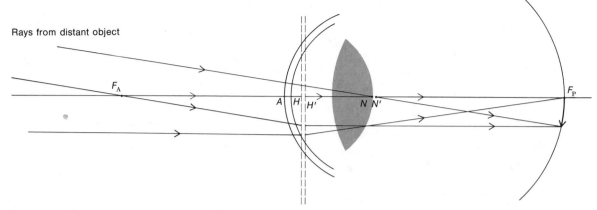

Rays from distant object

FIGURE 19-39
Image formation by the eye. A distant object is focused by the combined effect of the cornea and the eye lens. The eye is a thick-lens system.

TABLE 19-6
Optical distances in a typical eye in mm.

			Relaxed eye (mm)	Accommodation maximum (mm)
Front focal point (F)	to	Cornea (A)	$FA = 15.707$	12.397
Cornea (A)	to	principal point (H)	$AH = 1.348$	1.772
Front focal point (F)	to	principal point (H)	$FH = 17.055$	14.169
Cornea (A)	to	principal point (H')	$AH' = 1.602$	2.086
Cornea (A)	to	rear focal point (F')	$AF' = 24.387$	21.016
Principal point (H')	to	rear focal point (F')	$H'F' = 22.785$	18.930
Cornea (A)	to	nodal point (N)	$AN = 7.078$	
Coroea (A)	to	nodal point (N')	$AN' = 7.332$	

front principal point H. The nodal points for an un-deviated ray are 7.1 and 7.3 mm behind the cornea apex, just before and behind the rear surface of the lens. The rear focal distance F' is 22.79 mm behind the second principal plane P_2. The ratio of the front focal distance FH to the rear focal distance $F'H'$ is

$$\frac{F'H'}{FH} = \frac{22.79 \text{ mm}}{17.06 \text{ mm}} = 1.336$$

which is the index of refraction of fluid before and behind the lens. Table 19-6 lists the optical distances in the eye.

We can roughly approximate the optics of the eye by a single curved surface. In front of the surface is the object. The image is formed in a region of interocular fluid with index of refraction n. The image formation of the eye can be approximated by

$$P = \frac{1}{f} = \frac{1}{s_o} + \frac{n}{s_i} \qquad (19\text{-}12)$$

The effective focal length of the relaxed eye is typically $f = 17.1$ mm, the index of refraction $n = 1.336$. If the object is at infinity $s_o = \infty$, the image distance is

$$s_i = nf = (1.336)(17.1 \text{ mm}) = 22.8 \text{ mm}$$

which is typical of the distance from the nodal point N' to the retina. In this simple model, the image distance can be kept fixed and the focusing power of the cornea and lens adjusted for accommodation.

EXAMPLE

The eye is focused at $s_o = 0.25$ m. Assume an image distance to the retina $s_i = 22.8$ mm. With the simple model, determine the required focusing power.

$$P = \frac{1}{f} = \frac{1}{0.25 \text{ m}} + \frac{1.336}{0.0228 \text{ m}} = 4.0 + 58.6$$
$$P = 62.6 \text{ diopters}$$

Eye Defects

Most failures of the eye to form a focused image are caused by the length of the eyeball s_i being too great or too little although the refracting power of the lens is normal.

Myopia In nearsightedness, **myopia,** the eyeball is too long and the image of a distant point falls short of the retina. For the image to reach the retina, the object must be moved closer to the eye. There is a certain maximum distance that the nearsighted eye can see distinctly.

EXAMPLE

A nearsighted eye has an effective focal length of 17.1 mm but a rear focal distance $H'F$ of 25 mm. The interocular fluid has an index of 1.336.
(a) At what distance from the retina will the image of a distant point be formed?
(b) What is the furthest distance at which an object can be brought into focus?

The eye has normal refracting power, so objects at infinity will form an image at a distance $s_i = nf =$

22.8 mm. If the image must be formed at a distance $s_i = 25$ mm, then the object distance can be determined by the simple model of the eye:

$$\frac{1}{s_o} = -\frac{n}{s_i} + \frac{1}{f} = -\frac{1.336}{0.025 \text{ m}} + \frac{1}{0.0171 \text{ m}} = 5.04$$
$$s_o = 0.20 \text{ m}$$

A small lengthening of the eye can give significant nearsightedness.

Nearsightedness can be corrected by a negative meniscus lens (see Figure 19-40), which is thinner in the center than at the edges. (If you wear glasses, examine them to see if the lenses thicken or thin at the center. Your eyeglass prescription also gives the correcting power in diop-

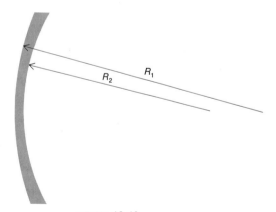

FIGURE 19-40
The diverging eyeglass lens.

ters.) A nearsighted eye, when relaxed, focuses at some short distance in front of the eye; distant objects are out of focus. The corrective lens brings the image of a distant object close to the eye so that it can be seen clearly: Parallel rays from a distant object will appear to be coming from the focal point of the negative corrective lens, as Figure 19-41 shows.

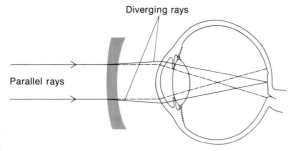

FIGURE 19-41
The nearsighted eye can be corrected by a diverging lens.
Courtesy of American Optical Corporation.

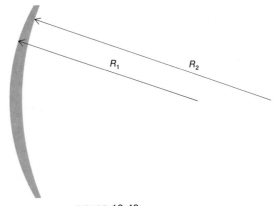

FIGURE 19-42
The converging eyeglass lens.

EXAMPLE

A nearsighted eye focuses clearly from a near point of 0.10 m to a far point of 0.80 m.
(a) What corrective lens is required to focus an object at infinity, $s_o = \infty$?
(b) What is the closest point that will be in focus with the corrective lens?

An object at infinity $s_o = \infty$ must form a virtual image in front of the negative lens, $s_i = -0.80$ m. The power of the corrective lens is

$$P = \frac{1}{f} = \frac{1}{s_o} + \frac{1}{s_i} = \frac{1}{\infty} + \frac{1}{-0.80 \text{ m}} = -1.25 \text{ diopters}$$

The near point is given by an image at $s_i' = -0.10$ m.

$$\frac{1}{s_o'} = \frac{1}{f} - \frac{1}{s_i'} = -1.25 - \frac{1}{-0.10 \text{ m}} = +8.75/\text{m}$$

$$s_o' = 0.11 \text{ m}$$

With the corrective lens, the eye can focus on a point from 0.11 m to ∞.

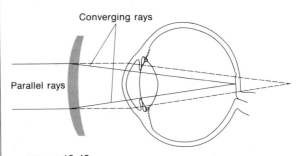

FIGURE 19-43
The farsighted eye can be corrected by a converging lens.
Courtesy of American Optical Corporation.

the retina of a relaxed eye, as shown in Figure 19-43. The accommodation of the eye will then focus closer objects.

Hypermetropia In farsightedness, **hypermetropia,** the image of a distant object in the relaxed eye falls behind the retina. The eye must accommodate even to focus on an object at infinity. Whereas a normal eye can focus to a near point of perhaps 20 cm, the near point of a farsighted eye at maximum accommodation is perhaps 5 m. Farsightedness can be corrected by a converging or positive meniscus lens (see Figure 19-42). The converging lens causes parallel rays from a distant object to be focused at

---

EXAMPLE

A farsighted eye focuses clearly from a near point of 4.0 m out to infinity. What corrective lens is required to focus on an object at 0.20 m?

A positive lens forms a virtual image of an object at 0.20 m at $s_i = -4.0$ m. The power of the lens is

$$P = \frac{1}{f} = \frac{1}{s_o} + \frac{1}{s_i} = \frac{1}{0.20 \text{ m}} + \frac{1}{-4.0 \text{ m}} = +4.75/\text{m}$$

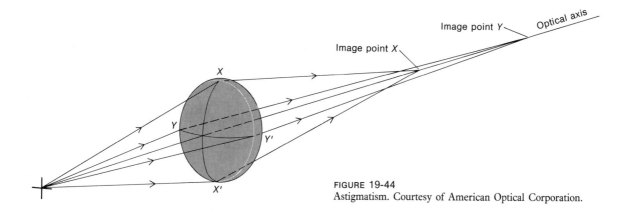

FIGURE 19-44
Astigmatism. Courtesy of American Optical Corporation.

Many people with otherwise normal eyes lose accommodating power with age as the maximum refraction of the eye decreases. For close work, special reading glasses are then required. If both near and far vision are needed, bifocal lenses, ground with two different corrections, or half-frame glasses are used, so that one can look through different parts of the glass for near and for far vision.

Astigmatism Thus far, we have considered only spherical lenses. But often the cornea or the retina is slightly aspherical. In one plane, say the horizontal, the cornea may have an outer radius of curvature of 7.7 mm; in the vertical, it may be 8.0 mm. Because the cornea is a primary refracting surface, rays passing through the vertical plane will be focused at a different point than those passing through the horizontal plane, as shown in Figure 19-44. The result is astigmatism: a blurred image that cannot be brought clearly into focus. The astigmatism, which can be oriented along any axis, is corrected by use of a cylindrical lens. Most eyeglasses have a combination of spherical and cylindrical correction. Some people whose eyes are otherwise fairly normal have significant astigmatism which requires correction.

Contact Lenses A small change in the curvature of the surface of the cornea, the primary refracting surface, gives significant correction. A **contact lens** is a small glass or plastic lens that is placed on the cornea. The interior radius of curvature closely matches that of the cornea. The space between the contact lens and the cornea is filled with moisture with an index of refraction of 1.33. The primary refracting surface, then, is the front surface of the contact lens. Contact lenses allow large corrections at no sacrifice of clear vision.

Figure 19-45 shows a nearsighted eye; total focusing power is 60 diopters: the cornea is 45 diopters and the eye lens, 15 diopters. Without correction (A), light from a distant object is focused in front of the retina because the eye is too long. With a contact lens, the focusing power of

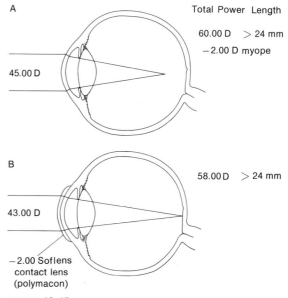

FIGURE 19-45
A nearsighted eye can be corrected with a flexible contact lens that reduces the focusing power of the cornea. A. A nearsighted eye with total power of 60 diopters and a length greater than 24 mm. B. The refraction of the cornea corrected with a −2 diopter contact lens. SOFLENS is a registered trademark of Bausch & Lomb, Inc. Courtesy of Bausch & Lomb.

the cornea is reduced to 43 diopters and light from a distant object is brought into focus on the retina (B). The improvement in vision with contact lenses can be striking, often much greater than is possible with eyeglass lenses.

Adapting to contact lenses entails some reshaping of the surface of the cornea. Returning to regular glasses requires a period of readaptation lasting several days.

Review

Give the significance of each of the following terms. Where appropriate give two examples.

Huygens' principle
particles
rays
geometrical optics
Newton's axioms
incidence
reflection
normal
Snell's law
index of refraction
dispersion
critical angle
internal reflection
light pipe
thin lens

converging lens
focal point
focal distance
lensmaker's equation
diverging lens
ray tracing
chief ray
nodal point
principal plane
real image
virtual image
thin-lens equation
linear magnification
diopters

chromatic aberration
spherical aberration
achromatic doublet
coma
astigmatism
objective lens
eyepiece lens
angular magnification
cornea
accommodation
myopia
hypermetropia
contact lens

Further Readings

R. L. Gregory, *Eye and Brain* (New York: McGraw-Hill, World University Library, 1966).
Ruth van Heyningen, "What Happens to the Human Lens in Cataract," *Scientific American* (December 1975).
Sir Isaac Newton, *Opticks* (New York: Dover Publications, 1952)

Problems

GEOMETRICAL OPTICS

1 A submerged scuba diver looks upward at the smooth surface of a lake and sees the sun at an angle of 30° to the normal. The index of refraction of water is 1.33. Where is the sun?

2 A man stands beside an empty swimming pool 4 m wide and 3 m deep. He can just see past the edge to the opposite bottom corner. If the pool were full of water, where would his line of sight strike the pool?

3 A fisherman is standing on the bank of a pond. A straight line from a fish to the fisherman makes an angle of 45° with the surface of the water (this is not the line of sight). Water has an index of refraction of 1.33. The fish sees the fisherman:
(a) above this line;
(b) below this line;
(c) only if the fish swims deeper; or
(d) only if the fish swims closer to the surface.

4 A person looks into a full swimming pool 2 m deep. The surface is smooth and the water has an index of refraction of 1.33. A coin at the bottom appears to be at an angle of 45° from the edge of the pool. Where is the coin?

5 Explain why light coming through an ordinary pane of window glass is not obviously refracted.

6 A ray is incident at an angle of 45° on a glass pane 2.5 cm thick with an index of refraction of 1.5. Trace the ray through the glass. Approximately how far will the ray be displaced?

7 The porpoise tank at an aquarium has a window 2.5 cm thick with an index of refraction of 1.5; the water has an index of refraction of 1.33. A porpoise is 2 m from the window. How far away does it appear to be?

8 A piece of optical glass has a critical angle of reflection of 40°. What is its index of refraction?

9 A diamond has an index of refraction of 2.44.
(a) What is the critical angle for internal reflection?
(b) How does internal reflection contribute to the sparkle of a diamond?

10 A glass prism with angles 45°–90°–45° is used for internal reflection. The glass has an index of refraction of 1.65.
(a) What is the critical angle for the prism in air?
(b) If the prism is placed in water, with an index of refraction 1.33, what is the critical angle? Will it still work as a reflecting prism for light incident normal to the prism?

11 Two plane mirrors intersect at an angle of 60°. A point is 10 cm from the intersection of the mirrors, halfway between them. Make a sketch showing the different images of the point.

12 Two plane mirrors intersect at 120°. A point is 20 cm from the point of intersection and 10 cm from one of the mirrors. Make a sketch.
(a) How many images are formed?
(b) Where are they?

13 A corner reflector consists of two mirrors intersecting at 90°. Show that any ray entering the corner mirror will be reflected along the direction of incidence. (A corner reflector consisting of many such small mirrors was left on the moon by an Apollo mission to serve as a target for bouncing laser beams off the surface of the moon.)

THIN LENSES

14 The cornea of the eye has a radius of curvature of 7.7 mm. The index of refraction of the cornea and of the medium behind the cornea is about 1.34. What is the approximate focal length due to refraction at the cornea?

15 Crown glass has an index of refraction of about 1.53. A plano-convex lens has a focal length of

33 cm. What is the radius of curvature of the convex surface?

16 Flint glass has an index of refraction of about 1.65. A double-convex lens has a focal length of 33 cm. If each surface has the same curvature, what is the radius of curvature?

17 A Leica M5 "35 mm" camera has a lens of 50 mm focal length and will focus to within 1.0 m.
(a) What is the image distance?
(b) What is the image height of an object 10 cm high at 1.0 m?

18 A Rollieflex SL 66 "$2\frac{1}{4} \times 2\frac{1}{4}$" single lens reflex camera has a lens of 80 mm focal length. An object 5 cm high is brought into focus at a distance of 16 cm in front of the lens.
(a) What is the image distance?
(b) How large is the image?

19 A 35 mm slide of dimension 24 mm × 36 mm is projected onto a screen at a distance of 10 m by a projector with a lens of 125 mm focal length.
(a) What is the distance from the slide to the lens?
(b) How big is the projected image?

20 A light bulb serves as the object for a lens with a focal length of 8.0 cm. The bulb is 36 cm from a screen.
(a) Where should the lens be placed to form a clear image?
(b) Is there more than one point of focus?

21 A lens has a focal length of 25 cm. At what object distance is the image size the same as the object size?

22 A 24 mm × 36 mm negative is placed in an enlarger with a lens of focal length 75 mm. The lens is a distance of 50 cm from the image plane.
(a) What is the distance from the negative to the lens?
(b) What is the size of the final negative image?
(c) What is the magnification of the enlarger?

23 A camera has a lens with a focal length of 75 mm. What is the power of the lens in diopters?

24 A closeup lens has a power of +4.0 diopters. What is the focal length?

25 An object is placed 20 cm behind a lens with a power of −5 diopters. Where is the image?

26 A nickel coin 2 cm in diameter is placed 5 cm in front of a lens with a focal length of 8 cm. Sketch the ray diagram showing the position of the image.
(a) Where is the image? Is it real or virtual?
(b) What is the magnification?
(c) If the nickel is moved 1.0 cm closer to the lens, does the image get larger or smaller? What is the new magnification?

OPTICAL INSTRUMENTS

27 An achromatic doublet magnifying lens has a 12x magnification with an image distance of 25 cm.
(a) What is the focal length of the lens?
(b) What is the working distance from the lens to the object?

28 A pocket magnifying glass has a focal length of 5 cm and an image at a distance of 25 cm.
(a) Where is the object plane?
(b) The reticule consists of a transparent screen with lines spaced 1.0 mm apart placed at the object plane and used to measure small objects. What is the apparent size of a 1.0 mm spacing when viewed through the magnifying glass?

29 A hand magnifying glass not only magnifies the image but places it at a greater distance from the eye. This is a definite benefit for older people who have become farsighted. A magnifying glass has a focal length of 10 cm and forms an image at a distance of 2 m.
(a) What is the distance to the object?
(b) What is the magnifying power of the lens under this condition?
(c) Make a sketch showing the formation of the final image.

30 A stamp collector's 5x magnifying glass produces an image at 25 cm. Make a sketch showing the object distance and image formation. What is the focal distance of the lens?

31 A microscope objective has a focal length of 4 mm. The image plane is 176 mm above the objective lens.
   (a) Where is the object plane?
   (b) What is the magnifying power of the objective?

32 A microscope objective has a focal length of 16 mm and a magnifying power of 10x.
   (a) Where is the image plane?
   (b) Where is the object plane?

33 A refracting optical telescope has a magnifying power of 150x, when the eyepiece of 30 mm focal length forms an image at a distance of 25 cm.
   (a) What is the focal length of the objective lens?
   (b) What is the distance between the objective and the eye lens?

34 A Galilean telescope is formed by an objective lens of 10 cm focal length and an eyepiece of -2.5 cm focal length. The eye lens forms a virtual image at 25 cm.
   (a) What is the separation of the lenses to focus an object at a distance of 20 meters?
   (b) What is the angular magnification?

35 A quarter is 24 mm in diameter.
   (a) If the quarter is held at arm's length, 30 cm, what is its angular size?
   (b) The moon has an angular diameter of about 0.5°. At what length must the quarter be held to be the same angular size as the moon?

36 When we look at the moon, which has an angular size of about 0.5°, we see the image on the retina of the eye. It appears large with considerable detail. If we take a photograph of the moon it appears surprisingly small.
   (a) If we take a photograph of the moon with a polaroid camera with a focal length of 117 mm, how large is the image?
   (b) If we look at the photograph at a distance of 25 cm, what is the angular size of this image?

37 Newton's telescope had a mirror with a radius of curvature of 63.5 cm. If the final image is viewed with an eyepiece of 2.5 cm focal length, what is the angular magnification of the telescope?

38 The Mt. Palomar telescope has a concave objective mirror of diameter 5.08 m, with a radius of curvature of 57 m.
   (a) What is the focal length of the objective mirror?
   (b) What is the angular magnifying power when used with an eyepiece of 24 mm focal length?

39 The Questar reflecting telescope has a mirror with an effective focal length of 1400 mm. When viewed through an eyepiece of 24 mm focal length, what is the angular magnification?

40 A convex spherical mirror causes light rays to diverge. Parallel rays will appear to be focused at a point behind the surface a distance $f = -R/2$. Like a diverging lens, it has a negative focal length. A round Christmas tree ornament has a diameter of 6.0 cm.
   (a) What is the effective focal distance of the spherical ornament?
   (b) A child looks into the ornament and sees its own image. Show by ray tracing that the image is virtual and behind the surface.

41 A close look reveals a light reflection in the cornea of the eye. The "apple of the eye" is the reflection of the sky. The cornea has a radius of curvature of 7.7 mm.
   (a) What is the focal length of the reflecting surface of the eye?
   (b) A 100 watt light bulb 5.5 cm across is 50 cm from the cornea. Where is the image? How big is the image?

OPTICS OF HUMAN VISION

42 The cornea of the eye has a radius of curvature of 7.7 mm and an index of refraction of 1.376. A person is swimming underwater with an index of refraction of 1.33.
   (a) What is the focusing power of the cornea underwater?
   (b) Can one see clearly underwater? Why or why not?
   (c) Explain why one can see clearly with a face mask.

43 An eye has an effective focal length of 20 mm. An optometrist prescribes eyeglasses with a correction of -2.0 diopters.
(a) Is the eye myopic or hypermetropic?
(b) What is the eye's effective focal length with the corrective lens?

44 A nearsighted eye will focus clearly at a distance of 1.5 m. What lens is required to correct the eye to see clearly at infinity?

45 A nearsighted eye can see clearly from distances of 6 cm to 20 cm.
(a) What lens is required so that the eye can see a distant object clearly?
(b) What is the closest object that can be seen clearly with the corrective lens?

46 A myopic eye has a normal refracting power of 58 diopters but requires a correction of -20 diopters to see distant objects clearly.
(a) Using the simple model for the eye optics (equation 19-12), what is the approximate image distance for the eye?
(b) What is the greatest distance the uncorrected eye can see clearly?

47 A person with normal vision can see clearly from 25 cm to infinity. The normal refracting power of the relaxed eye is 70 diopters. Using the simple model for the eye optics (equation 19-12), what is its accommodation power?

48 A farsighted person can see clearly at a distance of 2 m and beyond. What corrective lens is required to allow this person to see clearly at a distance of 0.25 m?

49 A person with normal eyes can see clearly at infinity. The normal refracting power of the relaxed eye is 58 diopters. The effective image distance is 23 mm. Because of age the accommodation power is reduced to 3 diopters. What is the closest distance the person can see?

ADDITIONAL PROBLEMS

50 An aquarium with parallel glass sides 3 mm thick and 20 cm apart front to back is filled with water. The glass and water have an index of refraction of 1.5 and 1.33 respectively. A ray of light is incident at an angle of 30° to the normal of the glass.
(a) What is the angle of refraction in the glass?
(b) What is the angle of refraction into the water?
(c) Approximately where does the ray leave the far side of the aquarium?
(d) With what angle does the ray leave the far glass?

51 An isosceles prism has an apex angle of 60°. The prism is made of high-dispersion crown glass as described in Table 19-2. A ray is incident at an angle so that yellow light is refracted at an angle of 30° to the normal.
(a) What is the angle of incidence to the first face?
(b) What is the total deviation of deep red light after two refractions?
(c) What is the total deviation of violet light after two refractions?

52 An eye with a cornea with a 7.7 mm radius of curvature has an optical refracting power of 70 diopters. The image distance of the eye is 25 mm and the interocular fluid has an index of refraction of 1.336. Using the simple model for the refraction of the eye,
(a) At what distance will the eye focus?
(b) What is the refracting power of the outer surface of the cornea?
(c) What refraction power is necessary for the eye to see clearly at infinity?
(d) If the refraction of the cornea is replaced with the refraction of the front surface of a contact lens with an index of refraction 1.40, what outer radius of curvature of the contact lens is required to give clear vision?

53 A closeup lens with a power of $+2$ diopters is placed in front of a camera lens with a focal length of 50 mm. The maximum distance between the lens and the film plane is 53 mm.
(a) What is the closest distance to which the camera lens can focus without the closeup lens?

(b) What is the combined focal length of the two lenses?

(c) What is the closest distance to which the camera lens can be focused with the closeup lens?

54 A sphere of 5.0 cm radius is made of glass with an index of refraction of 1.65. A point source of light at infinity produces an image because of refraction at the first surface. The image produced by the first surface becomes the object for refraction by the second surface. Make a sketch tracing the rays.

(a) Where is the first image?

(b) Where is the final image? Is it erect or inverted?

55 A guppie is at the center of a spherical fishbowl of radius 20 cm. The water has an index of refraction of 1.33 (neglect the refraction of the glass).

(a) Where is the image of the guppie?

(b) Is the image real or virtual?

56 A guppie at the center of a spherical fishbowl with thin walls is eyeing his sweetie at the center of an identical fishbowl situated close to but not touching his. To him she appears

(a) erect, or inverted?

(b) smaller than life, larger than life, or actual size? Make a sketch to justify your answer.

57 An aplanatic oil-immersion objective has an effective focal length of 1.8 mm. The lens forms an image at a distance of 180 mm from it. A red blood cell has a diameter of 8 μm.

(a) How large is the image of the red blood cell formed by the objective lens?

(b) If this image of the red blood cell is viewed with a 10x eyepiece, what is the apparent size of the final image?

20

Wave Optics

INTERFERENCE

Isaac Newton

Newton's geometrical optics were based on his particle theory of light and successfully explained the refraction and reflection of light. However, many optical phenomena that Newton described in his *Opticks* he could not adequately explain. He observed that light reflected from soap bubbles and from other thin films, such as oil on water and the oxide formed on metals when heated, showed the regular sequence of rainbow colors. Newton also observed unusual dark bands in light reflected from two glass surfaces.

If two optically flat glass plates (such as microscope slides) are placed on top of one another in contact at one edge, and separated by a small distance at the other edge (by, for instance, placing a hair or a fine sewing thread between them), a small wedge of air is formed, as is shown in Figure 20-1. If, say, yellow light is directed at the plates from above the pair, it is reflected from the two surfaces on either side of the wedge, but not at the point where the two surfaces touch; there a narrow dark band is observed. However, as the separation of the two surfaces increases, there is a small reflection from both the lower surface of the top plate and the upper surface of the bottom plate, and a series of bright and dark bands are observed as shown in Figure 20-1. If viewed in white light, a series of bands fringed with colors are observed.

Newton refined these observations by placing a large

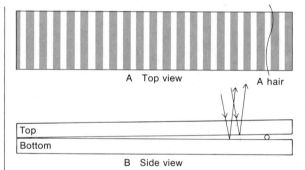

FIGURE 20-1
When viewed in reflected yellow light, two glass plates separated at one edge by the thickness of a fine hair show a series of bright and dark bands.

plano-convex lens in contact with an optically flat surface, both illuminated from above as shown in Figure 20-2. The reflected light, also viewed from above, shows a series of rings, **Newton's rings,** strikingly shown in Figure 20-3. The dark spot at the center is where the surfaces are in contact and no light is reflected.

Newton, believing that light acted as a particle (for if it were a wave, it would bend around obstacles and soften shadows), attempted to explain these puzzling phenomena: Perhaps the light particles excite vibrations in the refracting or reflecting surfaces. His explanation was not convincing.

Distant light source Observer

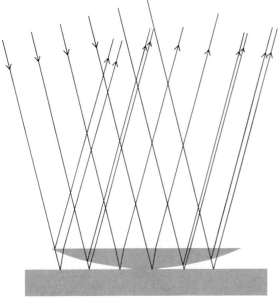

FIGURE 20-2
Newton's experiment for observing rings. A film of air lies between a plane surface and a convex lens surface. Light from a distant source is reflected by these two surfaces, reaches the observer, and interferes constructively or destructively, giving bright and dark bands.

FIGURE 20-3
Newton's rings formed by the interference in air between reflections from a convex and a plane surface. Courtesy of Bausch & Lomb.

Thomas Young

Although Newton could not give a satisfactory explanation to these phenomena, his logic and experimental approach were correct. And his *Opticks*, almost a hundred years after its publication, inspired the English physician, Egyptologist, and scientist Thomas Young to perform new experiments with light. Young's first experiment was one of Newton's: He studied a beam of sunlight that entered a darkened room through a pinhole. Observing the light striking a card held at different distances from the pinhole, Young found color fringing, both within the bright spot and into the shadow beyond the edge of the spot. Light did indeed bend around obstacles, but only slightly. Young postulated that the color fringing of the narrow beam of sunlight was produced by **interference** of two portions of light. Light was a wave: Different light

waves arriving out of phase would interfere with one another, canceling at some points and reinforcing at others.

Young's Double-Slit Experiment

The best-known and most convincing demonstration of the interference of light waves is **Young's double-slit experiment.** Parallel rays of light from a distant pinhole fall on a pair of narrow slits separated by some distance d. If light acts as a wave, each slit will act as a source of waves. The waves spread outward from each slit to a screen at a large distance D. Young postulated that the light rays from different slits would interfere with one another. Figure 20-4 shows a drawing similar to Young's representing the interference of waves from two slits.

FIGURE 20-4
Thomas Young explained interference among light waves by analogy with water waves. Where the wave crests from both slits travel together and are therefore in phase there is maximum intensity. At the right is a plot of the resulting wave intensity. From Gerald Feinberg, "Light." Copyright © 1968 by Scientific American, Inc.

As can be seen in Figure 20-5, light from the two slits travels slightly different distances to reach a point P on a screen. If the **path difference** is an integer number of wavelengths, the waves will arrive in phase and add, giving a maximum intensity, or *constructive interference*. If the path difference equals an odd number of half wavelengths, the waves arrive out of phase and cancel, giving a minimum intensity, or *destructive interference*. By simple trigonometry, the path difference Δs is

$$\Delta s = d \sin \theta$$

where d is the spacing of the slits.

The condition for *constructive* interference is that the path difference is an integer number of wavelengths:

$$\Delta s = n\lambda = d \sin \theta_n \qquad (20\text{-}1)$$

where $n = 0,1,2,3, \ldots$ The integer n is the **order** of the interference maximum and can be positive or negative (first order, $n = \pm 1$; second order $n = \pm 2$; etc.).

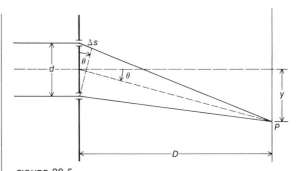

FIGURE 20-5
Young's double-slit experiment. The path difference between the two slits is $\Delta s = d \sin \theta$.

The condition for *destructive* interference is that the path difference is an odd number of half wavelengths:

$$\Delta s = \left(m - \frac{1}{2}\right)\lambda = d \sin \theta_m \qquad (20\text{-}2)$$

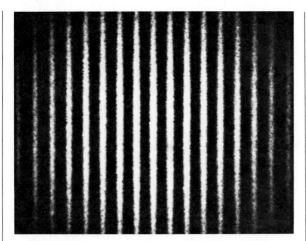

FIGURE 20-6
Interference fringes photographed from a double-slit experiment similar to Thomas Young's. Courtesy of Brian J. Thompson, the University of Rochester.

where $m = 1,2,3, \ldots$ Figure 20-6 shows a photograph of the interference pattern produced by two narrow slits, confirming Young's theory.

Newton's Rings

Young gave a detailed explanation of Newton's ring experiment based on the concept of interference. Figure 20-7 shows the Newton's ring experiment as viewed in cross section. Light is incident from above. Some of the incident light will be reflected both at the curved surface and, after crossing the space beneath that surface, at the plane surface. The distance between the convex glass, of radius of curvature R, and the plane glass at a distance r from the point of contact is just

$$\Delta s = R - (R^2 - r^2)^{1/2} = R - R(1 - r^2/R^2)^{1/2}$$

If $r/R \ll 1$, we can approximate the square root by

$$(1 - r^2/R^2)^{1/2} = 1 - \tfrac{1}{2}(r^2/R^2)$$

For small values of r, the distance between surfaces Δs becomes

$$\Delta s = r^2/2R$$

The radius of the convex lens R is typically 1 m, while the rings have a radius r of a few to several millimeters. The air gap between the surfaces is quite small.

A light wave from a distant source is reflected at two surfaces, as shown previously in Figure 20-2: The first reflected wave is produced at the lower surface of the convex lens at the glass-to-air boundary; the second reflected wave is produced at the plane surface. At the second reflection, the wave traveling through the air is reflected from the glass and is inverted (corresponding to a 180° phase shift), as shown in Figure 20-8.

Because of this phase shift, if the total path difference between the two reflected waves $2\Delta s$ is an integer number of wavelengths,

$$2\Delta s = m\lambda$$

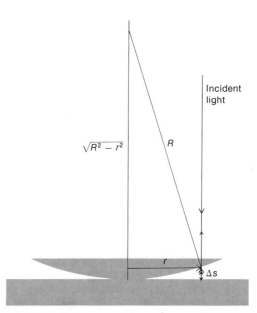

FIGURE 20-7
Newton's ring experiment. The thickness Δx depends on the radius of curvature R and the radius r from the center.

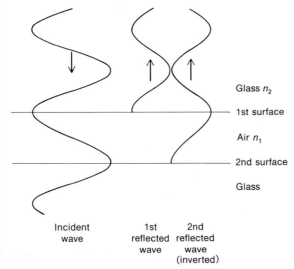

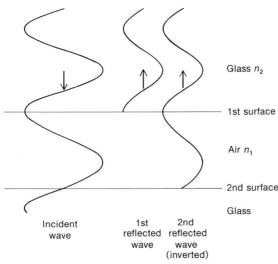

FIGURE 20-8
Destructive interference of two reflected waves. If the total path difference $2\,\Delta s = m\lambda$, the two reflections cancel, giving a dark band. Note that the second reflection going from air to glass is inverted.

FIGURE 20-9
Destructive interference of two reflected waves. If the total path difference $2\,\Delta s = (n + \frac{1}{2})\lambda$, the two reflections add, giving a bright band. Note that the second reflection going from air to glass is inverted.

the two reflections cancel exactly (destructive interference). In Figure 20-8, the path difference is one wavelength and the two reflected waves cancel, giving a dark band. At the point where the two glass surfaces are in contact, the path difference is zero; the two reflections cancel exactly, and the central point of contact is dark.

If the total path difference between the two reflected waves $2\Delta s$ is an odd number of half wavelengths, then the reflected waves arrive in phase and add (constructive interference). In Figure 20-9 the path difference is $\frac{3}{2}$ wavelength and the waves arrive in phase, giving a bright band:

$$2\Delta s = (n + \tfrac{1}{2})\lambda = 2r_n{}^2/2R$$

The radius of the nth bright ring is given by

$$r_n{}^2 = (n + \tfrac{1}{2})\lambda R \qquad (20\text{-}3)$$

If an optically flat surface and a convex lens of radius R are illuminated by a distant light source, we can determine the wavelength of light by counting and measuring the radii of the bright rings.

Thomas Young not only used the Newton ring experiment to validate the wave theory of light, but also, from Isacc Newton's own data, computed the wavelengths of different colors of visible light, obtaining 424 nm in the extreme violet to 705 nm in the extreme red. Although Young successfully extended Newton's investigations of light, his contemporaries saw his work as an attack on Newton and refused to accept his results. The wave theory of light was not to be accepted until, in 1816, Augustin Fresnel presented his own wave theory before the French Academy of Science. Fresnel's theory was challenged by the mathematician Simeon Poisson, who asserted that Fresnel's theory could not possibly be true, for it made a ridiculous prediction: If a circular obstruction is placed in the path of a beam of light, at a proper distance from a screen, then light is diffracted around the obstacle and a bright spot would appear in the exact center of the shadow. This seemed highly unlikely. Subsequently the experiment was performed by Dominique Arago (and presumably also by Fresnel) and the bright spot, shown in Figure 20-10, was observed. Fresnel's theory, developed to explain one set of phenomena, pre-

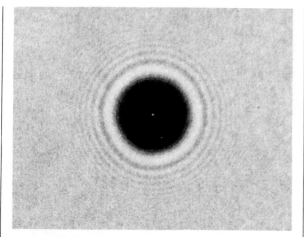

FIGURE 20-10
Because of the diffraction of light at a certain distance, the center of the shadow cast by an opaque disk shows a small bright spot (variously called Fresnel's, Arago's, or Poisson's spot). Courtesy of Brian J. Thompson, the University of Rochester.

dicted another, totally unsuspected and previously unrecorded phenomenon. Fresnel's work also conclusively established the wave nature of light.

DIFFRACTION

Multiple-Slit Diffraction

Diffraction is the bending of light around an obstacle. A series of narrow **multiple slits** will produce a diffraction pattern. If light illuminating the slits comes from a distance, and if the image produced by the slits is viewed at a large distance, the rays diffracted from each slit are parallel. Often the incident light, as Figure 20-11 shows, comes from a narrow slit at the focal point of a converging lens.

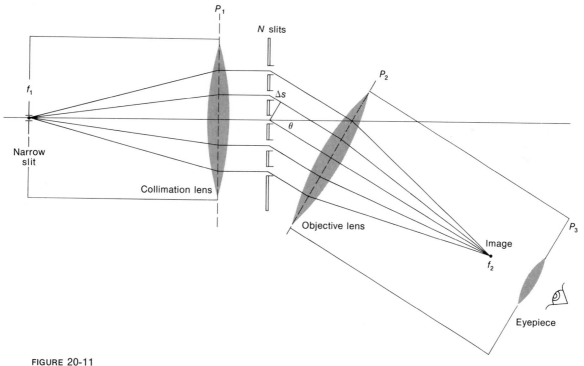

FIGURE 20-11
Fraunhoffer diffraction from multiple slits. A collimating lens gives parallel rays from a narrow slit. A telescope collects the diffracted rays to form the diffraction image of the slit. The greater the number of slits, the sharper is the diffraction image.

The final image is formed at the focal point of a second lens. At the focus, the light waves diffracted by all the slits interfere with one another.

If the path difference Δs of the light waves from adjacent slits is an integer number of wavelengths $n\lambda$, the condition for constructive interference is

$$\Delta s = n\lambda = d \sin \theta$$

where d is the spacing between adjacent slits, θ is the angle to the nth interference maximum, and n is the order of the maximum ($n = 0,1,2,\ldots$). The position of the first-order diffraction maximum ($n = 1$) is simply given by

$$\lambda = d \sin \theta$$

The condition for constructive interference in the multiple-slit experiment is the same as in Young's double-slit experiment. Although the positions of the maxima do not change, the multiple slits make each maximum much sharper. Light from the multiple slits cancels in the region between the diffraction maxima. Figure 20-12 is a photograph of the diffraction of a small beam of light produced by two through six slits respectively. As the number of slits increases, the diffraction image gets sharper as more of the light intensity is concentrated in narrow interference maxima. Figure 20-13 shows the light intensity of the multiple-slit diffraction image.

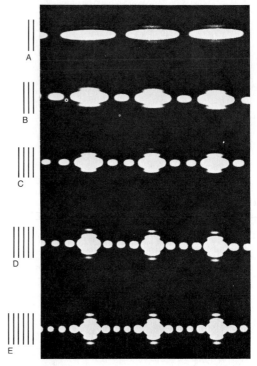

FIGURE 20-12
Multiple-slit interference pattern produced by 2, 3, 4, 5, and 6 slits. The greater the number of slits, the narrower are the principal diffraction maxima. Courtesy of Brian J. Thompson, the University of Rochester.

Diffraction Gratings

Any surface that has a set of identical, parallel, equally spaced slits, lines, or grooves can be used as a **diffraction grating.** In 1882, Henry A. Rowland, Professor of Physics at John Hopkins University, brought the art of ruling diffraction gratings to a state unsurpassed until recent years (see Figure 20-14). His gratings had 568 lines/mm. Modern high-precision diffraction gratings are ruled on carefully selected optical blanks, which are polished to be optically flat within $\frac{1}{10}$ the wavelength of green light, and then are coated with aluminum or gold. Precise rulings are made in the optical surface with a diamond stylus that has been carefully shaped. The spacing between the grooves of a precision grating is accurate to 10 nm. A grating with an area as large as 20 cm $\times$ 30 cm, with as many as 1200 lines/mm, may take as long as 4 weeks to rule. High resolution can be obtained by the use of diffraction gratings with very many ruled slits. Gratings are now commercially available with 20 lines/mm, 300 lines/mm, 1200 lines/mm, and even 3600 lines/mm.

Figure 20-15 shows a schematic profile of grooves ruled in a typical diffraction grating. Light incident at an angle α to the grating normal is diffracted into some angle β. If d is the spacing between the grooves, the difference in

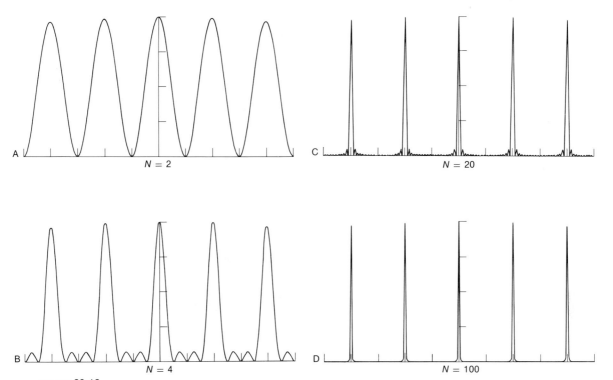

FIGURE 20-13
The intensity of the multiple-slit diffraction image as a function of distance from the central maximum, for $N = 2$, 4, 20, and 100 slits. The greater the number of slits, the sharper is the diffraction maximum. The position of the maximum does not change.

FIGURE 20-14
Henry A. Rowland with his ruling engine at Johns Hopkins University in the 1890s. Courtesy of The American Institute of Physics, Niels Bohr Library.

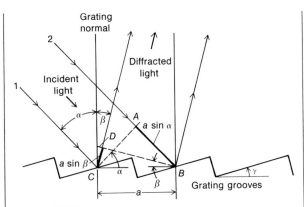

FIGURE 20-15
A ruled grating. The path difference between rays diffracted
from two adjacent grating grooves depends on the angle of
incidence α and of diffraction β, as shown. The grating
grooves are cut at an angle γ known as the blaze angle. Courtesy of Bausch & Lomb.

optical paths between light from adjacent slits is

$$\Delta s = d \sin \alpha - d \sin \beta = n\lambda \qquad (20\text{-}4)$$

Constructive interference (a diffraction maximum) occurs
when the path difference is equal to $n\lambda$.

Many modern diffraction gratings are ruled so that the
surface of the grooves makes an angle γ, the **blaze angle,**
to the surface of the grating itself. If the grating is turned
to the proper angle it will blaze—that is, it will strongly
reflect light toward the observer, increasing the amount of
light thrown into the diffraction image. If light is incident
normally to the diffraction grating, $\alpha = 0$, then the angle
of mirror reflection is $\beta = 2\gamma$. This angle will give a
diffraction maximum at a particular wavelength. Because
this angle corresponds to the mirror reflection, the image
will be quite bright, increasing the sensitivity of the
grating spectrometer.

EXAMPLE

Light is incident on a grating at normal incidence
$\alpha = 0$. The grating has 1200 lines/mm and a blaze
angle $\gamma = 17°$. For what wavelength is the first-order

diffraction maximum at an angle equal to twice the
blaze angle $\beta = 2\gamma$?

The condition for diffraction maximum is given by
equation 20-4:

$$n\lambda = d \sin 0 - d \sin 2\gamma$$

Because of the minus sign, we must choose for the
first-order maximum $n = -1$. The wavelength is then
given by

$$\lambda = \frac{1 \text{ mm}}{1200} \sin (2 \times 17°) = \frac{(10^{-3} \text{ m})(0.559)}{1200}$$

$$\lambda = 4.66 \times 10^{-7} \text{ m} = 466 \text{ nm}$$

At this wavelength, the first-order diffraction maximum
also corresponds to the bright reflection from the mirror surface.

Original gratings are quite expensive, and replica grating are quite acceptable for many applications. A replica
is made by applying a transparent resin to the original
grating, peeling it off, and mounting it on a glass plate.
Many replica gratings can be made from a single master
grating. Transparent replica gratings are usually used with
the light at normal incidence $\alpha = 0$, and the diffracted
light is transmitted through the grating at some angle θ,
just as with the multiple slits discussed previously. The
condition for a transmission grating diffraction maximum
is simply

$$n\lambda = d \sin \theta$$

where θ is the angle of diffraction maximum.

Czerny-Turner Spectrometer

Figure 20-16 shows a schematic of the widely used
Czerny-Turner spectrometer: a narrow entrance slit,
adjustable from 5 to 2000 μm in width; two concave
spherical mirrors, one collimating, one focusing; a diffraction grating; and an exit slit, behind it a light detector.

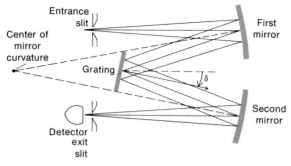

FIGURE 20-16
The geometry of the Czerny-Turner spectrometer. The entrance slit is placed at the focus of the first spherical mirror, parallel rays from which strike the grating. The parallel diffracted rays are collected by the second mirror and are focused on the exit slit. Light leaving the exit slit is detected. The grating angle determines the wavelength of the diffraction maximum.

The entrance slit is located at the focal distance of the collimating mirror, which focuses parallel light onto the grating (typically 10 cm by 10 cm and having from 50 lines/mm to 2400 lines/mm). The grating surface is at some angle δ to the axis of the spectrometer. The focusing mirror collects the parallel rays from the grating and focuses them on the exit slit, where the light then falls on a suitable light detector. Changing the grating angle δ changes the wavelength of the diffraction maximum at the exit slit. Typically, the grating can be rotated through an angle δ of 0° to 50°. Using a grating with 1200 lines/mm, a precision spectrometer with a 1.0 m optical path length can achieve a resolution of 0.01 nm in first order, making the Czerny-Turner spectrometer a sensitive analytical instrument.

DIFFRACTION FROM A SINGLE APERTURE

Single-Slit Diffraction

Light through a single slit is also diffracted. Both Newton and Young had observed the color fringing of light through a small pinhole. In **single-slit diffraction,** each point in the aperture is considered a source of waves. Viewed on a screen that is a large distance from the

aperture, the pattern is somewhat larger than the geometrical image of the aperture.

Figure 20-17 shows the path of diffracted rays from two different portions of a single slit. The path difference between a ray from the edge of the slit and a ray from the center of the slit of width w is

$$\Delta s = \frac{w}{2}\sin\theta$$

If the path difference is half a wavelength of light, the two waves cancel. This cancellation gives us the condition for the first diffraction minimum.

$$\Delta s = \frac{\lambda}{2} = \frac{w}{2}\sin\theta$$

This difference in paths applies for every pair of points across the aperture. The condition for the first minimum in single-slit diffraction is

$$\lambda = w\sin\theta \qquad (20\text{-}5)$$

Figure 20-18 shows the intensity of the single-slit diffraction pattern. And Figure 20-19 shows a photograph of the diffraction pattern produced at a large distance from a long, narrow single slit. Amplitude declines as distance from the central maximum increases.

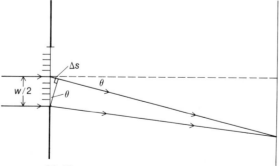

FIGURE 20-17
Diffraction from a single aperture. The path difference between one edge and the center of the slit is given by $\Delta s = \dfrac{w}{2}\sin\theta$.

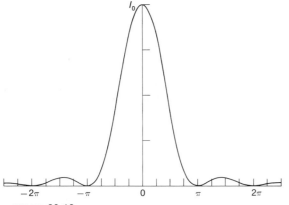

FIGURE 20-18
The single-slit diffraction pattern. The intensity is given by

$$I = I_0 \left(\frac{\sin \alpha}{\alpha} \right)^2 \text{ where } \alpha = \pi w \sin \theta / \lambda.$$

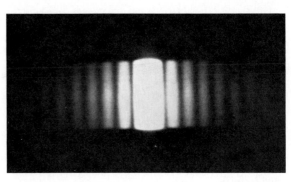

FIGURE 20-19
A single-slit diffraction pattern. Courtesy of Brian J. Thompson, the University of Rochester.

FIGURE 20-20
The combined effect of single-slit and multiple-slit diffraction. Intensity I as a function of $x = \pi w \sin \theta / \lambda$; $d/w = 4$.

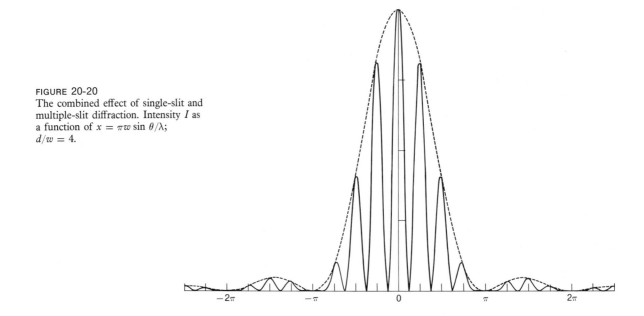

A diffraction grating consists of a set of N slits of width w and spacing d. The diffraction pattern produced by a multiple-slit grating consists of the N slit interference pattern superimposed on the single-slit diffraction pattern p produced by each slit. Figure 20-20 shows the intensity of the multiple-slit pattern, the single-slit pattern, and the resulting pattern caused by superposition.

The Rayleigh Criterion

The resolving power of a lens system such as of a telescope, a microscope, or the eye is limited by the diffraction pattern produced by light waves entering the aperture. Figure 20-21 shows light from a distant object such as a star incident on a telescope aperture. The diffraction

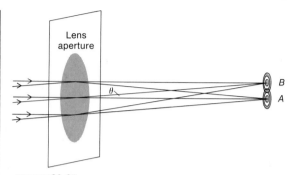

FIGURE 20-21
Diffraction images of two distant stars clearly resolved. Rayleigh's criterion gives the minimum angular separation θ for two images to be clearly resolved.

image of the star (neglecting the star's apparent motion, caused by atmospheric turbulence) is at point A. The image of a second star falls at some point B. If the two images are too close, they merge into a single spot. The angular separation needed for two point sources to be resolved clearly is given by **Rayleigh's criterion:** The diffraction maximum of the second image must fall no closer to the first image than the first diffraction minimum. If the diffraction images are too close together, as shown in Figure 20-22 A and B, they are unresolved; if separated as in Figure 20-22 D, they are clearly resolved. Figure 20-22 C shows the images just barely resolved as given by Rayleigh's criterion.

Light waves from each point of a distant source passing through a **circular aperture** will each arrive at a position on the screen with different phases, and consequently will interfere. The angular size to the first minimum—that is, the angular width of the central maximum—is given by

$$d \sin \theta = 1.22 \lambda \tag{20-6}$$

where d is the diameter of the opening. Figure 20-23 shows a photograph of the diffraction pattern produced by a circular aperture. The angular resolution of two point sources, as given by Rayleigh's criterion, is

$$\sin \theta = \frac{1.22 \lambda}{d}$$

EXAMPLE
The resolution of the eye is limited by the spacing of the photoreceptors on the retina. In the fovea of the retina, the receptors are separated by a distance of about 2 μm. The eye being about 17 mm long, this separation distance on the fovea corresponds to an angular separation of about 25 seconds of arc. To be resolved, two points of light should be separated by one minute of arc or greater. For light of wavelength 510 nm, at what pupil diameter will the Rayleigh criterion for resolution equal one minute of arc?

$$d = \frac{1.22 \lambda}{\sin \theta}$$

Since the angle is small, its sine is approximately equal to the angle in radian measure:

$$\sin \theta \cong \theta = (1') \frac{1°}{60'} \times \frac{\pi \, \text{rad}}{180°}$$

$$\sin \theta = 2.91 \times 10^{-4}$$

$$d = \frac{(1.22)(510 \, \text{nm})}{2.91 \times 10^{-4}} = 2.1 \times 10^{-3} \, \text{m}$$

For diameters less than 2 mm, the resolution of the eye becomes diffraction limited.

The eye can change the pupil aperture from 2 mm to 6 mm to accommodate varying brightnesses without sacrificing resolving power. At larger apertures there is greater chromatic and spherical aberration of the eye, but a narrower diffraction limit. The two effects tend to compensate and the resolution remains about constant at one minute of arc for the normal eye. The ability to resolve two points on an object is called visual acuity, commonly tested by eye charts displaying letters or symbols. The separations required to distinguish between letters subtend a visual angle of one minute of arc when viewed from a distance of 20 ft. A person with 20/40 vision can clearly see at 20 ft letter separations that subtend one minute of arc at a distance of 40 ft.

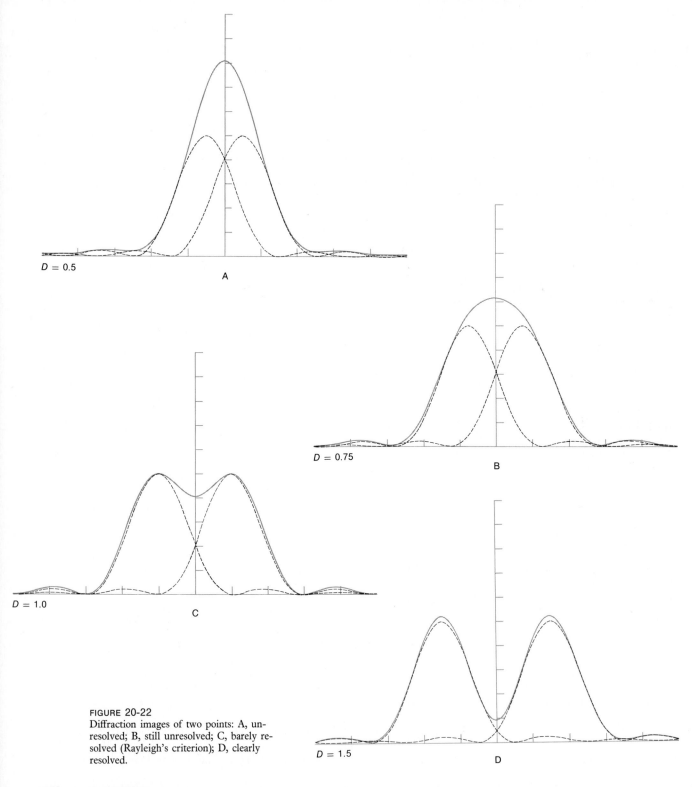

$D = 0.5$

A

$D = 0.75$

B

$D = 1.0$

C

$D = 1.5$

D

FIGURE 20-22
Diffraction images of two points: A, un-resolved; B, still unresolved; C, barely re-solved (Rayleigh's criterion); D, clearly resolved.

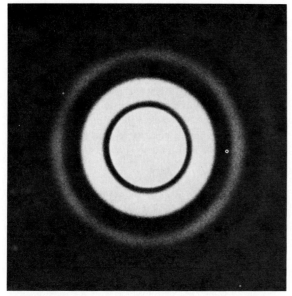

FIGURE 20-23
The diffraction pattern produced by a circular aperture viewed a large distance from the aperture. Courtesy of Brian J. Thompson, the University of Rochester.

Resolution of the Light Microscope

As Figure 20-24 shows, the diffraction pattern produced by the circular aperture of the light microscope's objective lens limits the microscope's resolving power. If the lens has an angular aperture 2α and a focal length f_o, the diameter of the aperture is

$$d = 2f_o \sin \alpha$$

The angular separation θ between two points O and O' near the focal point of the objective lens separated by a

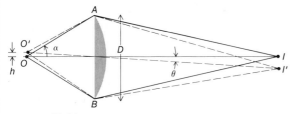

FIGURE 20-24
A microscope's resolving power is limited by the diffraction image produced by the objective aperture.

distance h, viewed through a medium with index of refraction n, is

$$\sin \theta = \frac{h}{f/n}$$

The minimum separation h is just

$$h = \frac{f \sin \theta}{n}$$

The Rayleigh criterion states that the minimum angular separation θ is given by

$$\sin \theta = \frac{1.22\lambda}{d}$$

Combining the last two expressions gives the minimum separation

$$h = \frac{f(1.22)\lambda}{nd} = \frac{1.22\,\lambda}{2n \sin \alpha}$$

The **numerical aperture,** NA, of a microscope is a measure of the light-gathering power of the objective lens, and is stated on each objective lens:

$$\text{NA} = n \sin \alpha$$

The minimum separation distance that can be resolved as two distinct points of light depends on the wavelength and numerical aperture:

$$h = \frac{1.22\,\lambda}{2\,\text{NA}}$$

Dry objective lenses have numerical apertures ranging from 0.12 to 0.85, as shown previously in Table 19-3. Aplanatic oil-immersion lenses have a numerical aperture of about 1.3. The region between the lens and the thin cover glass is filled with an oil with an index of refraction $n = 1.5$, the same as that of the lens. Because of its special design, as discussed in Chapter 19, the aplanatic objective also has a large angular aperture. The maximum resolution for two points is about one-half the wavelength of the light used. Because the eye is most sensitive at a wave-

FIGURE 20-25
Theoretical limits of light microscopy. The diatom *Amphipleura Pellucida,* the most difficult of all diatoms to resolve, is shown here photographed through an interference contrast microscope. This photomicrograph approaches the theoretical limit of the light microscope. Courtesy of Carl Zeiss, Inc., New York.

length of 510 nm, the best resolution obtainable with a visible-light microscope is about 250 nm. A red blood cell, having a diameter of about 10 000 nm, is clearly visible. However, cell walls and organelles have a thickness of about 7 nm and cannot be resolved in any detail. Magnification can be increased, but beyond some point the image simply gets larger with no increase in resolution. Figure 20-25 shows a photomicrograph of the diatom *Amphipleura Pellucida* at a resolution that approaches the theoretical limit of the light microscope.

EXAMPLE

A 100x oil-immersion objective lens has a numerical aperture of 1.25 and a 10x eyepiece. The object is viewed in yellow light with a wavelength of 590 nm. What is the minimum observable separation if the resolution is diffraction limited? How large will this separation appear?

The minimum separation is

$$h = \frac{(1.22)(590 \text{ nm})}{2(1.25)} = 343 \text{ nm}$$

The apparent size of the final image is

$$h'' = M_o M_e h = (100)(10)(343 \text{ nm})$$

$$h'' = 0.343 \text{ mm}$$

An image size of 0.34 mm at a distance of 25 cm is clearly distinguishable to the eye. The image could be magnified further but with no improvement in detail.

WAVE OPTICS

Index of Refraction

A light ray incident on the interface between two regions with different indexes of refraction is refracted, as shown in Figure 20-26. Snell's law gives the refraction.

$$n_1 \sin \theta_1 = n_2 \sin \theta_2$$

In the optically denser medium, the rays are refracted toward the normal. Huygens was the first to show that this refraction is caused by a change in wavelength as the ray enters the medium.

If parallel wave fronts are incident at an angle θ_1 in region one, the distance along the interface between the points of contact of successive waves s is given by

$$\lambda_1 = s \sin \theta_1$$

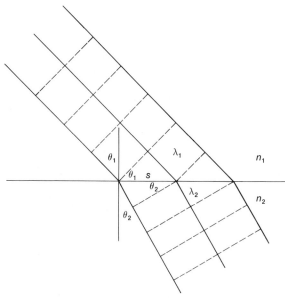

FIGURE 20-26
The velocity of propagation of light is slower in a region with a larger index of refraction n. $v = c/n$. In going from a less optically dense medium n_1 to a more optically dense medium n_2, where $n_2 > n_1$, the wavefronts are refracted toward the normal. The reflected waves are phase shifted by 180°.

In the second region this distance is given by

$$\lambda_2 = s \sin \theta_2$$

Eliminating the sines from the above equations leaves

$$n_1\lambda_1 = n_2\lambda_2$$

The wavelength in the optically denser medium is shorter, but the wave's frequency is unchanged in going from one medium to another. The velocity of propagation in vacuum, where $n_1 = 1.0$, is just c:

$$v_1 = \lambda_1 f = c$$

In the second region, the speed of the wave is

$$v_2 = \lambda_2 f = \frac{\lambda_1 f}{n_2} = \frac{c}{n_2}$$

The speed of light in the medium is equal to the speed of light in a vacuum divided by the **index of refraction.**

Polarization

Until the nineteenth century, proponents of the wave theory of light assumed that the wave motion was longitudinal; that is, the displacement of the medium was in the direction of propagation. In 1820, Fresnel suggested that light was a transverse wave; that is, the displacement was perpendicular to the direction of travel.

Recall, from Chapter 18, that James Clerk Maxwell was able to demonstrate that electromagnetic waves had all of the properties of visible light: reflection, refraction, and polarization. In a vacuum, the waves propagated with the speed of light. He made the strong inference, later conclusively demonstrated, that light was an electromagnetic wave with an oscillating electric field perpendicular to the direction of travel. The **polarization** of light, then, is in the direction of the oscillating *electric field vector*.

Light incident at some angle θ on a glass surface can be resolved into two components at right angles to the line of incidence. One component has the electric vector parallel to the surface. The circles and x's in Figure 20-27 represent the heads and tails of the parallel electric field vector. The arrows represent the other component of the oscillating electric field, a component perpendicular to both the line of incidence and the parallel component.

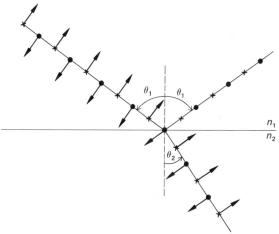

FIGURE 20-27
At Brewster's angle, the reflected light is polarized. The reflected ray is at right angles to the refracted ray.

Reflection of electromagnetic waves from a surface can be thought of as originating from the vibration of the electrons in the surface. In the direction of the parallel electric field component, this excitation and radiation readily takes place. For the perpendicular component, at a certain angle of incidence, known as **Brewster's angle,** the direction of motion of the electrons is along the line of the reflected ray. Electromagnetic waves cannot be excited in this direction and no reflection takes place; only the parallel component of light can be reflected. Thus, the two polarizations of light can be separated. By Snell's law,

$$n_1 \sin \theta_1 = n_2 \sin \theta_2$$

At Brewster's angle of incidence $\theta_1 = \theta_B$ the reflected and refracted rays are perpendicular,

$$\theta_1 + \theta_2 = 90°$$

therefore

$$n_1 \sin \theta_1 = n_2 \sin(90 - \theta_1) = n_2 \cos \theta_1$$

At Brewster's angle

$$\tan \theta_B = \frac{\sin \theta_B}{\cos \theta_B} = \frac{n_2}{n_1} \qquad (20\text{-}7)$$

Light reflected from a surface at Brewster's angle is polarized; light refracted through the surface is only partially polarized. Reflection is a common method of producing polarized or partially polarized light. Glare reflected from the shiny hood of an automobile or the surface of a lake can also be polarized. Consequently, sunglasses that selectively absorb one polarization of light can effectively reduce glare.

Dichroism

Some crystals transmit one polarization of light while strongly absorbing the other. If the crystal is thick enough, the beam that emerges is completely polarized in a selected direction as shown in Figure 20-28. Polarization by absorption is known as **dichroism.**

Among dichroic crystals are the naturally occurring tourmaline and the synthetic "herapathite" (iodoquinine

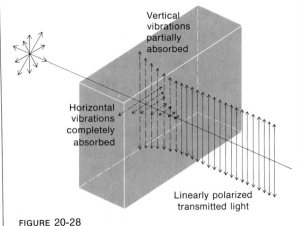

FIGURE 20-28
A dichroic filter transmits linearly polarized light by selectively absorbing one polarization. From Francis Sears and Mark Zemansky, *University Physics,* 4th ed., 1970, Addison-Wesley, Reading, Mass. Figure 42-9. Reprinted with permission.

sulfate). The long, needlelike herapathite crystal absorbs light polarized along its axis. However, the crystals were so fragile that there seemed to be no way of making practical use of them, until Edwin Land, in 1928 while still a college freshman, invented a plastic sheet that he called "Polaroid." After the herapathite crystals are imbedded in the plastic sheet, it is stretched; the crystals align themselves in the direction of stretch. In 1938, Land developed a molecular polarizing material consisting of long molecules of polyvinyl alcohol(PVA). The PVA molecules are also oriented by stretching the material, which, when stained with an ink containing iodine, becomes dichroic. The stretched PVA sheet is laminated to an acetate supporting sheet to form a polaroid filter.

Unpolarized light or randomly polarized light is orientated in no particular direction. Upon passing through a polarizing filter, only the components aligned with the axis of the filter are transmitted. Thus, a beam of light of a given polarization can be obtained. If a second polarizer, an analyzer, is placed at some angle θ, only the component of the electric field in the direction of the second polarizer passes through the analyzer, as shown in Figure 20-29.

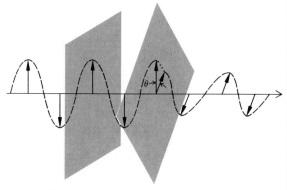

FIGURE 20-29
Polarizing filters. Only the component parallel to the filter's transmission axis passes through.

The component is given by

$$E_2 = E_1 \cos \theta$$

The intensity of the light depends on the square of the electric field vector, so that the intensity depends on

$$I_2 = I_1 \cos^2 \theta$$

If the polarizer and the analyzer are placed at 90°, almost no light is transmitted.

Polaroid filters do not absorb all wavelengths equally: When two such disks are crossed (that is, when the axes of the PVA molecules are perpendicular to those of a second disk), a small amount of light at the red and blue-violet ends of the spectrum is passed. White light passed through a single polaroid filter becomes lightly colored. However, the ease of making large polaroid filters and their modest cost more than compensate for these deficiencies.

Review

Give the significance of each of the following terms. Where appropriate give two examples.

Newton's rings
interference
Young's double-slit experiment
path difference
order
diffraction

multiple slits
diffraction grating
blaze angle
Czerny-Turner spectrometer
single-slit diffraction
Rayleigh Criterion

circular aperture
numerical aperture
index of refraction
polarization
Brewster's angle
dichroism

Further Readings

Isaac Newton, *Opticks* (New York: Dover Publications, 1952). In books two and three, Newton discusses his observations of the "Colours of thin transparent Bodies" and considers several interference phenomena.

Lasers and Light: Readings from Scientific American (San Francisco: W. H. Freeman and Co., 1969). Contains many articles which discuss the wave nature of light.

Problems

INTERFERENCE

1 Light from a helium-neon laser of wavelength 633 nm shines on a pair of slits 0.15 mm apart. The light falls on a screen 4.5 m away.
 (a) Make a sketch of the diffraction pattern.
 (b) Indicate the distance to the first five minimum points on each side of the central maximum.

2 A hair is placed at one edge between two optically flat plates 10 cm long. It is illuminated with sodium yellow light of wavelength 589 nm. Sixty-nine dark fringes are observed, counting the dark fringe at the point of contact of the two plates. How thick is the hair?

3 In a Newton's ring experiment, light is reflected from a spherical surface and a plane surface that are in contact. The radius of the spherical surface is 50 cm. The apparatus is illuminated from above with sodium light of wavelength 589 nm.
 (a) At which reflections will there be a 180° phase shift?
 (b) What is the condition for a bright ring in reflected light?
 (c) What is the radius of the fifth bright ring?

4 An argon laser shines on a pair of thin slits 0.25 mm apart. A double-slit interference pattern is observed on a screen 2.0 m away. The separation between the first interference maxima and the central maximum is 4.1 mm. What is the wavelength of the laser light?

5 In a ripple tank, parallel water waves with a wavelength of 2 cm are incident on a barrier that has two 1 cm holes separated by a distance of 5 cm, center to center. Describe the pattern that results beyond the barrier. Where are the minima and maxima?

DIFFRACTION

6 Light from a helium-neon laser of wavelength 633 nm passes through a transmission grating at normal incidence to form a pattern of dots on a screen 1.0 m from the grating. The second dot is at a distance of 1.0 m on either side of the central maximum.
 (a) What is the grating spacing?
 (b) What is the distance to the first dot from the central maximum?
 (c) How many orders will appear? Where is the third-order dot?

7 Yellow sodium light of wavelength 589 nm shines at normal incidence on a transmission grating with 1200 lines/mm. Th first-order maximum is observed through the eyepiece of a spectroscope. At what angle should it appear?

8 A transmission grating has 900 lines/mm. Light at normal incidence is diffracted in second order into an angle of 60°.
 (a) What is the spacing of the grating grooves?
 (b) What is the wavelength?
 (c) Where is the first-order image?

9 Light from a sodium lamp of wavelength 589 nm falls on a transmission grating at normal incidence. The first-order maximum is observed at 36.8°.
 (a) What is the grating spacing?
 (b) The lamp is replaced with a mercury vapor lamp and a first-order spectrum line is observed at an angle of 26.4°. What is the wavelength of the mercury line?

10 A transmission grating has 500 lines/mm and forms a spectrum on a flat screen 1.0 m away. The light is at normal incidence to the grating.

COLOR PLATES

SPECTRUM CHART

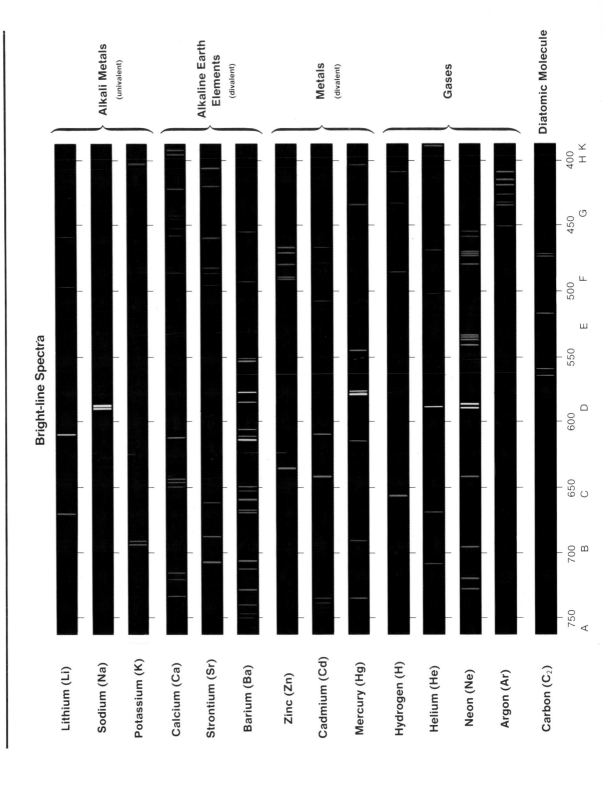

Bright-line Spectra

Alkali Metals (univalent)

Alkaline Earth Elements (divalent)

Metals (divalent)

Gases

Diatomic Molecule

Lithium (Li)
Sodium (Na)
Potassium (K)
Calcium (Ca)
Strontium (Sr)
Barium (Ba)
Zinc (Zn)
Cadmium (Cd)
Mercury (Hg)
Hydrogen (H)
Helium (He)
Neon (Ne)
Argon (Ar)
Carbon (C₂)

750 700 650 600 550 500 450 400
A   B   C   D   E   F   G   H K

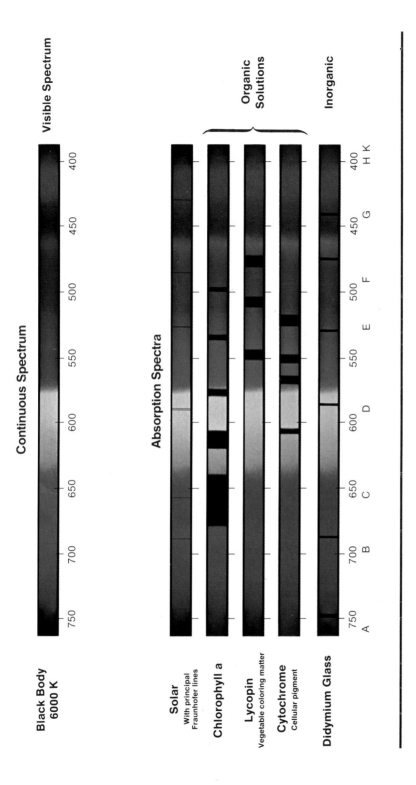

Continuous Spectrum

Visible Spectrum

Black Body 6000 K

Absorption Spectra

Solar
With principal Fraunhofer lines

Chlorophyll a

Lycopin
Vegetable coloring matter

Cytochrome
Cellular pigment

Organic Solutions

Didymium Glass

Inorganic

750   700   650   600   550   500   450   400
A   B   C   D   E   F   G   H K

COLOR PLATE 1

The visible spectrum (wavelength scale in nanometers).

Bright-line spectra of common elements. Only the most prominent lines are shown.

Continuous spectrum of visible light emitted from a black body at 6000 K.

Absorption spectra. The solar spectrum shows the principal Fraunhofer absorption lines. Also shown is absorption of solutions commonly used in laboratory work: The center of each band indicates the position of maximum absorption. Didymium glass is frequently used to calibrate photometers. Courtesy of Sargent-Welch Scientific Company.

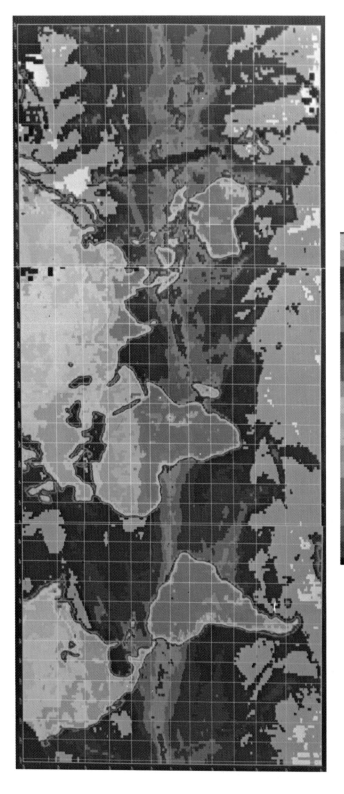

COLOR PLATE 2

Radiobrightness temperature of the world taken from Nimbus-5, a meteorological satellite launched in 1972, into an almost circular orbit 1100 km above the earth. A sensitive scanning microwave radiometer translates the power radiated from the earth's surface into radiobrightness temperature and measures it at a wavelength of 1.55 cm or a frequency of 19.35 GHz. At this frequency typical absorption coefficients are 1.0 for heavy vegetation; 0.9 for rough seas and 0.4 for calm seas; and 0.95 for dry soil and 0.75 for soil with 30% moisture. The microwave emission measurements reflect both the absolute temperature and surface conditions. Courtesy of Dr. William J. Webster, Jr., Geophysics Branch, Goddard Space Flight Center, National Aeronautics and Space Administration.

(a) How far from the central maximum will the first-order red (700 nm) be? the first-order blue (400 nm)?

(b) Repeat for the second-order spectra.

(c) In what order does the red spectrum begin to overlap with the previous order of the blue spectrum?

11 A given reflection grating at normal incidence gives a second-order maximum for a sodium line of 589 nm at $\sin 37° = 0.600$.

(a) What is the spacing of the grating lines?

(b) If the grating is illuminated with cadmium red light of wavelength 644.0 nm, what are the angles of the maxima at all orders?

12 If light is incident on a grating at twice the blaze angle $\alpha = 2\gamma$, then the angle of reflection is $\beta = 0$. If the grating has 600 lines/mm and a blaze angle of 8.6°, for what wavelength is this also a first-order diffraction maximum?

DIFFRACTION FROM A SINGLE APERTURE

13 A helium-neon laser with a wavelength of 633 nm shines on a single slit. At a screen distance of 1.5 m the first diffraction minimum is 2 cm from the central maximum. How wide is the slit?

14 A single slit 0.20 mm wide is illuminated with parallel light of wavelength 589 nm. The diffraction pattern is observed on a screen 50 cm away. How far from the central maximum is the fourth dark band on either side?

15 Sunlight with a wavelength of maximum visibility of 510 nm comes through a pinhole of 0.6 mm diameter. It crosses a room a distance of 2.0 m and strikes a screen. The sun is 0.53° in angular size.

(a) What is the geometrical size of the image of the sun as viewed through the pinhole?

(b) What is the diameter of the diffraction pattern to the first minimum? Can the diffraction be observed?

(c) What is the diameter of the diffraction pattern if the size of the hole is reduced to 0.1 mm? Can the diffraction be observed?

16 For a pupil diameter of less than 2 mm, the human eye is diffraction limited. Two tiny spots at a wavelength 589 nm are just barely resolved at a distance of 10 m.

(a) If the eye is diffraction limited and the pupil has a diameter of 1.0 mm, what is the minimum separation of the spots?

(b) In practice, the normal resolving power of a good eye is one minute of arc. How far apart must the spots be?

17 The Mt. Palomar telescope has an aperture of 5 m. If the image properties of the telescope are limited by the diffraction of the mirror aperture,

(a) What is the angular size separation in seconds of arc of two barely resolved points of light at a wavelength of 510 nm?

(b) At this resolution, if the telescope is pointed at the moon (a distance of 3.78×10^5 km), how far apart would two resolved points on the moon be?

18 Under typical atmospheric observing conditions, a star image will appear to be 1 to 4 seconds of arc in diameter. Under optimum conditions, star images seldom have diameters of less than $\frac{1}{4}$ second of arc. For a telescope to be diffraction limited under optimum seeing conditions at a wavelength of 510 nm, how large must the aperture be? (Large telescopes are designed to gather light, thus to record faint images, not to improve on resolving power. A relatively small orbiting telescope could produce clearer images than the largest earth-based telescopes.)

19 A microscope with a 40x objective lens and a numerical aperture 0.65 is used to observe a red blood cell of diameter 10 000 nm. The microscope has a 15x eyepiece and uses visible light at a wavelength 510 nm. What is the minimum separation that can be resolved by the microscope?

20 Ultraviolet light of wavelength 250 nm is used with a microscope with a 100x oil-immersion objective with numerical aperture of 1.25. Ultraviolet light is strongly absorbed by nucleic acids and protein, and thus specimens can be seen in good contrast.

(a) What is the minimum separation that can be resolved by the microscope?

(b) The bacteria E. Coli has a dimension of about 1000 nm. How large would the final virtual image appear at a distance of 25 cm through a 10x eyepiece?

WAVE OPTICS

21 Light from an argon laser has a wavelength 514 nm in air. The light travels through water with an index of refraction of 1.33.

(a) What is the wavelength of light in water?

(b) What is the frequency of the light in water?

22 Light from a helium-neon laser has a wavelength of 633 nm in air. The light travels through a 1 cm thickness of glass with an index of refraction of 1.52. What is the wavelength of the light in the glass?

23 Plane polarized light falls on a polarizing filter so that a maximum intensity I_o is transmitted. The polaroid is then rotated 45°. What is the intensity of the transmitted beam?

24 Plane polarized light falls on a polarizing filter so that a maximum intensity I_o is transmitted.

Through what angle should the polarizing filter be rotated to reduce the light intensity to

(a) $\frac{1}{2}$ the original intensity?

(b) $\frac{1}{4}$ the original intensity?

(c) $\frac{1}{8}$ the original intensity?

25 A piece of glass has an index of refraction 1.54. What is Brewster's angle for the glass?

26 A beam of light is reflected at an angle of 58° from a plane glass surface. The reflected beam is completely polarized. What is the index of refraction of the glass?

27 Sunlight is reflected off the surface of a still pool of water.

(a) What is the angle of reflection for polarized light?

(b) What is the plane of the electric field of the polarized light?

28 Glare is produced by reflection from water, metal, or other shiny horizontal surfaces if the light is incident at Brewster's angle.

(a) What is the orientation of the electric field of the polarized reflection?

(b) What is the orientation of the polarizing filter in polaroid sunglasses? Why?

21

Quantum Optics

THE SPECTRUM OF LIGHT

Line Spectra

In the eighteenth century, light was thought to be a particle, in the nineteenth century, a wave. The twentieth century settled on a quantum description of light—which, however, is rooted in the nineteenth century study of light emitted by the elements.

In 1814, a German optician, Joseph von Fraunhofer, testing prisms of his manufacture, allowed sunlight to pass through a fine slit and be collimated to make parallel rays. The light refracted through the prism was then collected by a telescope. Figure 21-1 shows a modern spectrometer that uses a grating instead of a prism. Fraunhofer observed some 600 dark lines, carefully noted their positions and spacing, and labeled the most prominent of these alphabetically. Fraunhofer's letters are still

FIGURE 21-1
The grating spectrometer is used to accurately measure the wavelength of spectrum lines. Courtesy of American Optical Corporation.

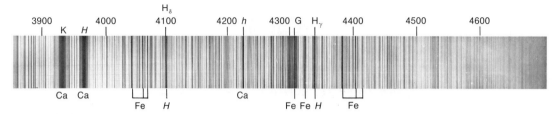

FIGURE 21-2
The Fraunhofer dark lines are prominent in the solar spectrum. The portion of the spectrum from 385 nm to 450 nm shows the *G* line of iron (Fe) as well as the *K* and *H* lines of calcium (Ca) and the hydrogen lines H_γ and H_δ. Courtesy of Hale Observatories.

used to identify them. These **Fraunhofer dark lines** can be clearly seen in the portion of the solar spectrum shown in Figure 21-2. Fraunhofer also observed dark lines in the spectra of stars (among them Sirus, the brightest star)—and found many, although not all, of these dark lines present in starlight. Fraunhofer concluded that "these lines and bands are due to the nature of sunlight and do not arise from diffraction, optical illusion, etc."

Fraunhofer had noticed that two bright yellow lines appeared in the spectrum of an ordinary candle flame. The same lines, also observable when common salt is placed in the flame of a burner, are associated with the element sodium. These bright-line spectra, or **emission spectra,** are observed with all elements. Plate 1 B, facing page 464, shows the emission spectra of several common elements. Fraunhofer had noticed that the two bright lines of the sodium spectra coincided with certain prominent, double dark lines of the solar spectra, the Fraunhofer D lines.

In 1859, Gustav Kirchhoff and Robert Bunsen (famous for his burner) proposed an explanation of the dark lines in the sun's spectra. They set up an experiment: They passed light with a continuous spectrum, as shown in Plate 1 A, through a sodium-laden burner flame and found dark lines at the same wavelength position as the sodium emission lines. When the continuous light was dimmed, the bright emission lines reappeared. They concluded that the dark lines were caused by absorption of the light corresponding to a particular wavelength. Thus, the same atoms that produced the bright-line emission spectra also produced the dark-line **absorption spectra.** Plate 1 C shows the absorption spectra of the sun and several materials. Kirchhoff and Bunsen concluded that

the bright D line always indicates the presence of sodium in a flame. Said Kirchhoff, "the dark D lines in the solar spectrum permit us to conclude that sodium is present in the sun's atmosphere."

Soon after, Kirchhoff announced his two fundamental laws, **Kirchhoff's laws of spectroscopy:**

1. Each chemical species has a characteristic emission spectrum.
2. Each element is capable of absorbing the radiation that it is able to emit.

From study of the dark lines of the sun's absorption spectrum or that of a distant star or planet, it can be deduced what elements are present in its outer atmosphere. In 1825, the French philospher Auguste Comte had suggested that the chemical composition of the stars is an example of knowledge that is beyond human reach. Thanks to Kirchhoff's laws of spectroscopy, more was known about the chemical composition of stars than of the moon until its exploration by the astronauts. Figure 21-3 shows a portion of the infrared spectrum of sunlight reflected from Venus. Also shown are the dark molecular bands caused by carbon dioxide in the planet's atmosphere.

In 1860, while testing certain minerals, Kirchhoff and Bunsen found strange spectral lines and began a search for unknown elements. One metal element chemically related to sodium and potassium, with a sky-blue spectral line, they named cesium (after the Latin for blue). Another alkali metal element, with a red line, was named rubidium (after the Latin for red).

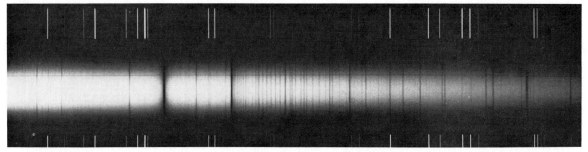

FIGURE 21-3
A portion of the spectrum of Venus in reflected sunlight from 860 to 883 nm in the infrared. The dark bands indicate the presence of carbon dioxide in the planet's atmosphere. The bright lines above and below are comparison spectra of neon and iron. Lick Observatory photograph.

Emission flame photometry, the measure of the intensity and wavelength of spectrum lines, remains a simple yet sensitive method for identifying and quantitatively analyzing easily excited elements, notably the alkali metals, the spectra of which are easy to produce. The solution containing the sample to be analyzed is aspirated directly into an oxygen-hydrogen, oxygen-acetylene, or air-hydrogen flame. The resulting excitation causes the atoms to emit the characteristic wavelength of light. Table 21-1 shows the flame spectra wavelength and detection limit of some common metals.

TABLE 21-1
Flame spectra wavelengths of some common metals, giving the most sensitive wavelength in nanometers (nm) and the detection limits in nanograms (ng) per milliliter (ml).

	Element	Wavelength (nm)	Detection limit (ng/ml)
Ca	calcium	422.7	6
Cu	copper	327.4	20
Fe	iron	372.0	100
Li	lithium	670.8	0.2
Mg	magnesium	285.2	40
K	potassium	766.5	10
Na	sodium	589.2	0.1

Atomic Absorption Spectroscopy

Spectroscopy has become a powerful technique for observing the presence of chemical elements. A particularly important technique for studying trace elements in environmental samples is known as **atomic absorption spectroscopy,** developed in the early 1950s by A. Walsh as a method to measure concentrations of elements, notably the heavy metals, the emission spectra of which are difficult to produce. Atomic absorption spectroscopy uses the absorption of the light of a particular wavelength emitted by a light source as it traverses a flame containing the free atoms of the sample under test. The atoms absorb only a very narrow band of wavelengths. Hollow cathode light sources with carefully selected spectrum lines, usually in the ultraviolet, illuminate the sample. The light source is modulated, usually by a physical light chopper, which interrupts the beam of light typically 285 times per second. After passing through the flame, the light passes through a spectrometer and is detected by the photomultiplier tube, as shown in Figure 21-4. Because the light is modulated, only the fluctuating part of the final detected signal contains the absorption information, eliminating the background effect of flame luminescence and stray light.

The sample to be studied is periodically aspirated into the flame, causing a sudden increase in the absorption of the source light. This absorption is measured. The *detection limit* is the concentration of an element in solution

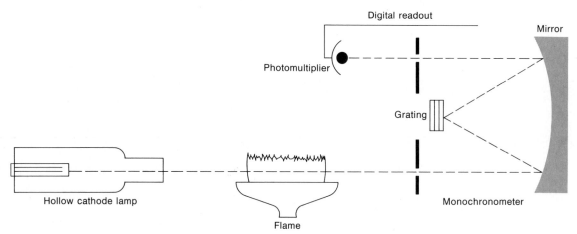

that gives a signal equal to twice the standard deviation of
the background signal with a blank sample (a measurement uncertainty of ±50%). Table 21-2 gives the detection limits for a commercial atomic absorption spectrometer.

The atomic absorption spectrometer, shown in Figure
21-5, is a valuable tool for measuring the concentration of

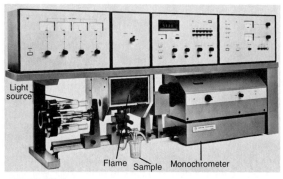

FIGURE 21-5
Atomic absorption spectrograph. At the left is the four-turret
changeable light source. Next is the flame with the sample to
be aspirated. At the right are the monochrometer and photomultiplier to measure the absorption at a particular wavelength. Courtesy of Varian Techtron, Palo Alto, Calif.

TABLE 21-2
*Atomic absorption spectroscopy giving the
detection limit in nanograms per milliliter and
principal wavelength for typical elements.*

Element		Wavelength (nm)	Detection limit ng/ml
Ag	silver	328.1	2
Al	aluminum	309.2	18
Ba	barium	553.6	20
Bi	bismuth	223.1	46
Ca	calcium	422.7	2
Cu	copper	324.7	2
Fe	iron	248.3	6
Hg	mercury	253.7	160
Li	lithium	670.8	1.5
Mg	magnesium	285.2	0.2
Ni	nickel	232.0	8
Pb	lead	283.3	15
Rb	rubidium	780.0	2
Sr	strontium	460.7	2
V	vanadium	318.4	50

TABLE 21-3
The Balmer series for hydrogen.

n	m	Line	Balmer's prediction (nm)	Ångström's value (nm)	difference (nm)
2	3	H_α	$\frac{9}{5}b = 656.208$	656.210	$+0.002$
2	4	H_β	$\frac{4}{3}b = 486.08$	486.074	-0.006
2	5	H_γ	$\frac{25}{21}b = 434.00$	434.01	$+0.01$
2	6	H_δ	$\frac{9}{8}b = 410.13$	410.12	-0.01

SOURCE: After Johann Jacob Balmer, "Note on the Spectral Lines of Hydrogen," in Jerry B. Marion, *A Universe of Physics* (New York: John Wiley, 1970), p. 169.

trace elements in biological and environmental samples. Atomic absorption spectroscopy has been used since 1960 for the clinical analysis of blood serum levels of calcium, magnesium, sodium, and potassium. This was soon followed by measurement of calcium in saliva and calcium, magnesium, lead, nickel, bismuth, and mercury in the urine. A technique for determining strontium levels in biological fluids was soon developed. (The interfering phosphate ion must first be removed.) For nearly 30 years, the potential of atomic absorption spectroscopy as an analytical procedure for clinical and biological research work has been fully appreciated.

The Balmer Series

The discovery of line spectra and their application to chemistry, and advances in the development of grating spectrometers set off a flurry of scientific measurement of the spectra produced by different elements. In 1862, the Swedish astronomer Anders Jonas Ångström used spectroscopic techniques to detect the presence of hydrogen in the sun and later made precise measurements of the hydrogen spectrum. In 1885, Johann Jacob Balmer found a simple numerical relationship between the wavelengths of the visible spectrum lines of hydrogen, the **Balmer series.** "The wavelengths of the first four hydrogen lines are obtained by multiplying the fundamental number $b = 364.56$ nm in succession by the coefficient $\frac{9}{5}$; $\frac{4}{3}$; $\frac{25}{21}$; and $\frac{9}{8}$."[1] His results closely agreed with Ångström's observations, as Table 21-3 shows. Balmer concluded that

[1] Quoted in *A Universe of Physics: A Book of Readings,* Jerry B. Marion, editor (New York: John Wiley and Sons, 1970), pp. 169–170.

these coefficients were part of a general mathematical series that is determined by integer values of m and n, where the wavelengths of the hydrogen spectrum are given by

$$\lambda = \frac{bm^2}{m^2 - n^2}$$

The four hydrogen lines correspond to values of $n = 2$ and $m = 3; 4; 5;$ and 6. This general series can be put in the form

$$\frac{1}{\lambda} = \frac{n^2}{b}\left[\frac{1}{n^2} - \frac{1}{m^2}\right] = R_H\left[\frac{1}{n^2} - \frac{1}{m^2}\right]$$

where the constant $R_H = n^2/b$ is the Rydberg constant. Balmer's value for the Rydberg constant was very close to the presently accepted value for hydrogen:

$$R_H = 10\ 967\ 758\ \text{m}^{-1}$$

The Balmer series is characterized by the spectrum with $n = 2$

$$\frac{1}{\lambda} = R\left[\frac{1}{2^2} - \frac{1}{m^2}\right] \tag{21-1}$$

This obviously leaves open the possibility for a series of hydrogen in the ultraviolet, $n = 1$, and in the infrared, $n = 3, 4,$ and 5. It was natural, then, to assume that other series exist. The correctness of this speculation was confirmed, in 1908, by Friedrich Paschen, who discovered the first two members of the infrared $n = 3$ series (the Paschen series). In 1906, Theodore Lyman observed the first member of the ultraviolet series (the Lyman series)

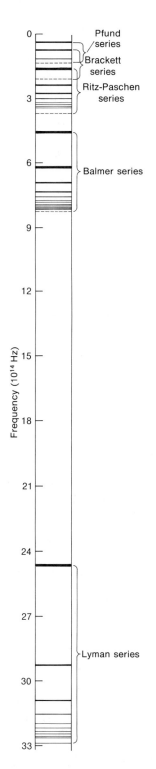

FIGURE 21-6
The hydrogen spectrum. The positions of the emission lines of hydrogen are represented as a function of frequency.

with $n = 1$ and $m = 2$, but not until 1913 was this line attributed to hydrogen. The Brackett series $n = 4$ and the Pfund series $n = 5$ in the far infrared were not observed until the 1920s. Figure 21-6 shows the Balmer series of hydrogen extending from about 660 nm in the red to 360 nm in the ultraviolet, as well as the other spectrum series plotted as a function of frequency.

BLACK-BODY RADIATION

We discussed the transfer of energy by radiation in some detail in Chapter 11. We now consider the spectrum of light emitted by a hot object. In 1809, P. Prevost stated a law of exchange: A body may lose heat by emission of radiation, and this heat can be transferred to a second body by absorption of the radiation. In our ordinary experience, an object on which light falls reflects certain wavelengths, or colors, absorbs some wavelengths, and can transmit others.

Kirchhoff's Emission Law

Kirchhoff sought to understand how energy from light or radiation is transferred to a body and is subsequently radiated. If light of a certain intensity in watts/m² at a certain frequency ν (the lowercase Greek letter nu) is incident on a surface, the energy absorbed A_ν, depends on the intensity of the light I_ν and the **absorption** a_ν at that frequency:

$$A_\nu = I_\nu a_\nu$$

The energy liberated from the surface at a particular frequency is the **emissivity** of the surface e_ν.

In 1859, Kirchhoff argued that the ratio of the emissivity e_ν to the absorption a_ν depends only on the temperature and frequency. If it depended on some other parameter, two objects placed in a chamber at a uniform temperature could transfer energy and one would heat up at the expense of the other in violation of the second law of thermodynamics. **Kirchhoff's emission law** states that the ratio of the emissivity to the absorption is a function

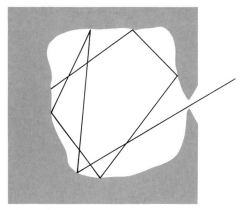

FIGURE 21-7
A cavity acts as a black body. Any radiation entering the cavity through its small aperture is trapped.

only of the frequency and temperature:

$$\frac{e_\nu}{a_\nu} = K(\nu, T)$$

Emission by a heated body is often compared to the radiation of a **black body.** An ideal black body absorbs all radiation that falls upon it at all frequencies, $a_\nu = 1$. We can approximate a black body by making a thermally isolated cavity with a small hole to the outside, as shown in Figure 21-7. If the cavity is large and irregular in shape, any light falling through the hole into it is absorbed. When the interior surface is heated, the cavity reaches an equilibrium temperature, T. The radiation within the cavity is in equilibrium with that radiated from the walls. The radiation coming from a hole in the cavity wall is independent of the material enclosing the cavity. For an ideal black body in equilibrium at temperature T, the radiation emitted at a particular frequency ν is a function only of the temperature and frequency.

$$e_\nu = K(\nu, T)$$

At the end of the nineteenth century, the explanation of black-body radiation became one of the central problems of theoretical physics. In 1893, Wilhelm Wien showed, by an extended theoretical argument treating the radiation in a cavity as an ideal gas, that the emission from a black body should be approximately

$$e_\nu = K(\nu, T) \propto \nu^3 e^{-a\nu/T}$$

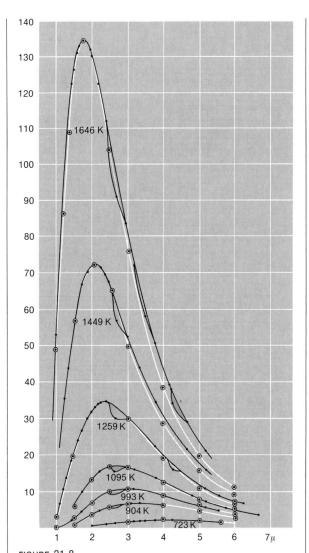

FIGURE 21-8
Black-body radiation data. The solid lines and plain crosses show Lummer and Pringsheim's experimental data from about 1899. The circled crosses and dotted lines represent Wien's formula, which agrees well at short wavelengths. The shaded areas show the absorption by water vapor and carbon dioxide. *Verhand lungen der Deutschen Physikalischen Gesellschaft.*

Figure 21-8 shows in detail the experimental data published in 1899. Wien's formula, shown by the dashed lines, agrees well with the experimental data at high frequencies (short wavelengths), but at lower frequencies (longer wavelengths), it gives too small a value.

Wien was able to demonstrate what is known as the Wien displacement law. The wavelength at maximum emission, the peak of the emission, decreases with increasing temperature, as was discussed in Chapter 11. The product of the wavelength at the maximum, λ_{peak} and the absolute temperature is constant

$$\lambda_{\text{peak}} T = 2.89 \times 10^{-3} \text{ m-K}$$

This result agrees well with the experimental data.

By about 1893, the black body radiation spectrum was almost explained, but not quite. The small discrepancy between the theory and experiment at low frequencies remained a major puzzle at the turn of the nineteenth century.

In 1900, Lord Rayleigh concluded that at low frequencies (long wavelengths), the emission from a black body should be proportional to

$$e_\nu \propto \nu^3(T/\nu) = \nu^2 T$$

This agreed well with the observed spectrum of cavity radiation at low frequencies. However, at high frequencies, Rayleigh's model indicated that an infinite amount of energy would be radiated. This flaw in the theory has been called the *ultraviolet catastrophe*. Lord Rayleigh explained the low-frequency end of the black-body radiation spectrum and Wien the high-frequency end, but a comprehensive theory was needed to explain the entire spectrum.

Max Planck

Max Planck, at the University of Berlin, considered the problem of black-body radiation from the very general perspective of the first and second laws of thermodynamics. By Maxwell's theory of electromagnetism, if a charged particle such as an electron oscillates at some frequency ν it radiates electromagnetic waves of the same frequency. The product of the wavelength λ and the frequency ν is the speed of light:

$$\lambda\nu = c$$

Planck considered the equilibrium between the electro-magnetic radiation in the cavity and the radiation coming from the oscillating charges in the wall at a temperature T. By assuming that the energy in the cavity is distributed as uniformly as possible through every frequency, Planck was able immediately to derive Rayleigh's result, which suffered from the so-called ultraviolet catastrophe.

Planck found a way to predict the correct black-body radiation spectra: He assumed that the energy that could be stored in the electromagnetic waves in the cavity at a particular frequency should be quantized (that is, considered to exist in finite increments) to some value ε_0. There could be one unit, $1\varepsilon_0$, or two units, $2\varepsilon_0$, or some integer number of units $n\varepsilon_0$.

$$U = \varepsilon_0, 2\varepsilon_0, \ldots, n\varepsilon_0$$

Calculating the average energy in the cavity at a particular frequency, Planck found that the emission from a cavity at a temperature T at a frequency ν should depend as

$$e_\nu = \frac{2\pi\nu^2\varepsilon_0}{c^2[e^{\nu_0/kT} - 1]}$$

For this result to agree with Wien's formula, the energy quanta ε_0 should be proportional to the frequency:

$$\varepsilon_0 = h\nu$$

The single constant h, **Planck's constant,** can be determined by fitting the experimental data. Planck's constant has a value

$$h = 6.63 \times 10^{-34} \text{ J-s.}$$

At high frequencies (short wavelengths),

$$\frac{h\nu}{kT} \gg 1$$

In this limit,

$$e^{h\nu/kT} \gg 1$$

and Planck's formula reduces to Wien's result:

$$e_\nu = K(\nu, T) = \frac{2\pi h\nu^3}{c^2}e^{-h\nu/kT} \qquad (21\text{-}2)$$

At low frequencies (long wavelengths),

$$\frac{h\nu}{kT} \ll 1$$

In this limit,

$$e^{h\nu/kT} \cong 1 + (h\nu/kT)$$

Planck's formula reduces to Rayleigh's result:

$$e_\nu = K(\nu, T) = \frac{2\pi\nu^2 kT}{c^2}$$

Planck's result not only gave the correct frequency dependence of the black-body radiation spectrum over the entire frequency spectrum, but also gave the correct value for the total energy radiated. It also changed the history of science, for Planck introduced the notion that electromagnetic energy exists in quanta: discrete increments, or bundles.

Satellite Infrared Measurements

In 1964, the Nimbus I meteorological satellite was launched into an elliptical orbit around the earth. It carried not only three advanced television cameras but also an infrared scanning detector to measure the infrared emission from the earth's surface at wavelengths from 3.4 μm to 4.2 μm. The theory of black-body radiation just discussed can be applied to calculate the energy that is radiated from the surface of the earth.

Because a good absorber of radiation is a good emitter, the energy radiated depends on the absorption coefficient of the surface, a_ν. The atmosphere is quite transparent in the infrared region beyond 3 μm wavelength, so the radiation observed by a satellite in the absence of cloud cover is from the earth's surface. Areas of open water, ground with high moisture content, or forests with heavy vegetation radiate nearly as a black body $a_\nu = 1$ in the infrared. Dry soils and mineral areas have an absorption of less than unity. At infrared wavelengths, Wien's result (equation 21-2) is valid. If the range of frequencies observed, $\Delta\nu = \nu_2 - \nu_1$, is small, then the total infrared emission from a surface is given by

$$\Delta I = a_\nu K(\nu, T)\Delta_\nu = \frac{a_\nu 2\pi h\nu^3 \Delta\nu}{c^2} e^{-h\nu/kT} \quad (21\text{-}3)$$

If the surface is assumed to be radiating as a black body, an effective radiation temperature can be calculated for the surface.

The earth's infrared emission is sufficient to be measured from a satellite. The intensity of the radiation from a 6 km square area of the surface is observed. The radiation is characteristic of the absolute temperature (in Kelvin) and the infrared absorption of the surface a_ν. If there is a cloud cover over the surface, the indicated temperature is the effective temperature at the tops of the clouds. The temperature of the atmosphere decreases with increase in altitude, and the approximate height of the cloud cover can be estimated to supply data for weather forecasting. Figure 21-9 shows the satellite schematically. Today, as part of its weather forecasting, the National Weather Service routinely monitors the infrared emission from the earth at wavelengths of 10.5 μm and 11.6 μm and in the visible region. Monitoring the earth in the infrared allows the movement of the cloud cover to be followed both day and night. Figure 21-10 shows a typical infrared weather photograph. Figure 21-11 shows microwave and infrared images taken of Hurricane Fifi in 1974. Plate 2, facing page 465, shows the microwave image of the world.

EXAMPLE

A weather satellite monitors the infrared emission in the wavelength range $\lambda_1 = 3.4$ μm to $\lambda_2 = 4.2$ μm. Suppose that the earth radiates as a black body at 280 K. Use the Wien result to determine the energy radiated from the ground over this narrow range.

The frequencies correspond to $\nu_1 = 8.82 \times 10^{13}$ Hz and to $\nu_2 = 7.14 \times 10^{13}$ Hz, and the frequency interval is $\Delta\nu = 1.68 \times 10^{13}$ Hz. The average frequency is about $\nu = 8.0 \times 10^{13}$ Hz with a photon energy

$$h\nu = (6.63 \times 10^{-34} \text{ J-s})(8.0 \times 10^{13} \text{ Hz})$$
$$= 5.29 \times 10^{-20} \text{ J}$$

$$h\nu = 0.33 \text{ eV}$$

The value of kT is 0.024 eV at 280 K, so that $h\nu/kT = 13.75 \gg 1$. Wien's result is valid. The intensity of radi-

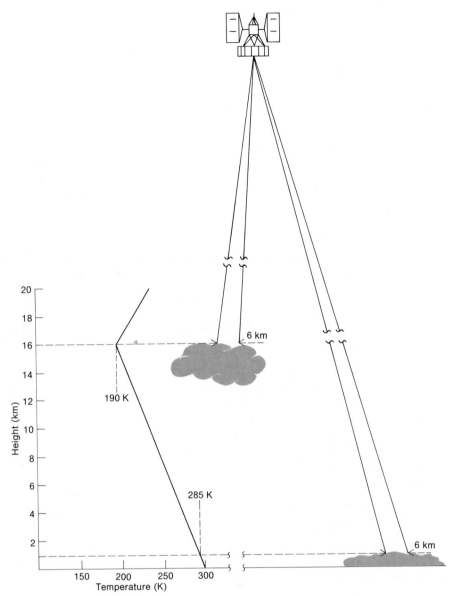

FIGURE 21-9
The Nimbus-I satellite measures the infrared emission from the surface, allowing determination of the effective emission temperature and the approximate height of cloud tops. From William Nordberg, "Geophysical Observations from Nimbus-I," *Science* (29 October 1965), p. 561. Copyright 1965 by the American Association for the Advancement of Science.

FIGURE 21-10
A typical infrared satellite photograph. Taken by a stationary meteorological satellite over the Pacific Ocean, it shows infrared brightness values from 184 K (white) to 330 K (black) observed at a wavelength of about 10.5 μm. The resolution of the photograph is 6.4 km². Taken at 11:45 Greenwich Mean Time on 8 February 1977, it shows the high, cold clouds of a winter storm traveling to the northeast, north of California into Canada. The clear region off the California coast is the Pacific high-pressure region that caused the winter drought. The northward movement of the storm track brought the severe cold to the eastern United States in the winter of 1977. Courtesy of National Environmental Satellite Service, U.S. Department of Commerce.

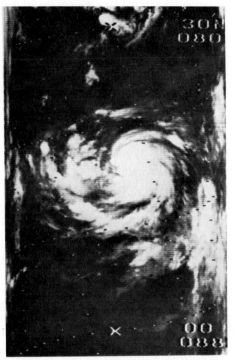

FIGURE 21-11
Hurricane Fifi, 18 September 1974. Microwave (left) and infrared (right) images obtained from Nimbus-5. The microwave picture shows principally the ground conditons; the infrared image shows the cloud cover. Courtesy of the Goddard Space Flight Center, National Aeronautics and Space Administration, Greenbelt, Md.

ation in this narrow frequency band is given by equation 21-3

$$\Delta I = \frac{a_\nu 2\pi h\nu^3 \Delta\nu}{c^2} e^{-h\nu/kT}$$

The temperature dependence occurs in the exponential. The energy radiated from the ground over this narrow spectrum is extremely responsive to the temperature kT. If the earth is treated as a black body, $a_\nu = 1$, the effective emission temperature is equal to the thermodynamic temperature.

Evaluating the above equation, we find that the black-body radiation in this frequency band at a temperature of 280 K is

$$\Delta I = \frac{(2\pi)(6.63 \times 10^{-34}\text{ J-s})(8.0 \times 10^{13}\text{ s}^{-1})^3}{(3.0 \times 10^8\text{ m/s})^2}$$
$$\times (e^{-13.75})(1.68 \times 10^{13}\text{ s}^{-1})$$

$$\Delta I = 0.43\text{ W/m}^2$$

In this narrow frequency band, at an effective temperature of 300 K, the emission increases to 1.23 W/m^2.

THE PHOTOELECTRIC EFFECT

Lenard's Experiment

In 1887, while conducting experiments on electromagnetic waves, the German physicist Heinrich Hertz happened to observe the **photoelectric effect:** A metal surface can emit electric charges when ultraviolet (short-wavelength) light shines on it.

In 1902, Philipp Lenard, who worked under Hertz, published a detailed account of his experimental study of the action of light on a metal surface. Lenard allowed light of different frequencies to fall on a clean metal surface in a vacuum, as shown in Figure 21-12. Electrons were emitted from the surface, the photocathode. A voltage placed between the photocathode and an electrode, the anode, slowed these electrons. The maximum energy of the electrons could be measured by measuring the voltage required to stop them. Lenard found the following:

1. The electron current emitted from the surface was proportional to the intensity of the light.

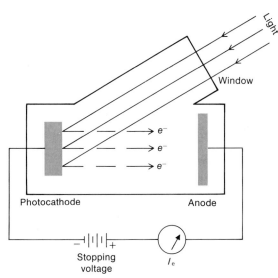

FIGURE 21-12
Lenard's experiment for observing the photoelectric effect. Light (photons) incident on a photocathode ejects electrons (photoelectrons) that are collected at the anode.

2. The energy of the electrons emitted was independent of the intensity of the light.

3. The energy of the emitted electrons depended on the type of metal.

According to the electromagnetic theory of light, the intensity of the light is proportional to the square of the oscillating electric field strength. A brighter light source would have a larger electric field, which would give greater acceleration to the electrons in the surface of a metal. From the electromagnetic theory, a more intense light source would be expected to give the emitted electrons a greater energy—time is required to absorb the kinetic energy necessary to eject the electron. However, the electrons are observed to be released from the surface instantaneously, even in very faint light. Thus, the observed photoelectric effect contradicts the predictions of electromagnetic theory.

Einstein's Explanation

In 1905, in the same volume of the *Annals of Physics* in which his papers on relativity and on Brownian motion appeared, Einstein offered an explanation for the photoelectric effect. He assumed that the energy quantization used by Planck,

$$\varepsilon = h\nu$$

is a universal characteristic of light. Light is composed of discrete bundles of energy, quanta, or **photons.** When light quanta strike a metal surface, their energy is transferred instantaneously to a single electron. Some of this energy is converted into potential energy as the electron escapes the metal surface; this is the **work function** W. The remaining energy appears as kinetic energy of the emitted electron. Einstein states:

In accordance with the assumption to be considered here, the energy of a light ray spreading out from a point source is not continuously distributed over an increasing space but consists of a finite number of energy quanta which are localized at points in space, which move without dividing, and which can only be produced and absorbed as complete units. . . .

According to the concept that the incident light consists of energy quanta of magnitude[2] $R\beta\nu/N$, however, one can conceive of the injection of electrons by light in the following way. Energy quanta penetrate into the surface layer of the body, and their energy is transformed, at least in part, into kinetic energy of electrons. The simplest way to imagine this is that a light quanta delivers its entire energy to a single electron; we shall assume that this is what happens.[3]

In modern notation,

$$\varepsilon = h\nu = \tfrac{1}{2}m_e v^2 + W \qquad (21\text{-}4)$$

The kinetic energy of the electron is then the difference between the photon energy and the work function of the metal:

$$KE = h\nu - W$$

The work function for a pure metal surface is well defined. However, any impurities of the surface can change the work function significantly. Table 21-4 shows the range of values of the photoelectric work function of different metals.

At any frequency above the threshold frequency, the reverse voltage required to stop the electrons increases directly with the frequency. The stopping voltage V_0 times the electron charge e is equal to the maximum kinetic energy:

$$eV_0 = KE = h\nu - W$$

At frequencies less than some **threshold frequency**, the photon does not have enough energy to liberate an electron from the metal. The threshold frequency is given by

$$\nu_0 = \frac{W}{h}$$

---

[2]In Einstein's expression for the light quanta, $R\beta\nu/N$, $\beta = h/k$, the ratio of Planck's constant h to the Boltzmann constant k; $k = R/N$. Thus, $R\beta\nu/N = h\nu$, which is the familiar form.

[3]Einstein's paper appears in translation in A. B. Arons and M. B. Pappard, "Einstein's Proposal of the Photon Concept," *American Journal of Physics 33* (May 1965), pp. 367–374.

TABLE 21-4
The work function for clean polycrystalline metal surfaces.

		Work function (eV)
Ag	silver	4.73
Al	aluminum	4.08
Au	gold	4.82
Fe	iron	3.9–4.7
Cs	cesium	1.81
Ni	nickle	5.01
Zn	zinc	3.08–4.3
Li	lithium	2.28–2.42
Na	sodium	2.28

The *threshold wavelength* is given by

$$\lambda_0 = \frac{c}{\nu_0} = \frac{hc}{W}$$

Robert Millikan, at the University of Chicago, undertook a series of detailed measurements of the threshold frequency and stopping potential as a function of frequency for a variety of light sources and metal surfaces. His results, published in 1916, confirmed every particular of Einstein's theory of the photoelectric effect. Figure 21-13 shows some of Millikan's data.

EXAMPLE

A photoelectric surface is made of cesium metal with a work function of 1.81 eV. What are the threshold frequency and threshold wavelength? What is the kinetic energy of the emitted electrons if the light has a wavelength of 400 nm?

The threshold frequency is such that

$$\nu_0 = \frac{W}{h} = \frac{(1.81 \text{ eV})(1.6 \times 10^{-19} \text{ J/eV})}{6.63 \times 10^{-34} \text{ J-s}}$$

$$\nu_0 = 4.37 \times 10^{14} \text{ Hz}$$

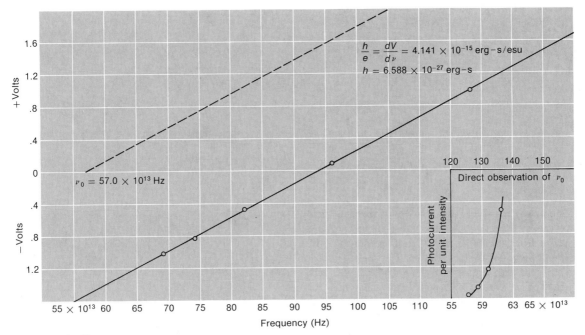

FIGURE 21-13
The photoelectric effect. Millikan's data for stopping potential as a function of frequency (in Hz) for light falling on a lithium metal surface. The dotted line is corrected for the contact potential at the copper lithium junction and shows the threshold frequency. The data fall on a straight line with a slope h/e as predicted by Einstein. From R. A. Millikan, "A Direct Photoelectric Determination of Planck's 'h'," *Physical Review* (1916), p. 377.

The threshold wavelength is just

$$\lambda_o = \frac{c}{\nu_o} = \frac{3 \times 10^8 \text{ m/s}}{4.37 \times 10^{14} \text{ Hz}} = 686 \text{ nm}$$

The kinetic energy of the emitted electrons is given by

$$KE = \frac{hc}{\lambda} - W$$

$$KE = \frac{(6.63 \times 10^{-34} \text{J-s})(3.0 \times 10^8 \text{ m/s})}{(400 \times 10^{-9} \text{ m})(1.6 \times 10^{-19} \text{J/eV})} - 1.81 \text{ eV}$$

$$KE = 3.11 - 1.81 \text{ eV}$$

$$KE = 1.30 \text{ eV}$$

The Image Orthicon Tube

Television is made possible by the **image orthicon tube,** the television camera's light-sensitive component that uses the photoelectric effect to convert an image to an electrical signal. The television camera's lens system focuses an image on the photocathode shown in Figure 21-14. The back of the photocathode emits electrons in proportion to the intensity of light hitting it. The electrons are accelerated in a straight line through a potential difference of some 300 V and strike a glass membrane only 2 μm thick. Each electron striking the membrane knocks out several electrons, leaving behind positive charge at each point on the glass membrane. The glass membrane thus holds a charge replica of the original image.

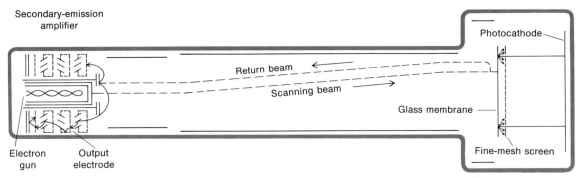

FIGURE 21-14
The image orthicon tube. The television camera lens focuses light on the sensitive photocathode. Photoelectrons are emitted from the left side of the photocathode and strike the glass membrane. A beam of electrons scans the membrane and carries off the image. The return beam is amplified to yield the electrical output of the tube. From R. Clark Jones, "How Images Are Detected." Copyright © 1968 by Scientific American, Inc. All rights reserved.

A focused electron beam deflected by a magnetic field scans the membrane 30 times a second. The negative electrons discharge the positive glass membrane. The electron beam is adjusted so that the electrons have just enough energy to return the glass membrane to zero potential. The electron current will just neutralize the maximum positive charge at any part of the glass plate. Where the membrane has less positive charge, some of the electrons are returned and strike the output electrode structure. The output electrode structure multiplies the return electron current by secondary emission of electrons from a series of metal accelerating plates. The output becomes the video signal of black and white television.

Response of the Eye

The spectral response of the human eye, shown in Figure 21-15, extends in wavelength from about 410 nm to 720 nm. The energy of a photon is given in terms of wavelength by

$$\varepsilon = \frac{hc}{\lambda} = \frac{(6.63 \times 10^{-34} \text{ J-s})(3 \times 10^8 \text{ m/s})}{(1.6 \times 10^{-19} \text{ J/eV})\lambda}$$

$$\varepsilon = \frac{1.24 \times 10^{-6} \text{ eV-m}}{\lambda} = \frac{1\,240 \text{ eV-nm}}{\lambda} \quad (21\text{-}5)$$

A photon with a wavelength of 1.00 nm has an energy of 1240 eV. The energy of the photons in the visible range extends from a low energy range of 1.72 eV at 720 nm to a maximum of 3.02 eV at the short wavelength end of the

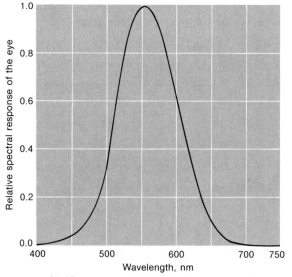

FIGURE 21-15
The spectral response of the human eye, based on measurements by the *Commission Internationale de l'Eclair* (CIE).

spectrum, 410 nm. The photon energy of 1.7 eV is sufficient to rearrange an organic molecule.

The ozone layer in the upper atmosphere absorbs the high-frequency, ultraviolet end of the solar spectrum. An ultraviolet photon has sufficient energy to cause damage to organic molecules. The eye is sensitive to wavelengths with an energy sufficient to cause rearrangement of organic molecules but not sufficient to cause molecular damage.

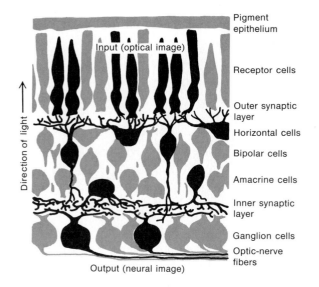

Pigment epithelium

Input (optical image)

Receptor cells

Outer synaptic layer

Horizontal cells

Bipolar cells

Amacrine cells

Inner synaptic layer

Ganglion cells

Optic-nerve fibers

Direction of light

Output (neural image)

FIGURE 21-16
Structure of the vertebrate retina, based on a micrograph of a retina. The vertically oriented receptor cells, the bipolar cells, and ganglion cells constitute the input-output pathway. Horizontally oriented cells and amacrine cells carry information across the retina at two distinct levels to relate activity in different parts of the visual field. Light enters the retina from below and must pass through several layers of cell bodies, nerve fibers and blood cells to reach the receptor cells. From Frank S. Werblin, "The Control of Sensitivity in the Retina." Copyright © 1972 by Scientific American, Inc. All rights reserved.

As was discussed in Chapter 19, the image formed by the cornea and lens of the eye falls on the retina. In the human eye the retina is complex, consisting of nine layers (see Figure 21-16). In mammals, as in nearly all vertebrate animals, the retina is inside out. Light striking the retina must pass through a web of blood vessels and nerve fibers before reaching the light-sensitive receptor cells. These cells are rods and cones, so named for their appearance under the microscope. Figure 21-17 shows a scanning electromicrograph of the rod cells. The **rod cells** function in low-level illumination and give us only shades of gray, *scotopic* vision; the **cone cells** function in bright light and give rise to color vision, *photopic* vision. Figure 21-18 gives the relative sensitivity of rods and cones, as a func-

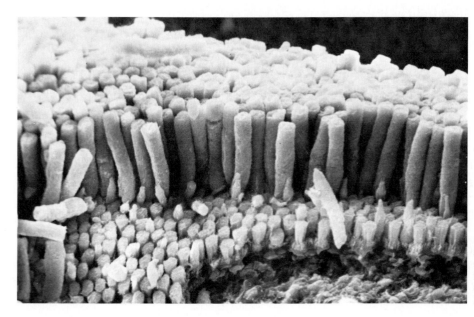

FIGURE 21-17
Scanning electronmicrograph showing the retina, rod cells, inner segments, and bipolar cells. Courtesy of Deric Bownds and Stan Carlson.

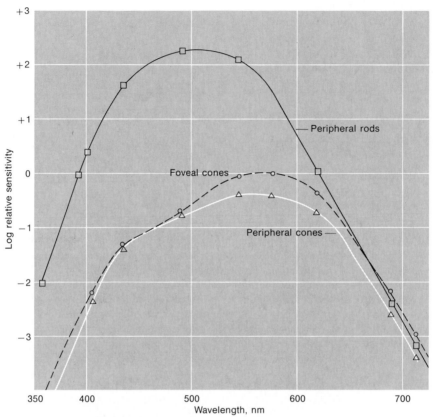

FIGURE 21-18

Logarithm of the relative spectral sensitivity of dark-adapted rods and cones. Cone vision peaks at a wavelength of 555 nm, rod vision at 507 nm, foveal cones, peripheral cones, peripheral rods. Replotted from George Wald, "Human Vision and the Spectrum," *Science, 101* (1945), p. 657.

tion of wavelength. Rod vision, which peaks at a wavelength of 507 nm, is a thousand times more sensitive than cone vision, which peaks at 555 nm.

Rod and cone cells contain photopigments and act as transducers, converting light energy into nerve signals. The pigment in the rod cells is **rhodopsin.** Rhodopsin in an eye adapted to the dark is a dark purple, called visual purple. To nullify rod vision, bright light bleaches the rhodopsin. The dim-light sensitivity of the rod cells as a function of wavelength is directly related to the photoabsorption of rhodopsin. Figure 21-19 is an electron

microscope photograph of the ultrastructure of part of a rod cell. The outer segment of the rod cell contains hundreds of thin rod sacs containing rhodopsin. The inner segment contains mitochondria which supply the cell with ATP as biochemical energy. The thin sacs containing rhodopsin are also seen in the cone cells in Figure 21-20.

In 1942, S. Hecht, S. Schlaer and M. H. Pirenne published a careful, elaborate quantitative study of the sensitivity of the eye. A very weak source of light was used that produced a single flash of 0.001 s duration with a known

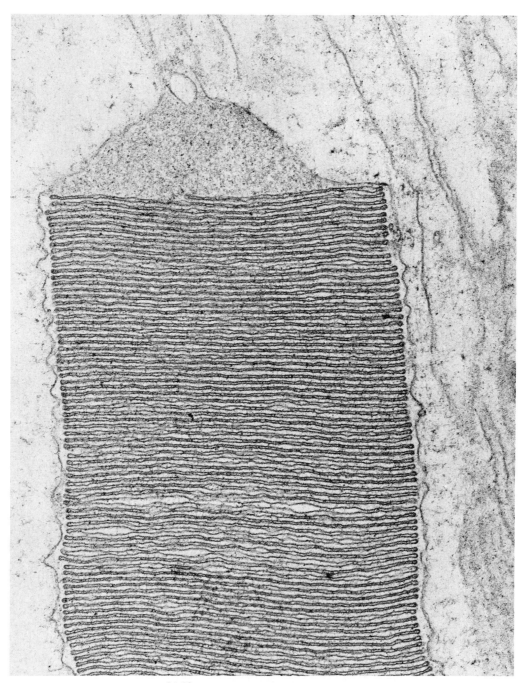

FIGURE 21-19
The end of an outer rod segment of a cow, magnification
80 000×. The rod sacs or disks appear to be freely floating.
Light enters the rod segment from below. Courtesy of
Arnold I. Goldman, National Eye Institute.

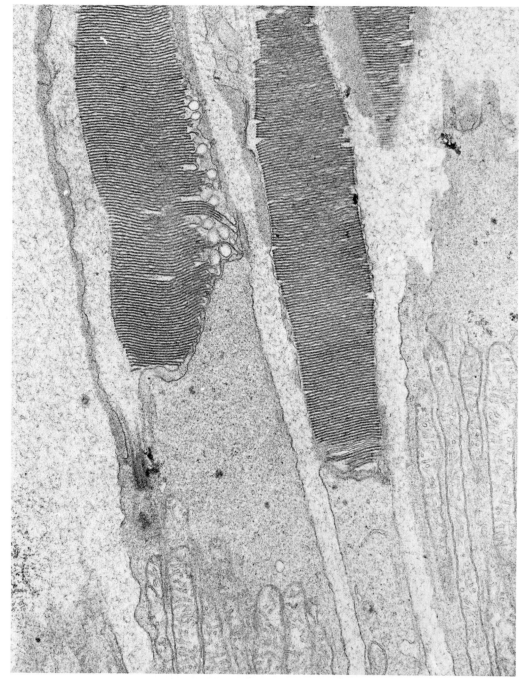

FIGURE 20-20
The base of a cone outer segment of a rhesus monkey, magnification 26 000X. The inner segment, with the mitochondria, can be seen at the bottom. Courtesy of Arnold I. Goldman, National Eye Institute.

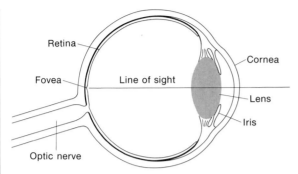

FIGURE 21-21
The location of the fovea of the eye. This tiny retinal depression is densely packed with nerve cells. Their organization and density make possible a high degree of visual acuity. From Ulric Neisser, "The Processes of Vision." Copyright © 1968 by Scientific American, Inc. All rights reserved.

amount of energy. Sensitivity was measured by recording the number of times a flash was actually seen by an observer with dark-adapted vision. The minimum recorded detectable energy was between 2.1 and 5.7×10^{-17} joules. At a wavelength of 500 nm, and allowing for 4% reflection at the cornea, 50% absorption in the light path in the eye, and 80% transparency of the retina, they arrived at the minimum observable light flash: ". . . in order for us to see, it is necessary for only one quantum of light to be absorbed by each of 5 to 14 retinal rods." [4]

Under bright light, the cones are the primary photoreceptors. As Figure 21-21 shows, not far from the optical axis of the eye, on the retina, is the fovea, giving us our remarkable ability to distinguish fine detail in good light. The central territory of the fovea, where the cones are almost uniformly thick and may total as many as 2000 in number, is approximately 100 μm across. The cones are about 2 μm apart. The typical angular separation of two cone cells is 24 seconds of arc. The cone cells of the fovea can discriminate between two points of light separated by only one minute of arc.

[4] Selig Hecht, Simon Sclaer, and Maurice H. Pirenne, "Energy, Quanta, and Vision," *Journal of General Physiology 25* (July 1942), pp. 819–840. This very readable paper, the basis of most textbook discussions of the quantum nature of seeing, remains the definitive study.

Perception of Color

Isaac Newton demonstrated that white light can be broken into colors, and that two colors combine to produce a third color. Our perception of color depends on the physiology of the retina, and on our subjective evaluation. What we perceive as white can range from a yellowish white to a bluish white depending on the time of day and the surroundings. Some persons also perceive color differently from most persons, for example, confusing red with green (less commonly they confuse green with blue). Such persons suffer from color blindness.

In the **tricolor theory** first put forward in 1801, Thomas Young showed that any color of the spectrum could be produced by the mixture of three colored lights of suitable wavelength, shown in Figure 21-22. Hermann von Helmholtz, in his experimental studies on human vision, amplified Young's work. Helmholtz's *Physiological Optics* remains one of the important studies of the human eye. What is known as the Young-Helmholtz theory postulates that the color-sensitive photoreceptors in the retina have different pigments sensitive to violet, green, or red portions of the spectrum.

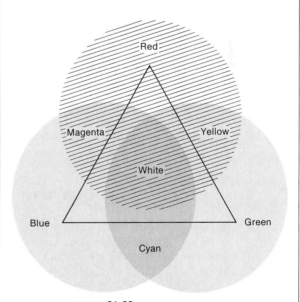

FIGURE 21-22
The tricolor triangle of mixed light.

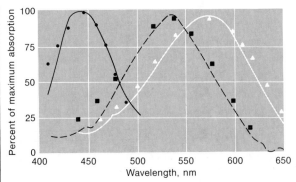

FIGURE 21-23

Color vision in primates is mediated by three cone pigments that sense light in the blue, green, and red portions of the spectrum. Their sensitivity as a function of wavelength is shown here. Although the red receptor peaks in the yellow, it extends far enough into the red to see red well. From Edward F. MacNichol, Jr., "Three-Pigment Color Vision." Copyright © 1964 by Scientific American, Inc. All rights reserved.

The tricolor theory is supported by measurements of the light-absorption spectrum of individual cone cells in the retinas of goldfish, rhesus monkeys, and humans. These researches were conducted, in the 1960s, by Edward MacNichol, Jr., and coworkers at Johns Hopkins University and by George Wald and coworkers at Harvard. Three pigments the spectral sensitivities of which peak at 447 nm (blue-violet), 540 nm (green), and 577 nm (yellow) have been observed. Figure 21-23 shows the absorption of the pigments.

Review

Give the significance of each of the following terms. Where appropriate give two examples.

Fraunhofer dark lines
emission spectra
absorption spectra
Kirchhoff's laws of spectroscopy
atomic absorption spectroscopy
Balmer series

absorption
emissivity
Kirchhoff's emission law
black body
Planck's constant
photoelectric effect
photons

work function
threshold frequency
image orthicon tube
rod cells
cone cells
tricolor theory

Further Readings

Sterling B. Hendricks, "How Light Interacts with Living Matter," *Scientific American* (September 1968). Reprinted in *Lasers and Light* (San Francisco: W. H. Freeman, 1969).

Edwin H. Land. "Experiments in Color Vision," *Scientific American* (May 1959).

Edward F. MacNichol, Jr., "Three-pigment Color Vision," *Scientific American* (December 1964), pp. 48–56.

Charles R. Michael, "Retinal Processing of Visual Images," *Scientific American* (May 1969).

George Wald, "Molecular Basis of Visual Excitation," *Science* (11 October 1968), pp. 230–239. Harvard professor of biology, George Wald was awarded the Nobel Prize in physiology in 1967 for his biochemical and physiological work on visual excitation. See also his early paper in *Science* (29 June 1945), p. 653.

George Wald, "Light and Life," *Scientific American* (October 1959). Reprinted in *Lasers and Light* (San Francisco: W. H. Freeman, 1969)

Frank S. Werblin, "The Control of Sensitivity on the Retina," *Scientific American* (January 1973), pp. 71-79.

Problems

THE SPECTRUM OF LIGHT

1 Using the general expression for the spectrum lines of hydrogen, determine the wavelengths of the first four lines of the Lyman series.

2 Using the general expression for the spectrum lines of hydrogen, determine the wavelengths of the first four lines of the Paschen series.

3 If the sensitivity of the eye extends from 700 nm to a short-wavelength limit of 400 nm.
   (a) How many Balmer lines can the eye see under ideal conditions?
   (b) What is the wavelength of the last observable line?
   (c) What is the wavelength of the next line (unobserved) in the series?

BLACK-BODY RADIATION

4 The standard of illumination, the candela, is defined in terms of the emission from an area of $1/60$ cm^2 at the freezing point of platinum metal (2063 K) at a pressure of one atmosphere. What is the wavelength of maximum emission?

5 The color of a hot metal object has long been used to estimate temperature. For the following wavelengths of maximum emission of black-body radiation, determine the temperature:

(a) A dull red with the maximum in the infrared at 800 nm.
(b) Red-hot with the maximum at the edge of visibility, 700 nm.
(c) Yellow-hot with the maximum at 600 nm.
(d) White-hot with maximum at the peak of eye sensitivity, 510 nm.

6 A radiation pyrometer is used to determine the emission temperature of a surface above 500°C, by comparing the color of the source with the color of a hot filament. A particular pyrometer has a useful range from 700°C to 1900°C.
   (a) What is the temperature in Kelvin? What is the wavelength of maximum emission over this range?
   (b) What is the wavelength of maximum emission of iron at its melting point of 1535°C?

7 The wavelength of maximum emission for a black body is given by the Wien displacement law. The mean kinetic energy of a molecule is given by $KE = \frac{3}{2}kT$. The ground has an absolute temperature of about 280 K.
   (a) What is the wavelength of maximum emission at this temperature?
   (b) What is the photon energy at this wavelength in eV?
   (c) What is the average kinetic energy of a molecule at this temperature in eV?

8 The top of the cloud cover 6 km above the earth's surface is at a temperature of about 265 K and radiates as a black body in the infrared. How much power is radiated in the wavelength range from 3.4 μm to 4.2 μm by a 1 km area of the cloud cover?

9 The surface of the ocean is at a temperature of 285 K and has an absorption coefficient of 0.45 in the frequency range 19.3 GHz to 19.4 GHz.
   (a) How much power is radiated by a 1 km square area of the ocean's surface in this frequency range?
   (b) If this power were radiated in this frequency range by a perfect black body, what would be the effective emission temperature?

PHOTOELECTRIC EFFECT

10 A mercury vapor flame has a spectral line of wavelength 334 nm shining on a sodium metal surface with a work function 2.46 eV. What is the maximum kinetic energy of the ejected electron in eV?

11 A helium-neon laser has a wavelength of 632.8 nm.
   (a) What is the frequency of the laser light?
   (b) What is the energy of a photon at this wavelength?

12 Ultraviolet light in the wavelength range 290 nm to 200 nm is harmful to plants. What is the range of photon energies?

13 An image orthicon used in a television camera has a maximum efficiency at a wavelength of 450 nm. The threshold wavelength is 720 nm.
   (a) What is the work function of the image orthicon surface?
   (b) What is the kinetic energy of a photoelectron emitted by a 450 nm photon?

14 An infrared-sensitive photomultiplier has a silver-cesium oxide (AgCsO) photocathode that has a threshold wavelength of 1.18 μm. What is the work function for the cathode material?

15 A clean zinc metal surface has a work function of 4.2 eV.
   (a) What is the threshold wavelength for the emission of photoelectrons?
   (b) If ultraviolet light of wavelength 200 nm falls on the surface, what is the maximum kinetic energy of the ejected photoelectrons?

16 A modern photomultiplier tube has a cesium-tin oxide (CsSbO) photocathode. The detector has a maximum efficiency at a wavelength of 400 nm. It has a threshold wavelength of 650 nm.
   (a) What is the work function of the material?
   (b) What is the kinetic energy of a photoelectron emitted by a 400 nm photon?

17 Sunlight reaches the earth's surface with a power of 800 W/m². If the light is composed of photons with an average wavelength of 510 nm, how many photons per second will strike a surface 1 cm square?

18 A 100 watt light bulb produces about 10% of its energy as visible light. If the light has an average wavelength of 510 nm, how many photons are given off per second?

19 Some of Millikan's original data of 1916, corrected for the contact potential, show the stopping potential V_o as a function of wavelength: $\lambda_1 = 254$ nm, $V_1 = 3.04$ V; $\lambda_2 = 312.5$ nm, $V_2 = 2.14$ V; $\lambda_3 = 365$ nm, $V_3 = 1.61$ V; and $\lambda_4 = 546$ nm, $V_4 = 0.48$ V.
   (a) Make a sketch of stopping potential as a function of frequency.
   (b) Estimate the threshold frequency. The work function.
   (c) From these data what is Planck's constant?

20 Hecht's experiments on the detection of a flash of light of wavelength 510 nm showed that a flash of duration 0.001 s and energy of 2.5×10^{-17} J incident on the cornea of the eye could be seen 60% of the time. How many photons is this?

21 The early Daguerrotype photographic materials required a light intensity per unit area of 8×10^{-1} J/m² to form an image. Modern high-speed materials require only 3×10^{-6} J/m². How many photons at 450 nm are required to produce an image 1 mm square on the two materials?

VI

THE ATOM

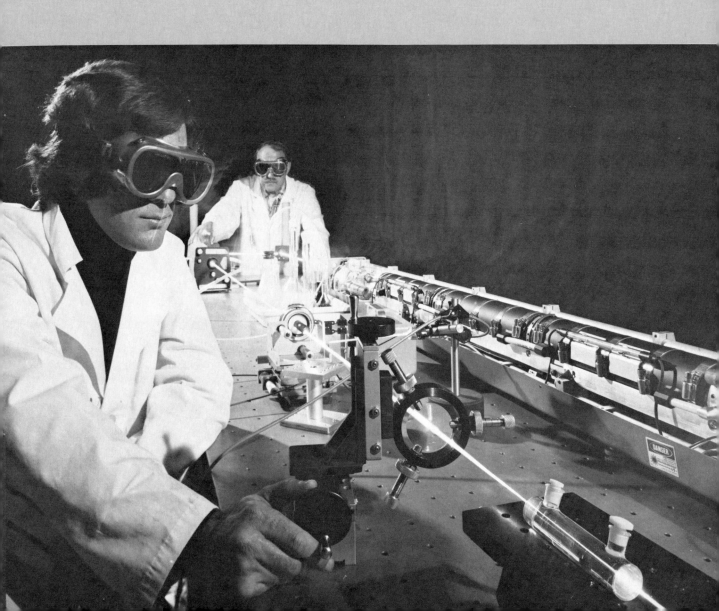

22

The Quantum Atom

The work of such nineteenth-century chemists as Joseph Dalton, Jöns Berzelius, Amedeo Avogadro, Dmitri Mendeleev yielded much knowledge about the chemical atom: electrical properties, atomic weights, how molecules are formed, and the like. The elements could be systematically arranged by weights and chemical properties in a periodic table. But although, after Kirchhoff's work, chemists and physicists had been able to observe emission spectra of the elements that showed sharply defined, widely spaced lines, there was no theory that could predict what the sequence of spectrum lines for a particular element would be.

The structure of the atom could not be accurately described until more was known about its constituents, the electrons and nucleus, and about how an atom gives off light. To explain the atom, quantum notions were needed. In this chapter we briefly examine the work and ideas of a few key scientists, Thomson, Roentgen, Becquerel and Rutherford, men whose work led to Bohr's quantum atom. Bohr's model of the atom was confirmed by Moseley's study of the x-ray spectra of atoms that predicted three new elements, which were subsequently discovered. Finally we shall apply energy levels and the emission of photons to modern lasers.

SETTING THE STAGE

In 1838, when Michael Faraday sent a current from an electrostatic generator through a partially evacuated glass tube, a purple glow extended from the *anode* (the positive electrode) to the *cathode* (the negative). The cathode was covered with a purple *cathode glow*. Between the cathode glow and the glow in the tube there was a dark space, the *Faraday dark space*. With better vacuum pumps to more completely evacuate the tube, the dark space became larger and **cathode rays,** rays traveling from the cathode, struck the end of the tube and caused the glass to fluoresce. A solid object placed between the end of the tube and the cathode cast a shadow. In 1879, the Englishman Sir William Crookes developed several highly evacuated tubes, **Crookes tubes,** to demonstrate cathode rays. Figure 22-1 shows one such tube with a Maltese cross that casts a shadow at the end of the tube, indicating that cathode rays travel in straight lines. Cathode rays, which occur no matter what the material of the cathode, are deflected by a magnetic field; they have sufficient energy to heat a thin metal foil inside the tube to red hot, and exert a force on an object inside the tube, such as a metal paddle, causing it to turn. If allowed to pass out of the

FIGURE 22-1
A Crookes tube with a Maltese cross, for studying the properties of cathode rays. Courtesy of Central Scientific Co.

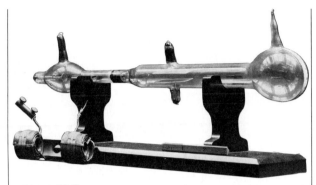

FIGURE 22-2
Thomson's apparatus for determining the charge-to-mass ratio of cathode rays. The cathode rays are produced at the left and pass through slits into a field-free region. An electric field can be applied across the metal-plate electrodes inside the tube. A magnetic field can be applied in the same region by field coils shown on the table. Lent to Science Museum, London, by J. J. Thomson, Trinity College, Cambridge.

vacuum through a window of thin glass, cathode rays travel only about a centimeter in air. Cathode rays carry a negative charge.

The Ratio of Charge to Mass

In 1897, J. J. Thomson, then director of the Cavendish Laboratory at Cambridge, England (a laboratory central to the development of twentieth-century physics), reported a series of experiments in which he measured the charge-to-mass ratio of cathode rays. Figure 22-2 shows Thomson's apparatus, and Figure 22-3 shows how it operates. A beam of cathode rays in the evacuated tube passes through two successive slits into the space between two metal plates. An electric field $\mathbf{E}$ can be applied across the plates. A magnetic field $\mathbf{B}$ can also be applied in the same region. The cathode ray particles travel in straight lines except when they are in the magnetic or electric fields. The magnetic field provides the centripetal force to deflect the particles in a circular path of radius R:

$$F = \frac{mv^2}{R} = qvB$$

The velocity of the particles is related to the charge per unit mass. Solving the preceding equation we get

$$v = \frac{q}{m} RB$$

The magnitude B of the magnetic field and the curvature R of the beam can be measured.

To determine q/m, we must determine the velocity v of the particles. To do so, Thomson had the energy of the cathode rays be deposited on a small metal target as heat and measured the temperature rise with a thermocouple. The energy in the beam can be determined by simple calorimetry. If each particle has a mass m and a velocity v, it has a kinetic energy $KE = \frac{1}{2}mv^2$. If N particles are collected, the total energy collected $U = N\frac{1}{2}mv^2$. If each particle has a charge q, the total charge collected is $Q = Nq$. By solving, both v and q/m can be determined. Using this method, Thomson found that

$$\frac{q}{m} = -1.0 \text{ to } 1.4 \times 10^{11} \text{ C/kg}$$

Thomson's values were a bit small. For the negative cathode ray particles, or **electrons** as we know them, the

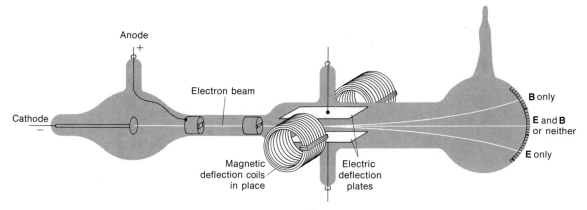

Anode
+

Cathode
−

Electron beam

Magnetic
deflection coils
in place

Electric
deflection
plates

B only

E and **B**
or neither

E only

FIGURE 22-3
A schematic of Thomson's experiment showing the path of the cathode rays.

present value is the ratio of charge to mass

$$\frac{q}{m} = -1.76 \times 10^{11} \text{ C/kg}$$

Discovery of the X-Ray

In December 1895, two memorable events occurred: the first commercial motion picture projector, *le cinematographe*, was introduced, and the discovery of x-rays was announced by Wilhelm Roentgen, a German professor of physics (see Figure 22-4). It would be hard to say which event has had a greater influence on modern life.

In November 1895 Roentgen observed that cathode rays from a Crookes tube caused certain minerals to fluoresce. He was working in a darkened room with the Crookes tube enclosed in a black lighttight box. When he turned on the tube, he noticed, much to his surprise, that a small fluorescent screen some distance away was glowing. He concluded that some unknown ray had penetrated the lighttight box and was causing the fluorescent glow. In the next six weeks he investigated these rays, which, to distinguish them from other rays, he called **x-rays.**

The x-rays were invisible to the eye, were not influenced by magnetic fields, and easily penetrated several meters of air, thick blocks of wood, rubber, and ordinary glass, and darkened photographic plates. They were stopped by a few millimeters of lead or of lead glass. The x-rays were produced at the point where the cathode rays struck the glass end of the Crookes tube.

Placing his wife's hand between his x-ray tube and a photographic plate, Roentgen made the first x-ray photo-graph, which revealed the bones and shadowy outline of her hand. News of his discovery caused an instant sensation in the press. The medical profession was quick to use x-rays to reveal broken bones, and shotgun pellets and other foreign objects lodged in the body.

THE NEW PHOTOGRAPHIC DISCOVERY.

THANKS TO THE DISCOVERY OF PROFESSOR RÖNTGEN, THE GERMAN EMPEROR WILL NOW BE ABLE TO OBTAIN AN EXACT PHOTOGRAPH OF A "BACKBONE" OF UNSUSPECTED SIZE AND STRENGTH !

FIGURE 22-4
Once discovered, x-rays quickly captured the public imagination. This cartoon appeared in the 25 January 1896 issue of *Punch*, only one month after Röentgen announced his discovery. Reproduced by permission of *Punch*.

Rays

The French physicist Henri Antoine Becquerel, who had been studying the fluorescent properties of certain uranium salts, soon heard news of Roentgen's x-rays. He supposed that the emission of x-rays was related to the fluorescence produced when the cathode rays struck the glass. To test his idea, he placed uranium crystals in the sunshine (to excite, so he supposed, fluorescent generation of x-rays) atop a covered photographic plate. The photographic plate was indeed blackened as if by x-rays. He repeated his experiment the following week, but because the sun was not shining, he placed his samples in a drawer. Developing the plates, he again found a dark image and concluded that the uranium salt emitted continuously some unknown ray. The task of chemically isolating the source of the rays he gave to one of his students, Marie Curie (whose work we discuss in Chapter 24).

Both Roentgen's and Becquerel's work quickly attracted the attention of J. J. Thomson and his young student Ernest Rutherford, who immediately undertook a study of the ionizing power of these new rays. They found that x-rays ionize air molecules and neutralize charged bodies, such as an electroscope, giving a quantitative method of measuring x-ray intensities. In studying the ionizing properties of Becquerel's rays, they found that two distinct types of rays were given off by uranium compounds: **Alpha rays,** which were easily absorbed by a sheet of paper or thin metal foil, were found to be helium atoms with two electrons removed, **alpha particles;** and **beta rays** penetrated 2 mm of aluminum, were deflected by a magnetic field, had a negative charge, and had approximately the same charge-to-mass ratio as the cathode ray electrons.

In 1899, the Dutch physicists H. Haga and C. H. Wind performed a single-slit diffraction of x-rays and concluded that they were waves like light with a very short wavelength of about 0.1 nm. In 1900, the Frenchman P. Villard found a third radiation coming from uranium compounds. The **gamma ray** had great penetrating power and behaved like an x-ray with a very short wavelength.

The five-year period from 1895 to 1900, then, was marked by the discoveries of x-rays and gamma rays, which behaved as very-short-wavelength light; and cathode rays, alpha rays, and beta rays, which behaved as particles. These new phenomena raised hundreds of questions that triggered a flurry of scientific activity and set the stage for the developments of twentieth-century physics.

A NEW ATOMIC MODEL

By 1910, from accumulating experimental evidence a new picture of the atom was slowly emerging. The atom was electrically neutral, but the primary forces within it were assumed to be electrical. The mass of the atom was associated with the positive charge. The negative charge was carried by electrons that had a mass of a $\frac{1}{1840}$ the mass of a hydrogen atom. J. J. Thomson had developed a classical model of the atom on the assumption that radiation from the atom should be given by the equations of electricity and magnetism, and that the dynamics of the atom should follow Newton's laws of motion. Being at rest within the atom, the electrons did not radiate. In 1902, Lord Kelvin had proposed a "raisin cake" model of the atom, so called because the electrons were thought to be embedded in the positive charge like raisins in a cake (see Figure 22-5). But still a piece was missing.

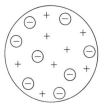

FIGURE 22-5
Lord Kelvin's raisin cake model of the atom. The negatively charged electrons (the raisins) are embedded in a uniform sphere of positive charge (the cake). However, electrostatic forces make this model unstable.

Ernest Rutherford and Hans Geiger

In 1907, Ernest Rutherford, now professor of physics at the University of Manchester in England, continued his researches, begun at McGill, on the scattering of alpha particles (see Figure 22-6). These particles, massive doubly ionized helium atoms, seemed to Rutherford to be ideal probes with which to study the structure of the atom. In 1908, his student Hans Geiger undertook a semi-quantitative study of **alpha scattering** by a gold foil, observing through a microscope the fluorescence produced as individual alpha particles struck a zinc sulfide screen. The experiments were tedious and the "sparks" of fluorescence were difficult to count for longer than two minutes at a time. Figures 22-7 and 22-8 show Geiger's

FIGURE 22-6
Hans Geiger and Ernest Rutherford in the Schuster Laboratory, at Manchester, at about the time of their researches on alpha particle scattering. Courtesy of the Department of Physics, the University of Manchester.

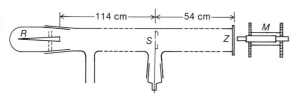

FIGURE 22-7
Geiger's early apparatus for observing alpha particles from source *R* scattering from a foil at *S*. Alpha particles striking the screen *Z* were observed by eye through a measuring microscope *M*. Adapted from *Proceedings of the Royal Society (London) A81* (1908), p. 174.

early apparatus and the results for alpha scattering. Curve *A* is the result of collimation of the alpha particle beam through a simple slit in vacuum. The other curves show the small scattering for one and two thin gold foils.

The scattering of alpha particles from a thin gold foil was usually small, less than one degree. As a small research project, Geiger and Rutherford asked a young student, E. Marsden, to see if he could observe any alpha particles being scattered backward from the foil at large angles. They did not think this occurrence likely, but much to their surprise, Marsden indeed found that a few alpha particles returned from the same side of the foil, having been scattered into a large angle of almost 180°.

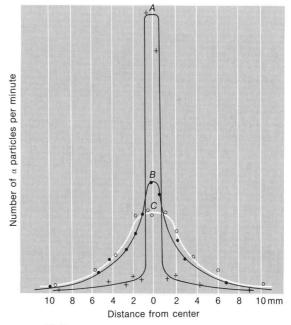

FIGURE 22-8
Geiger's measurements of alpha particle scattering. Curve A shows the collimation of the alpha particle beam in vacuum. Curves B and C represent the scattering due to one and two thin gold foils respectively. Adapted from *Proceedings of the Royal Society (London) A81* (1908), p. 176.

Rutherford Scattering

Rutherford worked out systematically the type of scattering that could explain the large-angle scattering of alpha particles by an atom of gold. Only Coulomb forces due to the positive charge of the atom could produce the scattering of a heavy alpha particle. The electrons were too light and would simply be deflected; they would contribute little to the scattering. The magnitude of the force between two charges is given by Coulomb's law:

$$F = \frac{k_e Q_1 Q_2}{R^2}$$

The force acts along the line of centers between the charges.

If a second charge Q_2 approaches a charge Q_1 from infinity to within a distance of R, the potential energy of the system increases:

$$PE = \frac{k_e Q_1 Q_2}{R}$$

The atom has a charge $Q_1 = +Ze$. For gold, $Z = 79$. The approximate distance between atoms of a gold foil, $2R_0$, can be determined from the density of gold $\rho = 19\ 300$ kg/m³, Avogadro's number N_A, and from the atomic weight of gold $M_A = 0.197$ kg/mol. The volume occupied by an atom of gold V_{atom} is approximately

$$V_{atom} = \frac{M_A}{N_A} = \frac{0.197 \text{ kg/mol}}{(6.023 \times 10^{23}/\text{mol})(19\ 300 \text{ kg/m}^3)}$$

$$V_{atom} = 17 \times 10^{-30} \text{ m}^3$$

The approximate spacing of the atoms is the cube root of the volume.

$$2R_0 = (V_{atom})^{1/3} = 2.6 \times 10^{-10} \text{ m} = 0.26 \text{ nm}$$

The radius of the atom R_0 is approximately 0.13 nm.

The charge on an alpha particle is $Q_2 = +2e$. Rutherford had determined that the alpha particle has a mass of 6.6×10^{-27} kg and a velocity of 1.6×10^7 m/s. The alpha particle had an initial kinetic energy of

$$KE_1 = \tfrac{1}{2}mv^2 = \tfrac{1}{2}(6.6 \times 10^{-27} \text{ kg})(1.6 \times 10^7 \text{ m/s})^2$$

$$KE_1 = 8.5 \times 10^{-13} \text{ J} = 5.3 \times 10^6 \text{ eV}$$

As the alpha particle approaches the gold atom, the Coulomb force converts kinetic energy to electrical potential energy.

We can determine the size of a charge distribution required to just stop an alpha particle in a head-on collision from conservation of energy. If the alpha particle is stopped, its kinetic energy goes to zero. The initial kinetic energy KE_1 is converted to potential energy PE_2:

$$KE_1 = PE_2 = \frac{k_e Q_1 Q_2}{R'}$$

We can at once calculate the radius of closest approach R':

$$R' = \frac{k_e Q_1 Q_2}{KE_1} = \frac{2k_e Z e^2}{KE_1}$$

$$R' = \frac{2(9 \times 10^9 \text{ N-m}^2/\text{C}^2)(79)(1.6 \times 10^{-19} \text{ C})^2}{8.5 \times 10^{13} \text{ J}}$$

$$R' = 4.3 \times 10^{-14} \text{ m}$$

To be stopped or to be scattered through a large angle, the alpha particle must approach within a distance of 4×10^{-14} m of the positive charge. The size of the atom is of the order of 10^{-10} m. Thus, to explain the results of Geiger's and Marsden's alpha-scattering experiment it must be assumed that the positive charge of the atom was concentrated in a small region of space, the **nucleus**. Rutherford considered the scattering of an alpha particle from a point charge and calculated the probability that a beam of particles is scattered into some angle θ, as shown in Figure 22-9. He demonstrated theoretically that the scattering into a particular angle should be proportional to

$$\frac{1}{\sin^4 \tfrac{1}{2}\theta}$$

An extensive series of experiments by Geiger and Marsden, published in 1913, confirmed this result in every particular. Scattering was firmly established both experimentally and theoretically, and little doubt remained that

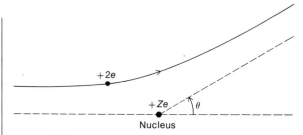

FIGURE 22-9
Alpha particle scattering. Rutherford calculated the scattering due to the Coulomb force between the alpha particle and the nucleus.

the positive charge and the mass of the atom were concentrated in a small nucleus.

Rutherford wasted no time in proposing a new, solar-system model of the atom in which the electrons as particles move in circular orbits at a radius of about 10^{-10} m around the small, massive nucleus, as shown in Figure 22-10. This model had one serious flaw: An electron rotating in such an orbit would continuously lose energy by electromagnetic radiation. This atom would in fact collapse in a time of less than 10^{-8} s! The model also explained nothing about the emission spectrum of atoms.

THE BOHR MODEL

The young Danish physicist Niels Bohr was a strong proponent of the new quantum notions by which Planck had explained black-body radiation and Einstein the photoelectric effect. In the spring of 1912, while spending four months with Rutherford in Manchester, Bohr "became convinced that the electric constitution of the Rutherford atom was governed throughout by the quantum of action." The theoretical collapse of Rutherford's solar-system model presented a challenge to the young theorist. By the time Bohr returned to Copenhagen in July 1912, he had formulated ideas that would bring revolution to physics.

Bohr based his own theory on the Rutherford atom. He first considered the energy of an electron of charge $-e$ in a uniform circular orbit of radius R around a nucleus of charge Ze, where Z is the atomic number of the element. The force, shown in Figure 22-11, between the electron and the nucleus, given by Coulomb's law, provides the centripetal force to keep the electron in a circular orbit:

$$F = \frac{m_e v^2}{R} = \frac{k_e Z e^2}{R^2}$$

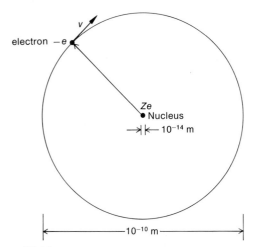

FIGURE 22-10
Rutherford's solar-system model of the atom. Unfortunately, classical theory predicts that this model will collapse in about 10^{-8} s.

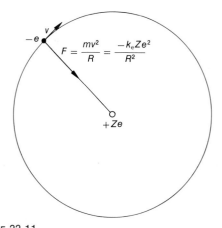

FIGURE 22-11
The Bohr model of the atom starts from the assumption that the electron is in a uniform circular orbit about the nucleus with a charge $+Ze$.

Assuming that the speed of the orbiting electron is small compared to the speed of light, its kinetic energy can be determined from the centripetal force equation:

$$KE = \tfrac{1}{2}m_e v^2 = \frac{1}{2}\frac{k_e Z e^2}{R}$$

Because removing the electron from the atom requires work, the potential energy is negative. The electrical potential energy of an electron at a radius R from the nucleus is

$$PE = -\frac{k_e Z e^2}{R}$$

The total energy of the electron in orbit is the sum of the kinetic and potential energy:

$$E = KE + PE$$

$$E = \frac{1}{2}\frac{k_e Z e^2}{R} - \frac{k_e Z e^2}{R}$$

The total energy is negative, indicating a bound orbit:

$$E = -\frac{1}{2}\frac{k_e Z e^2}{R}$$

This is the classical result using Newton's laws for a particle in a uniform circular orbit. It ignores the problem of radiation caused by the accelerating electron. By the theory of electromagnetic waves such a system radiates energy and the electron moves closer and closer to the nucleus, leading to collapse of the atom. To forestall this theoretical problem, Bohr assumes that "the energy radiation from an atomic system does not take place in the continuous way assumed in the ordinary electrodynamics, but that it, on the contrary, takes place in distinctly separate emissions."

The **Bohr model** of the atom, then, makes three assumptions:

1. The electron in an orbit around the nucleus can be treated as a classical system in a circular orbit.
2. The system settles down to certain stationary states that do not radiate.
3. The system radiates when the electron changes from one stationary state to another.

Bohr's interpretation of the condition for stable orbits was simple; the angular momentum of the electron in orbit around the nucleus is equal to an integer value of Planck's constant divided by 2π:

$$L = m_e v R_n = \frac{nh}{2\pi} \tag{22-1}$$

The quantum condition can be written

$$m_e v = \frac{L}{R_n} = \frac{nh}{2\pi R_n}$$

The kinetic energy can then be written

$$KE = \frac{(m_e v)^2}{2m_e} = \frac{1}{2m_e}\left[\frac{nh}{2\pi R_n}\right]^2 = \frac{k_e Z e^2}{2R_n}$$

Solving for the radius of the orbit, R_n is

$$R_n = \frac{n^2 h^2}{4\pi^2 m_e k_e Z e^2} = \frac{n^2 a_o}{Z} \tag{22-2}$$

For hydrogen, $Z = 1$. The lowest possible orbit is for $n = 1$. The radius of this lowest possible orbit is

$$a_o = \frac{h^2}{4\pi^2 m_e k_e e^2}$$

$$= \frac{(6.63 \times 10^{-34}\ \text{J-s})^2}{4\pi^2(9.1 \times 10^{-31}\ \text{kg})\left(9 \times 10^9\ \dfrac{\text{N-m}^2}{\text{C}^2}\right)(1.6 \times 10^{-19}\ \text{C})^2}$$

$$a_o = 5.291 \times 10^{-11}\ \text{m} = 0.05291\ \text{nm}$$

This is known as the **Bohr radius** for the hydrogen atom.

The total energy of an electron in a uniform circular orbit is

$$E_n = -\frac{k_e Z e^2}{2R_n} = -\frac{k_e Z e^2}{2a_o}\left[\frac{Z}{n^2 a_o}\right] = -\frac{k_e e^2}{2a_o}\left[\frac{Z^2}{n^2}\right]$$

$$E_n = -E_o \frac{Z^2}{n^2}$$

The constant E_o is the binding energy for the hydrogen atom, the energy required to remove the electron from the lowest energy level.

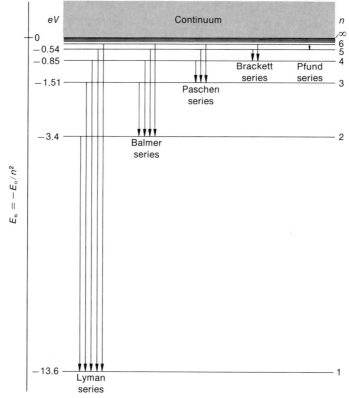

FIGURE 22-12

The energy levels for a bound electron of the hydrogen atom. The lowest energy level is often called the ground state. The free electron levels are called the continuum. From Audrey L. Companion, *Chemical Bonding* (New York: McGraw-Hill Book Co., © 1964).

$$E_o = \frac{k_e e^2}{2a_o} = \frac{(9 \times 10^9 \text{ N-m}^2/\text{C}^2)(1.6 \times 10^{-19} \text{ C})^2}{2(5.29 \times 10^{-11} \text{ m})}$$

$$E_o = 2.18 \times 10^{-18} \text{ J} = 13.6 \text{ eV}$$

The Bohr model predicts that stationary states can occur at different energy levels. These energy levels, characterized by the principal quantum number n, are given by

$$E_n = -\frac{E_o Z^2}{n^2} \qquad (22\text{-}3)$$

Figure 22-12 shows the energy levels for the hydrogen atom.

The energy of the lowest possible level, often referred to as the ground state, is given by $n = 1$.

Bohr immediately concluded that the atom radiates energy, as a photon, in going from one stationary orbit to a lower orbit. The photon energy is equal to the energy difference between the two levels. In going from energy level m to n, the photon energy is

$$h\nu = W_m - W_n$$

We can write the reciprocal of the wavelength:

$$\frac{1}{\lambda} = \frac{h}{c} = \frac{W_m - W_n}{hc}$$

$$\frac{1}{\lambda} = \left[\frac{E_o}{hc}\right]\left[\frac{1}{n^2} - \frac{1}{m^2}\right] \qquad (22\text{-}4)$$

For $n = 2$, this equation immediately yields the Balmer series for the hydrogen lines discussed in Chapter 21. For

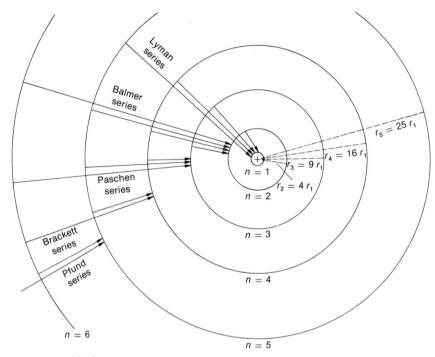

FIGURE 22-13
The Bohr orbits for hydrogen, showing the radii of the different stationary states and the transitions corresponding to different spectroscopic series. From Audrey L. Companion, *Chemical Bonding* (New York: McGraw-Hill Book Co., © 1964).

$n = 1$, we obtain the Lyman series. The transition between energy levels gives rise to the emission of spectrum lines. The constant

$$R_{\mathrm{H}} = \frac{E_{\mathrm{o}}}{hc} = \frac{2\pi^2 m_e k_e^2 e^4}{h^3 c} = 1.10 \times 10^7 \text{ m}^{-1}$$

is almost identical with the **Rydberg constant,** discussed in Chapter 20.

The above derivation assumes that the mass of the nucleus m_p is infinite. When a small correction is made for the finite mass of the nucleus, E_o/hc is identical with the Rydberg constant for hydrogen. One of the strongest confirmations of the Bohr model is its having correctly determined the spectrum lines of hydrogen and the value of the Rydberg constant on strictly theoretical grounds.

The Bohr model also predicts that the radius of a given stationary orbit increases as the square of the principal quantum number n. For the hydrogen atom, the radius of the nth orbit is

$$R_n = n^2 R_{\mathrm{o}}$$

Because of collisions, the higher n states are seldom populated in normal gas discharges. Figure 22-13 shows, approximately to scale, the relative sizes of Bohr orbits for the hydrogen atom.

The Bohr model was immediately extended for hydrogenlike ions; that is, nuclei with one electron, such as He^+, Li^{++}, and Be^{+++}. The energy levels of such ions are given by equation 22-3:

$$E_n = -\frac{Z^2 E_{\mathrm{o}}}{n^2}$$

The radii of the orbits are given by equation 22-2:

$$R_n = \frac{n^2 R_{\mathrm{o}}}{Z}$$

The electron is pulled closer to the nucleus, and the binding energy for a given energy level increases as Z^2. The energy required to remove the electron from the ground state of He^+ is four times as great as that from H. Simi-

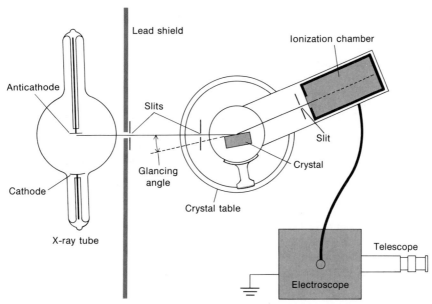

larly, the energy to remove the electron from the lowest state of Be^{+++} is 16 times as great as the ionization energy of hydrogen. However, when the Bohr theory was applied to even the simplest two-electron system, the helium atom with a nucleus and two electrons, it failed completely. It gave the wrong numerical result for the binding energy of helium, and it could not predict the observed helium spectra.

X-RAY SPECTRA

Bragg Diffraction

Further evidence that x-rays are waves came in 1912, when Max von Laue produced interference patterns by directing x-rays through a sodium chloride crystal, used as a three-dimensional diffraction grating, onto a photographic plate. The regular pattern of spots that appeared corresponded to different spacings between atoms of the crystal. Von Laue deduced that his x-ray beam had wavelengths of about 0.1 to 0.05 nm.

Almost immediately, Sir William Henry Bragg constructed a crystal spectrometer to further study the diffraction of x-rays (see Figure 22-14). His son William Lawrence Bragg found a convenient way to analyze the diffraction pattern. The x-rays are partially scattered by the electron distribution of the atoms in crystal. Heavier elements produce more scattering. By a simple Huygens wavefront argument, illustrated in Figure 22-15, Bragg showed that the angle of reflection of the x-rays was equal to the angle of incidence from a plane of atoms.

Waves scattered from two successive planes in the crystal lattice interfere with one another, as shown in Figure 22-16. If the path difference is an integer number of wavelengths $n\lambda$, the scattered waves will arrive in phase, giving an interference maximum. This, the condition for **Bragg diffraction,** is given by

$$2d \sin \theta = n\lambda \qquad (22\text{-}5)$$

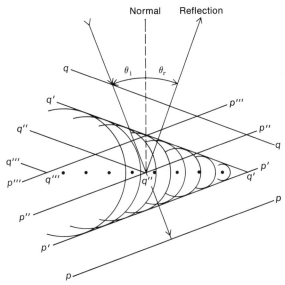

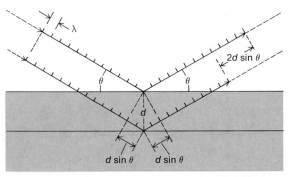

FIGURE 22-15
Coherent scattering of x-rays from a layer of atoms. The lines *p, p', p''* represent successive wave fronts. The scattering from the atoms, represented as dots, sets up waves radiating outward from the scattering center. The angle of incidence equals the angle of reflection.

FIGURE 22-16
Bragg condition for diffraction of x-rays from successive sheets of atoms in a crystal. From Sir Lawrence Bragg, "X-Ray Crystallography." Copyright © 1968 by Scientific American, Inc. All rights reserved.

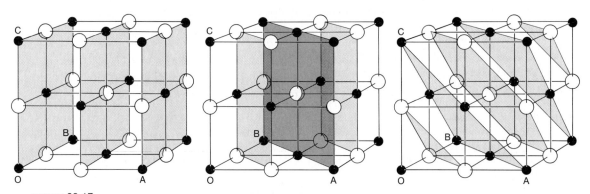

FIGURE 22-17
Bragg planes. Three possible arrangements of reflecting sheets of atoms in a crystal of sodium chloride. Each arrangement produces a diffraction maximum. From Sir Lawrence Bragg, "X-Ray Crystallography." Copyright © 1968 by Scientific American, Inc. All rights reserved.

For a well-ordered crystal there are different *Bragg planes* that lead to different diffraction maxima, as shown in Figure 22-17. (Both Braggs, father and son, were awarded the Nobel Prize in 1915 for their constribution to crystal analysis.)

As Figure 22-18 shows, x-rays are produced when electrons, accelerated up to energies of 20 keV or more, strike an anode containing the target metal. If the Bragg spectrometer is used with a known crystal, the wavelength spectrum of the x-rays produced in the anode target can

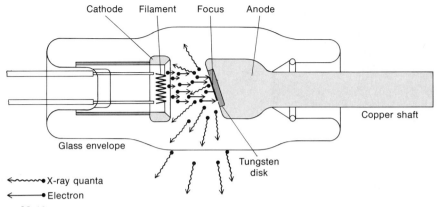

FIGURE 22-18
An x-ray tube with stationary anode. Electrons are emitted from the filament, are focused, and are accelerated through 30 kV to 100 kV. They strike the anode, which contains a tungsten target. About 99% of the energy of the electrons is converted to heat in the anode, and only 1% into x-ray energy. The copper shaft conducts heat away from the tungsten target. Courtesy of Siemens Corporation.

be determined. Figure 22-19 shows the energy spectrum, the intensity as a function of wavelength for two different target metals at a maximum electron energy of 35 keV.

The x-ray spectrum consists of a sharp line and a continuous spectrum. The continuous spectrum rises to a maximum and then falls quickly to zero with decreasing wavelength. The minimum wavelength corresponds to a photon, which has an energy equal to the maximum electron energy. This continuous spectrum, known as **bremsstrahlung,** or breaking radiation, results when the electron is deflected as it strikes the crystal lattice of the metal target. The electron is decelerated and emits electromagnetic radiation. The x-ray photon has a maximum energy equal to the maximum electron energy:

$$h\nu_{max} = eV_o$$

The minimum wavelength is

$$\lambda_{min} = \frac{c}{\nu_{max}} = \frac{hc}{eV_o}$$

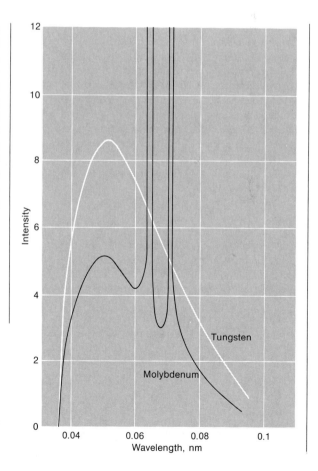

FIGURE 22-19
The x-ray spectrum for an electron energy of 35 keV. For a molybdenum target the characteristic line spectra are observed. For tungsten the electron energy is not sufficient to excite the characteristic x-ray lines.

FIGURE 22-20
X-ray tube with rotating anode. Use of a rotating anode for improved heat dissipation obtains higher x-ray intensities. The anode rotates typically at 150 Hz and can withstand power levels of 30 kW (300 mA at 100 kV) for a 5 s exposure time. The electron-emitting cathode is off-center and focuses the electron beam onto a narrow anode track. Courtesy of Siemens Corporation.

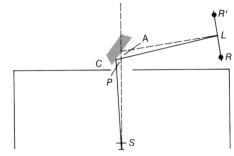

FIGURE 22-21
Moseley's arrangement for diffraction of x-rays by a crystal of potassium ferrocyanide. The x-ray source is mounted behind the narrow slit S. The crystal C, mounted vertically, contains the geometrical axis of the spectrometer. X-ray plates R-R' were placed equidistant from the crystal. In this arrangement the diffraction maximum position L is independent of the point where the x-rays strike the crystal. From H. G. J. Moseley, "Atomic Numbers," *Philosophical Magazine* (1913).

This gives a small-wavelength cutoff to the bremsstrahlung spectrum. If the target is tungsten, with 35 keV electrons, only the continuous spectrum appears. With a molybdenum target, strong x-ray lines appear. A considerable fraction of the electron beam energy in the target excites sharp x-ray spectrum lines known as **characteristic x-rays.** The wavelength is characteristic of the anode target metal. Figure 22-20 shows a modern x-ray tube.

Moseley's Law

Harry Moseley, one of Rutherford's students at Manchester, used the Bragg crystal spectrometer to systematically determine the characteristic x-ray wavelengths of as many elements as possible. Moseley developed a photographic technique shown in Figure 22-21, for recording the spectra; he used a potassium-ferrocyanide crystal, the crystal lattice spacings of which he determined by careful comparison with the diffraction patterns of sodium chloride. Examining the x-ray spectra of heavier elements, he observed decreasing angular displacement and thus decreasing wavelength of the characteristic x-ray lines (see Figure 22-22). Moseley noted in his first paper, published in 1913:

The prevalance of lines due to impurities suggests that this may prove a powerful method of chemical analysis. Its advantage over ordinary spectroscopic methods lies in the

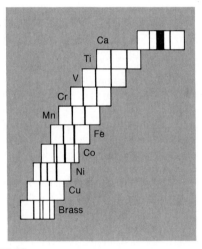

FIGURE 22-22
Moseley's x-ray spectra. Cobalt (Co) has nickel (Ni) as an impurity. Brass shows both copper (Cu) and zink (Zn) lines. From H. G. J. Moseley, "Atomic Numbers," *Philosophical Magazine* (1913).

simplicity of the spectra. . . . It may even lead to the discovery of missing elements, as it will be possible to predict the position of their characteristic lines.[1]

In his second paper (1914), Moseley showed that the square root of the frequency of the characteristic x-ray lines for different elements was in linear relation to an integer number Z. This is **Moseley's law.** The spectra shown in Figure 22-23 form two series. For the lighter

---

[1] Quoted in William and Margaret Dampier, *Readings in the Literature of Science* (New York: Harper and Row, 1959), p. 154.

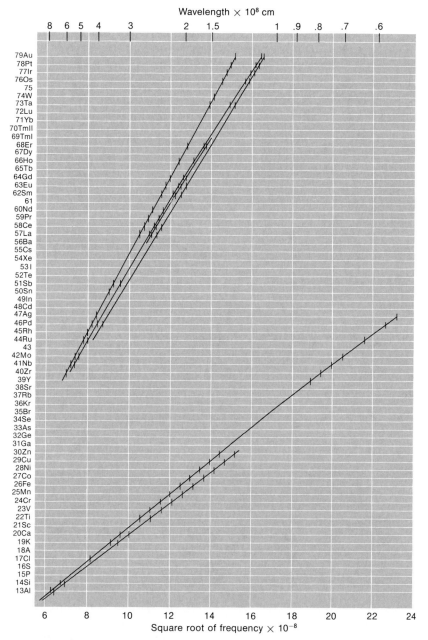

Wavelength × 10⁸ cm

FIGURE 22-23
Moseley's results for the atomic number Z as a function of the square root of the
frequency of the characteristic x-rays. From H. G. J. Moseley, "Atomic Numbers,"
Philosophical Magazine (1913).

elements, the strongest line K_α has the lower frequency; the weaker line K_β has a slightly higher frequency. For the heavier elements there is a series, noted by L_α, L_β, and L_γ, in order of increasing frequency.

Moseley pointed out that the frequency of the K_α spectrum line was closely related to the number Z by

$$\nu = \frac{3}{4}(Z - a)^2 \nu_0$$

where ν_0 is the Rydberg frequency $\nu_0 = R_0 c$, and a is a constant very nearly equal to 1.0. Moseley's relationship can be written in terms of the reciprocal of the wavelength. Because the fraction $\frac{3}{4}$ can be written as

$$\frac{3}{4} = \left[\frac{1}{1^2} - \frac{1}{2^2}\right]$$

Moseley's law for the reciprocal of the wavelength for the K_α lines becomes

$$\frac{1}{\lambda} = \frac{\nu}{c} = \left[\frac{1}{1^2} - \frac{1}{2^2}\right](Z - a)^2 R_0$$

This equation is remarkably similar in form to the sequence of the hydrogen spectra predicted by Bohr's model.

Because the sequence of atomic numbers of the chemical elements had not yet been definitely established, Moseley assigned integer numbers to the elements, starting with aluminum, $Z = 13$. To fit his data, he was required to leave vacancies for $Z = 43$, 61, and 75 for unknown elements. Thus, Moseley made the correct assignment of atomic numbers to the elements.

When World War I broke out, Moseley refused to be dissuaded from enlisting in the Army and was killed in action within less than a year. Rutherford commented that Moseley's work "ranks in importance with the discovery of the periodic law of the elements, and in some respects is more fundamental."

In 1917, Walther Kossel demonstrated that the x-ray spectrum could be interpreted by assigning different energy levels to shells of electrons around the nucleus, as shown in Figure 22-24. The innermost, K-shell electrons are closely bound and correspond to a Bohr $n = 1$ orbit. The next shell, the L-shell, corresponds to $n = 2$, and so

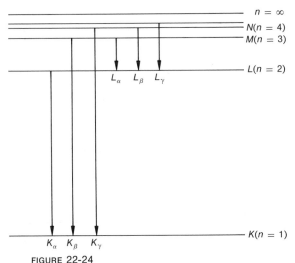

FIGURE 22-24
Energy levels and transitions for characteristic x-rays.

on. When an atom is struck by an energetic electron, an inner K-shell or L-shell electron is sometimes knocked loose. An electron from a higher shell drops down to fill the vacancy, losing energy and giving off a characteristic x-ray. The energy of the inner shells is approximately given by the Bohr model, if allowance is made for a certain small screening of the nucleus charge Ze by the presence of other electrons.

With modern x-ray detectors small concentrations of trace elements can be measured by observation of characteristic x-rays. In x-ray fluorescence, an intense x-ray source, shining on a sample, knocks out a K- or L-shell electron from some fraction of the atoms present. As the shell is filled, the atom immediately radiates its own characteristic x-ray. Modern detectors can measure x-ray energies precisely and thus one can distinguish between characteristic x-rays from different elements in the sample. In Figure 22-25 the x-ray fluorescence spectrum of dried whole blood, the detector count rate is a function of channel number, which is proportional to x-ray energy. At the high-energy end is a large peak caused by the exciting radiation. The iron (Fe) K_α and K_β lines are due to the iron in hemoglobin. This sample shows traces of

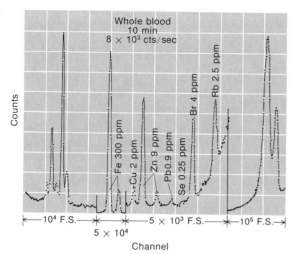

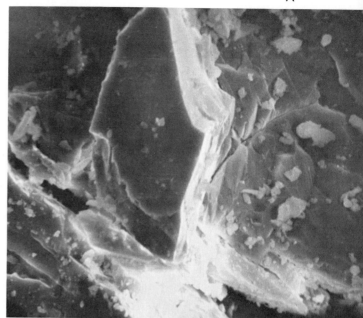

A

FIGURE 22-25

X-ray fluorescence spectrum of whole blood. The measured concentration of lead is several times normal, indicating lead poisoning. The channel number is proportional to the energy of the observed x-ray. Note that the full-scale number of counts (F.S.) changes over different parts of the spectrum. Courtesy of University of California, Lawrence Berkeley Laboratory.

lead (Pb), indicating possible lead poisoning, and several other trace elements.

In electron microscopy, the scanning electron beam is sometimes used to directly excite the sample's characteristic x-rays, which are then observed by an x-ray detector. The image produced by selecting a particular x-ray line is proportional to the concentration of a particular element as Figure 22-26 shows. This technique has been used to analyze the compositions of metal alloys and the elemental concentrations in microscopic biological samples.

Compton Scattering

In 1923, the American physicist Arthur Holly Compton, working at the University of Chicago, observed the scattering of x-rays from the electrons in a graphite target. X-rays from a molybdenum x-ray tube incident on a graphite target are scattered in all directions. When x-rays scattered at some particular angle, say 45° or 90°, were

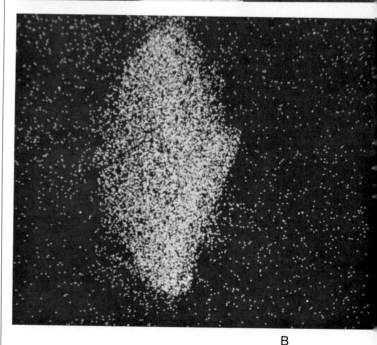

B

FIGURE 22-26

A grain of a moon rock. A. Scanning electron micrograph. B. Scanning x-ray image of the same grain showing a titanium-rich diamond-shaped area. Courtesy of Dr. Rolf Woldseth, Kevex Corporation.

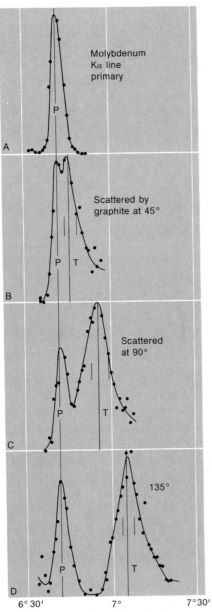

FIGURE 22-27
Compton's spectra of x-rays scattered from graphite at various angles. From *Physical Review, 22* (1923), p. 411.

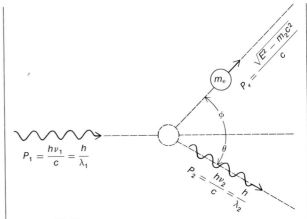

FIGURE 22-28
Compton scattering of x-ray photons. The photon will transfer momentum to an electron in a single collision.

collimated into a narrow beam and observed with a Bragg spectrometer, Compton found the usual characteristic molybdenum K_α x-ray line. However, as the scattering angle was increased, Compton observed a curious result: a second x-ray, shifted in wavelength (see Figure 22-27).

Normally, x-ray waves scatter in all directions without changing wavelength or frequency; this is *coherent scattering*. To explain the second shifted x-ray line, Compton assumed that the x-rays scatter as a particle, behavior now called **Compton scattering**, as shown in Figure 22-28. The x-ray photons possess a momentum:

$$p = \frac{E}{c} = \frac{h\nu}{c} = \frac{h}{\lambda}$$

Compton calculated the momentum that would be transferred, simply by conservation of momentum and energy, to a single electron if the photon is scattered into some angle θ. If the x-ray photon is treated as a single particle, classical particle mechanics can be used, although relativity effects must be included to obtain the correct result. In losing momentum, the x-ray photon is shifted to a longer wavelength. The shift in wavelength is given by

$$\lambda' - \lambda = \Delta\lambda = \frac{h}{m_e c}[1 - \cos\theta]$$

The constant $\lambda_o = h/m_ec = 0.00242$ nm is often referred to as the *Compton wavelength*. Compton scattering can occur with visible light but the effect is so small as to be unmeasurable. With x-rays with an energy of 25 keV, the wavelength is of the order of .05 nm and the effect of Compton scattering becomes clearly observable.

Compton's experiment confirmed that x-rays can scatter from electrons as particles. However, there is a remarkable ambiguity in the experiment, which Compton himself points out:

The crystal spectrometer measured a wavelike property, the wavelength, and measured it by means of a characteristically wavelike phenomenon, interference. But the effect of the graphite scatterer on the value of that wavelike property could be understood only in terms of a particlelike behavior.[2]

Thus x-rays behaved as both particles and waves in the same experiment! We shall examine the consequences of this wave-particle duality, both for photons such as light and x-rays and for particles such as electrons, when we examine the new quantum theory of wave mechanics in Chapter 23.

X-ray Crystallography

Bragg diffraction of x-rays from crystals, **x-ray crystallography,** is a widely used technique for studying the properties of materials. The application of x-ray diffraction has brought fundamental advances in inorganic and organic chemistry, metallurgy, and biology. Powder crystallography, for example, is a basic tool in the study of minerals (see Figure 22-29). A powdered sample is placed in a beam. The crystals are randomly oriented, some of them at a different Bragg diffraction maximum. If the sample is rotated, all angles of incidence are observed. The angle 2θ from the sample to the film image of the diffraction ring is a measure of the lattice spacings. From the Bragg condition, equation 22-5, the separations of the different lattice planes can be determined.

---

[2]Quoted in George Trigg, *Crucial Experiments in Modern Physics* (New York: Von Nostrand Reinhold, 1971), p. 103.

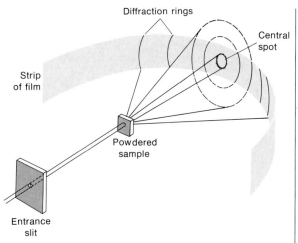

FIGURE 22-29
X-ray photographs are commonly used to study the properties of crystals. The samples are finely powdered and rotated in the x-ray beam so that all possible angles of incidence are observed. The resulting diffraction rings give a direct measure of the lattice spacings in the crystal.

$$d = \frac{n\lambda}{2\sin\theta}$$

From the spacing of the diffraction rings, the lattice constants of the crystal sample can be systematically determined.

X-ray crystallography is a powerful tool for determining the structure of molecules. Myoglobin, hemoglobin, and other organic molecules can be crystallized. The atoms in the crystal produce variations in electron density that scatter incident x-rays. Where heavy atoms such as iron are present, the scattering is greater. The diffraction spots that occur depend on the spatial periodicity of the molecules in the crystal, as shown in Figure 22-30.

Early x-ray work with human hair showed than an unstretched hair fiber has a molecular shape, a so-called α form. Stretching the fibers changes the molecular shape to a so-called β form. These two forms are quite common in biological structures. In 1951, Linus Pauling and Robert Corey, at the California Institute of Technology, determined that the α form is a chain of polypeptides in the shape of a helix.

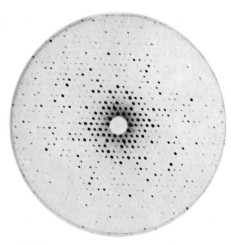

FIGURE 22-30
X-ray diffraction pattern of a single crystal of hemoglobin.
Each spot corresponds to a different periodicity of the crystal.
Courtesy of Dr. M. F. Perutz, Medical Research Council,
Laboratory of Molecular Biology, Cambridge, England.

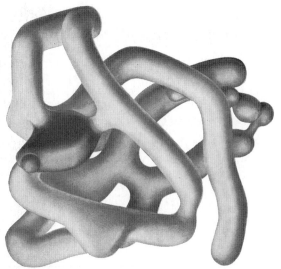

FIGURE 22-31
The myoglobin molecule as first reconstructed by John
Kendrew and his coworkers in 1957. The molecule is resolved
to 0.6 nm. From M. Perutz, "The Hemoglobin Molecule."
Copyright © 1964 by Scientific American, Inc. All rights re-
served.

In 1937, William Lawrence Bragg succeeded Rutherford as director of the Cavendish Laboratory, a major
center for x-ray research. Max Perutz, a principal research
assistant of Bragg's, had conducted x-ray studies of the
hemoglobin molecule since 1936. In 1948, he was joined
by a graduate student, John C. Kendrew, who began
working on the structure of myoglobin, a molecule in
muscle tissue. Shortly thereafter, Francis Crick and James
Watson joined Perutz at Cambridge. Maurice Wilkins and
Rosalind Franklin, doing crystallographic studies of DNA
at Kings College of the University of London, found, in
one of their x-ray photographs, a regular spacing of
0.34 nm between groups and a helical structure 2 nm in
diameter. After a period of intense work, Watson and
Crick produced their now-famous double helix model of
DNA.

In 1957, John Kendrew, using some 400 reflections,
made the first three-dimensional model of the myoglobin
molecule, shown in Figure 22-31, to a resolution of
0.6 nm. Later Kendrew and coworkers refined their work
to include some 10 000 reflections to give a 0.15 nm
resolution, calculating the electron density at some
100 000 positions in the molecules, as shown in Figure
22-32 Perutz and his colleagues later determined the
complete three-dimensional structure of the hemoglobin
molecule.

LASERS

In Bohr's atomic model, an electron in an atom has certain
stationary states or energy levels. When an electron
changes from one energy level to another, lower level, the
atom radiates a light photon. If the atom absorbs the
energy of a photon an electron goes from a lower level to
a higher. Although Bohr's model gives correct numerical
results only for hydrogen and for hydrogenlike atoms, the
concept of energy levels is quite useful.

One of the most fascinating inventions in the second
half of the twentieth century is the **laser** (standing for
light **a**mplification by **s**timulated **e**mission of **r**adiation).
Laser beams are used to align subway tunnels, to paint
moving-light sculptures, to precisely adjust integrated
circuit chips, to bore holes in steel, to perform surgery on
the retina of the eye, and to measure the distance to the
moon. Lasers are being developed to crush small pellets of
deuterium to liberate fusion power, and to carry thousands of telephone and many television signals over a
slender strand of glass.

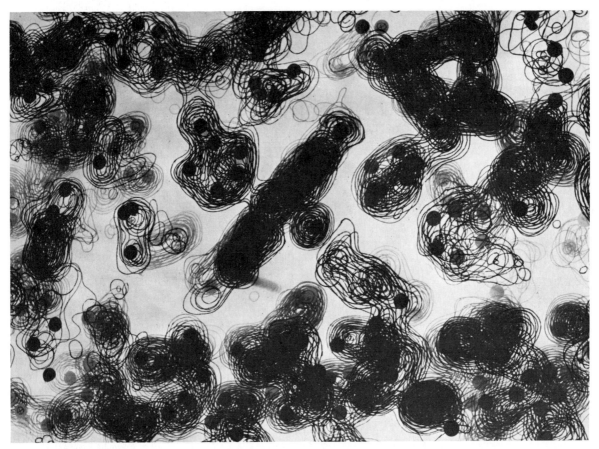

FIGURE 22-32
Electron density map of the region around the heme group of myoglobin viewed parallel to the heme group at a resolution of 0.15 nm. On one side (left) of the heme group can be seen the five-membered ring of the histidine residue that links the heme group to the rest of the protein molecule. On the other side (right), the isolated peak is a water molecule linked to the iron atom of the heme. Beyond this a second histidine ring is seen nearly edge on. Top right is an alpha-helix seen end-on. Courtesy of J. C. Kendrew and H. C. Watson.

Development of the Laser

The laser was preceded by the maser (**m**icrowave **a**mplification by **s**timulated **e**mission of **r**adiation). The first maser, using vibrations of molecules of ammonia (NH_3), was developed by James Gordon, H. J. D. Zeiger, and Charles Townes, in 1954. The ammonia molecules, pumped to an excited state, radiate electromagnetic waves (microwaves) at a precisely determined frequency (the same frequency used in the first atomic clock, and the basis of extremely sensitive microwave amplifiers for use by astronomers).

The first successful optical maser or laser was developed in 1960 by T. H. Maiman, using a ruby crystal. In the laser, the *active medium* consists of atoms which are "pumped" to an excited state. The outer electron is raised to an energy level above the ground state. In the ruby laser (see Figure 22-33), the chromium atoms in the ruby are pumped by absorbing intense light from a xenon flash tube.

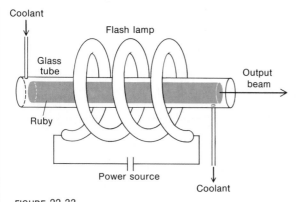

FIGURE 22-33
The ruby laser. Adapted from Arthur L. Schawlow, "Optical Masers." Copyright © 1961 by Scientific American, Inc. All rights reserved.

Figure 22-34 is a simple diagram of the energy levels of the chromium atoms in a ruby. The electrons are first in the ground state, or lowest energy level (A). The electrons absorb energy from the xenon flash lamp and are raised to one of two electron energy bands (B). Each electron gives up some of its energy to the crystal lattice and falls into a metastable level (C), an energy level of relatively long

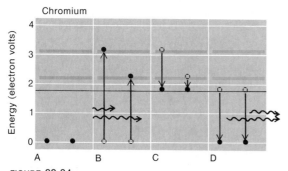

FIGURE 22-34
Energy levels of the ruby laser. A. Chromium atoms in the ground state absorb photons that pump them to one of two energy "bands." B. The atoms lose some of their energy to the crystal lattice and fall into a metastable energy level, C. When stimulated by other photons they fall to the ground state, D, emitting photons of a characteristic wavelength. Adapted from Arthur L. Schawlow, "Optical Masers." Copyright © 1961 by Scientific American, Inc. All rights reserved.

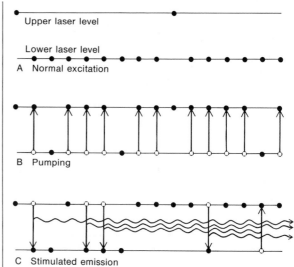

FIGURE 22-35
Laser energy levels: A. normal atomic excitation; B. pumping into the upper state, producing a population inversion; C. stimulated emission. From Arthur L. Schawlow, "Laser Light." Copyright © 1968 by Scientific American, Inc. All rights reserved.

duration. Thus, the electrons have been pumped into this upper laser level. When stimulated by an electromagnetic wave of the proper frequency from other chromium atoms, these electrons drop back into a lower laser energy level, emitting photons in phase with the electromagnetic wave (D). This amplifies the wave.

In a normal excited gas, most of the electrons are in the lower energy levels, with fewer and fewer electrons at increasing energy states (see Figure 22-35 A). By Kirchhoff's law of spectroscopy, an atom can absorb any light that it can radiate. An electron in the lower laser level can absorb a laser photon and jump back into the upper laser level. The trick to laser action is to have a **population inversion,** that is, to have many more atoms in the upper laser level than in the lower laser level, as shown in Figure 22-35 B. Once the upper state is populated, stimulated emission exceeds absorption by the lower state, and the laser electromagnetic wave can build up, as shown in Figure 20-35 C. (Even then, some of the lower states absorb energy.)

The active medium of the laser is placed in a reflecting cavity. The ends of the ruby are highly polished, optically flat surfaces. One end of the ruby is completely reflecting, while the second end has a partially reflecting mirror. In the Q-switched, or giant-pulse ruby laser, a fast-acting shutter is placed in the light path of the cavity. As long as the shutter is closed, the stimulated emission will not build up. The unexcited chromium atoms of the ruby (see Figure 22-36 A) are pumped into their upper laser levels by a xenon flash lamp indicated by the black arrows in Figure 22-36 B and C. The atoms begin to radiate at their characteristic frequency. However, coherent radiation will not take place. Opening the shutter starts an electromagnetic wave toward one end of the cavity, as shown in Figure 22-36 D. It will be partially reflected and grow in amplitude as chromium atoms are stimulated to emit in phase with the initial wave. At the fully reflecting end, it is again reflected. When it reaches the partially silvered mirror, some of the energy leaves the cavity as the laser beam and some is reflected to continue to stimulate emission. This process continues until no more atoms are in the upper laser level. When the shutter of the giant pulse laser is opened, the energy in the upper level is suddenly dumped in a short-giant pulse that can deliver a power as large as 50 megawatts for 10 nanoseconds.

There are several methods of preparing the excited state population, and there are many different active laser materials. Table 22-1 lists some of the principal active materials. In optical pumping, a light from a xenon flash-tube or from another laser excites the atoms or molecules to the upper laser level. Solid ruby or glass lasers are optically pumped. The rare-earth elements such as neodymium are used in the glass laser. The search for a tunable laser which can operate over a range of wavelengths has led to the use of chemical dies as the active material. Chemical dye lasers, tunable over a range of wavelengths, are usually optically pumped by another laser. Gas lasers such as helium-neon, argon, carbon monoxide, and carbon dioxide, and metal vapors such as helium-selenium, use an electrical discharge to excite the atoms to the upper laser level. In solid state lasers, the pumping energy comes directly from electrical work done across a special semiconductor diode junction such as gallium arsenide.

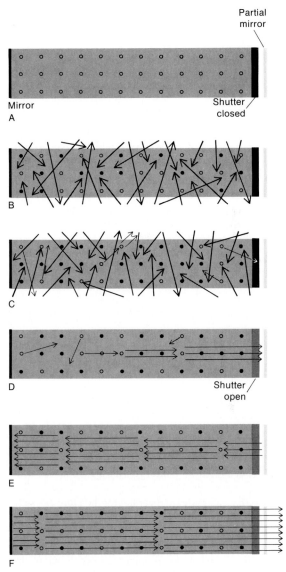

FIGURE 22-36
Giant pulsed ruby laser employs a shutter to delay laser action until a large population of atoms has been raised to an excited state. When the shutter is opened a photon cascade produces an intense pulse of laser light. From Arthur L. Schawlow, "Advances in Optical Masers." Copyright © 1963 by Scientific American, Inc. All rights reserved.

TABLE 22-1
Types of lasers

Active material	Principal wavelengths
Solid	
Chromium in aluminum oxide (ruby)	694.3 nm
Neodymium in glass	1064 nm
Erbium in glass	1540 nm
Gas	
Argon (Ar)	351.1, 363.8, 456.5, 457.9, 488, 514.5 nm
Krypton (Kr)	350.7, 356.4, 647.1, 530.9, 678.4 nm
Nitrogen (N_2)	337.1 nm
Helium-neon (He-Ne)	632.8 nm
Hydrogen fluoride (HFl)	4–6 μm
Carbon monoxide (CO)	5.0–5.5 μm
Carbon dioxide (CO_2)	9.6, 10.6 μm
Water vapor (H_2O)	118 μm
Hydrogen cyanide (HCN)	337 μm
Metal vapor	
Helium-cadmium	325 nm
Helium-selenium	460.4–1250 nm (30) lines
Organic dye lasers (tunable)	
Rhodamin 6G	500–650 nm
Coumarin 2	430–490 nm
Sodium fluorescein	540–580 nm
Carbostyril 165	415–490 nm
Nile blue-A perchlorate	690–780 nm
Solid state lasers	
Gallium arsenide	905 nm

The Helium-Neon Laser

Figure 22-37 shows a common helium-neon gas laser. The laser consists of a gas-discharge chamber filled with a mixture of helium and neon gas. An electrical current driven from a radiofrequency generator flows back and forth through the gas, ionizing a small fraction of the gas. Accelerated electrons colliding with the atoms excite the helium atoms into higher energy levels. Figure 22-38 shows the energy-level diagram for helium and neon. Collisions pump the helium electron from the ground state (*a*) into an excited level about 19.7 eV above the ground state (*b*). The collision of a helium atom with a neon atom transfers the excitation energy to one of four closely spaced neon energy levels (*c*). This populates the neon upper laser level. When stimulated by a photon at the proper frequency, the neon emits a photon, shown by the wavy arrow, as the neon electron falls into one of 10 closely spaced lower laser levels. The neon decays to the ground state by a series of photon emissions (which do not

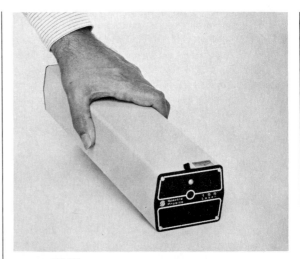

FIGURE 22-37
A typical helium-neon demonstration laser produces a power output of 0.5 mW at a wavelength of 632.8 nm. Courtesy of Spectra-Physics.

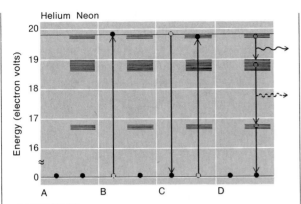

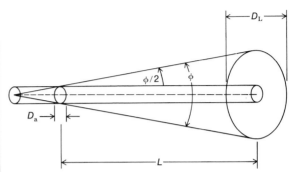

FIGURE 22-39
Beam divergence. The laser beam expands as it leaves the laser aperture.

contribute to the laser action). The helium-neon laser operates primarily at a wavelength of 632.8 nm, which corresponds to a transition energy of 1.96 eV. Because pumping the helium atoms requires about 19.7 eV of energy, 90% of the pumping energy is wasted. Helium-neon lasers are not very efficient. Lasers must be handled with care, for the alignment of the optical cavity's end mirrors is critical. If the laser is dropped or mishandled and the mirror alignment disturbed, it may stop operating.

Typical student lasers operate with a power range of less than a milliwatt. The laser light is in a narrow beam. Because a helium-neon laser typically has an angular spread of 0.5 to 1.5 milliradians, the beam travels great distances with little spreading. Figure 22-39 shows the beam divergence. If the laser output $P = 1.0$ mW over a beam diameter of $D_a = 1.5$ mm, with a beam divergence of $\phi = 1.0$ milliradians, we can calculate the spreading of the beam. The radiation at the aperture, the power per unit area is

$$P/A = \frac{P}{(\pi D_a^2/4)} = \frac{10^{-3}\text{ W}}{\pi(1.5 \times 10^{-3}\text{m})^2/4} = 570\text{ W/m}^2$$

This is comparable to the solar radiation at the surface of the earth on a clear day, typically 800 W/m³.

As Figure 22-35 shows, the laser beam diverges as it travels beyond the aperture. If the beam travels a distance L, its size increases. The diameter of the beam is given approximately by

$$D_L = D_a + L\,\phi$$

where ϕ is the *beam divergence* in radians. At a distance of 12 m, a reasonable distance across a classroom, the diameter of the laser beam is still

$$D_L = 0.0015\text{ m} + (12\text{ m})(10^{-3}\text{ rad})$$
$$D_L = 0.0015\text{ m} + 0.0120 = 0.0135\text{ m}$$

The irradiance of the beam has been reduced to 7 W/m² at this distance.

The Carbon-Dioxide Laser

In the carbon-dioxide laser, a major breakthrough in high-power lasers, the molecular energy levels are excited. (Figure 22-40 shows the energy levels for a typical atomic laser compared to those of a molecular laser.) The vibrational energy levels of the carbon-dioxide molecule are

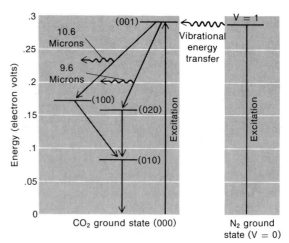

FIGURE 22-41
An energy-level diagram for carbon dioxide molecular laser. Collisional excitation of nitrogen can be used to pump the upper laser level. From C. K. N. Patel, "High-Power Carbon-Dioxide Lasers." Copyright © 1968 by Scientific American, Inc. All rights reserved.

excited. The excitation energy for carbon-dioxide being much smaller than for atomic levels, much less energy is wasted.

The carbon dioxide can be pumped directly by an electrical discharge or indirectly by excitation of vibrational modes of nitrogen, N_2. Figure 22-41 shows the energy levels of the carbon-dioxide laser excited by nitrogen. The nitrogen gas is excited and selectively pumps a particular upper level of the carbon dioxide molecule (001 shown on figure). Transition to the two lower laser levels gives rise to infrared radiation at 10.6 μm and 9.6 μm. Commercial carbon-dioxide lasers are available with a continuous power level of 1 to 500 watts. They are focused to a small spot for use in machining, cutting, and drilling. Researchers at Raytheon Company have constructed a folded carbon-dioxide laser some 180 m long, which has produced a continuous power of 8.8 kilowatts. A typical carbon-dioxide laser about 2 m in length can produce a continuous power output of about 150 watts.

Lasers in Medicine

In the past 10 years, lasers have become widely accepted in medicine, both for eye surgery and for bloodless inter-

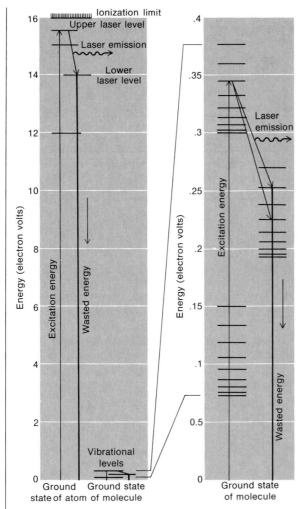

FIGURE 22-40
An energy-level diagram for a typical atomic system and a typical molecular system. The vibrational levels of the molecule are close to the ground state. The laser photon energy is a large fraction of the total excitation energy of the molecule. From C. K. N. Patel, "High-Power Carbon-Dioxide Lasers." Copyright © 1968 by Scientific American, Inc. All rights reserved.

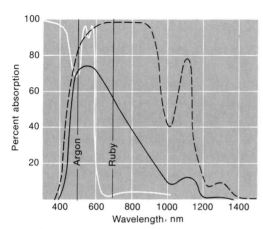

FIGURE 22-42
Ocular transmission (dashed line), combined absorption of pigment epithelium and choroid (solid line), and absorption of hemoglobin in 100 μm thickness of blood (dotted line) as a function of wavelength. Courtesy of Coherent Radiation.

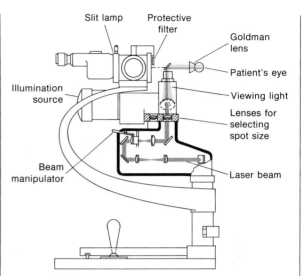

FIGURE 22-43
The argon laser photocoagulator uses a slit lamp system and a Goldman lens to overcome the refraction at the cornea. Courtesy of Coherent Radiation.

nal surgery. Other applications of lasers in medicine have been extensively developed, as in the treatment of skin cancer, and others still under study.

Eye surgery is the largest use of lasers in medicine. In the early 1950s, the xenon arc light **photocoagulator** was developed. Light is transmitted through the transparent cornea, lens, and ocular media to the retina. The pigment epithelium and the choroid absorb light reaching the retina, with resulting heating and subsequent tissue change. The xenon arc was replaced by a laser. Ophthalmologists have readily accepted the laser as a new technique for eye surgery, and as early as 1964, ruby lasers were being used experimentally in several hospitals.

In 1968, the argon laser was applied to laser surgery. The argon laser has specific beneficial properties for laser surgery. Its light is highly collimated in a narrow beam that can be focused to a very small, highly controllable spot. The physician can perform very specific surgery, and has the high power necessary to cause heating and structural changes in the eye's retina and blood vessels. Figure

22-42 shows the absorption of the pigment epithelium and hemoglobin as a function of wavelength, and transmission through the ocular medium. The argon and ruby wavelengths are also indicated. The argon laser wavelength is near the maximum in the epithelium absorption. Argon laser photocoagulators are among the most successful medical applications of lasers.

Figure 22-43 shows the slit-lamp system employed in the treatment of the eye. In one technique, a special "Goldman lens" is placed as a contact lens on the patient's cornea to remove the refraction at the cornea. A viewing light illuminates the patient's retina and allows the physician to look through the slit lamp to accurately position the laser spot, as shown in Figure 22-44. For precise surgery, the physician can control the spot size from 50 μm to 1 mm and the laser pulse duration from 20 ms to 5 s. Although the laser can do very little for a detached retina, it can do much for a torn retina that might later become detached. Laser photocoagulation has also been used to arrest the development of diabetic retinopathy.

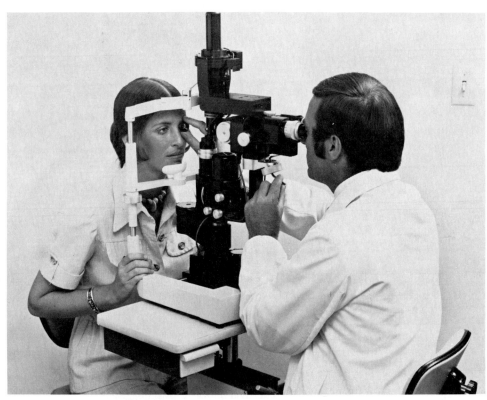

FIGURE 22-44
The argon laser photocoagulator. Courtesy of Coherent Radiation.

Laser Safety

The principal hazard in laser use is damage to the eye. The eye can focus the parallel rays from a laser to a small point, incurring small burns on the retina if the power is sufficient. A laser that can produce observable lesions on the retina is clearly a hazard. Exposure to laser power of 30 to 40 milliwatts in the visible for 100 milliseconds will produce a small lesion observable by an ophthalmologist. Yet the recipient of such a small lesion is usually unaware of any effect on the visual field.

The optics of the eye are not perfect, and the eye undergoes a variety of motions, including a tremor at 30 to 80 Hz. As a result the image of a point source of light makes a circle on the retina about 100 μm in diameter. The exposure time for a laser accident is limited to the blink reflex time, typically 12 milliseconds. A laser with a power output of 3 mW to 10 mW should be considered a *possible hazard* and used with caution—anything over 10 mW should be considered *dangerous*. Even a laser with a stated power output of 1 mW could in fact produce 2 or 3 mW, for the power rating often specifies the minimum, not the maximum, power output. Even lasers with low power rating should be respected.

Review

Give the significance of each of the following terms. Where appropriate give two examples.

cathode rays	alpha scattering	Moseley's law
Crookes tube	nucleus	Compton scattering
electrons	Bohr model	x-ray crystallography
x-rays	Bohr radius	laser
alpha rays	Rydberg constant	population inversion
alpha particles	Bragg diffraction	photocoagulator
beta rays	bremsstrahlung	
gamma rays	characteristic x-rays	

Further Readings

E. N. Andrade, *Rutherford and the Nature of the Atom* (New York: Doubleday, Anchor Science Study Series S35, 1964). A good account of Rutherford's life and work by a fellow physicist who worked with him at Manchester.

Max Born, *The Restless Universe* (New York: Dover Publications, 1951). This revision of a book first published in 1936 is a quite readable account of the fundamentals of modern physics.

Sir Lawrence Bragg, "X-Ray Crystallography," *Scientific American* (July 1968), reprinted in *Lasers and Light* (San Francisco: W. H. Freeman and Co., 1969).

George Gamow, *Thirty Years That Shook Physics: The Story of Quantum Theory* (New York: Doubleday, Anchor Science Study Series, S45, 1966). A very readable account by one who was there.

Banesh Hoffman, *The Strange Story of the Quantum* (New York: Dover Publications, 1959). An account for the general reader of the growth of the ideas underlying our present atomic knowledge.

John C. Kendrew, "The Three-Dimensional Structure of a Protein Molecule," *Scientific American* (December, 1961).

Lasers and Light: Readings from Scientific American (San Francisco: W. H. Freeman and Co., 1969). A comprehensive series of articles.

Ruth Moore, *Niels Bohr: The Man, His Science and the World They Changed* (New York: Alfred A. Knopf, 1966). The definitive biography of Bohr, capturing the man and the times.

Max F. Perutz, "The Hemoglobin Molecule," *Scientific American* (November, 1964).

Problems

SETTING THE STAGE

1 An electron traveling at 2.8×10^7 m/s has a charge-to-mass ratio of $q/m = 1.76 \times 10^{11}$ C/kg. What magnetic field is required to bend the electron beam in a radius of 10 cm?

2 An electron is accelerated to an energy of 2 keV. What magnetic field is required to bend the electron beam in a radius of 1.0 cm?

3 An electron is traveling at 2.5×10^7 m/s in a magnetic field of 2.0×10^{-3} N/A-m (T). What electric field would give the same deflection to an electron?

4 The target in Thomson's experiment has a heat capacity of 5×10^{-3} calorie/°C. The target collects electrons with energies of 2 keV, increasing temperature by 2.0°C.
 (a) How many electrons are collected?
 (b) How much charge is collected?

5 In Thomson's experiment, the target with a heat capacity of 5×10^{-3} calorie/°C increases by 2.0°C. A total charge of 2×10^{-5} C is collected. The electrons are deflected on a radius of curvature of 30 cm in a magnetic field of 5×10^{-4} T.
 (a) What is the speed of the electrons?
 (b) What is the charge per unit mass?

A NEW ATOMIC MODEL

6 Radium emits a 4.78 MeV alpha particle with mass of 6.6×10^{-27} kg.
 (a) What is the speed of the alpha?
 (b) What is the distance of closest approach to a gold atom with $Z = 79$?

7 A beryllium nucleus $Z = 4$ has a radius of about 5×10^{-15} m. What kinetic energy is required of an alpha particle to penetrate to within this distance of the beryllium nucleus?

8 What energy does an alpha particle require to get within a distance of 6.2×10^{-15} m of a nitrogen nucleus with $Z = 7$?

9 A Polonium-210 source gives off alpha particles with an energy of 5.3 MeV.
 (a) What is the speed of the alpha particles?
 (b) What is the distance of closest approach to a uranium nucleus with $Z = 92$?

NIELS BOHR

10 The Balmer series gives radiation transitions ending in the Bohr level $n = 2$.
 (a) What transition has the longest wavelength? What are the energy and wavelength of that transition?
 (b) What transition has the highest frequency? What are its wavelength and energy?

11 The Lyman series is determined by transitions ending in the ground state, $n = 1$
 (a) What are the energies of the first four Lyman lines?
 (b) What are the wavelengths of these transitions?

12 The spectrum of He$^+$ ($Z = 2$), a helium nucleus with one electron, can be determined from the Bohr model. Make a diagram comparing the energy levels of He$^+$ with H up to $n = 8$.
 (a) What transitions lie in the visible with wavelengths between 400 and 700 nm for He$^+$?
 (b) What are the wavelengths of these transitions?

13 What is the ionization energy for He$^+$, Li^{++} as determined from the Bohr model? Express each energy in eV and J/mol.

14 A bare proton can recombine with an electron, forming a neutral hydrogen atom. The binding energy must be liberated in photons.
 (a) What is the total energy liberated in eV and in joules in recombination of one atom?

(b) What is the energy liberated in joules and in calories in the recombination of one mole?

15 An electron is in a Bohr orbit with $n = 2$ around a hydrogen nucleus.
 (a) How many times per second does the electron orbit the nucleus?
 (b) If the atom were to radiate at this frequency, what wavelength would the light have?
 (c) If the electron dropped into the ground state, what wavelength would it emit?

16 An electron is in a Bohr orbit with $n = 3$ around a hydrogen nucleus.
 (a) How many times per second does the electron orbit the nucleus?
 (b) If the lifetime in the excited state is 10^{-8} s, how many orbits does it complete before radiating?

17 A small correction for the finite mass of the nucleus must be made to the results of the Bohr model. This requires replacing the electron mass m_e by the reduced mass $m_e' = m_e/(1 + m_e/m)$. With this correction, compare the wavelength of the Lyman H_α ($n = 2$ to $n = 1$ transition) line for hydrogen, $m = m_p$ and tritium $m = 3m_p$.

X-RAY SPECTRA

18 In a diffraction experiment, x-rays with a wavelength of 0.36 nm produce an interference maximum $n = 1$ at a total deflection $2\theta = 60°$. What is the spacing of the lattice planes?

19 Moseley determined that his potassium ferrocyanide crystal had a lattice spacing of $d = 8.45 \times 10^{-10}$ m. The copper x-rays have wavelengths $K_\alpha = 0.155$ nm and $K_\beta = 0.140$ nm. What is the angle of deflection 2θ for these lines in third order?

20 The distance between layers of atoms in a sodium chloride crystal are 0.281 nm. The iron K_α x-ray has an energy of 6.4 keV. What are the defraction angles for the interference maxima for $n = 1, 2$, and 3?

21 Characteristic K_α x-rays from molybdendum have an energy of 17.1 keV and are incident on a salt crystal. There is a first-order interference maximum

at an angle $2\theta = 14.7°$. What is the lattice spacing of the crystal?

22 The atoms of rock salt form a simple cubic array. Each molecule of salt (NaCl) consists of two atoms. The density of the crystal is 2.167 gm/cm³; molecular weight is 58.46 gm/mole.
 (a) What is the average spacing between the atoms in the crystal?
 (b) The interference pattern in first order from the chromium K_α x-rays occurs at an angle of $2\theta = 48°$. What are the wavelengths and energy of this characteristic x-ray?

23 Make a two-dimensional crystal out of 10 or more pennies.
 (a) What is the diameter of a penny?
 (b) With the pennies in a triangular array, what are the spacings of the first four different Bragg planes?
 (c) With the pennies in a rectangular array, what are the spacings of the first four different Bragg planes?

24 A sodium-chloride crystal is a regular cubic lattice with a perpendicular spacing of about 0.28 nm between adjacent atoms. Draw a two-dimensional picture of one of the layers of atoms. What are the spacings of the first four Bragg planes?

25 Electrons are accelerated to an energy of 40 keV and strike a tungsten target emitting bremsstrahlung radiation. What is the minimum wavelength of x-rays emitted?

26 A color television picture tube accelerates electrons through an electrical potential difference of 24 kV. When the electrons strike the picture tube, some x-rays are produced. What is the minimum wavelength of x-rays produced?

27 Use Moseley's law (with a screening constant $a = 1.0$) to estimate the energy of the K_α ($n = 1$, $m = 2$) x-ray for iron, $Z = 26$.

28 From Moseley's law for uranium, $Z = 92$, using a screening constant $a = 1$,
 (a) Estimate the energy of the K_α x-ray. (Measured value, 98 keV)
 (b) Estimate the binding energy of the K-shell electron. (Measured value, 116 keV)

29 An x-ray has an energy of 40 keV. By Compton scattering, the x-ray scatters from an electron at an angle of 90°,
 (a) What is the new wavelength?
 (b) How much energy is transferred to the electron?
 (c) What is the momentum transferred to the electron?

LASERS

30 A helium-neon laser has a power output of 1 mW, an aperture diameter of 2.0 mm, and a beam divergence of 0.5 milliradians.
 (a) How big is the beam spot 10 m from the laser?
 (b) 1 km from the laser?

31 An argon laser has a power output of 2 watts at 514.8 nm. The beam diameter is 1.4 mm and beam divergence 0.5 milliradians at the aperture.
 (a) What is the spot size at a distance of 1 km?
 (b) How does the power per unit area of the beam at that distance compare to the radiation of the sun (800 W/m²)?

32 A laser power of 30 mW for 100 ms is sufficient to cause a small retinal burn.
 (a) If the laser beam falls on a spot 100 μm in diameter, what is the power per unit area at the surface of the retina?
 (b) If the power is on for 100 ms, how much energy per unit area falls on the surface of the retina?

33 On a bright day, radiation from the sun is approximately 800 W/m².
 (a) If an eye with a pupil diameter of 1 mm stares at the sun, how much power enters the eye?
 (b) The sun's disk has an angular size of about 30 minutes of arc. This will form an image at the retina about 150 μm in diameter. What is the power per unit area at the retina?
 (c) How much energy per unit area falls on the retina in one second? How does this compare with the answer to part (b) of the previous problem?

34 A class III visible laser with a power output of 3 mW is accidentally focused on a spot 150 μm in diameter on the retina for a blink-response time of 25 ms.
 (a) What is the power per unit area at the retina?
 (b) How much energy falls on the retina per unit area?

23

Wave Mechanics

. . . there are not two worlds, one of light and waves, one of matter and corpuscles. There is only a single universe. Some of its properties can be accounted for by the wave theory, and others by the corpuscular theory.

*Nobel presentation speech honoring
Louis de Broglie, 1929*

MATTER WAVES

Bohr's theory of the atom, the so-called old quantum theory, was based on quantizing the orbits of the electrons in the atom. Because the theory had only limited success in explaining the spectra of atoms, in the 20 years after its publication, an entirely new quantum theory emerged that unified our view of waves and particles and gave firm principles for applying quantum conditions.

The history of the new quantum theory, or **wave mechanics,** is rich and complex. The student is referred to several excellent introductions listed at the end of this chapter. Table 23-1 lists the Nobel Prizes awarded for work, principally in the 1920s, that contributed to the development of the new quantum theory, and gives a brief outline of the major developments. In this chapter we discuss some of the central ideas of wave mechanics.

Louis De Broglie

In 1924, at Paris University, Louis de Broglie presented his doctoral thesis, "Researches on the Quantum Theory," which outlined his theory of propagating **matter waves.**

In some situations, light acts as a particle, a photon, with a clearly defined energy $E = h\nu$. That the energy of the photon was proportional to the frequency ν, a purely wavelike property, bothered de Broglie: Light seemed to act as both a particle and a wave. Electrons, on the other hand, were particles with clearly defined mass and charge. Yet the stable Bohr orbits of the electron in the hydrogen atom were determined by a series of integer numbers—in particular the radius of the Bohr orbits was proportional to the square of the quantum number n. Only two phenomena known to de Broglie depended on integer numbers: interference, such as in Newton's rings or multiple-slit diffraction; and standing waves, such as those on a vibrating string. De Broglie suspected, then, that the electron might not be simply a particle, but should also be assigned some wave property. But what?

De Broglie used an analogy with light. The energy of a photon is given by

$$E = h\nu$$

The momentum of a photon is given by

$$p = \frac{E}{c} = \frac{h\nu}{c} = \frac{h}{\lambda}$$

If light can exist as a wave and as a particle, it is only a short step in reasoning to speculate that a material particle could behave as a wave. The **de Broglie wavelength** is given by

$$\lambda = \frac{h}{p} = \frac{h}{mv} \qquad (23\text{-}1)$$

De Broglie applied the hypothesis of matter waves to the stationary states of the Bohr model, in which the circular orbits can be represented as waves traveling around the atom, as shown in Figure 23-1. We are familiar

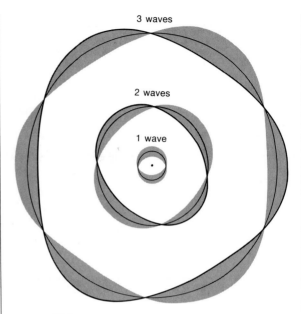

FIGURE 23-1
De Broglie's matter waves can be related to the stationary orbits of the Bohr model of the hydrogen atom. Adapted from *Thirty Years That Shook Physics* by George Gamow. Copyright © 1966 by Doubleday & Company, Inc. Reprinted by permission of the publisher.

with standing waves on a string. Perhaps there are other, similar standing waves for the electron in an atom. If the circumference of the Bohr orbit $2\pi R_n$ has n wavelengths, then we have

$$2\pi R_n = n\lambda_n$$

From the Bohr theory of the hydrogen atom, the radius of the nth orbit is, as discussed in Chapter 22:

$$R_n = \frac{n^2 h^2}{4\pi^2 k_e m_e e^2} = n^2 a_o \qquad (23\text{-}2)$$

The Coulomb force between the electron and the proton provides the centripetal force to hold the electron in the orbit:

$$F = \frac{m_e v_n{}^2}{R_n} = \frac{k_e e^2}{R_n{}^2}$$

We can solve this for $k_e e^2$

$$k_e e^2 = m_e v_n{}^2 R_n$$

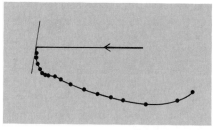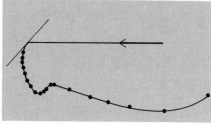

Scattering of 75 volt electrons from
a block of nickel (many small crystals)

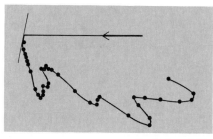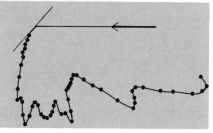

Scattering of 75 volt electrons from
several large nickel crystals

FIGURE 23-2
The results of the Davisson and Germer experiments. Electrons scattered from large nickel
crystals show a series of interference maxima. Adapted from *Physics Review, 30* (1927),
p. 706.

Substituting this into the previous equation, we can eliminate the electron charge. Equation 23-2 becomes

$$R_n = \frac{n^2 h^2}{4\pi^2 R_n m_e^2 v_n^2}$$

Solving for the orbit radius after some manipulation, we obtain

$$2\pi R_n = \frac{nh}{m_e v_n}$$

Comparing this to the standing waves around the nth Bohr orbit, we find that

$$2\pi R_n = n\lambda_n = \frac{nh}{m_e v_n}$$

or that the wavelength is related to

$$\lambda_n = \frac{h}{m_e v_n} = \frac{h}{p}$$

Thus, de Broglie was able to describe the stationary states of the Bohr model of the hydrogen atom as corresponding to standing electron waves. He established a correspondence between the classical trajectories of particles and the rays of wave propagation. But did these matter waves have a basis in reality? Could an electron be a wave?

Electron Diffraction

Clinton Davisson and Lester Germer, at the Bell Telephone Laboratories, and, independently, George Thomson, at the University of Aberdeen, in Scotland, observed the diffraction and interference effects of electron beams. Davisson and Germer had been studying the reflection of low-energy (50-75 eV) electrons from a block of nickel. Their target had by accident been exposed to air at high temperature and had become oxidized. To reduce the oxidized layer, they heated and cooled their target several times. When they continued their electron scattering experiments, the results had changed dramatically, as shown in Figure 23-2. Shortly thereafter, George Thomson observed the diffraction of an electron beam on passing through a thin metal foil.

The diffraction of electrons from a crystal is analogous to the diffraction of x-rays from a crystal. Waves scattered from two successive planes in the crystal lattice interfere with one another as was shown in Chapter 22, Figure 22-18 A. If the path difference is an integer number of wavelengths $n\lambda$, then the scattered waves arrive in phase, giving an interference maximum. This is the Bragg condition (also discussed in Chapter 22):

$$2d \sin \theta = n\lambda$$

EXAMPLE

In Davisson's and Germer's experiment, they observed the first two strong diffraction maxima from the metal crystal at angles of 18° and 38° from the incident beam. The nickel crystal lattice has a spacing of about 0.222 nm. What is the wavelength, calculated from the diffraction? The electrons had an energy of 75 eV. What is the de Broglie wavelength of the electrons? The condition for interference maxima is that

$$n\lambda = 2d \sin \theta$$

For the first maximum $n = 1$, the wavelength is

$$\lambda = 2(0.222 \text{ nm}) \sin 18°$$

$$\lambda = 0.137 \text{ nm}$$

For the second maximum $n = 2$, the wavelength is

$$2\lambda = 2(0.222 \text{ nm}) \sin 38°$$

$$2\lambda = 0.273 \text{ nm}$$

$$\lambda = 0.137 \text{ nm}$$

The de Broglie wavelength for the electron is

$$\lambda = \frac{h}{m_e v}$$

Because it is a low-energy electron, we can determine the momentum nonrelativistically

$$m_e v = (2m_e KE)^{1/2}$$

$$m_e v = 2(9.1 \times 10^{-31} \text{ kg})(75 \text{ eV})(1.6 \times 10^{-19} \text{ J/eV})^{1/2}$$

$$m_e v = 4.67 \times 10^{-24} \text{ kg-m/s}$$

The wavelength is

$$\lambda = \frac{6.63 \times 10^{-34} \text{ J-s}}{4.67 \times 10^{-24} \text{ kg-m/s}}$$

$$\lambda = 0.14 \text{ nm}$$

Within the limits of accuracy, the observed diffraction wavelength agrees with the de Broglie wavelength.

Erwin Schrödinger

A new mechanics must be developed
which is to classical mechanics
what wave optics is to geometrical optics.

Louis de Broglie, Nobel lecture, 1929

In the geometrical optics of Newton, light particles travel in straight lines, rays, unless they are incident on some surface. In the wave optics of Huygens, Young, and Fresnel, each point of the wave acts as a source of new waves as it propagates through space. In the classical mechanics of Newton, material particles travel in straight lines, unless they are acted on by forces. It was Erwin Schrödinger who would develop the new mechanics of matter, wave mechanics.

In early 1925, Erwin Schrödinger, a professor of physics at the University of Zurich, learned of de Broglie's ideas of electron waves in a footnote in one of Einstein papers. Schrödinger's development of the theory of matter waves was swift and sure. In January 1926 his first published paper hit the world of physics like a bombshell. Unlike the earlier pioneers in quantum theory, he introduced no strange new mathematics but instead relied on well-developed nineteenth-century mathematics. His first paper "Quantization as an Eigen-Value Problem" caused consternation because he simply wrote down a basic equation for matter waves without explaining its derivation. He then proceeded with no difficulty to solve several important problems. This was the first of a dramatic series of five papers published during the first half of 1926. In his second paper, he disclosed that his approach had been a natural extension of the ideas of de Broglie. At the heart of the matter is the quantum wave function Ψ (the Greek capital letter *psi*).

The Quantum Wave Function

The quantum wave function for an electron can be represented as a traveling wave. The dynamics of the quantum wave are described by the **Schrödinger equation,** a differential equation for the propagation of the electron wave in space. The solution of Schrödinger's equation forms the basis of wave mechanics. (We have already studied several different types of waves: In Chapter 8, we examined traveling waves on a string, sound waves, and the production of standing wave patterns; in Chapter 18, the propagation of electromagnetic waves; and in Chapter 20, the interference and polarization of light waves.) Although the statement and solution of the Schrödinger equation are beyond the scope of this course, we can treat in some detail the quantum wave function, for its properties are similar to other waves that we have studied.

The quantum wave function, which describes an electron with a kinetic energy E traveling in the x-direction with a momentum p, can be represented as a traveling wave. Although describing the wave function in fact requires use of complex numbers that have both real and imaginary parts, we can gain understanding by looking only at the real part of the wave. The real part of the wave can be represented as

$$Re\ \Psi(x, t) = \Psi_o \cos 2\pi(x/\lambda - \nu t)$$

The de Broglie wavelength of a particle with momentum p is

$$\lambda = h/p$$

The frequency of a de Broglie wave with an energy E is

$$\nu = E/h$$

The real part of the quantum wave takes the form

$$Re\ \Psi(x, t) = \Psi_o \cos \frac{2\pi}{h}(px - Et)$$

This is the fundamental quantum wave function.

The wave function $\Psi(x, t)$ represents a traveling wave. But what in fact is waving? In the case of light, the electromagnetic field $E(x, t)$ is oscillating and propagating through space. The intensity of light is proportional to the square of the electric field $E^2(x, t)$. The quantum wave function is simply a number that changes its value in space and time. The magnitude[1] of the quantum wave function at a point in space is given by the square of the number $\Psi^2(x, t)$. In 1926, Max Born correctly suggested that Ψ^2 represented the statistical probability of finding the electron at a particular point in space. If we have one electron and sum Ψ^2 over all space, the probability of finding the electron somewhere is unity. The quantum wave function, then, yields the probability of finding the electron in a particular volume of space.

CONFINED WAVES

Standing Waves in One Dimension

The ancient Greek philosopher Pythagoras discovered the relationship between vibrating strings and music: The length of a wave propagating on a string depends on the frequency of vibration. However, after a very short period of time, only certain vibrations remain on a plucked string: the fundamental frequency plus the harmonics, such that the length of the string L is an integer number of half-wavelengths:

$$L = n\frac{\lambda_n}{2}$$

These **standing waves** that are produced in one dimension are a consequence of the string's wave equation,

---

[1] The quantum wave function at a particular point in space $\Psi(x,t)$ is actually a complex number with a real and imaginary part. The magnitude of the wave function is given by $\Psi^*\Psi$ where Ψ is the complex conjugate. For our purposes we will treat Ψ as a real quantity.

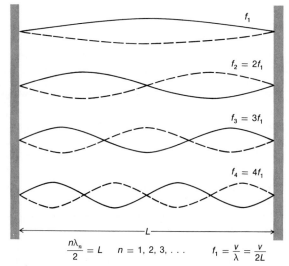

$$\frac{n\lambda_n}{2} = L \qquad n = 1, 2, 3, \ldots \qquad f_1 = \frac{v}{\lambda} = \frac{v}{2L}$$

FIGURE 23-3
One-dimensional standing waves on a string. Only certain values (eigen-values) of the frequency and wavelength are allowed.

which gives the propagation, and the boundary conditions. The standing wave pattern can in fact be represented as a displacement of the string:

$$y(x, t) = y_0 \sin \frac{2\pi nx}{2L} \cos 2\pi \nu t$$

These standing wave patterns vibrate in time. Figure 23-3 shows simple one-dimensional standing wave patterns. The equation governing the propagation of waves on a string resembles the Schrödinger equation. Anchoring the string at each end gives the "boundary conditions," permitting only certain values of the wavelength or of the frequency. In the language of modern physics, these certain values are known as *eigen-values*.

An Electron in a One-Dimensional Box

Suppose that an electron is to be confined in a one-dimensional box of some length L. The standing wave pattern of the electron must go to zero at the walls of the box. $\Psi(x) = 0$ at $x = 0$ and $x = L$. This implies that the

electron is reflected back and forth in the box to produce a standing wave. The standing wave function must then be of the same form as that of a vibrating string. This implies that only certain electron wavelengths are permitted, given by

$$\frac{n\lambda_n}{2} = L$$

This in turn implies that only certain energies are permitted

$$E_n = \frac{p_n^2}{2m_e}$$

where

$$p_n = \frac{h}{\lambda_n} = \frac{nh}{2L}$$

The permitted energy levels, known as the energy eigenvalues, are

$$E_n = \frac{n^2 h^2}{8 m_e L^2}$$

For a box of ordinary size, the permitted energy levels are so close together that they are in effect continuous. However, in atomic systems the permitted energy levels may be widely spaced.

EXAMPLE
Suppose an electron is confined in a one-dimensional box of length 10^{-10} m, the diameter of the lowest Bohr orbit. What is the energy of the lowest standing wave in eV?

$$E_1 = \frac{h^2}{8m_e L^2} = \frac{(6.63 \times 10^{-34} \text{ j-s})^2}{8(9.1 \times 10^{-31} \text{ kg})(10^{-10} \text{ m})^2}$$

$$E_1 = \frac{6.04 \times 10^{-18} \text{ J}}{1.6 \times 10^{-19} \text{ J/eV}}$$

$$E_1 = 37 \text{ eV!}$$

If we confine the electron in a box of the order of the size of the hydrogen atom, we find that the minimum energy is quite large. Figure 23-4 shows the quantum

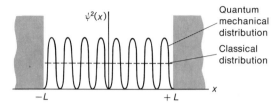

FIGURE 23-4
The quantum distribution of an electron in a one-dimensional box, for $n = 8$, compared to the classical distribution. The probability of finding the electron at any given position in the box depends on the square of the wave function.

distribution of an electron in a box with $n = 8$ compared to the classical distribution.

Standing Waves and Resonance

One-dimensional waves such as standing waves on a string are quite easy to visualize, as are two-dimensional standing waves such as the vibration of a drum head. But three-dimensional quantum waves associated with the hydrogen atom are difficult to visualize. Yet two-dimensional standing waves have many features quite similar to those of three-dimensional quantum waves. We now discuss two-dimensional standing waves and resonance.

A membrane such as a drum head or a soap film can be stretched over a circular hoop. If the membrane is under tension, the wave equation governing the motion is quite similar to the Schrödinger equation. If a drum head such as a tympani is struck, it resounds at certain resonant frequencies—the frequency eigen-values. Figure 23-5 dramatically shows three different possible modes of vibration of a two-dimensional membrane. We can classify these patterns by a number n, the number of nodal circles. The lowest mode of vibration has $n = 1$, a single nodal circle at the outer boundary.

The symmetric vibration of a drum head is not the only possible mode of vibration. Any point on a circular drum can be described by specifying the radius r from the center and the azimuthal angle ϕ defined from the x-axis. The standing wave pattern can be represented as the product of a function of radius $R(r)$, a function of the

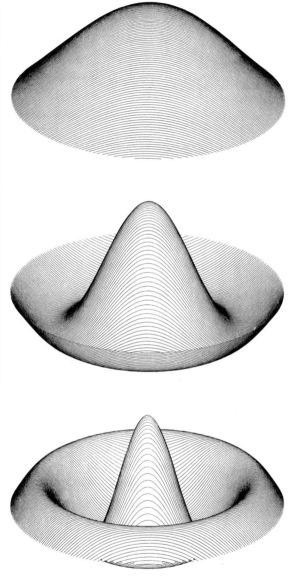

FIGURE 23-5
Symmetrical confined waves on a two-dimensional drumhead, showing different modes of vibration of the drumhead for $m = 0$. A circular drumhead does not have integer harmonics.
A. $n = 1$, the lowest mode, frequency f_1.
B. $n = 2$, frequency $f_{02} = 2.295f_1$.
C. $n = 3$, frequency $f_{03} = 3.599f_1$.
Courtesy of Ron Hipschman, San Francisco State University.

azimuthal angle $\Phi(\phi)$, and an oscillation in time $\cos 2\pi\nu t$. Figure 23-6 shows one possible mode of vibration of the drum head. If we go around the drum in angle Φ, the displacement behaves as

$$\Phi(\phi) = \cos m\phi$$

where $m = 1$. Because a nodal line runs through the origin, the displacement at the center of the drum is zero. Figure 23-7 shows the first three modes, $n = 1,2,3$, corresponding to $m = 1$.

Higher modes of vibration are also possible. Solutions with $n = 2,3$, etc., will have their corresponding radial functions. Figure 23-8 shows the standing wave patterns for circular membranes for azimuthal number $m = 0,1,2$, and 3 for one nodal line $n = 1$ and for two nodal lines $n = 2$. In each case, the radial dependence of the distribution is different. Also shown is the frequency of each node. The frequencies of the higher modes are not an even multiple of the fundamental frequency because of the nature of the solutions to the wave equation.

Figure 23-9 shows the modes with the higher number of nodal lines, corresponding to higher frequencies. For

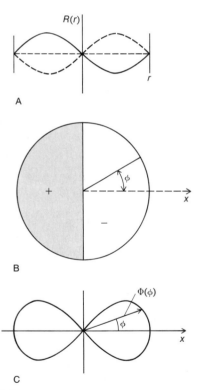

FIGURE 23-6
An asymmetrical standing wave on a drum. At different points around the drum the amplitude is proportional to the azimuthal factor $\Phi(\phi) = \cos m\phi$, for $m = 1$ when combined with the radial behavior, and the time variation produces a standing wave pattern which has a nodal line. From Herman Y. Carr and Richard T. Weidner, *Physics from the Ground Up* (New York: McGraw-Hill Book Co., © 1971).

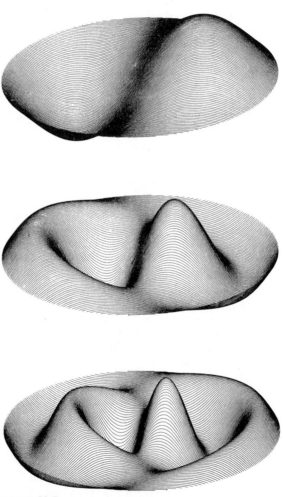

FIGURE 23-7
Asymmetrical vibration of a drumhead $m = 1$.
A. $n = 1$, frequency $f_{11} = 1.593f_1$.
B. $n = 2$, frequency $f_{12} = 2.917f_1$.
C. $n = 3$, frequency $f_{13} = 4.230f_1$.
Courtesy of Ron Hipschman, San Francisco State University.

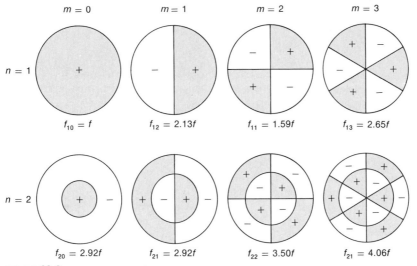

FIGURE 23-8
Standing wave patterns for a circular drum. The number of nodal circles is given by n. The number of times the pattern repeats itself in going around the drum once is given by m, and is equal to the number of nodal lines through the origin.

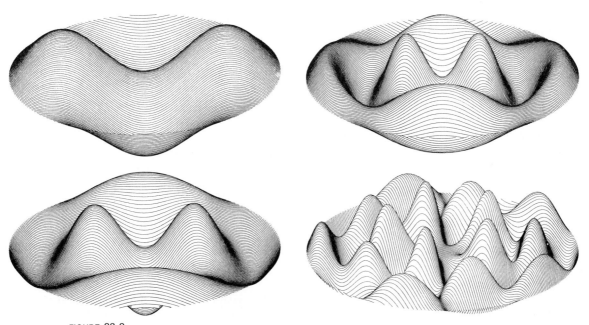

FIGURE 23-9
Higher modes of vibration for a circular drum. A. $n = 1$, $m = 2$, frequency $f_{21} = 2.1316f_1$. B. $n = 2$, $m = 2$, frequency $f_{22} = 3.500f_1$. C. $n = 3$, $m = 2$, frequency $f_{23} = 4.83f_1$. D. $n = 3$, $m = 5$, frequency $f_{53} = 6.528f_1$. Courtesy of Ron Hipschman, San Francisco State University.

the same amplitude, more energy is required to excite these modes. In the normal thumping of a drum, these higher modes die out rather quickly; after a short time, only the lowest-order modes remain. When a bass drum is struck, only higher frequencies are heard at first but after a short time only the boom of the fundamental is heard.

Whereas visualizing these two-dimensional confined waves is fairly easy, visualizing three-dimensional confined waves is difficult, but is required for the solutions to the wave equation for the hydrogen atom.

THE HYDROGEN ATOM

Confined Waves in Three Dimensions

The Schrödinger equation makes it possible to determine the standing wave patterns, the confined waves of an electron trapped in the vicinity of a positive nucleus. The total energy of an electron is the sum of the kinetic energy and the potential energy:

$$E = KE + PE$$

The total energy is

$$E = \frac{p^2}{2m_e} - \frac{k_e e^2}{R}$$

For the electron in a bound state, the total energy is negative; the electron wave is confined to the vicinity of the atom.

The Schrödinger equation can be solved for three-dimensional confined waves. Fortunately, Schrödinger did not have to solve this equation from scratch, for its type was well studied in the nineteenth century. Only certain solutions to the Schrödinger equation are permitted. These solutions can be characterized by three quantum numbers: n, ℓ, and m.

The Quantum Numbers

The **principal quantum number** n is associated with the energy level of the wave electron. Only certain dis-

crete values are permitted. These energy eigen-values give the frequencies of vibration of the different possible confined electron standing waves. For the hydrogen atom, the energy eigen-values are the same as those obtained by Bohr:

$$E_n = -E_o/n^2$$

where $E_o = 13.6$ eV.

The **orbital quantum number** ℓ is associated with the orbital angular momentum of the electron; ℓ is always less then n:

$$\ell < n$$

If a solution to Schrödinger's equation is to be obtained, the angular momentum of the electron must have a value:

$$L = [\ell(\ell + 1)]^{1/2} \frac{h}{2\pi} \qquad (23\text{-}4)$$

where h is Planck's constant. The solution for $\ell = 0$ is called an s-state. The p-state has $\ell = 1$; the d-state has $\ell \neq 2$; the f-state, $\ell \neq 3$; the g-state, $\ell = 4$; and so on. The possible states for various values of the quantum numbers n and ℓ are shown in Table 23-2.

TABLE 23-2
Possible quantum states for values of the quantum numbers n and ℓ.

n	ℓ: 0	1	2	3	4	5
1	$1s$					
2	$2s$	$2p$				
3	$3s$	$3p$	$3d$			
4	$4s$	$4p$	$4d$	$4f$		
5	$5s$	$5p$	$5d$	$5f$	$5g$	
6	$6s$	$6p$	$6d$	$6f$	$6g$	$6h$

There is also a **magnetic quantum number** m which is related to the z component of angular momentum:

$$L_z = m \frac{h}{2\pi} \qquad (23\text{-}5)$$

For a given value of the orbital quantum number ℓ, the magnetic quantum number m may take on any integer value from $-\ell$ to $+\ell$, which is $2\ell + 1$ possible values. These last two quantum numbers are related directly to the angular distribution of the confined wave.

Classically, any orientation of the angular momentum vector is possible. In quantum mechanics, however, only certain orientations are possible. These correspond to the z-component of the angular momentum, with discrete values given by the quantum number m, as shown in Figure 23-10. Table 23-3 shows the possible m values for a given ℓ value.

The spatial distribution of the quantum wave function about the z-axis is related to the magnetic quantum number m and corresponds exactly with the possible azimuthal modes of vibration as discussed previously and shown in Figure 23-6. The magnetic quantum number m deter-

TABLE 23-3
Allowed values of m *for a given* ℓ *value.*

State	ℓ	Allowed values (m)	$(2\ell + 1)$
s	0	0	1
p	1	$-1, 0, +1$	3
d	2	$-2, -1, 0, +1, +2$	5
f	3	$-3, -2, -1, 0, +1, +2, +3$	7
g	4	$-4, -3, -2, -1, 0, +1, +2, +3, +4$	9

mines how often the distribution repeats itself as the electron goes around the z-axis. For $m = 0$, the distribution is symmetric about the z-axis.

Angular Distribution

Because we can no longer properly define an electron orbit, the spatial distribution of the electron in an atom is often referred to as an **orbital.** The s-orbitals are spherically symmetric, and $\ell = 0$ and $m = 0$. Figure 23-11 shows the spatial distribution of the confined waves of the $1s$ and $2s$ orbitals.

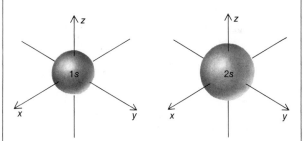

FIGURE 23-11
The s-orbitals are spherically symmetric. A. 1s-orbital. B. 2s-orbital. From Audrey L. Companion, *Chemical Bonding* (New York: McGraw-Hill Book Co., © 1964).

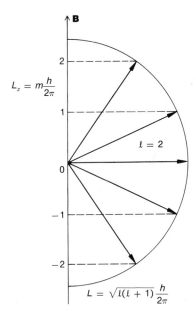

FIGURE 23-10
Projection of the angular momentum **L** along the z-axis determined by the magnetic quantum number m. $L_z = mh/2\pi$. $L = [\ell(\ell + 1)]^{1/2} h/2\pi$.

The p-orbitals correspond to $\ell = 1$. For $m = 0$, the distribution is symmetric about the z-axis. Figure 23-12 shows the angular distribution of the $m = 0$ p-orbital wave function. The p-orbitals with $m = \pm 1$ are given by

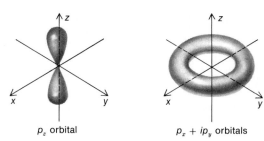

p_z orbital $p_x + ip_y$ orbitals

FIGURE 23-12
The p-orbitals used by the physicist. From Alan Holden,
Bonds Between Atoms (New York: Oxford University Press,
© 1971). Used with permission.

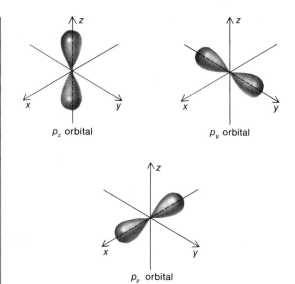

p_z orbital p_y orbital

p_x orbital

FIGURE 23-13
The spatial p-orbitals used by the chemist. From Audrey L.
Companion, *Chemical Bonding* (New York: McGraw-Hill
Book Co., © 1964).

distributions that are also symmetric about the z-axis.
Figure 23-12 shows the form that this angular distribution
takes.

The p-orbitals with $m = -1, 0$, and $+1$ are quite
satisfactory for the physicist. They are useful in explaining both atoms with more than one electron and certain
magnetic phenomena, such as the Zeeman effect. However, the chemist, working with spatial configurations of
molecules, uses orbitals to construct chemical bonds between atoms of a molecule. Fortunately, spatially oriented
configurations can be constructed from the physicist's
p-orbitals.

Figure 23-13 shows the angular distribution of the p_x,
p_y, and p_z orbitals. The p-orbitals form a set of electron
configurations aligned along the x-, y-, and z-axes. The
s-orbitals are spherically symmetric. The s- and p- orbitals
are of particular importance to chemistry.

The d-orbitals correspond to $l = 2$. There are 5 different possible distributions corresponding to $m = -2, -1$,
$0, +1$, and $+2$. The solutions to the Schrödinger equation can be combined to form a series of five spatial
d-orbitals, shown in Figure 23-14. The d_{xy} and $d_{x^2-y^2}$
orbitals lie primarily in the x-y plane. The d_{yz}-orbital lies
primarily in the yz plane rotated by 45°. The d_{xz}-orbital
lies primarily in the x-z plane rotated by 45°. The d_{z^2}-
orbital is symmetric about the z-axis.

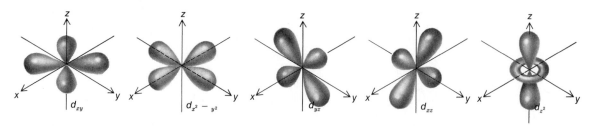

d_{xy} $d_{x^2 - y^2}$ d_{yz} d_{xz} d_{z^2}

FIGURE 23-14
The five spatial d-orbitals. From Audrey L. Companion, *Chemical Bonding* (New York: McGraw-Hill Book
Co., © 1964).

The Radial Distribution

The radial distribution of the confined electron wave depends on the quantum numbers n and ℓ. At a large distance from the nucleus, the wave function must go to zero, just as the displacement of a drumhead goes to zero at the edge of a vibrating drum. For higher n values, we have a larger number of nodal lines. The s-orbitals have spherical nodal surfaces. For higher angular momentum, the p-states, d-states, and so on, the wave function must also go to zero at the origin. Figure 23-15 shows the radial wave functions of hydrogen $R_{n\ell}(r)$. The probability that

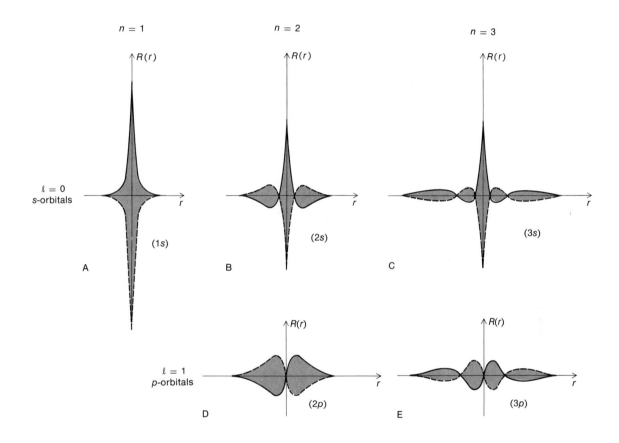

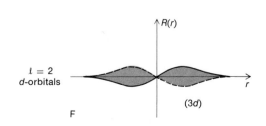

FIGURE 23-15
The radial wave function of hydrogen $R_{n\ell}(r)$. From Herman Y. Carr and Richard T. Weidner, *Physics from the Ground Up* (New York: McGraw-Hill Book Co., © 1971).

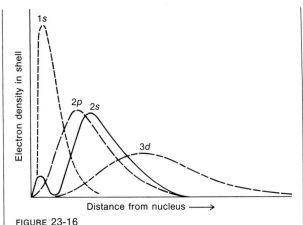

FIGURE 23-16
The radial distribution of the electron density of the hydrogen atom. The maximum of the electron distribution for the 1s, 2p, and 3d states corresponds closely to the radius of the Bohr orbit. From Audrey L. Companion, *Chemical Bonding* (New York: McGraw-Hill Book Co., © 1964).

the electron will be found at a certain distance from the nucleus is shown in Figure 23-16. The maximum of the electron distribution corresponds closely to the diameter of the Bohr orbit.

MANY-ELECTRON ATOMS

If anybody says he can think about quantum problems without getting giddy, that only shows he has not understood the first thing about them.[2]

Niels Bohr and his students and colleagues at the Institute for Theoretical Physics, in Copenhagen, struggled to understand the quantum atom. Bohr was convinced that the chemical elements could be arranged in a systematic sequence. Because the chemical characteristics of an element are always the same, the outer electrons of the atom of an element must always end up in the same state. Until about 1925, however, two key ideas were missing: spin and the exclusion principle.

---

[2] Niels Bohr, quoted in Ruth Moore, *Niels Bohr* (New York: Alfred A. Knopf, 1966), p. 127.

Spin

In the 1920s, two Dutch physicists, Samuel Goudsmit and George Uhlenbeck had been studying the experimental evidence on the **Zeeman effect,** the splitting of spectrum lines in a magnetic field. Hendrik Lorentz and Pieter Zeeman had received the 1902 Nobel Prize in physics for their research on how a magnetic field affects the radiation from atoms. In the *normal Zeeman effect*, a magnetic field splits the spectral lines into three parts: one of higher frequency, one of lower frequency, and one unchanged. The small frequency shift is proportional to the strength of the magnetic field. There was, however, an *anomalous Zeeman effect*, wherein the spectrum lines were split into more than three components. In 1925, Goudsmit and Uhlenbeck put forth a bold proposal. They showed that this anomalous effect could be explained if the electron possessed, in addition to its orbital angular momentum, an intrinsic angular momentum, or **spin.** The z-component spin of the electron could be either up or down and would have a value

$$s_z = \pm \frac{1}{2} \frac{h}{2\pi}$$

where h is Planck's constant. Spin was combined with orbital angular momentum in a complicated theory that was able to correctly predict the observed Zeeman effect.

The Pauli Exclusion Principle

Wolfgang Pauli, a legendary theoretician,[3] suggested that some physical principle prevents all electrons in an atom from being in the same state. Pauli suggested that matters could be settled if no two electrons can occupy the same

---

[3] Pauli is particularly known for the *Pauli effect*. Often when Pauli came near an experimental laboratory, something went wrong or broke down. George Gamow, in *The Thirty Years That Shook Physics*, tells one story concerning an ocurrence in the laboratory of Professor Franck in Göttingen, Germany:
One afternoon, without apparent cause, a complicated apparatus for the study of atomic phenomena collapsed. Franck wrote humorously about this to Pauli at his Zurich address and, after some delay received an answer in an envelope with a Danish stamp. Pauli wrote that he had gone to visit Bohr and at the time of the mishap in Franck's laboratory his train was stopped for a few minutes at the Gottingen railroad station. . . . You may believe this anecdote or not, but there are many other observations concerning the reality of the Pauli effect.

FIGURE 23-17
Wolfgang Pauli (left) and Niels Bohr fascinated by a spinning top. Could this be the source of their inspiration? Courtesy of American Institute of Physics, Niels Bohr Library, Margrethe Bohr Collection.

TABLE 23-4
The sequence of possible states of the atom defined from the quantum numbers n, ℓ, m, spin, and the Pauli exclusion principle, for the first four shells.

State	n	ℓ	Number of states	Total	Shell	Total
1s	1	0	2	2	K	2
2s	2	0	2			
2p	2	1	6	8	L	10
3s	3	0	2			
3p	3	1	6			
3d	3	2	10	18	M	28
4s	4	0	2			
4p	4	1	6			
4d	4	2	10			
4f	4	3	14	32	N	60

quantum state with all the same quantum numbers. The **Pauli exclusion principle** can be stated that no two electrons can have all the same quantum numbers; n, ℓ, m, and spin.

For a given quantum number n, the value of ℓ must be less than n. For a given value of ℓ, the magnetic quantum number m can take on any integer value from $-\ell$ to $+\ell$. There are two possible spin orientations for the electron. For a given value of ℓ there are $2(2\ell + 1)$ possible states. As Table 23-4 shows, this limits the number of possible states for an electron in the atom. The electron configuration of an atom can be built up by filling each shell in turn. The electronic configurations of the elements are shown in the Periodic Table in Appendix B. For $\ell = 0$, we have s-states. The notation $(1s)^2$ means that there are two s-electrons with $n = 1$. For $\ell = 2$, we have p-states. The notation $(3p)^6$ indicates there are six p-electrons with $n = 3$.

The elements hydrogen (1) through argon (18) consist of the sequential filling of the $(1s)$, $(2s)$, $(2p)$, $(3s)$, and $(3p)$ states. The $(4s)$ state fills before the $(3d)$ state. The outer electron configuration of calcium $(4s)^2$ is very similar to the outer electron configuration of magnesium $(3s)^2$, so they have similar chemical properties.

The $(3d)$ state fills with 10 electrons, scandium (21) through nickel (28), before any electrons are added to the $(4p)$ state. The unpaired $(3d)$ electrons, with their intrinsic spin and magnetic moment, are responsible for the magnetic properties of these elements. As the atomic number increases, the next state to be filled is the one with the lowest energy: The $(5s)$ level fills before the $(4d)$ level, the $(5p)$ and $(6s)$ levels fill before the $(4f)$. The higher angular momentum states such as $(5d)$ and $(4f)$ tend to have an electron distribution further from the nucleus and thus at a higher, less strongly bound energy level.

From lanthanum (57) through lutetium (71), the inner $(5d)$ and $(4f)$ levels fill. These rare-earth metals all have the same outer electron configuration $(6s)^2$, and are chemically similar. The chemically similar actinide series, starting with actinium (89), all have an outer-electron configuration $(7s)^2$ and differ only in the number of inner-shell $(5f)$ and $(6d)$ electrons.

WAVE-PARTICLE DUALITY

Uncertainty

By 1927, the quantum mechanics had unmistakably established its power to explain the atom in detail. Great advances were being made on all fronts. Yet the wave nature of electrons was troublesome. Whereas the notion that an electron is a little ball in a neat orbit around the nucleus is somehow comforting, wave motion is vague. Werner Heisenberg, working with Bohr, undertook to give an interpretation of the electron waves. He arrived at an interesting conclusion known as the **Heisenberg uncertainty principle.** This can be illustrated by considering the diffraction of a quantum wave by a single-slit aperture.

Suppose a beam of electrons is traveling to the right in the x-direction with a well-determined energy and momentum, as shown in Figure 23-18. We can determine the position of the electrons in space by having them pass through a single slit of width a. The uncertainty in the y-position is

$$\Delta y = a$$

The electrons with a momentum in the x-direction p_x have a de Broglie wavelength:

$$\lambda = \frac{h}{p_x}$$

The electron wave will be diffracted at the slit. The angular width of the diffraction pattern some distance from the slit is given by our law of single-slit diffraction, $\lambda = a \sin \alpha$. Using the wavelength and slit width this can be written

$$\sin \alpha = \frac{\lambda}{a} = \frac{h}{p_x \Delta y}$$

Where does the electron strike the screen? A rifle bullet would either pass through the slit or would not. It would strike the screen immediately behind the slit unless it were deflected off the edge of the slit. The attempt to localize the electron in space gives rise to an uncertainty in the direction of electron travel. The electron is diffracted into some angle between $+\alpha$ and $-\alpha$. For the electron to go through the slit and arrive at point Y_1 on a screen, it must have a component of momentum in the y-direction, p_y. There is, then, an uncertainty in the momentum, which is given by

$$\Delta p_y = p_x \sin \alpha$$

But combined with the angular size of the diffraction pattern, the uncertainty in momentum can be written as

$$\Delta p_y = p_x \sin \alpha = p_x \frac{h}{p_x \Delta y} = \frac{h}{\Delta y}$$

The uncertainty in the momentum of the particle depends on

$$\Delta y \Delta p_y = h$$

Because of diffraction effects caused by the wave properties of the electron, the product of the uncertainty of the

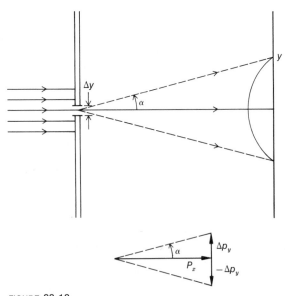

FIGURE 23-18
The Heisenberg uncertainty principle. Any attempt to localize the electron in space introduces uncertainty about the momentum.

position Δy times the uncertainty in the momentum Δp_y is limited by Planck's constant.

Because of the wave nature of matter, it is impossible to measure both the position and the momentum of a particle in a particular direction to a precision greater than some limit. As rigorously deduced from quantum mechanics, the theoretical limit is

$$\Delta y \Delta p_y \geq \frac{1}{2}\frac{h}{2\pi} \qquad (23\text{-}7)$$

where Δy and Δp_y represent the root-mean-square uncertainties of the observed value.

The uncertainty condition also relates the simultaneous measurement of energy and time

$$\Delta E \Delta t \geq \frac{1}{2}\frac{h}{2\pi} \qquad (23\text{-}8)$$

In quantum physics there is much discussion of hypothetical experiments. In principle, these "thought" experiments, or **gedanken experiments,** could be performed, but in practice they present real technical difficulties. Such gedanken experiments are at the heart of philosophical discussions directed toward clarifying the foundations of quantum theory and revealing hidden contradictions in it.

One such gedanken experiment is the double-slit experiment shown in Figure 23-19. In the experiment, a beam of electrons or photons passes first through a single slit to form a coherent beam and then through a double slit. The intensity of the electron or photon distribution is then observed on a screen. We have discussed the double-slit experiment in some detail in our discussion of wave optics in Chapter 20. If light illuminates the slits, the outcome of the experiment is clear. Because of its wave character, light passing through each slit spreads out and forms a single-slit diffraction pattern. Light from the two slits interferes, giving rise to a double-slit interference pattern superimposed on the single-slit diffraction pattern. (This was shown in Chapter 20, Figure 20-18.)

The intensity of the light reaching the screen can be found by combining amplitudes of the waves from the two slits. If the amplitude of the wave from slit one is represented by Ψ_1 and from slit two by Ψ_2, then the total amplitude reaching the screen will be

$$\Psi = \Psi_1 + \Psi_2$$

The intensity represents the probability that the photon wave will be observed at a particular point on the screen. The intensity, and thus the probability distribution, is given by the square of the wave amplitude:

$$P = (\Psi)^2 = (\Psi_1 + \Psi_2)^2$$

If the path difference between the waves is an odd number of half wavelengths, the waves will arrive out of phase and cancel (destructive interference). If the path difference between the waves is an integer number of wavelengths, the waves arrive in phase and add (constructive interference). The double-slit interference pattern will be observed even if the intensity is so small that only one photon is in the apparatus at a time. This double-slit pattern is essentially the same whether we are treating water waves, sound waves, light waves, or quantum waves. The pattern is determined by the slit width a, the slit spacing d, and the wavelength λ.

If we look at the double-slit experiment with reference to particles, we get a different result. Suppose the electrons are small particles analogous to bullets. A bullet going through slit one will arrive at the screen with some probability distribution P_1. We can allow for the bullet's scattering a bit from the edges of the slit, but we do not allow the bullet to break into pieces. The bullets going through the second slit will also arrive at the screen with some probability distribution P_2. The probability of the bullet reaching some position on the screen in passing

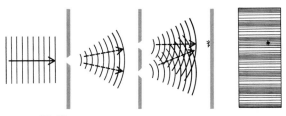

FIGURE 23-19
A double-slit experiment can in principle be done with either photons or electrons, which are observed as a particle (*).

through either slit will be the sum of the two probabilities.

$$P = P_1 + P_2$$

There are no interference effects with bullets; either the bullet arrives at a spot or it does not. How can one bullet cancel another?

When the light or the electron passes through the slits, it acts as a wave to produce an interference pattern. When detected, the light acts as a single bundle of energy or photon. When we detect the electron, we always observe a whole electron, never $1/2$ or $3/4$ electron. The probability of observing these particles is distributed like the intensity of a wave. Thus we have the **wave-particle duality.** Is it a wave or a particle? The answer is *neither* and *both*. This is a paradox of quantum theory.

Because the electron is observed as a particle in the double-slit experiment, it seems reasonable to ask which slit the electron goes through. To answer this question, we can perform another experiment: We can place a light source near the pair of slits so that whenever an electron goes through one slit or the other, we can look for a flash of light. However, to determine which slit the electron went through, the experiment must be sufficiently disturbed to destroy the interference pattern.

It is impossible to arrange the experiment so that we can both tell which slit the electron went through and have interference. Because of the wave-particle nature of both photons and electrons, we must give up trying to know too much. We can no longer predict exactly what will happen in an experiment. This uncertainty is built into the foundations of quantum theory.

Complementarity

Niels Bohr and his colleagues attempted to distill the philosophical significance of the advances of quantum theory. Bohr advocated a point of view which he called **complementarity:** The wave and particle pictures of matter and light are complementary to one another. Thus, in Young's double-slit experiment, light is emitted as photons, which can be diffracted through a double slit as waves. When the photons enter the detector they again act as particles. The wave description and the particle de-

scription are thus complementary. Any attempt to define the momentum and position of a particle is limited by the Heisenberg uncertainty principle, as is any attempt to precisely define the energy and time of the emission of a photon.

For 50 years, most physicists have accepted the statistical interpretation of quantum mechanics. The uncertainty principle has been accepted as a fundamental reality by new generations of physicists, including the author. The uncertainty principle states that our ability to observe nature is limited. A few notable exceptions, Einstein and Schrödinger among them, have attempted to find a more complete description of nature. Thus, the quantum description remains a subject of recurring discussion.

An uncertainty principle also applies to the study of living systems. The observer and the observed form a unity, and observation can have the effect of changing what is being observed. For example, studying the detailed structure of an organism often involves preparing microscope slides from frozen tissue samples. The study of the ultimate structure of organisms and the study of the dynamic processes *in vivo* are complementary in that they cannot both proceed at the same time. As Bohr comments,

In every experiment on living organisms there must remain some uncertainty about the physical conditions to which they are subjected, and the idea suggests itself that the minimal freedom we must allow the organism will be just large enough, so to say, as to hide its ultimate secrets from us. . . .

Understanding of the nature and structure of the atom has moved from the perfection of methods of calculation to the stage of grand philosophical discussion. Meanwhile, the swift-flowing stream of discovery was changing direction: As early as 1929, the Italian physicist Enrico Fermi had realized that the new atomic physics was essentially complete. The new fundamental discoveries and breakthroughs were to come in the study of the nucleus, which Fermi and his colleagues began to do. The Seventh Solvay Conference in 1933 would be devoted to the nucleus. Ominously, 1933 was the year in which Adolf Hitler came to power.

---

[4] Niels Bohr, *Atomic Physics and Human Knowledge* (New York: John Wiley and Sons, 1958), p. 9.

Review

Give the significance of each of the following terms. Where appropriate give two examples.

wave mechanics
matter waves
de Broglie wavelength
Schrödinger equation
standing waves
principal quantum number

orbital quantum number
magnetic quantum number
orbital
Zeeman effect
spin
Pauli exclusion principle

Heisenberg uncertainty principle
gedanken experiment
wave-particle duality
complementarity

Further Readings

Niels Bohr, *Atomic Theory and the Description of Nature* (Cambridge: Cambridge University Press, 1961). Four essays originally published in 1934. Also, *Atomic Physics and Human Knowledge* (New York: John Wiley & Sons, 1958). Seven essays from 1932 to 1957.

Louis de Broglie, *The Revolution in Physics: A Non-mathematical Survey of Quanta* (New York: The Noonday Press, 1953).

Albert Einstein and Leopold Infeld, *The Evolution of Physics* (New York: Simon and Schuster, 1938). The growth of ideas from early concepts to relativity and quanta.

George Gamow, "The Uncertainty Principle," *Scientific American* (January 1958).

Werner Heisenberg, *Physics and Philosophy: The Revolution in Modern Science* (New York: Harper and Row, Harper Torchbooks, 1958). A nonmathematical discussion of the historical roots of atomic science and the status of quantum theory.

Heinrich A. Medicus, "Fifty Years of Matter Waves," *Physics Today* (February 1974), pp. 38–45.

Erwin Schrödinger, *Science, Theory and Man* (New York: Dover Publications, 1957). Nonmathematical essays originally published in 1935.

Erwin Schrödinger, "What is Matter?" *Scientific American* (September 1953).

Problems

MATTER WAVES

1 An electron is accelerated to 32 keV. What is the electron's de Broglie wavelength?

2 An electron in a high-resolution electron microscope has a kinetic energy of 100 keV. Neglect relativity effects.
(a) What is the momentum of the electron?
(b) What is the de Broglie wavelength?

3 Calculate the quantum wavelength of the following:
(a) A 20 keV x-ray.
(b) A 20 keV electron.
(c) A 20 keV proton. ($m_p = 1.6 \times 10^{-27}$ kg)

4 A particle has a quantum wavelength of 0.1 nm. What is its momentum and energy if it is
(a) An electron?
(b) A photon?
(c) A proton?

5 A golf ball of a mass 0.047 kg is traveling at 5 m/s. What is its quantum wavelength?

6 In the Davisson-Germer experiment, electrons with an energy of 54 eV are incident on a monocrystalline nickel surface. The first interference maximum is at an angle of 50°. The crystal spacing is 0.215 nm.
(a) What is the wavelength from crystal diffraction?
(b) What is the de Broglie wavelength?

7 In the Thomson configuration, an electron beam of energy 1.6 keV is diffracted through a thin aluminum crystal. The first-order ($n = 1$) diffraction pattern occurs at an angle of 2°11′ ($\sin \theta = 0.038\ 1$). The spacing of the crystal lattice is 0.405 nm.
(a) What is the wavelength of the electron beam from the diffraction measurement?
(b) What is the momentum of the electron beam?
(c) Using these measurements, what is Planck's constant?

8 A commercial electron microscope has an electron energy of 50 keV and a numerical aperture of 0.02. The microscope has a guaranteed resolution of 0.8 nm. What is the resolution of the microscope if it is diffraction limited? (The resolution of a microscope was discussed in Chapter 21.)

CONFINED WAVES

9 An electron is placed in a one-dimensional box of 0.1 nm.
(a) Draw a sketch of the first three possible electron waves.
(b) What is the energy associated with these three waves?

10 Suppose that an electron were confined in a one-dimensional box of the dimension of the nucleus 10^{-14} m.
(a) Draw a sketch of the lowest possible electron wave.
(b) What is the electron energy associated with that wave?
(c) What is the energy associated with the lowest possible proton ($m_p = 1.67 \times 10^{-27}$ kg) wave?

THE HYDROGEN ATOM

11 The hydrogen atom has energy levels given by $E_o = -13.6\ eV/n^2$.
(a) Make an energy level diagram showing the possible energy levels through $n = 5$.
(b) Indicate which transitions correspond to the Balmer series.

MANY-ELECTRON ATOMS

12 An aluminum atom has 13 electrons surrounding the nucleus. What is the configuration of the electrons?

13 Phosphorus ($Z = 15$) and sulfur ($Z = 16$) have significantly different chemistries. What is the ground state arrangement of electrons in each atom?

WAVE-PARTICLE DUALITY

14 A visible-light microscope will resolve the position of a particle to within about 250 nm. A red blood cell has a diameter of about 10 000 nm and a mass of approximately 2×10^{-13} kg.
(a) What is the uncertainty in the momentum of the red blood cell?
(b) What is the uncertainty in the velocity given by the quantum limit?
(c) Is the observation of a red blood cell limited significantly by the Heisenberg uncertainty principle?

15 An electron is known to be localized somewhere within a space of dimension 0.10 nm.
(a) What is the uncertainty of the momentum?
(b) If the magnitude of the momentum is the same order of magnitude as the uncertainty in the momentum, what is the magnitude of the kinetic energy?

16 Before the discovery of the neutron, it was thought that perhaps electrons were trapped within the nucleus of dimension 10^{-14} m.
(a) What would be the uncertainty of the momentum?
(b) If the magnitude of the momentum is the same order of magnitude as the uncertainty in the momentum, what is the magnitude of the kinetic energy? Is it valid to use a nonrelativistic formula? Is it possible to trap an electron in the nucleus?

ADDITIONAL PROBLEMS

17 A circular drum has a fundamental frequency of 40 Hz in its lowest mode of vibration. The wavelength of this lowest mode is approximately 0.40 m.
(a) Draw a sketch of the first three higher modes.
(b) What are the frequencies of these modes?

18 A high-precision electron microscope has a rated resolution of $\Delta y = 0.50$ nm and an angular aperture of $\alpha = 0.20$ radians. The 30 keV electron beam is scattered from the target atom, which in turn recoils. The uncertainty in the momentum of the target atom is related directly to the uncertainty in the momentum Δp_y of the scattered electron. The electron is scattered into an angle $\pm \alpha$. The uncertainty of momentum is just

$$\Delta p_y = p \sin \alpha$$

(a) What is the momentum p of the electron beam?
(b) What is the uncertainty in the electron momentum Δp_y because of scattering?
(c) What is the product of the uncertainties in momentum and position?

19 The momentum of a particle of charge q, which makes a curved track of radius R in a magnetic field of strength B, is given by

$$p = mv = qBR$$

A positive pi-meson with a charge $q = +e$ leaves a track in a bubble chamber photograph with a radius of curvature of 0.325 m. The magnetic field of the bubble chamber has a value of 1.70 ± 0.07 T. The position of the particle can be defined to about the radius of the bubbles, or about 5×10^{-4} m.
(a) What is the momentum of the pion?
(b) What is the uncertainty in the pion's momentum that is due to lack of precise knowledge of the magnetic field strength?
(c) What is the product of the uncertainties in the pion's position and momentum?

20 In the photoelectric effect, electrons are emitted from the photocathode with energies ranging from zero to the maximum determined by the photon energy and the metal work function. Thus, the uncertainty in the electron energy is of the order of the observed maximum kinetic energy. Experiments indicate that the photoelectrons are emitted from the photocathode instantaneously, within an uncertainty in time of 3.0×10^{-9} s (3.0 ns). A potassium surface has a work function of 2.26 eV and is illuminated with light of wavelength 214 nm.

(a) What is the maximum energy of the emitted photoelectron?

(b) What is the product of the observed uncertainties in time and energy?

(c) Is the experiment significantly limited by Heisenberg uncertainty?

21 In the Mössbauer experiment, the emission of a 14.4 keV gamma ray has an extremely precise energy with an uncertainty of at most 2.9×10^{-8} eV! This is so sharp that in resonance absorption experiments, a Doppler shift caused by velocities as small as 1 mm/s can be observed. The lifetime, and thus the time uncertainty of the 14.4 keV excited state of the nucleus, is 1.4×10^{-7} s. What is the uncertainty product for the energy and time in the Mössbauer experiment?

VII

THE NUCLEUS

24

Fission

The discovery of nuclear fission and development of nuclear weapons have profoundly affected our world. In the 40 years since the first sustained nuclear reaction and the first nuclear detonation, humankind has had to live with the promise and the peril of nuclear energy. In this chapter, we discuss the central discoveries and concepts related to the atomic nucleus and examine the work of a few of its principal discoverers. As a survey of important discoveries Table 24-1 lists the Nobel Prizes awarded for work related to the nucleus.

THE EARLY WORK

Sometimes I had to spend a whole day mixing a boiling mass with a heavy iron rod nearly as large as myself. I would be broken with fatigue at the day's end. Other days, on the contrary, the work would be a most minute and delicate fractional crystallization, in the effort to concentrate the radium.[1]

Marie and Pierre Curie

Marie Sklodowsky arrived in Paris in 1891 to pursue her study of physics, after the University of Warsaw had refused her admission because she was a woman. She earned her doctorate in physics at the Sorbonne, and in

---

[1] Marie Curie, *Pierre Curie* (New York: Dover Publications, 1963), p. 92.

1895 married one of her teachers, Pierre Curie, with whom she collaborated in her research.

In 1896, Henri Becquerel, the discoverer of radioactivity, presented the young Madame Curie with a research problem: Some unknown element was causing *pitchblend*, an ore of uranium, to show more radioactivity than was expected from uranium. The Curies dropped all other research to find this element. Having virtually no research money and working in a makeshift laboratory in an abandoned shed, they started with a ton of pitchblend ore, dissolving, boiling down, and chemically separating its different components and then testing each for radioactivity.

In September 1897, their first daughter was born; soon after Marie was back to work. In July 1898, they announced the discovery of a new element, **polonium** (named after Marie's native Poland), which has chemical properties similar to those of bismuth. Further refinement by fractional crystallization, a technique using the slight differences in solubility of different salts, isolated a second new element, **radium,** chemically similar to barium. From the ton of pitchblend ore, they succeeded in isolating about 8 kg of barium and radium chloride, some 60 times more radioactive than a similar amount of uranium. In March 1902, they finally produced a sample of 0.12 gm of almost pure radium chloride and determined the approximate atomic weight of radium.

The Curies achieved fame and received the 1903 Nobel Prize in physics for their work. (The monetary prize, however, did little more than pay off the debts they had incurred in their research.) After Pierre's accidental death in 1906 (he was struck down by a horsedrawn wagon) Marie Curie succeeded him as a professor of physics, becoming the first woman faculty member at the Sorbonne. Figure 24-1 shows her at work in the laboratory. She continued her research and preparation of radium salts for cancer therapy in hospitals around the world until her death in 1934.

TABLE 24-1
Nobel Prizes awarded for work and discoveries related to the nucleus. (Prize awarded in chemistry.)*

1903 Henri Becquerel (1852-1908), for discovery of radioactivity; Pierre Curie (1859-1906) and Marie Curie (1867-1934), for research on radiation phenomena, including discovery of polonium and radium.

1908* Ernest Rutherford (1876-1937), for investigation of radioactive decay of elements such as thorium and the properties of alpha and beta particles. (He later discovered that protons are part of the nucleus.)

1916* Marie Curie (1867-1934), for chemical separation of radium.

1927 Charles Wilson (1869-1959), for development of the cloud chamber that makes the paths of electrically charged particles visible.

1935 James Chadwick (1891-1974), for discovery of the neutron.

1936 Carl Anderson (1905-   ), for discovery of the positron in cosmic rays.

1938 Enrico Fermi (1901-1954), for demonstration of new radioactive isotopes produced by neutron bombardment.

1939 Ernest O. Lawrence (1901-1958), for invention and development of the cyclotron.

1951 John Cockroft (1897-1967) and Ernest Walton (1903-   ), for the transmutation of atomic nuclei by artificially accelerated particles: the atom smasher.

1960 Donald Glaser (1926-   ), for invention of the bubble chamber.

1963 Eugene Wigner (1902-   ), for contributions to the theory of the nucleus; Marie Geopert-Mayer (1906-1972) and Hans Jensen (1907-1973), for discovery of nuclear shell structure.

1967 Hans Bethe (1906-   ), for theories of nuclear reactions and of energy production in stars.

1976 Aage Bohr (1922-   ), Ben Mottelson (1926-   ), and James Rainwater (1917-   ), for development of a theory of the nucleus based on collective and particle motion of nucleons.

FIGURE 24-1
Marie Curie showing the patience required to perform the delicate and tedious radium separations. (Photographer unknown.) Courtesy of American Institute of Physics, Center for the History of Physics.

Natural Radioactivity

By 1900, the naturally radioactive elements polonium, radium, thorium, and actinium had been isolated and their forms of radiation—alpha particles, beta particles, and gamma rays—were being studied. At McGill University, Rutherford discovered a new group of radioactive substances associated with thorium. A series of experiments, conducted with Frederick Soddy, showed that thorium, the **parent nucleus,** underwent **alpha decay,** the emission of an alpha particle, becoming a radioactive material thorium-x, the **daughter nucleus.** If thorium-x and the parent nucleus are chemically separated, it takes about

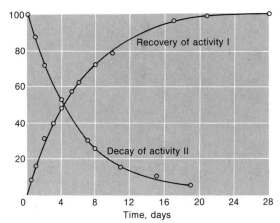

FIGURE 24-2
Radioactive emanation of thorium. Partial results of experiments conducted by Rutherford and Soddy. Curve I shows the recovery of activity of the thorium. Curve II shows the decay of activity of a chemically separable product, which they named thorium-x. From *Journal of the Chemical Society, Transactions*, 81 (1902), pp. 837–860.

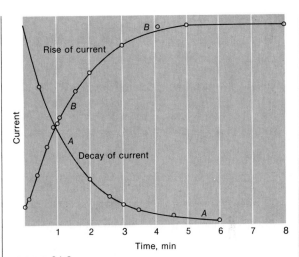

FIGURE 24-3
The decay current from radioactive thorium emanation as reported by Rutherford. From *Philosophical Magazine*, 5 (1900), 49: 1–14.

four days for thorium-x to lose about half its activity and for the parent thorium to recover its activity, as shown in Figure 24-2. It was also found that thorium-x would decay into the noble gas radon (chemically similar to such noble gases as helium and neon) and would lose about half its activity in about one minute, as shown in Figure 24-3.

One of Rutherford's principal discoveries was the law of **radioactive decay.** The radioactivity of a single substance, such as thorium-x, decreases with time in a geometric progression: 1, $\frac{1}{2}$, $\frac{1}{4}$, $\frac{1}{8}$, etc. The time in which activity diminishes by one-half is the radioactive **half-life.** Thorium-x has a half-life of about four days; radon about one minute. After 10 half-lives (about 10 minutes for radon), there is very little activity.

Rutherford gave a theoretical explanation of this process: The **activity** A, the number of radioactive decays per unit time, is proportional to the number of radioactive nuclei N present and a **decay constant** λ:

$$A = -\frac{\Delta N}{\Delta t} = \lambda N$$

where the minus sign indicates that the number of nuclei is decreasing in time. The decay constant gives the probability per unit time that a particular nucleus will undergo radioactive decay.

In our discussion of RC time constants in Chapter 16 (see also Appendix A), we found that when the rate of change in a quantity is proportional to the quantity, the result is exponential behavior. If initially N_0 nuclei are present at time $t_0 = 0$, then the number of nuclei present at a later time t is given by

$$N = N_0 e^{-\lambda t}$$

Figure 24-4 plots exponential decay as a function of time. After one half-life $T_{1/2}$, the number of radioactive nuclei has decreased by a factor of two. The decay constant λ is inversely proportional to the half-life, $\lambda T_{1/2} = \ln 2 = 0.693$ or

$$\lambda = \frac{0.693}{T_{1/2}}$$

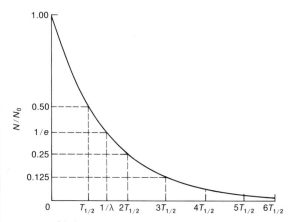

FIGURE 24-4
Radioactive decay as a function of time. For each half-life $T_{1/2}$, the relative activity falls by a factor of one-half.

If N radioactive nuclei are present, the activity of the sample is

$$A = \lambda N = \frac{0.693}{T_{1/2}} N \qquad (24\text{-}1)$$

If the initial activity of a sample is A_0 at $t_0 = 0$, the activity at time t is given by

$$A = A_0 e^{-\lambda t}$$

EXAMPLE
Thorium-x decays by alpha particle decay with a half-life of 3.64 days.
(a) If the initial activity of the sample is 10^4 counts per minute, how many nuclei are initially present?
(b) What is the activity after 7.0 days?
    The number of nuclei initially present N_0 can be found from the initial activity

$$N_0 = \frac{A_0}{\lambda} = \frac{T_{1/2} A_0}{0.693}$$

$$N_0 = \frac{(3.64 \text{ days})}{(0.693)} \frac{(24 \text{ hr})}{(1 \text{ day})} \frac{(60 \text{ min})}{(1 \text{ hr})} (10^4 / \text{min})$$

$$N_0 = 7.56 \times 10^7 \text{ nuclei}$$

The activity after 7.0 days is given by

$$A = A_0 e^{-\lambda t} = A_0 e^{-0.693t/T_{1/2}}$$

$$A = (10^4 / \text{min}) e^{-(0.693)(7.0 \text{ days})/(3.64 \text{ days})}$$

$$A = (10^4 / \text{min}) e^{-1.33}$$

The value of the exponential can be found from a sliderule, a scientific calculator, or logarithm tables: $e^{-1.33} = 0.264$

$$A = 2640 / \text{min}$$

The activity after 7 days is 2640 counts per minute.

Radioactive Decay

Natural radioactive decay occurs by emission of alpha particles or beta particles or gamma rays. Moseley's work on x-rays, discussed in Chapter 22, allows the unambiguous assignment of an **atomic number** Z to represent the charge on the nucleus of different chemical species. The mass spectrometer work of Aston, discussed in Chapter 16 shows that the same chemical element consists of **isotopes,** nuclei with the same Z but with different mass and **atomic mass number** A.

A particular isotope can be described by a chemical symbol X with a subscript representing the atomic number Z and a superscript representing the atomic mass number A:

$$_Z X^A$$

For instance, carbon-12, the common isotope of carbon, is $_6 C^{12}$. This notation for isotopes allows us to keep track of charge and mass in nuclear reactions.

By spectroscopic observation, Rutherford learned that the alpha particle was a helium nucleus produced by alpha decay. The alpha particle has an atomic number $Z = 2$ and a mass number $A = 4$:

$$\alpha = {}_2 He^4$$

In the process of $\boldsymbol{\alpha}$ **decay** (alpha decay), an alpha particle with a precise energy is emitted from the nucleus. A radioactive nucleus with atomic number Z and mass number A will change to atomic number $Z - 2$ and mass number $A - 4$. For instance, Rutherford's thorium-x (which is actually radium-224, $_{88}Ra^{224}$) is produced by the alpha decay of thorium-228:

$$_{90}Th^{228} \longrightarrow _{88}Ra^{224} + _2He^4 + 5.42 \text{ MeV}$$

with the release of 5.42 MeV of energy. In nuclear reactions, the atomic number ($90 = 88 + 2$) and the mass number ($228 = 224 + 4$) are conserved. By α decay, thorium-x in turn decays into radon-220, $_{86}Rn^{220}$:

$$_{88}Ra^{224} \longrightarrow _{86}Rn^{220} + _2He^4 + 5.68 \text{ MeV}$$

In the process of **β^- decay** (beta minus decay), the radioactive nucleus emits an electron with a charge $Z = -1$ and a mass number $A = 0$.

$$\beta^- = _{-1}e^0$$

If the nucleus decays by emitting only a single particle, the decay energy should be precisely divided between the β^- and the remaining nucleus. By conservation of momentum, if the momentum of the nucleus at rest before decay is zero, then the total momentum of the two particles after decay is also zero:

$$m_1v_1 + m_2v_2 = 0$$

If the electron leaves with a velocity v_1, then the nucleus will recoil with a velocity $v_2 = -m_1v_1/m_2$. The energy released in the decay is divided between the two particles:

$$E = \tfrac{1}{2}m_1v_1{}^2 + \tfrac{1}{2}m_2v_2{}^2$$

This, however, is not observed in β^- decay. The electron is emitted with energies ranging from zero to some maximum of 0.048 MeV, as shown in Figure 24-5.

In 1934, the Italian physicist Enrico Fermi theorized that in beta decay a third particle, a **neutrino,** ν, is emitted that has virtually no mass and no charge and is almost undetectable. By postulating the existence of the neutrino, Fermi developed a theory to explain the energy spectrum of beta decay. The energy of the nuclear decay is distributed among three particles: the β^-, the recoil nucleus, and the neutrino. For instance, radium-228 de-

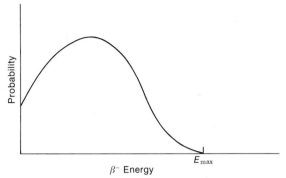

FIGURE 24-5
Beta decay. In beta-minus decay the electron is emitted with a range of energies from zero up to some maximum.

cays to actinium-228 with a half-life of 5.9 years. The reaction can be written

$$_{88}Ra^{228} \longrightarrow _{89}Ac^{228} + _{-1}e^0 + \nu + 0.048 \text{ MeV}$$

The existence of these ghostlike neutrinos was confirmed experimentally 22 years later, in 1956. Neutrinos have recently become an important factor in theories of solar evolution and energy production (as we discuss in Chapter 26).

In **gamma decay,** a radioactive nucleus emits a high-energy photon. Because photons possess no electrical charge or rest mass, the atomic number Z and the atomic mass number A of the nucleus remain unchanged. Some internal energy of the nucleus is radiated away. Gamma rays are radiated in many radioactive processes.

There are three naturally occurring radioactive decay series: the *uranium series*, starting with $_{92}U^{238}$; the *actinium series*, starting with $_{92}U^{235}$; and the *thorium series*, starting with $_{90}Th^{232}$. Each series consists of a chain of radioactive decays by α and β^- decays that start with an isotope of a very long half-life. For instance, the first step of the thorium series begins, as shown in Figure 24-6, with the α decay of $_{90}Th^{232}$ (with a half-life of 14 billion years) to radium-228, $_{88}Ra^{228}$, which is produced at a slow, constant rate. The remaining decay sequence proceeds rather quickly, each step with its characteristic half-life.

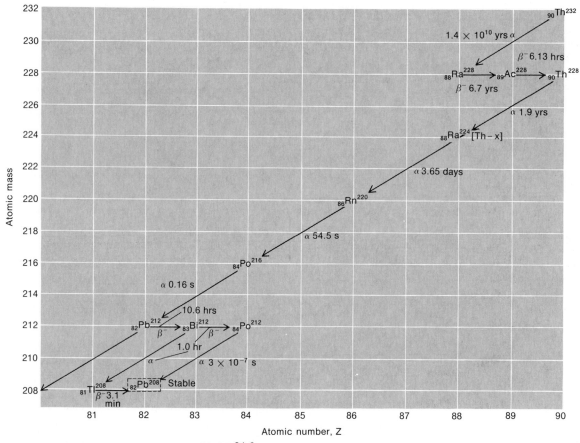

FIGURE 24-6
The decay scheme of the thorium series.

So complex is each of the three decay series that observing radioactive products and doing the necessary chemical separations to isolate the radioactive fractions took many years and occupied many investigators: the Curies and later their daughter Irene Joliot-Curie and her husband Frederic Joliot; Otto Hahn (who in 1905 spent time with Rutherford at McGill) and his coworkers Lise Meitner and Hans Strassmann.

In **β^+ decay** (beta plus decay), an unstable nucleus emits a **positron,** a particle with a mass equal to that of the electron but with a positive charge. When radioactivity was artificially produced (as discussed below) by bombarding nuclei with energetic particles, some nuclei were observed to decay by positron emission. For instance sodium-22 decays to neon-22 plus a positron and a neutrino

$$_{11}Na^{22} \longrightarrow {}_{10}Ne^{22} + {}_{+1}e^0 + \nu + 0.546\,MeV$$

Neutrinos are also associated with β^+ decay.

The existence of the positron was first predicted in 1928 by Dirac as a result of his efforts to reconcile the quantum wave equation with Einstein's special theory of relativity. A positron combined with an electron cancels the charge and liberates their rest mass energy as two gamma rays:

$$_{-1}e^0 + {}_{+1}e^0 \longrightarrow 2\gamma$$

Each gamma has 0.511 MeV of energy.

Dirac's prediction remained a subject of theoretical speculation until 1932, when Carl Anderson, studying the tracks left by cosmic rays in a *cloud chamber*, came upon tracks that resembled an electron with positive charge. (The cloud chamber, invented by Charles Wilson, is a closed chamber with a saturated vapor of alcohol or water under a slight pressure. Charged particles leave a trail of ionization behind in passing through the chamber. When the pressure in the chamber is reduced, the ions act as centers of condensation and small droplets form. A camera with flash lamp photographs the trail of particles and thereby records the passing particle.)

Working at the University of Chicago, Anderson was using a cloud chamber placed in a magnetic field to investigate cosmic ray "showers," bursts of rays from space that enter the earth's atmosphere. He observed a 63 MeV particle pass through 6 mm of lead and emerge with an energy of 23 MeV. The curvature of the particle's path indicated that it must be positive. This could be interpreted as a particle with the mass of an electron and a positive charge, the positron. Soon he observed that from gamma rays passing through aluminum produced pairs of positrons and electrons. Figure 24-7 shows a modern experiment in which an energetic gamma ray produces an electron positron pair or an electron positron pair plus a recoil electron.

NUCLEAR STRUCTURE

The Proton and Neutron

In 1919, using alpha particles to bombard the nucleus, Rutherford produced artificial disintegration of the nucleus. Alpha particles passing through a nitrogen gas caused many long-range particles to be emitted from the nitrogen nuclei. An alpha particle striking a nitrogen nucleus released a hydrogen nucleus, or **proton:**

$$_2\text{He}^4 + {_7}\text{N}^{14} \longrightarrow {_8}\text{O}^{17} + {_1}\text{H}^1$$

This was the first observation that the proton was a constituent of the nucleus. Thus, the positive charge of the nucleus is associated with protons in the nucleus.

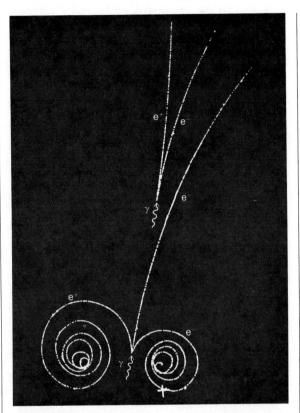

FIGURE 24-7
Electron-positron pair production. A high-energy gamma ray scatters off an electron, losing some of its energy and producing an energetic recoil electron and an electron-positron pair. A second gamma ray produces a second electron-positron pair. Courtesy of Lawrence Berkeley Laboratory, University of California.

Rutherford also suggested that the nucleus might also contain a neutral particle, but it was not until 1932 that James Chadwick, working with Rutherford, made the discovery. (In the 1930s, the Cavendish laboratory under Rutherford's leadership—see Figure 24-8—was a center for the study of the nucleus.) Chadwick bombarded beryllium metal with alpha particles and found an unknown radiation that would even penetrate lead. Like x-rays, it would darken a photographic film, and, like alpha parti-

FIGURE 24-8
Ernest Rutherford with his students at the Cavendish Laboratory during the early 1930s: " . . . one would receive occasionally . . . a more or less unannounced visit from Rutherford at one's own working bench. He would seat himself on a laboratory stool and put one through a quite searching examination: 'What, precisely, are you doing?'" Photograph by Paul Harteck. Courtesy of American Institute of Physics, Niels Bohr Library.

tracks of the recoil particles, Chadwick was able to determine a recoil velocity of a proton of $v_p = 3.7 \times 10^7$ m/s and of a nitrogen of $v_n = 4.7 \times 10^6$ m/s. From these observations he concluded that the mass of this neutral particle m_n must be approximately equal to the mass of the proton m_p. The new particle was named the **neutron.** Modern values give the neutron a mass m_n compared to the proton mass m_p as

$$m_n = 1.0014\, m_p$$

The reaction by which Chadwick produced the neutron is

$$_2\text{He}^4 + {}_4\text{Be}^9 \longrightarrow {}_6\text{C}^{12} + {}_0\text{n}^1$$

The discovery of the neutron provided a missing piece in the puzzle of nuclear structure. The nucleus consists of numbers of the two **nucleons:** protons and neutrons. The atomic mass number A corresponds to the number of nucleons in the nucleus; the atomic number Z corresponds to the number of protons.

The Atom Smashers

In April 1931, J. D. Cockcroft and E. T. S. Walton, shown with Rutherford in Figure 24-9, had developed a high-voltage machine capable of accelerating protons up to energies of 600 keV. Bombarding a lithium target with energetic protons produced alpha particles. The reaction that Cockcroft and Walton observed,

$$_3\text{Li}^7 + {}_1\text{H}^1 \longrightarrow {}_2\text{He}^4 + {}_2\text{He}^4$$

was the first artificial splitting of the atom with electrically accelerated particles. After this event, larger and larger particle accelerators were used to study the nucleus. High-energy physics had been born. By the end of 1932, Ernest O. Lawrence, at the University of California at Berkeley (see Figure 24-10), had succeeded in operating an 11 in cyclotron able to produce a beam of protons with a current of 10^{-9} A at an energy of 1.22 MeV. He immediately began bombarding various targets and obtained results similar to Cockcroft's and Walton's.

cles or protons, it would produce ionization in passing through a gas. These unknown radiated particles were undeflected by a magnetic field and therefore were electrically neutral.

The experimental problem faced by Chadwick was to either determine whether these neutral penetrating rays were actually photons with zero rest mass, or else to measure the mass of the particles. Using a Wilson cloud chamber, Chadwick observed the recoil of both hydrogen and nitrogen nuclei from such radiation. By measuring the

FIGURE 24-9
E. T. S. Walton (left) and John Cockcroft (right) with Rutherford in 1932, after their successful disintegration of nuclei with energetic protons from their accelerator. Photograph: UK Atomic Energy Authority. Courtesy of American Institute of Physics, Niels Bohr Library.

FIGURE 24-10
E. O. Lawrence and Robert Thornton discuss an early experiment with the 27 in cyclotron in 1934. Courtesy of Lawrence Berkeley Laboratory, University of California.

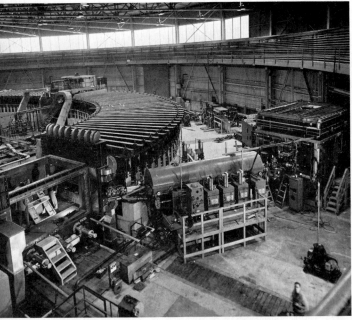

FIGURE 24-11

The bevatron as it nears completion in about 1954, before the shielding blocks are in place. Far right is the Cockcroft-Walton preaccelerator. The cylinder in the center contains the linear accelerator, which injects the particles into the main accelerator chamber. The workman standing in the beam pipe indicates the scale of the machine. Courtesy of Lawrence Berkeley Laboratory, University of California.

As larger accelerators at higher energies became operational, the inner secrets of the nucleus began to be revealed. In 1952, Donald Glaser, at the University of Michigan, showed that the track of a charged particle passing through a liquid could be revealed in small gas bubbles as the liquid boils, when the pressure in the chamber suddenly drops. Using this phenomenon, he constructed the first bubble chamber to observe and measure the tracks of high energy particles. Many different liquids, including hydrogen (which boils at 25 K), have been used in bubble chambers.

In 1954, the first of a new generation of particle accelerators became operational. The Bevatron in Berkeley, shown in Figure 24-11, began producing protons with an energy of 2 billion[2] electron volts (2 GeV). Development of new accelerators and of the bubble chamber brought a "golden age" to high-energy physics in the 1950s and 1960s.

Binding Energy of the Nucleus

In the mid-1930s, Hans Bethe and others accurately compared the masses of the initial and final products of nuclear reactions. They found that a small portion of the mass was converted into energy. When a nuclear reaction takes place, the total mass of the final products is less than the total mass of the initial products. This **mass difference** is converted into energy as given by Einstein's relationship:

$$\Delta E = \Delta M c^2$$

By measuring the mass and energy of the particles emitted in radioactive decay, and by measuring the masses of the initial and final atoms, they showed that energy is conserved.

The number of positive charges, protons, in the nucleus is given by the atomic number Z. The total number of protons and neutrons in the nucleus gives the atomic mass number A. The higher the position of an element in the sequence of elements, the greater is its number of protons and neutrons; the mass of the nucleus is slightly less than the sum of the proton and neutron masses. This small mass difference was liberated as energy when the elements were formed in stars early in the evolution of the universe. This mass difference gives rise to the nuclear **binding energy** B. The binding energy of a nucleus of mass M is

---

[2] In the United States and France, one billion represents a thousand million (10^9), in Great Britain and Germany, it represents a million million (10^{12}). Such confusion is intolerable in science. The prefix giga- (G-), pronounced as jig-a, was adopted to represent 10^9. Consequently the unit for energy for large particle accelerators is the giga-electronvolt (GeV). The proper name of the machine, the "Bevatron," was not changed, however. The prefix tera- (T-) has been adopted to represent 10^{12}. At present, superconducting magnets are being installed at the 500 GeV proton accelerator at Fermilab in Illinois. When operational, the machine will accelerate protons to 1000 GeV or 1 tera-electron volt (1 TeV). The machine will be called the "Tevatron."

given by

$$B = \Delta Mc^2 = [ZM_p + (A - Z)M_n - M]c^2 \quad (24\text{-}2)$$

The total binding energy increases with increasing atomic mass number.

Figure 24-12 shows the binding energy per nucleon B/A, and also shows an interesting effect. For the more

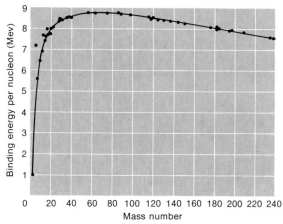

TABLE 24-2
Masses of atoms and of subatomic particles, in atomic mass units (amu).

Electron	$_{-1}e^0$	0.000 549
Positron	$_{+1}e^0$	0.000 549
Proton	$_1p^1$	1.007 277
Neutron	$_0n^1$	1.008 665
Hydrogen-1	$_1H^1$	1.007 825
Deuterium	$_1H^2$	2.014 102
Tritium	$_1H^3$	3.016 050
Helium-3	$_2He^3$	3.016 030
Helium	$_2He^4$	4.002 603
Lithium-6	$_3Li^6$	6.015 124
Lithium-7	$_3Li^7$	7.016 004
Beryllium	$_4Be^9$	9.012 186
Boron-11	$_5B^{11}$	11.009 305
Carbon	$_6C^{12}$	12.000 000

The amu are based on carbon-12 (= 12.000 000 amu), 1 amu = $1.660\ 566 \times 10^{-27}$ kg. All values are from the Atomic Mass Table in *Radiological Health Handbook* (Public Health Service, US HEW, Revised January 1970), pp. 51–64.

FIGURE 24-12
Binding energy per nucleon in MeV as a function of mass number.

Table 24-2 shows the masses of the proton, neutron, and electron and other light nuclei.

stable nuclei, with atomic mass numbers 20 to 190, the binding energy is greater than 8 MeV per nucleon (a proton or neutron). Iron, with an atomic mass number of 56, has a binding energy of 8.79 MeV/nucleon, making it one of the most stable nuclei. Above atomic mass number 60, the binding energy per nucleon decreases, because the Coulomb force between electrically charged protons in the nucleus tends to make the nucleus fly apart. The atomic mass scale (sometimes called the unified mass scale) is based on the mass of carbon-12, including the mass of the atomic electrons. The mass of carbon-12 is 12.000 000 atomic mass units (amu). One atomic mass unit is

$$1 \text{ amu} = 1.660\ 53 \times 10^{-27} \text{ kg}$$

EXAMPLE
The helium atom $_2He^4$ has a mass of 4.002 603 amu. What are the total binding energy and the binding energy per nucleon for helium?

The mass difference is given by
$$\Delta M = ZM_p + (A - Z)M_n - M$$

$$\Delta M = 2(1.007\ 825 \text{ amu}) + 2(1.008\ 665 \text{ amu})$$
$$- 4.002\ 603 \text{ amu}$$

$$\Delta M = 0.030\ 377 \text{ amu} = 5.0443 \times 10^{-29} \text{ kg}$$

The binding energy is

$$B = \Delta Mc^2 = (5.044\ 3 \times 10^{-29} \text{ kg})(2.997\ 9 \times 10^8 \text{ m/s})^2$$

$$B = 4.534 \times 10^{-12} \text{ J} = 28.296 \text{ MeV}$$

The binding energy is over 28 MeV. The binding energy per nucleon is

$$B/A = \frac{28.296 \text{ MeV}}{4 \text{ nucleons}} = 7.074 \text{ MeV/nucleon}$$

In the radioactive decay of heavy elements, energy is available because of the difference between the masses of the initial and final products. It is a matter of simple arithmetic to determine both the available energy and the possibility of a reaction.

EXAMPLE

Radium-226 decays to radon-222 by the emission of an alpha particle:

$$_{88}\text{Ra}^{226} \longrightarrow {}_{86}\text{Rn}^{222} + {}_2\text{He}^4$$

The mass difference between the final products and the initial radium is given by

$_{86}\text{Rn}^{222}$	222.017 61 amu
$_2\text{He}^4$	4.002 60 amu
products	226.020 21 amu
$_{88}\text{Ra}^{226}$	226.025 44 amu
mass difference	.005 23 amu

The energy available for this reaction from the conversion of mass to energy is

$$\Delta M c^2 = 4.87 \text{ MeV}$$

The alpha particle emitted by radium has an observed energy of 4.78 MeV.

The successful smashing of the atom by Cockcroft and Walton in England, the Joliot-Curies in France, and Lawrence in the United States verified that mass could be converted to energy. For very heavy elements, the binding energy per nucleon falls below 8 MeV nucleon, while elements of intermediate mass have binding energies of about 8.7 MeV/nucleon. If a heavy element such as lead or radium or perhaps uranium were split into two fragments, a single reaction might release millions of electron volts of energy.

NUCLEAR FISSION

Enrico Fermi

Enrico Fermi and his colleagues at the Physical Laboratory of the University of Rome studied the nucleus by means of neutrons, which, being electrically neutral particles, reach the nucleus unaffected by the Coulomb force. The early experiments were quite simple. The neutron source was a sealed glass tube about 6 mm in diameter and 15 mm long, containing radon-222 gas and beryllium powder. The radon-222, a daughter nucleus of radium-226, emits an alpha particle that strikes the beryllium, which gives off a neutron. About 2×10^7 neutrons per second are emitted with energies up to about 8 MeV. A substance to be investigated is shaped into cylindrical form and fitted first around the neutron source and then around a detector so that the decay products from the neutron-induced activity can be detected. Fermi's group systematically worked its way through the periodic table of the elements.

By the summer of 1934, the group had completed a survey of some 60 elements and found more than 40 product nuclei that were radioactive by beta decay. In several cases, sufficient radioactivity was obtained so that by special chemical techniques they could separate and identify the radioactive element formed. In three cases, the active isotope formed by bombarding an element (aluminum, chlorine, or cobalt) with atomic number Z resulted in an isotope with atomic number $Z - 2$. For instance,

$$_{13}\text{Al}^{27} + {}_0\text{n}^1 \longrightarrow {}_{11}\text{Na}^{24} + {}_2\text{He}^4$$

Sodium-24 subsequently decays by β^- decay with a half-life of 15 hr. In four cases, the bombardment of an element (phosphorus, sulfur, iron, or zinc) resulted in an isotope with atomic number $Z - 1$. For instance,

$$_{15}\text{P}^{31} + {}_0\text{n}^1 \longrightarrow {}_{14}\text{Si}^{13} + {}_1\text{H}^1$$

The silicon-31 subsequently decays by β^- emission with a 2.6 hr half-life. The nuclei of elements bombarded by neutrons undergo *transmutation*, changes in their chemical nature.

In still other cases (bromine or iodine), the radioactive product is an isotope of the bombarded element plus a gamma ray:

$$_{35}Br^{79} + {_0}n^1 \longrightarrow {_{35}}Br^{80} + \gamma$$

Bromine-80 subsequently decays by β^- decay with an 18 min half-life. The creation of radioactive materials by absorption of neutrons is known as **neutron activation.**

Uranium Fission

The results obtained by neutron bombardment of element 92, uranium, were puzzling and complicated. In 1934, Fermi announced that his group had probably produced element 93. Seeing reports of Fermi's work, Lise Meitner, Fritz Strassmann, and Otto Hahn, at the Kaiser Wilhelm Institute in Berlin, repeated the experiments and reached a similar conclusion. In a similar experiment in 1937, Irene Joliot-Curie and P. Savitch, in Paris, discovered a material chemically similar to lanthanum $Z = 57$.

Lise Meitner, an Austrian Jew, had worked with Otto Hahn in Berlin for 30 years (see Figure 24-13) when, in 1938, she moved to Stockholm after Hitler's invasion of Austria. Hahn and Strassmann continued the uranium work, repeating the Joliot-Curie experiments. Finding four isotopes, supposedly of radium, that precipitated out with barium, they came to a remarkable conclusion: "We come to the conclusion that our 'radium isotopes' have the properties of barium. As chemists we should actually state that the new products are not radium, but barium itself."[3] Bombarding uranium $Z = 92$ with neutrons produces relatively lightweight barium $Z = 56$. The uranium has undergone **fission;** that is, the nucleus splits in two.

---

[3] Otto Hahn and Fritz Strassmann, "Concerning the Existence of Alkaline Earth Metals Resulting from Neutron Irradiation of Uranium" *Naturwissenschaften* (January 1939), p. 11; translation quoted in Hans G. Graetzer and David L. Anderson, *The Discovery of Nuclear Fission* (New York: Van Nostrand, Momentum Book #20, 1971), pp. 44–47.

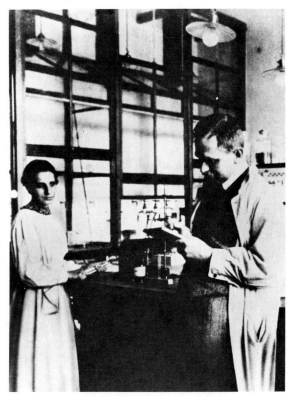

FIGURE 24-13
Lise Meitner and Otto Hahn at the Kaiser Wilhelm Institute for Chemistry, Berlin, Germany, in 1913. Photo courtesy of American Institute of Physics, Niels Bohr Library.

In late 1938, Hahn communicated his and Strassmann's findings to Lise Meitner, who shared them with the physicist Otto Frisch. Frisch and Meitner confirmed the theoretical possibility of uranium fission, and gave Niels Bohr a rough calculation of the energy released in the splitting of uranium. By early 1939, both Bohr and Enrico Fermi had arrived in the United States, where the news about fission spread quickly by word of mouth. By the time of the Washington Conference on Theoretical Physics, only 10 days after Bohr's arrival in New York City, fission had been experimentally confirmed at Columbia University, the University of California at Berkeley, Carnegie Institute in Washington, D.C., and at Johns Hopkins University in Baltimore.

Frisch and Meitner describe the process of fission:

On account of their close packing and strong energy exchange, the particles in a heavy nucleus would be expected to move in a collective way which has some resemblance to the movement of a liquid drop. . . . [see Figure 24-14] It seems therefore possible that the uranium

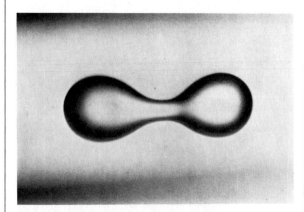

FIGURE 24-14
The liquid drop model. A water drop suspended in oil is deformed. If sufficient energy is imparted to the drop it will fission. In 1939 Niels Bohr and John A. Wheeler developed the analogy between the splitting of a drop of liquid by deformation and nuclear fission. Courtesy of Lawrence Berkeley Laboratory, University of California.

nucleus has only small stability of form, and may after neutron capture divide itself into two nuclei of roughly equal size (the precise ratio of sizes depending on finer structural features and perhaps partly on chance). These two nuclei will repel each other and should gain a total kinetic energy of [about] 200 MeV, as calculated from nuclear radius and charge. This amount of energy may actually be expected to be available from the difference in packing fraction [meaning binding energy] between uranium and the elements in the middle of the periodic system. The whole "fission" process can be thus described in an essentially classical way. . . .[4]

---

[4] Lise Meitner and Otto Frisch, "Disintegration of Uranium by Neutrons," *Nature* (11 February 1939), p. 239.

The theoretical and experimental confirmation of fission led immediately to the understanding that:

1. A large amount of energy is released during fission.

2. Neutrons are liberated, thus suggesting the possibility of a chain reaction.

3. Uranium-235 is the isotope responsible for fission.

4. Delayed neutrons are emitted.

5. Elements 93 and 94 would be produced if unfissionable uranium-238 absorbed neutrons.

6. The element 94 (plutonium) is fissionable.

EXAMPLE

Energy is liberated because the mass of the fission products is less than the initial mass. A typical fission reaction could be of the form

$$_0n^1 + {}_{92}U^{235} \longrightarrow {}_{92}U^{236*} \longrightarrow$$
$$_{56}Ba^{141} + {}_{36}Kr^{92} + 3\,_0n^1 + \text{energy}$$

where ${}_{92}U^{236*}$ represents the unstable compound nucleus formed for an instant. The initial masses are

$_0n^1$	1.008 665 amu
$_{92}U^{235}$	235.043 943 amu
M_i	236.052 608 amu

The final masses are

$_{56}Ba^{141}$	140.914 050 amu
$_{36}Kr^{92}$	91.897 3 amu
$3\,_0n^1$	3.025 995 amu
M_f	235.837 345 amu

The mass difference is converted to energy

$$\Delta M = M_i - M_f = 0.215 \text{ amu}$$

The energy difference $E = \Delta M c^2$, which is released in the reaction, is

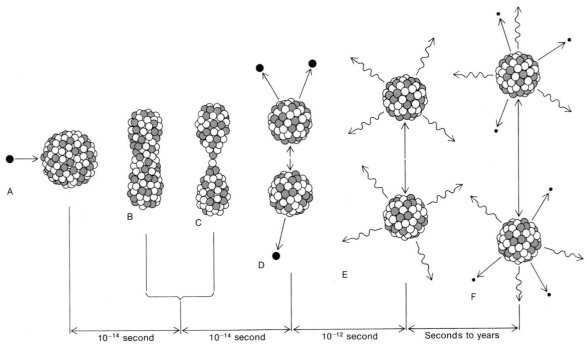

$$\Delta E = (0.215 \text{ amu})(1.66 \times 10^{-27} \text{ kg/amu})(3.0 \times 10^8 \text{ m/s})^2$$

$$\Delta E = 3.216 \times 10^{-11} \text{ J}$$

$$\Delta E = 200 \text{ MeV!}$$

The energy released in a single fission reaction is about a hundred million times the energy released in an ordinary chemical reaction.

Fission Fragments

In the process of fission, as shown in Figure 24-15, the uranium-235 nucleus breaks into two unequal pieces, or **fission fragments.** The most probable splitting consists of one mass of about 92 amu and the larger of about 141 amu, with the emission of about 3 neutrons. Of the many possible mass fragments for uranium-235, the larger fragment usually has a mass between 130 to 145 amu and the smaller fragment between 88 and 103 amu. The fission yield (see Figure 24-16) gives the percentage occurrence of different mass numbers in the fission.

Many of the newly formed fragments decay to a stable isotope within a few seconds by the emission of an electron (β^-). Others go through a chain of beta decays ending in a stable isotope. A few decay isotopes have a long half-life and remain radioactive for a long time. Figure 24-17 shows a possible uranium fission and beta decay chain.

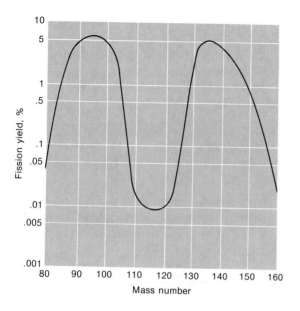

FIGURE 24-16
Yield from U-235 fission. The curve shows the percentage of
fission products with different mass numbers. The probability of
equal mass fragments $A = 119$ is quite small.

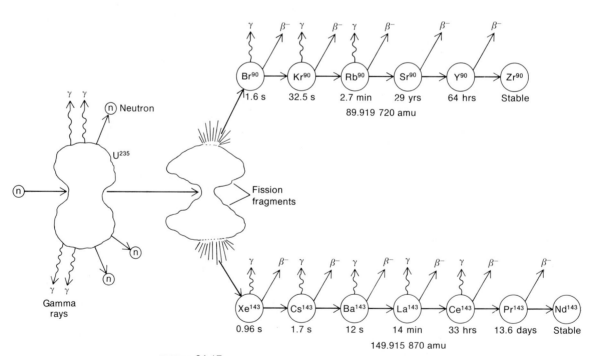

FIGURE 24-17
The fission fragments from U-235 fission undergo a series of
radioactive decays by beta-minus, β^-, emission.

Power from Fission

In December 1941, Fermi, shown in Figure 24-18, began commuting from New York to the University of Chicago, where, with Arthur Compton, he began work on construction of the first nuclear reactor, or "atomic pile," in a squash court at Stagg Field (see Figure 24-19). On 2 December 1942, the first sustained nuclear reaction of uranium was demonstrated (see Figure 24-20).

In August 1942, the project for development of military potential of uranium was reorganized with the establishment of the Manhattan District of the Army Corps of Engineers, mobilizing one of the greatest concentrations of scientific talent and financial resources in the history of this country. Massive installations were created at Oak Ridge, Tennessee, and Hanford and Richland, Washington, to concentrate the fissionable isotope U-235. Later the "atomic city," Los Alamos, was built outside Albuquerque, New Mexico. The first atomic bomb was detonated on the Trinity Flats in the New Mexico desert at dawn, 16 July 1945.

FIGURE 24-18
Enrico Fermi led the team of scientists who built the first nuclear reactor and the first nuclear bomb. Courtesy of U.S. Energy Research and Development Administration.

FIGURE 24-19
The first nuclear reactor or "atomic pile" was constructed in a squash court under the stands of Stagg Field at the University of Chicago. Courtesy of U.S. Energy Research and Development Administration.

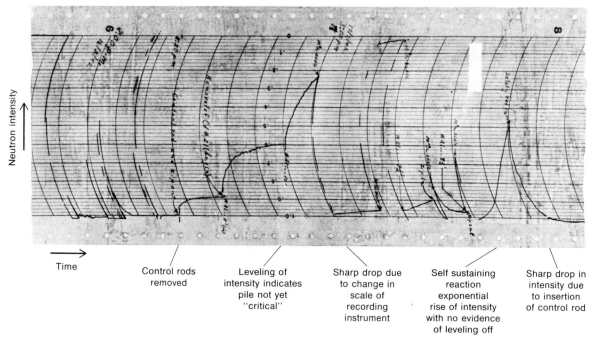

Neutron intensity

Time

Control rods removed

Leveling of intensity indicates pile not yet "critical"

Sharp drop due to change in scale of recording instrument

Self sustaining reaction exponential rise of intensity with no evidence of leveling off

Sharp drop in intensity due to insertion of control rod

FIGURE 24-20
The startup record of the first self-sustaining chain reaction on 2 December 1942.
Courtesy of U.S. Energy Research and Development Administration.

Light-Water Reactors

In a nuclear reactor, the fission of uranium-235 is the energy source for generating electric power. Figure 24-21 shows three systems for generating electric power. In a fossil fuel system, oil or coal is burned in air to produce heat that boils water to produce steam, which in turn drives a turbine that rotates an electric generator. The steam is condensed at the low-pressure end of the turbine by condenser cooling water. In the boiling water reactor system, heat from the fission of uranium in the reactor core boils water to produce the steam. In the pressurized water reactor system, also shown in Figure 24-21, heat is removed from the reactor by water in a primary loop kept under pressure to prevent boiling. The pressurized water goes into a steam generator, which transfers the heat to boil water in a secondary loop to produce steam.

The fuel for a light-water nuclear reactor is enriched in fissionable U-235 (most uranium is nonfissionable

U-238.) The reactor fuel is enriched to about 3% U-235 in large gaseous diffusion plants such as those operating near Hanford, Washington. U-235 and U-238 as uranium hexafluoride, UF_6, are gases. Because of the slight difference in atomic weight of the two isotopes, U-235 hexafluoride diffuses through a porous membrane slightly faster than U-238 hexafluoride. After many stages of diffusion, the uranium is enriched.

The enriched uranium hexafluoride is reduced to uranium oxide and formed into ceramic pellets that are sealed in zirconium alloy tubes about 3.8 m long. Some 60 tubes are bundled together to make up the fuel assembly shown in Figure 24-22. The core of a nuclear reactor has an array of about 700 fuel assemblies. The new uranium-oxide pellets and fuel assemblies are quite safe to handle, requiring only minimal precaution.

For the reactor core to undergo sustained fission reactions, there must be a *multiplication* of neutrons. For maximum effect in the fission of uranium-235, these

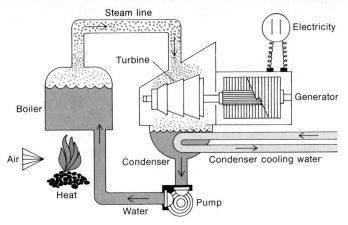

Fossil fuel power production

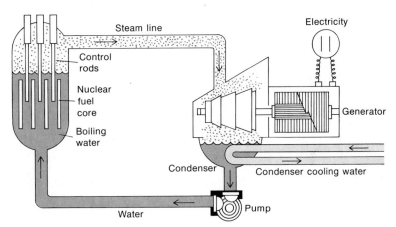

Boiling water reactor system

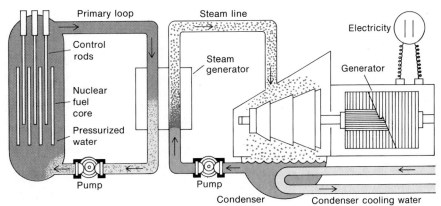

Pressurized water reactor system

FIGURE 24-21
Reactor systems. Fossil fuel power production, A, is compared with the boiling water reactor system, B, and the pressurized water reactor system, C. Courtesy of Pacific Gas and Electric Co.

FIGURE 24-22

A technician checks a nuclear fuel element assembly for a pressurized water reactor. The fuel assembly is about 4.0 m long and about 0.20 m square. In each fuel assembly are about 250 fuel pins containing 240 precision uranium dioxide fuel pellets. Each pellet can liberate the energy equivalent to 200 tonnes of coal. Water channels surrounding each fuel pin allow the uniform flow of cooling water. Until they are placed in the reactor, the uranium fuel assemblies are not particularly radioactive. Courtesy of Westinghouse Electric Corporation.

neutrons are slowed to thermal speeds by a *moderator*. The neutron multiplication of the reactor is regulated by *control rods*.

A neutron emitted by the fission of one uranium nucleus can have one of four possible outcomes. The neutron can:

1. Escape from the reactor.

2. Be absorbed by an impurity of other metal.

3. Be absorbed by U-238, which does not produce fission.

4. Be absorbed by U-235, giving rise to an additional fission.

If more neutrons are released by the additional fission than were lost or absorbed in a nonfission reaction, a chain reaction will sustain itself and will grow.

In the original Chicago stadium reactor and in later test reactors, graphite (carbon) was used as a moderator. Heavy water (D_2O or HDO) is also an effective moderator, but it is difficult to obtain. In light-water reactors using enriched fuels, ordinary water purified of minerals is used as both the moderating medium and the heat transfer fluid.

Reactor control rods are made of material, such as cadmium, boron, or hafnium, with a large probability of absorbing neutrons. By absorbing neutrons in nonfission reactions, they control neutron multiplication. These rods are inserted into or retracted from the reactor core to control its power level.

Figure 24-23 shows the massive stainless-steel upper support plate for a modern reactor. The control rods and fuel assemblies are inserted through this plate into the reactor core. Figure 24-24 is a schematic diagram of the boiling water reactor. Jet pumps force water downward to the bottom of the core. As it passes through the core, it is superheated and flashes to steam at the top of the reactor. In the modern pressurized water reactor, water is heated as it passes through the reactor core, but because it is kept under pressure it does not boil. A reactor that produces 1000 megawatts (MW) of electric power has an active core about 3.8 m long and 4.6 m in diameter, and it must produce a maximum of 3000 MW of thermal power. The core would melt in seconds if there were not a constant flow of water to carry off the heat.

In the boiling water reactor, the water in the secondary loop boils under pressure of about 85 atmospheres and at a temperature of about $T_U = 287°C = 560$ K. By expanding to do mechanical work, this high-pressure steam drives the power turbine. At the low-pressure end of the turbine, the steam is condensed, $T_L = 285$ K, to give a maximum pressure difference across the turbine. The waste heat is carried away by cooling water from a nearby lake, river, ocean, or cooling tower. The mechanical work is converted into electrical work in the generator. The efficiency of conversion of heat energy to electrical energy

FIGURE 24-23
The workman indicates the massiveness of the upper support plate of a pressurized reactor. The thick steel plate contains alignment holes bored to a tolerance of $\pm 50\ \mu m$ (0.002 in) by a computer-controlled milling machine. Courtesy of Westinghouse Electric Corporation.

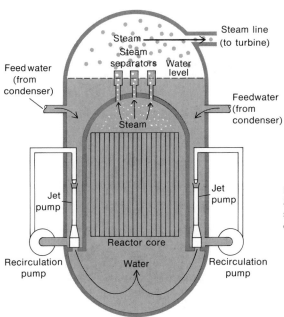

FIGURE 24-24
Schematic arrangement of the boiling water reactor. Water is superheated in the reactor core and flashes to steam at the top of the reactor.

is limited by the thermodynamic efficiency of the steam cycle (as discussed in Chapter 12):

$$\text{eff} = \frac{T_U - T_L}{T_U} = \frac{560 \text{ K} - 285 \text{ K}}{560 \text{ K}} = 49\%$$

The overall efficiency of a nuclear powerplant is about 33% after other heat losses are considered. The steam cycle of a nuclear powerplant differs from that of the fossil fuel plant only in having lower operating temperatures and steam pressure.

Reactor Energy Release

In the light-water reactor, the enriched uranium fuel is about 3% U-235. Fission of U-235 releases fission fragments that eventually poison the fuel. About 1% of the total fuel ($\frac{1}{3}$ of the U-235) can be burned before the fuel element must be recycled. The approximate atomic weight of the uranium fuel (mostly nonfissionable) is 238 gm/mole. Each fission reaction releases about 200 MeV of energy. The energy released in the 1% burning of 1 kg of fuel can be estimated:

$$E = \frac{\left[\begin{array}{c}(1 \text{ kg})(0.01)(6.02 \times 10^{23} \text{ nuclei/mol}) \\ (200 \times 10^8 \text{ eV})(1.6 \times 10^{-19} \text{ J/eV})\end{array}\right]}{0.238 \text{ kg/mol}}$$

$$E = 8.1 \times 10^{11} \text{ J/kg}_{\text{fuel}}$$

A 1000 MW electric powerplant operating at 33% thermal efficiency requires 3000 MW of thermal energy. The energy released at full power for one year is

$$W = (3000 \text{ MW})(3600 \text{ s/hr})(24 \text{ hr/day})(365 \text{ days})$$

$$W = 9.5 \times 10^{16} \text{ J}$$

The total fuel required to produce this power is

$$M = \frac{9.5 \times 10^{16} \text{ J}}{8.1 \times 10^{11} \text{ J/kg}_{\text{fuel}}} = 1.2 \times 10^5 \text{ kg}_{\text{fuel}}$$

of fuel per year of operation. This is only 120 tonnes of uranium fuel, 1% of which is burned up in one year's operation, producing 1.2 tonnes of fission fragments. (It is because of this intensely radioactive waste product that the commercial development of nuclear power has occasioned much concern and controversy.) To reclaim the unburned U-235, and the plutonium-239 produced by neutron capture, the spent fuel rods must be recycled. The intensely radioactive isotopes can also be separated out and concentrated for ultimate disposal.

The attractive feature of nuclear power plants is the small amount of fuel required. Only 120 tonnes of enriched uranium fuel will run a major power plant for a year. By comparison, 2.1 million tonnes of oil, which has an energy content of about 4.5×10^7 J/kg, would be required to produce the same amount of energy.

Review

Give the significance of each of the following terms. Where appropriate give two examples.

polonium	atomic number	proton
radium	isotope	neutron
parent nucleus	atomic mass number	nucleon
alpha decay	α decay	mass difference
daughter nucleus	β^- decay	binding energy
radioactive decay	neutrino	neutron activation
half-life	gamma decay	fission
activity	β^+ decay	fission fragments
decay constant	positron	

Further Readings

E. N. daC. Andrade, *Rutherford and the Nature of the Atom* (New York: Doubleday, Anchor Science Study Series S35, 1964). A history of Rutherford and his work written by one of his former students.

Marie Curie, *Pierre Curie* (New York: Dover Publications, 1963). Madame Curie's reflections on her late husband together with autobiographical notes.

Laura Fermi, *Atoms in the Family: My Life with Enrico Fermi* (University of Chicago Press, 1954). A family view of one of the founders of the nuclear age.

H. A. Bethe, "The Necessity of Fission Power," *Scientific American* (January 1976). A forceful statement by a proponent of nuclear power.

Harry Foreman, ed., *Nuclear Power and the Public* (New York: Doubleday, Anchor Science Study Series S68, 1972). A reasonable, balanced examination of the basic issues surrounding the full development of nuclear power.

Robert Jungk, *Brighter than a Thousand Suns: The Story of the Men Who Made the Bomb* (New York: Harcourt Brace Jovanovich, 1958). A dramatic history from the end of World War I, through Hiroshima and Nagaskai, to the proceedings against J. Robert Oppenheimer in 1954.

Lansing Lamont, *Day of Trinity* (New York: Atheneum, 1965). An account of the events leading to the first testing of the atomic bomb.

Problems

THE EARLY WORK

1 Technetium-99m has a half-life of 6.0 hr and an initial activity of 6.4×10^7/s. What is the activity after 6 hr? 12 hr? 24 hr?

2 Potassium-43 has a half-life of 22.4 hr. If the initial activity of a sample is 3.2×10^7/s, how long will the activity of the sample take to fall off to 1.0×10^6/s?

3 Phosphorus-32 has a half-life of 14.3 days.
(a) What is the decay constant?
(b) If there are initially 10^6 atoms of P^{32}, what is the initial activity?
(c) How many atoms are left after seven days?

4 Carbon-14, with a half-life of 5760 years, is produced in the upper atmosphere by cosmic ray bombardment. The concentration of C^{14} in the atmosphere probably has not changed significantly in thousands of years, being about one part in 10^{12}. Carbon-14 is fixed in the structure of living plants. When the plant dies, the carbon-14 decays. 1 gm of recently produced carbon-14 has an activity of about $15.3 \pm .1$/min. A 1 gm sample prepared from the remains of a tree buried in Wisconsin in the last glaciation is found to have a decay activity of 3.8/min. How old is the tree?

5 Strontium-90, with a half-life for a radioactive decay of 25 years, is a waste product of nuclear reactors. A sample of Sr^{90} has an initial activity of 10^9/s. What is the activity of the sample after 100 years? 1000 years?

6 Iodine-123 is used in thyroid function tests and has a half-life of 13.3 hr. The initial activity of the iodine is 1.5×10^7/s.
(a) What is the decay constant?
(b) How many I^{123} atoms are present initially?
(c) What is the activity of the iodine after 24 hr? after 7 days?

NUCLEAR STRUCTURE

7 Thorium-x, $_{88}Ra^{224}$, decays by emitting an alpha particle.
(a) What are the Z and A of the product nucleus?
(b) What element is it?

8 Carbon-14 (C^{14}) decays by β^- decay.
(a) What are the Z and A of the product nucleus?
(b) What element is it?

9 Madame Curie observed that a gram of pure radium generates about 80 calories of heat per hour. A gram of radium has an activity of 3.7×10^{10}/s. What is the average energy released in each disintegration in joules? in MeV?

10 Manganese, $_{25}Mn^{56}$, decays by the emission of an electron.
(a) What are the Z and A of the product nucleus?
(b) What is the element?

11 A neutron collides with lithium, $_3Li^6$. The reaction gives off an alpha particle, kinetic energy, and a product nucleus. What is the product nucleus?

12 Complete the following nuclear reaction:

$$_5B^{10} + _0n^1 \longrightarrow _2He^4 + \underline{\hspace{2cm}}$$

13 An important thermonuclear reaction involves deuterium, $_1H^2$, and tritium, $_1H^3$.
(a) Complete the reaction:

$$_1H^2 + _1H^3 \longrightarrow _2He^4 + \underline{\hspace{2cm}}$$

(b) How much energy, in MeV, is released?

Use the atomic masses listed in Table 24-2 for the following problems.

14 A positron collides with an electron and annihilates it, producing two gamma photons of equal energy. Show that the energy of each gamma photon is 0.51 MeV.

15 A free neutron is unstable and will decay into a proton and an electron with a half-life of 13 min.
(a) Write the reaction.
(b) What is the energy released in the reaction in MeV?

16 Verify that about 3.3 MeV is released in the nuclear reaction

$$_1H^2 + {}_1H^2 \longrightarrow {}_2He^3 + {}_0n^1$$

17 A neutron collides with lithium-6, $_3Li^6$. The reaction gives off an alpha particle plus another nucleus.
(a) Write the reaction and identify the other nucleus.
(b) Show that the energy liberated in the reaction is 19.5 MeV.

18 Boron-12, $_5B^{12}$ (12.014 354 amu), decays to carbon-12.
(a) What particle is emitted in the decay?
(b) Verify that the energy released is 13.4 MeV.

19 The first nuclear reaction from high-energy protons was observed by Cockcroft and Walton in 1932. Protons with a kinetic energy of 0.50 MeV were fired at lithium-7. In each reaction, two alpha particles were observed. What is the total kinetic energy of the two alpha particles?

20 Radium-226, $_{88}Ra^{226}$ (226.025 438 amu), decays by the emission of an alpha particle to radon-222, $_{86}Rn^{222}$ (222.017 610 amu).
(a) What is the binding energy per nucleon for radium-226?
(b) How much energy is released in the reaction?

21 Manganese, $_{25}Mn^{56}$ (55.938 910 amu), decays by beta minus emission to iron, $_{26}Fe^{56}$ (55.938 299 amu).
(a) What is the binding energy per nucleon for iron-56?
(b) What is the maximum energy of the beta particle?

22 An alpha particle striking beryllium produces carbon-12 plus a neutron:

$$_2He^4 + {}_4Be^9 \longrightarrow {}_6C^{13} \longrightarrow {}_6C^{12} + {}_0n^1$$

How much energy is released in the reaction?

23 A deuterium nucleus is formed by the capture of a nucleus neutron on hydrogen: $_0n^1 + {}_1H^1 \longrightarrow {}_1H^2$ What is the binding energy of the deuterium?

24 Carbon-12, $_6C^{12}$, consists of 6 protons, 6 neutrons, and 6 electrons, and has a mass of 12.000 000 amu.
(a) What is the total binding energy of carbon-12?
(b) What is the binding energy per nucleon?

FISSION

25 In a fission reaction, U-235 (235.043 933 amu) plus a neutron yield tin-134 (133.896 857 amu), yttrium-95 (94.912 540 amu), an alpha particle, three neutrons, and considerable energy:

$$_{92}U^{235} + {}_0n^1 \longrightarrow$$
$$_{51}Sb^{134} + {}_{39}Yt^{95} + {}_2He^4 + 3_0n^1 + energy$$

(a) How much mass is converted into energy?
(b) How much energy is released in the reaction?

NUCLEAR POWER

26 The world's annual energy consumption is estimated to be 3×10^{20} joules. To produce this energy from "burning" uranium in a light-water reactor,
(a) Assuming 1% of the fuel undergoes reaction, how much total fuel would be consumed?
(b) How much uranium would actually undergo fission and produce fission fragments?

27 The world supply of uranium is estimated to be about 10^6 tonnes. About 0.7% of this fuel is U-235.
(a) How much energy is available from the fission of U-235?
(b) If all of the uranium is used, how much energy is available?
(c) At a world energy consumption rate of 3×10^{20} J/year, how long would it last?

25

Ionizing Radiation

IONIZING RADIATION

The very process that releases energy in nuclear fission of uranium or plutonium also yields intensely radioactive fission fragments. This radioactivity is the cause not only of much of the apprenhension about nuclear warfare but also much of the debate surrounding the development and deployment of civilian nuclear power. Are nuclear reactors safe? What if there is an accident? In this chapter, we examine the source of radioactivity, its measurement, its biological effects and the health standards these impose. Finally, we examine its medical applications.

Alpha Particles

In alpha decay, a radioactive nucleus throws out a bare helium nucleus—two protons and two neutrons. The natural radioactive decay of uranium, radium, plutonium, and other heavy isotopes with an atomic number greater than 210 proceeds by alpha decay. The energy of the emitted alpha particle is sharply defined, and is characteristic of the particular emitting nucleus. The alpha particle has a positive electric charge $+2e$, and so, when passing through matter, interacts strongly with electrons. Alpha particles of modest energy can be easily stopped by only the thickness of this piece of paper.

The alpha particle will lose energy by causing ionization. Some of its kinetic energy will be lost by producing ion-electron pairs. It will lose energy all along its path.

Much radiation damage results from this production of ion-electron pairs. The **specific ionization** is the number of ion-electron pairs produced per unit length of the path. In air, the average energy loss required to produce one ion-electron pair is about 33.7 eV. For the 5.5 MeV alpha particle from the decay of americium-241, about 1.6×10^5 ion-electron pairs are produced in stopping. In air at one atmosphere pressure, a 5.5 MeV alpha particle will travel a distance of about 39 mm before stopping. It requires an alpha particle energy of at least 7.5 MeV to penetrate the protective layer of skin, which is about 0.07 mm thick.

The rate at which energy is being deposited along the particle track also has biological significance. The energy loss per unit path length is the **linear energy transfer,** LET. The LET is usually quoted in keV per micro-meter (μm) of path length. (Because LET indicates the rate at which energy is being deposited, it is particularly important in dose rate calculations in radiation therapy using charged particles.) Because most of the energy loss goes into ionization, LET is closely related to the specific ionization. For alpha particles and other heavy particles, energy loss per unit path length is inversely proportional to the energy of the particle over most of the path. Consequently, much of the ionization and energy loss occur near the end of the particle path. Figure 25-1 shows the LET for an alpha particle in air as a function of range. The maximum in the LET at the end of the range is known as the Bragg peak.

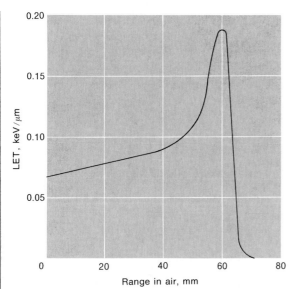

FIGURE 25-1
Linear energy transfer, LET, for 7.4 MeV alpha particles in air. Much of the energy loss and ionization occurs at the Bragg peak at the end of the range.

TABLE 25-1
Beta particle penetration in various materials.

Energy of beta particle Material	Maximum range for beta particles (mm)	
	1 MeV	4 MeV
Air (1 atmosphere pressure)	3050	19 000
Water	3.8	23.0
Lucite	3.0	19.0
Concrete	2.0	11.0
Aluminum	1.0	8.0
Iron	0.5	3.0
Lead	0.3	2.0

SOURCE: *Radiological Health Handbook* (Washington, D.C.: U.S. Department of Health, Education, and Welfare, 1970).

Beta Particles

A beta particle is emitted by an unstable nucleus, either as a positron (β^+) or an electron (β^-). Beta particles are emitted with a wide range of energy up to some maximum energy. In beta decay, some fraction of the total energy and momentum is ejected with a neutrino (as discussed in the previous chapter). Beta particles, having small mass, are elastically scattered by the electrons in the material and tend to wander around, losing only a small amount of energy in each scatter. The path of the beta twists and turns and is difficult to define (in contrast to the straight track of the alpha particle). On the average, the beta particle loses energy more slowly than the alpha particle, and consequently penetrates deeper into matter. Table 25-1 gives typical ranges of beta particles in matter. For example, it requires a beta with an energy of at least 70 keV to penetrate the 0.07 mm thick protective layer of skin.

Gamma Rays

A gamma ray, a high-energy photon, the most penetrating form of radiation, loses its energy by any or all of three processes:

1. scattering by electrons in the material (Compton scattering);

2. knocking out an electron from the material (photoelectric effect);

3. creating an electron-positron pair (pair production), if there is sufficient energy.

Being electrically neutral, a gamma ray interacts only weakly with matter, and therefore may penetrate very deeply before interacting and losing its energy. Where a particular gamma ray will lose its energy cannot be specified, but the probability of its being absorbed in a certain distance can be given. The number of photons absorbed

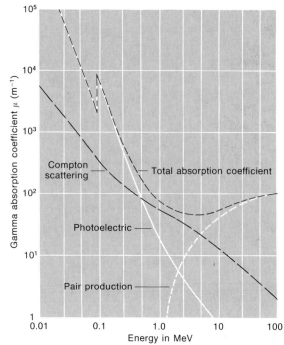

FIGURE 25-2
Gamma absorption coefficient μ in lead showing the contributions of the photoelectric effect, Compton scattering off electrons, and pair-production. Adapted from *Radiological Health Handbook*, U.S. Department of Health, Education, and Welfare, Public Health Service, 1970.

TABLE 25-2
Half-thickness for gamma absorption by common materials.

Energy of gamma ray Material	Half-thickness for absorption (mm) 1 MeV	5 MeV
Air (1 atmosphere)	84.3	195.0
Polyethelene	100.0	240.0
Muscle tissue	100.0	230.0
Water	98.0	230.0
Compact bone	55.0	126.0
Aluminum	42.0	90.0
Iron	15.0	28.0
Concrete	38.0	84.0
Lead	9.0	14.0

SOURCE: Derived from mass absorption coefficients in the *Radiological Health Handbook* (Washington, D.C.: U.S. Department of Health, Education, and Welfare, 1970)

ΔN in a thickness Δx depends on the number of photons present N and an absorption coefficient μ. The number of photons lost in a thickness Δx is given by

$$\Delta N = -\mu \Delta x N$$

The fractional loss $\Delta N/N$ depends on the distance Δx:

$$\frac{\Delta N}{N} = -\mu \Delta x$$

This gives an exponential behavior. If initially N_0 gamma photons are incident on the material, the number surviving after a thickness x is

$$N = N_0 e^{-\mu x}$$

The **gamma half-thickness** $x_{1/2}$ is just the thickness required to reduce the number of photons to half the initial value $\mu x/2 = \ln 2$:

$$x_{1/2} = \frac{\ln 2}{\mu} = \frac{0.693}{\mu}$$

The energy absorbed from gamma photons produces either recoil electrons or electron-positron pairs that in turn cause ionization along the particle path. Figure 25-2 shows the various contributions to the absorption coefficient for gamma ray photons in lead as a function of energy: At low energies (less than 0.5 MeV), the photoelectric effect dominates; at about 2 MeV the Compton scattering dominates; at high energies (above 5 MeV), positron-electron pair production dominates. Table 25-2 gives the half thickness for absorption of 1 MeV and 5 MeV gamma rays in common materials.

Neutrons

A nuclear reactor releases large numbers of both high-energy and low-energy neutrons, which are hazardous because they produce ionization in matter. Being electrically neutral, neutrons can travel large distances in matter undisturbed by the electrons until they collide with a hydrogen or other nucleus. They then transfer a large part of their kinetic energy in a single collision, producing a

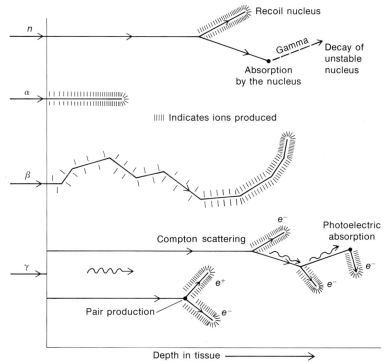

FIGURE 25-3

Schematic representation of tracks of an alpha particle, a beta particle, a gamma photon, and a neutron in tissue. Density of ionization increases as the particles slow. The alpha trail of ionization is dense; the beta particle wanders; the gamma photon and neutrons produce ionization by recoil particles. Adapted from *Biophysics: Concepts and Mechanisms* by E. J. Casey. © 1962 by Litton Educational Publishing, Inc. Reprinted by permission of Van Nostrand Rheinhold Company.

very energetic charged particle that rapidly loses its energy by ionization. Half of the high-energy neutrons will lose their energy and be stopped by a thickness of 10 cm of water or 70 cm of paraffin.

The second hazard associated with neutrons is *neutron activation*. The nuclei of many elements have a relatively high probability of capturing a neutron. This may produce an unstable nucleus, one subject to radioactive decay. In the intense neutron flux of a nuclear reactor, many common elements such as copper and sodium become intensely radioactive. In neutron activation analysis, samples are deliberately irradiated with neutrons as a means of determining the trace element concentration of biological or environmental samples. After irradiation, the sample is checked for the characteristic radiation of different elements. The chance activation of impurity nuclei in nuclear reactor cooling water or structures or fuel elements can pose a radiation hazard.

MEASUREMENT OF RADIOACTIVITY

Almost all methods of measuring radioactivity depend on the ionizing property of radiation. Alpha and beta particles and gamma rays produce ionization as they pass through matter; neutrons produce recoil particles that in turn produce ionization. Figure 25-3 shows schematically the ionization produced by radiation in tissue.

Gas-Filled Detectors

The oldest and simplest method for detecting ionizing radiation is the **gas-filled detector,** usually a hollow metal cylinder with a fine wire electrode extending the length of its axis, as shown in Figure 25-4. The cylinder is

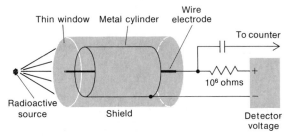

FIGURE 25-4
A gas-filled detector.

shielded by a metal can. Its end is closed off by a thin window. The air in the can is partially evacuated and replaced with a halogen or organic vapor. The central wire is kept at some positive voltage, the magnitude of which depends on the specific detector. The tube then acts as a capacitor with an effective detector capacitance C_{det}, with an electric field pointing radially outward. When an ionizing particle enters the detector through the thin window, it loses energy by creating positive ion-electron pairs along its path. Because of the electric field, the electrons are accelerated toward the center electrode, where they are collected as a pulse. The size of the pulse depends on the type and energy of the particle and on the size of the voltage across the detector tube.

At a small voltage, the device is an **ionization detector,** and measures only the ionization produced by the passage of the particle. If the voltage is too small, some ion-electron pairs actually recombine before being collected. Figure 25–5 shows the relative pulse size as a function of detector voltage. In region I, there is significant recombination. In region II, all of the ionization produced by the particle is collected. The amount of charge collected depends on the type and energy of the incoming particle. The three lines on the graph could represent alpha, beta, or gamma particles of the same energy, or they could represent alpha particles with three different energies.

If the detector voltage, and thus the detector electric field, is increased, the electrons are accelerated toward the center electrode and gain sufficient energy to cause further ionization, creating ion-electron pairs. The new electrons in turn are accelerated toward the center electrode, causing still further ionization. This process is known as avalanche multiplication. Each primary electron produced by the ionizing particle passing through the detector can produce several thousand electrons, which are collected at the central wire electrode. There, a significant pulse of negative charge ΔQ is collected, causing the voltage to momentarily drop by

$$\Delta V = \frac{\Delta Q}{C_{det}}$$

The electronic circuit records this voltage drop as a pulse.

Over a range of operation, region III in Figure 25-5, the output pulse is proportional to the energy of the incoming particle. The gas-filled detector is then functioning as a **proportional detector,** which can determine the energy of an incoming particle. The magnitude of the voltage pulse increases with the detector voltage, eventually reaching a maximum as indicated by region V. In this region, any incoming radiation produces a pulse of maximum size. The gas-filled detector is then functioning as a **Geiger detector,** which when combined with an electronic circuit to count pulses is known as a Geiger counter.

Operating a gas-filled detector as a Geiger counter requires that the detector voltage be correctly set. If too low, there is not enough avalanche multiplication and some pulses will be too small to be counted; if too high, as in region VI, the resultant continuous discharge may severely damage the detector tube. There is a systematic procedure for determining the proper operating voltage. If a known radioactive source in a fixed geometry is used, the count rate for different tube voltages can be measured and recorded. Figure 25-6 shows the count rate as a function of tube voltage for a typical Geiger detector. Below a threshold voltage, the count rate is zero. As the detector voltage is increased, the count rate increases in the proportional region, reaching a maximum. In the *plateau region,* the count rate is almost constant and independent of the detector voltage. The Geiger detector is operated in the middle of this plateau region.

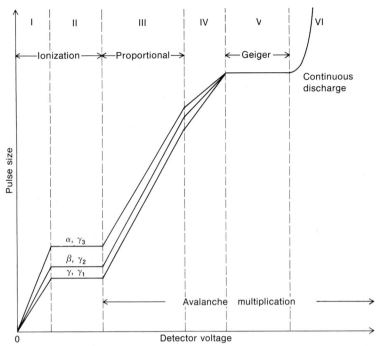

FIGURE 25-5.
Pulse size as a function of voltage for a gas-filled detector. The different curves show the response to different particles, or for different gamma rays of different energy levels. In region V the pulse size is independent of the type or energy of the particle entering the detector. Adapted from Victor Arena, *Ionizing Radiation and Life: An Introduction to Radiation Biology and Biological Radiotracer Methods*, St. Louis, Mo.: The C. V. Mosby Co., 1971.

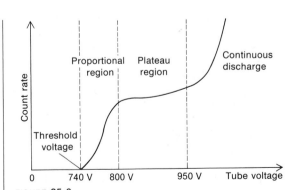

FIGURE 25-6
The Geiger count rate as a function of tube voltage. The normal operating voltage is set in the plateau region where the count rate is insensitive to tube voltage.

Units of Activity

The count rate measured by the Geiger counter is simply the number of α, β, or γ that enter the detector per given time interval. This determines the **activity** of a particular sample, measured by a unit, the **curie,** Ci. One curie of activity corresponds to 3.7×10^{10} disintegrations per second.[1] The curie is equal to the activity of 1 gm of pure radium. A 1 Ci source is very active; the units of millicurie (mCi) and microcurie (μCi) are commonly used.

Neither the type of emitted radiation nor its energy affects the activity of a sample. The activity is simply related to the number of counts on a Geiger counter or

---

[1] The SI unit of activity is the becquerel, Bq, which is equal to one disintegration per second: 1 Bq = 1/s. One curie = 3.7×10^{10} Bq. This unit was adopted by the CGPM in 1975 and has not yet been widely accepted. By strength of historical precedent, the curie will probably remain the unit of activity.

other sensitive detector, corrected for its geometry. A detector placed right on top of a sample will intercept no more than 50% of the radioactive decays.

Another measure of the strength of a radioactive material is the **specific activity;** the activity per unit mass. The specific activity of pure radium is by definition 1.00 Ci/gm. Table 25-3 gives the specific activity for pure radioactive isotopes. The specific activity of a radioactive

TABLE 25-3
Specific activity of common radioactive isotopes.

Isotope	Specific activity (Ci/gm)	Half-life
Radium-226	1.000	1602 yr
Uranium-238	3.3×10^{-7}	4.5×10^9 yr
Plutonium-239	6.1×10^{-2}	2.44×10^4 yr
Polonium-210	4.5×10^3	138.4 day
Iodine-131	1.24×10^5	8.05 day
Technetium-99m	5.28×10^6	6.0 hr
Cobalt-60	1.13×10^3	5.26 yr
Strontium-90	1.41×10^2	28.1 yr
Cesium-137	87.0	30.0 yr
Carbon-14	4.46	5730 yr

material depends on the probability that a particular nucleus will decay, which in turn depends on the half-life, and on the atomic weight of the atom and the extent to which the isotope is diluted by a carrier. Short-half-life isotopes have much higher specific activities.

The Geiger counter, the simplest type of radiation detector, gives us only very crude information about the activity of a radioactive sample. Thus, it can warn us of possible hazards in our environment, but cannot identify or indicate the energy of radiation.

Ionization Chambers

Biological radiation damage is caused by ionization. Whenever radiation passes through a cell, energy is absorbed, ionization is produced, and chemical bonds are broken and rearranged. Different types of radiation have different energies. Measuring the activity of a particular radioactive source in curies gives no indication of how much damage that radiation can do. We need a unit of radiation that measures ionizing power.

The **ionization chamber,** in its simplest form, consists of two parallel plates (for instance, 10 cm square) separated by some distance (perhaps one cm) that is filled with dry air at one atmosphere pressure. One cubic centimeter of dry air at $0°C$ has a mass of 0.001293 gm. Gamma rays or x-rays passing through this air create ion-electron pairs, which can be collected and measured with a sensitive electrometer. This measurement is used to define a unit of the ionizing power of radiation.

The **Roentgen unit,** R, is defined as the amount of gamma ray or x-ray radiation that produces a quantity of electron-ion pairs equal to a charge of one electrostatic unit in one cubic centimeter, or 0.001293 gm, of dry air. One electrostatic unit (1 esu) is equal to a charge of 3.34×10^{-10} coulomb. The radiation survey meter measures the ionization produced in an air chamber per unit time and electrically translates this reading to give a measure of the ionizing power of the radiation in R/hr, or mR/hr (mR, a milliroentgen, is 10^{-3} R). One R of gamma radiation will produce an ionization of 2.58×10^{-4} coulomb in one kilogram of dry air.

The ionization chamber of a radiation survey meter typically has a volume of 580 cm^3 and an end window of aluminized plastic film with a thickness of 6.4 μm. It will detect alpha particles with energies greater than 3.5 MeV; beta particles with energies greater than 70 keV; and gamma and x-rays with energies greater than 7 keV. A protective end cap can also be used that effectively stops all alpha and beta particles, and all gamma rays with energies less than about 30 keV. Figure 25-7 shows a common radiation survey meter.

Everyone who works around radioactive materials routinely wears a pocket dosimeter to monitor possible exposure to ionizing radiation. Most dosimeters measure a total exposure of 0 to 200 mR; however, workers may have a second dosimeter with a range of 0 to 10 R in case of accidental large exposure. The pocket dosimeter shown in Figure 25-8 consists of an electrically charged ionization chamber. A small fiber indicates the degree of charge. As radiation produces ionization in the chamber, it discharges and the needle records the accumulated radiation dosage, as shown in Figure 25-9.

FIGURE 25-7
The "cutie-pie" radiation survey meter measures the ioniza-
tion produced per unit time in a known volume of dry air.
Courtesy of Victoreen Instrument Co., Cleveland, Ohio.

FIGURE 25-8
A pocket dosimeter. The direct-reading dosimeter discharges
when exposed to ionizing radiation. It can be read immediately to determine if accidental exposure has occurred. Courtesy of Nuclear Associates, Inc.

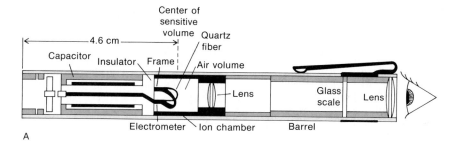

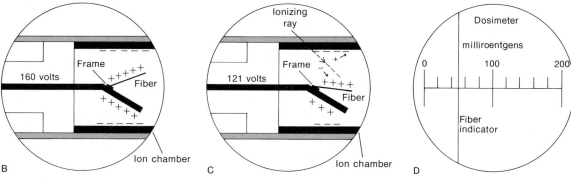

FIGURE 25-9
A. Schematic of the direct-reading dosimeter. The fiber and frame are attached to a capacitor and are initially electrically
charged. B. The fiber stands away from the frame by electrostatic repulsion. The size of the capacitor determines the sensitivity
of the dosimeter. Ionizing radiation causes ions to form in the chamber, which discharges the capacitor, causing the fiber to
move. C. The position of the fiber indicates the accumulated radiation dose and can be viewed on a scale, D, through the optical system. Courtesy of Nuclear Associates, Inc.

Other Detectors

The Geiger counter measures the activity of a radioactive material; an ionization detector, the total ionization produced in a given volume. But neither of these detectors tells us about the energy of the emitted particles. For this purpose, two other classes of detectors are commonly used in work with radioactivity: scintillation detectors and solid-state detectors.

In a **scintillation detector,** a scintillation material (a plastic, crystal, or liquid) is placed in a lighttight container next to a photomultiplier tube. Scintillation detectors are commonly used for measuring the energy of gamma rays. A scintillation material, such as a crystal of sodium iodide with a certain quantity of the element thalium present, absorbs the energy of an incident gamma ray, then re-emits the energy as visible light photons. This process is called scintillation. The photons given off, proportional in number to the energy of the incident gamma ray, strike the sensitive element of the photomultiplier tube, where electrons are emitted by the photoelectric effect. The photoelectrons are accelerated and multiplied to give a large pulse at the final electrode. The height of the final electron pulse, then, is proportional to the energy of the incident gamma ray. Figure 25-10 shows several sodium-iodide crystal detectors with the photomultiplier tube in an integrated unit. Two of the crystals, having a recessed cavity to receive the radioactive sample, are called *well detectors.*

Output pulses from the photomultiplier are electrically analyzed to determine the number of times a particular pulse height appears. Figure 25-11 shows the output from a pulse height analyzer for a cobalt-60 source. The output indicates the 1.17 MeV gamma peak at channel 124 and the 1.33 MeV gamma peak at channel 140.

Liquid scintillators are commonly used in biological assay work. The liquid scintillator is mixed directly with the sample being analyzed. The entire sample is then placed next to a photomultiplier tube, and the light output pulses are analyzed. Elaborate gamma photon-counting systems, now in routine clinical use, can automatically count several different gamma photon energies associated with radioactive labels used in clinical determinations.

A variety of solid-state detectors operate by absorbing the particle energy in a semiconductor crystal. Ionizing

FIGURE 25-10
Sodium-iodide crystal detectors are mounted directly on a photomultiplier tube. The assembly is surrounded by a chrome-finished magnetic and light shield. The diameter of the large detectors is 8.3 cm; of the small, 5.9 cm. Two detectors have recessed wells. Courtesy of Bicron Corporation.

radiation striking the crystals produces pairs of electrons and "holes." (Conduction of charge by holes in semiconductors was discussed in Chapter 14.) Because the crystal is a solid, the radiation is stopped in a relatively short distance. Surface barrier detectors are commonly used to detect charged particles. A 5 MeV alpha particle will travel only 30 μm in the crystal before losing its energy; a 1 MeV beta particle, only 1.5 mm. Large silicon or germanium crystals are used to detect gamma photons. A thicker crystal is required to stop the gamma rays. A crystal of nearly pure silicon or germanium is made electrically neutral by drifting lithium atoms into the crystal. To maintain the electrical properties, these detectors are usually cooled to liquid nitrogen temperatures (77 K). These detectors have exceptionally sharp energy resolution of the order of 1 keV, and are used in high-precision neutron activation and in x-ray fluorescence analysis.

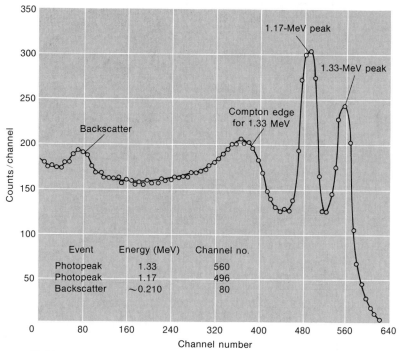

Event	Energy (MeV)	Channel no.
Photopeak	1.33	560
Photopeak	1.17	496
Backscatter	~0.210	80

FIGURE 25-11

Sodium-iodide pulse-height spectrum for cobalt-60. The 1.17 MeV and 1.33 MeV gamma ray peaks are shown. Also shown are the Compton edge resulting from scattering of gamma photons off electrons, and the backscatter peak resulting from gamma photons scattering into the detector. Courtesy of ORTEC Incorporated.

HEALTH STANDARDS FOR RADIATION EXPOSURE

The Rad

The roentgen unit R is a measure of the ionizing power of gamma photons but is not defined for alpha and beta particles. Because ionization is ultimately the result of energy loss, it is useful to define a unit of radiation related to the energy absorbed by tissue. The *radiation absorbed dose*, the **rad,** is defined as the amount of radiation required for 100 ergs (10^{-5} joules) of energy to be absorbed in one gram (10^{-3} kg) of tissue:[2]

$$1 \text{ rad} = 100 \text{ erg/gm} = 10^{-2} \text{ J/kg}$$

[2] The SI unit of absorbed dose is the gray, Gy, which is equal to a joule per kilogram: 1 Gy = 1 J/kg. This unit was also adopted by the CGPM in 1975 and has not yet been widely accepted. One rad = 10^{-2} J/kg = 10^{-2} gray.

An exposure of 1 R of gamma radiation absorbed in tissue gives a dose of 0.83 to 0.93 rad, depending on the energy of the gamma photons. Thus, radiation survey meters that measure the radiation exposure in R can give an estimate of the radiation dose of gamma rays in rad.

Many of the biological effects of radiation depend on the rate at which energy is lost in tissue, which is related to the linear energy transfer, LET. Radiation that is most easily stopped, such as alpha particles, has a greater ionization and energy loss along the path, and produces more damage to sensitive tissue. Thus, a *quality factor, QF,* has been defined that relates the relative biological effectiveness of various forms of radiation. Because changes in a charged particle's energy change the rate of energy loss, the quality factors can also vary with energy. For most purposes, the quality factor for gamma photons and x-rays and electrons is 1, for protons is 5, and for alpha particles is 10. Table 25-4 gives quality factors for different particles.

TABLE 25-4
Quality factors for various types of radiation.

Radiation type	Quality factor
Gamma rays, medium energy	1
Gamma rays > 4 MeV	0.7
X-rays	1
Beta particles > 30 keV	1
Beta particles < 30 keV	1.7
Thermal neutrons	4 or 5
Fast neutrons	10
Protons	10
Alpha particles	10
Heavy ions	20

TABLE 25-5
Units of radiation.

Quantity	Unit	Value
Activity	Ci	3.7×10^8 s^{-1}
Ionization in dry air	R	2.58×10^{-4} C/kg
Energy in tissue	rad	10^{-2} J/kg
Effect in humans	rem	rad $\times$ QF

The Rem

The effective dose of radiation exposure is given by the **rem,** the *roentgen equivalent man.* The radiation exposure in rem is given by the radiation absorbed dose in rad multiplied by the quality factor QF:

$$\text{rem} = \text{rad} \times QF$$

Estimating the biological effect of a particular radiation exposure on a nuclear worker or on a laboratory hamster requires some inference. The ionizing power of the radiation can be measured in roentgen by suitable instruments. If the type of radiation is known, the absorbed dose can be determined in rad. Estimating the quality factor, the time, and extent of exposure allows the effective dose received by the subject to be computed in rem.

Table 25-5 gives the four basic units for measuring exposure to radiation: the activity of the source (Ci), the ability to ionize air (R), the ability to deposit energy in tissue (rad), and the biological effect in humans (rem). Once units of radiation exposure are defined, the question is, What is safe (or, conversely, what is dangerous)? Figure 25-12 gives the percentage of mortality among rhesus monkeys 30 days after receiving a single whole-body

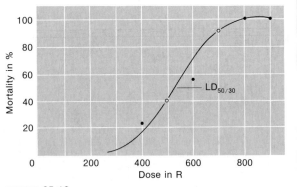

FIGURE 25-12
Mortality of rhesus monkeys at 30 days after total body exposure to x-rays. From U. Henschke, *American Journal of Roentgenology, 77* (1957), pp. 899–909.

exposure to x-rays at various dose levels. A single x-ray exposure of 500 R was sufficient to kill 50% of the monkeys within 30 days. The human data are quite similar—although of course they come from accidental exposures rather than systematic experiments. The dose has to be estimated after the fact, and the effects observed in the unfortunate patients.

An x-ray exposure of 125 R produces virtually no apparent symptoms in humans. If the dose is doubled to 250 R there will be evidence of radiation sickness, but there will also be a high probability of recovery within a few weeks. If however the whole body dose to humans exceeds 500 R there is less than a 50% chance of survival. A whole body radiation dose of 500 R of x-rays, or an equivalent dose of *500 rem is plainly dangerous.* Localized radiation exposure to parts of the body in cancer therapy will have a less severe effect on the organism than will whole body exposure. Local doses of 150 R per treatment with a total exposure over the therapy of as much as 5000 R are sometimes used in cancer treatment.

How Much Is Safe?

The U.S. federal government has established standards for the **maximum exposure limit** for civilian federal radiation workers (see Table 25-6). The government considers these exposure levels to be safe, for they have produced no clinically observable effects. Under normal conditions, a worker can be exposed to no more than 5 rem per year, 3 rem in any calender quarter, and 0.150 rem (150 mrem) in any week. Under emergency conditions these limits can be stretched a bit, but the absolute limit is that one should not exceed a total of 5 rem/yr for every year since the person's eighteenth birthday.

A radiation area is defined as any area in which a person in the area could get 10 mR of exposure for every hour (10 mR/hr) in the area. In only 15 hours, the weekly maximum limit would be reached, equal to about two

FIGURE 25-13
This radiation warning sign must be posted when the exposure level exceeds 10 mR/hr.

working days of continuous exposure. A radiation area must have controlled access and be posted with the familiar radiation warning sign as shown in Figure 25-13. A high-radiation area is one in which the exposure rate exceeds 100 mR/hr, which exposes a worker to the entire weekly permitted exposure to radiation in 1.5 hrs. Work in high-radiation areas is usually supervised by a radiation health physicist or other trained person who monitors personnel exposure.

General Population Standards

Three basic standards apply to the general population.

1. Whole-body exposure to an individual in the general population should not exceed 500 millirem per year (mrem/yr).

2. For a sample of the general population at risk, the average exposure must be less than 170 mrem/yr.

3. Geneticists recommend that exposure to radiation over one generation (30 years) should not exceed 10 rem.

TABLE 25-6
Federal government maximum exposure limits.

Occupational exposure limits:
 Whole-body:
  5 rem/year
  3 rem/calendar quarter
  0.150 rem/week
  $(N - 18) \times 5$ rem = maximum to age N.
 Skin exposure only:
  7.5 rem/calendar quarter
 Extremities (hands, forearms, feet, ankles):
  18.5 rem/calendar quarter

General population exposure from manmade sources
 An individual in the population:
  0.5 rem/year
 Average over general population:
  0.170 rem/year
 Average genetic dose over population:
  5.0 rem/30 years = 0.170 rem/year

(Since roughly half of this could result from natural exposure and medical uses, it has been recommended that exposure to manmade radiation sources should not exceed 5 rem over 30 years.)

Cosmic rays from outer space continually bombard the earth, so all life on earth has evolved in constant exposure to natural radioactivity. As Table 25-7 shows, because of absorption in the atmosphere, exposure to cosmic rays is greater at higher elevations. There is also environmental radioactivity, caused by naturally occurring uranium, radium, and thorium in granitic rocks, the type of house lived in (a concrete or brick house can increase the local background by as much as 40 mrem/yr), and global fallout from atmospheric testing of nuclear weapons in the 1950s. The average exposure to diagnostic x-rays is about 72 mrem/yr. As Table 25-7 also shows, each of us receives a background radiation exposure of from 179 to 324 mrem/yr. Thus, we are continuously exposed to a certain amount of radiation. For the general population, the federal maximum radioactive exposure level, 170 mrem/yr, is about the same as the natural background exposure.

TABLE 25-7
Estimated annual exposure to radiation in 1970 averaged over the U.S. population.

Source	Exposure (millirem/yr)		
	Low	High	Average
Cosmic rays	40	120	45
Terrestrial			
External gamma	30	95	60
Internal (dietary)	25	25	25
Global fallout	4	4	4
Medical			
Diagnostic radiology	72	72	72
Radiopharmaceuticals	1	1	1
Occupational	.8	.8	.8
Ingested global fallout	4	4	4
Nuclear power			
Gaseous effluents			.002
Fuel reprocessing			.0008
Miscellaneous sources	___	___	2.6
Total exposure	179	324	214

Source: Estimates of Ionizing Radiation Doses in the United States, 1960–2000 (Washington, D.C.: U.S. Environmental Protection Agency).

Dose Rate from Gamma Sources

The dose of radiation received from a radioactive source depends on the type of radiation, its energy and activity levels, and the person's distance from the source. Alpha and beta particles, even at high energies, are rapidly attenuated by air. They are biologically hazardous only on the surface of the skin, or if they are actually ingested. Gamma radiation, however, is highly penetrating. An unshielded intense gamma source can be very dangerous.

A physically small unshielded source of radiation radiates equally in all directions. At a distance s, the radiation spreads evenly over a sphere of area $A = 4\pi s^2$; at twice the distance, over four times the area. Thus, exposure falls off as $1/s^2$. If the radiation dose rate in rad/hr from a 1 Ci gamma source at a distance of 1 m is known, the exposure at other distances for other sources can be estimated. This **equivalent dose rate** the *Rhm*, has the units of

$$\frac{\text{rad-m}^2}{\text{hr-Ci}}$$

The dose rate, given in rad/hr, at a distance s m from a gamma source with a certain activity in Ci is given by

$$\text{dose rate} = \frac{(Rhm)(\text{activity})}{s^2}$$

Table 25-8 lists typical Rhm values for common gamma emitters as well as the energies of the principal gamma rays. Because of the strong absorption of alpha and beta particles in air, such analysis is not useful for alpha and beta radiation. Because strontium-90 does not emit gamma rays, its Rhm = 0, as shown in the table.

EXAMPLE

A radiation therapy unit used on cancers contains a 1250 Ci cobalt-60 source at a distance from the patient of 0.75 m. What is the dose rate at this distance?

$$\text{dose rate} = \frac{(Rhm)(\text{activity})}{s^2}$$

$$\text{dose rate} = \frac{\left(1.3\,\frac{\text{rad-m}^2}{\text{hr-Ci}}\right)(1250\text{ Ci})}{(0.75\text{ m})^2}$$

$$\text{dose rate} = 2890\text{ rad/hr} = 48\text{ rad/min}$$

TABLE 25-8

Equivalent dose rate, Rhm, for various isotopes. The Rhm is equivalent to the dose in rad/hr at a distance of 1.0 m from a 1.0 Ci source.

Isotope	Rhm $\left(\dfrac{\text{rad-m}^2}{\text{hr-Ci}}\right)$	Principal radiation energy (MeV)
Cobalt-60	1.3	1.17, 1.33
Strontium-90	0.0	beta decay—no gamma
Technetium-99m	0.059	0.14
Iodine-131	0.21	0.364, 0.637
Cesium-137	0.32	0.662
Radium-226	0.825	0.187

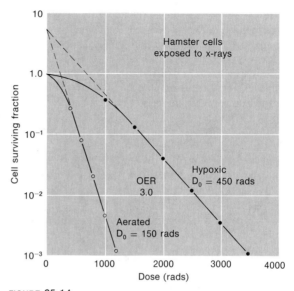

FIGURE 25-14
The oxygen effect. Survival curves for cultured hamster cells exposed to x-rays in the presence of oxygen (aerated radiation) and the absence of oxygen (hypoxic radiation). The modification of the surviving fraction is known as the oxygen effect. From Eric J. Hall, *Radiobiology for the Radiologist* (New York: Harper & Row, Publishers, Inc., 1973).

Biological Effects of Radiation

All forms of radiation affect biological systems by producing ionization and by excitation of molecules within the organism's individual cells. Primary, or **molecular, lesions** are caused by passage of this radiation through the cell from the outside, as in the case of exposure to gamma or neutron radiation, or by exposure to radiation from the decay of alpha and beta particle emitters that have been taken up by the organism. Unusual ions can be formed and chemical bonds can be broken at unusual points. An important cause of damage is the formation of free radicals—atoms or groups of atoms with unpaired electrons that are highly reactive and that can disrupt key molecular bonds within the cell.

The extent of cell damage depends significantly on the concentration of oxygen available at the time of the primary damage. Often ionizing radiation produces a hydrogen radical by breaking the hydrogen from a molecule. If this hydrogen is returned to the original molecule, the damage is repaired. However, if the broken molecule, reacts with any molecular oxygen present, forming an organic peroxide, the hydrogen cannot return, and the molecule suffers irreversible damage. This effect, known as the *oxygen effect,* has clinical implications during radiation therapy for cancer. Figure 25-14 shows the survival of typical mammalian cells exposed to x-rays in the presence and absence of oxygen: For a surviving fraction of

20%, about three times more radiation is required to produce the same biological effect in the absence of oxygen.

Alterations in cellular chemistry in turn alter cell metabolism, an effect called a *biochemical lesion.* These alterations can lead to observable damage, so-called *anatomical lesions,* and also cell deaths. The cell deaths give rise to **radiation sickness.**

If the whole-body dose to a person is of the order of 450 to 550 rem, there is greater than 50% chance that the person will die within about four weeks. The early symptoms of radiation sickness are nausea, vomiting, and pallor. After a subsequent latency period of about a week (during which symptoms subside), there appear symptoms of general malaise, fatigue, loss of appetite, skin redness, and susceptibility to infection. By the third week, the skin peels and there is internal bleeding. This stage is followed either by recovery or death.

If the whole-body dose is of the order of 100 to 250 rem, there is good chance of recovery. By the fourth week, the person begins to recover health, unless the weakened condition has precipitated some other illness. However, a person who recovers from such a large radiation dose stands a greater than normal risk of contracting leukemia or some other cancer after a latency period of as long as 20 years.

Radiation damage can alter a cell's genetic material; damage appears after several cell generations as cell mutations. This is particularly serious for the rapidly dividing cells in the bone marrow that produce red blood cells. Radiation cell damage in the gonads can damage genetic material, which damage can show up in future generations. One of the controversies surrounding manmade radiation is how to evaluate the long-range effects of radiation exposure.

NUCLEAR MEDICINE

Medical application of radiation has become routine hospital procedure. The two major uses of radiation in medicine are diagnosis and therapy.

Diagnostic Procedures

Scintillation detectors can detect radioactive isotopes in very low concentration. Isotopes in the body behave chemically exactly like their nonradioactive cousins. A wide variety of isotopes such as iodine-123, iodine-131, cobalt-58, and technetium-99m are used (see Table 25-9). The isotopes are used either directly in simple salt solutions or indirectly, to tag chemicals introduced into the body, as in the iodine-131 uptake test shown in Figure 25-15. In studies done *in vivo*, that is, in a living body, isotopes enter the normal body metabolism, are distributed by the bloodstream, and are taken up by specific organs or are excreted. Observing the disposition of the isotope in time allows a clinical evaluation of specific organ function to be made. Radioactive tracers should be used in human subjects only when the benefit to health outweighs the hazards of radiation exposure. Certain radioisotope diagnostic procedures are not used on pregnant women to avoid radiation exposure of the fetus.

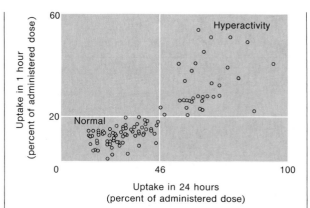

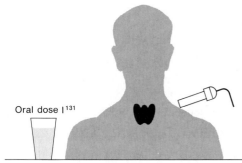

FIGURE 25-15
Iodine-131 uptake can be used as a screening test for hyperthyroidism. Courtesy of U.S. Energy Research and Development Administration.

Gamma emitters with a relatively short half-life are used to keep the radiation exposure to a minimum. Some isotopes, such as cobalt-57 and xenon-133, are rapidly cleared from the body and thus are said to have a short biological half-life.

Radioisotopes are used in clinical laboratory tests done *in vitro*, that is, in solution, in tests called **radioimmunoassay**, RIA. (In 1977, Rosalyn Yallow was awarded a Nobel Prize for her work on the development of radioimmunoassay.) A radioactive tracer is carried through a series of analytical procedures. The quantity of tracer in

TABLE 25-9
Radioactive isotopes used in medicine.

Isotope	Half-life biological	physical	Principal radiation energy in keV	Form of administration	Use
F^{18} fluorine-18	808 days	110 min	β^+-635, γ-511	Fluorine ion in saline 0.5-2.0 mCi	Bone imaging
K^{43} potassium-43	58 days	22.4 hrs	β^--820 γ-373, 619	Ion in saline 1 − 2 mCi	Myocardial imaging
Cr^{51} chromium-51	616 days	27.8 days	γ-320	Sodium chromate	Red blood cell studies
Co^{58} cobalt-58	9.5 days	71.4 days	β^+-480 γ-811, 864	Labeled vitamin B-12	Schilling test for pernicious anemia
Ga^{67} gallium-67	4.8 days	78 hrs	γ-93, 184	Gallium citrate	Imaging—Hodgkin's disease, lymphomas, malignant tumors
Rb^{81} rubidium-81	45 days	4.6 hrs	β^+-1400 γ-511, 190	Rubidium chloride	Myocardial imaging
Tc^{99m} technetium-99m	20 days	6.02 hrs	γ-140	Pertechneate ion in saline Human serum albumin	Brain scans, salivary imaging, blood flow studies, cardiac studies
In^{111} indium-111	10 hrs	2.8 days	γ-173, 247	DTPA (diethylene-triamine-penta-acetic acid)	Spinal and cranial fluid imaging
In^{113m} indium-113m	10 hrs	1.73 hrs	γ-390		Liver, lung, brain, blood pool scan
I^{123} iodine-123	138 days	13.3 hrs	γ-159	Sodium iodide 100 μCi–400 μCi	Thyroid function Thyroid imaging
I^{125} iodine-125	138 days	109.7 min	γ-635	Reagent	Radioimmunoassay T-3, T4, Digoxin
I^{131} iodine-131	138 days	8.07 days	β^--330 γ-364	Reagent	Radioimmunoassay T-3
				Sodium iodide 10 μCi	Thyroid uptake, thyroid cancer, metastases scan
				Iodonated human serum, albumin (IHSA)	Blood plasma volume, lung blood flow
				Sodium iodohippurate Bengal dye	Kidney function Liver function
Xe^{133} xenon-133	—	5.3 days	β^--346 γ-81	Gas 10 mCi	Lung ventilation, lung inhalation
Yb^{169} ytterbium-169	1000 days	32 days	γ-177, 198	DTPA (diethylene-triamine-penta-acetic acid)	Spinal and cranial fluid imaging
Au^{198} gold-198	280 days	2.7 days	β^--329 γ-412	Colloid	Liver structure scans
Hg^{203} mercury-203	14.5 days	46.9 days	β^--214 γ-279	Chlormerodin	Kidney scans

the final solution is then measured with a detector, and a quantitative determination of specific chemicals can then be made. Iodine-125 is widely used as a label in thyroid function tests that measure the levels of certain thyroid hormones in a blood sample. It is also used to measure digoxin in the blood, caused by the digitalis therapy commonly used to treat heart conditions. Radioimmuno-assay is much more sensitive than conventional chemical assay techniques. Other isotopes include chromium-51, used to label red blood cells, and cobalt-58, used to label vitamin B-12.

Radioisotope Imaging

The spatial distribution of radioisotope tracers in the human body yields valuable diagnostic information. Both time-resolved dynamic studies and studies of the ultimate disposition of materials are used. The distribution of radioiodine-labeled human serum albumin in the blood supply to the brain can locate regions with reduced blood supply, indicative of tumors. The distribution of radio-iodine absorbed by the thyroid can indicate growths or regions of hyperactivity. Scans of the blood supply to the lungs, using technetium-99m macroaggregated albumin or indium-113m ferric hydroxide complex, can indicate defects in pulmonary circulation. The size and function-ing of the liver can be observed by using gold-198 colloid or technetium-99m sulfur colloid.

Figure 25-16 shows a scintillation camera, first devel-oped by Hal Anger in 1957. The **Anger camera** forms an image of the brain. The blood supply carries a radio-isotope tracer to the brain. Gamma rays emitted by the tracer enter the detector through a lead collimator, which typically has 10,000 tiny holes that admit only gamma rays coming head-on. The admitted gamma rays enter a scintillation crystal, typically about 30 cm in diameter and about 1.3 cm thick, where they cause the usual flash of visible light. The detector of the Anger camera consists of an array of 19 or 37 photomultiplier tubes that observe the scintillation crystal. The pattern of flashes in the scintilla-tion crystal corresponds to the spatial distribution of the radioactive isotope in the patient. The photomultiplier tubes and electrical circuitry generate a dot on an oscillo-

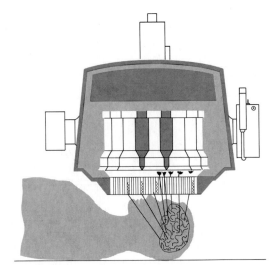

FIGURE 25-16
The Anger camera forms an image of a portion of the brain. Gamma photons enter the scintillation crystal through tiny holes in the collimating plate. An array of photomultiplier tubes detects the light flashes. The photomultiplier signals are processed to determine the spatial distribution of gamma rays. From Robert J. Shalek and Gordon L. Brownell, "Nuclear Physics in Medicine," *Physics Today* (August 1970). Used with permission.

scope screen corresponding to the position of each ob-served gamma ray. A physician analyzes a photograph of these dots, which becomes part of the patient's medical record. Figure 25-17 shows a type of Anger camera in clinical application.

Technetium-99m

Many radioactive tracers present two problems: First, they may initially decay by emitting beta particles, which can cause primary radiation damage within the body, and give no diagnostic benefit; second, many isotopes have a relatively long biological half-life, which can contribute more radiation exposure than is required for diagnosis.

In recent years, the radioisotope **technetium-99m** has become widely used in diagnostics. Tc-99m is produced

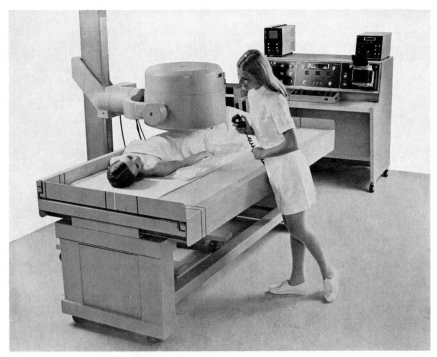

FIGURE 25-17
The Pho/Gamma Anger type scintillation camera in clinical
application. Courtesy of Searle Analytic, Inc.

by the radioactive decay of molybdenum-99, which has a
half-life of 67 hours (see Figure 25-18). Molybdenum-99
is produced by neutron activation of naturally occurring
molybdenum-98 in the neutron flux of a commercial
nuclear reactor, such as the General Electric facility at
Vallecitos, California. Tc-99m is in a relatively long-
lived, so-called metastable state that decays with a half-
life of 6.024 hours into the ground state Tc-99 by giving
off a 140.5 keV gamma ray. Having such a short half-life,
Tc-99m must be produced daily by a molybdenum-99
generator. In many regions, predawn couriers supply
hospitals their day's requirement of the isotope. Because
Tc-99m has a short half-life, the activity of a diagnostic
dose is reduced within 24 hours to $\frac{1}{16}$th of the initial
activity. The ground state, Tc-99, has a 210 000 year
half-life and produces no further radiation exposure before being eliminated from the body.

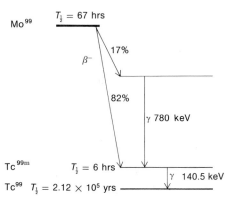

FIGURE 25-18
Scheme for production and decay of technetium-99m.

TABLE 25-10

Preparations using technetium-99m as an imaging agent; administered dose and typical radiation exposure.

Tc-99m preparation	Target organ	Administered dose (mCi)	Radiation exposure (rad)
Sulfur colloid	Liver	3	0.72–1.0
	Spleen		0.48–1.3
	Bone marrow		0.70–1.0
Bone agent	Bone	10	0.145 whole-body
			0.27 skeleton
			3.3 kidney
			6.8 bladder
Lung agent	Lungs	2	0.26 lungs
			0.14 liver
			0.24 kidney
Pertechneate	Brain	10	0.13 whole-body
			1.0 stomach
			2.7 thyroid

Sterile preparations of technetium-99m can be used for gamma ray imaging to visualize and diagnose abnormalities in the brain, the lungs, bone, the liver, the spleen, and other organs (see Figures 25-19 and 25-20). The simplest preparation is sodium pertechneate, given orally or intravenously and used primarily to observe the blood supply to the brain. A dose of some 10 mCi is given and scanning begins 30 to 45 minutes later, finding any extant lesions, blockages, or other irregularities.

Different preparations of scanning agents can be made to focus on different target organs: A bone-scanning agent looks at the blood flow to the bone to observe abnormal bone-forming activity; a lung preparation with micro-aggregated human serum albumin stays in the blood for a long time, revealing the blood flow to the lungs; a liver preparation is rapidly cleared from the blood and shows up liver function and any abnormal blood supply to it (as well as showing spleen and bone marrow function). With each of these preparations, there is a dose of radiation to various organs, as shown in Table 25-10.

Radiation Therapy

Gamma and beta radiation is used to treat certain medical conditions. In early work, radium-226 (as much as 10 mg) encapsulated in gold needles was used to treat cancer. The primary function of the radiation is to destroy abnormal tissue or to depress the activity of otherwise normal tissue. Wire containing cobalt-60, iridium-192, and tantalum-182 has also been used.

Radiation therapy can be from an intense external source such as gamma rays from radioactive cobalt-60 or cesium-137, high-energy x-rays, or particles from accelerators. Many major hospitals have large cobalt-60 teletherapy units (see Figure 25-21), which can contain a thousand curies of activity. Cobalt-60 has gamma rays with energies of 1.17 and 1.33 MeV. The intense high-energy gamma radiation requires thick shielding of the source. These high-energy gammas penetrate deep into tissue. Treatment not requiring thick penetration can use intense cesium-137 sources, with a gamma ray energy of

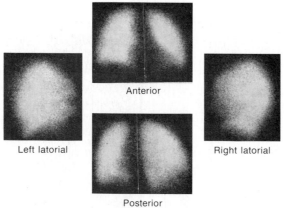

Anterior

Left latorial

Right latorial

Posterior

FIGURE 25-19
Lung images after intravenous injection of human albumin aggregated with Tc-99m. Courtesy of Medi+physics.

Meningioma (headaches 10 wks, seizures 4 wks)

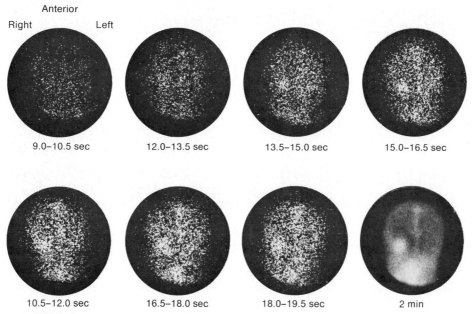

Anterior

Right　　　Left

9.0–10.5 sec　　12.0–13.5 sec　　13.5–15.0 sec　　15.0–16.5 sec

10.5–12.0 sec　　16.5–18.0 sec　　18.0–19.5 sec　　2 min

FIGURE 25-20
A rapid sequence of photographs showing scintillation of the head viewed from the front. The study, done by intravenous injection of 10 mCi of Tc-99m, follows the initial passage of the isotope from the heart into the head. Photographs at 1.5 s intervals follow passage of the isotope through the major neck arteries, cerebral arteries, cerebral capillaries, and the venous structures. 2000 to 4000 counts are observed in each 1.5 s exposure. This patient shows a meningioma, a lesion in the blood brain barrier. The lesion area fills rapidly with blood but drains slowly, as shown in the 2 min exposure. Courtesy of Medi+physics.

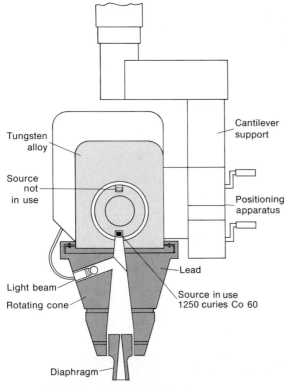

FIGURE 25-21
Cobalt-60 gamma ray teletherapy unit uses a 1250 Ci source. Courtesy of Energy Research and Development Administration.

Tungsten alloy

Cantilever support

Source not in use

Positioning apparatus

Light beam

Lead

Rotating cone

Source in use 1250 curies Co 60

Diaphragm

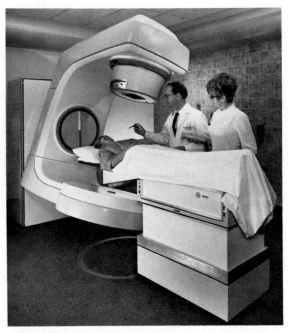

FIGURE 25-22
A 4 MeV x-ray unit, Cliniac-4, used for cancer therapy, can deliver an exposure of as large as 300 Rad/min. Courtesy of Varian Radiation Division, Palo Alto, Calif.

0.662 MeV, which require less shielding and are thus more portable.

In the past decade, technological advances have yielded very-high-energy x-ray machines. These x-rays are produced by accelerating a beam of electrons up to 6 MeV and having it strike a tungsten target, producing an intense energetic beam of x-rays. The 4 MeV Cliniac-4 shown in Figure 25-22 is capable of producing intense x-ray fields in the exposure range of 100 to 300 rad/min. Over 200 of these units are now in use at major hospitals for treatment of various forms of cancer.

Intense x-ray therapy has yielded excellent results, either by itself or in conjunction with surgery, to treat certain cancers (particularly when caught in early stages) such as certain brain tumors, throat cancer, and Hodgkin's disease of the lymphatic system. Intense x-ray therapy can also be used to ease pain in terminal cancer patients. Application of radiation therapy to medical treatment has turned hazardous radiation to human benefit.

RADIATION IN THE BIOSPHERE

On April 26, 1953, the city of Troy, New York, was drenched with a sudden cloud burst. As the rain fell, physicists in nearby university laboratories who were experimenting with radioactivity noticed a sudden surge in their "background" radiation counts. They soon discovered that the rain was highly radioactive and surmised that ·the cause was radioactive debris—fallout—from nuclear tests in Nevada . . . . inevitably strontium-90 would accompany calcium as it moved through the food chain, ultimately becoming concentrated, along with calcium, in vegetables, in milk, in the bones of people. . . .[3]

The Food Chain

Because biological organisms rapidly absorb many radio-isotopes, the release of radioactivity to the biosphere (the sphere of plant and animal life) poses a serious environmental problem. There is no way to decontaminate the biosphere. Once radioactive elements have been taken up by the food chain, there is no way to keep the radioactivity out of the human food supply. The radiation is passed from one organism to another, the concentration increasing with ascent up the food chain.

Early radiation health standards were directed to minimizing direct human exposure to radiation. For instance, the allowable concentration of radioisotopes in water was so low that one could swim all day in water and not receive appreciable radiation exposure. However, aquatic plants and animals constantly search out the elements they need for their metabolism. For instance, sea urchins, crustaceans, and fish concentrate zinc-65 many times over the amount present in the water (see Figure 25-23). The rapid increase in strontium-90 in cow's milk that followed intensive atmospheric testing of nuclear weapons in the early 1950s accentuated the worldwide concern about the hazards of radioactive contamination of the biosphere.

A similar direct food chain link to humans was discovered in Lapland: High concentrations of radiation discovered in native Laplanders were traced to their diet, which is mainly caribou. The caribou feeds mainly on lichen called caribou moss, which had absorbed and concentrated radioactive fallout. Once radioactive isotopes enter the food chain, humans can be exposed to radiation; it is therefore of critical importance that radioactive material be isolated from the biosphere.

Nuclear Power

Any industrial society needs energy resources. As oil and natural gas reserves have been depleted, nuclear power

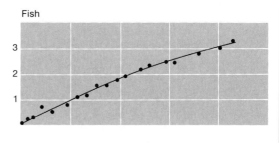
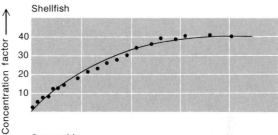
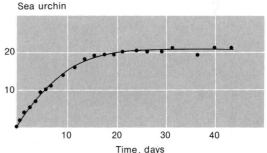

FIGURE 25-23
Uptake of zinc-65 by several marine organisms. From *Environmental Contamination by Radioactive Materials*, copyright 1969 by the International Atomic Energy Agency, Vienna, Austria. Used with permission.

has been offered as a major alternative. Nuclear reactor safety has thus been the focus of much public concern.

Nuclear reactor design incorporates three features to contain any radioactivity produced in its operation: First, the fuel elements holding the uranium pellets are sealed in a zirconium alloy cladding that confines the fission fragments produced in the release of energy. Second, in case of an accident, the reactor vessel itself is built to withstand a great pressure without rupture. And the reactor is designed to automatically shut itself down in case of an accident. (In case of accidental loss of cooling water, an emergency core-cooling system cools the reactor core

[3] Barry Commoner in *The Closing Circle* (New York: Alfred A. Knopf, 1971), pp. 45–48.

TABLE 25-11

Principal radioactive isotopes produced in a nuclear reactor; fp indicates a fission product, na indicates neutron activation of nonfuel material in fuel element, reactor coolant, or reactor vessel.

	Isotope	Half-life	Origin
I.	Very long	$T_{1/2} > 100$ yrs	
	Carbon-14	5700 yrs	na
	Technetium-99	5×10^5 yrs	fp
	Iodine-129	1.7×10^7 yrs	fp
	Plutonium-239	2.4×10^5 yrs	na
II.	Long	100 yrs $> T_{1/2} > 8$ yrs	
	Tritium-3	12.3 yrs	na
	Krypton-85	9.4 yrs	fp
	Strontium-90	25 yrs	fp
	Cesium-137	33 yrs	fp
III.	Medium	8 yrs $> T_{1/2} > 6$ mos.	
	Sodium-22	2.6 yrs	na
	Iron-55	2.9 yrs	na
	Cobalt-60	5.3 yrs	na
	Zink-65	0.7 yrs	na
	Ruthenium-106	1.0 yrs	fp
	Cerium-144	1.6 yrs	fp
	Prometheum-147	2.26 yrs	fp
IV.	Short	6 mos $> T_{1/2} > 5$ days	
	Phosphorus-32	14.3 days	na
	Chromium-51	27 days	na
	Iron-59	45 days	na
	Cobalt-58	71 days	na
	Strontium-89	54 days	fp
	Yttrium-91	58 days	fp
	Zirconium-95	65 days	fp
	Ruthenium-103	40 days	fp
	Iodine-131	8 days	fp
	Xenon-133	5.3 days	fp
	Barium-140	12.8 days	fp
	Cerium-141	32.5 days	fp
	Praseodynium-144	13.8 days	fp

before a "meltdown" condition can occur.) Third, the reactor vessel sits in a massive concrete well and is covered by a pressure dome (the characteristic visual feature of a nuclear power plant). In the improbable event of a breach of the reactor vessel, the containment dome will prevent radioactivity from reaching the environment.

Fission of uranium produces, as we know, large quantities of intensely radioactive fission fragments. When an operating reactor is shut down, after 15 minutes there is an activity of 3 Ci for every watt of power output. For a 1000 megawatt reactor this would be an activity of 3 billion Ci! Much of this is, however, from material with a very short half-life. Spent fuel elements removed from the reactor are stored underwater for 3 to 6 months simply to let their radioactivity subside, or "cool off." However, two years after shutdown, the activity of a reactor core is approximately 75 Ci for every megawatt day of operation. A 1000 megawatt reactor operating for a year would produce a long-term radioactivity of 27 million Ci!

Table 25-11 shows the principal radioisotopes produced in a nuclear reactor, classed according to length of half-life. These isotopes, which are either fission fragments or are produced by neutron activation of other nonfuel materials within the reactor, occur in such large quantities as to be very hazardous. Many of the chemical elements are biologically active.

After the reactor's spent fuel elements have cooled, they are sent to fuel reprocessing plants, where they are broken down. The unburned uranium-235 and plutonium are separated out and recycled for new fuel elements. The radioactive fission and activation products are chemically separated out, concentrated, and stored as dry solids or "high-level" (intensely radioactive) liquid wastes. If our nuclear power capacity reaches a projected 900 000 megawatts by 2000 A.D., we will be producing annually large quantities of this intensely radioactive waste. The public controversy over the development of civilian nuclear power has focused on a fear of an accident at a powerplant that would radioactively contaminate the surrounding community. Moreover, as a significant nuclear power industry develops, the problem of the ultimate disposal of long-lived radioactive wastes has become a major issue. Not until this problem is solved can civilian nuclear power be used on a large scale.[4]

Ultimate Disposal of Radioactive Waste

During the processing of the spent reactor fuel, the radioisotopes strontium-90 and cesium-137 are processed for storage for as long as 700 years. In addition, there are radioactive materials known as the actinides, which include isotopes of the elements actinium, thorium, uranium, neptunium, and plutonium, and are formed by neutron adsorption into the uranium fuel. The actinides emit alpha particles and generally have very long half-lives (for example, 24 000 years for plutonium-239).

At present, high-level liquid wastes are stored in large tanks such as those at the Department of Energy, Hanford Reservation, near Richland, Washington. Here they decay and cool and eventually are solidified in place or are removed for permanent storage elsewhere. Some 42 mil-

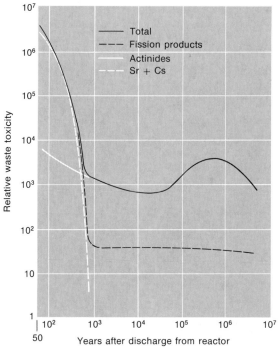

FIGURE 25-24
Relative toxicity of concentrated radioactive waste compared to uranium ore containing 1.4% U_3O_8 as a standard. From "Disposal of Nuclear Wastes," by A. S. Kubo and D. J. Rose, *Science, 182* (21 December 1973), pp. 1205–1211. Copyright 1973 by the American Association for the Advancement of Science.

lion gallons of this intensely radioactive waste are still in liquid form, much of it the remains of over 70 million gallons generated during the arms race, when thousands of kilograms of plutonium were produced and stockpiled for use in weapons. Each gallon of this waste has many curies of activity.

In a well-publicized radiation spill at Hanford in 1973, some 115 000 gallons of liquid waste dribbled out of a steel and concrete tank over a 51 day period. The total loss into the surrounding soil is estimated at 40 000 Ci of cesium-137, 14 000 Ci of strontium-90, 4 Ci of plutonium, and smaller amounts of other products. This material, now stored in the clay soil at Hanford, legitimates public concern over safe storage of such wastes.

Figure 25-24 shows the relative amount of toxicity of radioactive waste compared to uranium ore. The graph

---

[4]Christopher Hohenemser, Roger Kasperson, and Robert Kates, "The Distrust of Nuclear Power," *Science* (1 April 1977), pp. 25–34, report that

The only commercial reprocessing plant in West Valley, New York, closed in 1972 after 6 years of operation and is being redesigned and enlarged. A second plant, at Morris, Illinois, has been scrapped because of technical problems that would, among other things, have led to unacceptable occupational exposures. A third plant, at Barnwell, North Carolina, is under construction and is scheduled to open soon. As to the disposal of solid waste, it is still not clear what the product, and therefore the process, will be and where it will be stored. Most spent fuel is now stored at reactor sites in cooling ponds. Failure to resolve the waste disposal questions may delay opening and operation of reprocessing plants under construction, even if they are otherwise functional.

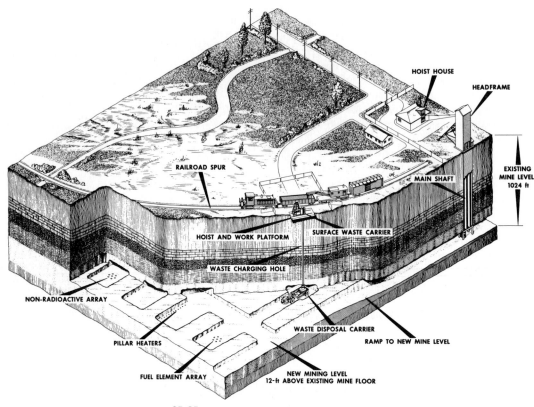

HOIST HOUSE

HEADFRAME

RAILROAD SPUR

MAIN SHAFT

EXISTING
MINE LEVEL
1024 ft

HOIST AND WORK PLATFORM

SURFACE WASTE CARRIER

WASTE CHARGING HOLE

NON-RADIOACTIVE ARRAY

PILLAR HEATERS

WASTE DISPOSAL CARRIER

RAMP TO NEW MINE LEVEL

FUEL ELEMENT ARRAY

NEW MINING LEVEL
12-ft ABOVE EXISTING MINE FLOOR

FIGURE 25-25
A radioactive waste repository. It may be feasible to store radioactive waste in deep mines. Courtesy of U.S. Energy Research and Development Administration.

starts 50 years after discharge from the reactor. In the first 50 years, isotopes with short half-lives will have disappeared, those with medium half-lives will show greatly reduced activity, and only those with long half-lives, such as strontium-90 and cesium-137, remain. Even after a thousand years, however, significant quantities of the actinide wastes will remain.

Many schemes for permanent storage of radioactive wastes have been proposed. Significant studies were made of a large, stable salt deposit near Lyons, Kansas, that seemed promising until a large quantity of water mysteriously disappeared from a test well in the vicinity, indicating a chance that radioactivity might escape into the environment. Salt beds in southern New Mexico and basalt rock formations near Hanford, Washington, have

also been studied. Figure 25-25 shows a possible deep mine storage site. Each proposed plan for permanent disposal has come under criticism from scientists, politicians, and nuclear critics.

An alternative to deep storage is to build storage vaults aboveground or near the surface. In such storage systems, periodic reprocessing would separate out certain wastes and reduce the total volume. If the actinides can be separated out, recycled, and burned in a nuclear reactor, the duration of the storage problem is reduced from a million years to 700 years. However, such long-term supervised storage would require remarkable institutional stability, lest these repositories fall into neglect or become objects of terrorist attack or political blackmail.

Review

Give the significance of each of the following terms. Where appropriate give two examples.

specific ionization
linear energy transfer
gamma half thickness
gas-filled detector
ionization detector
proportional detector
Geiger detector
activity

curie
specific activity
ionization chamber
Roentgen unit
scintillation detector
rad
rem
maximum exposure limit

equivalent dose rate
molecular lesion
radiation sickness
radioimmunoassay
Anger camera
technetium-99m

Further Readings

Peter Alexander, *Atomic Radiation and Life* (Baltimore, Md.: Penguin Books, 1965). A basic introduction.

Luther J. Clark, "Radioactive Wastes: Some Urgent Unfinished Business," *Science* (18 February 1977), pp. 661–666.

Christopher Hohenemser, Roger Kasperson, and Robert Kates, "The Distrust of Nuclear Power," *Science* (1 April 1977), pp. 25–34.

Arthur S. Kubo and David J. Rose, "Disposal of Nuclear Wastes," *Science* (21 December 1973), pp. 1205 ff. An interesting discussion of the dilemmas of waste disposal and the options open to us.

William D. Metz, "Reprocessing: How Necessary Is It for the Near Term?" *Science* (1 April 1977), pp. 43–45.

Gene I. Rochlin, "Nuclear Waste Disposal: Two Social Criteria," *Science* (7 January 1977), pp. 23–31.

David Rose and Richard Lester, "Nuclear Power, Nuclear Weapons, and International Stability," *Scientific American* (April 1978), pp. 45–57.

Problems

IONIZING RADIATION

1 In this chapter, we discussed the properties of different forms of radiation: alpha, beta, gamma, x-rays, and neutrons. (α, β^+, β^-, γ, x, n). Indicate which one or more types of radiation apply to the following statements.

(a) Results from an increase in the charge of a nucleus by $+e$.
(b) Results from a change in an atomic electron.
(c) Highly penetrating radiation.
(d) Results from a decrease in the charge of a nucleus by $+e$.
(e) Radiation easily stopped.
(f) A particle with a rest mass.
(g) A photon.
(h) A helium nucleus.
(i) Loses energy along its entire path through matter.
(j) Loses energy only when it strikes something.
(k) Appears to violate conservation of energy and momentum.
(l) Naturally occurring only for nuclei with atomic numbers > 210.
(m) A positron.
(n) A charged particle.
(o) An unstable particle.
(p) Can lose energy by creating an electron-positron pair.
(q) An electron.

2 A 1 MeV beta particle is effectively stopped after traveling a distance of about 1 mm in aluminum. If the betas traveled only through a very thin 0.1 mm aluminum foil,

(a) What is the qualitative effect on the number of betas passing through the foil?
(b) What is the qualitative effect on the energy of the betas passing through the foil?

3 The gamma half-thickness for stopping a 1 MeV gamma ray is about 42 mm in aluminum. If gamma rays travel only through 10 mm of aluminum,

(a) What is the qualitative effect on the number of gammas passing through the material?
(b) What is the qualitative effect on the energy of the gammas passing through the material?

MEASURE OF RADIOACTIVITY

4 The ionization chamber of a radiation survey meter has a thin aluminized Mylar window that allows α, β, and γ rays to enter the chamber and be observed, and has two shutters, one of thin (0.1 mm) aluminum foil, one of thicker (1.0 mm) aluminum. Say the meter is used to survey the radiation hazard associated with a chemical spill. With no shutter in place, the reading at the surface is 50 mR/hr; with the thin shutter in place, it drops to 48 mR/hr; and to 25 mR/hr with the thick shutter. What conclusions can you make about the type of radiation coming from the surface?

5 The average energy lost in producing one ion-electron pair in air is about 33.7 eV. A radiation dose of 1 R corresponds to 3.34×10^{-10} coulomb of electron charge per cm^3 of air. How much energy is lost in each cm^3?

6 A scintillation counter with 35% efficiency in counting iodine-131 measures 10 000 counts/minute. What is the activity of the sample in microcuries?

7 A sample of radioactive iodinated human serum albumin has an activity of 10 μCi.
(a) How many grams of I-131 are present?
(b) What is the activity of the sample in counts/minute?

(c) If the atomic weight of I-131 is 131 gm/mol, how many atoms of I-131 are present?

8 A Cs-137 source is placed close to a Geiger counter with 35% efficiency. The Geiger counter observes 1452 counts in one minute.
(a) What is the activity of the source in μCi?
(b) How much mass of Cs-137 is present?

9 A radiation survey meter observes gamma radiation of 150 mR/hr. The air-filled ionization chamber has a volume of 580 cm^3. If the electrons generated by the radiation in the chamber are swept out by an electric field, what is the current (the charge per second)?

10 A sample of plutonium-239 is observed to have an activity of 70 000 counts/minute in a detector that is 50% efficient.
(a) What is the activity of the source in μCi?
(b) How many μgm of plutonium are present?

11 A 10 μCi cobalt-60 source for student use is observed using a Geiger counter with an efficiency of 30%. What count rate can be expected?

12 A 5 mCi cobalt-57 low-energy gamma ray source is used in a student laboratory to do the Mössbauer experiment.
(a) If this source is counted by a detector with a 10% counting efficiency, how many counts per minute would be expected?
(b) This source has a half-life of 270 days. What would be the expected count rate with the same detector after 1.5 years? after 3.0 years?

HEALTH STANDARDS FOR RADIATION EXPOSURE

13 A 5 mCi cobalt-57 source is left exposed. The dose rate factor Rhm for this radiation is 0.093 $\frac{\text{rad-m}^2}{\text{hr-Ci}}$. How long would it take to get an exposure of 150 mrad at a distance of 1 m from the source?

14 Exposure to 1 rad of gamma rays would produce 100 ergs of absorbed energy per cm^3 of tissue. A dose of 1000 rad of gamma rays to the whole body would be fatal but would produce negligible heat.

(a) For a dose of 1000 rad, what is the energy absorbed per gram of tissue in joules? in calories? Assume tissue has the same density as water. (1 erg = 10^{-7} J.)
(b) Assuming the specific heat of tissue is the same as that of water (namely C_s = 1.0 cal/gm-°C), calculate the increase in temperature due to the absorbed energy.

15 In a classic experiment, performed in 1961 and 1962, to test the sensitivity of natural plant populations to ionizing radiation, a forest in Long Island, New York, was irradiated for an extended time from a radiation source equivalent to 9500 Ci of Cs-137. The Rhm for Cs-137 is 0.32 rad-m^2/hr-Ci.
(a) How many grams of Cs-137 are present?
(b) What would be the radiation exposure rate at a distance of 10 m from the source?
(c) What is the radiation exposure rate at a distance of 100 m?

16 A 1.0 μCi cobalt-60 source is used in a student laboratory. The Rhm for this gamma source is 1.3 rad-m^2/hr-Ci. What radiation exposure rate could be expected at a distance of 10 cm from the source?

ADDITIONAL PROBLEMS

17 Technetium-99m, delivered daily to hospitals, has a half-life of 6 hrs. The physician wants to administer a dose of 10 mCi of Tc-99m at 8:00 A.M., to do a brain scan. At 12:00 noon, the Tc-99m activity would be 20 mCi/cm^3. How much should the physician administer at 8:00 A.M.?

18 Fluorine-18 has a 110 min half-life. If the preparation has an activity of 2.0 mCi/cm^3 at 10:00 A.M., what is the activity at 8:00 A.M.? at 12:00 noon? at 10:00 A.M. the following day?

19 A Tc-99m generator contains 400 mCi of molybdenum-99, which has a half-life of 67 hrs. What is the activity of the generator after one week? two weeks?

20 A dose of 10 mCi of sodium pertechnetate Tc-99m, with a half-life of 6.0 hrs, is given to a patient for diagnostic purposes. After two days, almost all of the original Tc-99m has decayed to Tc-99, which has a half-life of 210 000 years.
   (a) How many atoms of Tc-99m are present in the original dose?
   (b) Assuming that all of the Tc-99m has decayed to Tc-99, what is the activity in μCi of the Tc-99 that remains in the patient? Hint: What is the decay constant?

21 Iodine-131, with a half-life of 8.0 days, is often used to study thyroid metabolism. If a particular thyroid has an observed activity of 1600 counts/minute today, neglecting the biological turnover, what will be the count rate in 8 days? in 64 days?

22 In a thyroid uptake study, 10 μCi of sodium-iodide-131 is given to a patient. The activity of the patient's thyroid and a sample identical to the original patient dose are then counted after 24 hrs to determine the percentage uptake. Normal range for the uptake is 16% to 42%. Assuming that the effective counting efficiency both for the patient and the identical sample are 10%, and neglecting background count rates;
   (a) What would be the initial activity of the original patient dose as observed by the detector?
   (b) What would be the observed activity of the identical sample after 24 hrs?

(c) If the patient thyroid count after 24 hrs was 4.25×10^5 counts/min, what is the percentage uptake by the thyroid?

23 In the Schilling test for pernicious anemia, a dose of 0.5 μCi of cobalt-58 labeled vitamin B-12 is administered. The detector used to determine activity has an efficiency of 40%.
   (a) What is the observed activity of this dose?
   (b) Cobalt-58 has a half-life of 71.4 days. What would be the observed activity of the administered dose after 24 hrs?
   (c) A 24 hr urine sample is taken. The activity of $\frac{1}{20}$th of the sample is 3.3×10^3 counts/min. What percentage of the cobalt-58 labeled vitamin B-12 is excreted? An excretion of less than 7% of the vitamin is suggestive of pernicious anemia.

24 A sealed 10 mg radium source is used in cancer therapy. What is the exposure rate at a distance of 1.0 cm from this source? What is the exposure rate at a distance of 1.0 m?

25 A cesium-137 needle for cancer therapy is 2.2 cm long and 1.6 mm in diameter and has an activity of 25 mCi.
   (a) What would be the radiation exposure rate to hospital personnel 1 m from the needle?
   (b) What is the radiation exposure rate 1.0 cm from the needle?

26

Fusion

ENERGY PRODUCTION IN STARS

High in the summer sky, warming the earth, stands the sun. Its heat drives the winds and the currents of the seas, and by photosynthesis, the sun causes the plants to grow. When temperature, as indicated by color, is compared with absolute photographic magnitude or brightness, most stars fall in a narrow sequence from cool, weakly luminescent stars to hot, very bright stars. This series of stars is called the main sequence. Main sequence stars, of which our sun is one, generate energy primarily by the **thermonuclear fusion** of hydrogen to helium.

Astrophysicists have developed theories of star formation. Some stars were formed very early in the evolution of the galaxy, other stars are thought to be forming even today, by a process that takes millions of years. Very early in the formation of a star, great clouds of hydrogen gas are contracted by gravitational attraction. As the gas is compressed, its temperature rises. When it reaches 6000 K, all of the hydrogen molecules are broken up into atoms. At a temperature of 10^4 K, the hydrogen is ionized. As the ionized gas is compressed further, the temperature reaches 10^6 K. Before the star begins its stable thermonuclear burning stage, nuclear collisions break any deuterium (D), lithium (Li), beryllium (Be), or boron (B) present into hydrogen. Once the temperature of the star has reached at least 5×10^6 K or as high as 3×10^7 K, the hydrogen nuclei have enough kinetic energy to overcome the Coulomb repulsion because of their charge. The nuclei can then interact.

Hydrogen Fusion

In 1939, Hans Bethe achieved the first step toward understanding the hydrogen-burning process in stars. Before then, research into reactions in the sun had been at an impasse, because all of the reaction products between protons and helium nuclei form unstable products:

$$p + p \longleftrightarrow {}_2He^2 \text{ (unstable)}$$

$$p + {}_2He^4 \longleftrightarrow {}_3Li^5 \text{ (unstable)}$$

$$_2He^4 + {}_2He^4 \longleftrightarrow {}_4Be^8 \text{ (unstable)}$$

Bethe was the first to realize that during a reaction between protons, a proton could be converted into a neutron, producing a deuterium nucleus. Two hydrogen atoms interact to form deuterium plus a positron and a neutrino. Atomic masses usually include the mass of the atomic electron. In the reaction, one atomic electron is left over:

$$_1H^1 + {}_1H^1 \longrightarrow {}_1D^2 + e^+ + [e^-] + \nu + 0.418 \text{ MeV}$$

The positron and the electron annihilate one another, giving rise to two 0.51 MeV gamma rays. The final products are then

$$_1H^1 + {}_1H^1 \longrightarrow {}_1D^2 + 2\gamma + \nu + 1.44 \text{ MeV}$$

The neutrino, a chargeless particle with zero rest mass,

was discussed in Chapter 24. It will carry about 0.26 MeV of the energy away from the star.

Once deuterium is introduced, a series of reactions convert hydrogen into helium. Table 26-1 shows the **proton-proton reactions,** so-called PP I reactions, which result in the synthesis of helium from hydrogen.

TABLE 26-1
Proton-proton reaction, PP I.

$$_1H^1 + {}_1H^1 \longrightarrow {}_1D^2 + e^+ + [e^-] + \nu + 0.418 \text{ MeV}$$
$$_1D^2 + {}_1H^1 \longrightarrow {}_2He^3 + \gamma + 5.49 \text{ MeV}$$
$$_2He^3 + {}_2He^3 \longrightarrow {}_2He^4 + {}_1H^1 + {}_1H^1 + 12.86 \text{ MeV}$$

Net reaction:

$$6_1H^1 \longrightarrow {}_2He^4 + 2_1H^1 + 2e^+ + 2\nu + 2\gamma + 26.72 \text{ MeV}$$

Because the resulting products have less mass than the initial products, a great deal of energy is released in each reaction. The net result of the PP I reactions is that six protons react to form a helium nucleus, two protons, two positrons, two gamma rays, two neutrinos, and 26.7 MeV of energy.

EXAMPLE

Two hydrogen atoms interact to form deuterium and a positron and leave behind an atomic electron.
(a) How much energy is released in the reaction? Table 24-2, in Chapter 24, lists the masses of the particles.
(b) The positron and electron annihilate each other. What is the total energy liberated?

Initial mass: $M_i =$	2_1H^1	= 2.015 650 amu
	$_1D^2$	2.014 120 amu
	e^+	0.000 549 amu
	e^-	0.000 549 amu
Final mass: $M_f =$		2.015 200 amu
Mass difference:	$\Delta M = M_i - M_f = 0.000\ 454$ amu	

The energy released is given by

$$\Delta E = \Delta Mc^2 = (0.000\ 454 \text{ amu})(931 \text{ MeV/amu})$$

$$\Delta E = 0.418 \text{ MeV}$$

The electron and positron annihilate one another, liberating the rest mass energy as two gamma rays. The energy released in electron-positron destruction is

$$\Delta E' = \Delta Mc^2 = 2(0.000\ 549 \text{ amu})(931 \text{ MeV/amu})$$

$$\Delta E' = 1.022 \text{ MeV}$$

The total energy liberated in the reaction is 1.44 MeV.

Once a star contains significant quantities of helium, more complicated reactions building up to boron-8 proceed. Table 26-2 lists these proton reactions, called PP II reactions.

TABLE 26-2
Other proton reactions prevalent in stars, PP II.

$$_2He^3 + {}_2He^4 \longrightarrow {}_4Be^7 + \gamma + 1.59 \text{ MeV}$$
$$_4Be^7 + e^- \longrightarrow {}_3Li^7 + \nu + 0.86 \text{ MeV}$$
$$_3Li^7 + {}_1H^1 \longrightarrow {}_4Be^8 + \gamma \longrightarrow 2_2He^4 + \gamma + 17.35 \text{ MeV}$$
$$_4Be^7 + {}_1H^1 \longrightarrow {}_5B^8 + \gamma + 0.133 \text{ MeV}$$
$$_5B^8 \longrightarrow {}_4Be^8 + e^+ + \nu + 17.98 \text{ MeV}$$
$$_4Be^8 \longrightarrow 2_2He^4 + 0.10 \text{ MeV}$$

Boron-8 is unstable and spontaneously decays to beryllium-8, which is very unstable and breaks up in 10^{-16} s into two helium nuclei. While proton reactions can produce elements heavier than helium, these quickly react or decay, forming only helium. It is these and similar reactions that provide the principal energy source for the main sequence stars, including our sun.

EXAMPLE

Lithium-7 interacts with a proton to form two helium atoms, releasing energy. How much energy is released?

	$_3Li^7$	7.016 004 amu
	$_1H^1$	1.007 825 amu
Initial mass:	M_i	= 8.023 829 amu
Final mass:	$M_f = 2_2He^4$	= 8.005 206 amu
Mass difference:	$\Delta M = M_i - M_f = 0.018\ 623$ amu	

The energy released is

$$\Delta E = \Delta Mc^2 = (0.018\ 623\ \text{amu})(931\ \text{MeV/amu})$$
$$\Delta E = 17.3\ \text{MeV}$$

Carbon-Nitrogen Cycle

In 1938, Hans Bethe and Carl von Weizacker suggested that the reactions of protons with carbon and nitrogen nuclei would also give rise to fusion of hydrogen into helium. As shown in Figure 26-1, the carbon and nitrogen nuclei act only as catalysts for the reaction. Table 26-3 shows the sequence of reactions of the **carbon-nitrogen cycle,** the net result of which is

$${}_6C^{12} + 4{}_1H^1 \longrightarrow$$
$${}_6C^{12} + {}_2He^4 + 2e^+ + 2\nu + 26.73\ \text{MeV}$$

(In 1967, Bethe received the Nobel Prize in physics for his contributions to the understanding of energy

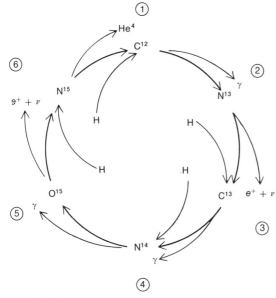

FIGURE 26-1
The carbon-nitrogen fusion cycle.
$$4\ {}_1H^1 + {}_6C^{12} \to {}_6C^{12} + 2\ e^+ + 2\ \nu + 3\ \gamma + {}_2He^4 + \text{energy}$$

TABLE 26-3
The carbon-nitrogen cycle for fusion of hydrogen (carbon-12 acts simply as a catalyst in the reaction).

① ${}_6C^{12} + {}_1H^1 \longrightarrow {}_7N^{13} + \gamma + 1.94\ \text{MeV}$
② ${}_7N^{13} \longrightarrow {}_6C^{13} + e^+ + \nu + 2.22\ \text{MeV}$
③ ${}_6C^{13} + {}_1H^1 \longrightarrow {}_7N^{14} + \gamma + 7.55\ \text{MeV}$
④ ${}_7N^{14} + {}_1H^1 \longrightarrow {}_8O^{15} + \gamma + 7.29\ \text{MeV}$
⑤ ${}_8O^{15} \longrightarrow {}_7N^{15} + e^+ + \nu + 2.76\ \text{MeV}$
⑥ ${}_7N^{15} + {}_1H^1 \longrightarrow {}_6C^{12} + {}_2He^4 + 4.97\ \text{MeV}$

Net reaction:

$${}_6C^{12} + 4{}_1H^1 \longrightarrow {}_6C^{12} + {}_2He^4 + 2e^+ + 2\nu$$
$$+ 3\gamma + 26.73\ \text{MeV}.$$

processes in stars.) In smaller main sequence stars, the carbon-nitrogen cycle contributes less than 10% of the total energy production; however, in main sequence stars larger than about four solar masses, it is most important.

SYNTHESIS OF THE ELEMENTS

In 1939, George Gamow proposed a **"big bang" theory** of the universe: Some 10 to 20 billion years ago, the entire universe consisted of an enormous ball of primordial matter-energy concentrated in a sphere about the size of the solar system. This primeval fireball expanded and cooled from a temperature of 10^{12} K in the first microsecond of creation, to 10^9 K after 5 minutes, and to 10^7 K after one day.

In the 1940s, Gamow, R. A. Alpher, and R. C. Herman attempted to explain the formation of heavy elements in the first 30 to 60 min of creation. Their predictions on the production of deuterium, helium-3, helium-4, and lithium-7 in the first hour of the big bang are in agreement with cosmic abundances. They were unable to explain the buildup of elements beyond atomic number 4 because beryllium-8 and boron-8 are unstable, and because no nuclei with atomic number 5 can be produced.

The details of the hydrogen reactions were worked out in a famous 1948 paper by Alpher, Bethe, and Gamow, on

"The Origin of the Chemical Elements." This theory successfully synthesized all the elements up to helium and accounted for energy production in the sun and stars by **hydrogen burning,** the fusion of hydrogen.

An essential key to the synthesis of heavier elements was first worked out, in 1952, by E. Salpeter, who found that carbon-12 could be produced from the interaction of helium with beryllium-8. Even though beryllium-8 is unstable, a small equilibrium quantity of it is produced at reasonable stellar temperatures:

$$_2He^4 + {_2}He^4 \longleftrightarrow {_4}Be^8$$

This in turn can react with alpha particles, producing carbon-12, which is stable:

$$_4Be^8 + {_2}He^4 \longrightarrow {_6}C^{16} + \gamma + 7.65 \text{ MeV}$$

Once carbon could be synthesized, the production of the remaining elements in stars was soon worked out. All modern theories of stellar evolution start from the classic paper on "The Synthesis of the Elements in Stars," published in 1957 by Geoffrey Burbidge, William A. Fowler, and Fred Hoyle. Although many of the details of stellar evolution and synthesis of the elements (nucleosynthesis) are still the subject of research and controversy, the basic process of stellar energy production and the formation of the elements is now clear.

Hydrogen and Helium Burning

By the time the gravitational contraction of a mass of hydrogen has raised the core temperature to about 5×10^6 K, the fusion of hydrogen, hydrogen burning, can begin. The energy released in the fusion reactions produces a pressure that stops the star's contraction. The rate of energy release from the hot core balances the radiation loss from the star. The star is relatively stable and will burn for hundreds of millions of years, slowly depleting the supply of hydrogen. The proton-proton reactions dominate in first-generation, pure-hydrogen stars. Second-generation stars with carbon-12 present will derive some fraction of their energy from the Bethe carbon cycle.

Eventually the star's central core will exhaust its supply of hydrogen, even though hydrogen burning may continue in the outer regions of the star. Gravitational contraction will again set in; as the star compresses, the interior temperature increases. Eventually the core reaches temperatures of the order of 10^8 K and densities of the order of 10^8 kg/m³. At these conditions, **helium burning** begins.

Two helium nuclei will form unstable beryllium-8. At these temperatures and densities, an equilibrium population of Be-8 builds up and interacts with helium to form carbon-12, which in turn interacts with helium to form oxygen-16:

$$_6C^{12} + {_2}He^4 \longrightarrow {_8}O^{16}$$

Helium burning in stars results in comparable production of both carbon-12 and oxygen-16.

A much smaller mass fraction (0.07%) is converted to energy in helium burning than in hydrogen burning. Thus, the helium burning phase is relatively short. During most of a star's life, energy production is by hydrogen and helium fusion and very little synthesis of the heavy elements takes place.

Carbon, Oxygen, and Silicon Burning

Once the helium in the core has been exhausted, leaving only carbon and oxygen, gravitational compression will again raise temperatures. Late in a star's life, the core may be carbon and oxygen, helium burning may be occurring in a shell near the core, and the outer shell may still be in the hydrogen burning phase.

When the temperature of the core reaches a range of 1.4×10^9 K and a density of 5×10^8 kg/m³; **carbon burning** may begin. The principal reactions are

$$_6C^{12} + {_6}C^{12} \longrightarrow (50\%) {_{10}}Ne^{20} + {_2}He^4 + 4.62 \text{ MeV}$$
$$\longrightarrow (50\%) {_{11}}Na^{23} + {_1}H^1 + 2.24 \text{ MeV}$$

These reactions not only produce sodium and neon but also make available protons and alpha particles which can react with the neon and sodium to produce magnesium-24, -25, -26, aluminum-27, and silicon-28. White dwarf stars are thought to be remnants of small stars that never reached sufficiently high temperatures to "burn" carbon.

The next burning stage is reached at a core temperature of 1.8×10^9 K. The oxygen that has survived the carbon burning phase reacts, forming silicon, $_{14}Si^{28}$.

The final phase of a stable star is silicon burning. The different reactions build the elements with masses between 28 and 57. With the production of iron, which has the highest binding energies (about 8 MeV per nucleon), the star will have run out of particle reactions that can supply it with energy. The universal abundance of iron-56 testifies to its being the end point for the fusion reactions in stars throughout the universe.

Once a star's silicon burning phase is complete, the star's core will compress further, increasing the temperatures and densities. The star's ultimate evolution from this point depends on its total mass. If the mass is sufficiently small, the star will ultimately cool as a mass of iron and heavier elements; if the mass is larger, gravitational compression can lead to photodisintegration of the nuclear matter. The electrons will be absorbed into the matter, forming an extremely dense neutron star. If the photodisintegration process is unstable, the star may explode, becoming a *supernova* and distributing its elements throughout its region of the galaxy. If the mass is sufficiently large, and if our present theories are correct, the ultimate collapse of nuclear matter gives rise to a black hole, from which nothing can escape.

The elements of the universe, including the oxygen and nitrogen of our atmosphere, and the carbon, calcium, sodium, and iron of our bodies has been synthesized in stars. The hydrogen is left over from the creation of the universe. All of the atoms on earth were present during the formation of our solar system around a second-generation star some 5 billion years ago. We are in fact just a little bit of stardust.

SOLAR NEUTRINOS

There is something fascinating about science. One gets such wholesale returns of conjecture out of such a trifling investment of fact.

Mark Twain

In addition to providing energy to keep a star burning, every four proton-proton reactions PP I listed in Table 25-1 give rise to two solar neutrinos. Unfortunately, these neutrinos are at a low energy and cannot be detected. The PP II proton reactions, involving heavier elements, as listed in Table 25-2, in particular produce boron-8. Boron-8 decays by the emission of a positron (beta plus) and a neutrino with an energy up to 14 MeV in energy

$$_5B^8 \longrightarrow \ _4Be^8 + e^+ + \nu + 14.06 \text{ MeV}$$

The existence of neutrinos was first proposed by Wolfgang Pauli in the early 1930s and was used by Enrico Fermi in his theory of the energy of beta decay, as discussed in Chapter 24. Neutrinos possess no rest mass or charge. They are thought to propagate at the speed of light. In 1958, William Fowler described the possibility of detecting the neutrinos emitted by the decay of boron-8 in the center of the sun. Because these **solar neutrinos** interact only very weakly with matter, they will emerge from the sun's interior. They go through tremendous thickness of material without interacting, and will travel from the core of the sun to its surface and through the reaches of space to the earth without being absorbed. It is, however, barely possible to detect neutrinos with energies greater than 0.81 MeV.

Since 1955, Raymond Davis at Brookhaven National Laboratory has studied the use of an inverse reaction to detect solar neutrinos. By capturing an electron, argon-37 will decay, with a half-life of 35 days and giving chlorine-37 and a neutrino:

$$e^- + Ar^{37} \longrightarrow Cl^{37} + \nu$$

If chlorine-37 captures a neutrino, the reverse reaction can occur:

$$Cl^{37} + \nu \longrightarrow Ar^{37} + e^-$$

Chlorine-37 absorbs a neutrino, giving an excited state of chlorine which then decays by the emission of an electron, leaving argon-37. The probability of neutrino capture is extremely small but sufficient to do an experiment.

In 1967, the Brookhaven Solar Neutrino observatory began experiments at the Homestake Gold Mine in Lead, South Dakota (see Figure 26-2). The neutrino detector consists of a 400 000 liter tank of perchloroethylene, $Cl_2C = CCl_2$, a common commercial dry-cleaning sol-

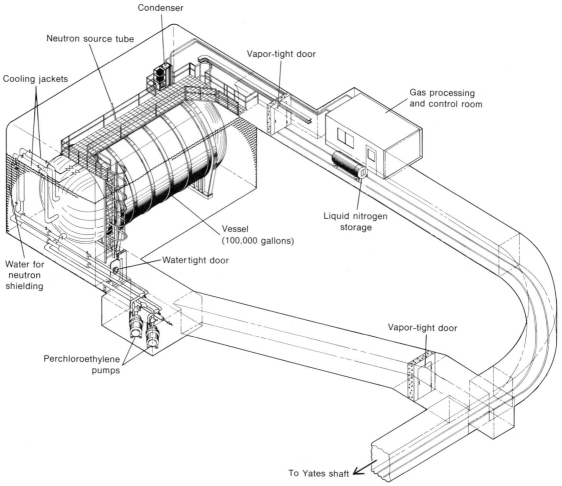

FIGURE 26-2
The neutron observatory in the Homestake Mine. The tank, containing about 400 000 liters of a commercial dry-cleaning solvent, is at a depth of 1500 m. Courtesy of Raymond Davis, Jr., Brookhaven National Laboratory.

vent. To minimize production of argon-37 by cosmic rays, the detector is buried in the mine about 1500 m below the surface. The alleged solar neutrinos will react with the chlorine-37 to produce argon-37. After about 100 days, the perchloroethylene tank is purged with helium gas, driving off about 90% of the argon-37 produced by neutrino capture. If the standard solar model were correct, one would expect to observe about 50 argon-37 atoms in the 400 000 liters after 100 days. The argon is condensed with liquid nitrogen and the radioactive decay of argon-37 is observed with particle detectors. The average observed production of argon-37 over all runs through 1975 is 0.32 ± 0.08 argon atoms per day—slightly higher than the production by cosmic rays that penetrated the 1500 m of rock. There is also considerable fluctuation from one experimental run to the next. Figure 26-3 shows the observations since 1970. Also shown is the theoretical prediction of the expected argon-37 production caused by the flux of neutrinos from the sun, which is about four times the observed value.

The low solar neutrino flux observed in these experiments has cast into question contemporary models of solar and stellar processes. The neutrino flux from the sun depends sensitively on the temperature assumed in the

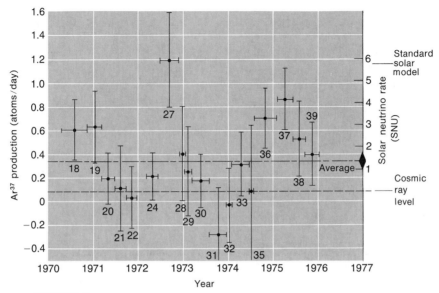

FIGURE 26-3

Summary of solar neutrino observations. These results (and more recent data) have been consistently below the prediction of the standard solar model. Courtesy of Raymond Davis, Jr., Brookhaven National Laboratory.

solar model. This sensitively changes the concentration of boron-8, one of the chief emitters of energetic neutrinos. Many hypotheses have been suggested to explain this unexpectedly low production of argon-37 in the solar neutrino experiment. Perhaps in traversing the eight light-minutes from the sun to the earth, neutrinos change their character in some way. Or perhaps additional or unknown nuclear processes take place in the sun and influence neutrino production. Or perhaps these results indicate some systematic error in the experiment.

The low solar neutrino flux observed in the Brookhaven experiments excites one of the more interesting scientific controversies of today, keeping theoreticians and experimentalists, astronomers and physicists busy in the enterprise of science. The present status of this controversy is well stated by John Bahcall and Raymond Davis, Jr.:

The attitude of many physicists toward the present discrepance is that astronomers never really understand astronomical systems as well as they think they do, and the failure of the standard theory in this simple case just proves that physicists are correct in being skeptical of the astronomers' claims. Many astronomers believe, on the other hand, that the present conflict between theory and observation is so large and elementary that it must be due to an error in the basic physics, not in our astrophysical understanding of stellar evolution.[1]

CONTROLLED THERMONUCLEAR FUSION

Since thermonuclear fusion was first demonstrated by the explosion of the first hydrogen bomb, in 1954, there have been great hopes that the enormous energy available from the fusion of hydrogen will find peaceful uses. Engineers and scientists, however, have found conditions suitable for controlled fusion difficult to attain. By 1963, temperatures required for fusion had been obtained briefly in the theta pinch experiments in Los Alamos, New Mexico, and by the late 1960s, much of the basic physics was well understood and had been verified by experiments.

[1] John N. Bahcall and Raymond Davis, "Solar Neutrinos: A scientific puzzle," *Science* (23 January 1976), p. 267.

Fusion Reactions

Fusion processes in stars involve the reaction of hydrogen. The most promising reactions for controlled thermonuclear fusion involve the isotopes of hydrogen, deuterium, $_1D^2$, and tritium, $_1T^3$. Deuterium is relatively plentiful, one part in 6700 occurring in the hydrogen in ordinary water. Tritium can be readily produced by the interaction of neutrons with lithium. Table 26-4 and Figure 26-4 give the principal reactions of deuterium and tritium.

Interaction of deuterium and tritium instantaneously forms a compound nucleus, which then flies apart, with an alpha particle and a neutron going in opposite directions. The mass difference between initial and final nuclei gives the total energy released in the reaction. A large fraction of the energy will be released as kinetic energy of the lighter neutron. With conservation of energy and momentum, the distribution of energy between two particles can be determined.

If m_1 is the mass of the alpha and m_2 of the neutron, the kinetic energy of each is given by

$$KE_1 = \frac{1}{2}m_1v_1{}^2; \; KE_2 = \frac{1}{2}m_2v_2{}^2$$

The total energy is $E_T = KE_1 + KE_2$.

The velocities of the two particles can be determined by using conservation of momentum (discussed previously in Chapter 3). In the center of mass of the compound nucleus, the momentum is zero:

$$m_1v_1 + m_2v_2 = 0$$

TABLE 26-4
Deuterium and tritium reactions, and the energy released in reaction. Neutrons carry off much of the energy.

$_1D^2 + _1T^3 \longrightarrow {}_2He^4[3.52 \text{ MeV} + {}_0n^1[14.06 \text{ MeV}]$

$_1D^2 + _1D^2 \xrightarrow{(50\%)} {}_2He^3[0.82 \text{ MeV}] + {}_0n^1[2.45 \text{ MeV}]$

$\xrightarrow{(50\%)} {}_1T^3[1.01 \text{ MeV}] + {}_1H^1[3.03 \text{ MeV}]$

$_1D^2 + _2He^3 \longrightarrow {}_2He^4[3.67 \text{ MeV}] + {}_1H^1[14.67 \text{ MeV}]$

$_1T^3 + _1T^3 \longrightarrow {}_2He^4 \qquad + 2_0n^1[11.32 \text{ MeV}]$

SOURCE: F. L. Ribe, "Fusion Reactor Systems," *Reviews of Modern Physics* (January 1975).

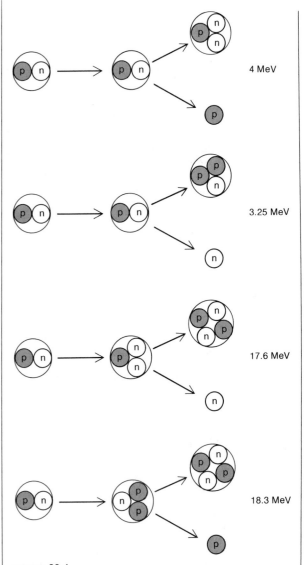

4 MeV

3.25 MeV

17.6 MeV

18.3 MeV

FIGURE 26-4
Deuterium and tritium reactions that may be useful for fusion power. From Richard F. Post, "Fusion Power." Copyright © 1957 by Scientific American, Inc. All rights reserved.

The velocity of the alpha v_1 is

$$v_1 = \frac{-m_2}{m_1} v_2$$

The minus sign indicates their travel in opposite directions. Since m_1 is greater than m_2, the neutron will be traveling faster than the alpha particle. We can solve for the kinetic energy of the neutron KE_2: The total energy of the alpha particle is

$$KE_1 = \frac{1}{2} m_1 v_1^2 = \frac{1}{2} m_1 \left(\frac{-m_2 v_2}{m_1} \right)^2$$

$$KE_1 = \frac{m_2}{m_1} \left(\frac{1}{2} m_2 v_2^2 \right) = \frac{m_2}{m_1} KE_2$$

The total energy is

$$E_T = KE_1 + KE_2 = \left(\frac{m_2}{m_1} \right) KE_2 + KE_2$$

$$= \left[\frac{m_2}{m_1} + 1 \right] KE_2$$

Since m_2 is less than m_1, the energy of the alpha KE_1 is less than that of the neutron KE_2.

$$KE_2 = \frac{E_T}{m_2/m_1 + 1}$$

EXAMPLE

Deuterium and tritium interact to produce an alpha particle and a neutron. What is the total energy released? What is the kinetic energy of the neutron?
The total energy released in the reaction can be calculated from the mass difference:

$_1D^2$	2.014 102 amu
$_1T^3$	3.016 050 amu
Initial mass: $M_i =$	5.030 152 amu
$_2He^4$	4.002 603 amu
$_0n^1$	1.008 665 amu
Final mass: $M_f =$	5.011 268 amu
Mass difference:	$\Delta M = M_i - M_f = 0.018\,884$ amu

The energy released in the reaction is

$$\Delta E = \Delta M c^2 = (0.01888 \text{ amu})(931 \text{ MeV/amu})$$
$$\Delta E = 17.58 \text{ MeV}$$

This energy will be distributed between the two particles. The deuterium and tritium form a compound nucleus; the neutron and alpha then fly out in opposite directions. The kinetic energy of the neutron is given by

$$KE_2 = \frac{\Delta E}{m_2/m_1 + 1} = \frac{17.58 \text{ MeV}}{1/4 + 1} = 14.06 \text{ MeV}$$

Most of the energy is carried off by the lighter neutron.

For two nuclei to interact, they must be close together, and the Coulomb force between the positive charges must be overcome. They must approach one another with considerable kinetic energy, which is converted to Coulomb potential energy. The total energy of two approaching particles is

$$E_T = \frac{1}{2} m_1 v_1^2 + \frac{1}{2} m_2 v_2^2 + \frac{k_e Q_1 Q_2}{R}$$

At the large distance, the Coulomb potential energy is unimportant and the total energy is just the kinetic energy of the two particles. At the distance of closest approach, the kinetic energy is zero. The total energy is then

$$E_T = \frac{k_e Q_1 Q_2}{R}$$

The two particles must approach one another within a distance of the order of 10^{-14} m to interact. At this distance, the energy of the system is

$$E_T = \frac{k_e Q_1 Q_2}{R} = \frac{(9 \times 10^9 \text{ N-m}^2/\text{C}^2)(1.6 \times 10^{-19} \text{ C})^2}{(10^{-14} \text{ m})}$$

$$E_T = 2.3 \times 10^{-14} \text{ joules} = 144 \text{ keV}$$

To overcome the Coulomb barrier, the two particles must approach one another with an initial kinetic energy of about 144 keV. Two deuterium ions must each have a kinetic energy of about 70 keV. Although these are modest energies for particle accelerators, such accelerators cannot produce enough particles per second to generate significant energy.

The energy required to ionize atomic hydrogen, deuterium, or tritium is 13.6 eV. At thermonuclear temperatures, the energy of the particles is 10 keV or higher. Collisions among particles will cause the gas to become totally ionized. The gas in this ionized state is known as a plasma and is electrically neutral, having equal numbers of positive and negative charges. Although the behavior of a plasma can often be analyzed as the motions of individual particles, the collective effect of the many charges often dominates behavior. In many ways a plasma behaves like a fluid. The individual motions of the ions are modified by collective effects.

A gas at a temperature T will have a distribution of molecular speeds given by the Maxwell-Boltzmann dis-

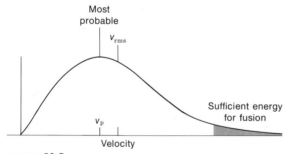

FIGURE 26-5
Particle distribution as a function of speed. The most probable speed is related to the temperature by $\frac{1}{2}mv_p^2 = kT$. If the most probable speed corresponds to a kinetic energy of 10 keV, a small fraction of the particles will have sufficient energy to undergo fusion reactions.

tribution shown in Figure 26-5 (and discussed in Chapter 10). The *most probable speed* for a particle in a gas is given by

$$\frac{1}{2}mv_p^2 = kT$$

where k is the Boltzmann constant:

$$k = 1.38 \times 10^{-23} \text{ J/K}$$

If the most probable speed of an ion in a thermonuclear plasma is about 10 keV, some small fraction of the ion distribution will have energies of 70 keV, sufficient to overcome the Coulomb barrier and undergo fusion. If the most probable ion energy is 10 keV, this would correspond to a temperature

$$T = \frac{\frac{1}{2}mv_p^2}{k} = \frac{(10^4 \text{ eV} \times 1.6 \times 10^{-19} \text{ J/eV})}{1.38 \times 10^{-23} \text{ J/K}}$$

$$T = 1.16 \times 10^8 \text{ K}$$

The temperature is 116 million Kelvin!

Figure 26-6 shows the energy production in W/cm^3 as a function of temperature for a typical deuterium plasma. To achieve the same power level, a deuterium-deuterium (D-D) plasma requires much higher temperatures than a deuterium-tritium (D-T) plasma requires. Figure 26-7 shows the ratio of nuclear power from fusion compared to the energy radiated from the plasma. To achieve ignition and spontaneous heating of the plasma, more energy must be liberated in fusion than is lost by radiation. The **ignition temperature** for D-D fusion is about 3×10^8 K, or a most probable kinetic energy of about 25 keV. For D-T fusion, the ignition temperature is about 10^8 K, or a kinetic energy of about 10 keV.

Lawson Criterion

Production of useful fusion energy requires that fusion reactions release more energy than is required to assemble the plasma. To proceed at all, fusion requires an ignition temperature of the order of 10^8 K. The density of the plasma determines the power generated. Plasma density n is expressed in ions/cm³. Densities of 10^{14} to 10^{15} ions/cm³ are typical. By comparison, the density of air at one atmospheric pressure and 0°C is 2.69×10^{19} molecules/cm³. The amount of energy liberated depends

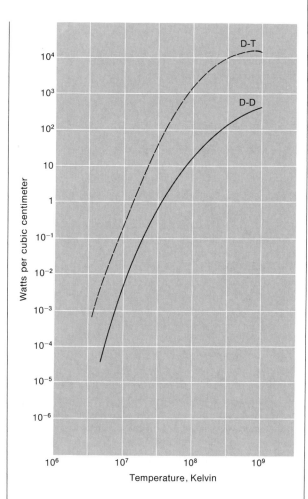

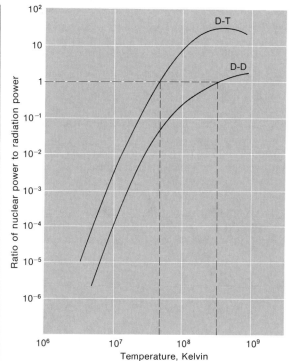

FIGURE 26-7
The ignition temperature, where the nuclear power produced in fusion equals the radiation power of the plasma, is shown for both D-D and D-T reactions. From Richard F. Post, "Fusion Power." Copyright © 1957 by Scientific American, Inc. All rights reserved.

FIGURE 26-6
Fusion power output at a plasma density of about $3 \times 10^{15}/\text{cm}^3$ plotted as a function of temperature for deuterium-deuterium, D-D, and for deuterium-tritium, D-T, reactions. From Richard F. Post, "Fusion Power." Copyright © 1957 by Scientific American, Inc. All rights reserved.

on how long the plasma is held together: the containment time τ.

The breakeven point for energy production in a deuterium-tritium plasma is given by the **Lawson criterion.** The product of the plasma density and the containment time $n\tau$ must be greater than 10^{14} s/cm³:

$$n\tau > 10^{14} \text{ s/cm}^3$$

Controlled thermonuclear fusion will become scientifically feasible only when both ignition temperature and the Lawson criterion for density and containment time are achieved.

Magnetic Confinement

Because of the high temperatures, no physical wall can be used to confine the plasma. A charged particle traveling in a magnetic field will travel in a curved path, as discussed in Chapter 15. This phenomenon is the basis of **magnetic confinement.** A particle has components of velocity perpendicular and parallel to the magnetic field. The parallel component will simply move the particle along the field line. Because of the component of velocity perpendicular to the magnetic field there is a force on the charged ion or electron. The force, perpendicular to both the direction of the particle and the direction of the magnetic field, is given by

$$\mathbf{F} = q\mathbf{v} \times \mathbf{B}$$

This provides a centripetal force, causing the particle to go in a circular orbit of radius R

$$F = qv_\perp B = \frac{mv_\perp^2}{R}$$

The radius of the orbit, the *cyclotron radius*, is

$$R = \frac{mv_\perp}{qB}$$

Figure 26-8 shows the rotation of a positive ion and an electron about the magnetic field lines. The mass of the ion being so much larger than that of the electron, the ion orbit is much larger. The cyclotron radius of ions in a plasma is usually small compared to the plasma's overall dimensions. In the absence of collisions among particles, the magnetic field traps the plasma ions and electrons and keeps them from moving across the field.

The **tokamak** confinement system derives from a series of Russian experiments in the late 1960s. As Figure 26-9 shows, a tokamak has a fat donut-shaped magnetic field. Coils wrapped around the torus provide a magnetic field that closes on itself forming a magnetic bottle. The plasma forms a single-turn secondary of an iron-core transformer that is used to heat the plasma. A current in the primary winding produces in the plasma a very large current that heats the plasma to perhaps as high as 7 keV. In addition, the large current produces a magnetic field that causes the plasma to compress.

The Princeton Large Torus, PLT, shown in Figure 26-10, began operation in December 1975 after three years of construction. Present theories predict that plasma confinement times will increase as the devices become larger. The toroidal magnetic field coils produce a magnetic field of 4.5 tesla at the center of the plasma.

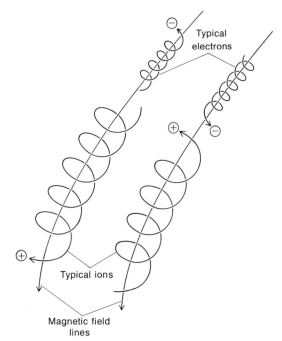

Typical electrons

Typical ions

Magnetic field lines

FIGURE 26-8
Ions and electrons are trapped around magnetic field lines.

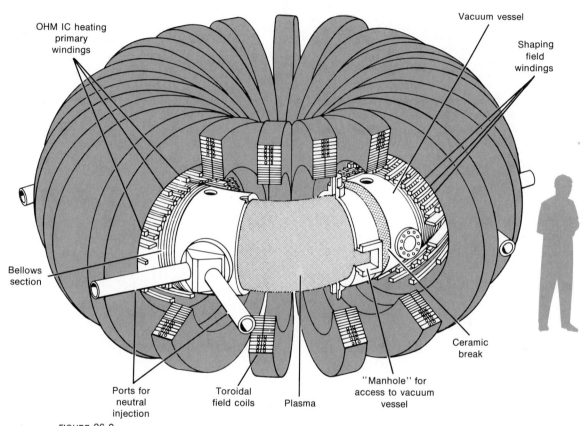

OHM IC heating
primary
windings

Vacuum vessel

Shaping
field
windings

Bellows
section

Ceramic
break

Ports for
neutral
injection

Toroidal
field coils

Plasma

"Manhole" for
access to vacuum
vessel

FIGURE 26-9
Schematic of a tokamak confinement system, the Princeton Large Torus. A toroidal magnetic field supports
the plasma inside the vacuum vessel, which is 1 m in height and almost 4 m in diameter. The toroidal coils
produce a magnetic field of 4.5 tesla at the center of the plasma. Courtesy of Princeton Plasma Physics
Laboratory.

There have been many plasma experiments in the attempt to achieve the conditions necessary for thermonuclear fusion. Figure 26-11 shows the research progress in
fusion power toward the goal of thermonuclear fusion as
defined by the Lawson criterion. The figure plots the
density-confinement time product $n\tau$ versus the burn rate
for D-T fuel. Also shown are quantitative data as well as
the goals of devices operating or under design or construction. Figure 26-12 shows the progress in fusion
research with time. Experimental reactors in the mid-
1980s should demonstrate the scientific feasibility of fusion power.

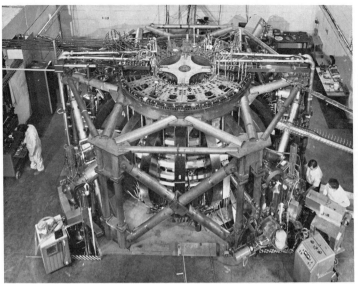

FIGURE 26-10
Princeton Large Torus (PLT) began operation in 1975. It is designed to further test plasma theory predictions that increased size of the plasma donut will yield increased plasma confinement times. Plasma heating methods are also under study. The complicated structure must support a weight of 150 tonnes and must hold three magnetic-coil systems against large magnetic forces. Courtesy of Princeton Plasma Physics Laboratory.

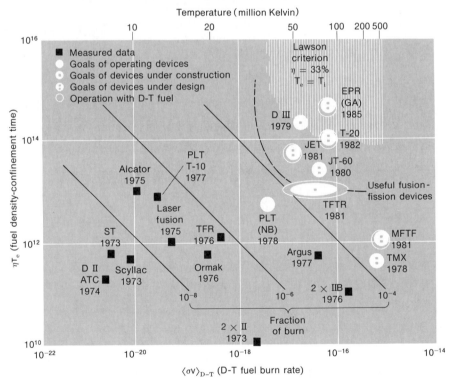

FIGURE 26-11
Research progress in fusion power, showing the performance of existing and planned experiments. The upper right-hand corner shows the goal as given by the Lawson criterion. The vertical scale gives the product of density-confinement time, $n\tau$. The horizontal scale is a measure of the D-T burn rate, and shows the approximate temperatures in million Kelvin. Courtesy of Electric Power Research Institute.

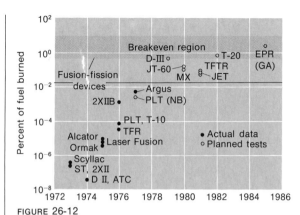

FIGURE 26-12
Fusion research progress as a function of time. The horizontal scale indicates the percentage of fuel undergoing fusion reactions. Breakeven conditions are expected in the early 1980s. Courtesy of Electric Power Research Institute.

Laser Fusion

If a deuterium-tritium pellet can be heated and compressed to great temperature and density, the entire fusion reaction can take place before the pellet can fly apart at thermal speeds. This process is called **inertial confinement.** Inertial confinement schemes using high-power lasers or intense electron or ion beams are under study. Recent developments in laser fusion have caused considerable excitement.

The goal in laser fusion is to implode fuel to extreme densities of about 10^{26} particles/cm^3. The effective containment time, the time required for them to fly apart, is slightly more than 10^{-12} s. Figure 26-13 shows the implosion of a symmetrically heated D-T pellet. A laser pulse of some 10 kilojoules of energy will heat the pellet's surface. As material flies from the surface, reaction forces establish a shock front that compresses the pellet. Once the surface has reached the thermonuclear ignition temperature, a thermonuclear burn front propagates inward and the pellet's density and fusion reaction rate increase by a factor of 10^3. If sufficiently large pellets can be compressed to ignition conditions, more energy will be released in the reaction than is required to compress the

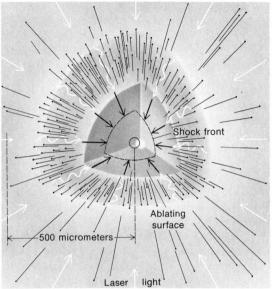

FIGURE 26-13
Implosion of a pellet of deuterium-tritium by symmetrically focused laser beams. Maximum absorption of light takes place at the critical surface, a narrow region in the low-density atmosphere surrounding the pellet. Hot energetic electrons transport energy inward to heat and boil off the pellet's surface. The surface explodes outward and the reaction force compresses the pellet. Courtesy of Lawrence Livermore Laboratory, University of California. From John Emmett, John Nuckolls, and Lowell Wood, "Fusion Power by Laser Implosion." Copyright © 1974 by Scientific American, Inc. All rights reserved.

pellet. Small glass microspheres about 10^{-4} m in diameter are filled with deuterium-tritium gas at 10 to 100 atmospheres. They have been compressed to about 20 times solid density with a 30 joule laser system pulsed for 250 picoseconds (10^{-12} s) and 14.1 MeV neutrons have been observed, indicating that thermonuclear fusion has been achieved.

The Fuel Cycle

The deuterium-tritium fusion yields a total of 17.58 MeV, 14 MeV as an energetic neutron that immediately leaves the reaction region and strikes the wall. Lithium-7, which is 92% of natural lithium, will absorb energetic neutrons, breeding tritium:

$$_0n^1 (2.47 \text{ MeV}) + {}_3Li^7 \longrightarrow {}_1T^3 + {}_2He^4 + n$$

Some of the energy is absorbed in driving the reaction. The low-energy neutron produced either decays to an electron and positron or is absorbed on lithium-6:

$$3 \text{ Li}^6 + {}_0n^1 \longrightarrow {}_2He^4 + {}_1T^3 + 4.80 \text{ MeV}$$

Thus, tritium is produced in the lithium blanket surrounding the reactor.

If deuterium-deuterium fusion is possible, the total energy resource from fusion becomes enormous. The net result of deuterium fusion is that six deuterium nuclei fuse to give two alpha particles, two protons, two neutrons, and 43.1 MeV of energy:

$$6 \, {}_1D^2 \longrightarrow 2 \, {}_2He^4 + 2 \, {}_1H^1 + 2 \, {}_0n^1 + 43.1 \text{ MeV}$$

About 7 MeV are liberated per deuterium atom.

Deuterium is relatively abundant, occurring to about one part in 6700 hydrogen atoms. Heavy water, HDO, is about 7% heavier than ordinary water, H_2O, and has a slightly higher boiling point. Concentration of heavy water is a relatively simple process.

EXAMPLE

One cubic meter of water has a mass of 1000 kg. Water has a molecular weight of 0.018 kg/mol.
(a) How many hydrogen atoms are present?
(b) If $1/6700$ of these are deuterium, how many deuterium atoms are present?
(c) How much energy would be released by the fusion of these nuclei?
The number of hydrogen nuclei in 1000 kg of water is

$$N_H = \left[\frac{1000 \text{ kg}}{0.018 \text{ kg/mol}}\right] \times \left[6.02 \times 10^{23}/\text{mol}\right]\left[\frac{2 \text{ H}}{\text{molecule}}\right]$$

$$N_H = 6.7 \times 10^{28} \text{ H}$$

The number of deuterium nuclei is

$$N_D = \frac{N_H}{6700} = \frac{6.7 \times 10^{28} \text{ H}}{6700 \text{ H/D}}$$

$$N_D = 10^{25} \text{ D}$$

The energy released in the fusion of the deuterium nuclei in 1 m^3 of water is

$$E = (10^{25} \text{ D})\left(\frac{7 \text{ MeV}}{D}\right)\left(\frac{10^6 \text{ eV}}{\text{MeV}}\right)\left(\frac{1.6 \times 10^{-19} \text{ J}}{\text{eV}}\right)$$

$$E = 1.2 \times 10^{13} \text{ joules!}$$

One barrel of oil has an energy content of about 6×10^9 joules. Thus, the fusion energy from deuterium in 1 m^3 of seawater is equivalent to about 2000 barrels of oil. The energy content in the deuterium in 1 km^3 of water is equivalent to 2×10^{12} barrels of oil, which is equal to the earth's known oil reserves. The oceans of the world have a volume of about 1.5×10^9 km^3. The potential resource from fusion energy is virtually unlimited.

PROSPECTS FOR THE FUTURE

A Fusion Reactor

Controlled thermonuclear fusion is not yet feasible. Yet there is interest in what a hypothetical 1000 megawatt fusion reactor would look like. It is assumed that the heating and confinement of the plasma will proceed according to known physical laws. The confinement time will increase as reactors increase in size. These will be large machines; much research will be done with smaller machines before any full-size machines are ever built.

An immediate problem in all approaches is the wall of the reactor vessel itself. The plasma in the center is at a temperature of 100 million K. The fusion reactions will expose the wall to an intense x-ray bombardment and a neutron flux of greater than 10^{13} neutrons/cm^2. A lithium blanket is required to absorb the neutrons, and to breed tritium. However, the wall material must physically with-

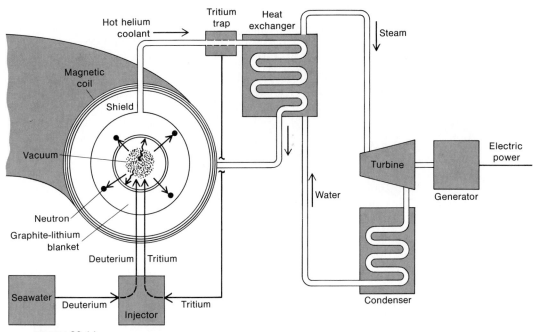

FIGURE 26-14
A conceptual drawing of a fusion powerplant. A heat exchanger extracts heat from the graphite lithium blanket. The tritium is recycled. The heat runs a conventional steam turbine. Courtesy of Energy Research and Development Administration.

stand the pressure difference between the outside and the vacuum of the plasma region. It must not react with the hot-liquid (1300 K) lithium blanket material. Rare metals such as niobium have been suggested for the inner wall of the vessel.

Figure 26-14 is a schematic diagram of power generation by fusion reactor. To remove the heat of fusion reactions, a coolant circulates past the lithium blanket and transfers the heat to produce steam to drive the turbine and generate electric power.

The Future

Laboratory research into controlled thermonuclear fusion is proceeding on many fronts, with the near-term goal of demonstrating the scientific feasibility of net energy production. As plasma experiments get larger, so do costs. A major American effort is the Tokamak Fusion Test Reactor, comparable in size to the Russian Tokamak T-20. The TFTR (see Figure 26-15) has a major radius of 2.7 m and a plasma diameter of 1.9 m; it is scheduled for operation in 1981. A large (10 kJ) laser fusion experiment, Shiva, underway at the Lawrence Livermore Laboratory in California, has 20 laser beams focused on a single pellet. Shiva will be superseded by the 200 to 300 kJ Nova laser system (see Figure 26-16) scheduled for operation by 1985. Results of these new experiments will influence the design of experimental power reactors in the late 1980s. Although creating fusion temperatures, densities, and confinement times poses formidable scientific and engineering problems, researchers believe that controlled thermonuclear fusion is indeed possible and has an excellent chance to become a practical power source early in the next century, creating an era of abundant energy.

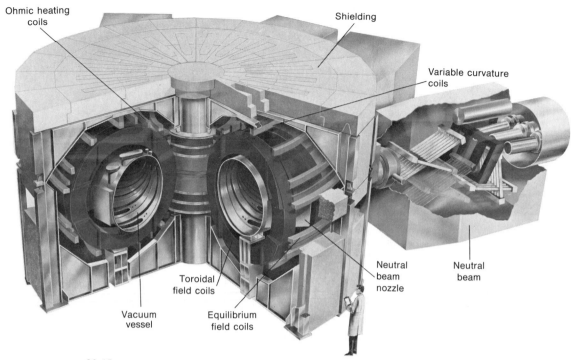

Ohmic heating coils

Shielding

Variable curvature coils

Neutral beam nozzle

Neutral beam

Toroidal field coils

Vacuum vessel

Equilibrium field coils

FIGURE 26-15
Tokamak Fusion Test Reactor, TFTR, is scheduled for completion in June 1981 at a cost of $228 million. TFTR is designed to demonstrate production of energy by deuterium-tritium fusion, to allow study of the basic plasma conditions and to provide design data for a future prototype fusion power reactor. Courtesy of Princeton Plasma Physics Laboratory.

FIGURE 26-16
A model of the 10 kilojoule Shiva laser facility operating at Lawrence Livermore Laboratory in California. The Nova laser system scheduled for operation by 1985 will yield 200 to 300 kilojoules. Courtesy of John Nuckolls, Lawrence Livermore Laboratory, University of California.

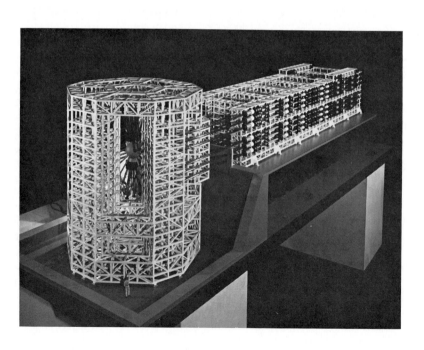

Review

Give the significance of each of the following terms. Where appropriate give two examples.

thermonuclear fusion
proton-proton reactions
carbon-nitrogen cycle
big bang theory
hydrogen burning

helium burning
carbon burning
solar neutrinos
ignition temperature
Lawson criterion

magnetic confinement
tokamak
inertial confinement

Further Readings

John N. Bahall and Raymond Davis, Jr., "Solar Neutrinos: A Scientific Puzzle," *Science* (23 January 1976), pp. 264–267.

John L. Emmett, John Nuckolls, and Lowell Wood, "Fusion Power by Laser Implosion," *Scientific American* (June 1974), pp. 24–36.

William A. Fowler, "The Origin of the Elements," *Scientific American* (September 1956).

George Gamow (all titles, New York: Viking Press), *A Star Called Sun* (1964); *The Creation of the Universe* (1956); *Biography of the Earth* (1941); *The Birth and Death of the Sun* (1940, revised 1952). Although these fascinating and delightful books do not reflect the latest theories, they present a significant discussion of the cosmological thinking that prevailed into the early 1960s.

Richard F. Post, "Fusion Power," *Scientific American* (December 1957); "Prospects for Fusion Power," *Physics Today* (April 1973), pp. 31–39.

Steven Weinberg, *The First Three Minutes: A Modern View of the Origin of the Universe* (New York: Basic Books, 1977). Cosmology as viewed by a major contemporary theoretical physicist.

Problems

See Table 24-2 (Chapter 24) for masses of atoms.

ENERGY PRODUCTION IN STARS

1 In hydrogen burning, four hydrogen atoms are converted into helium plus two positrons. The two positrons anihilate the two extra atomic electrons.
(a) How much mass is converted into energy?
(b) What fraction of the initial mass is this?
(c) How much energy is released?

2 In the Bethe carbon cycle, nitrogen-14 is converted into oxygen-15 by reaction with hydrogen; 7.29 MeV of energy is liberated in the reaction.
(a) Write the reaction.
(b) If the mass of $_7N^{14}$ = 14.003 074 amu, what is the mass of oxygen-15?

3 Oxygen-15 decays by beta plus emission.
(a) What is the product nucleus?
(b) Oxygen-15 has a mass of 15.003 070 amu; 2.76 MeV is liberated in the reaction. What is the mass of the product nucleus?

SYNTHESIS OF THE ELEMENTS

4 In helium burning, three helium-4 atoms combine to produce carbon-12.
(a) How much mass is converted into energy?
(b) What fraction of the total mass is converted into energy?
(c) How much energy is released?

5 In helium burning, carbon-12 plus helium yield oxygen-16; O^{16} has a mass of 15.994 915 amu.
(a) How much mass is converted into energy?
(b) What fraction of the total mass is this?
(c) How much energy is released?

6 Beryllium-8 has a mass of 8.005 308 amu; $_4Be^8$ is unstable and will fly apart in 10^{-16} s into two alpha particles. How much energy is liberated in this dissociation?

7 In carbon burning, two carbon atoms interact to yield sodium-23 and a proton:

$$_6C^{12} + {_6C^{12}} \longrightarrow {_{11}Na^{23}} + {_1H^1} + \text{energy}$$

Na^{23} has a mass of 22.989 771 amu.
(a) How much mass is converted into energy?
(b) What fraction of the total mass is this?
(c) How much energy is released?

SOLAR NEUTRINOS

8 In neutrino capture on chlorine-37, the neutrino supplies the necessary energy to make the reaction go:

$$Cl^{37} + \nu \longrightarrow Ar^{37} + e^-$$

How much energy must be supplied to make the reaction go? (Hint: What is the mass before and after?) Cl^{37} has a mass of 36.965 899 amu; Ar^{37} a mass of 36.966 772 amu.

CONTROLLED THERMONUCLEAR FUSION

9 A cup has about 200 cm³ of water with a molecular weight of 18 gm/mol and a density of 1 gm/cm³.
(a) How many hydrogen atoms are in the cup of water?
(b) Deuterium makes up one part in 6700 of hydrogen. How many deuterium atoms are in the cup?

(c) How much energy could be liberated by the fusion of the deuterium in the cup?

10 In one possible fusion reaction, deuterium reacts with deuterium to form a neutron and another nucleus.
 (a) Write the reaction. What is the other nucleus?
 (b) What fraction of the total mass is converted into energy?
 (c) How much energy is released?
 (d) How much of that energy is liberated with the neutron?

11 Lithium-6 absorbs a neutron, producing helium and another nucleus.
 (a) Write the reaction. What is the other nucleus?
 (b) How much energy is released?
 (c) How much of the energy is released with the helium nucleus?

12 The Adiabatic Toroidal Compression experiment has a deuterium ion temperature equivalent to a kinetic energy of 750 eV and a magnetic field of 3.5 tesla.
 (a) What is the ion's most probable speed?
 (b) What is the ion's cyclotron radius?

(c) If an electron has a kinetic energy of 2500 eV, what is its cyclotron radius?

13 The Princeton Large Torus Experiment will have a magnetic field of 4.5 tesla. If the deuterium ion temperature reaches 5 keV,
 (a) What is the deuterium ion's most probable speed?
 (b) What is the ion's cyclotron radius?

14 Energy consumption in the United States in 1985 will be an estimated 1.2×10^{20} J.
 (a) How many deuterium nuclei would be needed to produce this much energy?
 (b) What volume of water would contain this much deuterium?

15 A deuterium-tritium pellet is compressed with a laser pulse of 10^4 joules of energy.
 (a) How much energy does a single deuterium-tritium fusion liberate in joules?
 (b) How much energy would the fusion of 1 gm of deuterium-tritium liberate?
 (c) What mass of deuterium-tritium would be required to produce 10^6 joules?

Appendixes
and
Answers

Appendix A

Basic Mathematics

"I see trouble with algebra."

FIGURE A-1
Drawing by W. Steig: © 1963 The New Yorker Magazine,
Inc.

ALGEBRA

Rules of Algebra

In physics, concepts, theories, and mathematical models are often cast in the language of algebra. We can rearrange mathematical statements as long as we use the rules of algebra, brief descriptions of which follow.

The commutative principle: Addition or multiplication of real numbers is independent of order:

$$a + b = b + a$$
$$c \cdot d = d \cdot c$$

The associative principle: In a series of additions or of multiplications, the result is independent of order:

$$(a + b) + c = a + (b + c)$$
$$(b \cdot c)(d) = (b)(c \cdot d)$$

The distributive principle: Multiplication is distributive with respect to addition:

$$(a)(b + c) = a \cdot b + a \cdot c$$

Properties of zero and one: For any real number x:

$$x + 0 = x$$
$$x \cdot 0 = 0$$
$$x \cdot 1 = x$$

We can manipulate equations as long as we subtract, add, multiply, or divide both sides of the equation by the same quantity; and as long as we don't multiply by infinity or divide by zero.

Exponents

Often a quantity must be multiplied by itself. The area of a circle is πr^2, the volume of a sphere is $\frac{4}{3}\pi r^3$. If a quantity x is multiplied by itself n times, x is raised to the nth power:

$$x \cdot x \cdot x \cdot (\ldots) \cdot x = x^n$$

The multiplication of the same number raised to different exponents is given by

$$(x^n)(x^m) = x^{n+m}$$

For instance, $10^3 \times 10^6 = 10^{3+6} = 10^9$.
A negative exponent implies a reciprocal:

$$(x^n)(x^{-n}) = x^{n-n} = x^0 = 1$$

Dividing by x^n, we get

$$x^{-n} = \frac{1}{x^n}$$

We can raise an exponent to a power. Let $y = x^n$. We can raise y to the mth power:

$$y^m = (x^n)^m = x^{n \cdot m}$$

Fractional exponents are also permitted

$$y^{1/m} = (x^n)^{1/m} = x^{n/m}$$

If $n = m$, we have $x^{n/n} = x^1 = x$.
If $y = x^{1/2}$, this implies that $y^2 = x$, or y is the square root of x.

Logarithms

Common logarithms are based on powers of ten:

$$y_1 = \log x_1; \qquad y_2 = \log x_2$$

implies that

$$x_1 = 10^{y_1}; \qquad x_2 = 10^{y_2}$$

The product $z = x_1 \cdot x_2$ can be found from

$$z = x_1 \cdot x_2 = 10^{y_1} \cdot 10^{y_2} = 10^{(y_1 - y_2)}$$

$$\log z = \log (x_1 \cdot x_2) = y_1 + y_2 = \log x_1 + \log x_2$$

We can perform multiplication (or division) by addition (or subtraction) of logarithms. In this age of scientific pocket calculators, this may seem like a hard way to do multiplication. (Have you ever done longhand multiplication of two six-digit numbers?)

Logarithms are also useful for raising a number to a power:

Let $z = x^m = (10^y)^m = 10^{y \cdot m}$

$$\log z = m \cdot y = m \log x$$

For instance, let $z = 2^{100}$

$$\log z = 100 \log 2 = (100)(0.30103) = 30.103$$

$$z = 10^{\log z} = 10^{30.103} = 10^{(0.103+30)}$$

$$z = 10^{0.103} \times 10^{30}$$

$$z = 1.268 \times 10^{30}$$

Natural logarithms are based on powers of the number e.

$$e = 2.718\ 218\ 828$$

If

$$y = \log_e x = \ln x$$

then

$$x = e^y$$

For example,

$$\ln 1 = 0, \ \ln 2 = 0.693, \ \ln e = 1.0, \ \ln 10 = 2.30$$

Exponential functions based on e appear in nature whenever the rate at which a quantity is changing depends on the quantity present. For example, the voltage on a capacitor decays as

$$V = V_o e^{-t/RC}$$

In a time of one half-life, $T_{1/2}$, the voltage falls to one-half the initial value $V = V_o/2$

$$\frac{1}{2} = e^{-T_{1/2}/RC}$$

Taking the natural logarithm of both sides, we get

$$\ln\left(\tfrac{1}{2}\right) = \ln\left(e^{-T_{1/2}/RC}\right) = -T_{1/2}/RC$$
$$-\ln 2 = -T_{1/2}/RC$$

Solving for the half-life, we get

$$T_{1/2} = (\ln 2)RC = 0.693\, RC$$

Taking the common logarithm of both sides of the above expression accomplishes the same result:

$$\log\left(\tfrac{1}{2}\right) = \log(e^{-T_{1/2}/RC})$$
$$-\log 2 = (-T_{1/2}/RC)(\log e)$$
$$\log 2 = 0.30103; \ \log e = 0.4343:$$

$$T_{1/2} = (RC)\frac{(\log 2)}{(\log e)} = (RC)\frac{(0.30103)}{(0.4343)} = 0.693\, RC$$

The Quadratic Equation

The quadratic equation of the form

$$ax^2 + bx + c = 0$$

has two solutions. The roots x are given by

$$x = \frac{-b \pm (b^2 - 4ac)^{1/2}}{2a}$$

provided that $(b^2 - 4ac) > 0$

Binomial Expansion

The binomial expansion is of the form

$$(a + b)^n = a^n + \frac{na^{n-1}b}{1!} + \frac{n(n-1)a^{n-2}b^2}{2!}$$
$$+ \frac{n(n-1)(n-2)a^{n-3}b^3}{3!} + \cdots.$$

A commonly used expansion is $(1 + x)^{-1}$, where $x \ll 1$.

$$(1 + x)^{-1} = 1 + \frac{(-1)x}{1} + \frac{(-1)(-2)x^2}{1 \cdot 2}$$
$$+ \frac{(-1)(-2)(-3)x^3}{1 \cdot 2 \cdot 3}$$

$$(1 + x)^{-1} = 1 - x + x^2 - x^3 + \cdots \cong 1 - x$$

The factor $\gamma = (1 - v^2/c^2)^{-1/2}$, from relativity theory, can be expanded by use of the binomial expansion:

$$\gamma = 1 + \frac{(-\tfrac{1}{2})(-v^2/c^2)}{1} + \frac{(-\tfrac{1}{2})(-\tfrac{3}{2})(-v^2/c^2)^2}{1 \cdot 2} + \cdots.$$

$$\gamma = 1 + (\tfrac{1}{2})(v^2/c^2) + (\tfrac{3}{8})(v^4/c^4)$$

For nonrelativistic speeds, $v^2/c^2 \ll 1$, so that

$$\gamma = (1 - v^2/c^2)^{-1/2} \cong 1 + (\tfrac{1}{2})(v^2/c^2)$$

TRIGONOMETRY

Angles

As Figure A-2 shows, an angle in radians is defined by the

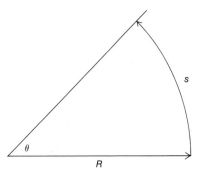

FIGURE A-2
An angle is defined as the ratio of the arc length to the radius.

ratio of the length of the arc s to the radius of the arc R:

$$\theta = \frac{s}{R}$$

One complete circle is an angle of 2π radians, where $\pi = 3.141\,592\,654.\ldots$ A complete circle is also $360°$. One radian corresponds to an angle of $\frac{180}{\pi} = 57.30°$.

The trigonometric functions of sine, cosine, and tangent are defined by use of the sides of a right triangle, as shown in Figure A-3. The length of the hypotenuse, the longest side c, is given in terms of the other two sides, by the Pythagorean theorem:

$$c^2 = a^2 + b^2$$

The sine of an angle is equal to the ratio of the side opposite the angle, opp, to the hypotenuse, hyp:

$$\sin \theta = \frac{\text{opp}}{\text{hyp}} = \frac{b}{c}$$

The cosine of an angle is equal to the ratio of the side adjacent to the angle, adj, to the hypotenuse, hyp:

$$\cos \theta = \frac{\text{adj}}{\text{hyp}} = \frac{a}{c}$$

The tangent of an angle is equal to (1) the ratio of the sine of the angle to the cosine, and (2) the ratio of the side opposite to the side adjacent:

$$\tan \theta = \frac{\sin \theta}{\cos \theta} = \frac{\text{opp}}{\text{adj}} = \frac{b}{a}$$

For the 3-4-5 right triangle shown in Figure A-3, $\sin \theta = 0.600$ and $\cos \theta = 0.80$. The angle $\theta = 36.87° \cong 37°$. The complementary angle ϕ is given by

$$\theta + \phi = 90°$$

For this triangle, $\phi = 53.13° \cong 53°$.
Figure A-4 shows a 30-60 right triangle. The hypotenuse

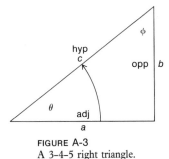

FIGURE A-3
A 3–4–5 right triangle.

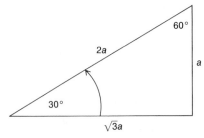

FIGURE A-4
A 30–60 right triangle.

2a is twice the side opposite the 30° angle. The side adjacent is $\sqrt{3}a$. Figure A-5 shows a 45-45 right triangle.

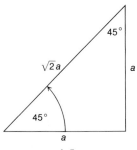

FIGURE A-5
A 45-45 right triangle.

The hypotenuse is $\sqrt{2}a$. Table A-1 lists the sines and cosines of common angles; Table A-2 lists the angles in radians and the sine, cosine, and tangent for angles from 0° to 90°.

TABLE A-1
Sines and cosines of common angles.

Angle	Sine	Cosine
0°	0.000	1.000
30°	0.500	0.866
37°	0.600	0.800
45°	0.707	0.707
53°	0.800	0.600
60°	0.866	0.500
90°	1.000	0.000

Trigonometric Identities

A fundamental identity derives from the Pythagorean theorem for a right triangle:

$$\sin^2 \theta + \cos^2 \theta = 1$$

The double angle formulas are also quite useful.

$$\sin (A \pm B) = \sin A \cos B \pm \cos A \sin B$$

$$\sin 2A = 2 \sin A \cos A$$

$$\cos (A \pm B) = \cos A \cos B \pm \sin A \sin B$$

$$\cos 2A = \cos^2 A - \sin^2 A = 2 \cos^2 A - 1$$

Figure A-6 shows a more general triangle, with angles α, β, and γ opposite sides a, b, and c respectively. The law of sines states that

$$\frac{\sin \alpha}{a} = \frac{\sin \beta}{b} = \frac{\sin \gamma}{c}$$

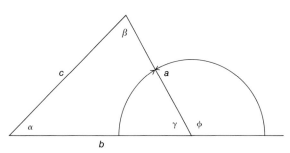

FIGURE A-6
A more general triangle.

The law of cosines gives that

$$c^2 = a^2 + b^2 - 2ab \cos \gamma$$

In terms of the complementary angle ϕ, the law of cosines states

$$c^2 = a^2 + b^2 + 2ab \cos \phi$$

The law of cosines is a generalization of the Pythagorean theorem.

VECTORS

Vector Addition

The laws of vector addition and subtraction resemble those of algebra. The commutative law states that the order of addition (or subtraction) of vectors is unimportant (see Figure A-7):

$$\mathbf{C} = \mathbf{A} + \mathbf{B} = \mathbf{B} + \mathbf{A}$$

TABLE A-2
Sines, cosines, and tangents.

Angle Degree	Angle Radian	Sine	Cosine	Tangent	Angle Degree	Angle Radian	Sine	Cosine	Tangent
0°	.000	0.000	1.000	0.000					
1°	.018	.018	1.000	.018	46°	0.803	0.719	0.695	1.036
2°	.035	.035	0.999	.035	47°	.820	.731	.682	1.072
3°	.052	.052	.999	.052	48°	.838	.743	.669	1.111
4°	.070	.070	.998	.070	49°	.855	.755	.656	1.150
5°	.087	.087	.996	.088	50°	.873	.766	.643	1.192
6°	.105	.105	.995	.105	51°	.890	.777	.629	1.235
7°	.122	.122	.993	.123	52°	.908	.788	.616	1.280
8°	.140	.139	.990	.141	53°	.925	.799	.602	1.327
9°	.157	.156	.988	.158	54°	.942	.809	.588	1.376
10°	.175	.174	.985	.176	55°	.960	.819	.574	1.428
11°	.192	.191	.982	.194	56°	.977	.829	.559	1.483
12°	.209	.208	.978	.213	57°	.995	.839	.545	1.540
13°	.227	.225	.974	.231	58°	1.012	.848	.530	1.600
14°	.244	.242	.970	.249	59°	1.030	.857	.515	1.664
15°	.262	.259	.966	.268	60°	1.047	.866	.500	1.732
16°	.279	.276	.961	.287	61°	1.065	.875	.485	1.804
17°	.297	.292	.956	.306	62°	1.082	.883	.470	1.881
18°	.314	.309	.951	.325	63°	1.100	.891	.454	1.963
19°	.332	.326	.946	.344	64°	1.117	.899	.438	2.050
20°	.349	.342	.940	.364	65°	1.134	.906	.423	2.145
21°	.367	.358	.934	.384	66°	1.152	.914	.407	2.246
22°	.384	.375	.927	.404	67°	1.169	.921	.391	2.356
23°	.401	.391	.921	.425	68°	1.187	.927	.375	2.475
24°	.419	.407	.914	.445	69°	1.204	.934	.358	2.605
25°	.436	.423	.906	.466	70°	1.222	.940	.342	2.747
26°	.454	.438	.899	.488	71°	1.239	.946	.326	2.904
27°	.471	.454	.891	.510	72°	1.257	.951	.309	3.078
28°	.489	.470	.883	.532	73°	1.274	.956	.292	3.271
29°	.506	.485	.875	.554	74°	1.292	.961	.276	3.487
30°	.524	.500	.866	.577	75°	1.309	.966	.259	3.732
31°	.541	.515	.857	.601	76°	1.326	.970	.242	4.011
32°	.559	.530	.848	.625	77°	1.344	.974	.225	4.331
33°	.576	.545	.839	.649	78°	1.361	.978	.208	4.705
34°	.593	.559	.829	.675	79°	1.379	.982	.191	5.145
35°	.611	.574	.819	.700	80°	1.396	.985	.174	5.671
36°	.628	.588	.809	.727	81°	1.414	.988	.156	6.314
37°	.646	.602	.799	.754	82°	1.431	.990	.139	7.115
38°	.663	.616	.788	.781	83°	1.449	.993	.122	8.144
39°	.681	.629	.777	.810	84°	1.466	.995	.105	9.514
40°	.698	.643	.766	.839	85°	1.484	.996	.087	11.43
41°	.716	.656	.755	.869	86°	1.501	.998	.070	14.30
42°	.733	.669	.743	.900	87°	1.518	.999	.052	19.08
43°	.751	.682	.731	.933	88°	1.536	.999	.035	28.64
44°	.768	.695	.719	.966	89°	1.553	1.000	.018	57.29
45°	.785	.707	.707	1.000	90°	1.571	1.000	.000	00.00

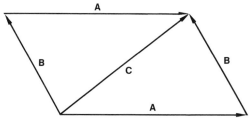

FIGURE A-7
The commutative law. The order of vector addition is unimportant: $\mathbf{C} = \mathbf{A} + \mathbf{B} = \mathbf{B} + \mathbf{A}$.

The associative law states that the sequence of addition of a series of vectors is unimportant (see Figure A-8):

$$\mathbf{D} = \mathbf{A} + \mathbf{B} + \mathbf{C} = (\mathbf{A} + \mathbf{B}) + \mathbf{C} = \mathbf{A} + (\mathbf{B} + \mathbf{C})$$

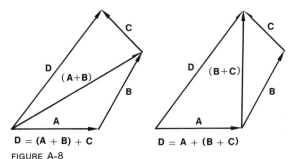

FIGURE A-8
The associative law. The order of a series of vector additions is unimportant: $\mathbf{D} = (\mathbf{A} + \mathbf{B}) + \mathbf{C} = \mathbf{A} + (\mathbf{B} + \mathbf{C})$

The negative of a vector is shown in Figure A-9. Adding the negative vector (-$\mathbf{A}$) to the positive vector ($\mathbf{A}$) gives zero:

$$\mathbf{A} + (-\mathbf{A}) = 0$$

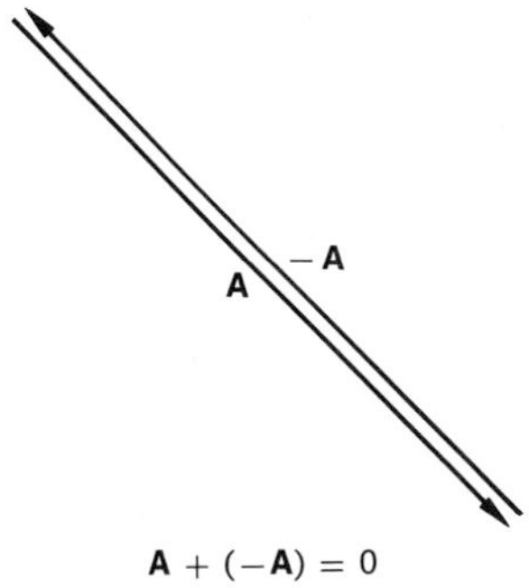

$$\mathbf{A} + (-\mathbf{A}) = 0$$

FIGURE A-9
The negative vector.

The displacement vector $\Delta\mathbf{R}$ is the difference between two position vectors:

$$\Delta\mathbf{R} = \mathbf{R}_2 - \mathbf{R}_1$$

Vector addition can be accomplished by the graphic method—drawing the vectors of lengths proportional to their magnitudes in their proper directions. They are then added head to tail (as shown in the previous three figures). Any vector can also be resolved into components, as shown in Figure A-10. The vectors $\mathbf{i}$, $\mathbf{j}$, and $\mathbf{k}$ represent

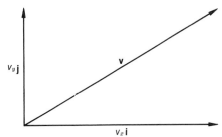

FIGURE A-10
A vector can be resolved into components: $\mathbf{v} = v_x\mathbf{i} + v_y\mathbf{j}$

unit vectors that point in the x-, y- and z-directions respectively.

Vector Multiplication

There are three cases of multiplication involving vectors:
1. multiplication of a vector by a scalar to produce a vector;
2. multiplication of a vector by a vector to produce a scalar quantity, the vector dot product or scalar product; and
3. multiplication of a vector by a vector to produce a vector result, the vector cross product.

A scalar quantity has a magnitude and units. If a vector is multiplied by a scalar, the magnitude and units change but the vector's spatial orientation does not. The force acting on an object, as given by Newton's second law, is in the same direction and proportional to the acceleration. This is really a vector equation:

$$\mathbf{F} = m\mathbf{a}$$

The inertial mass m is a scalar quantity.

The vector dot product, or scalar product, is the multiplication of two vectors to produce a scalar quantity:

$$C = \mathbf{A} \cdot \mathbf{B} = AB \cos \theta$$

The dot product is equivalent to the magnitude of vector A multiplied by the component of vector B that lies in the direction of A (see Figure A-11):

$$C = AB_\| = A(B \cos \theta)$$

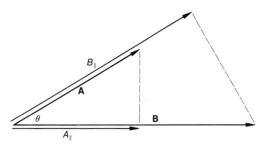

FIGURE A-11
The dot product. $\mathbf{A} \cdot \mathbf{B} = A_\| B = AB_\| = AB \cos \theta$.

This is also equivalent to the magnitude of B times the component of A that lies in the direction of B.

$$C = A_\| B = (A \cos \theta)B$$

The order of multiplication of the dot product is not important:

$$C = \mathbf{A} \cdot \mathbf{B} = \mathbf{B} \cdot \mathbf{A}$$

Any vector dotted into itself is the square of the magnitude of the vector. Suppose we have a velocity $\mathbf{v}$, which we can represent as two components v_x and v_y:

$$\mathbf{v} = v_x\mathbf{i} + v_y\mathbf{j}$$

The dot product of the vector with itself is

$$(\mathbf{v} \cdot \mathbf{v}) = (v_x\mathbf{i} + v_y\mathbf{j}) \cdot (v_x\mathbf{i} + v_y\mathbf{j})$$

$$= v_x v_x\mathbf{i} \cdot \mathbf{i} + v_x v_y\mathbf{i} \cdot \mathbf{j} + v_y v_x\mathbf{j} \cdot \mathbf{i} + v_y v_y\mathbf{j} \cdot \mathbf{j}$$

Because $\mathbf{i}$ and $\mathbf{j}$ are unit vectors at right angles, $\mathbf{i} \cdot \mathbf{i} = 1$, $\mathbf{j} \cdot \mathbf{j} = 1$, and $\mathbf{i} \cdot \mathbf{j} = \cos \theta = 0$

$$(\mathbf{v} \cdot \mathbf{v}) = v_x^2 + v_y^2 = v^2$$

This is just the square of the magnitude.

We can use the dot product to derive the law of cosines. Suppose that a vector $\mathbf{C} = \mathbf{A} + \mathbf{B}$, where $\mathbf{A}$ and $\mathbf{B}$ are at an angle ϕ as shown in Figure A-12. The magnitude of C is found from

$$C^2 = \mathbf{C} \cdot \mathbf{C} = (\mathbf{A} + \mathbf{B}) \cdot (\mathbf{A} + \mathbf{B})$$

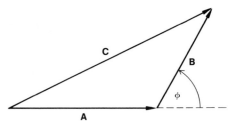

FIGURE A-12
Addition of vectors $\mathbf{C} = \mathbf{A} + \mathbf{B}$. The magnitude of $\mathbf{C}$ can be determined by using the dot product.

This reduces to

$$C^2 = \mathbf{A} \cdot \mathbf{A} + \mathbf{B} \cdot \mathbf{B} + 2\mathbf{A} \cdot \mathbf{B}$$

This is just

$$C^2 = A^2 + B^2 + 2AB \cos \phi$$

which is the law of cosines.

A vector $\mathbf{A}$ can be multiplied by a vector $\mathbf{B}$ to produce a vector $\mathbf{C}$ by using the vector cross product:

$$\mathbf{C} = \mathbf{A} \times \mathbf{B}$$

Any two vectors $\mathbf{A}$ and $\mathbf{B}$, if they are not parallel, will lie in a plane that is uniquely defined in space. We define the resultant vector $\mathbf{C}$ to be perpendicular to that plane, as shown in Figure A-13. The magnitude of the vector product is given by

$$C = AB \sin \theta$$

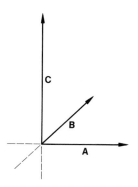

FIGURE A-13
Vector cross product. $\mathbf{C} = \mathbf{A} \times \mathbf{B}$. The vector $\mathbf{C}$ is perpendicular to the plane formed by $\mathbf{A}$ and $\mathbf{B}$.

The direction of the resultant vector is given by the right-hand rule. If the first finger of the right hand is $\mathbf{A}$ and the second finger is $\mathbf{B}$, then the thumb points in the direction $\mathbf{C}$. A number of physical quantities are defined in terms of the vector cross product, including torque, $\tau = \mathbf{R} \times \mathbf{F}$; angular momentum $\mathbf{J} = \mathbf{R} \times m\mathbf{v}$; and force due to a magnetic field, $\mathbf{F} = q\mathbf{v} \times \mathbf{B}$.

The vector cross product does not obey the commutative law; the order of multiplication is important:

$$\mathbf{C} = \mathbf{A} \times \mathbf{B} = -\mathbf{B} \times \mathbf{A}$$

If a vector is crossed into itself, the resultant is zero.

$$\mathbf{D} = \mathbf{A} \times \mathbf{A} = 0$$

Further Readings

Arthur Beiser, *Essential Math for the Sciences: Algebra, Trigonometry, and Vectors* (New York: McGraw-Hill, 1968). A good review.

Clarence Bennett, *Physics Problems and How to Solve Them* (New York: Barnes and Noble, College Outline Series, 1958). Many worked problems and examples.

J. B. Marion and R. C. Davidson, *Mathematical Preparation for General Physics* (Philadelphia: W. B. Saunders, 1972). Quite useful, with many applications.

G. Polya, *How to Solve It* (Princeton, N.J.: Princeton University Press, 1973). A comprehensive discussion of strategy for solving problems.

Clifford E. Schwartz, *Used Math for the First Two Years of College Science* (Englewood Cliffs, N.J.: Prentice-Hall, 1973). Contains some calculus and is quite thorough. By the same author, *Theory and Problems of College Physics* (New York: McGraw-Hill, Schaum's Outline Series). Many worked problems and examples.

Appendix B

Useful Information

The Greek alphabet.

Name	Capital	Lowercase
Alpha	A	α
Beta	B	β
Gamma	Γ	γ
Delta	Δ	δ
Epsilon	E	ϵ
Zeta	Z	ζ
Eta	H	η
Theta	Θ	θ
Iota	I	ι
Kappa	K	κ
Lambda	Λ	λ
Mu	M	μ
Nu	N	ν
Xi	Ξ	ξ
Omicron	O	o
Pi	Π	π
Rho	P	ρ
Sigma	Σ	σ
Tau	T	τ
Upsilon	Υ	υ
Phi	Φ	ϕ
Chi	X	χ
Psi	Ψ	ψ
Omega	Ω	ω

TABLE B-2
Physical Constants.

Mechanical

Universal Gravitation $\quad G = 6.6720 \times 10^{-11}$ N-m^2/kg^2

Thermodynamic

Boltzmann constant	$k = 1.3807 \times 10^{-23}$ J/K
Gas constant	$R = 8.314$ J/mol-K
Avogadro's number	$N_A = 6.022 \times 10^{23}$ /mol
Stefan-Boltzmann constant	$\sigma = 5.670 \times 10^{-8}$ W/m^2-K^4

Electrical

Electric constant	$k_e = 8.9876 \times 10^9$ N-m^2/C^2
Magnetic constant	$k_m = 1.000 \times 10^{-7}$ N/A-m
Electric charge	$e = 1.602 \times 10^{-19}$ C
Faraday constant	$\mathcal{F} = N_A e = 9.648 \times 10^4$ C/mol
Speed of light	$c = 2.9979 \times 10^8$ m/s

Atomic and Nuclear

Planck's constant	$h = 6.626 \times 10^{-34}$ J-s
Bohr radius	$a_o = 5.292 \times 10^{-11}$ m
Atomic mass unit	1 amu $= 1.6606 \times 10^{-27}$ kg
	1 amu $= 931.5$ MeV/c^2
Electron mass	$m_e = 9.1095 \times 10^{-31}$ kg
	$m_e = 0.5110$ MeV/c^2
Proton mass	$m_p = 1.6726 \times 10^{-27}$ kg
	$m_p = 938.28$ MeV/c^2
Neutron mass	$m_n = 1.6750 \times 10^{-27}$ kg
	$m_n = 939.57$ MeV/c^2
Electron volt	1 eV $= 1.602 \times 10^{-19}$ J

Conversion of units. The equivalent metric unit is shown to four significant figures.

Acre (U.S. survey)	$4.047 \times 10^3 \, \text{m}^2$	Inch	$2.540^* \times 10^{-2} \, \text{m}$
Angstrom (Å)	$1.000 \times 10^{-10} \, \text{m}$	Langley (ly)	$4.184^* \times 10^4 \, \text{J/m}^2$
Astronomical unit (AU)	$1.496 \times 10^{11} \, \text{m}$	Light year	$9.461 \times 10^{15} \, \text{m}$
Atmosphere (standard)	$1.013 \times 10^5 \, \text{N/m}^2$	Liter† (l)	$1.000^* \times 10^{-3} \, \text{m}^3$
Barrel (42 gal)	$1.590 \times 10^{-1} \, \text{m}^3$	Mile (statute)	$1.609 \times 10^3 \, \text{m}$
British thermal unit (Btu)	$1.054 \times 10^3 \, \text{J}$	Mile (international nautical)	$1.852 \times 10^3 \, \text{m}$
Bushel (U.S.)	$3.524 \times 10^{-2} \, \text{m}^3$	Ounce (avoirdupois)	$2.835 \times 10^{-2} \, \text{kg}$
Calorie	$4.182 \, \text{J}$	Ounce (troy)	$3.110 \times 10^{-2} \, \text{kg}$
Carat (metric)	$2.000 \times 10^{-4} \, \text{kg}$	Ounce (U.S. fluid)	$2.957 \times 10^{-5} \, \text{m}^3$
Cup (U.S.)	$2.366 \times 10^{-4} \, \text{m}^3$	Pound (mass, lbm)	$4.536 \times 10^{-1} \, \text{kg}$
Day (mean solar)	$8.640 \times 10^4 \, \text{s}$	Pound (force, lbf)	$4.448 \, \text{N}$
Dyne	$1.000 \times 10^{-7} \, \text{N}$	Quart (U.S. liquid)	$9.464 \times 10^{-4} \, \text{m}^3$
Erg	$1.000 \times 10^{-7} \, \text{J}$	Slug	$1.459 \times 10^1 \, \text{kg}$
Fathom	$1.829 \, \text{m}$	Tablespoon	$1.479 \times 10^{-5} \, \text{m}^3$
Foot (ft)	$3.048^* \times 10^{-1} \, \text{m}$	Ton (long, 2240 lb)	$1.016 \times 10^3 \, \text{kg}$
Gallon (U.S. liquid)	$3.785 \times 10^{-3} \, \text{m}^3$	Ton (metric), tonne	$1.000^* \times 10^3 \, \text{kg}$
Gauss	$1.000^* \times 10^{-4} \, \text{T}$	Ton (short, 2000)	$9.072 \times 10^2 \, \text{kg}$
Grain (avoirdupois)	$6.480 \times 10^{-5} \, \text{kg}$	Torr (mmHg, 0°C)	$1.333 \times 10^2 \, \text{N/m}^2$
Horsepower	$7.457 \times 10^2 \, \text{W}$	Yard	$9.144 \times 10^{-1} \, \text{m}$

* Indicates an exact equivalence.

† The liter was originally defined in 1901 as the volume of 1 kg of water at maximum density. Because a cubic decimeter (1000 cm³) of water at maximum density has a mass of 0.999 972 kg, there was a discrepancy of 28 parts in 10⁶ between the cubic decimeter (1000 cm³) and the liter (1000 milliliter). In 1964, the Twelfth General Conference "*declares* that the word 'litre' may be employed as a special name for the cubic decimeter," thus removing the discrepancy. This difference affects only reports of highly accurate work before 1964.

SI UNITS

The International System of Units (SI) has seven base units and two supplementary units, from which all other SI units are derived. Multiples and submultiples are expressed in a decimal system. The seven base units are:

meter (m): defined as 1 650 763.73 wavelengths in vacuum of the red-orange line of the spectrum of krypton-86.

kilogram (kg): the standard for the unit of mass, and the only base unit still defined by an artifact: a cylinder of platinum-iridium alloy kept by the International Bureau of Weights and Measures, in Paris, and a duplicate kept by the National Bureau of Standards as the U.S. standard of mass.

second (s): defined as the duration of 9 192 631 770 cycles of the radiation associated with a specified transition of the cesium-133 atom. It is realized by tuning an oscillator to the resonance frequency of cesium-133 atoms as they pass through a system of magnets and a resonant cavity into a detector.

ampere (A): defined as that current that, if maintained in each of two long, parallel wires separated by 1 m in free space, would produce a force between the two wires (due to their magnetic fields) of 2×10^{-7} newton for each meter of length.

kelvin (K): defined as the fraction $1/273.16$ of thermodynamic temperature of the triple point of water. The temperature 0 K is called absolute zero.

mole (mol): the amount of substance of a system that contains as many elementary entities as there are atoms in 0.012 kilogram of carbon-12. When the unit of mole is used, the elementary entities must be specified (that is, atoms, molecules, ions, electrons, other particles, or specified groups of such particles).

candela (cd): defined as the luminous intensity of $1/600\,000$ of a square meter of a black body at the temperature of freezing platinum (2045 K).

Table B-4 lists the seven base units as well as the two supplementary units, the plane angle in radian (rad) and the solid angle in steradian (sr). Table B-5 lists the derived SI units.

TABLE B-4

Base Units; SI units: seven well-defined base units and two supplementary units.

Quantity	Unit	Symbol
Length	meter (metre)	m
Mass	kilogram	kg
Time	second	s
Electric current	ampere	A
Thermodynamic temperature	kelvin	K
Amount of substance	mole	mol
Luminous intensity	candela	cd
Plane angle	radian	rad
Solid angle	steradian	sr

TABLE B-5

Derived SI units with special names.

Quantity	Unit	Symbol	Other units
Frequency	hertz	Hz	s^{-1}
Force	newton	N	$kg\text{-}m/s^2$
Pressure	pascal	Pa	N/m^2
Energy	joule	J	N-m
Power	watt	W	J/s
Electric charge	coulomb	C	A-s
Electric potential	volt	V	$N\text{-}m/C = W/A$
Capacitance	farad	F	C/V
Electric resistance	ohm	Ω	V/A
Conductance	siemens	S	$A/V = \Omega^{-1}$
Magnetic flux	weber	Wb	V-s
Magnetic field	tesla	T	$N/A\text{-}m = Wb/m^2$
Inductance	henry	H	$Wb/A = V\text{-}s/A$
Luminous flux	lumen	lm	cd-sr
Illuminance	lux	lx	lm/m^2
Activity (radionuclides)	becquerel	Bq	s^{-1}
Absorbed dose	gray	Gy	J/kg

30	Atomic number
Zn	Symbol
65.37	Atomic weight
Zinc	Element
$[Ar]3d^{10}4s^2$	Electron structure

1 **H** 1.00797 Hydrogen $1s^1$																	2 **He** 4.0026 Helium $1s^2$
3 **Li** 6.939 Lithium $1s^2 2s^1$	4 **Be** 9.0122 Beryllium $1s^2 2s^2$											5 **B** 10.811 Boron $1s^2 2s^2 2p^1$	6 **C** 12.01115 Carbon $1s^2 2s^2 2p^2$	7 **N** 14.0067 Nitrogen $1s^2 2s^2 2p^3$	8 **O** 15.9994 Oxygen $1s^2 2s^2 2p^4$	9 **F** 18.9984 Fluorine $1s^2 2s^2 2p^5$	10 **Ne** 20.183 Neon $1s^2 2s^2 2p^6$
11 **Na** 22.9898 Sodium $[Ne]3s^1$	12 **Mg** 24.312 Magnesium $[Ne]3s^2$											13 **Al** 26.9815 Aluminum $[Ne]3s^2 3p^1$	14 **Si** 28.086 Silicon $[Ne]3s^2 3p^2$	15 **P** 30.9738 Phosphorus $[Ne]3s^2 3p^3$	16 **S** 32.064 Sulfur $[Ne]3s^2 3p^4$	17 **Cl** 35.453 Chlorine $[Ne]3s^2 3p^5$	18 **Ar** 39.948 Argon $[Ne]3s^2 3p^6$
19 **K** 39.102 Potassium $[Ar]4s^1$	20 **Ca** 40.08 Calcium $[Ar]4s^2$	21 **Sc** 44.956 Scandium $[Ar]3d^1 4s^2$	22 **Ti** 47.90 Titanium $[Ar]3d^2 4s^2$	23 **V** 50.942 Vanadium $[Ar]3d^3 4s^2$	24 **Cr** 51.996 Chromium $[Ar]3d^5 4s^1$	25 **Mn** 54.938 Manganese $[Ar]3d^5 4s^2$	26 **Fe** 55.847 Iron $[Ar]3d^6 4s^2$	27 **Co** 58.933 Cobalt $[Ar]3d^7 4s^2$	28 **Ni** 58.71 Nickel $[Ar]3d^8 4s^2$	29 **Cu** 63.54 Copper $[Ar]3d^{10} 4s^1$	30 **Zn** 65.37 Zinc $[Ar]3d^{10} 4s^2$	31 **Ga** 69.72 Gallium $[Ar]3d^{10} 4s^2 4p^1$	32 **Ge** 72.59 Germanium $[Ar]3d^{10} 4s^2 4p^2$	33 **As** 74.922 Arsenic $[Ar]3d^{10} 4s^2 4p^3$	34 **Se** 78.96 Selenium $[Ar]3d^{10} 4s^2 4p^4$	35 **Br** 79.909 Bromine $[Ar]3d^{10} 4s^2 4p^5$	36 **Kr** 83.80 Krypton $[Ar]3d^{10} 4s^2 4p^6$
37 **Rb** 85.47 Rubidium $[Kr]5s^1$	38 **Sr** 87.62 Strontium $[Kr]5s^2$	39 **Y** 88.905 Yttrium $[Kr]4d^1 5s^2$	40 **Zr** 91.22 Zirconium $[Kr]4d^2 5s^2$	41 **Nb** 92.906 Niobium $[Kr]4d^4 5s^1$	42 **Mo** 95.94 Molybdenum $[Kr]4d^5 5s^1$	43 **Tc** (98) Technetium $[Kr]4d^5 5s^2$	44 **Ru** 101.07 Ruthenium $[Kr]4d^7 5s^1$	45 **Rh** 102.905 Rhodium $[Kr]4d^8 5s^1$	46 **Pd** 106.4 Palladium $[Kr]4d^{10} 5s^0$	47 **Ag** 107.870 Silver $[Kr]4d^{10} 5s^1$	48 **Cd** 112.40 Cadmium $[Kr]4d^{10} 5s^2$	49 **In** 114.82 Indium $[Kr]4d^{10} 5s^2 5p^1$	50 **Sn** 118.69 Tin $[Kr]4d^{10} 5s^2 5p^2$	51 **Sb** 121.75 Antimony $[Kr]4d^{10} 5s^2 5p^3$	52 **Te** 127.60 Tellurium $[Kr]4d^{10} 5s^2 5p^4$	53 **I** 126.904 Iodine $[Kr]4d^{10} 5s^2 5p^5$	54 **Xe** 131.30 Xenon $[Kr]4d^{10} 5s^2 5p^6$
55 **Cs** 132.905 Cesium $[Xe]6s^1$	56 **Ba** 137.34 Barium $[Xe]6s^2$	57 **La**★ 138.91 Lanthanum $[Xe]5d^1 6s^2$	72 **Hf** 178.49 Hafnium $[Xe]4f^{14} 5d^2 6s^2$	73 **Ta** 180.948 Tantalum $[Xe]4f^{14} 5d^3 6s^2$	74 **W** 183.85 Wolfram $[Xe]4f^{14} 5d^4 6s^2$	75 **Re** 186.2 Rhenium $[Xe]4f^{14} 5d^5 6s^2$	76 **Os** 190.2 Osmium $[Xe]4f^{14} 5d^6 6s^2$	77 **Ir** 192.2 Iridium $[Xe]4f^{14} 5d^7 6s^2$	78 **Pt** 195.09 Platinum $[Xe]4f^{14} 5d^9 6s^1$	79 **Au** 196.967 Gold $[Xe]4f^{14} 5d^{10} 6s^1$	80 **Hg** 200.59 Mercury $[Xe]4f^{14} 5d^{10} 6s^2$	81 **Tl** 204.37 Thallium $[Xe]4f^{14} 5d^{10} 6s^2 6p^1$	82 **Pb** 207.19 Lead $[Xe]4f^{14} 5d^{10} 6s^2 6p^2$	83 **Bi** 208.980 Bismuth $[Xe]4f^{14} 5d^{10} 6s^2 6p^3$	84 **Po** (210) Polonium $[Xe]4f^{14} 5d^{10} 6s^2 6p^4$	85 **At** (210) Astatine $[Xe]4f^{14} 5d^{10} 6s^2 6p^5$	86 **Rn** (222) Radon $[Xe]4f^{14} 5d^{10} 6s^2 6p^6$
87 **Fr** (223) Francium $[Rn]7s^1$	88 **Ra** (226) Radium $[Rn]7s^2$	89 **Ac**★★ (227) Actinium $[Rn]6d^1 7s^2$	104														

★	58 **Ce** 140.12 Cerium $[Xe]4f^2 5d^0 6s^2$	59 **Pr** 140.907 Praseodymium $[Xe]4f^3 5d^0 6s^2$	60 **Nd** 144.24 Neodymium $[Xe]4f^4 5d^0 6s^2$	61 **Pm** (147) Promethium $[Xe]4f^5 5d^0 6s^2$	62 **Sm** 150.35 Samarium $[Xe]4f^6 5d^0 6s^2$	63 **Eu** 151.96 Europium $[Xe]4f^7 5d^0 6s^2$	64 **Gd** 157.25 Gadolinium $[Xe]4f^7 5d^1 6s^2$	65 **Tb** 158.924 Terbium $[Xe]4f^9 5d^0 6s^2$	66 **Dy** 162.50 Dysprosium $[Xe]4f^{10} 5d^0 6s^2$	67 **Ho** 164.930 Holmium $[Xe]4f^{11} 5d^0 6s^2$	68 **Er** 167.26 Erbium $[Xe]4f^{12} 5d^0 6s^2$	69 **Tm** 168.934 Thulium $[Xe]4f^{13} 5d^0 6s^2$	70 **Yb** 173.04 Ytterbium $[Xe]4f^{14} 5d^0 6s^2$	71 **Lu** 174.97 Lutetium $[Xe]4f^{14} 5d^1 6s^2$
★★	90 **Th** 232.038 Thorium $[Rn]5f^0 6d^2 7s^2$	91 **Pa** (231) Protactinium $[Rn]5f^2 6d^1 7s^2$	92 **U** 238.03 Uranium $[Rn]5f^3 6d^1 7s^2$	93 **Np** (237) Neptunium $[Rn]5f^4 6d^1 7s^2$	94 **Pu** (242) Plutonium $[Rn]5f^6 6d^0 7s^2$	95 **Am** (243) Americium $[Rn]5f^7 6d^0 7s^2$	96 **Cm** (247) Curium $[Rn]5f^7 6d^1 7s^2$	97 **Bk** (247) Berkelium $[Rn]5f^9 6d^0 7s^2$	98 **Cf** (249) Californium $[Rn]5f^{10} 6d^0 7s^2$	99 **Es** (254) Einsteinium $[Rn]5f^{11} 6d^0 7s^2$	100 **Fm** (253) Fermium $[Rn]5f^{12} 6d^0 7s^2$	101 **Md** (256) Mendelevium $[Rn]5f^{13} 6d^0 7s^2$	102 **No** (254) Nobelium $[Rn]5f^{14} 6d^0 7s^2$	103 **Lw** (257) Lawrencium $[Rn]5f^{14} 6d^1 7s^2$

FIGURE B-1
Periodic table of the elements, based on the carbon-12 mass scale.

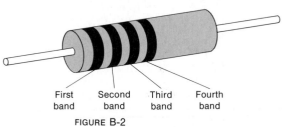

First band   Second band   Third band   Fourth band

FIGURE B-2
Color code bands of a resistor.

TABLE B-6
Color code for resistors. The resistors are coded with four bands. The first two represent a two-significant-digit number from 01 to 99. The third band represents a multiplier, some power of ten. If the third band is gold it represents a multiplier 0.1. The fourth band represents the tolerance: gold $\pm5\%$, silver $\pm10\%$, no band $\pm20\%$.

Color	Digit	Multiplier
Black	0	10^0
Brown	1	10^1
Red	2	10^2
Orange	3	10^3
Yellow	4	10^4
Green	5	10^5
Blue	6	10^6
Violet	7	10^7
Gray	8	10^8
White	9	10^9
Gold	—	10^{-1}

Appendix C

Well-Known Functions

In physics, the relationships among physical quantities are also described in the language of algebra. Many relationships take the form of easily recognized or well-known functions. If a variable y is related to a variable x so that for each possible value of x there is a unique value of y, then y is a **function** of the variable x, or $y(x)$. Thus, representative pairs of numbers (x,y) can be plotted on a graph, if we desire a visualization of the relationships between quantities. The possible values of the independent variable x make up the **domain** of x; the possible values of the dependent variable y make up the **range** of y.

Often the graph of the data shows a straight line or a parabola. The relationship, however, may be more complicated. The sine of an angle or the logarithm of a number gives us a number. But we can also consider the sine and the logarithm as functions—$y = \log x$, or $z = \sin \theta$—where all possible values of x or θ over some domain may be considered. Which types of functions you consider well-known depends on your mathematical sophistication. Very many functions have been studied by analytic mathematicians over the past 200 years. Here we are concerned with only a few: sine and cosine functions, exponential functions, logarithmic functions, and power laws.

Straight Line

The equation $y = mx + b$ over some domain of x, such as $0 < x < 10$, is a linear function. If we plot representative pairs of points, we obtain a graph representing the function as shown in Figure C-1. The constant m is the slope

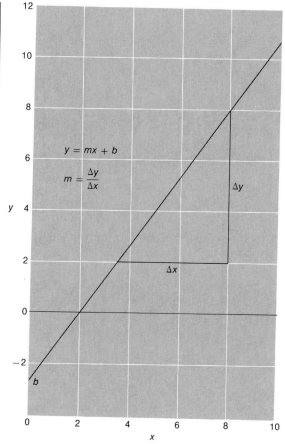

FIGURE C-1
The graph of a straight line $y = mx + b$. The slope of the line is given by $m = \dfrac{\Delta y}{\Delta x}$.

of the graph, or the rate of change of the function y with respect to x:

$$m = \frac{\Delta y}{\Delta x}$$

The constant b is the y-intercept of the function at $x = 0$. The x-intercept at $y = 0$ is $x = -b/m$. The equation of the straight line is characterized by the slope m and the y-intercept b. The simplest straight line has the y-intercept $b = 0$.

$$y = mx$$

Quadratic

Equations containing x^2 are considered quadratic. The equation $y = ax^2 + c$ over some domain, such as $-3 < x < 3$, is a parabola, as shown in Figure C-2. Such a function describes the trajectory of a ball when air

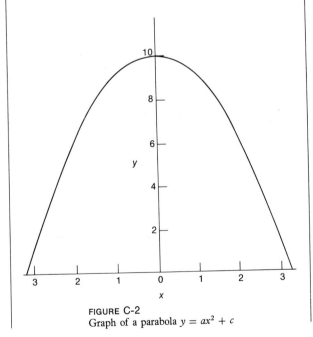

FIGURE C-2
Graph of a parabola $y = ax^2 + c$

resistance is neglected. Because this graph goes negative for large values of x, the constant a must also be negative. The graph has two x-intercepts, where $y = 0$. These are the roots of the equation.

The parabola is a special case of a more general quadratic function $y = ax^2 + bx + c$. The roots of the general quadratic equation, the x-intercepts, are given by the quadratic equation discussed in Appendix A.

TRIGONOMETRIC FUNCTIONS

The trigonometric relationships $\sin x$ and $\cos x$ and $\tan x$ can be treated as functions over some domain, where x may be in radian measure or in angle. Figure C-3 shows the sine, cosine, and tangent functions over the domain $0 < x < 3\pi$. Many wave and vibration phenomena can be characterized by sine and cosine functions. The amplitude of a standing wave of a vibrating string can be described as a sine function in space and as a cosine function in time:

$$y = y_0 \sin(2\pi x/L) \cos(2\pi f t)$$

where L is the length of the string and f is the frequency. Sine and cosine functions arise whenever a system has an acceleration that is opposite to and proportional to the system's displacement.

EXPONENTIAL FUNCTIONS

Exponential relationships arise when the rate of change of a quantity is proportional to the quantity present. For example, exponential relationships apply to compounded interest on savings accounts, inflation in the cost of living, the growth of bacteria cultures, the discharge of a capacitor, radioactive half-life, and penetration of radiation through matter. Positive exponential functions increase at an ever-increasing rate:

$$y = y_0 e^{+kx}$$

Negative exponentials decrease at an ever-decreasing rate:

$$y = y_o e^{-kx}$$

Figure C-4 shows increasing and decreasing exponential functions.

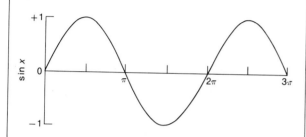

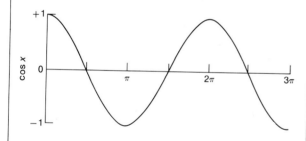

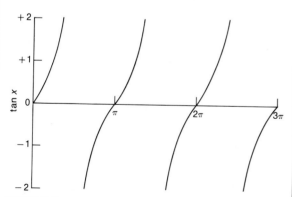

FIGURE C-3
Graphs of the sine, cosine, and tangent functions over a domain $0 < x < 3\pi$. A. The sine function $y = \sin x$. B. The cosine function $y = \cos x$. C. The tangent function $y = \tan x$.

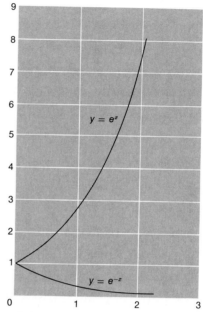

FIGURE C-4
Graph of the exponential functions $y = e^x$ and $y = e^{-x}$.

LOGARITHMS

The inverse of the exponential function is the logarithm (as discussed previously). Common logarithms are based on powers of ten:

$$z = 10^y; \text{ or } y = \log z$$

The natural logarithms are based on powers of $e = 2.718\ldots$

$$z = e^y; \text{ or } y = \log_e z = \ln e$$

We can treat the logarithm as a function if we specify the domain of z for positive values of z. Many phenomena are described by a logarithmic function, including the measure of acidity (pH), the intensity of sound in decibels (dB), stellar magnitudes, and the electrical potential from a line source. The logarithmic function is often used when a quantity changes over a large range of values. For instance, the intensity of sound can range from 10^{-12} W/m², at the threshold of hearing, to 1 W/m² or greater, and is described by the decibel scale, which is proportional to the logarithm (base 10) of the sound intensity.

Semi-Log Graphs

The logarithm is the inverse of the exponential. When an exponential relationship between physical quantities is anticipated, it is often useful to plot the logarithm of the dependent variable as a function of the independent variable. This is a **semilog** or log-linear graph. Suppose we anticipate that the decay of voltage across a capacitor is some function

$$V = V_0 e^{-t/RC}$$

a decaying exponential in time. If we take the logarithm (base 10) of each side, we obtain

$$\log V = \log(V_0 e^{-t/RC})$$

$$\log V = \log V_0 - (t/RC)\log e$$

where $\log e = 0.4343$. If we plot $\log V$ as a function of time t, we obtain the equation of a straight line:

$$\log V = \log V_0 - (0.4343/RC)t$$

The $\log V$ intercept is the constant $\log V_0$. The slope of the graph is $(-0.4343/RC)$. Figure C-5 A shows the linear plot of voltage as a function of time. When the log voltage is plotted, as shown in Figure C-5 B, the graph is a straight line, indicating exponential decay. Because

semilogarithmic plots are so common, special semilog graph paper is often used (see Figure C-5 C). The value of V is plotted against a logarithmic scale that automatically gives the logarithm of the number.

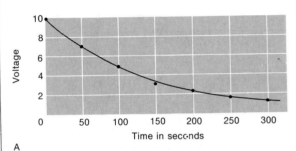

A

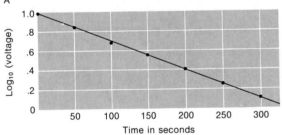

B

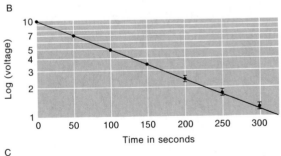

C

FIGURE C-5
Exponential decay of voltage as a function of time. A. Linear graph. B. Log-linear graph; log (voltage) as a function of time. C. Log-linear graph on semi-log graph paper.

Power Laws

When the relationship between physical quantities depends on a power law such as

$$y = ax^n$$

a plot of $\log y$ as a function of $\log x$ will be a straight line. Taking the logarithm of both sides, we get

$$\log y = \log (ax^n)$$

$$\log y = \log a + n \log x$$

The intercept $\log x = 0$ is the value of the constant $\log a$. The slope of the graph is the power n. Such a graph, known as log-log plot, is a valuable experimental technique for determining functional relationships between physical quantities. Figure C-6 shows the linear and log-log plots of several power laws.

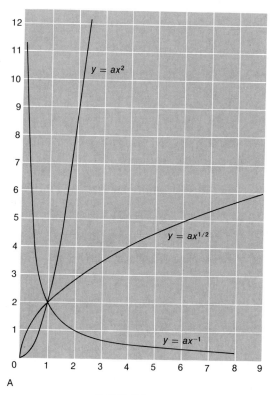

A

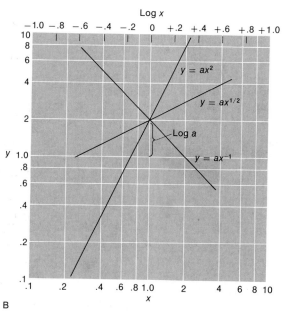

B

FIGURE C-6

Power laws $y = ax^n$. A. Linear graph. B. Log-log graph; $\log y$ as a function of $\log x$.

Appendix D

Calculators

"It hurts when I do square roots."

FIGURE D-1
Drawing by Irwin Caplan: © 1960 Look Magazine.

Widespread use of pocket calculators in recent years has revolutionized how we handle numbers. On one hand, we have become spoiled (Who among us can balance a checkbook without resort to a calculator?); on the other hand, we have available a great deal of numerical information and computational power.

The arithmetic calculator has at least four functions— $+$, $-$, $\times$, and $\div$—and may have one or more memory locations. Basic scientific calculators generally include the trig functions sin, cos, tan and their inverses; logarithms, exponentials, $1/x$, x^2, and perhaps π, y^x, as well as several storage locations for intermediate results. Advanced scientific calculators can be programed and have special function keys such as polar coordinates and standard deviation; they are particularly useful for repetitive calculations.

SETTING UP PROBLEMS

The most important consideration in using any calculator to solve scientific problems is to *set up the problem clearly*. It is hopeless to start punching numbers into the calculator, trying to hit upon the correct result. *Think the problem out carefully on paper first*. Systematic step by step problem solving is very important:

1. First read the text

2. Read the problem carefully and picture the situation.

3. Draw a sketch.

4. List the given information on the sketch.

5. Determine the unknown.

6. Determine the relevant equation.

7. Solve algebraically and insert the numbers.

Once the proper equation or equations have been determined and solved by algebra, we can write in the numerical values. We should verify that the units work out properly. Then and only then should we pick up the calculator to do the arithmetic. We have several options at this point.

(a) We can do longhand multiplication on paper and use tables for trig functions and logarithms.

(b) We can use a slide rule and tables with pencil and paper for intermediate results.

(c) We can use an arithmetic calculator with tables.

(d) We can use a scientific calculator.

The basic procedure for setting up the problem remains the same.

Significant Figures

Measurement of a given physical quantity is limited in precision (as was discussed in Chapter 1). In a given measurement, the last significant digit is usually estimated, and is a rough indication of the precision of the measurement.

When two three-digit numbers are multiplied, the result will have five or six digits:

$$123 \times 987 = 121\ 401$$

If the original numbers have only three significant figures, then the result should be rounded back to three significant figures. The matter is even worse with division: If 20.0 is divided by 70.0, each with three significant figures, the result is

$$\frac{20.0}{70.0} = 0.285\ 714\ 285\ 714\ldots.$$

which is a repeating decimal limited only by the capacity of our calculator. All rules concerning significant figures *must* be applied when using the calculator. The numerical quantities used to evaluate equations represent measure-

ments with limited accuracy. Before being written down, final results must be rounded back to the proper number of significant figures.

EXAMPLE

An automobile is traveling at $v_o = 30$ m/s when its brakes are suddenly applied and it decelerates at $a = -5.0$ m/s^2 for a distance of $s = 15$ m. How long does it take to travel the distance? The knowns are v_i, a, and s. The unknown is the time t. The correct relationship is:

$$s = v_o t + \tfrac{1}{2}at^2$$

This can be put in the quadratic form:

$$\tfrac{1}{2}at^2 + v_o t - s = 0$$

The solution is given by:

$$t = \frac{-v_o + \sqrt{(v_o{}^2 + 4(\tfrac{1}{2}a)(s)}}{2(\tfrac{1}{2}a)}$$

where only the positive root has been retained. This simplifies to:

$$t = \frac{-v_o + \sqrt{v_o{}^2 + 2as}}{a}$$

We can now write in the numbers and evaluate the expression.

CALCULATOR LOGIC

There are two basic logic systems for scientific calculators: the algebraic notation, and the reverse Polish notation. People have their staunchly defended preferences, but each is equally useful.

Algebraic Notation

The algebraic entry notation, or "equals" notation, allows the problem to be keyed as it would be stated in algebra. Depending on the sophistication of the calculator, intermediate results must be written on a piece of paper and reentered, or must be stored in a memory location and recalled, or are automatically stored and retrieved. If the calculator has a sum-of-products register, the first product can be stored while a second is being calculated. Evaluating complex expressions usually requires working from the inside out. Texas Instruments and most other calculator manufacturers except Hewlett Packard use the algebraic notation. Table D-1 shows the entry, operation, and display for evaluating the example above.

Some algebraic calculators have built-in parentheses so that the problem can be keyed as it would be stated in a computer language such as Fortran or Basic. In sophisticated calculators, the parentheses may be nested four deep. In parenthetical notation (see Table D-2), our previous example would be solved by stating the problem as:

$$(-v_o + (v_o^2 + 2as)^{1/2}) \div a = t$$

TABLE D-1
Algebraic entry notation (10 operations).

Quantity	Entry	Operation	Display	Quantity
v_o	30	x^2	900	v_o^2
		+	900	
	2	×	2	
a	−5	×	−10	$2a$
s	15	=	750	$v_o^2 + 2as$
		$\sqrt{\ }$	27.386	
		−	27.386	
v_o	30	=	−2.614	
		÷		
a	−5	=	0.5227	t

TABLE D-2
Parenthetical algebraic notation (12 operations).

Quantity	Entry	Operation	Display	Quantity
		(	0	
$-v_o$	−30	+	−30	
		(	−30	
v_o	30	x^2	900	
		+	900	
	2	×	2	
a	−5	×	−10	
s	15	)	750	
		$\sqrt{\ }$	27.386	
		)	−2.614	
		÷	−2.614	
a	−5	=	0.5227	$= t$ answer

Reverse Polish Notation

The reverse Polish notation, RPN, was named after its inventor, the Polish mathematician Jan Lukasiewicz. RPN, or "enter" notation, performs operations on two numbers after they have been entered. Calculators using RPN store intermediate results in a stack, or series of registers, where they are automatically available for further use. RPN is used principally in Hewlett Packard pocket calculators and in those of a few other manufacturers.

There are four steps to reverse Polish notation.

1. Start at the left side of the problem and key in the first number or next number.

2. Determine if any operations are possible. If so, do all operations possible.

3. If no operation is possible, press *Enter* to save the number for future use.

4. Repeat steps 1, 2, and 3 until the calculation is completed.

Table D-3 shows the entry, operation, and display for a calculator using reverse Polish notation.

TABLE D-3
Reverse Polish notation (10 operations).

Quantity	Entry	Operation	Display	Quantity
$-v_o$	-30	Enter	-30	
		x^2	900	
		Enter	900	
	2	Enter	2	
a	-5	$\times$	-10	
s	15	$\times$	-150	
		$+$	750	
		$\sqrt{}$	27.386	
		$+$	-2.614	
a	-5	$\div$	0.523	t answer

Appendix E

Uncertainty in Measurement

In any measurement there is uncertainty, which can be analyzed.

STANDARD DEVIATION

In any set of repeated measurements of some quantity, such as length, there is some deviation in the data. The following represents 10 measurements of a length:

92.3 cm	92.4 cm	92.3 cm	92.5 cm	92.3 cm
92.4 cm	92.4 cm	92.4 cm	92.6 cm	92.5 cm

Most of them lie between 92.3 and 92.5 cm, or 92.4 ± 0.1 cm. This is an estimate of the uncertainty.

These measurements, $(x_1, x_2, \ldots, x_{10})$ can be represented as a set of measurements $\{x_i\}$ where $i = 1$ to 10. The *mean value* of a set of N measurements is given by:

$$\bar{x} = \frac{1}{N} \sum_{i=1}^{N} x_i = 92.41 \text{ cm}$$

The *square deviation* of the measurement is given by:

$$\overline{\delta x^2} = \sum_{i=1}^{N} (x_i - \bar{x})^2 = \sum_{i=1}^{N} (\Delta x_i)^2$$

The square deviation can also be written:

$$\overline{\delta x^2} = \sum_{i=1}^{N} (x_i^2 - 2x_i\bar{x} + \bar{x}^2)$$

$$= \sum x_i^2 - 2\bar{x} \sum x_i + \bar{x}^2 \sum 1$$

The various sums are:

$$\sum x_i^2 = x^2 \qquad \sum x_i = N\bar{x} \qquad \sum 1 = N$$

The square deviation can then be written:

$$\overline{\delta x^2} = \bar{x}^2 - N\bar{x}^2$$

For the data sample, $N = 10$, $\bar{x} = 92.41$, and $\bar{x}^2 = 85\,396.17$.

The square deviation is:

$$\overline{\delta x^2} = 0.089 \text{ cm}^2$$

From a series of N measurements, we can estimate the mean square deviation of an infinite series of measurements. This is the *variance* of our sample:

$$\sigma^2 = \frac{1}{N-1} \sum (\overline{\delta x^2}) = \frac{1}{N-1} \sum_{i=1}^{N} (x_i - \bar{x})^2$$

The variance of our sample is

$$\sigma^2 = \frac{0.089 \text{ cm}^2}{10 - 1} = 0.0099$$

The *standard deviation* is just the square root of the variance:

$$\sigma = \left[\frac{1}{N-1} \sum_{i=1}^{N} (x_i - \bar{x})^2\right]^{1/2}$$

$$\sigma = 0.10 \text{ cm}$$

If a set of values $\{x_i\}$ has a normal distribution, then 68% of our data should lie within one standard deviation $\pm\sigma$. The normal distribution has the shape (see Figure E-1):

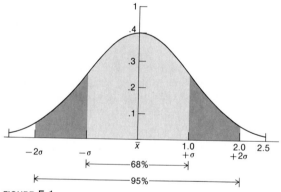

Normal probability distribution: 68% of the values lie within $\pm\sigma$; 95% lie within $\pm 2\sigma$.

$$f(x - \bar{x}) = \frac{1}{\sqrt{2\pi}} e^{\frac{-(x - \bar{x})^2}{2\sigma^2}}$$

95% of the data should fall within two standard deviations $\pm 2\sigma$.

The average value is better defined than any of the values contributing to it. The *standard deviation of the mean* is given by:

$$\sigma_{\bar{x}} = \frac{\sigma}{\sqrt{N}}$$

For our data set $\sigma_{\bar{x}} = \frac{0.10}{\sqrt{10}} = 0.03$. The mean value is

$$\bar{x} \pm \sigma_{\bar{x}} = 92.14 \pm 0.03 \text{ cm}.$$

PROPAGATION OF ERRORS

The uncertainty of a measurement can be estimated to be about the same size as the finest division on the scale of the instrument. The smallest division on a meter stick is 0.1 cm. If we measure a distance with a meter stick, then the uncertainty is ± 0.1 cm, or perhaps ± 0.05 cm if we are careful. If we determine the average of many measurements, then the standard deviation is perhaps ± 0.03 cm. When we calculate a quantity from measured values, we must propagate our estimated uncertainties to the final result. The following simple procedure is recommended for ordinary laboratory work.

When measurements are added or subtracted, make the largest uncertainty in the measurements the uncertainty of the result.

EXAMPLE

If $x_1 = 23.4 \pm 0.1$ cm and $x_2 = 76.5 \pm 0.2$ cm, then the difference is

$$\Delta x = x_2 - x_1 = 53.1 \pm 0.2 \text{ cm}$$

since 0.2 cm is the largest uncertainty.

When measurements are multiplied or divided, the largest fractional uncertainty in the measurements becomes the fractional uncertainty of the result.

We may wish to calculate the gravitational potential energy $U = mgh$, where $m = 0.1000 \pm 0.0005$ kg, $g = 9.80$ m/s^2 exactly, and $h = 0.703 \pm 0.002$ m. The fractional uncertainties are respectively:

$$\delta m/m = 0.0005 \text{ kg}/0.100 \text{ kg} = 5 \times 10^{-3}$$

$$\delta g/g = 0$$

$$\delta h/h = 0.002 \text{ m}/0.703 \text{ m} = 2.8 \times 10^{-3}$$

The potential energy is $U = mgh = 0.6894$ joules. The largest fractional uncertainty is

$$\delta U/U = 5 \times 10^{-3} \text{ or } \delta U = 0.00345 \text{ joules}$$

To three significant figures, the result is

$$U = 0.689 \pm 0.003 \text{ joules}$$

If a number enters a calculation as a square, its fractional uncertainty is doubled. For more complicated expressions such as $\cos \theta$, it is perhaps easiest to calculate extreme values. If $\theta = 10 \pm 2°$, then $\cos 8° = 0.990$ and $\cos 12° = 0.978$; $\cos \theta = 0.985 \pm 0.006$.

If several quantities entering a calculation have comparable uncertainties, then the uncertainty must be added in **quadrature:** The resulting uncertainty is the square root of the sum of the squares of the other uncertainties. Addition in quadrature is a statistical result based on the assumption that errors in different quantities are statistically unrelated. If we are adding or subtracting two quantities, then the uncertainties add as squares.

EXAMPLE

The uncertainty in $\Delta x = x_2 - x_1$ is given by

$$\delta x = (\delta x_1{}^2 + \delta x_2{}^2)^{1/2}$$

$$\delta x = ((0.1 \text{ cm})^2 + (0.2 \text{ cm})^2)^{1/2} = (0.05 \text{ cm}^2)^{1/2}$$

$$\delta x = 0.22 \text{ cm}.$$

This is very close to the largest uncertainty, 0.2 cm.

In the case of multiplication or division, the fractional uncertainties add in quadrature.

EXAMPLE

The uncertainty in $U = mgh$ is

$$\delta U/U = ((\delta m/m)^2 + (\delta g/g)^2 + (\delta h/h)^2)^{1/2}$$

$$\delta U/U = ((5 \times 10^{-3})^2 + 0 + (2.8 \times 10^{-3})^2)^{1/2}$$

$$\delta U/U = (3.28 \times 10^{-5})^{1/2} = 5.7 \times 10^{-3}$$

This is close to the largest fractional uncertainty.

Neglecting all but the largest uncertainties in an experiment slightly underestimates the experimental error.

Uncertainty and Discrepancy

The distinction must be made between **uncertainty** in a result and a **discrepancy** in a result. The final result in an experiment should agree with the predicted, or theoretical, value within the experimental uncertainty. By estimating the uncertainties in each measurement, we should be able to account for all sources of uncertainty in our procedure. If the final result and the predicted result disagree, there is a **discrepancy.** The discrepancy is caused by **error** in the experiment. There could be systematic errors because of improper calibration or misuse of instruments; or experimental errors because of improper experimental technique. Computational errors and blunders do not count, for they should be found and corrected at once. An experiment that has very small uncertainty in the result is said to have high **precision.** But a precise experiment can still give the wrong answer because of other systematic errors. An experiment that has very small systematic errors is said to have high **accuracy.**

COUNTING STATISTICS

Statistical fluctuations occur in counting random events such as radioactive decay. Although probability that a particular radioactive nucleus will decay is very, very small, the number of nuclei that might decay is very, very large. On the average, a certain number a will decay in a unit of time. The probability of observing n counts in a period of time is given by a Poisson distribution:

$$P(n) = \frac{a^n e^{-n}}{n!}$$

For events described by such a distribution, the expected standard deviation is equal to the square root of the number expected

$$\sigma = a^{1/2}$$

When such random events as radioactive decay are counted, there is considerable scatter in the data. It is useful to take the square root of the number of observed counts to verify that it is consistent with the scatter. If 100 counts are observed, the uncertainty will be ± 10, or a fractional uncertainty of 0.10. If 900 counts are observed, the uncertainty will be ± 30, or a fractional uncertainty of 0.033. If 2500 counts are observed, the uncertainty will be ± 50 and the fractional uncertainty is 0.02. In counting experiments, a large count yields a more precise result.

Further Readings

Hugh D. Young, *Statistical Treatment of Experimental Data* (New York: McGraw-Hill, 1962). A comprehensive introduction.

Answers to
Odd-Numbered Problems

Chapter 1 / Space and Time

1 (a) 1.1518×10^{12} $ (dollars);
  (b) 1.1518×10^{6} M$ (megadollars);
  (c) 1.1518×10^{3} G$ (gigadollars).

3 (a) $C = 28$ in, $A = 250$ in²; (b) $C = 7.98$ cm,
  $A = 5.07$ cm²; (c) $C = 60$ mils $= 0.060$ in,
  $A = 300$ mil² $= 3.0 \times 10^{-4}$ in²;
  (d) $C = 2.490 \times 10^{4}$ mi $= 4.07 \times 10^{4}$ km,
  $A = 4.934 \times 10^{7}$ mi² $= 1.278 \times 10^{8}$ km².

5 Typical suitcase is 20 cm $\times$ 50 cm $\times$ 70 cm $=$
  7×10^{4} cm³. Golf ball radius 2.2 cm. Minimum separation
  $4.4 \sin 60° = 3.8$ cm. Maximum separation 4.4 cm. Golf
  ball will fit in a cube of approximately 4 cm to a side:
  Vol $= (4 \text{ cm})^{3} = 64$ cm³. Total number of balls is of the
  order of $N = V/\text{Vol} = 7 \times 10^{4}$ cm³$/64$ cm³ $= 10^{3}$ golf
  balls, depending on the size of the suitcase.

7 How big is bucket? A cylinder 20 cm diameter and 30 cm
  high. $V = \pi r^{2}h = 9 \times 10^{3}$ cm³. How big is a grain of
  sand? $d \cong 0.05$ cm? (0.02 cm?) Volume of grain
  $V \cong d^{3} \cong 10^{-4}$ cm³ (10⁻⁵ cm³). Number of grains is
  approximately $N = 9 \times 10^{3}$ cm³$/10^{-4}$ cm³ $= 10^{8}$ grains

9 (a) How big is the balloon? (12 inches in diameter?)
  Vol $\cong 9 \times 10^{2}$ in³ $= 1.5 \times 10^{4}$ cm³

$$N = \frac{(1.5 \times 10^{4} \text{ cm}^{3})(6.0 \times 10^{23} \text{ molecules/mole})}{2.24 \times 10^{4} \text{ cm}^{3}/\text{mol}}$$

  $\cong 4 \times 10^{23}$ molecules

(b) A classroom $\cong$ 4 m $\times$ 5 m $\times$ 6 m. Vol $\cong$ 120 m³.

$$N \cong \frac{(120 \text{ m}^{3})(10^{6} \text{ cm}^{3}/\text{m}^{3})(6 \times 10^{23} \text{ molecules/mol})}{2.24 \times 10^{4} \text{ cm}^{3}/\text{mol}}$$

  $\cong 3 \times 10^{27}$ molecules.

Accuracy depends on knowing the actual size of the
classroom.

11 (a) 2.16 m; (b) 169 m; (c) 381 m; (d) 8847.7 m;
  (e) 72 803.6 m.

13 (a) 100 m $=$ 109 yd; (b) 1 mile $=$ 1609 m;
  (c) 440 yd $=$ 409 m.

15 A pace is typically 5 ft $\cong$ 1.5 m. At 90 paces/min one
  would walk about 8 km/hr. It would take about 7.5 min to
  walk 1 km, or 12 min to walk 1 mi. Your numbers may
  vary.

17 (a) 8.4×10^{16} m; (b) 5.6×10^{8} AU.

19 2 mi/(186 000 mi/s) $= 1.1 \times 10^{-5}$ s $= 11$ μs.

Chapter 2 / Motion

1 (a) Between frames 19 and 20; (b) between frames 21
  and 22; (c) $\Delta s = 0.33$ m, $\Delta t = 0.02$ s, $v = 16$ m/s.

3 (a) 1st trial: 274.5 m/s $= 988.7$ km/hr $= 617.6$ mi/hr,
  2nd trial: 282.2 m/s $= 1016.1$ km/hr $= 635.1$ mi/hr;
  (b) 2nd trial.

5 10 s.

7 4.0 min.

9 (a) 524 km; (b) 42° east of south.

11 (a) 57.4 km; (B) 36.9 km.

13 8.57 km, 44.6° south of east.

15 (a) 10 km/hr = 2.78 m/s; (b) 75 m; (c) 45 s.

17 750 km/hr, 58° west of north.

19 (a) 14.5° west of north; (b) 775 km/hr.

21 (a) 2.9 m/s²; (b) 132 m.

23 (a) 15.0 s; (b) 338 m.

25 (a) −4.4 m/s (downward); (b) 30.9 m; (c) 31.9 m.

27 (a) 1.08 m/s²; (b) 33 s.

29 52 mi/hr.

31 38 mi/hr or 6 mi/hr.

33 102 ft.

35 (a) 1st: 2.9 m/s², 2nd: 1.8 m/s², 3rd: 0.76 m/s²;
   (b) 1st: 23.2 m, 2nd: 80.5 m, 3rd: 347 m, Total: 451 m.

37 (a) 19.8 ft/s²; (b) 225 ft, 437 ft.

39 (a) 27.25 s, 412.5 m to start and to stop;
   (b) 2675 m, 88.4 s;
   (c) 3500 m/163 s = 21.5 m/s = 77.4 km/hr.

41 (a) 91.3 km/hr = 25.4 m/s;
   (b) 65.7 s, 8.83 m/s = 31.8 km/hr.

43 (a) 2.22 m/s; (b) 3.70 s.

Chapter 3 / Newton's Laws of Motion

1 (a) 2.1 m/s²; (b) 3060 N.

3 (a) 2.1 m/s²; (b) 4320 N.

5 (a) 2.4 × 10⁵ N; (b) 2260 m.

7 (a) 44 600 N; (b) 7280 N; (c) 1.8 m/s²; (d) No: Weight
   on earth is greater than thrust of engine.

9 4 m/s².

11 (a) 3.27 m/s²; (b) 2.45 N.

13 (a) 111 N; (b) 184 N.

15 (a) 130 lbf; (b) 150 lbf.

17 84 ft.

19 121 cm/s.

21 (a) 11 kg-m/s; (b) 1380 N.

23 2.17 m/s.

25 0.009 m/s = 9 mm/s.

27 (a) +3.2 kg-m/s; (b) −3.2 kg-m/s; (c) 0.2 kg.

29 (a) 4.9 m/s²; (b) 9.9 m/s.

31 2.6 m/s.

33 (a) 55.4 ft/s up, 50 ft/s horizontal, 74.6 ft/s at 48° angle;
   (b) 1.73 s; (c) 1.73 s up, 1.80 s down, distance 176.7 ft.

35 (a) 9.8 m/s²; (b) 2.21 m/s; (c) 0.25 m; (d) The mass
   cancels out.

37 (a) 104 m/s²; (b) 4.3 m.

39 (a) Yes. The ball is 62 ft above ground when it crosses the
   fence. (b) 81.5 ft/s.

Chapter 4 / Forces

1 7900 m/s.

3 (a) 1.47 hr; (b) 7.8 × 10³ m/s.

5 (a) 7.9 × 10⁷ N/m²; (b) 20%.

7 (a) If cable stretches only 0.2%, F = 2.1 × 10⁵ N;
   (b) 2.9 × 10⁵ N.

9 12.6 m/s².

11 (a) 3.05 × 10³ N = 312 kg-wt; (b) 4.11 × 10⁻³ m.

13 (a) 8.5 kg; (b) 242 kg.

15 (a) 0.10 m²; (b) 7.2 × 10⁸; (c) 90 m².

17 tan θ = 0.02, 1.14°.

19 0.76.

21 (a) 11.3°; (b) 2.5 N; (c) 1.2 m/s².

23 (a) 4436 N; (b) 3113 N.

25 1.9 × 10²⁷ kg.

27 0.82 m.

Chapter 5 / Energy

1 2350 J.

3 2.60 × 10⁵ ft-lbf = 3.5 × 10⁵ J.

5 (a) 2.35 × 10⁴ J; (b) 2.35 × 10⁴ J, 20 m/s; (c) 204 N,
   115 m, 2.35 × 10⁴ J.

7 3.8 m/s.

9 (a) 14 ft; (b) 300 ft-lbf.

11 (a) 7.8 km/s; (b) −3.2 × 10¹² J; (c) 1.6 × 10¹² J;
   (d) −1.6 × 10¹² J.

13 ΔPE = 0.84 × 10⁹ J, ΔKE = 6.6 × 10⁹ J.
   Total = 7.44 × 10⁹ J.

15 (a) 2.78 × 10⁶ J; (b) 28.9 eV.

17 (a) 5.4 × 10⁸ J; (b) 10⁷ W = 1.34 × 10⁴ hp.

19 (a) 5210 J; (b) 1110 W = 1.5 hp.

21 (a) 2.4 m in 100 m (2.4%);
   (b) 2.9 × 10⁶ J = 6.4 × 10⁵ cal = 640 kcal.

23 (a) 8.4 × 10⁵ J; (b) 1710 m.

25 (a) 1.7 × 10²¹ J; (b) 1.1 × 10¹² J; (c) 1.7 × 10²¹ J.

27 (a) 4800 W; (b) 5.2 × 10⁷ J = 52 MJ; (c) $0.078, $0.31.

29 4.5 × 10¹⁴ J/day = 1.6 × 10⁴ tonnes/day =
   5.8 × 10⁶ tonnes/year.

31 (a) 14.3 m/s = 32 mi/hr = 52 km/hr; (b) 0.17 gal/hr,
   185 mi/hr = 295 km/hr.

33 (a) 2.4 × 10⁵ J, 3.2 × 10⁴ J, 1.5 × 10⁵ J, 2.65 × 10⁴ J;
   (b) 20.5 m/s, 13.5 m/s, 20.8 m/s;
   (c) −27.0 m/s² (downward), +0.2 m/s² (upward),
   −33.4 m/s² (downward).

Chapter 6 / Relativity

1 3 × 10⁵ m, 300 m, 0.3 m.

3 2.8 × 10²⁰ m.

5 (a) 2.14 hr, 1.69 hr; (b) 3720 km; (c) [1 − v²/c²]^{1/2} =
   1 − 2.1 × 10⁻¹², 16 ns, 9.8 × 10⁻⁶ m.

7 (a) 0.28 hr; (b) 0.32 hr; (c) 6.93 km.

9 36 min.

11 (a) [1 − v²/c²]^{1/2} = 1 − 1.2 × 10⁻¹², 12 ns; (b) 64 ns;
   (c) 1.4 ns.

PART II / FORMS OF MOTION

Chapter 7 / Angular Motion

1 (a) See Figure 7-1 C; (b) $21\underline{6}0$ N-m; (c) $14\underline{4}0$ N.

3 (a) See Figure 7-21. Wall exerts only a horizontal force on ladder. Ground must exert both horizontal and vertical forces.

5 (a) See Figure 7-22; (b) $F = 1019$ N, $P = 1411$ N.

7 (a) 7.3×10^{-5}/s; (b) 465 m/s $= 16\underline{7}0$ km/hr;
  (c) 329 m/s.

9 (a) 628/s; (b) 77.2 m/s; (C) 4955 g.

11 (a) The end of the meterstick will accelerate at
  $a = L\alpha = \frac{3}{2}g$; (b) The penny will fall behind.

13 (a) 4.6 kg-m^2; (b) 9.4/s; (c) 204 J.

15 (a) 0.21 kg-m^2; (b) 524/s; (c) 2.86×10^4 J.

17 (a) 7.3×10^{-5}/s; (b) 9.8×10^{37} kg-m^2; (c) 2.6×10^{29} J.

19 (a) 0.066 m to right of center of gravity;
  (b) See Figure 7-14: $a_1 = 0.07$ m, $a_2 = 0.11$ m,
    $a_3 = 0.176$ m, $a_T = 0.33$ m;
  (c) $M = 2405$ N at $70°$, $P_x = M \cos 70 = 822$ N,
    $P_y = F - W_L + M \sin 70 = 3090$ N.

21 (a) $R = 2540$ N at $43°$ from vertical, $R_x = 1845$ N,
    $R_y = 1750$ N;
  (b) $R = 30\underline{9}0$ N at $82°$ from vertical, $R_x = 30\underline{6}0$ N,
    $R_y = 4\underline{4}0$ N.

23 (a) $\tau = mgR \sin \theta$; (b) $mgR \sin \theta = I\alpha = \frac{3}{2}mR^2\alpha$;
  (c) $a = \alpha R = 2g \sin \theta /3$.

Chapter 8 / Harmonic Motion

1 (a) 19.6 N/m; (b) 0.63 s.

3 (a) 49 N/m; (b) 0.63 s.

5 (a) 22.4 m/s (56.0 m/s); (b) 112 Hz (147 Hz).

7 For a mass 0.010 kg (a) 676 N; (b) 260 m/s, 520 m/s.

9 (a) 1.2 m for all three; (b) 264 m/s, 475 m/s, 792 m/s;
  (c) 453 N, 527 N, 376 N.

11 1530 m/s, 1498 m/s.

13 10^{-9} W/m^2.

15 10^{-4} W/m^2.

17 (a) 73 dB; (b) 10 instruments, 100 instruments.

Chapter 9 / Fluid Motion

1 (a) 294 N; (b) 14.7 m/s^2; (c) 60%.

3 (a) 18.6 N; (b) 18.2 N.

5 (a) 9.8×10^4 N/m^2; (b) 7.25×10^4 N/m^2;
  (c) 7.74×10^4 N/m^2.

7 (a) 9.6×10^5 N/m^2; (b) 4.8×10^5 N $= 1.1 \times 10^5$ lbf.

9 (a) 3.50×10^3 N/m^2; (b) 1.22×10^4 N/m$^2 = 91$ torr.

11 (a) 0.68 m/s; (b) 0.17 m/s.

13 (a) $v = 2.88$ cm/s; (b) 1105 cm/s.

15 (a) 2.06 liter/min $= 3.44 \times 10^{-5}$ m^3/s;
  (b) 108 N/m$^2 = 0.81$ torr.

17 2.5×10^3 m/s (This is unrealistically large. The flow around the golf ball is turbulent, giving a much larger drag force.)

19 (a) 0.98 J; (b) 1.14 W $= 68.6$ J/min.

21 (a) 16 J; (b) 10.7 W $= 640$ J/min.

23 (a) 6.75×10^3 N/m; (b) 1.05×10^5 N/m.

25 (a) 3.2×10^2 N/m^2; (b) 16 N/m.

27 (a) $11\underline{8}0$ liter/min; (b) 5.1 m; (c) 980 W.

29 (a) 7.9×10^{-7} m/s; (b) 1.8×10^{-3} m/s;
  (c) 1.27×10^4 s $= 3.53$ hrs, 5.62 s.

31 (a) 3.93 cm^3/s; (b) at 2 cm $A = 8.82 \times 10^{-2}$ cm^2,
    at 4 cm $A = 6.26 \times 10^{-2}$ cm^2,
    at 10 cm $A = 3.97 \times 10^{-2}$ cm^2.

PART III / THERMAL PHYSICS

Chapter 10 / Thermal Energy

1 $37.0°$C, 310.16 K.

3 $-40°$F $= -40°$C.

5 $\Delta L = 1.44 \times 10^{-2}$ m $= 1.44$ cm.

7 $\Delta h = 1.9$ cm.

9 (a) $35\underline{0}0$ cal/s; (b) 14.6 kW.

11 (a) 268 cal/s; (b) $85\,000$ cal; (c) 318 s $= 5.3$ min.

13 2.25×10^8 cal $= 9.42 \times 10^8$ J.

15 $89.7°$C.

17 (a) 7.27×10^7 cal/day; (b) $12\,000$ Btu/hr; (c) 3.52 kW.

19 (a) 137 atmo; (b) 95.7 moles.

21 (a) $31\underline{2}0$ moles; (b) 90.5 kg;
  (c) 1.13×10^5 cal $= 4.73 \times 10^5$ J.

23 (a) (N$_2$) 502 m/s, (O$_2$) 478 m/s;
  (b) 6.07×10^{-21} J for both; (c) $36\underline{5}0$ J for both.

25 (a) 2.40×10^6 cal, 1.00×10^7 J, 9.52×10^3 Btu;
  (b) 11.1¢; (c) 2.7¢.

Chapter 11 / Thermal Energy Transfer

1 (a) 0.58 m^2; (b) 14.5 W $= 3.47$ cal/s;
  (c) $12\,\underline{5}00$ cal $= 156$ gm of ice.

3 (a) 36.75 W/m^2; (b) 73.5 W/m^2.

5 (a) $39\,\underline{0}00$ W; (b) 161 W.

7 750 Btu/hr, 220 W.

9 $41.6°$C.

11 $11\underline{4}0$ W.

13 (a) 1.2×10^9 J/s;
  (b) 3.58×10^4 kg/s $= 1.29 \times 10^8$ liter/hr.

15 3.66 m.

17 5.2 watts.

Chapter 12 / Thermodynamics

1 (a) 1.00 atmosphere $= 1.0129 \times 10^5$ N/m^2; (b) $21\,\underline{2}00$ J.

3 (a) 397.9 K $= 124.8°$C; (b) 0.99 atmosphere;
  (c) 6.75×10^4 J.

5 (a) 61.5%; (b) 1.63×10^9 W; (c) 0.63×10^9 W.

Chapter 13 / Electric Charges and Fields

1 40 N.

3 5.6×10^{-11} N/C $= 5.6 \times 10^{-11}$ V/m downward.

5 1.14×10^7 m/s to the right (electric field points to left, accelerating the electron).

7 (a) 2.32×10^4 J/C $= 23.2$ kV; (b) 1.3×10^{16} V/m;
  (c) 3.7×10^{-15} J $= 2.32 \times 10^4$ eV $= 23.2$ keV.

9 (a) 1.0×10^{-5} C; (b) 3.0×10^5 V $= 300$ kV.

11 (a) 2.22×10^{-7} C; (b) 2.0×10^5 V/m.

13 (a) inner surface -10^{-9} C, outer surface $+10^{-9}$ C;
  (b) Assuming the sphere is very thin: Just inside
  $E = 3.6 \times 10^3$ V/m, just outside $E = 3.6 \times 10^3$ V/m.

15 (a) 720 V/m; (b) 0.5 m from Q_1 and 1.0 m from Q_2.

17 $E_x = 0$, $E_y = 15.6 \times 10^3$ N/C (upward).

19 (a) $\mathbf{E} = -1.1 \times 10^{10}$ V/m $\mathbf{j}$ (downward);
  (b) $\mathbf{E} = -0.82 \times 10^{10}$ V/m $\mathbf{j}$ (downward);
  (c) $\mathbf{E} = +3.99 \times 10^9$ $\mathbf{i}$ + 2.42 $\times 10^9$ $\mathbf{j}$ V/m (upward to right).

21 (a) 2.78×10^{-9} F $= 0.00278$ μF; (b) 8.33×10^{-7} C;
  (c) 3×10^6 V/m.

23 (a) 1.06×10^{-6} C; (b) 2.66×10^{-6} C.

25 11 μF (in parallel).

27 (a) 60 V each, 1800 μC, 1200 μC; (b) 50 μF;
  (c) 5.4×10^{-2} J, 3.6×10^{-2} J.

29 (a) 7.07 kV; (b) 0.113 C.

Chapter 14 / Electric Currents

1 (a) 4.32×10^5 C; (b) 2.7×10^{24} electrons.

3 (a) 2.9×10^8 C; (b) 2.9×10^5 s $= 80$ hrs $= 3.3$ days.

5 (a) 4.5×10^{-4} m/s; (b) 4.4×10^4 s $= 12.3$ hrs.

7 (a) 3.5×10^5 C; (b) 4.1×10^6 J $= 4.1$ MJ.

9 (a) 2.10×10^5 C; (b) 2.52×10^6 J $= 2.52$ MJ, 300 W.

11 3.2 W.

13 29 Ω.

15 (a) 0.104 Ω; (b) 0.170 Ω.

17 (a) 1.67 A; (b) 72 Ω.

19 6.25 Ω.

21 (a) 15 V; (b) 120 V. The voltage drop across the other sockets is zero because no current flows.

23 (a) 6 Ω; (b) 2 A; (c) 0.67 A.

25 (a) 2 A; (b) 0.5 V.

27 300 Ω resistance in series. Total resistance 400 Ω.

29 A shunt with resistance 0.025 Ω $= 25$ mΩ.

31 Solution 1:

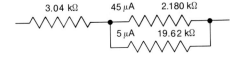

Solution 2:

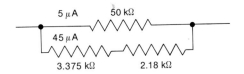

33 (a) 6.70×10^4 J; (b) 280 W; (c) 2.32 A; (d) 52 Ω.

35 (a) As in Figure 14-28; (b) 600 μA, 6.0 mA, 60 mA, 600 mA.

37 (a) -85 mV (interior negative); (b) $I_{\mathrm{L}} = 85$ nA (in), $I_{\mathrm{K}} = 85$ nA (out), $I_{\mathrm{Cl}} \cong 0$.

Chapter 15 / Magnetic Fields

1 1.13×10^{-2} N.

3 39 μA.

5 0.15 N.

7 (a) 0.245 A; (b) 0.49 kg-wt $= 4.8$ N.

9 (a) 1.58 T; (b) 1.12 T.

11 (a) 3.11×10^{-2} T; (b) (electrons are negative) left $= B$ down, up $= B$ left.

13 (a) 4.0×10^{-3} T; (b) 4×10^{-4} T; (c) counterclockwise.

15 6.23 mA.

17 (a) east; (b) 1.41×10^{-4} N.

19 9.4 m, to the left and down at a 20° angle to horizontal.

21 (a) 0.90 V; (b) 9 mA; (c) 2.7×10^{-2} N.

Chapter 16 / Electrical Signals

1 (a) 16.7×10^{-3} s $= 16.7$ ms; (b) 170 V peak.

3 (a) 10.9 Ω; (b) 15.6 A peak, 11 A rms; (c) 15.6 A peak, 11 A rms. Current is independent of frequency.

5 (a) T $= 200$ μs, f $= 5.0 \times 10^3$ Hz; (b) 6.2 V peak.

7 (a) 1.2×10^6 Ω $= 1.2$ MΩ; (b) 1.2 MΩ; (c) capacitive; (d) 100 μA rms, $\phi = 90°$.

9 (a) $X_c = 1.326$ MΩ, $Z = 1.330$ MΩ; (b) capacitive; (c) 127 μA peak $= 90$ μA rms, $\phi = 85.7°$.

11 (a) 137 s; (b) 13.7 MΩ.

13 (a) 10^{-3} s $= 1.0$ ms; (b) after several cycles it reaches a maximum of 0.73 and a minimum of 0.27.

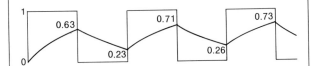

15 $Z = 1.88$ MΩ, $X_c = 1.59$ MΩ, $R = 1$ MΩ, $V = 51$ μV.

17 (a) $V_{\mathrm{Cl}} = -90.8$ mV, $V_{\mathrm{K}} = -107.5$ mV, $V_{\mathrm{Na}} = +34.8$ mV; (b) -90.8 mV, inside negative.

Chapter 17 / Electric Power and Safety

1 30 V peak.

3 (a) 177 V av; (b) 5.63 N-m; (c) 1.77 kW.

5 (A) 17 000 turns; (b) 2.0 A rms.

7 (a) $8.64; (b) $1.30; (c) $0.144.

9 (a) 0.072 Ω; (b) between 2 and 3 gauge;
   (c) The resistivity of aluminum is greater; (d) 24 kW, 720 W.

11 (a) 6 mA; (b) painful but not life-threatening.

13 (a) 120 mA; (b) serious possibility of ventricular fibrillation, life-threatening.

15 5 mV.

17 (a) 40 kW; (b) 1.4 kV; (c) 28 A.

19 75 mA. It can be life-threatening; current path through heart.

21 $X_c = 120$ kΩ, $C = 0.022$ μF.

23 (a) The safety ground is not solidly connected to ground. The safety ground is broken. (b) The neutral (or safety ground!) is hot. The wiring is crossed.
   (c) There is a resistive path between one side of the power line and the case (usually a short in a motor coil). It is potentially very dangerous.

Chapter 18 / Electromagnetic Waves

1 For $f = 2$ MHz and $C = 6.3$ pF, $L = 1.0 \times 10^{-3}$ H.

3 2.9 m.

5 1.7 m.

7 0.131 m.

PART V / OPTICS

Chapter 19 / Geometrical Optics

1 41.7° from vertical.

3 (a) above the line.

5 There is refraction at both surfaces.

7 1.503 m + 0.017 m = 1.520 m.

9 (a) 24.2°; (b) Light entering the diamond is internally reflected many times.

11 There will be five images at the same distance from the intersection separated by 60 degrees.

13 First deviation = $180 - 2\theta$. Second angle of incidence = $(90 - \theta)$. Second deviation = $180 - 2[90 - \theta] = +2\theta$. Total deviation = $(180 - 2\theta) + 2\theta = 180°$.

15 0.175 m.

17 (a) 52.6 mm; (b) 5.26 mm.

19 (a) 126.6 mm; (b) 1.90 m × 2.84 m.

21 50 cm.

23 13.3 diopters.

25 −10 cm (behind the lens).

27 (a) 2.27 cm; (b) 2.08 cm.

29 (a) 0.0952 m; (b) 21 ×; (c) See Figure 19-23.

31 (a) 4.09 mm; (b) 43 ×.

33 (a) 4.5 m; (b) 4.5 m + 0.027 m = 4.527 m.

35 (a) 4.6°; (b) 2.75 m.

37 12.7 ×.

39 58.3 ×.

41 (a) 3.85 mm; (b) −3.82 mm (behind the cornea), 0.42 mm.

43 (a) myopic (nearsighted); (b) 20.83 mm (48 diopters).

45 (a) −5 diopters (f = −0.20 m); (b) 8.6 cm.

47 Power of lens accommodates 4 diopters—from 70 to 74 diopters.

49 0.343 m.

51 (a) $\theta_i = 48.48°$; (b) $\theta_1 = 30.13°$, $\theta_2 = 29.87°$, $\theta_3 = 48.00°$, $\theta_D = \theta_i + \theta_3 - 60° = 36.48°$;
   (c) $\theta_1 = 29.73°$, $\theta_2 = 30.27°$, $\theta_3 = 49.53°$, $\theta_D = 38.01°$.

53 (a) 0.883 m; (b) 0.0454 m; (c) 0.319 m.

55 (a) −8.58 cm; (b) virtual.

57 (a) 0.79 mm; (b) 7.92 mm.

Chapter 20 / Wave Optics

1 (a)

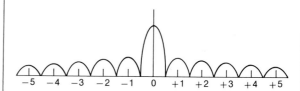

   (b) $x_1 = 19$ mm; $x_2 = 38$ mm; $x_3 = 57$ mm; $x_4 = 76$ mm; $x_5 = 95$ mm.

3 (a) from the lower spherical surface;
   (b) $2\Delta s = (n + \frac{1}{2})\lambda$; (c) $r_n = \sqrt{(n + \frac{1}{2})\lambda R} = 1.27$ mm.

5 0° max, 11.5° min, 23.5° max, 36.9° min, 53.0° max, 90° min. The waves from the two slits produce an interference pattern. Each slit acts as a point source of waves.

7 45°.

9 (a) 983 nm; (b) 437 nm.

11 (a) 1.96 μm; (b) $\theta_0 = 0$, $\theta_1 = 19.1°$, $\theta_2 = 41.0°$, $\theta_3 = 79.8°$.

13 47 μm.

15 (a) 18.5 mm; (b) 4.15 mm; (c) 25 mm.

17 (a) $0.26'' = 1.24 \times 10^{-7}$ radians; (b) 47 m.

19 0.48 μm.

21 (a) 386 nm; (b) 5.84×10^{14} Hz (same as in air).

23 $I_0/2$.

25 57°.

27 (a) 53°; (b) horizontal.

1 121.6 nm, 102.6 nm, 97.3 nm, 95.0 nm.

3 (a) 4; (b) $n = 2, m = 6, \lambda = 410.1$ nm; (c) $n = 2$,
$m = 7, \lambda = 397.1$ nm.

5 (a) 36$\underline{1}$0 K; (b) 41$\underline{3}$0 K; (c) 48$\underline{2}$0 K; (d) 56$\underline{7}$0 K.

7 (a) 10.3 μm; (b) 0.12 eV; (c) 0.036 eV.

9 (a) $I = \dfrac{a_\nu \, 2\pi\nu^2 \, kT\Delta\nu}{c^2} = 4.63 \times 10^{-9}$ W/m² = 4.6 mW/km²,

   (b) $T' = a_\nu T = 128$ K.

11 4.74×10^{14} Hz; (b) 1.96 eV.

13 (a) 1.72 eV; (b) 1.03 eV.

15 (a) 295 nm; (b) 2.0 eV.

17 2.1×10^{17} photons/s.

19 (a) Plot is a straight line with slope $\dfrac{\Delta V}{\Delta\nu} = \dfrac{2.56 \text{ V}}{0.631 \times 10^{15}/\text{s}}$

   $= 4.06 \times 10^{-15}$ V-s

   (b) Intercept $\nu_o = 0.31 \times 10^{15}$ Hz, $W = h\nu_o = 1.28$ V.

   (c) $h = e\dfrac{\Delta V}{\Delta\nu} = 6.49 \times 10^{-34}$ J-s.

21 1.8×10^{12} photons, 6.8×10^6 photons.

PART VI / THE ATOM

Chapter 22 / The Quantum Atom

1 1.6×10^{-3} T.

3 5.0×10^4 N/C = 5.0×10^4 V/m.

5 (a) 2.79×10^7 m/s; (b) 1.86×10^{11} C/kg.

7 2.3 MeV.

9 (a) 1.6×10^7 m/s; (b) 5.0×10^{-14} m.

11 (a) 10.2 eV, 12.1 eV, 12.8 eV, 13.1 eV; (b) 122 nm,
103 nm, 97 nm, 95 nm.

13 He⁺, 54.4 eV/ion = 5.24 MJ/mol; Li⁺⁺,
122.4 eV/ion = 11.8 MJ/mol.

15 (a) 5.8×10^{14} Hz; (b) 516 nm; (c) In ground state
$n = 1$ the electron would orbit at a frequency of
4.65×10^{15} Hz and classically radiate at wavelength
64.5 nm. The first Lyman line $m = 2$ to $n = 1$ has a
wavelength 122 nm.

17 $\dfrac{1}{\lambda} = R_H\left(\dfrac{m'}{m_e}\right)$;

   $R_H = \dfrac{2\pi^2 m k_e^2 e^4}{h^3 c} = \dfrac{2\pi^2(8.9876 \times 10^9)^2(1.602 \times 10^{-19})^4}{(6.626 \times 10^{-34})^3 \ (2.9979 \times 10^8)} m^{-1}$
   $= 10\,969\,589 \ m^{-1}$. This problem should be worked to five
   significant figures. For hydrogen $\dfrac{m'}{m_e} = \dfrac{1}{1.000\,5498}$,

   $\lambda = 121.61$ nm; for tritium $\dfrac{m'}{m_e} = \dfrac{1}{1.000\,1833}$,

   $\lambda = 121.37$ nm.

19 31.9°, 38.8°.

21 0.283 nm.

23 (a) 19 mm; (b) 16.5 mm, 19 mm, 33 mm, 38 mm;
   (c) 13.4 mm, 17.0 mm, 19 mm, 26.9 mm ($1/\sqrt{2}$, $2/\sqrt{5}$,
   1, $\sqrt{2}$).

25 0.031 nm.

27 6.375 keV.

29 (a) 0.03342 nm (37.1 keV); (b) 2.9 keV;
   (c) 2.9×10^{-23} J-s.

31 (a) 0.50 m; (b) 10 W/m².

33 (a) 6.28×10^{-4} W; (b) 3.56×10^4 W/m²;
   (c) 3.56×10^4 J/m², about one-fifth as much energy.

Chapter 23 / Wave Mechanics

1 6.86×10^{-12} m.

3 (a) 6.2×10^{-11} m; (b) 8.7×10^{-12} m; (c) 2.0×10^{-13} m.

5 2.8×10^{-33} m.

7 (a) 0.309×10^{-11} m; (b) 2.16×10^{-23} J-s/m;
   (c) 6.66×10^{-34} J-s.

9 (a) Standing waves as in Figure 23-3. $\lambda_1 = 0.2$ nm,
   $\lambda_2 = 0.1$ nm, $\lambda_3 = 0.67$ nm;
   (b) 3.7 eV, 151 eV, 340 eV.

11 (a)

13 Phosphorus $(1s)^2(2s)^2(2p)^6(3s)^2(3p)^3$
   Sulfur $(1s)^2(2s)^2(2p)^6(3s)^2(3p)^4$

15 (a) $\Delta p \geq \dfrac{1}{2}\dfrac{h}{2\pi\Delta x} = 5.3 \times 10^{-25}$ J-s/m;
   (b) 1.53×10^{-19} J = 0.95 eV.

17 (a) See Figure 23-8. $n = 1$, $m = 1, 2, 3$; (b) 63.6 Hz,
   85.2 Hz, 106 Hz.

19 (a) 8.84×10^{-20} J-s/m; (b) $\Delta p = \pm 3.64 \times 10^{-21}$ J-s/m;
   (c) N_o 1.8×10^{-24} J-s $\gg \dfrac{1}{2}\dfrac{h}{2\pi}$.

21 6.5×10^{-34} J-s $> \dfrac{1}{2}\dfrac{h}{2\pi} = 5.9 \times 10^{-35}$ J-s; Mössbauer
   effect is almost limited by Heisenberg uncertainty.

PART VII / THE NUCLEUS

Chapter 24 / Fission

1 3.7×10^7/s, 1.6×10^7/s, 4×10^6/s.

3 (a) $4.84 \times 10^{-2}/\text{day} = 5.60 \times 10^{-7}/\text{s}$;
(b) $4.85 \times 10^{4}/\text{day} = 0.56/\text{s}$; (c) 0.71×10^{6}.

5 $6.25 \times 10^{7}/\text{s}$, $9.1 \times 10^{-4}/\text{s}$.

7 $Z = 86$, $A = 220$, radon, Rn.

9 2.51×10^{-12} J, 15.7 MeV.

11 $_1\text{H}^3$, tritium.

13 (a) $_0\text{n}^1$; (b) 0.0189 amu, 17.6 MeV.

15 (a) $_0\text{n}^1 \rightarrow {}_1\text{p}^1 + {}_{-1}\text{e}^0$; (b) 0.000839 amu = 0.78 MeV.

17 (a) $_0\text{n}^1 + {}_3\text{Li}^6 \rightarrow {}_2\text{He}^4 + {}_1\text{H}^3$, tritium; (b) 0.00514 amu, 4.8 MeV.

19 17.8 MeV. $_1\text{H}^1 + {}_3\text{Li}^7 \rightarrow {}_2{}_2\text{He}^4$. Note: Electron masses are included with atomic masses.

21 (a) 0.525 amu/56 = 8.73 MeV/nucleon;
(b) 0.000611 amu = 0.57 MeV.

23 0.00239 amu = 2.22 MeV.

25 (a) 0.215 amu; (b) 200 MeV = 3.2×10^{-11} J.

27 (a) 5.7×10^{20} J; (b) 8.1×10^{22} J; (c) 270 years.

Chapter 25 / Ionizing Radiation

1 (a) β^-; (b) x; (c) n, γ, x; (d) β^+; (e) α, β^+, β^-;
(f) α, β^+, β^-, n; (g) x, γ; (h) α; (i) α, β^+, β^-;
(j) n, γ, x; (k) β^+, β^-; (l) α; (m) β^+; (n) α, β^+, β^-;
(o) n, β^+; (p) γ; (q) β^-.

3 (a) Most of the gammas will get through; (b) unchanged.

5 1.13×10^{-8} J/cm^3.

7 (a) 8.1×10^{-11} gm; (b) $2.22 \times 10^{7}/\text{min}$;
(c) 3.7×10^{11} atoms.

9 8.1×10^{-12} A = 8.1 picoamperes (pA).

11 $1.11 \times 10^{5}/\text{s}$.

13 323 hr.

15 (a) 109 gm; (b) 30.4 rad/hr;
(c) 0.304 rad/hr = 304 mrad/hr.

17 0.315 cm^3.

19 70.4 mCi after 1 week, 12.4 mCi after 2 weeks.

21 800/min, 6.25/min (8 half-lives).

23 (a) $7.40 \times 10^{3}/\text{s}$; (b) $7.33 \times 10^{3}/\text{s}$; (c) 15%.

25 8 mrad/hr, 80 rad/hr.

Chapter 26 / Fusion

1 (a) 0.0287 amu; (b) 7.1×10^{-3}; (c) 26.7 MeV.

3 (a) $_8\text{O}^{15} \rightarrow {}_7\text{N}^{15} + {}_{+1}\text{e}^0 + {}_{-1}\text{e}^0 +$ Energy, nitrogen-15.
There is an extra electron that, with the positron, is annihilated, giving energy included in the 2.76 MeV.
(b) 15.000 105 amu.

5 (a) 0.00769 amu; (b) 4.8×10^{-4}; (c) 7.16 MeV.

7 (a) 0.0024 amu; (b) 10^{-4}; (c) 2.24 MeV.

9 (a) 1.34×10^{25} atoms; (b) 2.0×10^{21} atoms;
(c) 2.3×10^{9} J.

11 (a) $_3\text{Li}^6 + {}_0\text{n}^1 \rightarrow {}_2\text{He}^4 + {}_1\text{T}^3$, tritium; (b) 0.005136 amu;
(c) 3.27 MeV; (d) 2.45 MeV.

13 (a) 6.9×10^{5} m/s; (b) 3.2 mm.

15 (a) 2.81×10^{-12} J; (b) 3.39×10^{11} J;
(c) 2.95×10^{-6} gm.

Index

Index

and convection, 247, 248
defined, 193
density of, 194, 195
electricity passed through, 298, 299
energy conservation and, 202, 203
flow continuity, 201, 202
in heat engines, 260
in motion, 201–212
Pascal's principle, 196
pressure of, 196, 197
at rest, 193–201
Reynolds number, 209, 210
Stokes' law, 210, 211
surface of, 212–216
velocity of fluid flowing in pipe, 207–209
viscosity, 206, 207
Fluorine ions and ultraviolet rays, 250
Fluxions, 49
FM radio, 396, 397
Focal distance, 416
Focal length of lens, 420
Focal point, 416
in diverging lens, 417
Food chain, radiation in, 595
Force
buoyant, 193
centripetal, 63, 78, 79
Coulomb, *see* Coulombs
defined, 50
gravity, *see* Gravity
normal, 89–91
reaction, 65, 66
restoring, 167
unit of, 51
Forced convection, 247
Fossil fuels, 107, 108
consumption of, 114–116
power production, 567
Foucault, Jean, 125
Fourier's theorem, 176–178
Fovea, 487
Fowler, William A., 606, 607
Fractional crystallization, 549
Frames of reference, 25, 122–124
Franklin, Benjamin, 276, 277, 296
and charges, 281, 282
Franklin, Rosalind, 512
Fraunhofer, von Josef, 467, 468
Fraunhofer dark lines, 451, 467, 468
Free body diagram, 150, 152, 154
Free expansion, 268, 269
Freezing point, 233
of mercury, 227
Frequency, 347
and black-body radiation, 472–474
and capacitors, 351

cutoff frequency, 353
intermediate frequency, 395
of LC resonant circuit, 392
and Moseley's law, 506–509
resonant frequency, 359, 392
spectrum of, 398–401
threshhold frequency, 480, 481
waves, characterization of, 173
Frequency modulation, 396, 397
for television, 397
Fresnel, Augustin, 450
Friction, 88–92
and heat, 230
kinetic, 91, 92
rolling, 91, 92
static, 89–91
Frisch, Otto, 561, 562
Fuels
deuterium-tritium fusion, 617, 618
fossil, *see* Fossil fuels
Fulcrum, 145
Fundamental wave pattern, 174, 175
Fusion, 112
"big bang" theory, 605, 606
carbon, burning, 606, 607
carbon-nitrogen cycle, 605
controlled thermonuclear, 609–620
deuterium and tritium reactions, 610–612
fuel cycle, 617, 618
helium burning, 606
hydrogen, 603–606
inertial confinement, 617
laser, 617
Lawson criteria, 612, 613
magnetic confinement, 614–616
reactors, 618–620
research progress in, 616, 617
thermonuclear, 603
tokamak confinement system, 614, 615

Galen, 435
Galilean telescope, 430
Galilean transformation, 123
Galileo, 10, 11, 14, 31, 49
inertial reference frame, 122–124
scaling laws, 86
Galvanometer, 323, 324
and Faraday's law, 336
Gamma, 135
Gamma rays, 401, 496, 575, 576
decay, 553
half-thickness, 576
radiation, dose rate from, 586, 587
Gamow, George, 605
Gas-filled detectors, 578, 579
Gas lasers, 515